Frontiers of Quality Electronic Design (QED)

Ali Iranmanesh
Editor

Frontiers of Quality Electronic Design (QED)

AI, IoT and Hardware Security

Editor
Ali Iranmanesh
Silicon Valley Polytechnic Institute
Los Altos Hills, CA, USA

ISBN 978-3-031-16346-3 ISBN 978-3-031-16344-9 (eBook)
https://doi.org/10.1007/978-3-031-16344-9

This Springer imprint is published by the registered company Springer Nature Switzerland AG
The registered company address is: Gewerbestrasse 11, 6330 Cham, Switzerland

Preface

This book intends to explore the latest trends in topics related to electronic design, with emphasis on hardware security and application of AI. It is formatted as a collection of articles, in the form of chapters, each written by one or several authors with expertise in their field.

I was motivated to create this book through my close association with many technical conferences, especially the events organized by the International Society for Quality Electronic Design (ISQED), a non-profit international organization devoted to innovation and advancement in electronic design, and engineering education, which I founded 23 years ago. Since inception, the organization has spanned over 30 events in various topics related to the field of electronic design, covering topics such as IC design, IoT, Sensors, smart power grid, solar energy technologies, and engineering education.

Successful implementation of these events has resulted in a wealth of technical articles to the engineering community. Today, the majority of these articles are available through ISQED proceedings and IEEE digital library, Xplore.

Last year, I approached several ISQED authors to see if there is any interest to start a book series around the same topics, where authors can get around conference's page limitations and delve deeper into technical subjects, and so the seed of this book was planted.

Initially, my goal was to cover the entire electronic design cycle, from design concept to manufacturing, and application and, in the process, highlight use of emerging fields such as artificial intelligence (AI) and machine learning (ML) and their application in the design process. This turned out to be a major undertaking, and at the end, I had to be content with a smaller set of topics, which are presented in this book. I hope I am able to continue the task and create follow-up books to cover a few other important topics, which were not covered in the present volume.

I would like to thank all authors for contributing to this book. Biographies of each chapter's authors are shown at the end of the chapter. Moreover, I would like to acknowledge many others, who have volunteered to review the manuscripts, and provide valuable feedback to authors.

Los Altos Hills, CA, USA
July 2022

Ali Iranmanesh

Acknowledgements

Special thanks to following chapter reviewers:

Prof. Omid Kavehei, University of Sydney
Prof. Ahmedullah Aziz, University of Tennessee, Knoxville
Prof. Siddhartha Nath, University of California, San Diego
Prof. Masoud Zabihi, University of Minnesota
Prof. Xinfei Guo, University of Michigan
Prof. Bassel Soudan, College of Computing and Informatics
Prof. Divya Akella Kamakshi, University of Virginia
Prof. Alvaro Cintas Canto, Marymount University
Prof. Bhasin Shivam, Nanyang Technological University
Prof. Amin Rezaei, California State University, Long Beach
Prof. Chidhambaranathan R, Utah State University
Dr. Deepashree Sengupta, Synopsys Inc.

Contents

About the Authors

Monther Abusultan received his B.Sc. degree with highest honors in CE with a minor in EE (2008) and his M.S. degree (2010) from Montana State University, Bozeman. He received his Ph.D. degree (2017) from Texas A&M University, College Station. He joined the Advanced Design team at Intel Oregon in 2017 where he worked on the development of Intel's future process technology nodes, then joined Microsoft Silicon Engineering and Solutions in 2021 to work on the development of Microsoft's custom microprocessors. He has published in many areas of VLSI including low power FPGA design, GPGPUs, flash-based digital design, logic synthesis, and optimization.

Ali Afzali-Kusha received the B.Sc. degree from the Sharif University of Technology, Tehran, Iran, in 1988, the M.Sc. degree from the University of Pittsburgh, Pittsburgh, PA, USA, in 1991, and the Ph.D. degree from the University of Michigan, Ann Arbor, USA, in 1994, all in electrical engineering. Dr. Afzali-Kusha has been with the University of Tehran since 1995, where he is currently a professor in the School of Electrical and Computer Engineering. His current research interests include low-power high-performance design methodologies from the physical design level to the system level, efficient hardware implementation of neural networks, in-memory computing.

Shaahin Angizi is currently an assistant professor in the Department of Electrical and Computer Engineering, New Jersey Institute of Technology (NJIT), Newark, NJ, USA, and the director of the Advanced Circuit-to-Architecture Design Laboratory. He completed his doctoral studies in Electrical Engineering at the School of Electrical, Computer and Energy Engineering, Arizona State University (ASU), Tempe, AZ in 2021. His primary research interests include ultra-low-power in-memory computing based on volatile & non-volatile memories, in-sensor computing for IoT, brain-inspired (neuromorphic) computing, and accelerator design for deep neural networks and bioinformatics.

Hongyu An is an assistant professor in Electrical and Computer Engineering at Michigan Technological University. He obtained his doctoral degree in Electrical Engineering at Virginia Tech. Prior to Virginia Tech, he received his B.S. and M.S. degrees in electrical engineering at Missouri University of Science and Technology and Shenyang University of Technology. He is the awardee of the Bill and LaRue Blackwell Thesis/dissertation award in 2020. His research interests include neuromorphic computing and engineering, neuromorphic electronic circuit design for artificial intelligence systems, and artificial intelligence for robotics and medical devices.

Kunal Bharathi received his Bachelor of Engineering degree from the PES Institute of Technology (now PES University) in Bangalore, India, and an M.S. in ECE from the University of Iowa. He is currently pursuing another graduate degree at Texas A&M University in College Station, Texas. His areas of interest include system software, networks, security, and machine learning.

Alberto Bosio received the Ph.D. in Computer Engineering from the Politecnico di Torino, Italy in 2006. From 2007 to 2018 he was an Associate Professor at LIRMM - University of Montpellier in France. He is now a Full Professor at the INL – Ecole Centrale de Lyon, France. His research interests include Approximate Computing, In-Memory Computing, Test and Diagnosis of Digital circuits and systems and Reliability. He co-authored 1 book, 3 patents, 35 journals, and over 120 conference papers. He is the chair of the ETTTC. He is a member of the IEEE.

Shao-Wei Chu received his B.S. degree in electrical engineering from National Taiwan University, Taipei, Taiwan, in 2019. He is currently pursuing the Ph.D. degree at Texas A&M University, College Station, TX, USA. His research interests include logic synthesis, hardware and software verification, and CAD frameworks for security (design-for-security).

Pierre d'Hondt received his Engineering degree from the Institut Supérieur de l'Electronique et du Numérique (ISEN), Lille, France, in 2019. He is currently a Ph.D. student at the University of Montpellier, and works with STMicroelectronics, Crolles, France. His main focus of research lies in Characterization for Test and Diagnosis of Digital circuits.

Wafi Danesh obtained his Master of Science degree from the South Dakota School of Mines and Technology in 2013. He is currently in the final year of his Ph.D. in Electrical and Computer Engineering at the University of Missouri – Kansas City. His research focuses on the nexus of hardware security, adversarial machine learning, and IoT network security. He has extensively reviewed numerous conferences and journals such as Elsevier Journal of Systems Architecture, GLVLSI, ICAIIC 2021, Springer Journal of Hardware and Systems Security (HASS), Elsevier DSP, and World Forum on Internet-of-Things (WF-IoT), among others.

Sujay Deb received MS degrees from the Indian Institute of Technology, Kharagpur, India, in 2007 and Ph.D. degree from Washington State University, USA, in 2012. He is currently an associate professor in the Department of Electronics and Communication Engineering, Indraprastha Institute of Information Technology, Delhi, India. His broader research interests in the design of novel interconnect architectures for multicore chips, heterogeneous system architectures. He is a member of IEEE.

Jaya Dofe received a Master of Science and Ph.D. in Electrical and Computer Engineering from the University of New Hampshire, New Hampshire, USA, in 2018. She is an assistant professor in the Computer Engineering Department at California State University, Fullerton, California, USA. Before joining CSUF, she was a faculty at Florida International University, Florida, USA. Her research focuses on hardware security, including design obfuscation, side-channel analysis of encryption algorithms, fault attack analysis, and emerging technologies, including 3D hardware security. She is also interested in engineering education research on active learning and equitable pedagogy. Dr. Dofe is guest editor of the Special Issue "3D Technology for Hardware Security: Opportunities and Challenges" of the Electronics journal. She is a technical committee member of IEEE conferences – ISVLSI, iSES, VDAT, VDEC. She chaired three panels for women in the engineering forum at the 7th IEEE International Symposium on Smart Electronic Systems and was a program chair for the Workshop for Women in Hardware and Systems Security (WISE), 2020.

Deliang Fan is currently an associate professor in the School of Electrical, Computer and Energy Engineering, Arizona State University, Tempe, AZ, USA. Dr. Fan's primary research interests include Machine Learning Circuits and Algorithms at Edge, In-Memory Computing Circuits and Architecture, Adversarial AI Security and Trustworthy. He has authored and co-authored around 140+ peer-reviewed international journal/conference papers in those areas. He is the recipient of NSF Career award, best IP paper award of DATE 2022, best paper award of GLSVLSI 2019, ISVLSI 2018 and 2017. His research works were also nominated as best paper candidate of ASPDAC 2019, ISQED 2019.

Raviv Gal joined IBM Haifa Research Lab in 2011 and is the manager of the Hybrid Cloud Quality Technologies department, leading projects utilizing data analytics including machine learning for all levels in the Hybrid Cloud, including hardware verification. Before this, Raviv worked 12 years in Marvell Israel where he was a verification leader, responsible for integration of several methodologies and technologies to the verification flow. He holds B.A. in Mathematics and C.S. and M.A. in C.S. from Tel-Aviv university. Besides HW verification, data, and AI, Raviv also likes to run, bike, and hike.

Georgios Georgakilas Circuits and Systems Lab UTH.

Patrick Girard received a Ph.D. degree in Microelectronics from the University of Montpellier, France, in 1992. He is currently Research Director at CNRS (French National Center for Scientific Research) and works in the Microelectronics Department of the Laboratory of Informatics, Robotics and Microelectronics of Montpellier (LIRMM) – France. His research interests include all aspects of digital testing and memory testing, with emphasis on critical constraints such as timing and power. Reliability and fault tolerance are also part of his research activities. Patrick Girard is a Fellow of the IEEE.

Wesam Ibraheem is the manager of the Analytics & Quality Technologies Group at IBM Research - Haifa. He holds a B.Sc. degree in Computer Science from the Technion – Israel Institute of Technology. Since joining IBM Research in 2006, Wesam has been involved as a manager, leader, and a developer in several projects concerning hardware verification and analytics tools. These included: a constraint satisfaction (CSP) engine, a pre silicon generator (IBM Genesys), a post Silicon exerciser (IBM Treadmill), a verification analyzing and dashboarding tool (Verification Cockpit), and more.

Alexander A. Ivaniuk received the M.Eng. degree from Belarusian State University of Informatics and Radioelectronics (BSUIR), Minsk, Republic of Belarus. He also received the Ph.D. degree (1999) and the Dr.S. degree (2010) from BSUIR. Since 2014, he is a professor in the Computer Science Department in BSUIR and Research Fellow at SK Hynix memory solutions Eastern Europe (SKHMSE). His areas of research interest are Design and Testing of Digital Devices and Systems, Hardware Synthesis and Simulations, Reconfigurable Hardware Design, and Physical Cryptography. Professor Ivaniuk is coauthor of 8 books, 50 journal papers, 80 international conference papers, and 6 patents.

Mehdi Kamal received the B.Sc. degree from the Iran University of Science and Technology, Tehran, Iran, in 2005, the M.Sc. degree from the Sharif University of Technology, Tehran, in 2007, and the Ph.D. degree from the University of Tehran, Tehran, in 2013, all in computer engineering. He was an Associate Professor with the School of Electrical and Computer Engineering at the University of Tehran. He is currently a research scientist at the Institute for Future of Computing at the University of Southern California, USA. His current research interests include approximate computing, neuromorphic computing, embedded systems design, low-power design, and security.

Sunil P. Khatri received his B.Tech. degree from IIT Kanpur, his M.S. degree from UT Austin, and his Ph.D. degree UC Berkeley. He is currently a Professor of Electronics and Communication Engineering at Texas A&M University, College Station, TX, USA. He has co-authored more than 275 peer-reviewed publications. Among these, five received a best paper award, while seven others received best paper nominations. His current research interests include VLSI IC/system-on-a-chip design (including energy efficient design of custom ICs and FPGAs, variation tolerant design, clocking), algorithm acceleration (FPGA, GPU as well as custom IC based), and interdisciplinary extensions of these topics to other areas.

Vipin Kumar Kukkala received his B.Tech. degree in Electronics and Communications Engineering from Jawaharlal Nehru Technological University, India, in 2013 and his Ph.D. in Electrical Engineering from Colorado State University, USA, in 2022. He is currently working as a senior high-performance computer architect at NVIDIA. His research interests include the design of next-generation automotive networks, security in cyber-physical systems, machine learning-based anomaly detection, and large-scale heterogeneous system design. He has co-authored multiple book chapters and published in several top-tier international peer-reviewed journals and conferences.

Aymen Ladhar received the Ph.D. degree in Electrical Engineering from the University of Sfax, Tunisia, in 2010. He is currently a test & yield engineer at STMicroelectronics Crolles, France where he is currently the responsible of the logic diagnostic activity for ST products and test vehicles. His research interests include VLSI testing, fault diagnosis, layout analysis, defect extraction and simulation.

Cheng-Yen Lee received his B.S. and M.S. degrees in ECE from the Department of Electrical and Computer Engineering, National Yang Ming Chiao Tung University (NYCU), Hsinchu, Taiwan, in 2017 and 2019, respectively. He is currently pursuing his Ph.D. degree in the Department of Electrical and Computer Engineering, Texas A&M University, College Station, Texas. His research areas focus on VLSI and mixed-signal IC designs using emerging technologies (multi-gate devices).

Konstantinos Liakos is a research assistant at Circuits and Systems Lab UTH.

Usha Mehta a Professor and PG Coordinator (VLSI Design) at EC department, Institute of Technology, Nirma University, Ahmedabad. She received her bachelor's degree in EC in 1994 and master's and Ph.D. degree in VLSI Design in 2005 and 2011, respectively. She has more than 22 years of experience of teaching She has guided more than six Ph.D. students and M. Tech by Research students. She has one Patent on her credit. She has been the Principal Investigator for research projects of ISRO and GUJCOST. She has published more than 55 research papers in the area of VLSI Design and Testing.

Khaled Ahmed Nagaty received the Bachelor of Science degree in statistics from Cairo University in 1982. He received the Ph.D. degree in Digital Image Processing from Cairo University in 1999. He was Associate Professor of Computer Science in the Faculty of Informatics & Computers at Ain Shams university in Cairo from 2007 until 2016. He is currently a professor with the faculty of Informatics & Computer Science at the British University in Egypt in partnership with London South Bank

University, UK since 2008. His current research interests include machine vision, pattern recognition, distributed systems, wireless network sensors, cryptography, tomography, biometrics, machine learning and optimization.

Ziv Nevo is a lead technical staff member in the Formal Quality Technologies Group at IBM Research - Haifa. He holds B.Sc. and M.Sc. degrees in Computer Science from the Technion – Israel Institute of Technology. Since joining IBM Research in 2002, Ziv was involved as a leader, architect and developer in several projects concerning hardware verification (static and dynamic) as well as cloud configuration analysis and synthesis.

Fabiha Nowshin received her B.S. and M.S. degrees in electrical engineering at Virginia Tech, Blacksburg, Virginia in 2019 and 2021, respectively. She is currently pursuing her Ph.D. degree in electrical engineering with the Bradley Department of Electrical and Computer Engineering (ECE) at Virginia Tech. Her research interests include emerging memory technologies for artificial intelligence applications as well as very large scale integrated (VLSI) circuits and system design for neuromorphic computing.

Sudeep Pasricha received his Ph.D. in computer science from the University of California, Irvine in 2008. He is currently a Professor at Colorado State University. He is Director of the Embedded, High Performance, and Intelligent Computing (EPIC) Laboratory. His research focuses on the design of innovative software algorithms, hardware architectures, and hardware-software co-design techniques for energy-efficient, fault-tolerant, real-time, and secure computing. He has co-authored multiple books, book chapters, and published close to 300 research articles in peer-reviewed journals/conferences, workshops, magazines, and books. He is a Senior Member of the IEEE and Distinguished Member of the ACM.

Massoud Pedram obtained his B.S. degree in Electrical Engineering from the California Institute of Technology in 1986. Subsequently, he received M.S. and Ph.D. in Electrical Engineering and Computer Sciences from the University of California, Berkeley in 1989 and 1991, respectively. In September 1991, he joined the Ming Hsieh Department of Electrical Engineering of the University of Southern California where he currently is the Charles Lee Powell Professor of Electrical Engineering and Computer Science in the USC Viterbi School of Engineering. His current research interests include energy-efficient computing, machine learning hardware, superconductive electronics, and homomorphic computing.

Fotis Plessas Circuits and Systems Lab UTH.

Jayesh Popat has done his Ph.D. in the area of "Hardware Security and VLSI Testing" under the guidance of Dr. Usha Mehta. He is currently working as a DFT engineer Microcircuits Technology, Ahmedabad. Having knowledge and experience of both academia and industry, he has a total 6 years of academic experience along with 4 years of industrial experience. He worked at different semiconductor industries like Intel India Pvt Ltd., AMD India Pvt. Ltd, Broadcom Ltd. etc. He has published more than 15 research papers in international journals and conferences. His research interest in Hardware security, cryptography hardware implementation and Design for Testability (DFT) techniques.

Sadhana Rai is a Research Scholar in the Department of Computer Science & Engineering at National Institute of Technology Karnataka(NITK), India. Her area of research is Hybrid Main Memories using Non-volatile memories. She works in SPARK(Systems, Parallelization and Architecture Research at NITK) Lab, under the supervision of Basavaraj Talawar.

Arman Roohi is currently an assistant professor with the School Computing, University of Nebraska-Lincoln, USA. Before joining UNL in 2020, he was a postdoctoral research fellow with UT Design Automation Laboratory, the University of Texas at Austin. He received the Ph.D. degree in Computer Engineering at the University of Central Florida, Orlando, FL, USA, in 2019. His research interests span the areas of design of cross-layer (device/ circuit/ architecture) co-design for implementing complex machine learning tasks secure computation, including hardware security, and the security of artificial intelligence, reconfigurable and adaptive computer architectures, and beyond CMOS computing, with emphasis on Spintronics.

Sidhartha Sankar Rout received the BTech degree in EEE from NIST, Odisha, India, in 2008 and the MTech degree in ECE from IIIT Delhi, India, in 2014. He is currently working toward the Ph.D. degree in Electronics and Communication Engineering from Indraprastha Institute of Information Technology Delhi, New Delhi, India. His research interests include system validation and hardware security. He is a student member of the IEEE.

Bilal Saleh is a research scientist at the IBM research lab in Haifa. His current role, in the Analytics & Quality Technologies group, is focused on leading the Coverage Directed Generation (CDG) project. Previously, Bilal worked on other projects at IBM research, including the Template Aware Coverage (TAC) project and the Hardware-Defects Triaging project. Bilal received his B.Sc. and Ph.D. degrees in Computer Science, from Haifa university, in 2010 and 2014. His Ph.D. thesis is about DNA packaging and condensation inside the living cell. Particularly, it deals with the extraction of Nucleosome's positioning patterns along Chromosomes.

Kyler R. Scott received his B.S. and M.S. degrees in Electrical and Computer Engineering from Texas A&M University (TAMU) in College Station, Texas. He is currently pursuing a Ph.D. degree at TAMU. He has held internships at Intel Corporation, where he worked on digital design verification, and Amazon

Web Services (AWS), where he worked on physical design for machine learning (ML) accelerators. His research interests include neuromorphic and mixed-signal computing, ML acceleration, and ML at the edge.

Gian Singh received his B.Tech. degree in E&CE from the National Institute of Technology (NIT-H), Hamirpur, India, in 2017. He has been pursuing a Ph.D. in Computer Engineering at Arizona State University (ASU), USA since Fall 2018. He has held internships at Maxlinear Inc., Qualcomm Technology Inc., and Micron Technology Inc., where he worked in Digital CAD, RTL design, and Memory architecture design teams. His current research interest includes the design of artificial neurons, in-memory computing, and near memory processing enabling high throughput and energy-efficient systems for data-intensive applications.

Mitali Sinha received the MTech degree from National Institute of Technology Agartala, India, in 2016 and Ph.D. degree from Indraprastha Institute of Information Technology Delhi, India, in 2022. She is currently working as a researcher in IMEC, Belgium. Her research interests include design space optimization and security analysis of accelerator-rich heterogeneous system-on-chips. She is a student member of the IEEE.

Sandeep Sunkavilli is a research assistant at the University of New Hampshire.

Basavaraj Talawar is an assistant professor in the CSE Department at NITK where he heads the SPARK (Systems, Parallelization and Architecture Research at NITK) lab. He has a Ph.D. from the Indian Institute of Science, Bangalore. His research interests are in the broad areas of Computer Architecture. He is a recipient of the Visvesvaraya Young Research Fellowship from the Govt. of India and a faculty award from IBM. His research is supported through grants from the DST, IBM, and Intel.

Sooryaa Vignesh Thiruloga received his M.S. degree in Computer Engineering from Colorado State University, USA in 2022. He is currently a Data Scientist at Hewlett Packard Enterprise, USA. His research interests include the design of scalable, efficient, lightweight deep learning architectures, and leveraging advanced artificial intelligence techniques for anomaly detection in automotive cyber-physical systems.

Shaghayegh Vahdat received the B.Sc., M.Sc., and Ph.D. degrees in electrical engineering from the University of Tehran, Tehran, Iran, in 2014, 2016, and 2021, respectively, where she is currently working as a postdoctoral fellow with the Tehran University of Medical Sciences. She will join the School of Electrical and Computer Engineering of the University of Tehran as an assistant professor from Oct. 2022. Her current research interests include hardware implementation and reliability enhancement of neural systems, mixed signal computations using emerging non-volatile memories, and low-power high-performance design of digital arithmetic units.

Arnaud Virazel received the Ph.D. degree in Microelectronics from the University of Montpellier, France, in 2001. He is currently Professor at the University of Montpellier – LIRMM (Laboratory of Informatics, Robotics and Microelectronics of Montpellier) where he is responsible for the TEST ("Test and dependability of microelectronic integrated Systems") team. He has published 7 books or book chapters, 50 journal papers, and more than 160 conference and symposium papers spanning diverse disciplines, including DfT, reliability, power-aware and memory testing. He is the head of the electrical engineering department (about 450 students in B.Sc. and M.Sc. programs) at the University of Montpellier.

Sarma Vrudhula (M'85-SM'02-F'16) is Professor of Computer Science and Engineering at Arizona State University, and the Director of the NSF I/UCRC Center for Embedded Systems. He received the B.Math. degree from the University of Waterloo, Waterloo, ON, Canada, and the M.S.E.E. and Ph.D. degrees in electrical and computer engineering from the University of Southern California, Los Angeles, CA, USA. His work spans several areas in design automation and computer aided design for digital integrated circuits and systems, focusing on low power circuit design, and energy management of circuits and systems.

Ankit Wagle (M'17) received the B.S. degree in Electronics and Telecommunication from the University of Pune, Maharashtra, India, in 2013, and the M.S. degree in VLSI Design from Vellore Institute of Technology, Vellore, TN, India, in 2015. He has been pursuing the Ph.D. degree with the School of Computing and Augmented Intelligence (SCAI), Arizona State University, Tempe, AZ, USA since 2016. His current research interests include new circuit architectures and design algorithms using threshold logic gates, and their applications to the design of energy efficient digital application-specific integrated circuits, field-programmable gate arrays, and neural network accelerators.

Yang Yi is an associate professor in the Bradley Department of Electrical Engineering and Computer engineering at the Virginia Tech. Her research interests include very large scale integrated (VLSI) circuits and systems, computer-aided design (CAD), neuromorphic architecture for brain-inspired computing systems, and low-power circuits design with advanced nano-technologies for high-speed wireless systems.

Chunxiu Yu is an assistant professor in the Department of Biomedical Engineering at Michigan Technological University. She received a Ph.D. degree in neurobiology from the Weizmann Institute of Science in Israel. She joined Duke University to study the neural mechanisms underlying reward-guided behaviors and University of North Carolina at Chapel Hill for research on the neural signal processing and network dynamics of visual attention. She was a research scientist in Neural Engineering and Neural Prostheses Laboratory at Duke University. Her research interests include neural signal processing, brain stimulation, optogenetics, brain and behavior, and brain-machine learning.

Qiaoyan Yu is a professor at the University of New Hampshire.

Siarhei S. Zalivaka received the B.Eng. (Hons.) degree and M.Eng. degree from Belarusian State University of Informatics and Radioelectronics (BSUIR) in 2012 and 2013, respectively. He also received his Ph.D. degree from Nanyang Technological University (NTU), Singapore, in 2018. He worked as an associate professor at BSUIR from 2018 to 2021. From 2018, he is working as a Research Fellow at SK hynix memory solutions Eastern Europe. His area of research interests are Hardware Security and Trust, Reconfigurable Computing, and NAND Flash Memory Devices. Dr. Zalivaka co-authored 1 book chapter, 8 journal and 24 conference papers, and 6 patents.

Yan Zhang is a research assistant professor in the Department of Biological Sciences at Michigan Technological University. She received her bachelor's degree in pharmacy from Sichuan University and master's degree in pharmacology from Peking Union Medical College in China. She received a Ph.D. degree in pharmacology and toxicology from University of Missouri-Kansas City. She worked at the Jared Grantham Kidney Institute at Kansas University Medical Center as a postdoctoral researcher. She received the Postdoctoral Fellowship Award from the Polycystic Kidney Disease Foundation. Her research interests include inflammation, dysregulated cellular signaling pathways in renal diseases, and neuromorphic computing.

Noah Zins is a graduate student in the Electrical and Computer Engineering Department at Michigan Technological University. He received a B.S. in Computer Engineering and Mathematical Sciences. His current research is on neuromorphic computing in robotics applications. His research interests include neuromorphic computing, signal and image processing, processor architectures, robotics, and embedded systems.

Avi Ziv is a Research Staff Member in the Hybrid Cloud Quality Technologies Department at the IBM Research – Haifa lab. Since joining IBM in 1996, Avi participated and led research projects that developed methodologies, technologies, and tools for various aspects of hardware functional verification including stimuli generation, checking, and functional coverage. In recent years, his focus is on utilizing data analytics techniques in general and machine learning specifically in functional verification. Avi holds a B.Sc. degree in Computer Engineering from the Technion – Israel Institute of Technology and M.Sc. and Ph.D. degrees in Electrical Engineering from Stanford University.

NAND Flash Memory Devices Security Enhancement Based on Physical Unclonable Functions

Siarhei S. Zalivaka and Alexander A. Ivaniuk

1 Multimode Physical Unclonable Function as an Entropy Source for Generating True Random Bits

1.1 Introduction

True random number generators (TRNGs) are used in a wide range of applications (e.g., cryptography, statistical sampling, simulation, computer games, etc.) [51]. TRNG can be implemented as a part of NAND flash memory device controller and used to support Trusted Computing Group (TCG) standard [1]. The main advantage of TRNGs comparing to pseudorandom number generators (PRNG) is the uniqueness and unpredictability of their produced output values. TRNG is a device or a part of a device that generates random numbers based on some intrinsic physical process. One of the possible ways of extracting random data from electronic devices is to implement physical unclonable functions (PUFs) (e.g., [50, 52]).

Nowadays physical unclonable functions (PUFs) are becoming ubiquitous cryptographic primitives as an alternative to classical cryptographic algorithms in compact digital devices [2]. Main semiconductor manufacturers actively introduce them into their IoT solutions [3], cutting-edge field programming gate array (FPGA) chips [4], authentication protocols [5], etc. In general, PUF can be represented as a mapping of external inputs (challenges) to the outputs (responses). This mapping is called challenge-response pair (CRP) set, which is unique for each integrated circuit

S. S. Zalivaka (✉)
SK hynix memory solutions Eastern Europe, Minsk, Republic of Belarus
e-mail: sergey.zalivako@sk.com

A. A. Ivaniuk
Belarusian State University of Informatics and Radioelectronics, Minsk, Republic of Belarus
e-mail: ivaniuk@bsuir.by

A. Iranmanesh (ed.), *Frontiers of Quality Electronic Design (QED)*,
https://doi.org/10.1007/978-3-031-16344-9_1

(IC) containing a PUF block even if the design and layout are the same [6]. This can be explained by intrinsic manufacturing process variations introduced during fabrication. Since physical properties of an IC may vary depending on temperature or voltage, some of the PUF response values are unstable. As a result, CRP set can be split into stable and unstable subsets and can be utilized for identification and random number generation, respectively.

PUF designs can be based on different physical phenomena, e.g., delay values [7], threshold voltages [8], operating frequencies [9], image sensor noise patterns [10], etc. Another subset of PUFs is utilizing memory to extract uniqueness from IC, e.g., SRAM PUF [11], DRAM PUF [12], Butterfly PUF [13], SR-Latch PUF [14], etc. NAND flash memory devices can be also successfully used to implement a PUF because some intrinsic effects, e.g., threshold voltages, erase times, bad block characteristics, program/read disturb, etc., uniquely characterize a memory device [15].

The proposed PUF design is based on using an inverter and a D-Latch which are controlled by enable (EN) signal. This circuit can operate in four modes, namely, initial memory, ring oscillator, metastability, and latch modes. All these modes can be used for different purposes, i.e., generating a unique identifier in initial memory mode, generating random numbers in ring oscillator or metastability modes, and storing generated ID or random value in latch mode. Thus, the proposed PUF design supports both PUF routines in a single device. One of the main challenges in TRNG design is consumed area and performance (rate of random bit generation). The proposed TRNG design is compact as it consumes a latch and an inverter gate and fast as ring oscillator mode operates on high frequency.

1.2 General Description of a Circuit

The proposed entropy source for random bit generation includes two elements, namely, Latch D-type (LD) and Inverter (INV). As shown in Fig. 1, INV is connected to LD and forms a negative feedback loop. The operation of this circuit is controlled by enable (EN) signal.

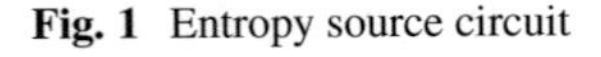
Fig. 1 Entropy source circuit

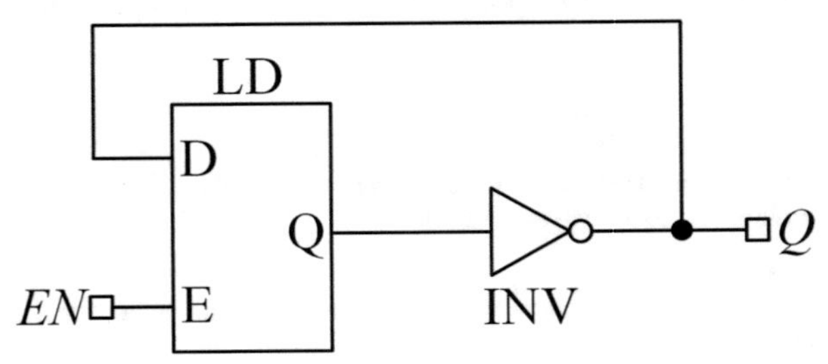

The proposed PUF supports four modes of operation:

1. **Initial memory**. This mode works only during start-up and EN="0". This is equivalent to SRAM PUF as the LD can generate either stable "0" or stable "1" or metastable value. According to SRAM PUF research [16], the output Q has a 10% chance of generating metastable value. If Q is stable, it can be used as a bit of a unique device ID.
2. **Ring oscillator (RO)**. If enable signal is kept as EN="1", the PUF will produce a meander signal with a unique frequency F_i (i is an index of individual entropy source) which is utilized for random bit generation similarly to RO PUF [17].
3. **Metastability**. Since LD is asynchronous and the value on data input (D) of LD is unpredictable, changing EN signal value from "1" to "0" can violate timing parameters of LD. In this case, LD may fall into a metastable state and the output Q can be either "0" or "1".
4. **Latch**. If enable signal is kept as EN="0", the LD stores random bit and the output Q value is stable.

As a result, the proposed PUF design can be used to generate unique stable ID bit (mode 1) or random bits (modes 2 and 3) or store generated ID or random bit (mode 4).

1.3 Operation of the Entropy Source

The circuit mentioned above (see Fig. 1) can be represented on gate level as shown in Fig. 2. The proposed circuit is named ROLD as a combination of different PUFs, i.e., ring oscillator and Latch D-type.

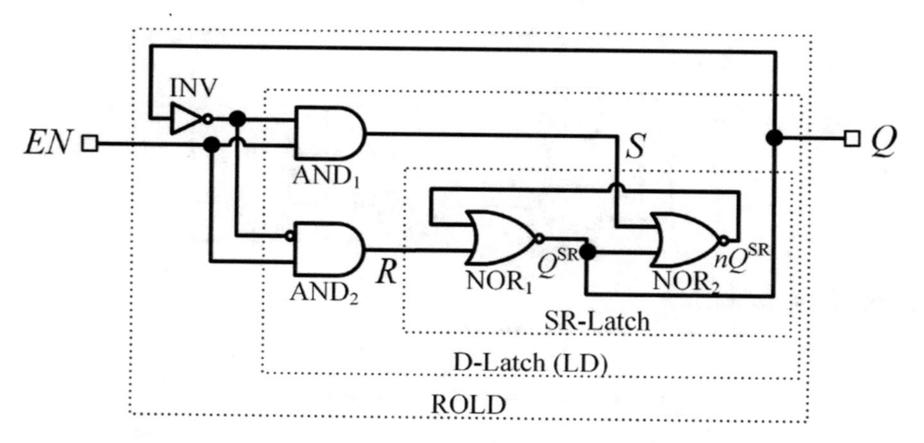

Fig. 2 Gate level of the entropy source circuit

The D-Latch component consists of basic SR-Latch circuit which has S (set), R (reset) inputs, and two complementary data outputs Q^{SR} and nQ^{SR}. In the case when SR-Latch is designed on NOR2 gates (NOR_1 and NOR_2 in Fig. 2), it has four operation modes: Setting "1" (when S="1" and R="0"), Resetting "0" (when S="0" and R="1"), Storing value (when S="0" and R="0"), and Forbidden mode (when S="1" and R="1"). The transaction from Forbidden state to Storing mode may cause generating metastable value on outputs Q^{SR} and nQ^{SR}. The D-type Latch is designed on the base of SR-Latch in such a way to prevent the occurrence of Forbidden mode by keeping S and R inputs in opposite values. The Storing mode is provided by additional input EN of enable signal and two additional AND2 gates (AND_1 and AND_2 in Fig. 2).

Let us describe equivalent circuits which are operating during four modes.

1.3.1 Initial Memory

When EN="0", the proposed circuit is equivalent to SR-Latch in storing mode (S="0", R="0"), which is shown in Fig. 3.

In this mode, AND elements (AND_1 and AND_2) generate constant "0" value and can be omitted for analysis of this circuit. NOR elements (NOR_1 and NOR_2) operate as inverters. Therefore, circuit in this mode operates as a bistable element as shown in Fig. 4.

During initialization (power-up) stage, the default value v is unknown due to manufacturing process variations (possible asymmetry of NOR gates NOR_1 and NOR_2 and connection wires between them). Therefore, unique ID values can be obtained from this PUF during power-up similarly to SRAM cells which are also based on bistable elements [16].

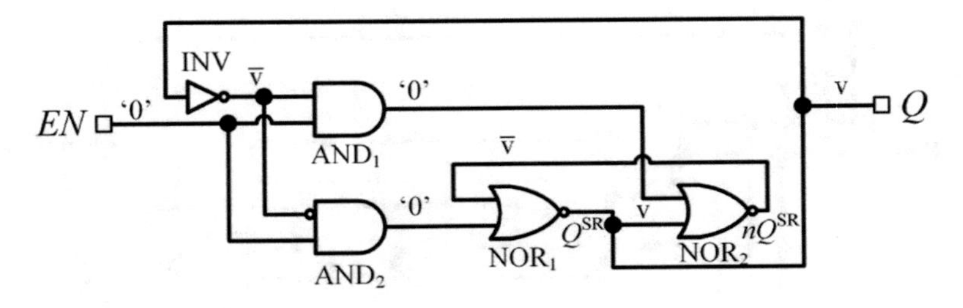

Fig. 3 ROLD circuit (EN="0")

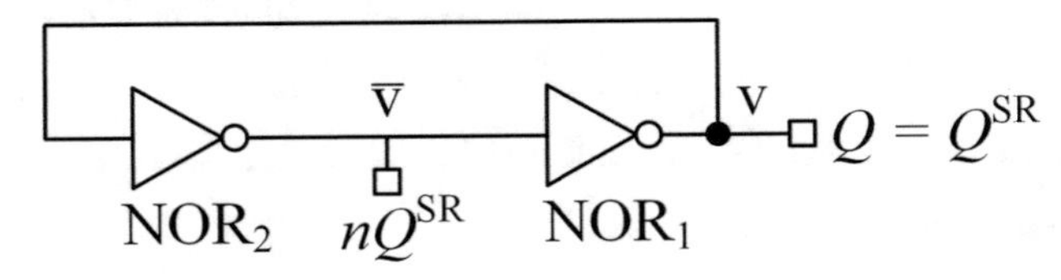

Fig. 4 Bistable element

1.3.2 Ring Oscillator

When EN="1", SR-Latch switches between Setting (S="1", R="0") and Resetting (S="0", R="1") modes based on the value obtained from inverter INV output as shown in Fig. 5.

In this mode, AND_1 element operates as a buffer repeating v or $\overline{v}$ values, NOR_2 element works as a constant "0" value generator, AND_2 and NOR_1 are identical to two inverters. This mode of operation is equivalent to the ring oscillator circuit with three inverters as shown in Fig. 6.

Thus, meander signal ($v \rightarrow \overline{v} \rightarrow v \rightarrow \ldots$) with unique frequency F_i appeared on the output Q. F_i is also determined by manufacturing process variations which make negative feedback loop delay unpredictable.

1.3.3 Metastability

Timing diagram in Fig. 7 shows three output values y_0, y_1, and y_2 from the output Q.

There are two possible ways how metastable state can appear on the output Q. First, initial value $y_0 \in \{v, X, \overline{v}\}$ (period of time from t_0 to t_1 as shown in Fig. 7) can be either stable zero, stable one, or a metastable state (X). In this case, metastability means the value with unknown stability, i.e., from time to time zero or one value appears on the output Q with different nonzero probability.

The second case is more complicated as it is based on SR-Latch phenomenon [18] which causes a high-frequency oscillation in addition to three values $\{v, X, \overline{v}\}$ in the first case. When both inputs S and R are fed with "1" value (forbidden state) for a short period of time and at this moment EN signal changes from "1" to "0", SR-Latch is trying to store forbidden state and generating damped

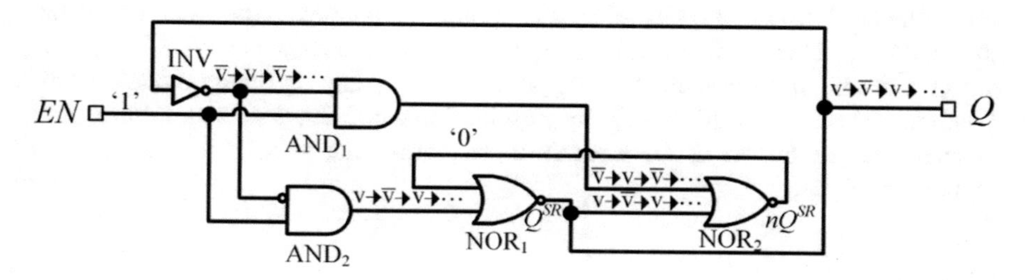

Fig. 5 ROLD circuit (EN="1")

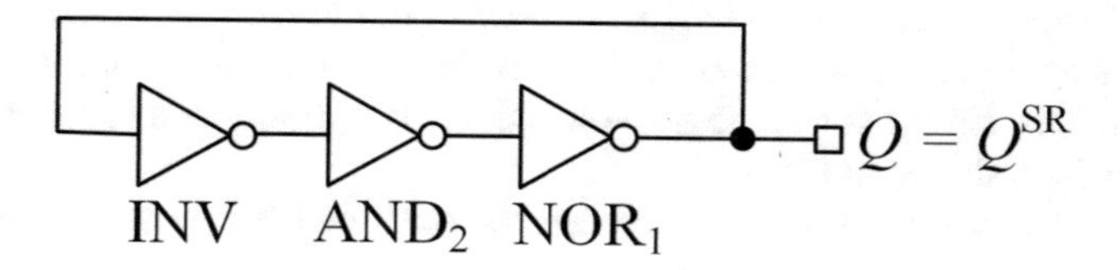

Fig. 6 Ring oscillator

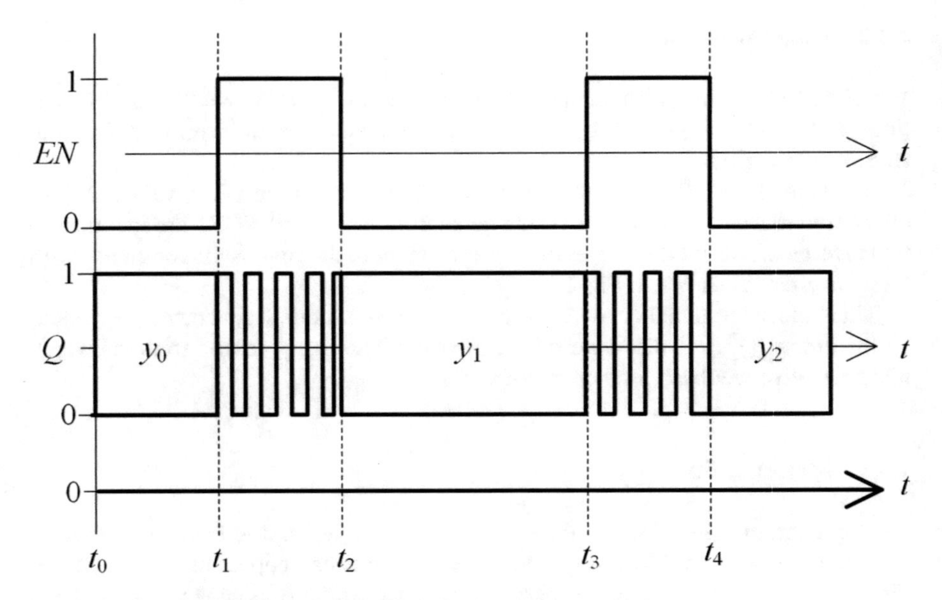

Fig. 7 SR-Latch timing diagram depending on changing EN signal

high-frequency oscillation. Metastable oscillation also dumps to the stable zero or one value after some time. So values y_1 (time period from t_2 to t_3) and y_2 (time period after t_4) will eventually get to stable zero or stable one value with or without metastable oscillation. This phenomenon is based on unique voltage and timing characteristics of SR-Latch and determined only after manufacturing.

Two mentioned scenarios of generating metastable value are shown in Fig. 8.

Possible values in the first case are shown in Fig. 8a and in second case (see Fig. 8b). Oscillation in the second case is eventually damped to the value v or $\overline{\text{v}}$, but the final value Q is more uncertain comparing to the first case.

As a result, transition of EN signal from "1" (ring oscillator mode) to "0" (latch mode) may cause high-frequency oscillation which leads to metastability state observed on the output Q. As a result, metastability can be used to generate true random numbers.

1.3.4 Latch

When EN signal is set into "0" value, it enables the possibility to store generated random values after initialization or ring oscillator mode or metastability caused oscillation. The circuit for storing N-bit unique ID (mode 1) or random number (mode 2 or 3) is shown in Fig. 9.

Thus, the proposed entropy source can be used for both purposes, unique bits producing and storing generated data.

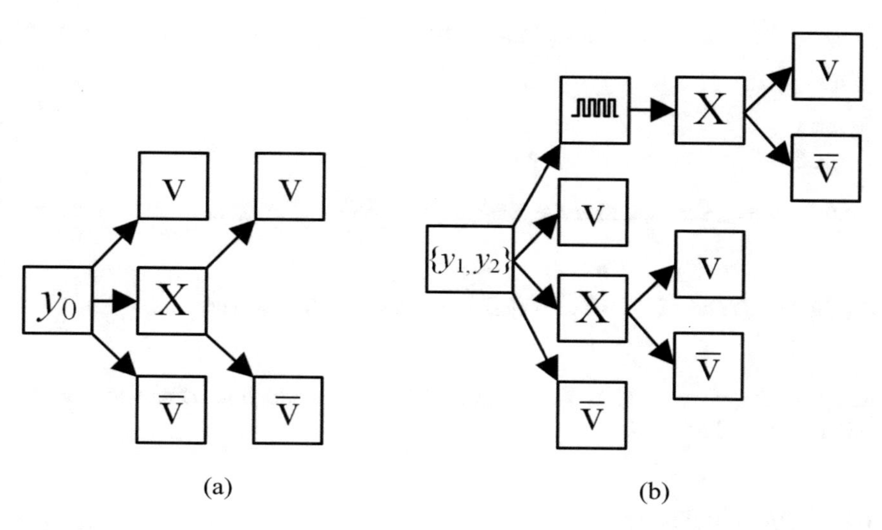

Fig. 8 Possible output values Q. (**a**) Period of time from t_0 to t_1. (**b**) Period of time from t_2 to t_3 and after t_4

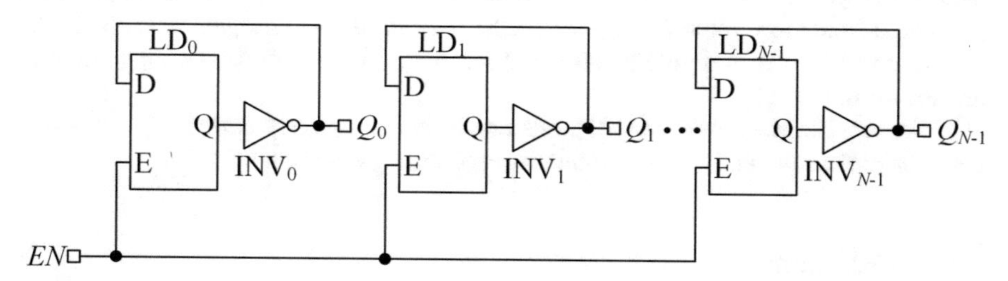

Fig. 9 Multi-bit latch for storing unique ID or random value

1.4 Experimental Results

The proposed entropy source has been implemented in Nexys 4 Xininx Artix-7 FPGA prototyping board [19], and characteristics for each mode have been collected.

1.4.1 Initial Memory

The total number of 128 entropy sources has been synthesized and implemented in FPGA. During $E = 100$ tests, each of the elements generated values shown in Fig. 10.

The distribution of probabilities of generating "1" value ($P_i^1(E)$) is the following: 61 elements with $P_i^1(E) = 0.0$, 56 elements with $P_i^1(E) = 1.0$, and 11 elements

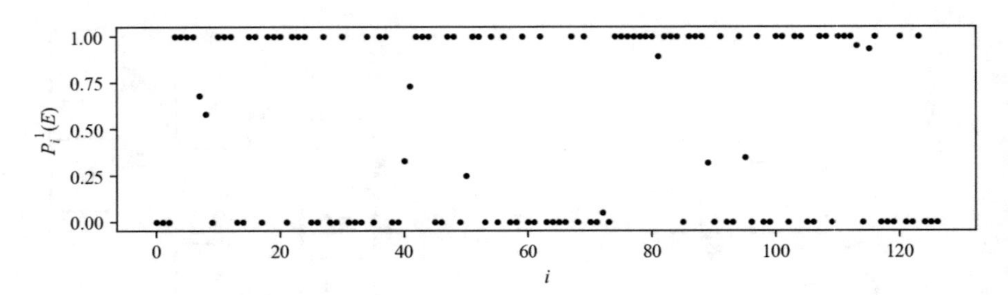

Fig. 10 Probabilities of "1" value ($P_i^1(E)$, $E = 100$) in initialization mode

with $0 < P_i^1(E) < 1$. Thus, reliable, unique, and reproducible ID can be generated using proposed method.

1.4.2 Ring Oscillator

These 128 generators have been tested in RO mode to show the uniqueness of generated frequency value F_i ($1 \leq i \leq 128$). The simulation frequency is 350 MHz (red line in Fig. 11); individual estimated frequency values F_i for entropy sources are shown in Fig. 11.

This experiment has been demonstrated that frequency value F_i for each generator is unique and unpredictable for each entropy source.

1.4.3 Metastability

Also the same 128 entropy sources have been tested $E = 100$ times in metastability mode (EN switches from "1" to "0"). The probabilities $P_i^1(E, k)$ of generating "1" value after k system clocks in RO mode ($EN =$ '1') for each element are shown in Fig. 12.

In contrast to initialization mode, the generated values have low reproducibility as all probabilities of generating "1" value ($P_i^1(E, k)$) are above 0.2 and below 0.8. Thus, this mode is more suitable for generating true random values.

1.4.4 Latch

To estimate the quality of random values produced by entropy sources, 128 elements have been utilized. As a result, a million 128-bit values have been generated by changing EN signal from "1" to "0". The duration of EN signal in "1" state is $k = 32$ system clocks. Each 128-bit value has been split into four 32-bit values. The histogram of the approximate distribution of generated four millions of 32-bit values is shown in Fig. 13.

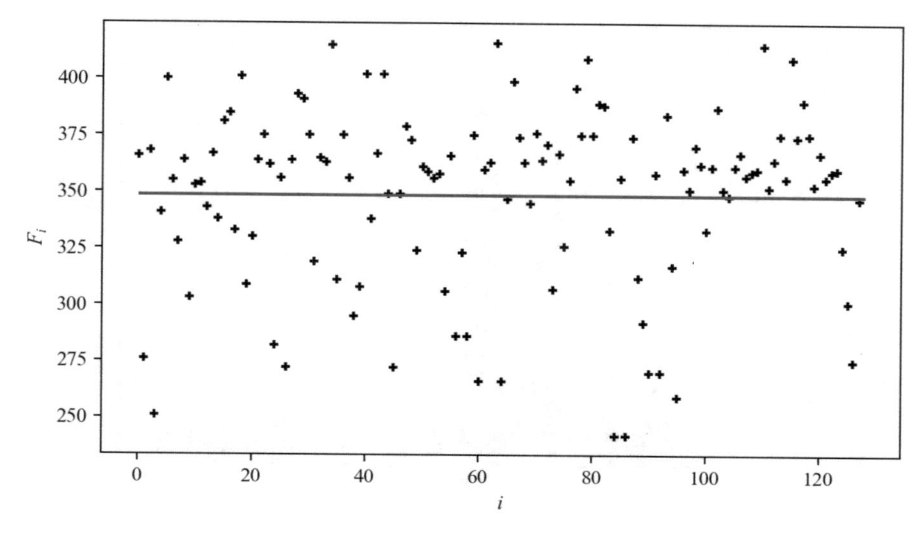

Fig. 11 Estimated frequencies F_i in RO mode

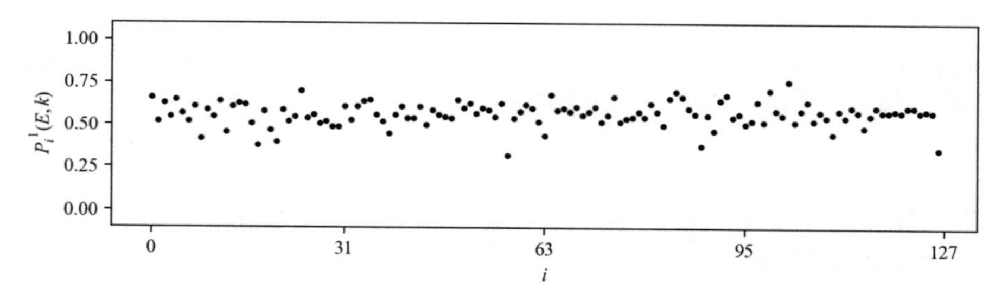

Fig. 12 Probabilities "1" value ($P_i^1(E, k)$, $E = 100$, $k = 32$) in metastability mode

The x-axis corresponds to the generated numerical value ranging from 0 to $\approx 4 \times 10^9$; the y-axis shows the estimated frequencies (the data is split into 100 bins) for each value. The generated values are truly random but not uniformly distributed. Therefore, the random sequence has to be post-processed in order to achieve required characteristics of randomness and be compliant with NIST standard [20].

1.5 Conclusion

The multimode physical unclonable function is presented. This design can be used for producing either stable unique ID or unpredictable random bit generation. The proposed design occupies smaller area comparing to classical PUF designs (e.g., Arbiter PUF, RO PUF, SRAM PUF) and can be used as an entropy source in

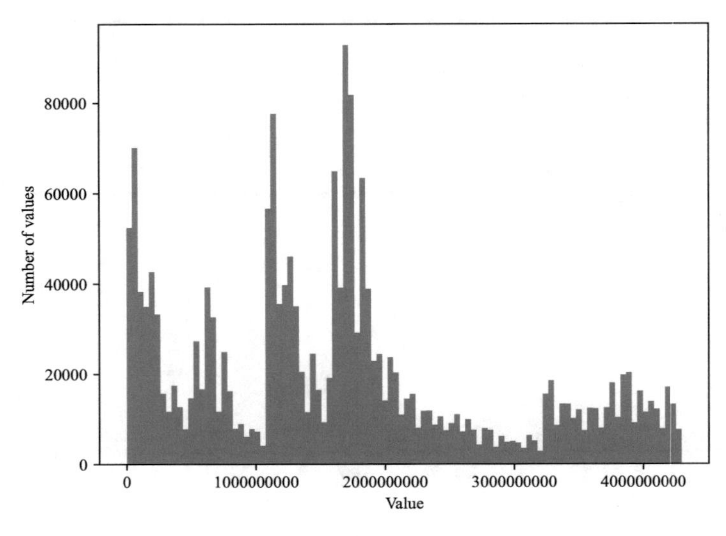

Fig. 13 Distribution of 32-bit random values ($k = 32$)

cryptography applications. For NAND flash memory devices, it can be utilized for entropy generation in encryption process and also for on-drive simulation purposes.

2 Raw Read-Based Physical Unclonable Function for TLC NAND Flash

2.1 Introduction

The increasing capacity of a single flash memory cell (SLC $\rightarrow$ MLC $\rightarrow$ TLC $\rightarrow$ QLC) has led to reliability issues with NAND-based storages [21]. This downside can be used for the opposite purpose, i.e., faults in blocks and pages can be utilized as a source of uniqueness for both chip identification and true random number generation. Modern TLC NAND flash memory devices have massive error correction code (ECC) engines which negotiate the effect of intrinsic NAND instability [22]. However, disabling ECC and scrambler modules during the read and write operation allows extracting less stable bits and using them to generate uniformly distributed random bits. As a result, one block of NAND can be separately used to generate a random number sequence during the read operation. The proposed flash memory operation is consistent with the definition of PUF, i.e., it provides a way to extract unique randomness characteristics from the physical super-high information content (SHIC) system [23]. Depending on the reliability of the obtained noise values, it can be used as random values (low reliability and high uniqueness) or unique identifiers (high reliability and high uniqueness). As a result, flash memory cells can be used as an entropy source for TRNG which does not require additional circuitry for its implementation and random numbers can be extracted during the read operation in the raw mode. The proposed method does not require a redesign of the existing NAND flash controller and can be used directly from the firmware level.

2.2 *Control of the Entropy Source*

The proposed entropy source is controlled by a two-stage algorithm. The first stage is enrollment, i.e., the positions of noisy bits are located during the read operation. The second stage is generation, i.e., read noisy bits from the positions are determined during the enrollment stage.

2.2.1 Enrollment

1. Choose a block from the reserve area.
2. Erase the whole block.
3. Write all zeros pattern to the block in the raw mode, i.e., ECC and scrambler are disabled during this operation.
4. For every page p_i $(0 \le i \le P-1)$, perform read operation in the raw mode R times. P is the number of pages in the block.
5. Calculate noise characteristic (Ψ) for each bit b_j $(0 \le j \le B)$ within all P pages. If $\Psi = 0$–bit b_j is stable and if Ψ is bigger, it means that the chosen bit b_j is more random. B is the number of bits in a page.
6. Bits with highest Ψ scores should be chosen as a source of true random number sequence.

As a result, the page can be represented as shown in Fig. 29 (the heatmap shows Ψ scores for each bit b_j within page p_i).

Note: The Ψ scores should be stored offline to the array A containing $P \times B$ elements.

2.2.2 Generation

1. Determine size L of a register R_{TRNG} to store a random number.
2. Store the information about noisy bits from A with the highest Ψ scores to the special data structure shown in Fig. 14.

   Bit $p'_k : b'_l$ $(0 \le k \le K-1, 0 \le l \le L-1)$ corresponds to a Ψ score $A[i][j]$ of some bit b_j from page p_i. K is the number of pages chosen for random number generation.
3. Initialize index $k = 0$ for cyclic iteration.
4. Read page p'_k.

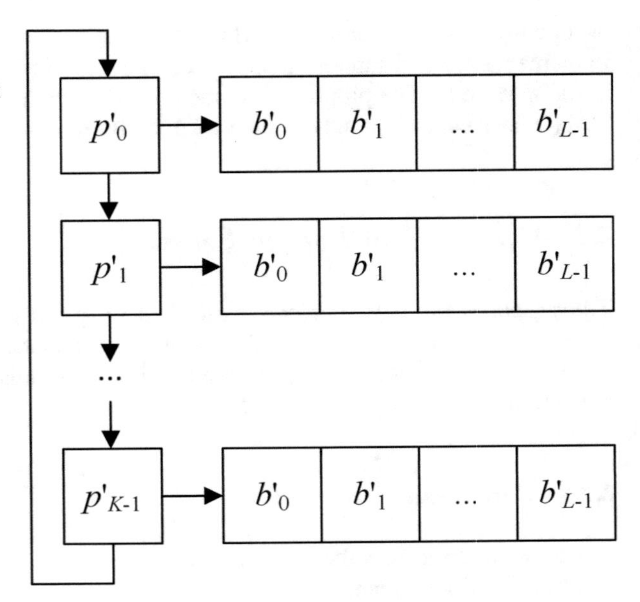

Fig. 14 The data structure for storing noisy bits for different pages within a block

5. Extract L bits $p'_k{:}b'_0 \ldots p'_k{:}b'_{L-1}$ and store them to the R_{TRNG} register.
6. Increment k by modulo K. Go to Step 4.

2.3 Experimental Results

SK hynix S72 512GB SSD drives have been tested in order to prove randomness of the proposed PUF design.

2.3.1 Enrollment

1–3. Block $0x84$ has been randomly chosen, erased, and written with zeros in the raw mode.

4. Read operation has been repeated in the raw mode for $R = 1000$ times.
5. For example, the randomness of each bit can be estimated as follows:

   Calculate two metrics for each bit, namely, uniformity (U) and bit flipping rate (BFR):

$$U = 1 - 2 \times |\frac{R_1}{R} - 0.5| \tag{1}$$

   R_1 is the number of bits with the value of "1".

For example, if there were 5 read operations and the values obtained were (1, 1, 0, 1, 0), then $U = 1–2 \times |3/5–0.5| = 1–2 \times 0.1 = 0.8$:

$$BFR = \frac{\sum_{i=0}^{B-2} b_i \oplus b_{i+1}}{B-1} \tag{2}$$

Based on U and BFR noise characteristic, Ψ can be calculated for each bit as follows:

$$\Psi = \alpha \times U + \beta \times BFR \tag{3}$$

α, β—tunable parameters which determine the importance of either uniformity or the bit flipping rate.

The example is summarized in Table 1.

Thus, increasing the importance of uniqueness sequence (0, 0, 0, 1, 1, 1) can be considered more random than (1, 1, 0, 1, 0, 1). However, usually, BFR is more important and correlated with uniqueness. Therefore, the third case is more realistic.

6. Array A has been computed based on the information obtained in Step 5.

   For example, a page with index $0x42$ has been chosen to demonstrate the uniqueness [24] of the noisy bit locations. Figure 30 shows the Ψ scores for the pages with index $0x42$ within block $0x84$ for different SSD samples.

2.3.2 Generation

1. RTRNG size is set to $L = 32$.
2. To estimate the number of noisy bits per page, all data has been aggregated, and average Hamming distances between reads for all pages have been computed.

   The graph for the chosen block is shown in Fig. 31.

   As shown in Fig. 31, different pages have various Hamming distances (HD) between reads. The value of HD shows the number of noisy bits per page. Therefore, pages with a bigger value of HD are to be stored in the data structure.

   For example, pages $0x120$ and $0x1c5$ have the highest HD among all pages (see Fig. 31). The data structure containing these pages is shown in Fig. 15.
3. $k = 0$ $(K = 2)$.
4. Read $p'_0 = 0x120$.

Table 1 Example of tuning α, β

Sequence	U	BFR	$\Psi, \alpha = 1, \beta = 1$	$\Psi, \alpha = 1, \beta = 0.1$	$\Psi, \alpha = 0.1, \beta = 1$
1 1 0 1 0 1	0.66	0.8	1.44	0.74	0.866
0 0 0 1 1 1	0.8	0.2	1.0	0.82	0.28

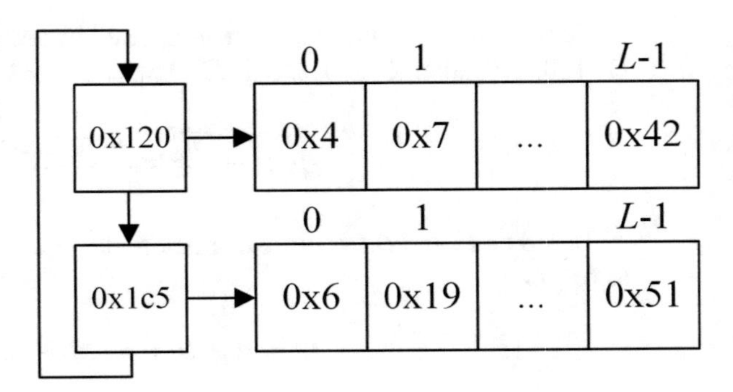

Fig. 15 Example of the data structure for noisy bits

5. $L = 32$ bits are extracted from the page p'_0 on the positions $0x4, 0x7, \ldots, 0x42$. $R_{TRNG} = (1, 0, \ldots, 1)$.
6. $k = 1$.
4. Read $p'_1 = 0x1c5$.
5. $L = 32$ bits are extracted from the page p'_1 on the positions $0x6, 0x19, \ldots, 0x51$. $R_{TRNG} = (1, 1, \ldots, 0)$.
6. $k = 0$.
4. Read $p'_0 = 0x120$.
5. $L = 32$ bits are extracted from the page p'_1 on the positions $0x4$, $0x7$, ..., $0x42$. $R_{TRNG} = (\mathbf{0}, \mathbf{1}, \ldots, 1)$.
6. $k = 0$.

The sequence of 800,000 bits has been obtained from SK hynix S72 SSD sample. The generated sequence contains 400188 zeros (50.02%) and 398812 ones (49.98%). The experiment confirmed the hypothesis of uniform distribution of noisy bits in TLC NAND.

2.4 Conclusion

The TLC NAND structure can be successfully utilized to extract uniqueness from the memory device. Existing NAND-based storage is quite unreliable for the write and read operations conducted without scrambling and ECC. Therefore, this disadvantage can be used to generate a true random number sequence. The proposed method is based on physical unclonable function (PUF) which is implemented using existing firmware functions.

The presented entropy source design has the following advantages:

- It does not require additional circuitry (hardware overhead) for its implementation.
- It cannot be reproduced on the different instance of the same device even knowing its configuration.
- It can be reconfigured using parameters L and K.

- Ψ metric can be tuned for particular requirements.
- It can generate true random numbers required for security protocols implementation using only firmware functions.

Thus, the proposed PUF-based entropy source can be utilized to enhance the security of the memory device without additional hardware cost and using only internal firmware commands.

3 Flash Memory Device Identification Based on Physical Unclonable Functions

3.1 Introduction

The memory cells of NAND flash devices have quite a low reliability, which leads to using error correction codes (ECC) with high correcting capability, e.g., BCH or LDPC code, in the data path [25]. On the other hand, excluding ECC from the data path creates a possibility of generating unique and unpredictable bits from the NAND memory cells. Thus, comparing the number of bits with one value between different pages is proposed as a source of unique and unpredictable identifiers.

The proposed ID generation method is based on the read operations, which bypass ECC and scrambling in the data path (raw read operations). The first stage (enrollment) includes erasing a block of NAND flash memory and writing an all-zero pattern to all pages within the block. Then, during multiple raw read operations, each page is characterized by an average number of ones obtained during the read operations. The second stage (uniqueness extraction) is using page statistics computed during enrollment to generate a sequence of page addresses (the number of pages is equal to the doubled ID length). Then, during the final third stage (ID generation), comparing the number of ones from the chosen pages allows generating unique ID bits, i.e., for two compared pages, if the first page has less ones than the second one during the raw read operation, zero is generated; otherwise, one is generated.

3.2 ID Generation Algorithm

A page is a minimal reading unit in the NAND flash memory, and it can be characterized by a number of bits which flip their values during the read operation. To easier highlight flipping bits, an all-zero pattern should be programmed in the page. Then, after multiple raw read operations (bypassing ECC and scrambling), the average number of ones obtained during the read operation can characterize the page. These statistics are obtained during the enrollment stage, which contains four steps:

1. Erase a block of memory.
2. Program in raw mode an all-zero pattern to all pages of a block.
3. Read in the raw mode each page N_r times.
4. Compute the average number of ones during N_r raw read operations.

For example, statistics for two blocks of memory with randomly chosen addresses 0xBE0 and 0x2F0 is shown in Fig. 16 ($N_r = 100$).

The distribution of the average number of ones in pages p_i^{avg} ($1 \leq i \leq N_p$, N_p– the number of pages in a block of memory) is unique for every block in the device. Therefore, the subtle intrinsic difference in this distribution can be utilized to design a NAND flash memory-based physical unclonable function. The block diagram for a proposed PUF design for ID generation is shown in Fig. 17, which has a similar principle as RO-PUF [17]. Instead of frequency comparison, the proposed algorithm compares the number of ones during raw read operations.

To generate a single response bit R, it is required to compare the number of ones during the raw read operation from two different pages p_i and p_j ($i \neq j$, $1 \leq i, j \leq N_p$), which are chosen based on challenge value $C = (i, j)$. C is an ordered pair of page addresses i and j which takes one of possible $\binom{N_P}{2}$ values. If $p_i < p_j$, $R = 0$; otherwise, $R = 1$. This PUF is able to generate $\binom{N_P}{2}$ possible response bits based on challenge value C. To generate an L-bit ID, identification server has to generate L challenges ($2L$ page addresses) and send them to the device. As a result, the device produces L response bits, which uniquely identify it.

Due to intrinsic NAND instability, values p_i and p_j may have different values from one read operation to another. This leads to instability of generated response

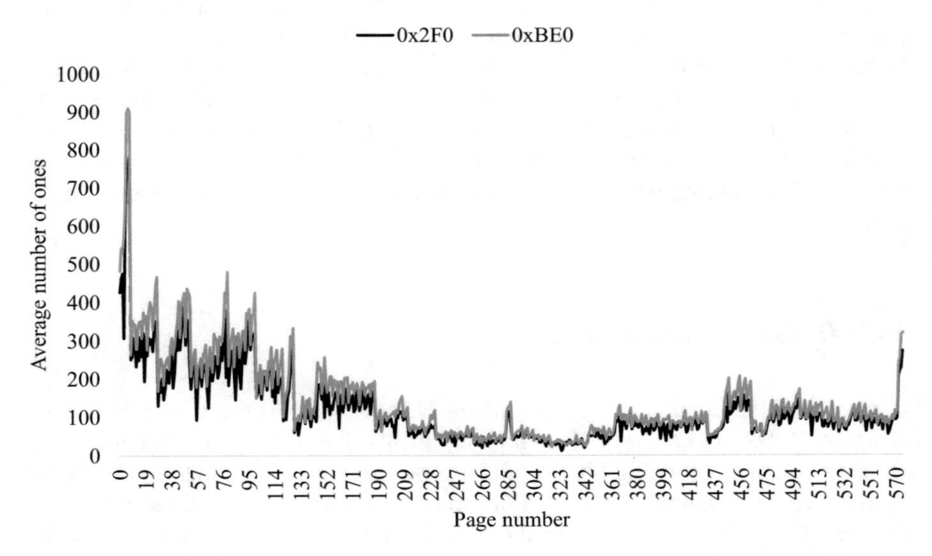

Fig. 16 The average number of ones obtained during raw read operations for two blocks of memory

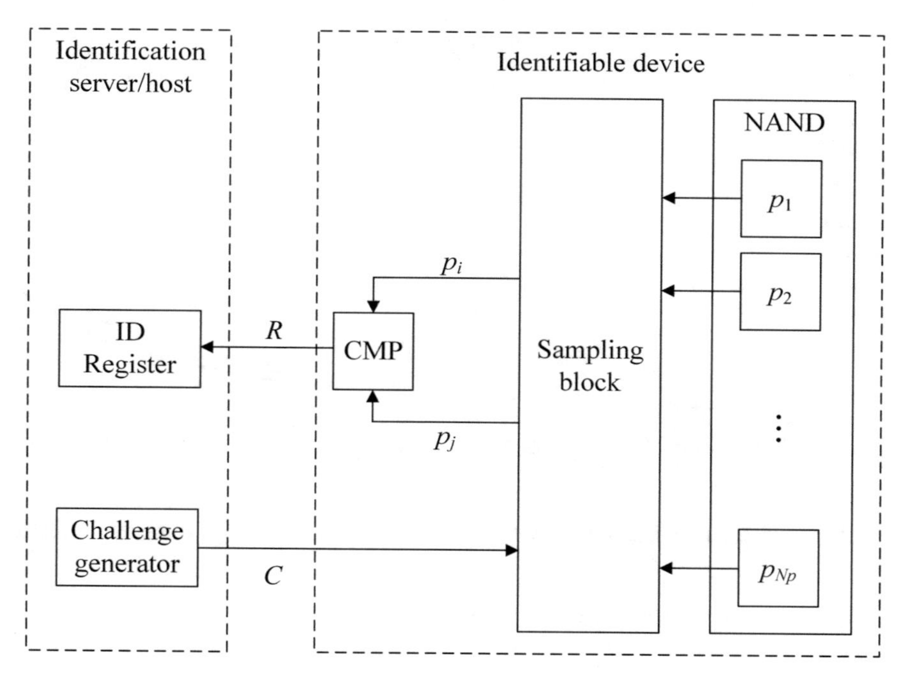

Fig. 17 ID generation based on NAND PUF

values R as during different read operations for the same address values i and j, the order of values p_i and p_j can be also different ($p_i < p_j$ or $p_i > p_j$). Thus, to provide a reliable identification, the subset of challenge values, which provide stable responses, has to be found.

If the average number of ones values obtained during the enrollment stage (see Fig. 16) are sorted, they can be separated into two groups, with lower and higher values of p_i^{avg}. Sorted values are shown in Fig. 18.

As shown in Fig. 18, the higher the difference between the average number of ones obtained for two pages p_i^{avg} and p_j^{avg} (e.g., $p_i^{avg} > p_j^{avg}$, $i \neq j$), the higher the probability to keep the order between the number of ones obtained during an arbitrary read operation ($p_i > p_j$). It also can be confirmed based on experimental data obtained from block 0x2F0. The data of 10-th and 100-th reads together with average values is shown in Fig. 19.

The value of the difference between two pages p_i (taken from pages with higher p_i^{avg} values) and p_j (taken from pages with lower p_i^{avg} values) ($p_i - p_j$) may change its value, but sign value ($p_i - p_j$) will be the same with a high probability for all read operations from 1 to at least 100. Therefore, to generate L-bit identifier, L challenges $C_k = (i, j)(1 \leq k \leq L)$ should be chosen based on enrollment data. There are multiple ways of doing this. For example, it can be done as shown in Fig. 32 in four steps:

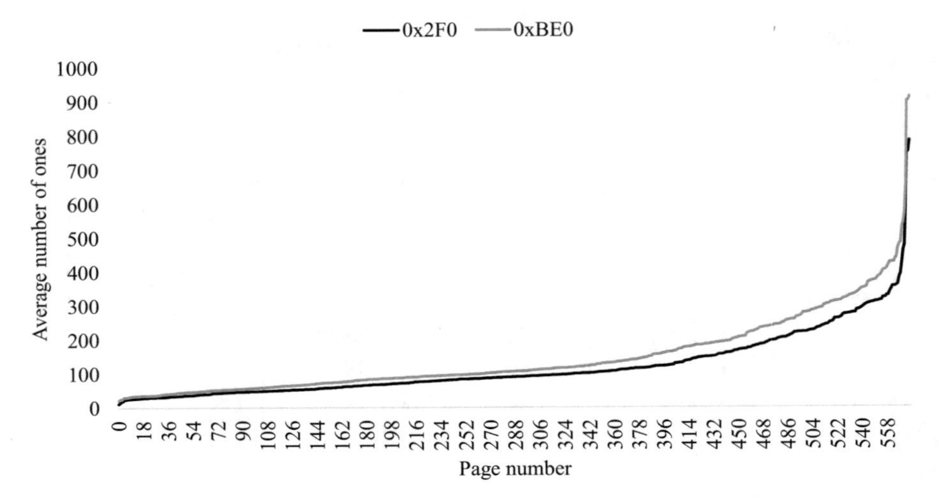

Fig. 18 The average number of ones (sorted) obtained during raw read operations for a block of memory

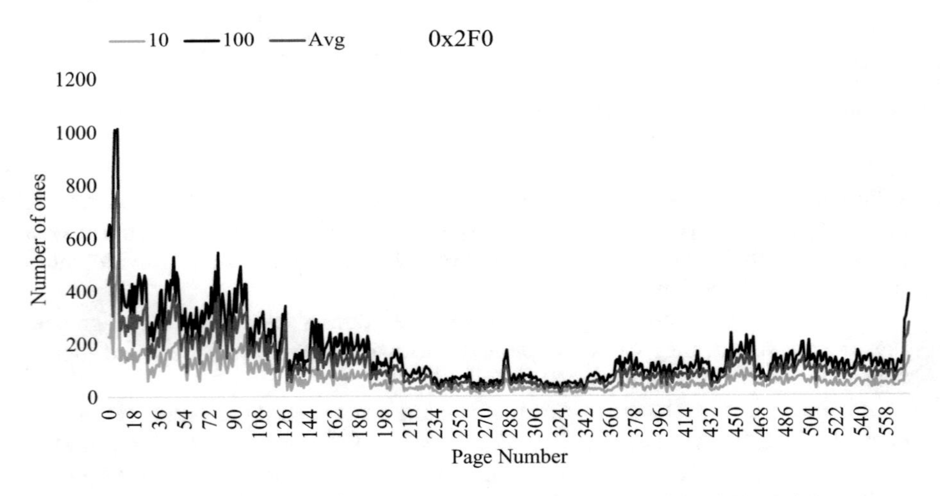

Fig. 19 The number of ones obtained during 10-th, 100-th raw read operations, and the average value

1. Sort pages by the average number of ones values obtained during raw read operations (p_i^{avg}) in ascending order. As a result, the sequence of page addresses corresponding to the sorted values can be represented as $A_1, A_2, \ldots, A_{N_p}$;
2. Split sequence into two L-element subsequences, namely, $A_{\text{low}} = (A_1, A_2, \ldots, A_L)$ with a lower value of p_i^{avg} and $A_{\text{high}} = (A_{N_p-L+1}, A_{N_p-L+2}, \ldots, A_{N_p})$ with a higher value of p_i^{avg};

3. To generate k-th bit of identifier, form an unordered pair of addresses $\{A_k, A_{N_p-L+1+k}\}$ $(1 \le k \le L < N_p)$, A_k is in A_{low} and $A_{N_p-L+1+k}$ is in A_{high}. If A_k and $A_{N_p-L+1+k}$ are chosen from A_{low} and A_{high} correspondingly, there is a high probability that $p_{A_k} < p_{A_{N_p-L+1+k}}$. Therefore, the unordered pair should be converted to the ordered pair (challenge value C_k) by some unique characteristics.
4. Each unordered pair $\{A_k, A_{N_p-L+1+k}\}$ can be converted to the challenge value $C_k = (A_k, A_{N_p-L+1+k})$ or $C_k = (A_{N_p-L+1+k}, A_k)$. This can be done based on the unique sequences of addresses in A_{low}:
   (a) Consider k-th element of $A_{\text{low}}(A_k)$ and the next one (A_{k+1}).
   (b) If $A_k < A_{k+1}$, unordered pair $\{A_k, A_{N_p-L+1+k}\}$ is converted to $C_k = (A_k, A_{N_p-L+1+k})$.
   (c) Otherwise, unordered pair $\{A_k, A_{N_p-L+1+k}\}$ is converted to $C_k = (A_{N_p-L+1+k}, A_k)$.
   (d) If $k = L$, A_{L+1} element is taken from a full sequence of sorted values.

The algorithm above is given for an exemplary purpose and can be changed to other ones in order to choose the most stable responses.

The final stage (ID generation) is to perform a raw read operation L times from two pages each time. To generate k-th bit, values p_{A_k} and $p_{A_{N_p-L+1+k}}$ are compared. If the pair of addresses is $(A_k, A_{N_p-L+1+k})$ in most cases, 0 value will be generated. If the pair of addresses is $(A_{N_p-L+1+k}, A_k)$ in the most cases, 1 value will be generated.

As a result, L-bit identifier can be generated using $2L$ raw read operations. The set of challenges $C_k = (A_k, A_{N_p-L+1+k})$ or $C_k = (A_{N_p-L+1+k}, A_k)$ can be either stored in the device memory for better reliability or generated by choosing L pairs from possible $\binom{N_P}{2}$ options.

3.3 Example of ID Generation

The results of the enrollment stage are shown in Fig. 16 for block 0x2F0. The uniqueness extraction stage is completed as follows:

1. The list of page addresses sorted by p_i^{avg} values is formed as follows:
   324, 325, 266, ..., 1, 5, 7 (576 addresses in total);
2. To generate $L = 128$ bit identifier, the sequence can be split into two groups:
   $A_{\text{low}} = (A_1, A_2, A_3, \ldots, A_{126}, A_{127}, A_{128}) = (324, 325, 266, \ldots, 254, 301, 242)$—128 addresses;
   $A_{\text{high}} = (A_{449}, A_{450}, A_{451}, \ldots, A_{574}, A_{575}, A_{576}) = (30, 159, 179, \ldots, 1, 5, 7)$—128 addresses;

3–4. These groups are merged into the sequence:

- The unordered pair $\{A_1, A_{449}\} = \{324, 30\}$ is converted to $C_1 = (A_1, A_{449}) = (324, 30)$ as $A_1 < A_2(324 < 325)$;
- The unordered pair $\{A_2, A_{450}\} = \{325, 159\}$ is converted to $C_2 = (A_{450}, A_2) = (159, 325)$ as $A_2 > A_3(325 > 266)$;
- The unordered pair $\{A_3, A_{451}\} = \{266, 179\}$ is converted to $C_3 = (A_3, A_{451}) = (266, 179)$ as $A_3 < A_4(266 < 314)$;
- ...
- The unordered pair $\{A_{126}, A_{574}\} = \{254, 1\}$ is converted to $C_{126} = (A_{126}, A_{574}) = (254, 1)$ as $A_{126} < A_{127}(254 < 301)$;
- The unordered pair $\{A_{127}, A_{575}\} = \{301, 5\}$ is converted to $C_{127} = (A_{575}, A_{127}) = (5, 301)$ as $A_{127} > A_{128}(301 > 242)$;
- The unordered pair $\{A_{128}, A_{576}\} = \{242, 7\}$ is converted to $C_{128} = (A_{128}, A_{576}) = (242, 7)$ as $A_{128} < A_{129}(242 < 110)$.

ID generation stage is based on the sequence generated during the second stage:

- $\text{ID}_1 = 0$ as $p_{324} < p_{30}$;
- $\text{ID}_2 = 1$ as $p_{159} > p_{325}$;
- $\text{ID}_3 = 0$ as $p_{266} < p_{179}$;
- ...
- $\text{ID}_{126} = 0$ as $p_{254} < p_1$;
- $\text{ID}_{127} = 1$ as $p_5 > p_{301}$;
- $\text{ID}_{128} = 0$ as $p_{242} < p_7$.

3.4 *Experimental Results*

3.4.1 Reliability

The 128-bit IDs were generated from two different samples (10 blocks each with the same addresses)—total 20 IDs.

Reliability shows how stable is generated ID during T tests (repeated generations) [26]. It can be computed as follows (HD, Hamming distance; ID_t, ID generated during t-th test):

$$R = 1 - \text{BER} = 1 - \frac{1}{T}\sum_{t=1}^{T} \text{HD}(\text{ID}, \text{ID}_t) \tag{4}$$

The ideal value of reliability is 1.0, i.e., that generated ID is stable and does not change its value during repeated generations.

All IDs generated in the experiment have $R = 1.0$ except three of them which have 0.980, 0.989, and 0.990.

3.4.2 Uniqueness

Uniqueness shows the difference between IDs generated from different samples (inter-die uniqueness) or different blocks within the same sample (intra-die uniqueness) [26]. The ideal value of uniqueness is 0.5 which is the biggest distance from both 0 (no difference) and 1 (each bit of the vector is flipped).

Intra-die uniqueness for m IDs can be computed as follows:

$$U_{\text{intra}} = \frac{2}{m(m-1)} \sum_{u=1}^{m-1} \sum_{v=u+1}^{m} \text{HD}(\text{ID}_u, \text{ID}_v) \tag{5}$$

For $m = 10$ IDs (each sample) $U_{\text{intra}} = 0.502$ for sample 1 and $U_{\text{intra}} = 0.498$ for sample 2.

Inter-die uniqueness for m IDs situated at the same address in different two samples can be computed as follows:

$$U_{\text{inter}} = \frac{1}{m} \sum_{i=1}^{m} \text{HD}(\text{ID}_i^1, \text{ID}_i^2) \tag{6}$$

$U_{\text{inter}} = 0.518$ for two identical samples ($m = 10$ for each sample).

Also, the algorithm has been stress tested by 10,000 erases. The ID was generated after each erase for five times. Therefore, 50,000 IDs were generated during the test. Only 16 of them had single bit flip, and the rest 49,984 were the same (without bit flips).

Thus, the proposed algorithm can be used to generate a unique, reliable, unpredictable, and unclonable ID for flash memory devices.

3.5 *Conclusion*

This section describes the method of generating stable unique ID based on NAND flash memory. Produced IDs have high reliability (0.99) and uniqueness (0.502) and also survived after erase stress testing without losing their characteristics. The proposed method does not require additional hardware overhead in devices having onboard flash memory. One block of memory provides more than 500 unique IDs.

4 Design of Data Scrambler with Enhanced Physical Security

4.1 Introduction

Modern NAND flash memory devices [47] usually contain three parts, namely, host, controller, and NAND memory cell array as shown in Fig. 20. The host usually communicates with the device using high-speed interface and generates workload

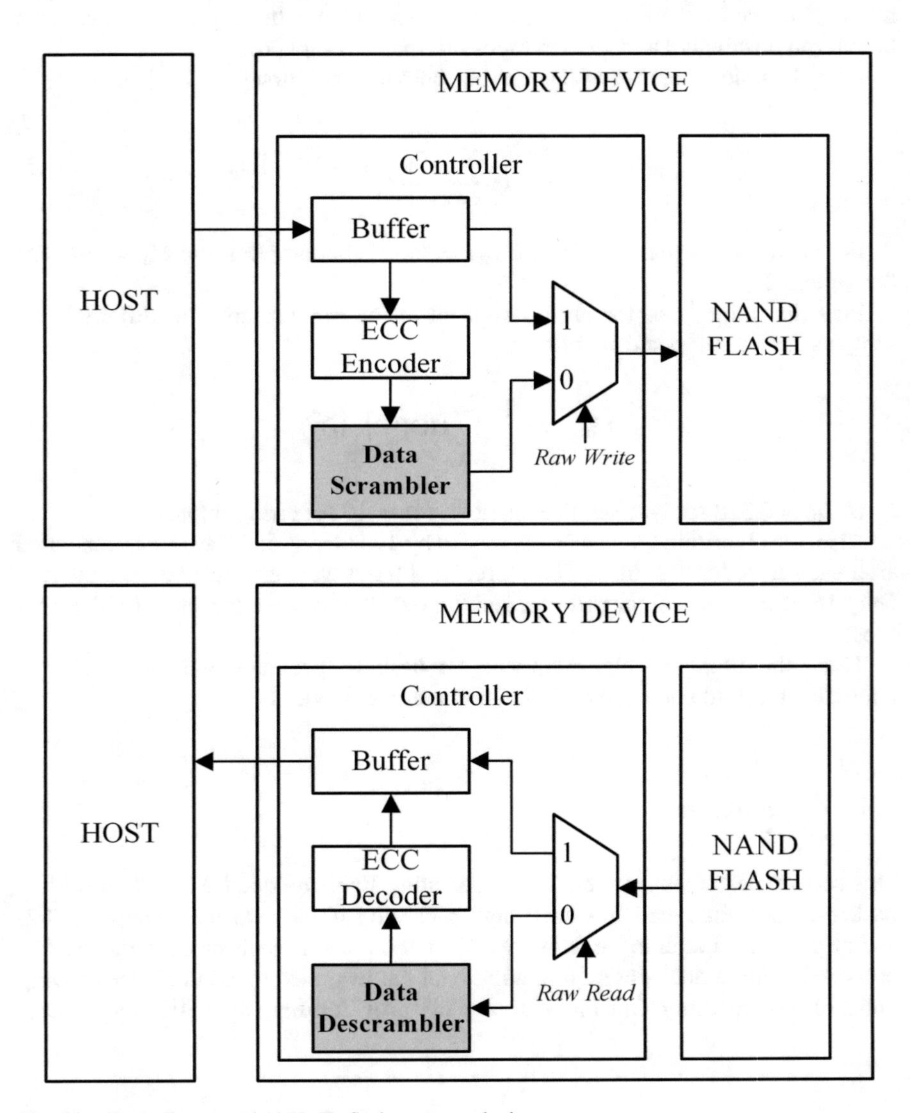

Fig. 20 Block diagram of a NAND flash memory device

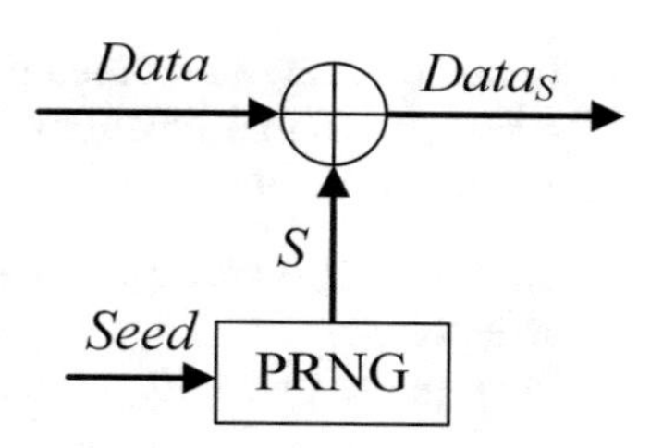

Fig. 21 Typical design of a data scrambler

for the controller. The data from the host is stored in buffer (usually DRAM) and then encoded by error correction codes (ECC). The encoding is required because basic reliability of NAND memory cells is quite low and this kind of memory introduces multiple errors during read and write operations [27].

One of the important blocks in NAND flash memory device is a hardware implementation of a scrambler (randomizer) which improves the reliability of memory cells (see Fig. 20). This can be achieved by transforming data patterns sent from the host to the uniformly distributed data [28]. The typical block structure of a data scrambler is shown in Fig. 21.

The scrambler usually contains a pseudorandom number generator (PRNG) block which is usually seeded by some value (e.g., logical block address (LBA) or a physical page number (PPN)). This block generates a uniformly distributed sequence S which is XORed with $data$ sent from the host. As a result, $Data_S =$ $Data$ XOR S is programmed to NAND memory cells.

This way of data scrambling has a vulnerability which gives an attacker a chance to degrade the reliability of NAND memory cells [29]. Since scrambler processes the data using a pseudorandom number sequence, the attacker can collect enough outputs ($Data_S$) and restore the configuration data of the PRNG block (e.g., polynomial coefficients) [30]. Then, the attacker is able to build a mathematical model of a scrambler and obtain output values for any input data patterns.

For example, if an attacker wants to program some particular data pattern (D_p) to NAND, he/she processes this sequence using the mathematical model to obtain $D_x = (D_p$ XOR $S)$ value and sends the output D_x to the device. As a result, $Data_S = (D_p$ XOR $S)$ XOR $S = D_p$ will be programmed to the memory device. Thus, the attacker is able to get any data patterns (worst-case data patterns, e.g., all zeros) in order to degrade NAND reliability. Since many memory devices are manufactured with the same circuit design, the attacker can take advantage of using the same mathematical model of a scrambler (obtained from a single device) to degrade reliability of other devices.

The reliability of the NAND memory cells can be also degraded using the same data pattern programming [29]. For example, if same data pattern ($Data$) is sent from the host to the same LBA or PPN (the same seed value for PRNG) multiple times, it is transformed to the same data pattern on NAND ($Data_S$). As a result, memory cells are programmed with the same value, and this leads to increasing of bit error rate (BER).

In this chapter, a modified design of the data scrambler is proposed. The use of physical unclonable functions (PUFs) [2] as an additional data processing before scrambling provides a way to:

1. Significantly decrease vulnerability to building a mathematical model of a scrambler.
2. Encrypt the data without hardware costly algorithms (e.g., AES), which are not used in mobile flash and IoT (Internet of Things) devices [49].
3. Increase the reliability of the NAND by avoiding programming the same data patterns [29].
4. The use of PUF as an additional block for scrambler data encryption provides additional security against cold-boot attacks [31] as PUF response for the same challenge changes its value after each restart.

The proposed design of a data scrambler is based on adding PUF circuit to the data path of a flash memory device. This provides enhanced security to the existing scrambler design as it encrypts the data using unique PUF-generated key. It also requires much smaller hardware overhead comparing to the classical encryption algorithms (e.g., AES). Since PUF adds unique signature to the data, it becomes much harder for an attacker to mathematically model scrambler and send worst-case data patterns, which degrade the reliability of NAND memory cells. Furthermore, even if the attacker managed to know the configuration of a PRNG block for a single device, it does not give him/her the advantage for the other devices as PUF responses are unique for every device. The presented solution has two possible options of implementing the PUF:

1. Implementation of a PUF remains noisy which does not require hardware for stabilization. However, NAND ECC engine has to be strengthened in order to provide correction capability for errors brought by both NAND memory cells and PUF response.
2. Design two separate ECC engines, a stronger one for NAND errors and a weaker one for correcting errors added to data by PUF. According to experimental data, the first option requires more hardware for implementation because it utilizes NAND ECC engine with bigger correction capability.

4.2 Proposed Scrambler Circuit Operation

The usual data path of NAND flash memory device consists of ECC encoder (decoder) and scrambler, which can be placed in different order (ECC before scrambler and vice versa) depending on the design of a memory controller. Without loss of generality, consider block design of a data path shown in Fig. 22.

In this case, ECC encoder is located before scrambler. Also PUF component is added to the data path in order to provide lightweight encryption for user data. PUF block is seeded by the same value as scrambler and generates a signature R which

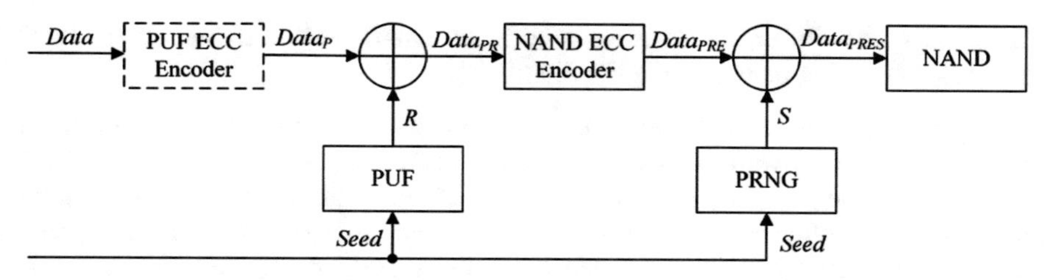

Fig. 22 Block diagram of the write data path including proposed scrambler design

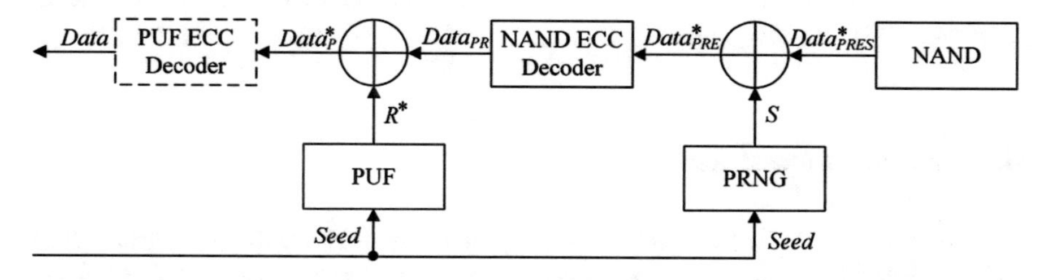

Fig. 23 Block diagram of the read data path including proposed scrambler design

is unique for every memory device even if it has completely same design. Since PUF output R can be noisy, optional small ECC engine (PUF ECC encoder) is added to the design. This engine encodes host $Data$ and converts it to $Data_P$. The PUF output R is XORed with encoded $Data_P$; encrypted data $Data_{PR}$ is further processed by NAND ECC encoder block in order to get a code word $Data_{PRE}$. The encoded and encrypted data is scrambled in a standard way by XORing with PRNG-generated value S as shown in Fig. 22. As a result, scrambled data $Data_{PRES}$ is to be programmed to the NAND. PUF ECC encoder is optional block which is used to protect data from PUF errors without modification of NAND ECC engine.

Decoding process is similar to the previously shown encoding (see Fig. 22) but performed in the opposite order. The decoding scheme is shown in Fig. 23.

First, scrambled data $Data^*_{PRES} \neq Data_{PRES}$ is read with errors as NAND-based storage usually produces multiple errors during read operation. Then, $Data^*_{PRES}$ is descrambled using value S generated by PRNG as $Data^*_{PRE}$. NAND ECC decoder corrects errors in $Data^*_{PRE}$ and produces $Data_{PR}$. Then, $Data_{PR}$ is decrypted to $Data^*_P$ using $R^* \neq R$ value produced by PUF block. As a result, $Data^*_P$ will be corrupted by noise from PUF which is basically not stable. Therefore, data sent to host is to be corrected by PUF ECC decoder as $Data$. In case of omitting PUF ECC decoder block, NAND ECC Decoder should be placed after XORing with PUF response as it has to correct both PUF and NAND noise.

Since the basic structure of the PUF can add errors to the data during decoding stage, the capability of ECC should be enlarged. This can be done using two techniques:

1. Enlarge the correcting capability of the NAND ECC engine.

2. Correct data after PUF using additional small ECC engine (PUF ECC decoder) [32] or enhancing PUF reliability [2].

Both techniques require additional hardware overhead for correcting unstable PUF outputs. This overhead is smaller than utilization of cryptographic algorithms (e.g., AES). The proposed design also decreases vulnerability to the same pattern programming [29] (because PUF response R is not stable) and to changing data pattern for every write operation. However, the presented implementation of the scrambler is still vulnerable to machine learning modeling attacks. For example, this issue can be addressed by adding obfuscating techniques to the challenge generator [33].

4.3 Experimental Results

Assume that NAND ECC engine can be implemented as BCH code, additional PUF ECC as Reed-Solomon code [34], and hardware overhead is estimated as FPGA LUT and flip-flop units. Host transmits 1023 bits of data and PUF also generates 1023-bit response:

1. PUF is noisy, and BER (bit error rate) is 0.01, i.e., that PUF generates around 11 errors in 1023-bit response.
2. NAND produces maximum 70 errors, and this can be corrected with BCH $[n = 1023, k = 323, t = 70]$ code.
3. NAND ECC overhead for this implementation consists of 5441 flip-flop and 17413 LUT blocks (Xilinx Artix-7 FPGA [35]).

4.3.1 Option 1

Since PUF response should not be corrected, NAND ECC correction capability should be increased to $t = 81 = 70 + 11$. As a result, BCH $[n = 1023, k = 213, t = 81]$ is to be implemented instead. Final hardware overhead for new ECC engine is 6512 flip-flop and 20840 LUT blocks. Therefore, additional hardware cost for PUF correction will be around 19.7%. However, the proposed approach can be used to improve reliability against same data pattern issue because PUF response is unpredictable.

4.3.2 Option 2

To correct errors brought by PUF responses, smaller PUF ECC engine (e.g., Reed-Solomon $[n = 1023, k = 1002, t = 11]$) is to be implemented. Therefore, it will require additional 624 flip-flop and 672 LUT blocks, which is less than 11%

of additional hardware cost. Furthermore, this approach includes additional latency overhead for PUF noise correction.

The estimation of hardware overhead is done in one of the possible ways (FPGA). It is not restricted to other technologies of scrambler implementation (e.g., ASIC).

Real implementation of a scrambler should be a trade-off between Option 1 and Option 2 in terms of hardware overhead and performance. Thus, the decision on a final implementation can be made based on constraints of a particular NAND flash memory device. Despite additional hardware cost, security and reliability enhancements are the benefits of implementation scrambler in the proposed way.

4.4 Conclusion

This section presents a new approach to designing a scrambler in NAND flash memory devices. The proposed design enhances physical security of data stored in a flash memory device and also provides better reliability comparing to the existing approaches. Scrambling algorithm has been implemented in Xilinx Artix-7 FPGA in order to compare with existing encryption schemes as AES which is usually not used in mobile NAND flash and/or IoT devices. In terms of hardware overhead, this approach is at least three times more efficient than existing encryption engines.

5 Physical Unclonable Function-Based Error Detection Algorithm for Data Integrity

5.1 Introduction

Data is usually stored in computer memory using many different representations (e.g., binary numbers, strings, compressed formats, etc.). The attribute-value pair format can be distinguished among the existing ones as it is widely used to represent data (e.g., header, email, query; string, URL; metadata, data, database entries, Internet messages, JSON objects, etc. [36]). Due to limitations of available memory space, some of these pairs can be stored externally on another device. For example, the general scheme of transmitting attribute-value pairs to the untrusted party is shown in Fig. 24. The memory controller extracts the data and generates an attribute (X) and value (Y) pair. This pair is further encoded by error correction codes (ECC) in order to avoid data losses during transmission. As a result, the encoder generates the value of X_e and Y_e and sends it via an untrusted channel to the untrusted party which stores the pair (X_e, Y_e) until requested by the device. The data should be sent back to the device and decoded to the original attribute-value pair ($X_d = X$, $Y_d = Y$).

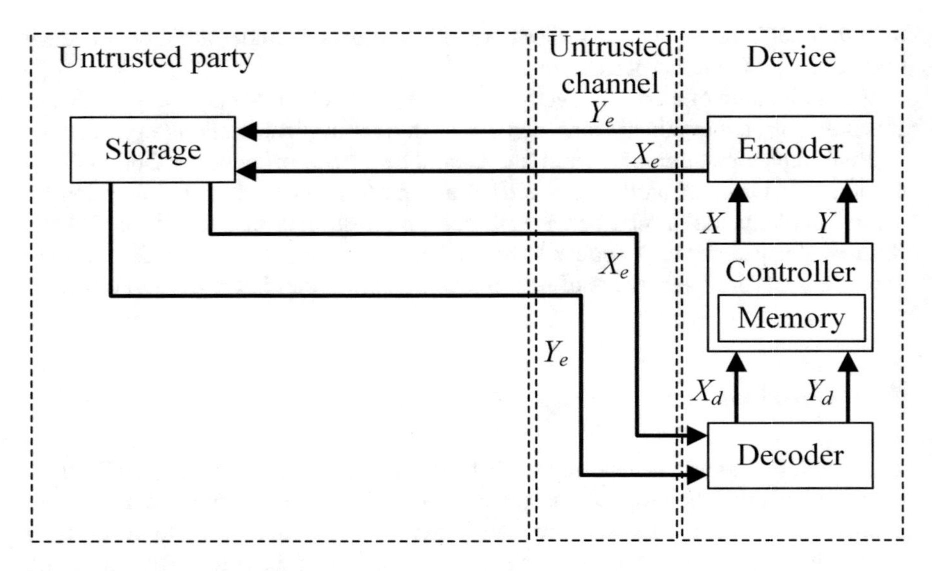

Fig. 24 General structure of data transmitting

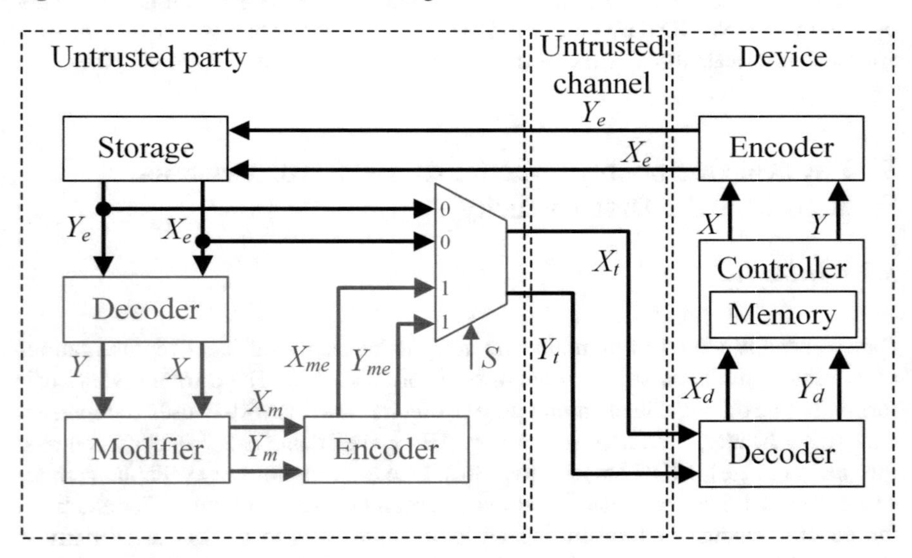

Fig. 25 Structure of untrusted party for the attack

However, since the data is stored on the untrusted party side, an attacker can observe and modify both the untrusted party and the channel [37]. Figure 25 shows one of the possible attack scenario implementations.

Since the ECC engine is used to encode the data from the device, it is possible to clone decoder and encoder blocks on the untrusted party side. Therefore, the original attribute-value pair (X, Y) can be modified by an attacker in order to reveal the information or modify and send it back to the device to degrade performance or data integrity. The untrusted party can operate in two modes, namely, ordinary mode ($S =$ "0") when data is not modified and attack mode ($S =$ "1") when pair (X, Y) is transformed to (X_m, Y_m) and encoded to (X_{me}, Y_{me}), which is sent back to the device. Thus, if $S =$ "1", a pair (X_t, Y_t) is decoded to the $(X_d, Y_d) \neq (X, Y)$.

This way of data transmitting causes the following problems:

1. The attacker has access to the data sent via an untrusted channel as he can decode it knowing the ECC algorithm.
2. The attacker also can modify X or Y value or both values at a time in order to modify critical data on the device, degrade performance, corrupt the data transmitted, etc.
3. Encryption can prevent these problems, but it usually requires significant memory and hardware resources to be utilized as a part of the controller.

For the problem described above, physical unclonable functions (PUFs) [2] can be efficiently utilized to protect the data against unauthorized modification and prove that pair (X, Y) is generated by a particular device. PUF is a hardware security primitive which maps external input (challenge) into an output (response). This mapping is unique, unpredictable, and unclonable for the particular chip which has a PUF instance. In addition to hashing capability, PUF also extracts unique intrinsic features of an integrated circuit. This property is used for making pair (X, Y) protected against illegal access and modification.

The proposed method is based on using two PUF instances implemented on the same circuit. The first PUF is utilized to generate a hash value (R_x) for the attribute value (X) in order to use it as a key for masking linked value (Y). The encryption process can be as complicated as possible, but for simplicity and for the sake of hardware overhead reduction, the generated hash value R_x can be simply XORed with Y. The result of encryption (Y^*) is further hashed by second PUF instance, and the response of the PUF is used to check whether the unique pair (X, Y) is generated on a particular device. The PUFs utilized in this chapter should be stable (Reliability value $\approx$1.0) and strong (the number of challenge-response pairs should be exponentially large). For example, Arbiter PUF design with enhanced reliability is a good candidate for the proposed method [38]. Using PUF for data integrity is beneficial for the following reasons:

1. The attacker is not able to reproduce hash values generated by PUFs as he doesn't have access to the internals of the original device.
2. The generated response values can be used to check whether the pair (X, Y) is generated by a particular device or never existed before.
3. The proposed algorithm is more hardware-efficient than the existing encryption engines in terms of utilized chip area and power consumption.
4. Furthermore, encoding pair (X, Y) using PUF instances also allows detecting errors even if they were not injected by an attacker. So it can be also utilized instead of error detection engines.

Points 2 and 4 provide data integrity based on PUF usage for both errors brought by an attacker and errors caused by the noise in the channel and untrusted party.

5.2 Proposed Data Path Design

The proposed algorithm can be implemented by three modifications of the scheme shown in Figs. 24 and 25.

A modified encoder is shown in Fig. 26.

In order to obfuscate the value of Y and the explicit connection between X and Y, the following steps are to be completed:

1. The value of Y should be obfuscated using cryptographic salt value S produced by salt generator. The generator can be implemented as a PUF or pseudorandom number generator (PRNG). As a result, the value of Y_s is obtained as an XOR operation of Y and S ($Y_s = Y$ XOR S).
2. Obtain hash value $R_X = \text{PUF}_0(X)$ (the response of PUF_0 on challenge X).
3. Encrypt the value of Y_s by XORing it with hash value R_X ($Y^* = Y_s$ XOR R_X).
4. Obtain a hash value $R_c = \text{PUF}_1(Y_s)$. This value is used to prove that pair (X, Y) is generated on this device.
5. The values (X, Y^*, R_c) should be encoded by the same ECC engine as used in Figs. 24 and 25 in order to obtain values (X_e, Y_e^*, R_{ce}).
6. The values (X_e, Y_e^*, R_{ce}) are to be sent via an untrusted channel to the untrusted party. Thus, the code word is changed by adding extra hash value R_{ce}.

The enhanced encoder requires two strong and stable PUFs and one multi-input XOR gate in addition to the ECC engine previously used.

A modified decoder is shown in Fig. 27.

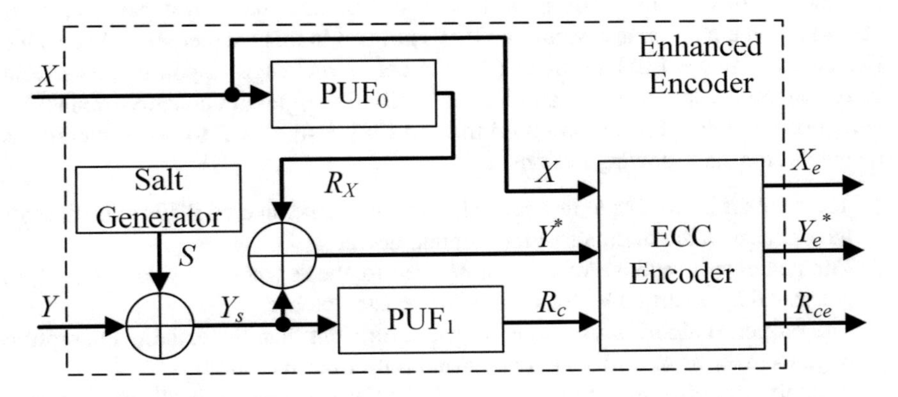

Fig. 26 Block diagram of enhanced encoder

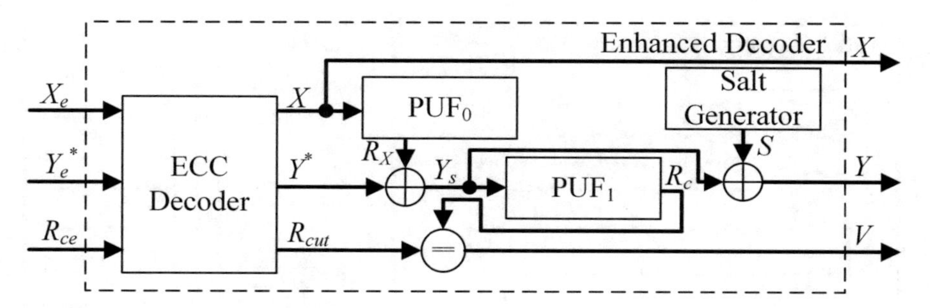

Fig. 27 Block diagram of enhanced decoder

Similarly to the encoding process, enhanced decoder utilizes two additional PUF circuits and XOR gate. To compare received hash value R_{cut} with genuine R_c value, an additional comparator is used:

1. Decode the received (X_e, Y_e^*, R_{ce}) values to get values of (X, Y^*, R_c) using the same decoding ECC engine as previously used.
2. Obtain hash value $R_X = \text{PUF}_0(X)$.
3. Decrypt the value of Y_s by XORing the value of Y^* with hash value R_X ($Y_s = Y^*$ XOR R_X).
4. Deobfuscate the value of Y_s into a value of Y by XORing with salt value S.
5. Obtain hash value $R_c = \text{PUF}_1(Y_s)$.
6. Compare the received hash value under test (R_{cut}) with the value of R_c. As a result, flag value V is generated ($V =$ "1" if received pair (X, Y) has been generated ($R_c = R_{cut}$) by this device and $V =$ "0" otherwise ($R_c \neq R_{cut}$)).

The code word transmitted via an untrusted channel should be transformed from (X_e, Y_e) to (X_e, Y_e^*, R_{ce}).

The changes in the communication process are shown in Fig. 28.

Modified communication protocol also includes pool of shared resources which consists of cryptographic primitives utilized by both enhanced encoder and decoder. Since PUF_0, PUF_1, and salt generator are the same, the keys are consistent for encoding and decoding processes.

As shown in Fig. 28, the attribute value (X) can be accessible by the untrusted party, because it is encoded only by the ECC engine. This does not give an advantage to the attacker as only the knowledge of the pair (X, Y) gives the possibility to observe the data stored on the device side.

Salt generator should change the value of S from time to time, e.g., based on timer (e.g., every 10 min), the number of exchanged pairs (X, Y), etc. It is used to prevent the attacker from taking advantage of functional dependency between X and Y. If X and Y do not depend on each other, this block can be omitted.

Furthermore, an attacker will not be able to modify the message as it is impossible to create a copy of PUFs to reproduce both encryption and encoding. Even if an attacker modifies the data, this fact will be detected by a decoding scheme

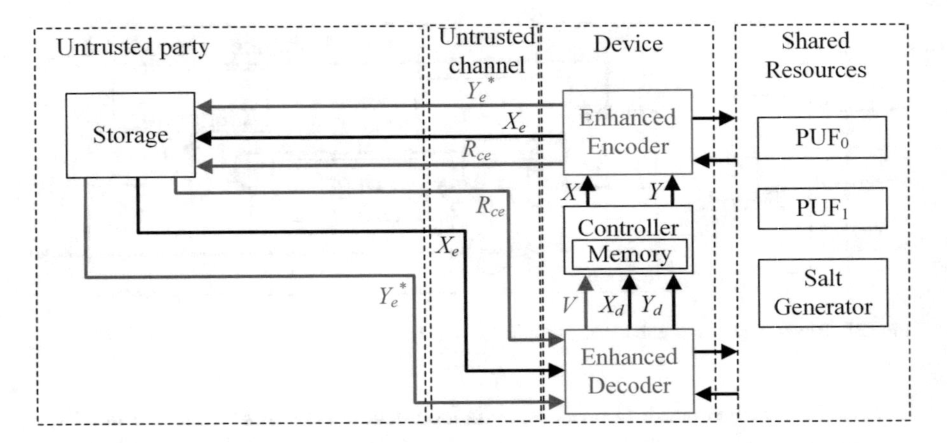

Fig. 28 Changed structure of data transmitting

based on the unique value of (Y_e^*, R_{ce}). The proposed approach also protects against errors caused by the noise on an untrusted channel and the untrusted party side.

5.3 *Example of Usage in Mobile NAND Flash Devices*

The proposed algorithm can be efficiently utilized in Host-aware Performance Booster (HPB) feature widely used in mobile flash devices [39] which is considered the same as Host Memory Buffer (HMB) used in SSD drives [40]. The block diagram of the proposed HPB algorithm enhancement is shown in Fig. 33.

Host stores HPB entries in the following format: LBA_e, PPN_e^*, R_{ce}. LBA_e is a logical block address encoded by ECC engine, i.e., it can be used by the host as a plaintext. PPN_e^* is a physical page number encrypted by enhanced encoder as shown in Fig. 26. In this case, LBA_e corresponds to X_e and PPN_e^* to Y_e^*. R_{ce} is a hash value of the PPN value.

The operation of a proposed modification of HPB algorithm can be described as follows:

1. A pair (LBA, PPN) is created by controller and stored in NAND as L2P table.
2. In order to use host memory as an external cache, NAND I/F (Interface) sends the pair (LBA, PPN) to enhanced encoder which encodes it into a triplet (LBA_e, PPN_e^*, R_{ce}) according to the encoding algorithm shown in Fig. 26.
3. If host decides to use this HPB entry (LBA_e, PPN_e^*, R_{ce}), it sends it back to the device.
4. Device controller decodes HPB entry (LBA_e, PPN_e^*, R_{ce}) into LBA, PPN_{HPB} (decoded PPN which could have been modified by host). The decoding scheme is shown in Fig. 27.

5. Enhanced decoder generates a value of V (validity of received HPB entry, V = "1" when the entry is valid and V = "0" otherwise). LBA is also checked in Dirty Map in order to ensure that (LBA, PPN) pair was not invalidated. Dirty Bitmap returns a validity value VD (VD = "1" when pair (LBA, PPN) is not invalidated and VD = "0" otherwise).
6. If both V and VD values are equal to "1", NAND I/F uses the received value and fetches the data by PPN_{HPB} address. Otherwise, it has to search the LBA and fetch the corresponding PPN from L2P table in NAND.
7. The proposed encoding and decoding algorithm provides a way to guarantee that a pair (LBA, PPN) is created by a unique NAND flash memory device as it utilizes PUF which is irreproducible by an attacker even if he knows the exact design of the encryption algorithm.

5.4 Conclusion

The encoding and decoding algorithm is proposed for attribute-value data which is transmitted via an untrusted channel. The algorithm utilizes strong and stable PUFs to prove that the attribute-value pair received from the untrusted party was generated by an authentic device. Furthermore, the algorithm is also used as an error detection method which can detect errors caused by the noise in the channel. The algorithm appends an initial code word with an additional hash value which proves the authenticity of the sent pair.

The advantages of the algorithm are listed below:

- Protection of the transmitted data from modifications by an attacker if the channel is untrusted.
- Detection of the errors caused by both noise in the channel (if ECC engine cannot correct all errors) and an attacker.
- Less additional hardware is required for the algorithm implementation compared to the encryption engines.

The proposed method can also be used as a part of HPB (HMB) algorithms in order to protect HPB entries and detect errors caused by the channel or injected by the attacker. Thus, this algorithm can be used simultaneously for error detection and security in NAND flash devices [48].

6 Conclusion

This chapter presents research results of SK hynix memory solutions Eastern Europe in area of physical security for NAND flash memory devices. Compact multimode PUF has been developed in order to be used as an entropy source with an identification feature. The proposed TRNG can be used within the existing NAND flash controller in order to be utilized by security protocols. Randomness also has been extracted directly from NAND flash memory by using read operations without ECC protection [41]. NAND flash memory is also a source of unique identification of the device which can generate more than 500 IDs utilizing only 1 block of memory of 2 MB. Classical PUF designs have been used in order to improve key-value pair transmission in HPB and HMB protocols [42]. Also scrambling engine has been enhanced in order to provide more secure and reliable way of randomizing data before sending it to the NAND memory cells [43].

Proposed solutions show the high potential of using NAND flash memory as an entropy source for cryptography and statistical simulation applications. Also classical PUF designs improve the security and reliability of data storage and transmitting protocols. Thus, presented PUF-based security solutions can be implemented in the areas with strict security and safety requirements (e.g., medical devices [44], avionics [45], critical firmware [46], etc.).

Appendix

See Figs. 29, 30, 31, 32, and 33.

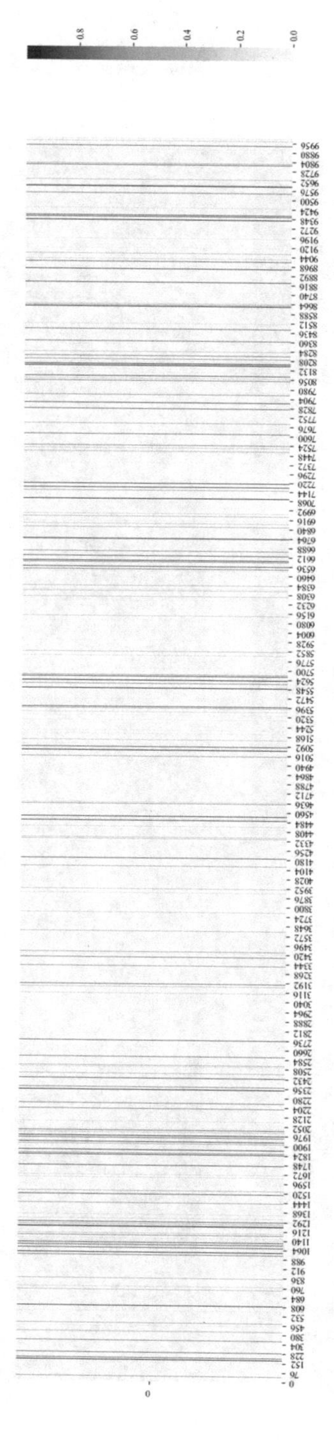

Fig. 29 Heatmap of flipping bits within a single TLC page after $R = 1000$ reads

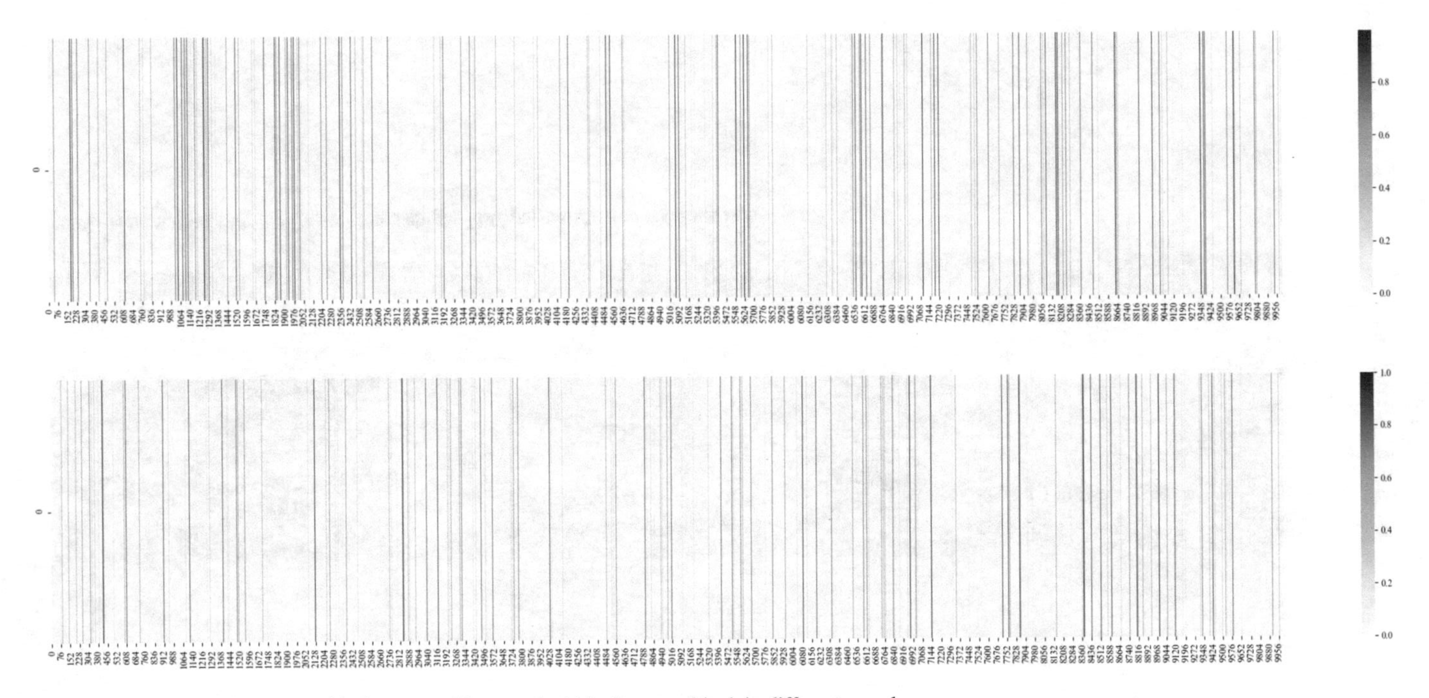

Fig. 30 Ψ scores for the pages with the same address and within the same block in different samples

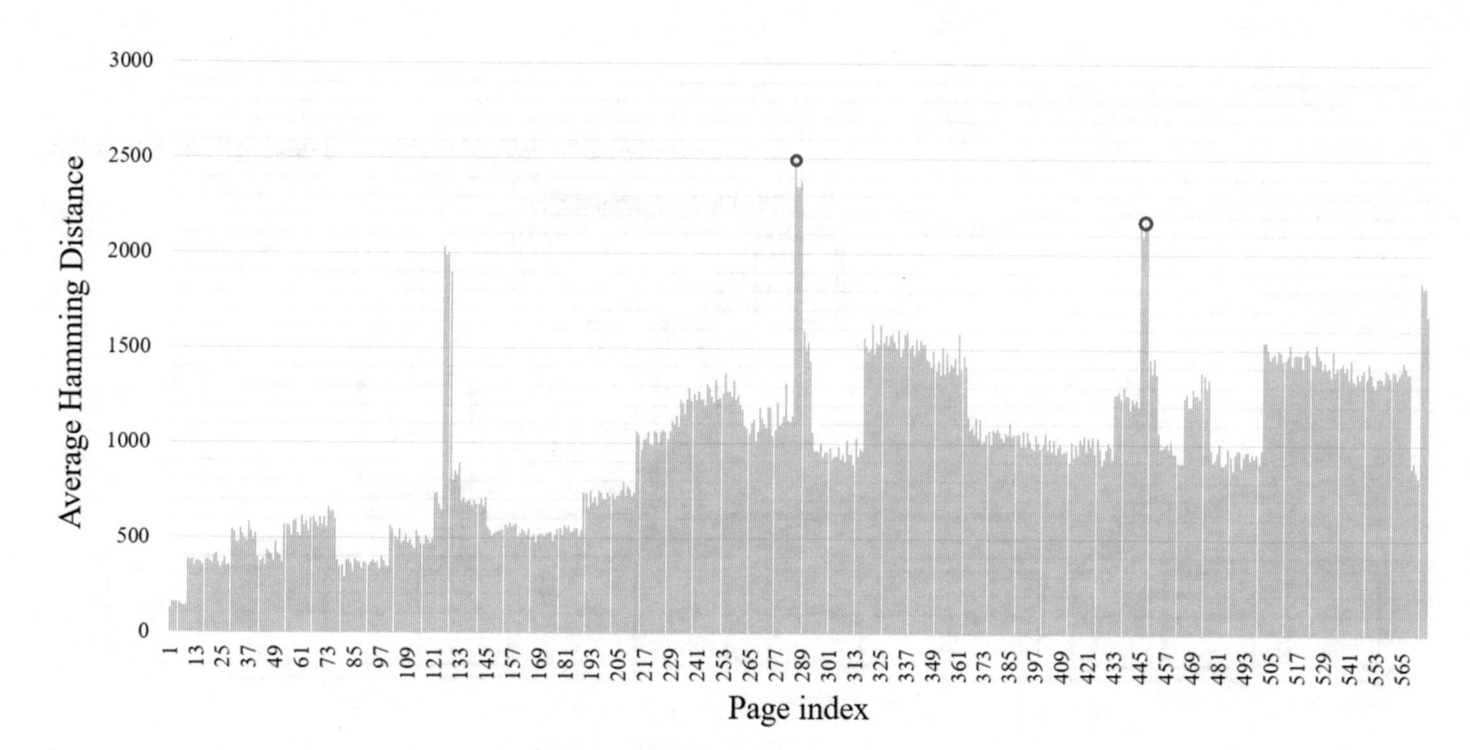

Fig. 31 Average hamming distances between reads for different pages within a block

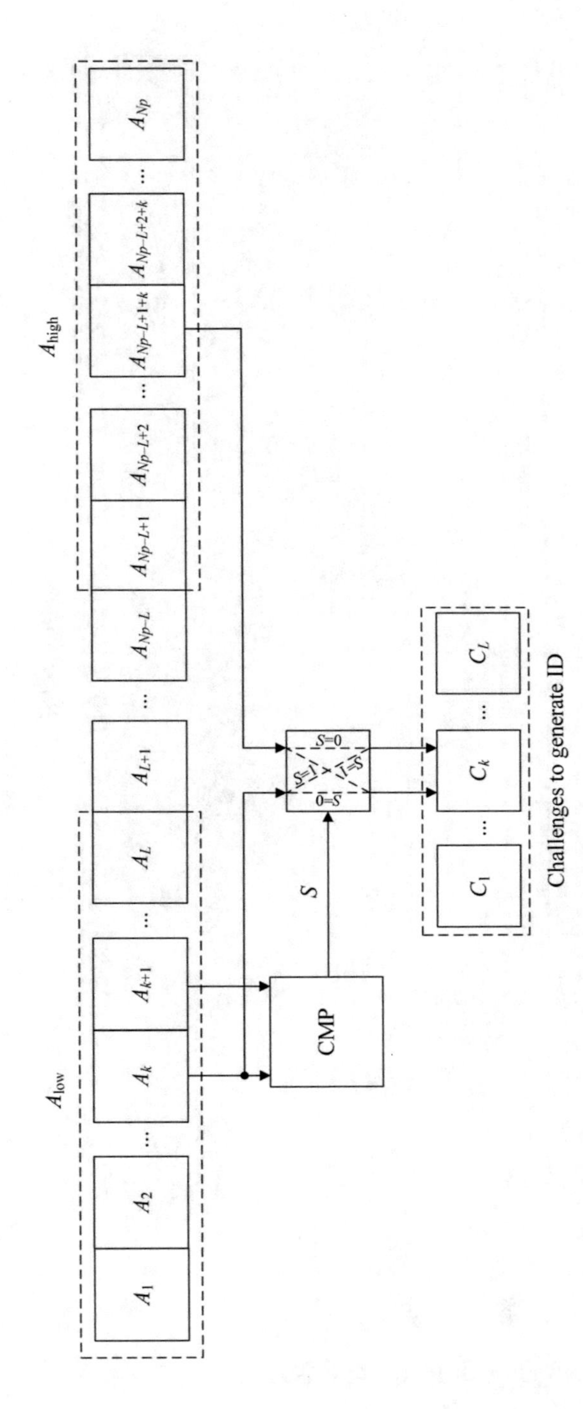

Fig. 32 Block diagram of challenges forming algorithm for ID generation

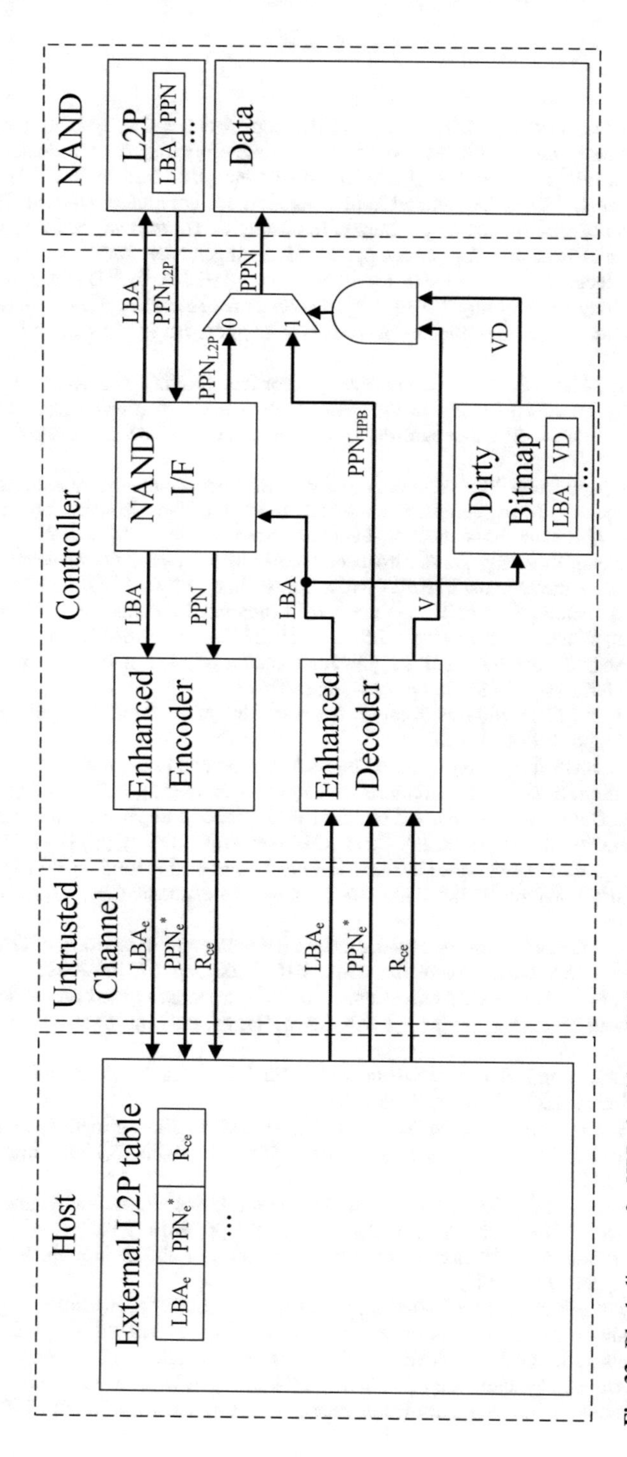

Fig. 33 Block diagram for HPB algorithm utilizing proposed encoding and decoding method

References

1. Trusted Computing Group: TCG Storage Workgroup: Storage Certification Program. In: Trusted Computing Group (2021) https://trustedcomputinggroup.org/wp-content/uploads/Storage_Certification_Program_Rev_1_33-Published-Copy.pdf. Cited 14 Oct 2021
2. Zalivaka, S.S., et al. (2016). Design and implementation of high-quality physical Unclonable functions for hardware-oriented cryptography. In Chang, C.-H., Potkonjak, M. (eds.) Secure System Design and Trustable Computing, pp. 39–81. Springer, New York
3. Samsung Introduces Exynos i T100 for Secure and Reliable IoT Devices with Short-Range Connectivity. In: Samsung (2019) . https://news.samsung.com/my/samsung-introduces-exynos-i-t100-for-secure-and-reliable-iot-devices-with-short-range-connectivity. Cited 14 Oct 2021
4. Lu, T., Kenny, R., Atsatt, S.: Secure device manager for Intel Stratix 10 devices provides FPGA and SoC security. In: Intel (2018). https://www.intel.com/content/dam/www/programmable/us/en/pdfs/literature/wp/wp-01252-secure-device-manager-for-fpga-soc-security.pdf. Cited 14 Oct 2021
5. Toshiba Develops A New PUF Technology for Solid-State Authentication of IoT Equipment. In: Istanbulpost (2018). https://www.istanbulpost.com.tr/toshiba-develops-a-new-puf-technology-for-solid-state-authentication-of-iot-equipment/. Cited 14 Oct 2021
6. Chang, C.H., Zheng, Y., Zhang, L.: A retrospective and a look forward: fifteen years of physical unclonable function advancement. IEEE Circ. and Syst. Mag. **17**(3), 32–62 (2017)
7. Lee, J., et al: A technique to build a secret key in integrated circuits for identification and authentication applications. Int. Symp. VLSI Circ. (VLSI'04), pp. 176–179 (2004)
8. Sehwag, V., Saha, T.: TV-PUF: a fast lightweight analog physical unclonable function. Int. Symp. Nanoel. Inf. Syst. (iNIS'16), pp. 182–186 (2016)
9. Gassend, B., et al: Silicon physical random functions. In: ACM Conf. Comput. and Comm. Secur. (CCS'02), pp. 148–160 (2002)
10. Cao, Y., et al.: CMOS image sensor based physical unclonable function for coherent sensor-level authentication. IEEE Trans. Circuits Syst. I Regul. Pap. **62**(11), 2629–2640 (2015)
11. Holcomb, D.E., Burleson, W.P., Fu, K.: Initial SRAM state as a fingerprint and source of true random numbers for RFID tags. In: Int. Conf. RFID Secur. (RFID'07), pp. 1–2 (2007)
12. Tehranipoor, F., et al: DRAM-based intrinsic physically unclonable functions for system-level security and authentication. IEEE Trans. Very Large Scale Integr. Circ. **25**(3), 1085–1097 (2017)
13. Kumar, S.S., et al: Extended abstract: the butterfly PUF protecting IP on every FPGA. In: IEEE Int. Worksh. on Hardw.-Orient. Secur. and Trust. (HOST'08), pp. 67–70 (2008)
14. Yamamoto, D., et al: Uniqueness enhancement of PUF responses based on the locations of random outputting RS latches. In: Int. Worksh. Crypt. Hardw. and Emb. Syst. 2011 (CHES'11), pp. 390–406
15. Jia, S., et al: Extracting Robust Keys from NAND Flash Physical Unclonable Functions. Int. Conf. on Inf. Secur. (ISC'15), pp. 437–454 (2015)
16. Holcomb, D.E., Burleson, W.P., Fu, K.: Power-up SRAM state as an identifying fingerprint and source of true random numbers. ieee T. Comp. 58(9), 1198–1210 (2009). https://doi.org/10.1109/TC.2008.212
17. Suh, G.E., Devadas, S: Physical unclonable functions for device authentication and secret key generation. In: ACM/IEEE Des. Autom. Conf. (DAC'07), pp. 9–14 (2007)
18. Kacprzak, T.: Analysis of oscillatory metastable operation of an RS flipflop. IEEE J. Solid State Cir. **23**(1), 260–266 (1988)
19. Digilent: Nexys 4 FPGA board reference manual. In: Digilent Inc. (2016). https://digilent.com/reference/_media/reference/programmable-logic/nexys-4/nexys4_rm.pdf. Cited 02 Nov 2021
20. Barker, E., Kelsey, J.: Recommendation for random bit generator (RBG) constructions. In: National Institute of Standards and Technology (NIST) (2016). https://csrc.nist.gov/CSRC/media/Publications/sp/800-90c/draft/documents/sp800_90c_second_draft.pdf. Cited 14 Oct 2021

21. Papandreou, N., et al.: Open block characterization and read voltage calibration of 3D QLC NAND flash. In: IEEE Int. Rel. Phys. Symp. (IRPS'20), pp. 1–6 (2020)
22. Papandreou, N., et al.: Reliability of 3D NAND flash memory with a focus on read voltage calibration from a system aspect. IEEE Non-Vol. Mem. Tech. Symp. (NVMTS'19), pp. 1–4 (2019)
23. Ruhrmair, U., Solter, J., Sehnke, F.: On the foundations of physical unclonable functions. In: Cryptology ePrint Archive (2009). https://eprint.iacr.org/2009/277.pdf. Cited 21 Feb 2022
24. Vijayakumar, A., Patil, V.C., Kundu, S.: On testing physically unclonable functions for uniqueness. In: IEEE Int. Symp. on Qual. El. Des. (ISQED'16), pp. 244–249 (2016)
25. Cai, Y., et al.: Error characterization, mitigation, and recovery in flash-memory-based solid-state drives. Proc. IEEE **105**(9), 1666–1704 (2017)
26. Hori, Y., et al.: Quantitative and statistical performance evaluation of arbiter physical unclonable functions on FPGAs. In: Int. Conf. Reconf. Comp. FPGA (ReConFig'10), pp. 298–303 (2010)
27. Micheloni, R., Crippa, L., Marelli, A.: Inside NAND Flash Memories, 582 p. Springer, New York (2010)
28. Cha, Y., Kang, S.: Data randomization scheme for endurance enhancement and interference mitigation of multilevel flash memory devices. ETRI J. **35**(1), 166–169 (2013)
29. Cai, Y., et al.: Vulnerabilities in MLC NAND flash memory programming: experimental analysis, exploits, and mitigation techniques. In: IEEE Int. Symp. on High-Perf. Comp. Arch. (HPCA'17), pp. 49–60 (2017)
30. Van Zandwijk, J.P.: A mathematical approach to NAND flash-memory descrambling and decoding. Digital Invest. **12**, 41–52 (2015). https://doi.org/10.1016/j.diin.2015.01.003
31. Heninger, N.: Cold-boot attacks. In: Encyclopedia of Cryptography and Security (2011). https://link.springer.com/referenceworkentry/10.1007/978-1-4419-5906-5_124. Cited 21 Feb 2022
32. Maes, R., Van Herrewege, A., Verbauwhede, I.: PUFKY: A fully functional PUF-based cryptographic key generator. In: Crypt. Hardw. and Emb. Syst. (CHES'12), pp. 302–319 (2012)
33. Zalivaka, S..S., Ivaniuk, A.A., Chang, C.H.: Reliable and modeling attack resistant authentication of arbiter PUF in FPGA implementation with trinary quadruple response. IEEE Trans. Inf. Forens. Secur. **14**(4), 1109–1123 (2019)
34. Tomlinson, M., et al.: Error-Correction Coding and Decoding, 522 p. Springer, New York (2017)
35. Xilinx: Artix-7 FPGAs Data Sheet: DC and AC Switching Characteristics. In: Xilinx Inc. (2021). https://www.xilinx.com/support/documentation/data_sheets/ds181_Artix_7_Data_Sheet.pdf. Cited 11 Oct 2021
36. Amazon: What is a key-value database? In: Amazon Inc. (2019). https://aws.amazon.com/nosql/key-value/. Cited 13 Oct 2021
37. Blahut, R.E.: Cryptography and Secure Communication, 587 p. Cambridge University Press, Cambridge (2014)
38. Zalivaka, S..S., et al.: Multi-valued arbiters for quality enhancement of PUF responses on FPGA implementation. In: Asia and South Pacific Des. Autom. Conf. (ASP-DAC'19), pp. 533–538 (2019)
39. Jeong, W., et al.: Improving flash storage performance by caching address mapping table in host memory. In: USENIX Worksh. on Hot Topics in Stor. and File Syst. (HotStorage'17), pp. 19–24 (2017)
40. Dorgelo, J.: Host memory buffer (HMB) based SSD system. In: Proc. Flash Memory Summit (FMS'15) (2015). https://www.flashmemorysummit.com/English/Collaterals/Proceedings/2015/20150813_FJ31_Chen_Dorgello.pdf. Cited 13 Oct 2021
41. Zalivaka, S.S., Ivaniuk, A.A.: Raw read based physically unclonable function for flash memory. US patent application (US20210055912A1). https://patents.google.com/patent/US20210055912A1 Cited 13 Oct 2021
42. Zalivaka, S.S., Ivaniuk, A.A.: Encoder and decoder using physically unclonable functions. US patent (US11394529B2). https://patents.google.com/patent/US11394529B2 Cited 05 Oct 2022

43. Zalivaka, S.S., Ivaniuk, A.A.: Data scramblers with enhanced physical security. US patent application (US20210326490A1). https://patents.google.com/patent/US20210326490A1 Cited 03 Nov 2021
44. Rodriguez, C.A.: Safeguard smart medical devices for enhanced patient safety. In: Maxim Integrated (2020). https://www.maximintegrated.com/en/design/blog/safeguard-smart-medical-devices-for-enhanced-patient-safety.html. Cited 22 Feb 2022
45. O'Neill, K., et al.: Protecting flight critical systems against security threats in commercial air transportation. In: Dig. Avion. Syst. Conf. (DASC'16), pp. 1–7 (2016)
46. Protection against Reverse-Engineering, Counterfeiting/Cloning and Overbuilding. In: Intrinsic ID (2021). https://www.intrinsic-id.com/firmware-ip-protection/. Cited 22 Feb 2022
47. McIntyre, D.: Annual flash controller update. In: Proc. Flash Memory Summit 2019 (FMS'19). https://www.flashmemorysummit.com/Proceedings2019/08-06-Tuesday/20190806_CTRL-102A-1_McIntyre.pdf. Cited 11 Oct 2021
48. Tyson, M.: Researchers find "pattern of critical issues" in SSD encryption. In: HEXUS.net (2018). https://hexus.net/tech/news/storage/123986-researchers-find-pattern-critical-issues-ssd-encryption/. Cited 11 Oct 2021
49. Pickering, P.: NAND rises to the occasion in data-heavy IoT applications. In: Electronic Design (2021). https://www.electronicdesign.com/technologies/iot/article/21807634/nand-rises-to-the-occasion-in-dataheavy-iot-applications. Cited 11 Oct 2021
50. Rajendiran, K.: Using PUFs for random number generation. In: Intrinsic ID (2021). https://semiwiki.com/ip/intrinsic-id/303704-using-pufs-for-random-number-generation/. Cited 21 Feb 2022
51. Neustadter, D.: True random number generators for heightened security in any SoC. In: Synopsys Inc (2021). https://www.synopsys.com/designware-ip/technical-bulletin/true-random-number-generator-security-2019q3.html. Cited 21 Feb 2022
52. Intrinsic ID: Zign RNG. In: Intrinsic ID (2021). https://www.intrinsic-id.com/products/zign-rng/. Cited 21 Feb 2022

ReRAM-Based Neuromorphic Computing

Fabiha Nowshin and Yang Yi

1 Introduction

As the complementary metal-oxide semiconductor (CMOS) continues to scale, it is becoming increasingly difficult to meet the energy and power requirements of processors [1]. The cost of data transmission will heavily affect the energy requirements of the cloud and therefore Internet of Things (IoT) devices [2]. As IoT continues to advance, it is becoming impossible to shuttle data back and forth for computation and analysis. Edge computation is necessary in this case to allow processing the relevant data that can be sent to the cloud [3, 4]. The edge devices that will be developed will need to be energy and power efficient to be able to preprocess the data before it can be transmitted [4]. Vector-matrix multiplications are a key operation in edge computing, and they can be carried out using resistive random access memory (ReRAM) crossbars [5]. ReRAMs are two-terminal nonvolatile memory devices that follow the switching mechanism of a memristor [5]. These ReRAMs are used to create accelerators for several deep learning applications. ReRAMs are emerging nonvolatile memories (eNVM)s that have been rigorously researched over the years and have emerged as invaluable to in-memory computations in the development of neuromorphic hardware. The traditional von Neumann computing system suffers from an issue known as memory bottleneck due to the CPU and memory being separate units shared by a bus [6]. The shuttling of data back and forth leads to increased energy consumption and latency that is worsened by CMOS scaling. ReRAM-based crossbars allow for in-memory or near-memory computations that can tackle the issues experienced by the von Neumann computing systems [7]. Furthermore, because of their ability

F. Nowshin (✉) · Y. Yi
Virginia Tech, Blacksburg, VA, USA
e-mail: fabiha27@vt.edu; yangyi8@vt.edu

A. Iranmanesh (ed.), *Frontiers of Quality Electronic Design (QED)*,
https://doi.org/10.1007/978-3-031-16344-9_2

to directly process analog signals, ReRAM-based accelerators will greatly benefit edge computing and IoT devices. ReRAMs will be crucial in replacing power- and energy-consuming analog-to-digital (ADC) and digital-to-analog (DAC) converters [8].

The traditional memories include static random access memory (SRAM), dynamic random access memory (DRAM), and flash. These typical memory technologies rely on the charge storage phenomenon, where in DRAM the charges are stored at the cell capacitor, at SRAM the charges are stored at the inverter nodes which are cross-coupled, and in flash the floating gate of the transistor is responsible for the charge storage [9]. The main disadvantage that arises from technology scaling is that these stored charges tend to get lost and introduce noise and reliability issues. Some key requirements for eNVMs are that they have to be scalable, possess nonvolatile storage, operate on low voltage, provide a long retention time, have high endurance, hold many synaptic strength levels, have a simple framework, and demonstrate synaptic learning capability. The popular eNVMs that were developed based on these properties include phase-change random access memory (PCRAM), spin-transfer-torque magnetic random access memory (STT-MRAM), and ReRAM that are not based on this charge storage mechanism. Compared to its eNVM counterparts, ReRAMs have received immense popularity in neuromorphic computing hardware due to its compatibility with CMOS technology, scalability, low power consumption, and analog conductance modulation properties [8–11]. Therefore, in this chapter we will focus on the use of ReRAMs in the area of neuromorphic computing.

In this topic we will cover the basic characteristics of a memristor device and its detailed switching mechanism. An overview of existing eNVMs will be discussed including the differences, advantages, and disadvantages of each technology compared to conventional memory structures. The use of ReRAMs as neurons and synapses will be discussed in this chapter to provide readers with more details about their use in neuromorphic computing. The process of writing and reading from crossbars will be covered as well, along with how they can be used to carry out in-memory computing operations. The different types of neural networks that can be built with the help of ReRAMs will be covered in detail in this chapter including the construction of multilayer perceptron (MLP), spiking neural networks (SNN), convolutional neural networks (CNN), recurrent neural networks (RNN), and in-memory computing architectures.

2 The Memristor

The memristor was developed as the fourth basic circuit element by Professor Leon Chua from the University of Berkeley, in the year 1971 [12]. The three fundamental circuit elements that previously existed are the resistors, capacitors, and inductors. There are four basic variables in the field of electrical engineering which are current, voltage, charge, and flux. As depicted in Fig. 1, the relationship between

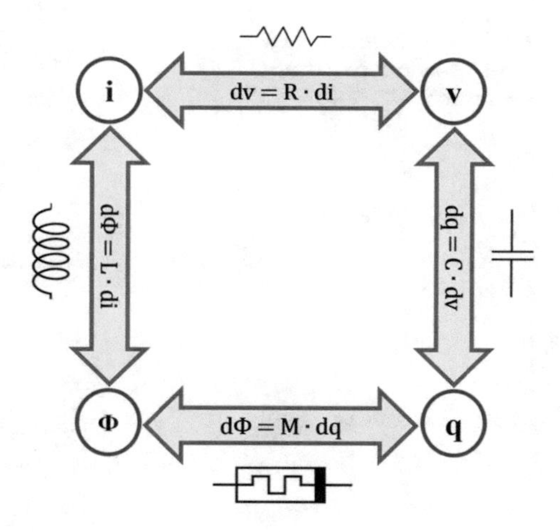

Fig. 1 Relationship between the four key electrical components

the fourth element, the memristor, can be used to explain the missing link between the variables charge and flux. From the figure, the equation for the memristor is given by:

$$d\varphi = M \cdot dq, \quad (1)$$

where the M is the memristance of the device and φ and q are the flux and the charge stored, respectively.

The memristor is a two-terminal nonvolatile passive device, also known as a memory resistor. The resistance of the memristor will increase when electric charge flows through the device in one direction and will decrease when it flows in the other direction. When no voltage is applied, the memristor retains its previous state of resistance which gives it the memory property. The current voltage or the I-V characteristic curve is the most important property of a memristive device. This curve is represented by a pinched hysteresis loop. The current and voltage are both zero at the origin. As the frequency increases, the hysteresis loop becomes thinner, and eventually with infinite frequency, the memristor starts to behave like a resistor, having a linear relationship between current and voltage as portrayed in Fig. 2 [13, 14]. This specific I-V characteristic curve is essential in demonstrating the on and off state of the memristor. The details on memristor implementation using ReRAMs are discussed in the next section which will show how this hysteresis curve is necessary for the switching mechanism and memory property of the device.

3 ReRAM: Implementation of the Memristor

Memristors can be implemented using resistive random access memory (ReRAM). The typical structure of a memristor or ReRAM is a metal-insulator-metal structure,

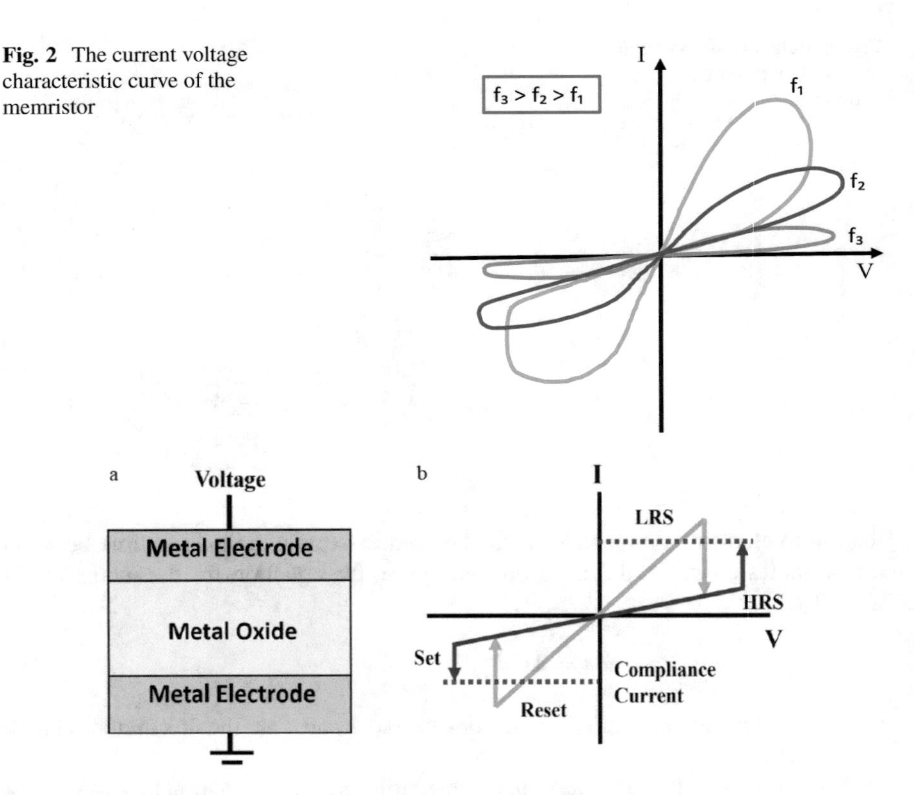

Fig. 2 The current voltage characteristic curve of the memristor

Fig. 3 (**a**) The structure of a ReRAM. (**b**) The operation of a ReRAM

an oxide layer sandwiched between two metal electrodes at the top and bottom as portrayed in Fig. 3a [7]. ReRAMs have a high resistance state (HRS) and a low resistance state (LRS). When switching from HRS to LRS, as shown in Fig. 3b, the process is termed as the set process, and the process of switching from LRS to HRS is termed as the reset process [7]. When the ReRAM is in the initial state, it has to go through an electroforming process where a voltage greater than the set voltage is necessary to allow the device to demonstrate its resistive switching property.

Two types of switching exist in ReRAMs, known as the unipolar and bipolar switching. The switching demonstrated in Fig. 3b is the unipolar switching behavior. In unipolar switching, the transition from HRS to LRS depends on only the amplitude of the voltage applied. In bipolar switching, the transition depends on both the amplitude and the polarity of the applied voltage. Compliance current on the other hand is a current value provided by the semiconductor parameter analyzer to prevent permanent dielectric breakdown in the set process.

ReRAMs in general can be classified into two categories, oxide-RAM (OxRAM) and conductive bridge RAM (CBRAM) [9]. The basic structure of the OxRAM and CBRAM are shown in Fig. 4a, b. The main difference between OxRAM and CBRAM is that in OxRAM the filament has oxygen vacancies in the metal oxide

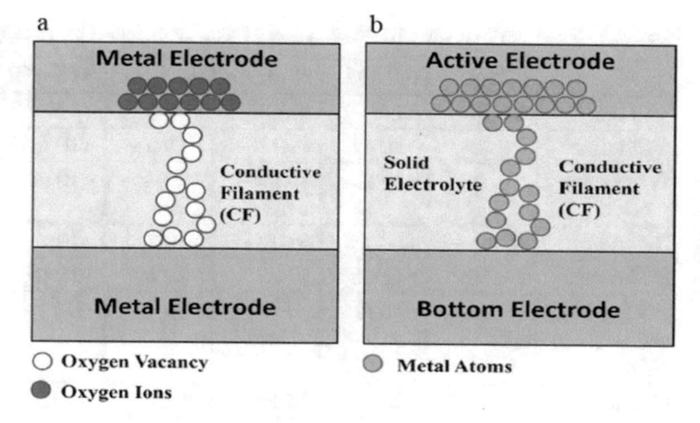

Fig. 4 (**a**) The structure of an OxRAM. (**b**) The structure of CBRAM

layer and in CBRAM the filament has mental atoms in the oxide layer that is created by the metal ions when they move into the solid electrolyte. While OxRAM and CBRAM have similar characteristics, the major difference is that OxRAM has a smaller on and off resistance ratio and higher endurance compared to CBRAM.

4 Comparison of ReRAMs with Other Memory Technologies

All the existing eNVMs including PCRAM, STT-MRAM, and ReRAM have a similar characteristic of being a two-terminal device with a nonvolatile property. These devices all switch from HRS to LRS to demonstrate their memory property that is made possible by applying a voltage or current signal, for instance, at one of the terminals of the device. The switching mechanism is different from one device to the other [9]. In PCRAM, the switching is based on chalcogenide materials which allow the device to switch between the crystalline phase which is the LRS and the amorphous phase which is the HRS under the application of an electric signal. STT-MRAM on the other hand is constructed by having a thin tunneling insulator layer sandwiched between two ferromagnetic layers. The parallel and antiparallel configuration demonstrate the HRS and LRS, respectively. The ReRAM, as discussed previously, bases its switching mechanism on the formation and the destruction of the conductive filament in the oxide layer between the two metal electrodes.

Table 1 demonstrates the comparison between the traditional memory technologies of SRAM, DRAM, NOR, and NAND and the eNVMs, STT-MRAM, PCRAM, and ReRAM [9]. Because of the different materials and type of switching in the eNVMs, PCRAMs, STT-MRAMs, and ReRAMs are used in different types of applications. STT-MRAM compared to the other eNVMs have low operating voltage and fast read and write time as well as a longer endurance which makes it suitable for replacing embedded DRAM or SRAM. On the other hand, PCRAM and

Table 1 Summary of the characteristics of traditional and emerging memory technologies

	SRAM	DRAM	NOR	NAND	STT-MRAM	PCRAM	ReRAM
Voltage	<1 V	<1 V	>10 V	>10 V	<1.5 V	<3 V	<3 V
Read time	~1 ns	~10 ns	~50 ns	~10 μs	<10 ns	<10 ns	<10 ns
Write time	~1 ns	~10 ns	10 μs–1 ms	100 μs–1 ms	<10 ns	~50 ns	<10 ns
Retention	N/A	~64 ms	>10y	>10y	>10y	>10y	>10y
Endurance	>1E16	>1E16	>1E5	>1E4	>1E15	>1E9	>1E66~1E12
Energy to write (J/bit)	~fJ	~10 fJ	~100 pJ	~10 fJ	~0.1 pJ	~10 pJ	~0.1 pJ
Cell area	>100 F^2	6 F^2	10 F^2	<4 F^2	6–50 F^2	4–30 F^2	4–12 F^2

ReRAM can be seen as an alternative for NOR and NAND flash memories due to their faster read and write times and low programming voltage.

5 Use of ReRAMs as Synapses

Human brains contain many neurons and synapses in the nervous system. As depicted in Fig. 5a, the neurons are connected to many other neurons in the human body. The synaptic weight changes based on the stimulus that passes from the presynaptic neuron to the postsynaptic neuron [15]. Figure 5b shows that as the input signals are applied to the dendrites of the neurons, they pass through the axon and accumulate; when the action potential exceeds a certain threshold, a spike is fired. The weight of the synapses is altered by the flow of signals through the neurons. In the synapse, the amplitude of the signal sent from the presynaptic neuron to the postsynaptic neuron can be increased or decreased which gives synapses the property of plasticity.

The function of the synapse can be demonstrated by a ReRAM device when developing neural networks. This is demonstrated in Fig. 5c, where if we use two CMOS neurons to carry out the pre- and post-neuron functions, the memristor can act as the synapse between them. The ReRAM can be used to implement the plasticity of the synapses which is a very popular training algorithm in ReRAM-based neuromorphic systems called spike time-dependent plasticity (STDP). This plasticity is a crucial aspect to mimic the memory aspect of the brain. Hardware implementations of these neural networks pose a particularly challenging problem in this field, and it becomes nearly impossible to map large-scale neural networks using CMOS circuitry which also adds to increased area and power consumption. However, with ReRAMs' compatibility property with CMOS technology and their ability to act as synapses, this aids in simpler hardware implementation of neuromorphic computing architectures and IoT devices.

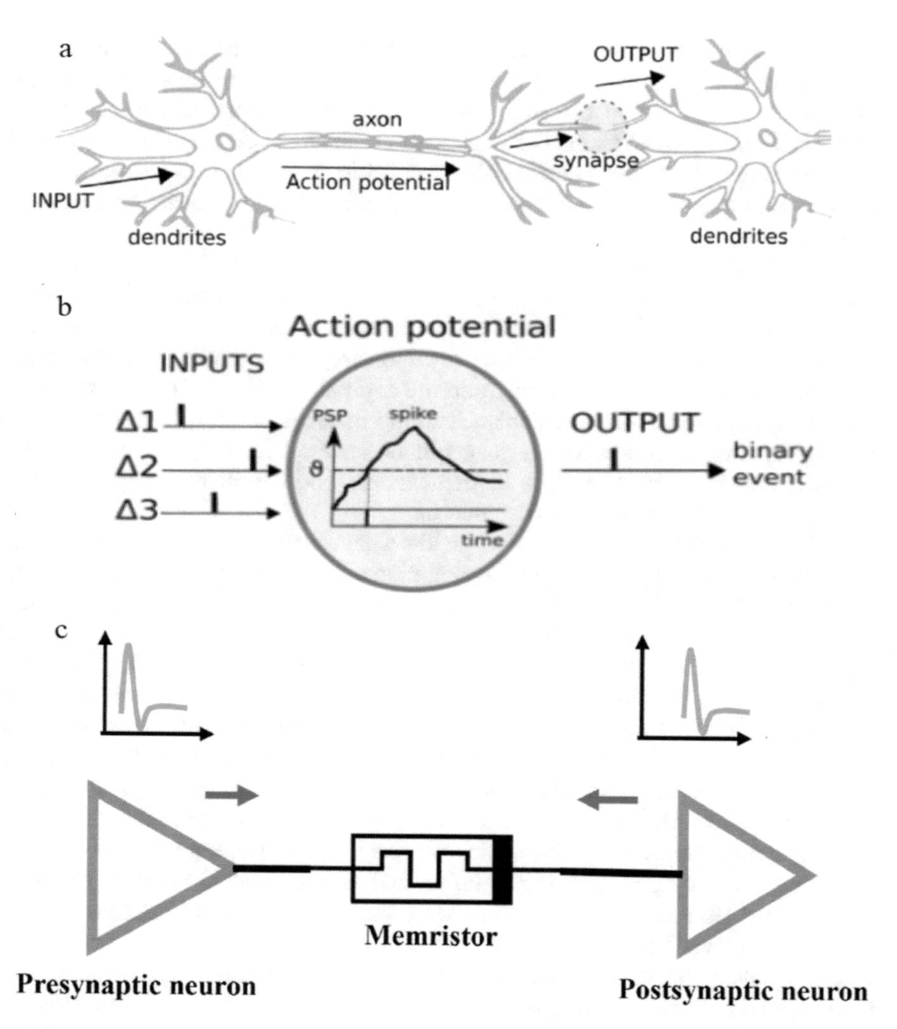

Fig. 5 (**a**) Biological neuron model. (**b**) Artificial neuron design. (**c**) Synapse design based on the memristor

6 Use of ReRAMs as Neurons

Aside from being used as synapses, ReRAMs can also be used as neurons. Leon Chua also predicted that there exists chaotic behavior in between regions in a memristor and exploiting these chaotic regions would lead to the development of devices that exhibits artificial neuronal behavior [16]. Recent research has shown that they can be used to model some of the dynamics of the neuron [17–20]. In ReRAM-based neurons, when pulses are applied, the devices accumulate the signal until the neuron generates a spike and therefore changes the conductance of the memristor. The Hodgkin-Huxley neuron model, for instance, can be implemented by ReRAMs [17]. Some research has also shown that cortical neurons can be mimicked using ReRAMs. In [18] a memristor emulator is discussed that is capable of producing spikes and demonstrates the firing mechanism of neurons. Their proposed emulator can imitate regular spiking, fast spiking, chattering, and intrinsic spiking.

A ReRAM-based neuron model is discussed in [19] which uses the unipolar operation of the ReRAM to demonstrate the Carillo-Hoppensteadt model. It further discusses neuromorphic pulse coding with the help of both ReRAM synapses and neurons. Their developed neuron has shown to carry out coincidence detection and pre-stimulus inhibition when pulse trains are applied. This is useful in pulse coding that is required for STDP, a Hebbian learning mechanism used in the process of training spiking neural networks that will be covered later in this chapter. The ReRAM-based neuron was able to demonstrate both oscillatory and excitatory modes, a characteristic of biological neuron.

In the year 2020, the first device to act like a neuron was developed, known as the Mott memristor [20]. Previous implementations of neurons included second-order elements that could demonstrate some neuromorphic properties like exhibiting periodic spiking and oscillations. However, in order to fully emulate a neuron, complete neuromorphic action potential functionality is necessary that include phasic and periodic spiking, bursting, chaos and threshold dynamics, and oscillatory behavior. The third-order Mott memristor combines both the resistor and the capacitor which changes the resistance based on a change in the temperature of the device, as depicted in Fig. 6. It essentially utilizes niobium oxide NbO_2 to which when a DC voltage is applied, the oxide switches from an insulating mode to conduction mode. As temperature drops, the device switches to back to the insulating mode, triggering a spiking current signal that is similar to that of a biological neuron. Two non-monotic and Boolean logic operations were implemented using a simple network consisting of these Mott memristors.

These developments of ReRAM-based neurons could eventually pave the way for transistor-less neuromorphic systems. With the nanoscale property of ReRAMs, the development of these ReRAM-based neurons could eventually lead to more brain-like energy and power-efficient systems. Research on the different oxide materials is still ongoing, and future work involves implementing these ReRAM devices on large-scale systems to carry out complex operations.

7 ReRAMs in Neuromorphic Computing

A simplified structure of a von Neumann computing system is shown in Fig. 7a. The CPU and the memory unit are separated in the structure and large amount of energy is spent in the fetching and storing of data to and from the memory unit to the CPU [21]. Machine learning applications rely on significant amount of data movements

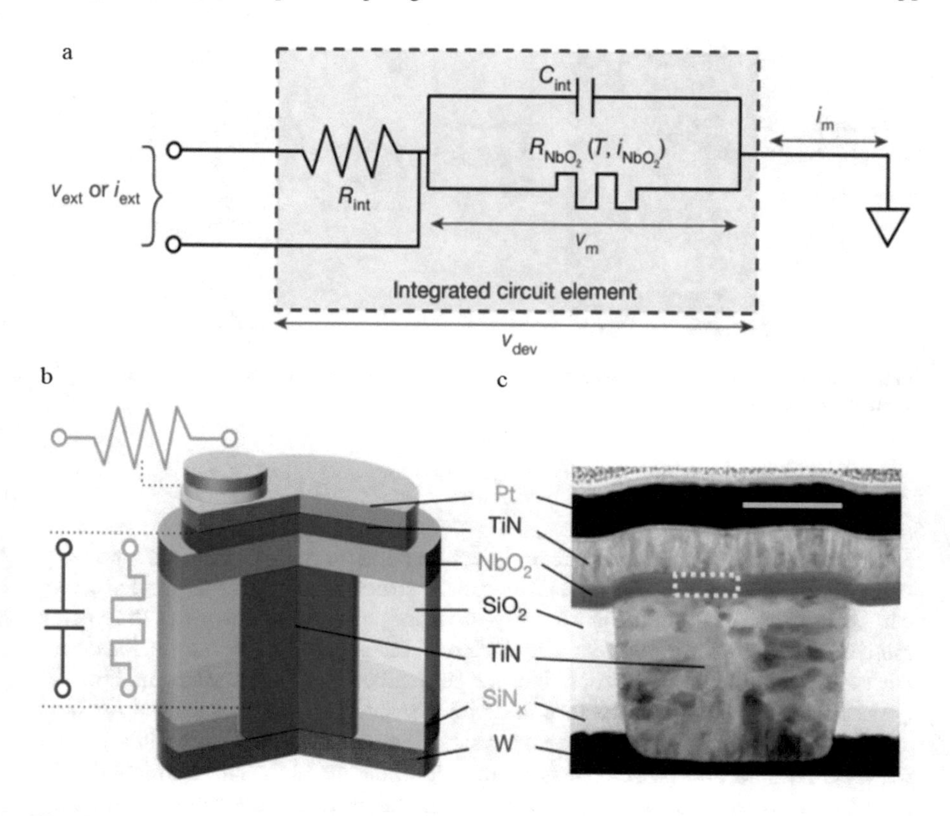

Fig. 6 (**a**) Circuit model of the Mott memristor. (**b**) Schematic of the memristor with different materials. (**c**) Actual cross-section of the memristor [20]

because of shuttling weights stored in the memory. This makes it very challenging for such applications to run on von Neumann architectures. The performance of the system degrades due to this data transfer, and the interconnect parasitic further contributes to the higher energy consumption of the system as technology scales. Furthermore, because of technology scaling, Moore's law is reaching a plateau where computers cannot double their performance every 18 months [1]. This is where neuromorphic computing comes in, an idea proposed by Dr. Carver Mead, which discusses the idea of using very large-scale integrated (VLSI) circuits to emulate biological nervous system [22].

An example of a typical neuromorphic computing architecture is shown in Fig. 7b. They have a highly parallelized structure where the input and output layers are connected by layers of neurons [23]. They mimic the biological system where the neurons have synaptic connections between them. As discussed in the previous section, this is where the synaptic property of the ReRAMs can be applied and hence can be used to build many types of neuromorphic computing architectures.

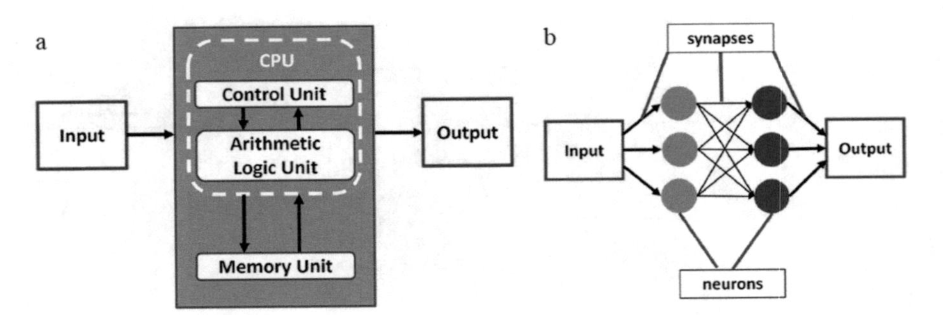

Fig. 7 (**a**) Architecture of a von Neumann computing system. (**b**) Architecture of a neuromorphic computing system

8 ReRAM Crossbars

Vector-matrix multiplications are a critical operation in IoT devices and edge computing. Due to the simple and compact structure of ReRAM devices, they can be used to integrate a high-density crossbar structure to carry out the vector-matrix multiplication operations. In ReRAM crossbars, the device is located at the crosspoint of two nanowires that are perpendicular to each other. In Fig. 8, a typical ReRAM crossbar structure is demonstrated. The crossbar allows for large amount of in-memory computation operations like vector-matrix multiplication and reduces the area and power consumption significantly [24]. The horizontal lines of the crossbar are the wordlines and the vertical lines are the bitlines. In neural networks the weights can be mapped into the conductance state of the ReRAMs in the crossbar. For a crossbar with i rows and j columns, when the input voltages are applied to the ith row of the crossbar, the output accumulated current can be calculated from the jth bitline as

$$I_j = \sum\nolimits_{i=0}^{n} G_{ij} V_i, \tag{2}$$

where G_{ij} is the conductance of the memristor and V_i is the voltage applied. To map the weights into the ReRAM crossbar, the values of the conductance in the crossbar can be written one by one [25]. For instance, from Fig. 8, to write the value G_{13} onto that specific ReRAM, a particular amplitude and writing pulse is applied to the ReRAM from the horizontal top wire, and the bottom wire is set to the ground voltage. Besides that specific ReRAM, the rest of the horizontal and vertical wires have half of the voltage applied to them. Since there is no voltage drop across the other devices, their conductances cannot be changed. The targeted ReRAM will experience the change in the conductance value with the application of the input voltage V that needs to be above the threshold of the ReRAM device. For the read operation, if, for instance, the desired ReRAM to read the voltage from is located at the first row and third column, a voltage of V_1 is applied to the first row while

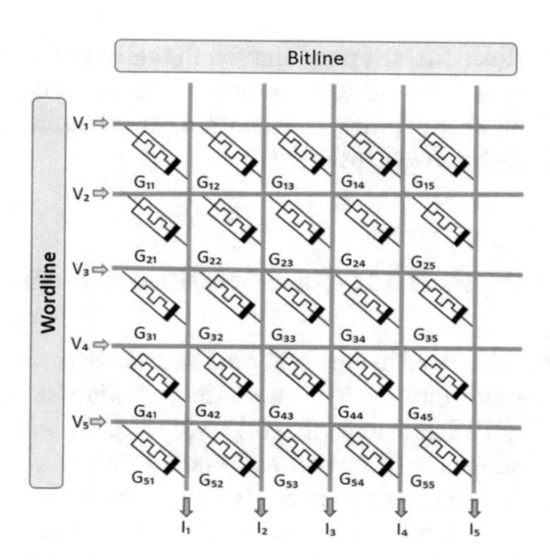

Fig. 8 A ReRAM crossbar

all other rows and columns are grounded. The output current can be calculated from the bitline using Eq. (2).

A major issue during the readout process of ReRAM crossbar is the problem of sneak path current. This is when the current leaks to the neighboring unselected ReRAM cells [25, 26]. Sneak path issue is being studied by researchers since it leads to increased energy consumption and reduces the read margins. To solve this problem, several access devices have been implemented with ReRAMs to form structures such as the one transistor-one memristor (1T1R) structure, one diode-one memristor (1D1R) structure, or one selector-one memristor (1S1R) structure. Using transistors as a selector device is most common among ReRAMs, giving them the 1T1R structure. They not only provide the selector function but also allow to control the switching behavior like the compliance current adjustment in ReRAMs. Using the 1T1R structure, each transistor can be used to select and update each ReRAM individually. Other common two terminal selectors for high-density crossbars include nonlinear devices, volatile switches, or rectifying diodes [27]. These configurations can be used to construct the crossbar and reduce the sneak path issue as well as facilitate its compatibility with CMOS technology.

ReRAMs have shown promising results in being integrated with CMOS technology in 3D configurations, leading to densely packed memory structures. Semiconductor/nanowire/molecular integrated circuits (CMOL) architecture is an example of a densely packed ReRAM crossbar array integrated with CMOS computational units [24, 28]. In the CMOL structure, the ReRAM crossbar array is slanted to the alignment of the CMOS neurons. The input and output of the CMOS neurons are connected to the top and bottom nanowires respectively by the use of the ReRAM device, as shown in Fig. 8. In simulation level, the ReRAM-based crossbar architectures PRIME and RESPARC have shown to achieve energy savings of more than 10^3 when compared to CMOS neural processor units [26]. CBRAM memristors integrated with CMOS neurons have been developed experimentally where these

ReRAMs are programmed digitally [29]. There has also been an implementation of a 3D large-scale crossbar of five layers of 100 nm crossbars [30]. CrossNets is another popular neuromorphic architecture that uses 3D integration of neurons and ReRAM synapses [31].

9 ReRAM-Based Spiking Neural Network

Compared to the traditional neural networks, SNNs mimic biological neuron models more closely by transmitting spiking signals. They are the third generation of artificial neural networks [32–34]. Similar to biological neurons, in SNNs, the signals accumulate, and once a threshold is exceeded, a spike is fired. They are more power and energy efficient due to the binary nature of spikes.

In SNNs, a common neuron model is the leaky integrate and fire (LIF) neuron model [35]. This model can be described using the following differential equations:

$$\tau_{\text{mem}} \frac{dV_i}{dt} = (V_i - V_{\text{rest}}) + RI_i - S_i(t)\,(V_i - V_{\text{rest}}) \tag{3}$$

$$\tau_{\text{syn}} \frac{dI_i}{dt} = -I_i(t) + \tau_{\text{syn}} \sum_j W_{ij} S_j(t) \tag{4}$$

From the equations, V_{rest} and V_i are the resting potential and the membrane potential of the neurons and τ_{syn} and τ_{mem} are the membrane time constants with I_i and R as the synaptic current and the input resistance. The input current is an integration of the spikes weighted together, and the Dirac delta function can be used to describe the spike train where the spiking time of the jth neuron is t_j^k:

$$S_j(t) = \sum_k \delta\left(t - t_j^k\right) \cdot t_j^k \tag{5}$$

Since ReRAM crossbar structures can be used to compute vector-matrix multiplications as discussed in the previous section, the weight matrix W_{ij} can be mapped into crossbar. In SNNs, the spike values can either be 0 or 1, and hence the output of the crossbar is an integrated sum of the input spikes.

10 Spike Time-Dependent Plasticity

SNNs have a major issue in terms of accuracy compared to other ANNs. Training SNNs are fairly difficult due to their spiking nature. The synapse between the two neurons carries the weight, and the main purpose of the training algorithm is to update the weights during training to be used for an application. Supervised and unsupervised learning are the two types of learning algorithms [35]. Supervised learning is where the datasets are labeled to correctly the classify the outputs. Unsupervised learning is where machine learning algorithms are used to group unlabeled datasets based on the similarities and the dissimilarities between them. Supervised backpropagation learning algorithms have been shown to be effective in ANNs to solve complex problems [35]. But for SNNs, supervised training methods do not exist due to their spiking nature. Backpropagation training techniques do not necessarily work well with SNNs because the spiking events become nondifferentiable while backpropagating and the information that was stored in spike timings gets lost.

One option explored by researchers was to train ANNs using static input images and then with the use of encoding to map the ANN to SNN [35–37]. Encoding schemes have received great attention in recent years in order to convert the inputs into spike events. There have been issues with this regarding efficiency and loss of information, for instance, when the information is encoded to spikes, they have values of either 0 or 1 and this may cause important information to vanish. Among the developed encoding schemes, the two common ones are rate and temporal encoding schemes. In rate encoding scheme, information is encoded in the frequency of the spikes, while in temporal encoding the information is encoded in the specific timing of the spikes [38, 39]. In all these networks, the training is carried out on the benchmark datasets like MNIST and CIFAR-10 which means they convert static images to spiking events. This indicates that the network is not a full SNN and it cannot take advantage of spatiotemporal information.

A very popular unsupervised learning technique in SNNs is the Hebbian learning rule, spike timing-dependent plasticity (STDP) [40–42]. This learning technique was initially developed in 1940 based on the dependence between the presynaptic and the postsynaptic spikes. The weight of the synapse increases if the presynaptic spike appears before the postsynaptic spike, also known as long-term potentiation (LTP), and the weight decreases if the presynaptic spike appears after the postsynaptic spike, also known as long-term depression (LTD) [40]. This STDP curve is illustrated in Fig. 9. The learning function for STDP is described in the following equation:

$$\text{STDP}\,(\Delta t) = \Delta w = |x| = \begin{cases} A^{-}e^{\Delta t/\tau-}, & \Delta t < 0 \\ A^{+}e^{-\Delta t/\tau+}, & \Delta t \geq 0 \end{cases} \tag{6}$$

where the terms A^{-} and A^{+} are the constants of the potentiation and the depression of the STDP curve for the time difference Δt between the presynaptic and the postsynaptic spikes and the terms $\tau+$ and $\tau-$ determine the gradient of the curve. The weight change is limited by the inequality $w_{\max} \geq w \geq w_{\min}$, and the weight adaptation speed is controlled by the weight change rate σ given by the

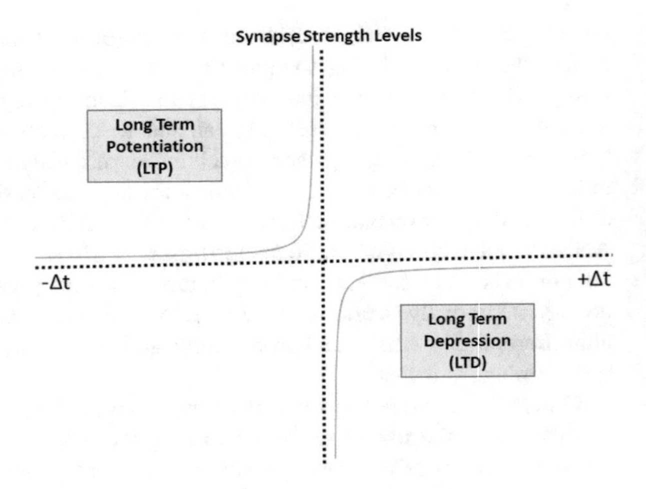

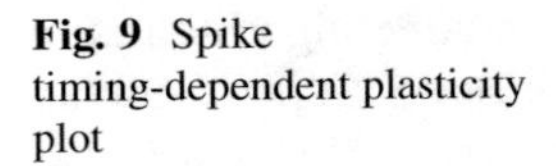
Fig. 9 Spike timing-dependent plasticity plot

following equation:

$$w_{\text{new}} = \begin{cases} w_{\text{old}} + \sigma \Delta w \left(w_{\max} - w_{\text{old}} \right), \Delta w > 0 \\ w_{\text{old}} + \sigma \Delta w \left(w_{\text{old}} - w_{\min} \right), \Delta w \leq 0 \end{cases} \tag{7}$$

11 STDP Functionality in ReRAMs

ReRAMs have shown to implement STDP characteristics as discussed previously. When voltage is applied on electrodes of these ReRAM devices, the conductance level of the device changes based on the timings of the applied voltage spikes. Recent research has shown several experimental demonstrations of this STDP functionality in binary ReRAM devices involving material of TiO_x, Al_2O_3/TiO_2, WO_x, TaO_x/Ta_2O_5, HfO_2, and CeO_x [43–48]. In this chapter we will demonstrate the STDP results from the $Pt/Al_2O_3/TiO_{2\text{-}x}/Ti/Pt$ ReRAMs that were fabricated in a 12 × 12 crossbar circuit to demonstrate the STDP mechanism in ReRAM-based materials [45]. In this work the authors used three different dependencies of weight change between the presynaptic spikes and the postsynaptic spikes to show the STDP behavior in weight updates, as depicted in Fig. 10g, h and i. The initial conductance value of the ReRAM was set to 33μS, and the presynaptic and the postsynaptic pulses applied to the electrodes of the selected ReRAM are the ones depicted in Fig. 10a, b, and c with a specific delay time of Δt between them. The memristor's new conductance value was measured afterward and calculated. The different pulse shape produces the different STDP curves. Using the waveform from Fig. 10a, the STDP function was measured for the different conductance values from G = 25 μS, 50 μS, 75 μS, and 100 μS. By changing the initial conductance value, each ReRAM develops its own dynamic range and hence produces a different STDP

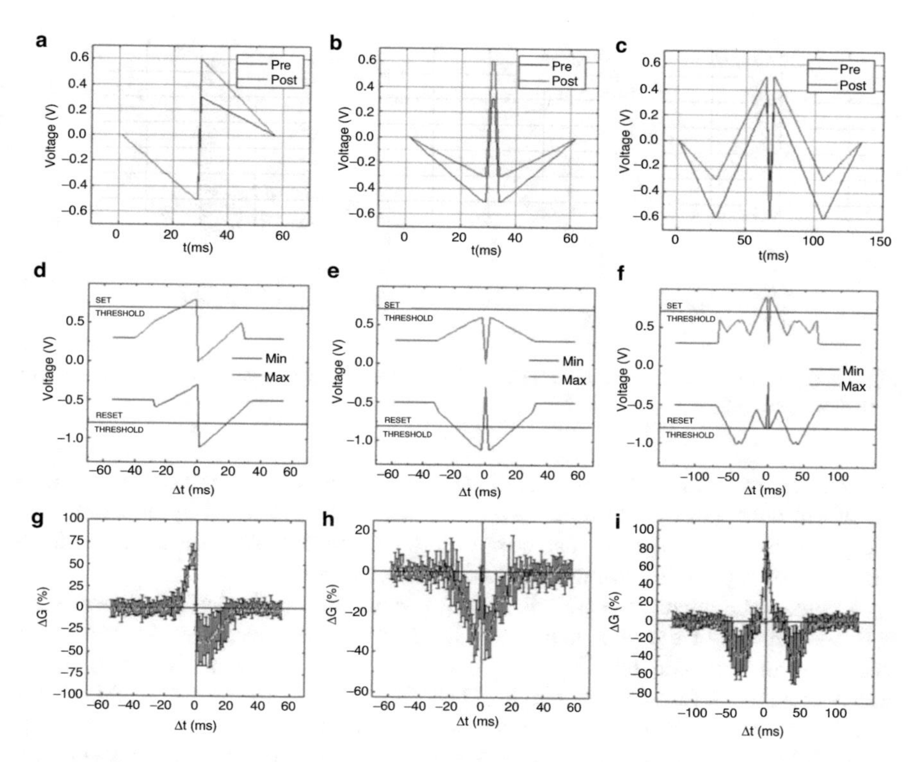

Fig. 10 STDP implementation emulating the biological synapses in Layer 5 and Layer 4 of the neurocortex, left and middle column respectively, and in the GABAergic synapses in the right column. (**a-c**) The applied pre synaptic and post-synaptic voltage pulses. (**d-f**) The applied time maxima and minima of the net voltage given as functions of the time interval between the pre and post-synaptic pulses that are applied to the memristor. (**g-i**) The measured STDP window with the red points showing the averages and the black error bars showing the standard deviations for 10 experiments for each time interval [45]

curve, as shown in Fig. 11. From the plot it can be observed that when the initial conductance value is nearer to the minimum value, the gradient of the LTD is very low, while when the initial conductance value is nearer to the maximum value, the gradient of the LTP is very low.

The STDP mechanism is demonstrated in another ReRAM device of the Ag/TiO_2:Ag/Pt configuration developed in [49]. This device shows that the pulse width that needs to be applied to carry out the STDP learning is in the order of nanoseconds, 10^5 times quicker than the human brain and significantly faster than other oxide-based memristors. This means there will be a large change in conductance as spike times are closer to each other. Hence this specific structure can be used to speed up neuromorphic applications. It can be seen that two different STDP learning rules can be implemented using this device. A biomorphic action potential-like waveform was used to obtain the asymmetric Hebbian learning rule.

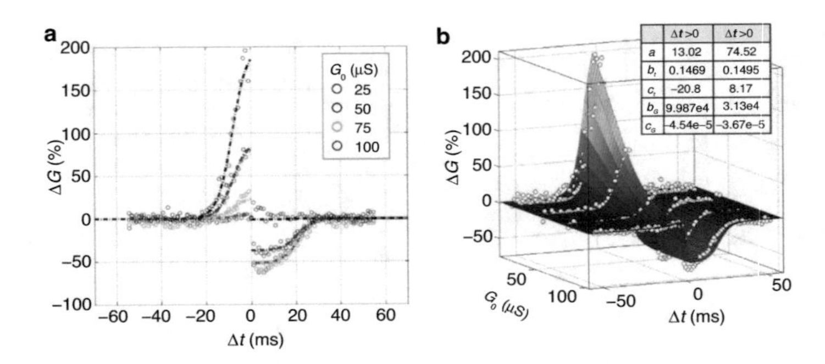

Fig. 11 (**a**) Experimentally measured STDP for different conductances. (**b**) 3D sur-face plot of the STDP curve [45]

Meanwhile, the anti-asymmetric Hebbian learning rule was obtained by using a waveform with a different order. Two STDP plots were obtained by injecting pre- and post-spiking pulse pairs that were preprogrammed in different time windows starting from 6 μs to 200 ns, demonstrating its rapidness.

A major challenge in STDP is that the synaptic weight update depends on both the spiking of the presynaptic and the postsynaptic signals, and when using ReRAMs, they should be adaptable to changes in conductance based on these signals. To successfully implement the STDP functionality, ReRAMs should be able to achieve different stable states depending on the applied signals, demonstrating their synaptic behavior. There have been several demonstrations of STDP on ReRAM devices, making them a suitable candidate for AI accelerators. Their ability to mimic the memory functions as well as the biological learning capabilities of synapses will allow ReRAMs to confront the von Neumann bottleneck problems in neuromorphic computing systems.

12 ReRAM-Based SNN Architectures

With SNNs being the third generation of ANNs, there have been several implementations of ReRAM-based SNNs. They take advantage of both the spatiotemporal capabilities of SNN and the synaptic properties of ReRAMs and use them as the vector-matrix multiplication blocks in the crossbar units. One such example is the ReRAM-based SNN architecture that exploits the computing-in-memory (CIM) property of ReRAMs [50, 51]. The designed architecture consists of four major parts: the inter-spike interval (ISI) encoded input layer, the fabricated memristor crossbar array to carry out the vector-matrix multiplication, hidden stages to transmit signals from one stage to the next, and a final output layer with a time-to-first-spike (TTFS) decoding scheme. The overall architecture is shown in Fig. 12a. From the architecture it can be seen that the inputs applied (images shown in this case) are converted to spiking signals and the information is encoded in the times between the spikes that are the ISI encoded signals. It is a temporal encoding scheme where the inputs are encoded in two dimensions, both the timing of the spikes and the time between the spikes. This encoding scheme is carried out by developing input preprocessing units using CMOS 180 nm technology. The units are based on a transconductance amplifier voltage-to-current conversion module, two LIF neuron units to generate the spiking signals, and a charge pump-based extractor unit. The matrix-vector multiplications are carried out in the intermediate stages and the signal is converted back to an ISI encoded signal. In the final output stage, the classification is done using a TTFS decoding scheme where the output from the ReRAM crossbar is converted to a spiking signal. This is done via a current amplifier to amplify the outgoing current from the crossbar and then feeding the signal to an LIF neuron module to generate spikes. The outputs are then classified based on whichever output neuron spikes first.

For the evaluation of the network, 5×4 images of the digits 0–9 were used and inputted to the system, as shown in Fig. 12b. For each clock cycle of 0.5 MHz, each image is processed, and the network is able to successfully identify all the digits. The designed ReRAM-based SNN architecture shows competitive results compared to the state-of-the-art neuromorphic architecture with a power consumption of 2.9 mW in evaluating images from 0–9 with an inference speed of 2 μS. A large-scale three-layer software model is developed to test the network against the benchmark handwritten dataset of MNIST. The software model uses the conductance parameters and models the memristor crossbar for the vector-matrix multiplication layers and adds the ISI and TTFS layers for the input and output layers, respectively. Based on the simulation results, using the ISI encoding scheme with the ReRAM crossbar provides the highest accuracy of 87%.

Another work realizes a novel SNN with ReRAM-based inhibitory synapses to demonstrate the lateral inhibition and homeostasis [52]. The advantage is that it can reduce the number of connections that are needed for lateral inhibition from N^2 to N and reduces the hardware complexity. The schematic of the fully connected and convolutional SNN is shown in Fig. 13. There is an inhibitory neuron and the ReRAM-based inhibitory synapses that were added to obtain homeostasis and lateral inhibition. The inhibitory neurons receive spikes through the ReRAM synapse from the excited neurons in the learning layer, and then this inhibitory neuron sends spikes to the connected excitatory neurons to carry out the task of lateral inhibition. Software simulations were carried out against the benchmark MNIST dataset, and the demonstrated method shows two times higher accuracy than unsupervised STDP-based SNNs. In these results, the nonideal characteristics of the ReRAM devices were also considered when carrying out the simulations such as limited resistor states, device variations, and open devices.

Although SNNs are faster and more power and energy efficient than traditional artificial neural networks, they suffer from the memory bottleneck issue of von Neuman computing systems. As we move into the future, the increase in the demand for data is becoming tremendous, and with the use of ReRAMs in SNNs for CIM operations, the issue of memory wall can be mitigated. The computations can be

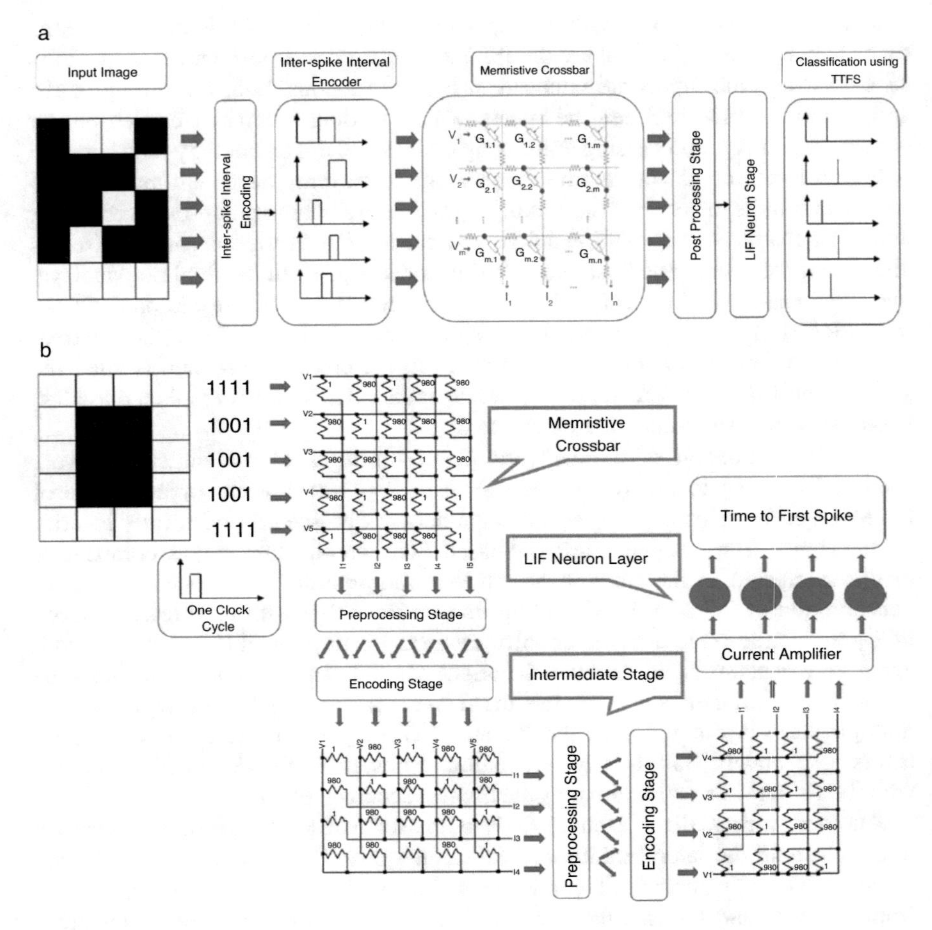

Fig. 12 (**a**) Architecture of the memristor-based spiking neural network. (**b**) Hardware simulation setup for the architecture

done inside the memory, using these ReRAM devices as storage. With developments in ReRAM-based SNN architectures, we can move one step closer to high-speed and compact IoT devices.

13 Other ReRAM-Based Neural Networks

Apart from being used in SNNs, ReRAMs have also been used in other types of neural networks like CNNs and RNNs. CNNs are mainly useful for large data analysis in the field of computer vision and consist of alternating convolution

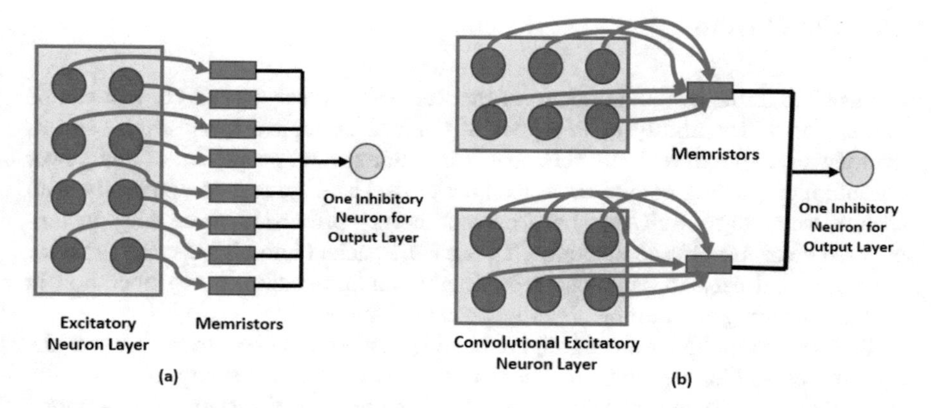

Fig. 13 The schematic of the proposed (**a**) fully connected and (**b**) convolutional SNNs for lateral inhibition and homeostasis

and pooling layers for feature extraction from images and then finally fully connected layers for classification [53, 54]. ISAAC, PipeLayer, and AtomLayer are some popular ReRAM-based CNN accelerators [55–57]. The initially developed pipelined architecture ISAAC demonstrated 14.8×, 5.5×, and 7.5× improvements in throughput, energy, and computational density compared to the DaDianNao architecture. While ISAAC does not consider weight updates and is susceptible to pipeline bubbles, PipeLayer uses highly pipelined execution of both training and testing [56]. On the other hand, AtomLayer uses atomic layer computation to carry out one network at a time to address the issues arising from the pipelined architectures and achieves a high power efficiency of 1.1× in inference for ISAAC and 1.6× in training for PipeLayer [57].

Another type of neural network where ReRAMs are used are RNNs. Derived from feedforward neural networks (FNNs), RNNs contain internal loops inside the hidden layers, giving them the recurrent connection property and allowing the information to stay in the network for a certain time [58]. They also mimic biological neurons more closely due to their dynamic property that allows them to process information in both spatial and temporal domains. In [59] a ReRAM-based processing-in-memory (PIM) architecture is designed that is able to accelerate RNN computation. The system throughput is increased using an RNN-friendly pipeline, and the architecture achieves a 79× higher computing efficiency compared to GPU baselines. A hardware implementation of RNN using ReRAM crossbar is demonstrated in [60] that uses feedforward and feedback matrices. Based on the simulation results, it was shown that compared to CMOS implementations, ReRAM synapses show significant improvements in terms of energy and computing efficiency. Furthermore, using ReRAMs for RNNs showed greater improvement in terms of speed compared to FNNs.

14 Conclusion

In this chapter the overall structure of the ReRAM is discussed and how it can be implemented. The ability of ReRAMs to emulate biological synapses as well as neurons are highlighted in this chapter. With their several properties of low power consumption, analog conductance modulation, scalability, and compatibility with CMOS technology, ReRAMs have become a very promising technology in bio-inspired neuromorphic applications. They can be packed into high-density crossbar structures and used to carry out vector-matrix multiplication, a key operation in machine learning and artificial intelligence applications.

ReRAM-based SNNs are highlighted in this chapter to demonstrate how both the event-based spiking networks can be combined with ReRAM synapses, especially since SNNs are the closest neural networks to the biological systems. Furthermore, with their STDP learning capability, ReRAMs pave the way for efficient training in SNNs, addressing the major issue of accuracy in these types of neural networks. The use of ReRAMs has also been seen in other neural networks like CNNs and RNNs. While ReRAMs still suffer from certain issues like sneak path current, unstable resistance states, and reliability issues, they have great potential in replacing CMOS circuitry and analog-to-digital and digital-to-analog converters and providing significant improvements in power, area, and energy consumptions. Ongoing research on ReRAMs and ReRAM-based neural networks are paving the way for efficient neural network accelerators.

References

1. Haron, N.Z., Hamdioui, S.: Why is CMOS scaling coming to an END? In: 2008 3rd International Design and Test Workshop, pp. 98–103 (2008). https://doi.org/10.1109/IDT.2008.4802475
2. Gubbi, J., Buyya, R., Marusic, S., Palaniswami, M.: Internet of Things (IoT): a vision architectural elements and future directions. Future Gener. Comput. Syst. **29**(7), 1645–1660 (2013)
3. Yocam, E.W.: Evolution on the network edge: intelligent devices. IT Professional. **5**(2), 32–36 (2003). https://doi.org/10.1109/MITP.2003.1191790
4. Li, C., et al.: Analogue signal and image processing with large memristor crossbars. Nat. Electron. **1**(1), 52–59 (2018)
5. Chua, L.: Memristor-the missing circuit element. IEEE Trans. Circuits Theory. **18**(5), 507–519 (1971). https://doi.org/10.1109/TCT.1971.1083337
6. Backus, J.: Can programming be liberated from the Von Neumann style? A functional style and its algebra of programs. Commun. ACM. **21**, 613–641 (1978)
7. Wong, H.-S.P., et al.: Metal–oxide RRAM. Proc. IEEE. **100**(6), 1951–1970 (2012)
8. Upadhyay, N.K., Jiang, H., Wang, Z., Asapu, S., Xia, Q., Yang, J.J.: Emerging memory devices for neuromorphic computing. Adv. Mater. Technol. **4**(4) (2019)
9. Yu, S., Chen, P.: Emerging memory technologies: recent trends and prospects. IEEE Solid-State Circuits Mag. **8**(2), 43–56 (2016). https://doi.org/10.1109/MSSC.2016.2546199
10. Xie, Y., Zhao, J.: Emerging memory technologies. IEEE Micro. **39**(1), 6–7 (2019). https://doi.org/10.1109/MM.2019.2892165

11. Park, J.: Neuromorphic computing using emerging synaptic devices: a retrospective summary and an outlook. Electronics. **9**(9), 1414 (2020)
12. Keshmiri, V.: A Study of the Memristor Models and Applications (2014)
13. Strukov, D.B., Snider, G.S., Stewart, D.R., Williams, R.S.: The missing memristor found. Nature. **453**(7191), 80–83 (2008)
14. Williams, S.R.: How we found the missing memristor. Spectrum, IEEE. **45**(12), 28–35 (2008)
15. Gerstner, W., Kistler, W.M.: Spiking Neuron Models. Cambridge Univ. Press, Cambridge (2002)
16. Moore, S.: Memristor breakthrough: first single device to act like a neuron. IEEE Spectrum. (2020)
17. Mehonic, A., Kenyon, A.J.: Emulating the electrical activity of the neuron using a silicon oxide RRAM cell. Front. Neurosci. **10**, 57 (2016)
18. Babacan, Y., Kaçar, F., Gürkan, K.: A spiking and bursting neuron circuit based on memristor. Neurocomputing. **203**, 86–91 (2016)
19. Nakada, K.: Neural pulse coding using ReRAM-based neuron devices. IEICE Tech. Rep. **117**(415), 63–68 (2018)
20. Kumar, S., Williams, R.S., Wang, Z.: Third-order nanocircuit elements for neuromorphic engineering. Nature. **585**(3474), 518–523 (2020)
21. Zhirnov, L., Cavin, R., Gammaitoni, L.: Minimum energy of computing fundamental considerations. In: ICT-Energy-Concepts Towards Zero-Power Info. and Commun. Technology, vol. 7, (2014)
22. Mead, C.: Neuromorphic electronic systems. Proc. IEEE. **78**(10), 1629–1636 (1990)
23. Walczak, S., Narciso, C.: Artificial neural networks. In: Encyclopedia of Physical Science and Technology, 3rd edn, pp. 631–645 (2003)
24. Huang, A., et al.: Memristor neural network design. In: Memristor and Memristive Neural Networks, pp. 1–35 (2018)
25. Shevgoor, M., Muralimanohar, N., Balasubramonian, R., Jeon, Y.: Improving memristor memory with sneak current sharing. In: 2015 33rd IEEE International Conference on Computer Design (ICCD), pp. 549–556 (2015)
26. Camunas-Mesa, L.A., Linares-Barranco, B., Serrano-Gotarredona, T.: Neuromorphic spiking neural networks and their memristor-CMOS hardware implementations. Materials. **12**(17) (2019)
27. Chen, Y.-C., Lin, C.-C., Hu, S.-T., Lin, C.-Y., Fowler, B., Lee, J.: A novel resistive switching identification method through relaxation characteristics for sneak-path-constrained selectorless RRAM application. Sci. Rep. **9**(1), 1–6 (2019)
28. Likharev, K.K., Strukov, D.B.: Introducing Molecular Electronics. Springer-Verlag, New York (2004)
29. Kim, K., et al.: A functional hybrid memristor crossbar-array/CMOS system for data storage and neuromorphic applications. Nano Lett. **12**(1), 389–395 (2011)
30. Li, C., Han, L., Jiang, H., Jang, M.-H., Lin, P., Wu, Q., et al.: Three-dimensional crossbar arrays of self-rectifying Si/SiO_2/Si memristors. Nat. Commun. **8**, 719–813 (2017)
31. Likharev, K.K.: CrossNets: neuromorphic hybrid CMOS/nanoelectronic networks. Sci. Adv. Mater. **3**(3), 322–331 (2011)
32. Ghosh-Dastidar, S., Adeli, H.: Spiking neural networks. Int. J. Neural Syst. **2009**, 295–308 (2009)
33. Wu, Y., Deng, L., Li, G., Zhu, J., Shi, L.: Spatio-temporal backpropagation for training high-performance spiking neural networks. arXiv preprint arXiv:1706.02609. (2017)
34. W. Maass, "Networks of spiking neurons: the third generation of neural network models," 1997.
35. Fouda, M., Kurdahi, F., Eltawil, A., Neftci, E.: Spiking neural networks for inference and learning: a memristor-based design perspective. arXiv preprint arXiv:1909.01771. (2019)
36. Diehl, P.U., Neil, D., Binas, J., Cook, M., Liu, S.-C., Pfeiffer, M.: Fast-classifying high-accuracy spiking deep networks through weight and threshold balancing. Proc. Int. Joint Conf. Neural Netw. **2015**, 2933–2940 (2015)

37. Rueckauer, B., Lungu, I.-A., Hu, Y., Pfeiffer, M., Liu, S.-C.: Conversion of continuous-valued deep networks to efficient event-driven networks for image classification. Front. Neurosci. **11**, 682 (2017)
38. Zhao, C., Wysocki, B.T., Liu, Y., Thiem, C.D., McDonald, N.R., Yi, Y.: Spike-time-dependent encoding for neuromorphic processors. ACM J. Emerg. Technol. Comput. Syst. **12**(3), 23–46 (2015)
39. Yu, Q., Tang, H., Tan, K.C., Yu, H.: A brain-inspired spiking neural network model with temporal encoding and learning. Neurocomputing. **138**, 3–13 (2014)
40. Iakymchuk, T., Rosado-Muñoz, A., Guerrero-Martínez, J.F., Bataller-Mompeán, M., Francés-Víllora, J.V.: Simplified spiking neural network architecture and STDP learning algorithm applied to image classification. EURASIP J. Image Video Process. **2015**(1), 4 (2015)
41. Shuai, Y., Pan, X., Sun, X.: Spike-timing-dependent plasticity in memristors. In: Memristor and memristive neural networks. IntechOpen, London (2017. [Online]. Available: https://www.intechopen.com/chapters/56763). https://doi.org/10.5772/intechopen.69535
42. Frohlich, F.: Network Neuroscience. Academic Press, Cambridge, USA (2016)
43. Seo, K., Kim, I., Jung, S., Jo, M., Park, S., Park, J., et al.: Analog memory and spike-timing-dependent plasticity characteristics of a nanoscale titanium oxide bilayer resistive switching device. Nanotechnology. **22**, 254023 (2011)
44. Tan, Z.-H., Yang, R., Terabe, K., Yin, X.-B., Zhang, X.-D., Guo, X.: Synaptic metaplasticity realized in oxide memristive devices. Adv. Mater. **28**(2), 377–384 (2015)
45. Prezioso, M., Merrikh-Bayat, F., Hoskins, B., Likharev, K., Strukov, D.: Self-adaptive spike-time-dependent plasticity of metal-oxide memristors. arXiv Preprint arXiv:1505.05549. (2015)
46. Hsieh, C.-C., et al.: A sub-1-volt analog metal oxide memristive-based synaptic device with large conductance change for energy-efficient spike-based computing systems. Appl. Phys. Lett. **109**(22), 223501 (2016)
47. Kim, S., Choi, S., Lu, W.: Comprehensive physical model of dynamic resistive switching in an oxide memristor. ACS Nano. **8**(3), 2369–2376 (2014)
48. Matveyev, Y., et al.: Crossbar nanoscale HfO2-based electronic synapses. Nanoscale Res. Lett. **11**(1), Dec (2016)
49. Yan, X., et al.: Memristor with Ag-cluster-doped TiO2 films as artificial synapse for neuroinspired computing. Adv. Funct. Mater. **28**(1), 1705320 (2017)
50. Nowshin, F.: Spiking neural network with memristive based computing-in-memory circuits and architecture. M.S. Thesis, Bradley Department of Electrical and Computer Engineering, Virginia Tech, VA (2019)
51. F. Nowshin, Y. Yi, "Memristor-based deep spiking neural network with a computing-in-memory architecture", n 2022 23rd International Symposium on Quality Electronic Design (ISQED), pp. 1-6. IEEE, 2022
52. Zhao, Z., et al.: A memristor-based spiking neural network with high scalability and learning efficiency. IEEE Trans. Circuits Syst. II Exp. Briefs. **67**(5), 931–935 (2020)
53. Kamencay, P., Benco, M., Mizdos, T., Radil, R.: A new method for face recognition using convolutional neural network. Digit. Image Process. Comput. Graph. **15**(4), 664–672 (2017)
54. Albawi, S., Mohammed, T.A., Al-Zawi, S.: Understanding of a convolutional neural network. In: Engineering and Technology (ICET) 2017 International Conference on, pp. 1–6. IEEE (2017)
55. Shafiee, A., et al.: ISAAC: a convolutional neural network accelerator with in-situ analog arithmetic in crossbars. In: Proc. ISCA, pp. 14–26 (2016)
56. Song, L., Qian, X., Li, H., Chen, Y.: PipeLayer: a pipelined ReRAM-based accelerator for deep learning. In: 2017 IEEE International Symposium on High Performance Computer Architecture (HPCA), pp. 541–552 (2017). https://doi.org/10.1109/HPCA.2017.55
57. Qiao, X., et al.: Atomlayer: a universal reram-based cnn accelerator with atomic layer computation. In: DAC (2018)
58. Schmiduber, J.: Deep learning in neural networks: an overview. Neural Netw. **61**, 85–117 (2015)

59. Long, Y., Na, T., Mukhopadhyay, S.: ReRAM-based processing-in-memory architecture for recurrent neural network acceleration. IEEE Trans. Very Large Scale Integr. VLSI Syst. **26**(12), 2781–2794 (2018). https://doi.org/10.1109/TVLSI.2018.2819190
60. Long, Y., Jung, E.M., Kung, J., Mukhopadhyay, S.: ReRAM crossbar based recurrent neural network for human activity detection. In: 2016 International Joint Conference on Neural Networks (IJCNN), pp. 939–946 (2016). https://doi.org/10.1109/IJCNN.2016.7727299

Flash: A "Forgotten" Technology in VLSI Design

Sunil P. Khatri, Sarma Vrudhula, Monther Abusultan, Kunal Bharathi, Shao-Wei Chu, Cheng-Yen Lee, Kyler R. Scott, Gian Singh, and Ankit Wagle

1 Chapter Summary

In this chapter, we begin in Sect. 2 with a background in floating gate technologies. In Sect. 3, we describe how flash transistors can be used to realize ASIC designs with significantly improved power, delay, and area metrics. Features such as the ability to control speed binning, mitigation of transistor aging, and delay tuning are quantified. In Sect. 4, we describe how flash can be used to realize both digital and analog convolutional neural network (CNN) accelerators. Section 5 described a flash-based processing in-memory approach, while Sect. 6 describes flash-based analog circuit design for DACs and LDO (low drop-out) DC-DC converters.

2 Technology Overview

In this section, we present a background on flash technology that is necessary for understanding the promise of flash—in processing in non-volatile memory (NVM), digital and analog circuit designs, machine learning accelerators, and more. In Sect. 2.1, we describe the structure of a basic flash transistor, and the methods used to adjust its device threshold voltage, V_{th}. Next, we present a survey of existing

S. P. Khatri (✉) · M. Abusultan · K. Bharathi · S.-W. Chu · C.-Y. Lee · K. R. Scott
Texas A&M University, College Station, TX, USA
e-mail: sunilkhatri@tamu.edu; abusultan@tamu.edu; kunal-bharathi@tamu.edu; shaowei22@tamu.edu; cylee@tamu.edu; kylerrscott@tamu.edu

S. Vrudhula · G. Singh · A. Wagle
Arizona State University, Tempe, AZ, USA
e-mail: vrudhula@asu.edu; gsingh58@asu.edu; awagle1@asu.edu

A. Iranmanesh (ed.), *Frontiers of Quality Electronic Design (QED)*,
https://doi.org/10.1007/978-3-031-16344-9_3

flash devices. Finally in Sect. 2.2, we show how "pseudo-flash" transistors can be constructed from regular MOSFETs.

2.1 Flash Transistors

As depicted in Fig. 1, a flash transistor is a field-effect transistor (FET) with two gates: a control gate, which is similar to the control gate of a CMOS transistor, and a buried and uncontacted *floating gate*, which acts as a capacitor that can store electric charge. By applying voltage pulses to the transistor terminals, one can *program* or *erase* the device. This changes the voltage of the floating gate, V_{fg}, and is used to set the transistor threshold voltage, V_{th}. During programming, electrons are forced through the oxide and become trapped on the floating gate, thereby decreasing V_{fg} and increasing V_{th}. The movement of electrons through the oxide is referred to as *tunneling*. During erasure, electrons are removed from the floating gate, which increases V_{fg} and decreases V_{th}. Erasure will result in a V_{th} that is negative. When done correctly, programming will result in a desired V_{th}.

There are two primary methods for inducing electron tunneling: Fowler-Nordheim (FN) tunneling [1] (Fig. 2a) and hot-electron injection (HEI) [2] (Fig. 2b). In FN tunneling, erasure is achieved by driving the control gate to GND, floating the source and drain terminals, and driving the substrate (bulk) to a voltage of 10–20 V [3–8]. For area efficiency reasons, several flash transistors share a common bulk, so erasure is performed on all of them simultaneously. In flash memory designs, erasure is applied to an entire block. Programming via FN tunneling is achieved by driving the source, drain, and bulk terminals to GND while applying voltage pulses to the control gate. Each pulse will tunnel some electrons through the oxide and onto the floating gate, causing an increase of the transistor threshold voltage, ΔV_{th}. The magnitude of ΔV_{th} for each voltage pulse is determined by the height and duration of the pulse, governed by the Incremental Step Pulse Programming (ISPP) model [9, 10]. Under the ISPP model, the device undergoes several program-verify cycles, until the desired V_{th} is achieved. The V_{th} after N_s programming pulses is given by $V_{th} = V_{th}^{init} + \beta \Delta V_{pp} N_s$, where β is a material-dependent constant and ΔV_{pp} is the pulse step increment. The ISPP can be used to achieve a very precise V_{th} with a

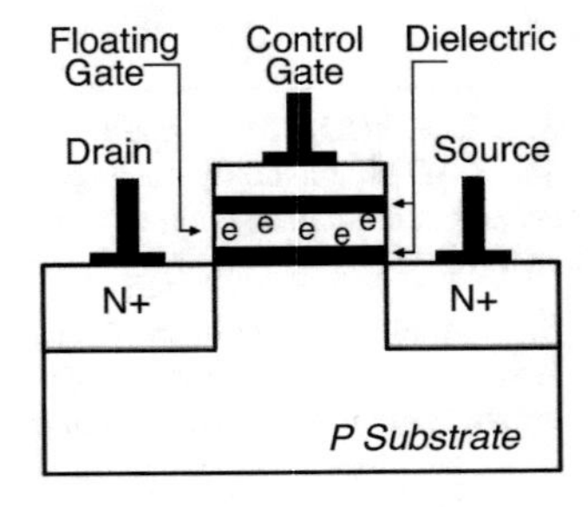

Fig. 1 Cross section of a flash transistor

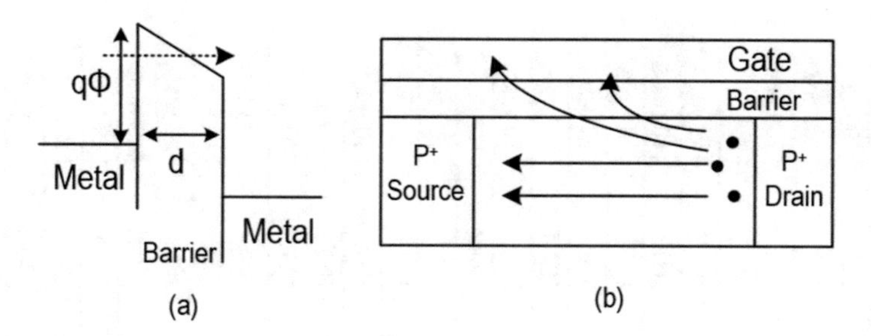

Fig. 2 Schematic diagram of (**a**) FN tunneling and (**b**) hot-electron injection

granularity approaching ΔV_{pp} [9, 11], at the cost of increased N_s. Once electrons are trapped in the floating gate, they remain trapped for several years [12, 13], or until removed by an erase operation.

When using HEI, electrons can be tunneled onto the floating gate, reducing V_{fg} and increasing V_{th}. A large gate-source voltage (V_{gs}) is applied. A high drain-source voltage (V_{ds}) is applied in pulses. Each pulse creates a high electric field, and causes electrons to gain energy as they travel through the field. Electrons are pulled toward the floating gate due to the high V_{gs}. If the energy of electrons is higher than the barrier of the oxide, some electrons tunnel through the oxide and are trapped on the floating gate. The amount of electrons tunneled with each pulse, and therefore ΔV_{th}, is determined by the duration and amplitude of the V_{ds} pulses.

Different flash technologies have been developed and have matured over the years to suit various applications. Table 1 summarizes the technology, characteristics, and durability of various embedded flash devices over the years. For each device, we present technological details (such as process node, charge storage material, additional mask layers required, and operating voltages) as well as performance characteristics such as endurance and retention.

2.2 *Pseudo-Flash Transistor*

As discussed in Sect. 2.1, flash transistors need additional steps in the fabrication flow to form a second gate for storing electrons. As a result, previous works have proposed floating gate transistors that use a vanilla CMOS process. We call these devices "pseudo-flash" transistors. A pseudo-flash transistor is a device created from multiple MOSFETs. Some of these MOSFETs act as capacitors, and they electrically isolate a node that will act as the floating gate. Like embedded flash transistors, charge can be stored on this floating gate to choose the transistor V_{th}.

The works of [42–53] propose a pseudo-flash transistor shown in Fig. 3. It consists of a PMOS transistor (M_{fg}), a poly-poly capacitor (C_g), and a MOS capacitor (C_{tun}). The gate input of the PMOS transistor is coupled by two capacitors. There is no DC path that allows electrons to escape from the node V_{fg}. To increase

Table 1 Summary of embedded flash devices

Type	Ref	Technology				Electrical characteristics			Durability	
		Node (nm)	Charge storage material	Extra masks	Storage levels	Vprog	Verase	Voperating	Endurance	Retention
EEPROM	[14]	2500	Polysilicon	1	4	13 V	13 V	5 V	10 K	10 years
HiV EEPROM	[15]	2000	Polysilicon	0	2	16 V	16 V	10 V	10 K	1000 h
C-Flash	[16]	180	Polysilicon	3	2	5 V	5 V	1.8 V	1 K	100 years
CMOS	[17]	180	Polysilicon	0	2	4.75 V	4.75 V	1.8 V	1 K	100 years
EEPROM	[18]	180	Polysilicon	0	2	4 V	4.5 V	1.8 V	10 K	NA
MONOS	[19]	180	Nitride	7	4	4.5 V	4.5 V	1.8 V	100 K	10 years
CMOS	[20]	130	Polysilicon	NA	2	12 V	12 V	5 V	10 K	10 years
EEPROM	[21]	130	Polysilicon	NA	4	8 V	14 V	3.3 V	10 K	10 years
Embedded flash	[22]	130	Polysilicon	2	2	7 V	7 V	1.2 V	NA	10 years
SONOS	[23]	130	Nitride	0	4	5 V	5.5 V	3 V	100 K	10 years
Split-gate	[24]	130	Polysilicon	NA	2	8 V	8 V	4 V	NA	10 years
1T MONOS	[25]	90	Nitride	NA	2	PHV	NHV	3.3 V	100M	NA
2T-SONOS	[26]	90	Nitride	NA	2	7 V	8 V	3.3 V	1 K	10 years
Charge trapping	[27]	90	Nitride	NA	2	10 V	17 V	NA	100 K	NA
eNVM	[28]	90	Polysilicon	0	2	8 V	8 V	3.3 V	500	10 years
eNVM	[29]	65	Polysilicon	0	2	8.8 V	8.8 V	2.5 V	10 K	486 h
Split-gate	[30]	65	Polysilicon	1	2	11 V	13 V	3.3 V	10 K	10 years
SONOS Pchannel	[31]	50	Nitride	NA	2	12 V	12 V	1.8 V	10 K	10 years
Split-gate	[32]	45	Polysilicon	NA	2	NA	NA	1.8 V	1 M	1000 h
SG MONOS	[33]	40	Nitride	NA	2	PHV	NHV	1.25 V	10M	20 years
SONOS	[34]	40	Nitride	0	2	4 V	4 V	0.6 V	100 K	10 years
Split-gate	[35]	40	Polysilicon	0	2	10.5 V	11.5 V	1.1/2.5 V	200 K	10 years
Charge trapping	[36]	32	Polysilicon	NA	8	NA	NA	NA	1M	NA
Charge trapping	[37]	22	HiK dielectric HfO2	0	2	2 V	2 V	1 V	10-1 K	10 years
SG MONOS	[38]	20	Nitride	NA	8	PHV	NHV	1 V	10 K	NA
FinFET eNVM	[39]	16	HiK dielectric	0	2	2 V	2 V	0.8 V	NA	10 years
FinFET CTT	[40]	14	HiK dielectric HfO2	NA	2	2 V	2 V	VDD	10 K	10 years
FinFET MONOS	[41]	14	Nitride	3	2	PHV	NHV	VDD	250 K	10 years

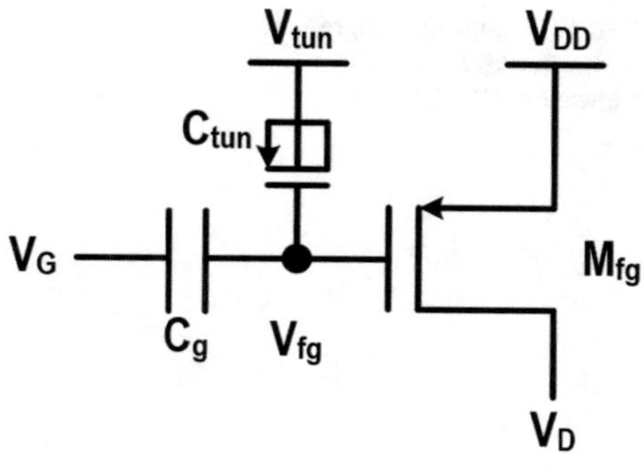

Fig. 3 The circuit diagram of a pseudo-flash transistor [42–53]

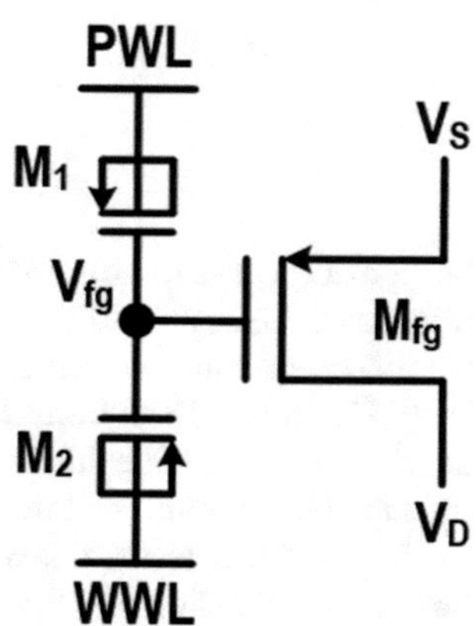

Fig. 4 The circuit diagram of a pseudo-flash transistor [54–56]

V_{fg}, FN tunneling is used to remove electrons from node V_{fg}. By increasing the tunneling voltage (V_{tun}), the electric field across the C_{tun} is increased, reducing the effective oxide thickness. This results in an increased probability of electron tunneling through the potential barrier. To decrease V_{fg}, HEI is used to inject electrons into node V_{fg}. A large voltage is applied across the drain and source of M_{fg}, creating a high electric field region between the drain and source terminals. Electrons gain energy when passing through the high electric field region. If the energy of electrons is higher than the barrier of the oxide, some electrons cross into the oxide and are trapped in the node V_{fg}. By using these two effects, the V_{fg} terminal of the pseudo-flash transistor can be set to different voltage levels, effectively changing the V_{th} of M_{fg}.

In [54–56], three transistors are used to form a pseudo-flash transistor, as shown in Fig. 4. M_{fg} is the read/program device, and M_1 and M_2 are the coupling device and the tunneling device, respectively. The size of M_1 needs to be larger compared to the size of M_{fg} and M_2 to cause a strong coupling between substrate terminal of M_1 and node V_{fg}. M_1, M_2, and M_{fg} block any DC path and form a node V_{fg} to store electrons. FN tunneling is used for the programming to add/remove electrons to/from node V_{fg}. To remove electrons, a boosted voltage is applied to write wordline (WWL), and program wordline (PWL) is set to 0 V. This creates a

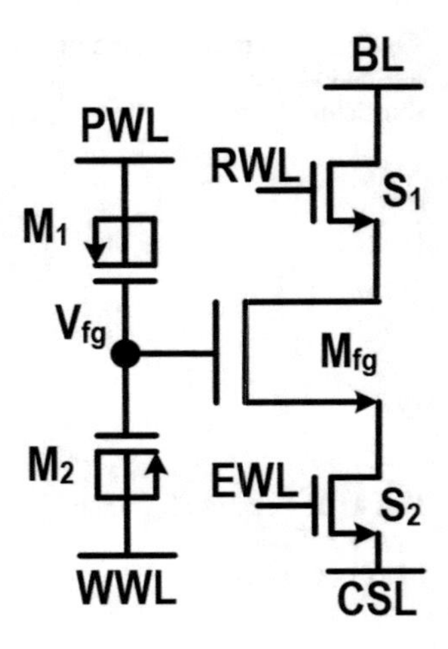

Fig. 5 The circuit diagram of a pseudo-flash transistor [57–59]

large electric field across M_2, resulting in a reduction in the effective oxide layer thickness. Therefore, the probability of the electrons to exit node V_{fg} increases. Similarly, to add electrons, WWL is set to 0 V, and a boosted voltage is applied to PWL causing electrons to be inserted into node V_{fg}. One downside of this architecture is that it exhibits an interference problem in the unselected pseudo-flash transistors, when the selected pseudo-flash transistors are programmed.

To avoid this issue, the works of [57–59] use five transistors to form a pseudo-flash transistor shown in Fig. 5. M_{fg} is the read/program device, M_1 is the coupling device, M_2 is the tunneling device, and S_1 and S_2 are switches. M_1, M_2, and M_{fg} form a node V_{fg} to store electrons. FN tunneling is used for programming, to add/remove electrons to/from node V_{fg} as well. WWL and PWL are set to high voltage and 0 V respectively to create a high electric field to remove electrons from node V_{fg}. To add electrons, both WWL and PWL are set to high voltage to allow electrons to be inserted into node V_{fg}. By turning on the switches, the corresponding pseudo-flash transistors are programmed. The switches of the rest of the pseudo-flash transistors are turned off to prevent them from being programmed.

3 ASIC Replacement

Historically, flash transistors have been used only for non-volatile memory (NVM) applications. However, recent works demonstrate their promise for use in analog and digital circuit designs as well. In this section, we describe two methods for implementing digital logic using flash transistors. These methods exhibit benefits to circuit delay, area, and power, design tunability, and security.

In Sect. 3.1, we describe one method for implementing digital logic using flash devices, using programmable logic array (PLA) device style structures. We demonstrate the benefits of this approach, including an ability to negate effects such as chip aging. Next in Sect. 3.2, we present flash threshold logic (FTL) cells, a method for implementing standard cells that realize digital threshold logic functions using flash transistors. We show that the use of FTL cells also significantly reduces circuit delay, power, and area. Finally in Sect. 3.3, we present a SAT-based solution for identifying threshold logic functions to be replaced by FTL cells in a larger digital logic design, and we demonstrate the security benefits of using FTL cells to protect circuit designs from IP theft.

3.1 *Digital Circuit Design Using Flash Transistors*

Flash transistors are primarily used for memory applications today. These applications include SD cards, USB flash drives, and SSDs. In this work, we explore the use of flash transistors to implement digital designs [60]. Flash transistors can be used to implement logic functions, replacing a section of a CMOS netlist. Because their threshold voltage (V_{th}) can be tuned with high precision, there are several benefits to using flash transistors in digital circuits. First, speed binning at the factory can be controlled with precision. Second, an IC can be reprogrammed in the field, to negate effects such as chip aging. And third, flash transistor V_{th} can be tuned to multiple levels, unlike MOSFETs which only have one threshold voltage level. This gives flash transistors the ability to encode more symbols than a regular MOSFET, enabling the implementation of multivalued logic natively. In this work, we evaluate digital circuit designs that have been implemented using flash transistors and compare their performance to traditional CMOS standard cell-based implementations. For these designs, we also demonstrate the control of speed binning and the tuning of circuit delay, power, and energy through reprogramming of flash transistor V_{th}. Our results show that, averaged over 20 circuit designs, our flash-based designs yield an improvement of $0.84\times$ the delay, $0.35\times$ the power, $0.3\times$ the energy, and $0.54\times$ the area of the equivalent circuit implemented using a CMOS standard cell-based design.

3.1.1 Flash-Based Digital Circuit Implementation

We present the implementation details for flash-based digital circuit designs, and compare the proposed implementation to a standard cell-based approach. Our implementation consists of a cluster of flash transistors arranged in a NAND configuration and programmed to implement the desired logic function. It is important to note that our proposed structure is *not* a programmable structure similar to an FPGA. Rather, we target an ASIC design flow. Unlike an FPGA, our implementation is not fully programmable because the metalization of the design is fixed—interconnects are hardwired and not adjustable after fabrication. We also note that the flash fabrication process is inherently compatible with the CMOS fabrication process. In fact, it is

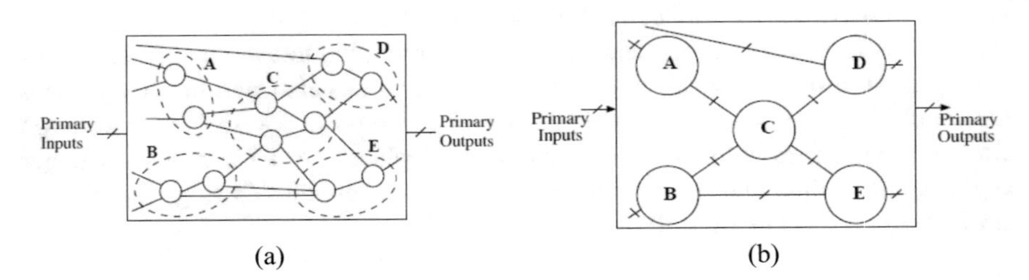

Fig. 6 Converting a logic netlist into a flash-based design. (**a**) Logic netlist. (**b**) Flash-based digital circuit

common for both flash and CMOS devices to be present on the same die, because flash memory designs use both flash and CMOS transistors simultaneously.

Netlist Conversion As depicted in Fig. 6, we convert a logic netlist into an equivalent (dynamic) flash-based digital circuit design. Starting with the logic netlist, we cluster the circuit nodes into multi-input, multi-output structures, with up to m inputs and n outputs. These clusters are shown as dotted circles in Fig. 6a. Later in Sect. 3.1.4, we describe this conversion process in more detail.

The flash-based digital circuit implements the logic functions of each CMOS cluster as a flash cluster (FC). FCs are shown as solid circles in Fig. 6b. Each FC implements a function of up to m inputs and n outputs, which we refer to as $F_{m,n}$. We choose the number of inputs ($m = 6$) and the number of outputs ($n = 3$) after much experimentation, because we found this choice to give the best trade-off of delay, area, power, and energy. Each FC in Fig. 6b has the same functionality and connectivity as its corresponding cluster in Fig. 6a. We next discuss the FC design details.

Flash Clusters (FCs) An FC is a generic circuit structure that is capable of implementing any logic function with up to m inputs and n outputs ($F_{m,n}$). FCs are also equipped with logic for programming the V_{th} of their constituent flash transistors. As shown in Fig. 7a, an FC is driven with the signals that are used for device programming ($mode$, row_select, col_select, $prog\_control$, and $prog\_pulse$) as well as the signals used during normal operation (*primary inputs* and *primary outputs*). In Fig. 8, we show the layout of a representative FC which has 35 pull-down stacks.

Flash Logic Arrays (FLAs) The FC consists of multiple flash logic arrays (FLAs) and an output generation circuit. An FLA is a group of pull-down stack structures made from flash transistors arranged in a NAND configuration. Each stack implements a logic cube of $F_{m,n}$, so that each FLA implements a group of input cubes that correspond to an output minterm of $F_{m,n}$. There are $2^n - 1$ FLAs in every FC, and only one FLA output pulls down when any input is applied to the FC. The outputs of the FLAs are connected to the output generation circuit in the

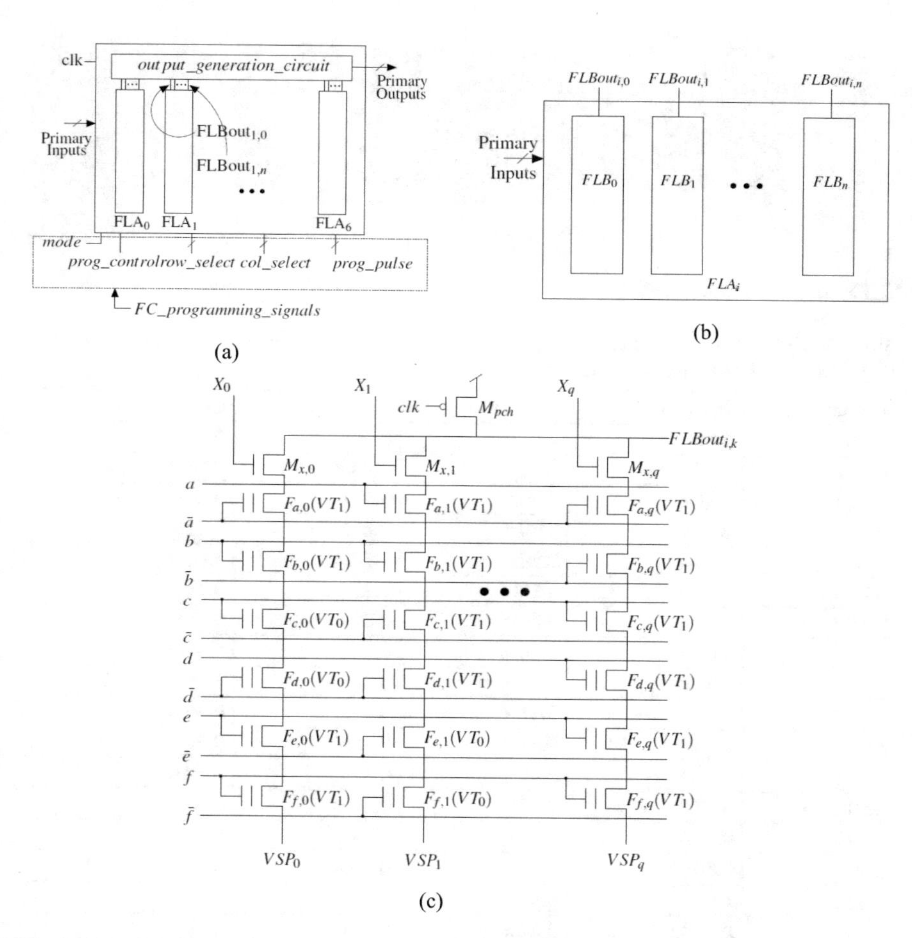

Fig. 7 Structure of flash clusters (FCs). (**a**) Flash cluster (FC). (**b**) Flash logic array i (FLA_i). (**c**) Flash logic bundle i, k $(FLB_{i,k})$

FC. Note that the FC is a dynamic circuit, so its default (precharged) output state is $2^n - 1$ (or <111> for $n = 3$). We therefore only need to implement $2^n - 1$ FLAs.

Flash Logic Bundles (FLBs) As shown in Fig. 7b, each FLA consists of multiple flash logic bundles (FLBs), and as shown in Fig. 7c, each FLB consists of a number of flash pull-down stacks which share the same output. Each pull-down stack implements one cube of $F_{m,n}$, and is made from m flash transistors and 1 regular NMOS transistor. The flash transistors are programmed to implement cubes. The transistor $M_{x,q}$ is used for evaluation and during programming. The shared transistor M_{pch} is used to precharge $FLBout_{i,k}$ before evaluation.

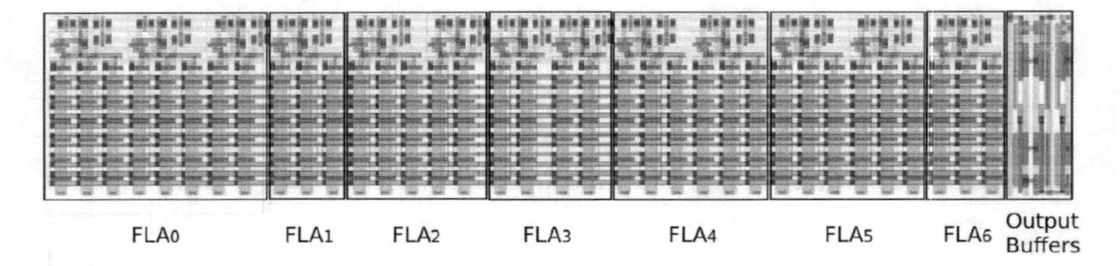

Fig. 8 Layout view of an FC

Table 2 Delay, power, energy, and cell area ratios of flash-based digital circuits relative to their CMOS standard cell-based counterparts

Circuit	D_{max} ratio	P_{avg} ratio	Eng ratio	Cell area ratio
des00	0.81×	0.34×	0.28×	0.50×
des01	0.75×	0.31×	0.24×	0.50×
des02	0.81×	0.35×	0.28×	0.59×
des03	0.74×	0.39×	0.28×	0.51×
des04	0.89×	0.38×	0.34×	0.62×
des05	0.71×	0.33×	0.23×	0.48×
des06	1.04×	0.34×	0.35×	0.58×
des07	0.83×	0.36×	0.30×	0.58×
des08	0.80×	0.35×	0.28×	0.56×
des09	0.87×	0.31×	0.27×	0.49×
des10	0.93×	0.38×	0.35×	0.54×
des11	0.87×	0.40×	0.35×	0.50×
des12	0.92×	0.38×	0.35×	0.53×
des13	0.89×	0.38×	0.34×	0.58×
des14	0.80×	0.33×	0.26×	0.51×
des15	1.01×	0.40×	0.40×	0.53×
des16	0.88×	0.34×	0.30×	0.59×
des17	0.77×	0.34×	0.27×	0.56×
des18	0.83×	0.34×	0.28×	0.55×
des19	0.69×	0.36×	0.24×	0.52×
Average	0.84×	0.35×	0.30×	0.54×

3.1.2 Flash-Based Implementation Results

In this section we present the delay, power, energy, and physical area of flash-based digital circuits. The results we present are a comparative study over 20 randomly generated functions (des00 to des19) implemented in both a CMOS standard cell-based approach and our flash-based digital circuit. The FC used to implement the logic functions was configured to have FLBs with three stacks. Table 2 shows the delay, power, energy, and physical area ratios of these 20 randomly generated logic functions implemented using our FC compared to a CMOS standard cell-based implementation. We note that the FC improves over the CMOS standard cell-based implementation in *all* four metrics.

Delay The measured delay D_{max} is the maximum delay of any transition seen at any primary output of the circuit, and it accounts for the precharge delay in our flash-based implementation. The delay of the flash-based circuits ranges from 0.69× to 1.04× of the CMOS standard cell-based circuit delay, with an average of 0.84×.

Power We also report the power dissipation (average of 0.35× of CMOS). It is well known that dynamic designs consume more power than static CMOS designs. Yet, our flash-based design consumes less power for several reasons. First, the number of nodes being precharged is less than in a CMOS approach. Also, the long transistor stacks result in smaller evaluation currents, and only one FLB pulls down during every evaluation resulting in low switching activity. Another reason for low power consumption is that the I_{ds} of a flash transistor is lower than that of a MOSFET.

Energy On average, energy utilization of the flash-based approach is about 0.3× of the CMOS approach.

Area On average, flash-based digital circuits use 0.54× of the area of CMOS-based designs.

3.1.3 Tuning Delay, Power, and Energy

The ability to change threshold voltages *after* fabrication in flash-based circuits enables adjustment of the design's speed to compensate for circuit aging. To demonstrate this, we perform 10,000 Monte Carlo simulations which model process variation (W and L) with a standard deviation of 5% of the nominal value. In Fig. 9 we show histograms of the maximum delay D_{max} of a CMOS design (CMOS), a

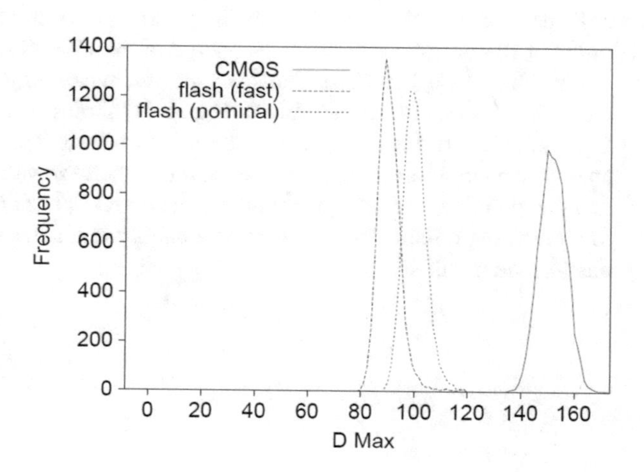

Fig. 9 Delay of CMOS and flash-based designs (before and after aging compensation)

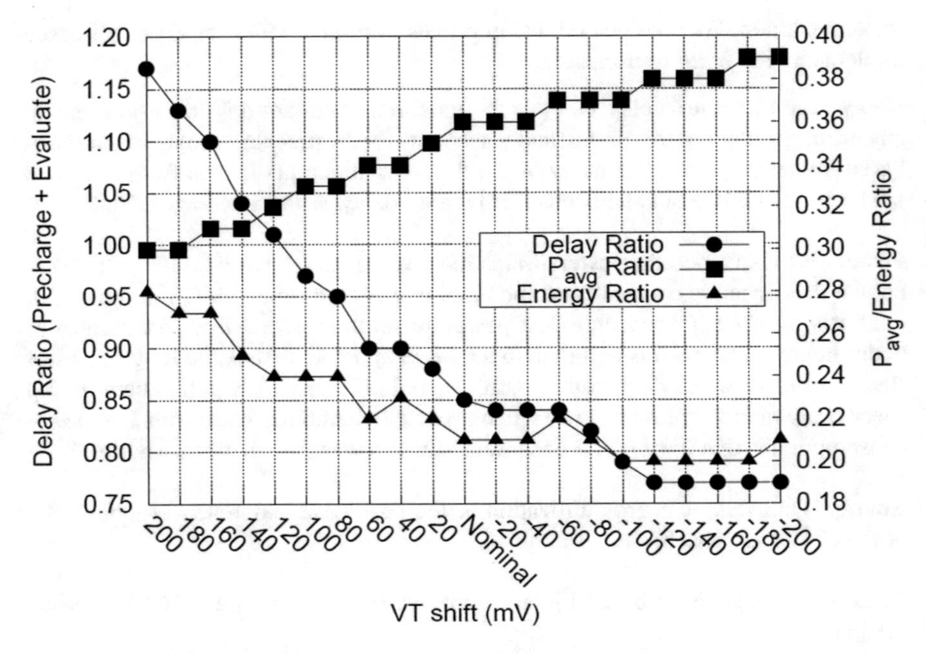

Fig. 10 Delay, power, and energy of the flash-based designs as V_{th} is shifted

flash-based design with nominal V_{th} (flash (nominal)), and the same flash-based design subsequently programmed with a lower V_{th} (flash (fast)). The lower V_{th} value was 50 mV lower than the nominal V_{th} value. Figure 9 demonstrates that in-field compensation of aging effects can be achieved by programming the flash-based design to have a lower V_{th} to decrease its delay.

In Fig. 10, we demonstrate the ability of tuning the delay, power, and energy of flash-based designs by shifting the V_{th} of flash transistors. The x-axis of the plot indicates the V_{th} shift in mV, and the y-axis shows the delay, power, and energy ratios of the flash-based design compared to the CMOS standard cell-based design as the value of V_{th} is shifted. The delay, power, and energy ratios are averaged across all of the designs listed in Table 2. Figure 10 shows that by reducing the value of V_{th}, the delays of the designs decrease and the dissipated power increases,while the energy consumption decreases. Conversely, when the value of V_{th} is increased, the delay and energy increase while power decreases. These results confirm our ability to control the circuit delay, power, and energy characteristics by tuning the V_{th} of flash-based circuits.

3.1.4 Flash-Based Design Conversion

In this section, we discuss the flow that we use to convert a CMOS digital circuit design into an equivalent (dynamic) flash-based implementation. Figure 6 illustrates this conversion. Starting with a technology-independent logic circuit (Fig. 6a), we cluster the circuit nodes into multi-input, multi-output structures with up to m inputs and n outputs (the dotted circles in Fig. 6a). We replace each logic cluster with an FC (the solid circles in Fig. 6b), and program the FC to implement the function of the cluster it replaced, $F_{m,n}$. As stated in Sect. 3.1.1, the FCs labeled A, B, C, D, and E in Fig. 6b have the same connectivity as the clusters labeled A, B, C, D, and E in the circuit in Fig. 6a.

There are several steps in the CAD flow to convert an input logic netlist. First, the input netlist is clustered into FCs (where FC_i implements $F^i_{m,n}$) with the goal of minimizing the wiring between FCs. This yields a multi-level netlist of interconnected FCs. Next, the layout of each FC is generated. All our designs are extremely regular in their physical layout, making them amenable to the on-the-fly physical synthesis flow that we use. Based on the fanout load of the jth output of FC_i, additional buffers are added for that output.

FC-Based Clustering Algorithm 1 outlines our clustering strategy, given an arbitrary logic netlist η. We first decompose η into a network of nodes with at most p inputs. We choose $p < m$ so that nodes can be grouped into FCs later. In particular, we found that $p = 3$ yielded good results. Next η is sorted in a depth-first manner, and placed into an array L.

Next, the logic in each FC is greedily constructed by successively grouping nodes from L such that the resulting implementation of the grouped nodes FC^* does not violate the input or output cardinality constraints of the FCs. The *get_next_element* procedure preferentially returns nodes in the fanout of the nodes

---

Algorithm 1 Clustering a logic netlist into a multi-level network of FCs

η = decompose_network(η, p)
L = dfs_and_levelize_nodes(η)
$FC^* = 0$
$\eta^* = 0$
while get_next_element(L) != NIL **do**
  $FC^* = FC^* \cup$ get_next_element(L)
  if (num_input(FC^*) $\leq m$) && (num_output(FC^*) $\leq n$) **then**
    continue
  else
    Q = remove_last_element(FC^*)
    $\eta^* = \eta^* \cup FC^*$
    $FC^* = Q$
  end if
end while
η^* = wiring_recovery(η^*)

---

of FC^*, in order to reduce the wiring between FCs. At every step of the construction of the result η^*, we verify that the induced graph is acyclic.

After the clustering step is complete, the routine *wiring_recovery* will attempt to reduce the wiring between FCs. This routine attempts to realize a wiring gain by moving individual nodes in L to a different FC other than their currently assigned FC. On average, the *wiring_recovery* routine is able to reduce wiring by about 9.6%.

On-the-Fly Layout Synthesis Once the multi-level netlist of FCs is generated, we next generate the layout of each $FC^i \in \eta^*$. First, we construct a table for all the 2^n output minterms o_p and their corresponding input cubes $C_p = \Sigma c_{p,q}$. The set of cubes C_p form a partition of the points in B^m, where $B = \{0, 1\}$.

This table is constructed from the truth table of $F^i_{m,n}$ by grouping all the input minterms for each output minterm. Next, the input minterms for each output minterm are minimized using Espresso [61]. The output minterm which has the largest number of input minterms is not implemented, and is mapped to the default output of the FC when it is precharged.

3.1.5 Performance of Our Conversion Flow

We evaluated our flash-based digital circuit design approach for larger digital designs by implementing a set of 12 of the largest benchmarks from ISCAS89 [62], ITC99 [63], and EPFL [64] benchmark suites. We compare the delay, power, and layout area metrics of the flash-based approach to a CMOS standard cell-based implementation of the benchmarks in Table 3. The delay, power, and area ratios of the flash-based designs are relative to their CMOS standard cell-based counterparts.

We observe that on average, each FC is equivalent to about 7.6 CMOS standard cells. The FCs have an average of 5.6 inputs and 2.4 outputs, which are close to the maximum number of inputs (6) and outputs (3).

Table 3 shows that the flash-based approach is ∼41% faster and consumes ∼65% lower power on average, compared to a traditional CMOS standard cell-based approach. This is a significant improvement, and results in an energy improvement of ∼5× over the standard cell-based approach.

The key reasons for the reduced delay are:

- Flash FET gate capacitance is ∼20× lower than MOSFET gate capacitance
- CMOS standard cells have increased parasitics due to the use of both NMOS and PMOS devices
- The use of shared diffusions in the NAND stack reduces parasitics significantly
- The FCs used to implement the benchmarks have a small number of input cubes (5.7 on average), which reduces input capacitive loads

Table 3 Delay, power, and cell area ratios of flash-based digital circuits relative to their CMOS standard cell-based counterparts

	CMOS	Flash				Flash	Flash	Flash
Benchmark	# Stdcells	# FCs	m_{Avg}	n_{Avg}	Avg # cubes	Delay ratio	Power ratio	Cell area ratio
b17	47,500	6658	5.61	2.32	5.49	0.90×	0.49×	0.67×
b20	22,983	2901	5.60	2.32	6.02	0.50×	0.18×	0.60×
b21	23,324	2963	5.62	2.32	6.03	0.49×	0.18×	0.61×
b22	34,693	4362	5.61	2.33	5.97	0.47×	0.19×	0.60×
s13207	2828	460	5.60	2.38	4.83	0.43×	0.91×	0.67×
s15850	3735	594	5.50	2.30	4.90	0.67×	0.62×	0.65×
s35932	10,661	1290	5.24	2.60	6.62	0.29×	0.22×	0.51×
s38417	10,771	1593	5.68	2.22	5.04	0.89×	0.43×	0.60×
s38584	13,895	2077	5.59	2.16	4.93	0.90×	0.61×	0.60×
Multiplier	46,363	5821	5.67	2.48	6.56	0.47×	0.11×	0.57×
Voter	22,453	2708	5.38	2.53	6.10	0.58×	0.14×	0.56×
Square	38,009	5109	5.63	2.49	5.61	0.49×	0.16×	0.58×
Average	23,101	3045	5.56	2.37	5.68	0.59×	0.35×	0.60×
Stdev	15582.94	2038.51	0.13	0.13	0.64	0.20×	0.26×	0.05×

3.1.6 Conclusion

Flash transistors are very important for non-volatile data storage applications today, and our work demonstrates their usefulness for the implementation of digital logic as well. This section presented an approach for using flash transistors to implement digital circuits by using stacks of flash transistors in a NAND configuration to implement cubes of a logic function. The V_{th} of flash transistors can be tuned with fine granularity, which yields several advantages, and we demonstrate two of these advantages in this section. First, speed binning at the factory can be performed with high precision. Second, an IC can be reprogrammed in the field to negate the effects such as chip aging. We present details of the circuit topology that we use in our flash-based approach. Our results show that, averaged over 20 digital circuit designs, our approach yields an improvement of 0.84× the delay, 0.35× the power, 0.30× the energy, and 0.54× the area of the equivalent circuit implemented using a CMOS standard cell-based design. When implementing a set of 12 of the largest benchmarks from ISCAS89, ITC99, and EPFL, our approach yields an improvement of 0.59× the delay, 0.35× the power, 0.20× the energy, and 0.60× the area versus CMOS standard cell-based implementations.

3.2 *Perceptron Hardware: Flash Threshold Logic Cells*

This section describes a new design for threshold logic gates (binary perceptrons). This new structure, called the flash threshold logic (FTL) cell [65], uses floating gate (flash) transistors to implement the weights associated with the threshold function.

A function $f : B^n \rightarrow B$ is called a *threshold function* iff $f(x_1, x_2, \cdots x_n) = 1 \Leftrightarrow \sum_{i=1}^{n} w_i x_i \geq T$, where $\sum$ is the arithmetic sum, w_i are weights and T is a threshold. The function can be equivalently expressed as $f \equiv (\mathbf{w} \cdot T) = (w_1, w_2, \cdots w_n; T)$.

In an FTL cell, the threshold voltage of the flash transistor acts as a proxy for the weights. An FTL cell is equivalent to a multi-input edge-triggered flip-flop that calculates a threshold function at the clock edge. This section focuses on the structure and properties of FTL cells. The FTL cell is designed using 40 nm technology as a standard cell, and simulations are done using layout-extracted parasitics. Results show improvements in area (79.7%), power (61.1%), and performance (42.5%) as compared to equivalent implementations having the same functionality that are built using conventional static CMOS design. This section also describes how to use FTL cells to correct post-fabrication setup and hold timing errors.

An FTL cell of n inputs can realize any *threshold* function having up to n variables. The input-output behavior of an FTL cell is equivalent to an *edge-triggered, multi-input* flip-flop that computes a threshold function.

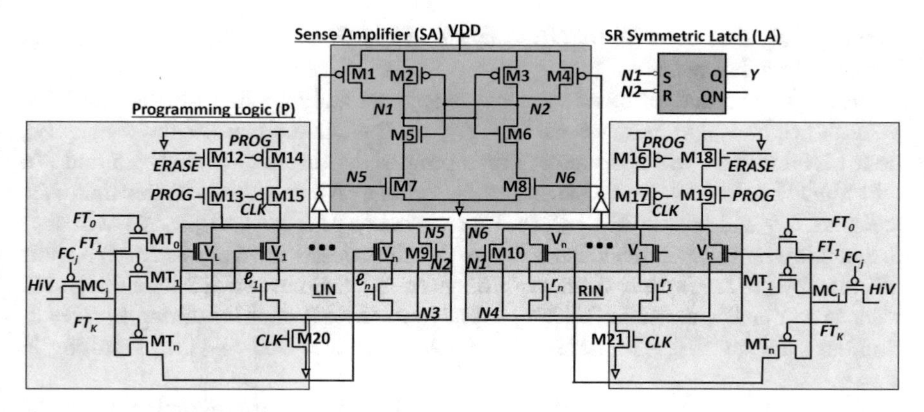

Fig. 11 FTL cell architecture

3.2.1 Flash Threshold Logic (FTL) Cell Architecture

The architecture of an FTL cell consists of five main components (shown in Fig. 11), namely, the left input network and right input network (LIN and RIN), a sense amplifier (SA), an output latch (LA), and a flash transistor programming logic (P). The LIN and RIN consist of two sets of inputs $(\ell_1, \cdots, \ell_n)$ and $(r_1, \cdots, r_n)$, respectively, with each input connected to NMOS devices in series with a flash transistor. In an FTL cell, $\ell_i = \overline{r_i}$ for all i. The state of the inputs and the threshold voltages of the flash transistors control the conductivity of these two networks. Signals are assigned to the LIN and RIN to guarantee that there is a sufficiently large conductivity difference across all minterm pairs (m_i, m_j) such that $f(m_i) \neq f(m_j)$. Two differential signals, $N1$ and $N2$, are used as inputs to an SR latch in the FTL cell. The latch is set (reset) when $[N1, N2] = [0, 1]$ ($[1, 0]$) and the output $Y = 1(0)$. In the formulation of a threshold function, the magnitudes of the two sides of the inequality are mapped to the LIN's *conductance* G_L and the RIN's *conductance* G_R, resulting in $[N1, N2] = [0, 1] \Leftrightarrow G_L > G_R$ and $[N1, N2] = [1, 0] \Leftrightarrow G_L < G_R$. As previously stated, the threshold voltages of flash transistors serve as a proxy for the weights of the threshold function. The threshold voltage of the flash transistor will be reduced to increase the corresponding weight. This nonlinear monotonic connection is **learnt** using a modified perceptron learning algorithm for a given threshold function.

There are three modes in an FTL cell: *regular*, *erase*, and *programming* mode. The $\boldsymbol{V_{th}}$ values of the flash transistors are set in the programming mode and erased in the erase mode. The evaluation takes place in regular mode.

FTL Regular Mode In this mode PROG = ERASE = 0. Assume that the $\boldsymbol{V_{th}}$'s of the flash transistors have been set to appropriate values corresponding to the weights of the threshold function, and their gates are being driven to 1 by setting HiV to VDD, FC_j to 0 V, and all FT_i to 0 V. When $CLK = 0$, the circuit is reset.

In this phase, the nodes $N5 = N6 = 0$, and $N1 = N2 = 1$. Therefore, the output Y remains unchanged.

Now suppose that an onset minterm is appropriately applied to LIN and RIN. With the right V_{th} values, we will get $G_L > G_R$. At the start of the evaluation, both LIN and RIN will conduct. As time progresses, the state of both $N5$ and $N6$ will transition from $0 \rightarrow 1$. Assuming $G_L > G_R$, $N5$ transitions faster than $N6$, and thus $N5$ will enable $M7$ before $N6$ can enable $M8$. As a result, $N1$ will start discharging before $N2$. Once the value of $N1$ falls below the V_{th} of $M6$, $M6$ shuts off, thereby preventing any further discharge of $N2$, and turn on $M3$ to recharge $N2$ back to 1. Finally, the state [N1,N2] = [0,1] sets the SR latch, resulting in $Y = 1$. Similarly, for an off-set minterm, $G_L < G_R$, and $[N1, N2] = [1, 0]$ resulting in $Y = 0$.

We introduce a new programming interface for off-chip programming, to set the V_{th} values of any FTL cell. This interface uses control bits FC_j to select the jth FTL cell and control bit FT_i signal to select the ith flash transistor of that FTL cell.

FTL Programming Mode (ERASE=0, PROG=1, CLK=0, FT$_i$=0, FC$_j$=0, HiV=20 V). The ERASE and PROG signals turn on M12 and M13 and turn off M14. The source of the flash transistor is floating. Meanwhile, the drain and the bulk terminals are grounded. By enabling appropriate high-voltage transistors MC_j and MT_i, a path from HiV to the gates of the flash transistors is opened (through FC_j and FT_i). High positive voltages are then sent using the HiV line, to inject charge in the floating gates till the desired threshold voltage (V_{th}) is set.

FTL Erase Mode (ERASE=1, PROG=1, CLK=0, FT$_i$=0, FC$_j$=0, HiV=−20 V). The ERASE signal turns off M12. Both the source and the drain terminals of the flash transistors are floating. The bulk is connected to the ground. High negative pulses are applied to the gate of the flash transistor to extract the buried charge in the floating gate, thereby erasing the flash transistor.

3.2.2 Related Work

Recently, [66] demonstrated the construction of a threshold logic gate as a standard cell, along with the flow needed to integrate these gates into ASICs in 65 nm using industry-standard tools. The demonstration included simultaneous improvements in the area and power of ASICs due to the integration of the threshold gates [67]. However, their design used CMOS devices exclusively, and their gate was limited in terms of the number of threshold functions that could be realized. This is because each of the weights w_i of the associated input x_i was implemented using w_i transistors, each driven by signal x_i. As a result, their circuit suffered from severe fan-in limitations. As compared to the threshold gate presented in [66], which could implement only 11 of the 5-input threshold functions, an FTL cell can realize all the 117 5-input threshold functions. Furthermore, since flash transistors allow fine control over the programmability of the weights of an FTL cell, FTL cells are far more robust than other threshold gates available in the literature. The increased robustness also allows the FTL cells to scale to geometries far lower than other (CMOS device-based) threshold gate architectures available in the literature.

3.2.3 Experimental Results

A 5-input FTL cell was designed as a standard cell layout using TSMC 40 nm LP technology. There are a total of 117 distinct threshold functions of 5 or fewer variables. We use the same indexing scheme for the threshold functions as used in [68]. Each of the 117 threshold functions was implemented as an FTL cell and then compared against a CMOS equivalent counterpart of the same functionality. Synthesis and analysis of both the circuits were performed using Cadence Genus©, and placement and routing was performed using Cadence Innovus©. The CMOS equivalent counterpart used TSMC 40 nm LP standard cells. Both circuits were simulated at 25 °C, and the static power analysis was performed with 20% input switching activity. The results of this comparison are shown in Fig. 12. This figure clearly shows that FTL cells have substantially improved area (79.5%), power (61.1%), and delay (42.5%) characteristics as compared to their respective CMOS equivalents, for all the 117 functions.

Experiments comparing the distributions of delays of FTL and CMOS implementations were also carried out. The results for the threshold function $F115 = [\mathbf{W;T}] = [4, 1, 1, 1, 1; 5]$ are shown as an example in Fig. 13. $[P, V, T] = [TT, 0.9V, 25\,°\text{C}]$ was the PVT corner setting for this experiment. For both the FTL cell and its CMOS equivalent, 100 K Monte Carlo instances were generated. For both circuits, the function of each of the 100 K FTL instances was checked against the truth table to ensure that it was correct. Even in the presence of process variations, the graph demonstrates the FTL cell's delay advantage over its CMOS counterpart. Furthermore, the difference in their standard deviation is also small. Note that the FTL instances with large delays can be *reprogrammed* to further reduce the delay. This capability is not possible for circuits built using conventional standard cells.

Figure 14 is used to demonstrate how timing violations can be corrected on a data path containing an FTL cell. The data path in the example consists of combinational delay (D2D), clock-to-Q (C2Q) delay, and the setup (DFF_setup) and hold (DFF_hold) times of a DFF. Timing violations are introduced into the example circuit by skewing the clock for the DFF by an appropriate amount Δ. For instance, Fig. 15a shows how the data launched from FTL cell X misses the target clock edge at DFF Y and causes a setup time violation. Reducing the C2Q of FTL X by reprogramming the flash transistors fixes this violation. Similarly, Fig. 15b shows how the data launched at FTL X gets captured by DFF Y one cycle early, thereby overwriting the old value at DFF Y and causing a hold time violation. This violation can be fixed by increasing the C2Q of FTL cell X, which allows the old value at

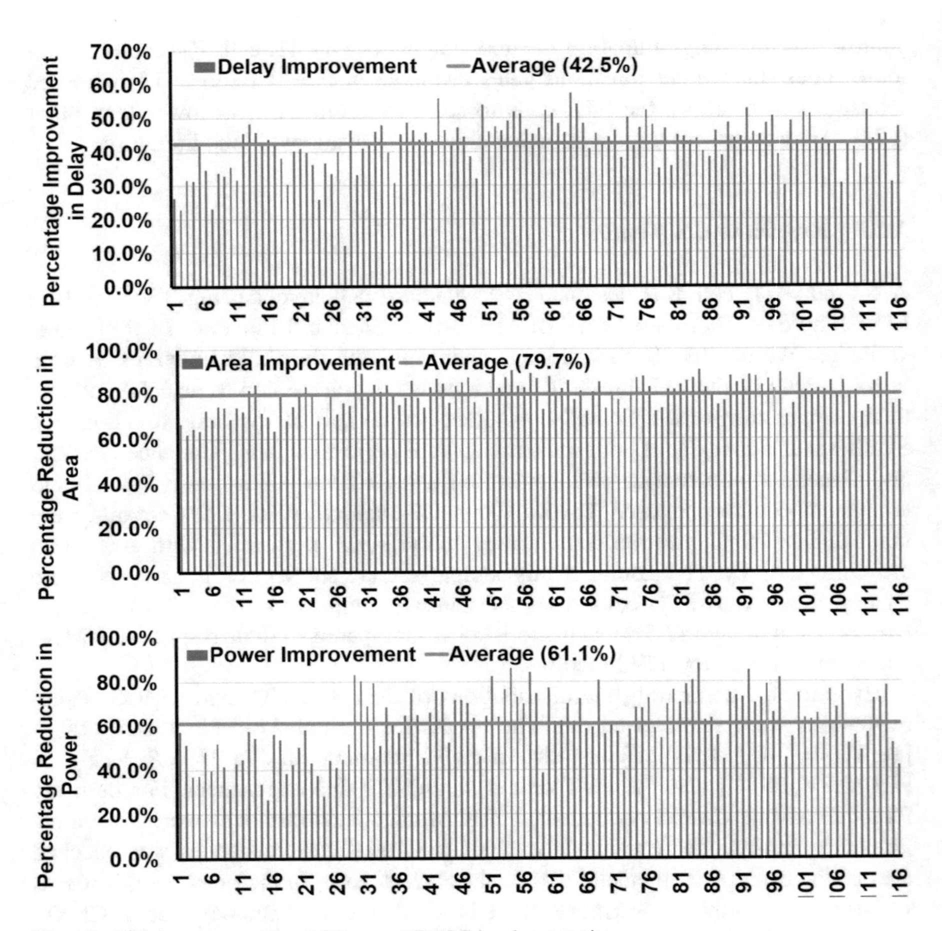

Fig. 12 PPA improvements of FTL over CMOS implementations

the input of Y to retain for a longer time. Since the FTL cells are programmed post-fabrication, the delay can also be modified after fabrication.

3.2.4 Summary

This section presented a new threshold logic cell (FTL) with integrated flash transistors. Significant performance improvements in FTL cell area (79.7%), power (61.1%), and performance (42.5%) have been demonstrated compared to traditional 40 nm standard cell-based designs that have the same functionality. We also showed that FTL cells can be used for fixing setup and hold time violations post-fabrication.

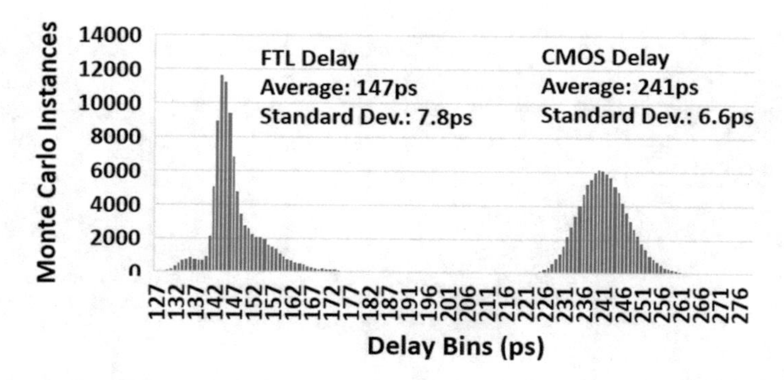

Fig. 13 Delay histogram of an FTL cell and its CMOS equivalent with 100 K Monte Carlo simulations. $PVT = [TT, 0.9V, 25\,°C]$

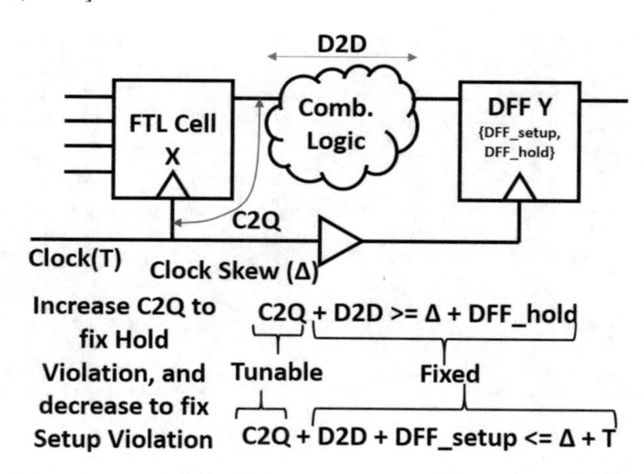

Fig. 14 Datapath to demonstrate post-fab timing corrections

3.3 Input and Output Hybridization for Enhanced Security in ASIC Circuits

ASIC circuits can be made secure against IP theft and/or malicious modification using FTL cells. A malicious agent cannot determine the functionality of an FTL cell without knowing its weight and threshold values, which are programmed by the IP owner, before the IC is deployed. Hybridization is defined as the process of identifying and replacing logic cones in ASIC circuits with FTL cells. Candidate logic cones are identified by analyzing the circuit backward from the primary outputs (*output hybridization*), and forward from the primary inputs (*input hybridization*). We present a SAT-based solution to identify threshold functions for input and output hybridization, and the corresponding weights needed to implement the threshold function. This information is then stored in a hash table for fast lookup during synthesis. The key in the hash table is the threshold logic function, and the value is the weights needed to realize it. NPN equivalence is accounted for in the

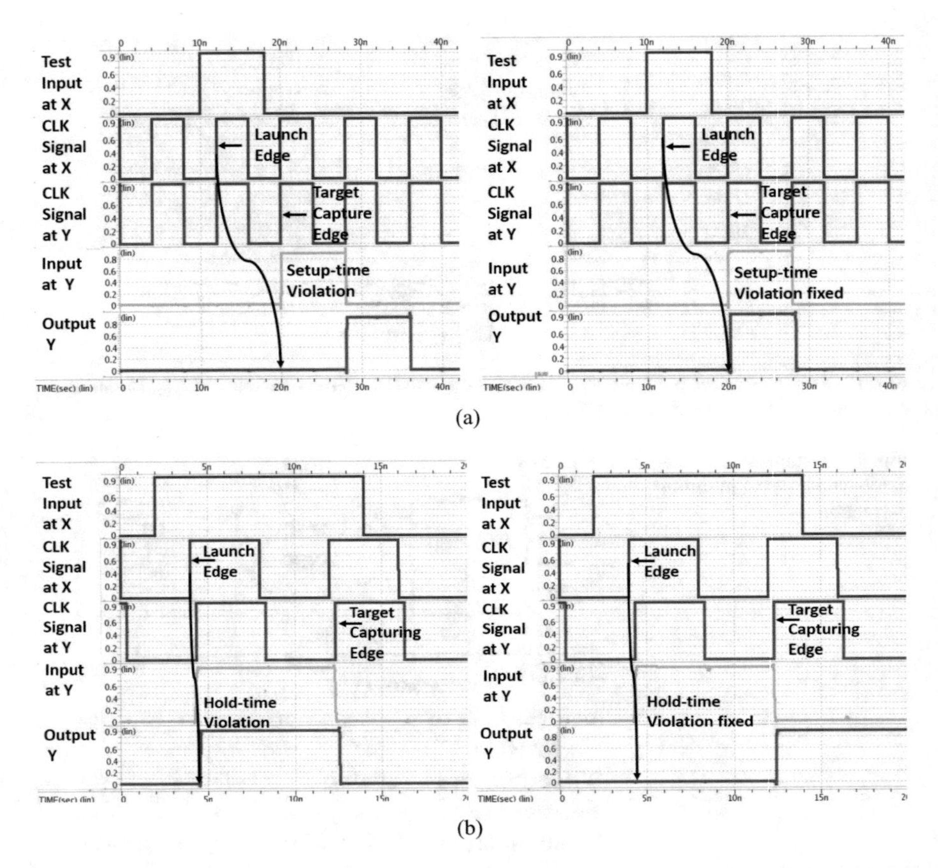

Fig. 15 Correcting setup and hold time violations with an FTL cell after fabrication. (**a**) Correcting setup time violation with an FTL cell after fabrication. C2Q of FTL cell reduced from 180 ps to 142 ps. (**b**) Correcting hold time violation with an FTL cell after fabrication. C2Q of FTL cell increased from 142 ps to 180 ps

hash table. A logic cone in the ASIC circuit is checked against this hash table to test if a function is a threshold function (and can be realized by an FTL cell). From the various hybridization choices available, we select those that maximize security and maximally improve delay, power, and area. We implemented our algorithms in ABC (an open-source synthesis tool), and our approach is compatible with existing ASIC design flows.

3.3.1 Security and Hybridization

Today, with the global nature of supply chains, IP theft and/or malicious modification of chips is a serious concern, with revenue losses estimates in the 10s of billion

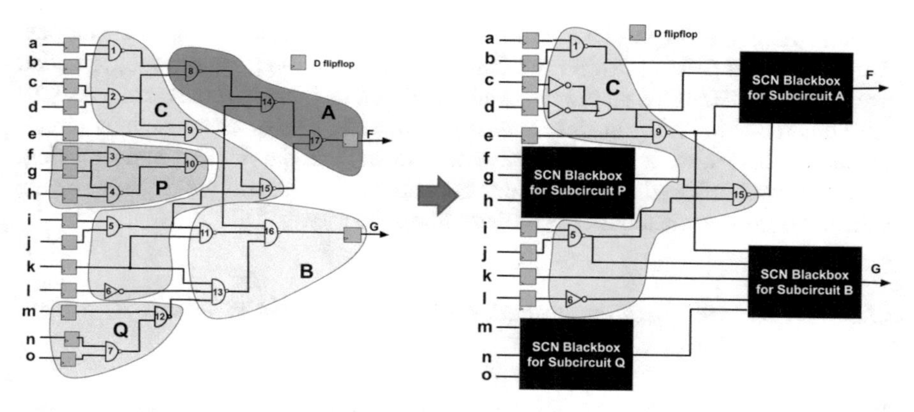

Fig. 16 Substituting FTL cells into ASIC circuits

dollars [69]. Hybridization leverages the benefits of FTL cells and helps designers protect their IP, both from discovery of chip function in the foundry and against reverse engineering by end users. Hybridization replaces logic cones in the original ASIC circuit with FTL cells. Figure 16 shows how logic cones (or "subcircuits") can be replaced by FTL cells. Replacing subcircuits A and B is an example of *output hybridization*. Similarly, replacing subcircuits P and Q is an example of *input hybridization*. This section will focus on the identification of logic cones in ASIC circuits that are suitable candidates for hybridization. The main contributions of this section are:

1. A SAT-based formulation to test if a function is threshold, and generation of the corresponding threshold value and weights. Creation of a hash table [70] with threshold function and weight information for efficiently testing if a logic function can be replaced by an FTL cell during synthesis.
2. Input and output hybridization is an open-source tool (ABC [71]). Our approach is designed to maximize security of the final circuit.
3. Results using a set of benchmark circuits that validate our approach.

3.3.2 Previous Work

Logic locking [72] is a well-known hardware security technique that protects against untrusted foundries as well as malicious end users. Gates are strategically added in the original circuit, such that the inputs to these new gates are a combination of the original nets in the circuit and a set of *key* bits. The outputs of the added gates are utilized by the circuit to generate the final circuit outputs. Without the correct *key* bits, the output of this new circuit will be incorrect (*output corruption*). Various attacks have been devised against logic locking (e.g., sensitization attacks [73], SAT-based attacks [74]), and in turn, logic locking methods have evolved accordingly [75] (SLL [73], WLL [76], SARLock [77]). However, there is always a trade-off in the resilience of logic locking to the various attacks that have been developed. For example, in SFLL [78], output corruption is quantified using a metric called error rate (ER). High ER makes the resultant circuit resilient to removal attacks [79] but vulnerable to SAT attacks. Similarly, low ER makes the resultant circuit resilient to SAT attacks but vulnerable to removal attacks. Unlike logic locking, our hybridization-based approach does not add new gates, and yields better delay, area, and power characteristics while providing security.

3.3.3 SAT-Based Threshold Function Identification and Weight Generation

In this section we will explore how we identify if a function is a threshold function. If a function is threshold, we are also able to generate the corresponding weights and the threshold value. Figure 17 provides an overview of our solution. We first devise a circuit that encapsulates the mathematical relationship between the inputs, weights, and threshold value. We then cast this circuit as a SAT problem, whose solution helps us identify threshold functions.

Circuit Construction If a function is threshold, for any onset minterm, we have $\sum_{i=0}^{k-1} W_i * X_i \geq T$, where k is the number of inputs to the threshold function, W_i are weights of the threshold function and X_i are the inputs of the function. We translate this relationship into a circuit for every minterm of a k input function. Our circuit has 2^k *planes* (the first plane corresponds to minterm m_0) as indicated in Fig. 17. Each plane has an adder that computes the weighted sum of the input bits and weight bits. The output of the adder, denoted as "SUM" in Fig. 17, is compared with the threshold value "T" using a comparator. The comparator has two outputs, one to indicate if SUM >= T (the "YES" branch) and another to indicate if SUM < T (the "NO" branch). The two comparator outputs are then used as inputs to a MUX, whose select line is the bit mi_0. Here, mi_0 is a bit that is set if the corresponding minterm (m_0) is an onset minterm, else it is unset. We call these MUX select bits as *minterm-indicator* bits. The weights and threshold value are shared across all planes, since they are independent of minterms. We can express a threshold function in terms of its 2^k minterms and set the minterm-indicator bits accordingly. For a threshold function, with the correct weight and threshold, the output of all the MUXes will be

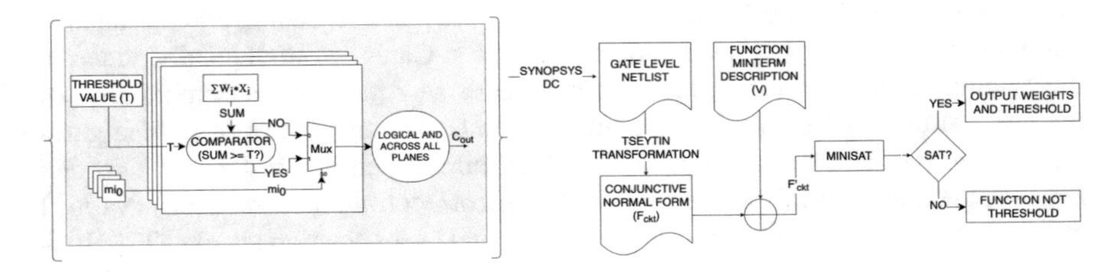

Fig. 17 SAT-based threshold check

set. We perform a logical AND across the planes to check this condition. The circuit as described in this section was expressed in Verilog, and Synopsys DC was used to synthesize the Verilog to a gate-level description of the circuit. Next, we converted this gate-level circuit to a SAT [80] instance which serves as a threshold function checker and weight generator.

Tseytin Transformation and SAT Any circuit composed of logic gates can be converted to a Boolean function in conjunctive normal form (CNF) using the Tseytin transformation [81] (*TS*). Applying TS to our circuit results in a CNF Boolean function (F_{ckt}). We also add a clause to F_{ckt} to force C_{out} (the output of F_{ckt}) to always be set. Next, to test if a given Boolean function F_{test} is threshold, we generate a vector V that encodes the onset and offset minterms for F_{test}. If F_{test} is a k-variable function, V has 2^k bits, where the ith bit is set if m_i is an onset minterm. We use V to construct clauses that force the minterm-indicator bits in F_{ckt} to represent F_{test}, and call it F'_{ckt}. In F'_{ckt} the only free variables are the weights and the threshold. Calling MiniSat [82] on F'_{ckt} has two possible outcomes. If F_{test} is a threshold function, F'_{ckt} is SAT, and MiniSat returns a set of weights and a threshold value corresponding to F_{test}. Otherwise F_{test} is not a threshold function and F'_{ckt} is UNSAT. Therefore, we can use this method to test if a function is threshold and, if it is, also find the weights and threshold value.

Optimizations Making repeated calls to a SAT solver is a computationally expensive process. The number of threshold functions is known for five variables or less. Therefore, we create a hash table [70] of all known five variable threshold functions along with their corresponding weights and threshold values. This data is generated using our SAT-based solution. Storing the data in a hash table makes the threshold function lookups during hybridization quick, allowing us to avoid repeated expensive calls to the SAT solver. In the next section, we will see how the SAT-generated weights are mapped to electrical weights in a FTL cell.

3.3.4 Results

To validate hybridization as an effective security option, we tested our algorithms on a set of benchmark circuits. Our entire suite of benchmark circuits consists of 28 combinational circuits of varying size from different sources, like ISCAS85, ISCAS99 [83], and EPFL [64]. In this section, we present our findings on a subset of circuits for brevity. When checking for hybridization choices, we consider all combinations of the input negation, output negation, and input permutation for each candidate cone. Table 4 consists of the seven large circuits from our benchmark suite. We report a summary for the output hybridization opportunities at each of the POs for each circuit. On average, there are approximately 791 different ways a PO can be hybridized among these 7 circuits. While a particular PO can be hybridized in many different ways, eventually we need to select a particular function. The choice of this function must maximize security. In order to achieve this, we use a "MAX

Table 4 Output hybridization—opportunities vs security

Output hybridization									
				Opportunities			% PO hybridized		
Name	# Gates	PIs	POs	Mean	Stdev	Total	Max reuse=1	Max reuse=3	Runtime(s)
div	57,247	128	128	967.25	143.25	123,808	100.00	100.00	17.256
log2	32,060	32	32	770	70.6	24,640	100.00	100.00	39.353
Multiplier	27,062	128	128	1130.94	197.03	144,760	100.00	100.00	78.54
sqrt	24,618	128	64	869.5	145.51	55,648	100.00	100.00	4.034
b22_C	18,450	766	757	707.53	250.45	535,600	36.46	99.60	2.837
Arbiter	11,839	256	129	619.72	42.26	79,944	100.00	100.00	0.18
C7552	7552	207	108	473.93	566.97	51,184	57.41	57.41	0.212
Average				791.27			84.84	93.86	20.34

Table 5 Output and input hybridization opportunities

Hybridization opportunities					
				Output	Input
Name	# Gates	PIs	POs	Mean	Mean
C3540	3540	50	22	853.45	218.08
i8	3310	133	81	1139.85	90.41
C1355	1355	41	32	856	20.29
C7552	7552	207	108	473.93	20.37
C5315	5315	178	123	497.17	29.48
apex2	4523	39	3	1096	516.92
C1908	1908	33	25	400.96	27.88
Average				759.62	131.92

REUSE" constraint that limits the number of times a particular threshold function is chosen to hybridize any PO. This may cause some of the POs to not be hybridized. For example, consider the circuit b22_C in Table 4. When MAX REUSE=1, we are able to only hybridize 36% of the POs, before running out of choices of threshold functions. When we relax this constraint and allow a threshold function to be used up to three times, we are able to hybridize over 99% of POs. Therefore, we are able to hybridize a large fraction of POs even with a relatively low MAX REUSE constraint. In addition, the longest runtime of our algorithm for finding these hybridization opportunities is just under 40s. Table 5 shows a summary of the output and input hybridization opportunities for seven other circuits selected from our benchmark suite. On average each PO can be hybridized 759 ways and each PI can be selected in one of the 131 hybridizations. For all circuits, all PIs are part of some input hybridization cone, across all the choices available for each circuit. These results demonstrate that there are abundant hybridization opportunities at the POs and PIs of an ASIC circuit. In addition, there is variety in the choice of threshold functions, making it hard for an adversary to guess the functionality of an FTL cell. Therefore, we propose hybridization as a practical method of securing circuits and protecting against IP theft.

3.4 *Benefits of Flash-Based ASIC Design*

The actual logic function realized by the ASIC approach of Sect. 3.1 or the FTL cell of Sect. 3.2 is *programmed after the circuit is manufactured.* This is made possible by the *floating gate* [84] transistors that are integrated alongside conventional MOSFETs within the FTL cell. Furthermore, in contrast to several *emerging technologies* [85–88], the approaches of Sects. 3.1 and 3.2 are built using mature technologies that are currently available in the industry. These designs can be used to tackle several problems that are currently present in the field of chip design:

1. **Superior Performance**: The designs of Sects. 3.1 and 3.2 deliver significantly better area, power, and delay boosts over conventional standard cells.
2. **IP Protection**: Since the functionality of the approaches of Sects. 3.1 and 3.2 is programmed post-fabrication, the functionality of the ASIC that uses these approaches cannot be reverse engineered by the foundry. This is because the functionality of the logic cells is unknown at the time of manufacturing.
3. **Correcting Timing Errors**: Post-fabrication programming enables precise speed binning, and setup (and hold) timing correction after the circuit is manufactured. This is not possible in traditional CMOS design.
4. **Mitigating Aging Effects**: Aging effects such as reduction in the operating speed of the circuit can be mitigated by reprogramming the flash-based logic cells. This too is not possible in standard CMOS designs.
5. **High Endurance**: Unlike the flash transistors that are used in memories, flash transistors used in our cells do not suffer from endurance issues, as very few program-erase cycles (far fewer than the flash endurance limits of 1–100 K)[12, 13] are needed to program our flash transistors with the correct functionality.

4 Neural Network Accelerators

Beginning in the early 2010s, deep neural networks (DNNs) have demonstrated state-of-the-art performance in a wide range of tasks including image classification [89], object detection [90], and text recognition [91]. Modern DNNs have significant resource requirements, making them unsuitable for use in edge computing contexts such as mobile and embedded devices. Consequently, there are many recent works that attempt to reduce the resource consumption of DNNs using various strategies like network pruning [92, 93], compact network designs [94, 95], and low-bit quantization [96, 97]. Quantization remains the most effective technique. It involves reducing the representation of network weights and activations to a small number of bits, yielding a quantized neural network (QNN). A binarized neural network (BNN) [98] is the extreme case of quantization, where network weights and activations are binary values. BNNs can achieve up to a 58× speedup and a 32× reduction in memory bandwidth requirements [96] on CPUs versus full-precision networks.

In a QNN, most computation is performed on binary values, or values with a low bit width. This makes them ripe for acceleration using novel mixed-signal methods. Primarily, these methods consist of using NVM devices—typically in a crossbar array structure—as the synapses for neurons in the DNN, resulting in a significant reduction of inference delay, power, and energy, and chip area. NVM devices are especially useful in this context, because in addition to storing network weight parameters, they can be used to perform some computation natively (in-memory computation). NVM technologies that have been used as synapses include flash transistors, resistive RAM (RRAM, also known as a memristor), phase-change memory (PCM), and conductive-bridging RAM (CBRAM). Unlike these other NVM technologies, flash is a well-understood, high-yielding technology, making it very practical for use in neural network accelerators.

In this section, we present two architectures for neural network acceleration, which are suitable for use in edge devices and which both utilize flash technology. In Sect. 4.1 we present TULIP, a digital BNN accelerator that is able to achieve very low power, energy, latency, and chip area. In Sect. 4.2, we present an analog QNN accelerator that uses flash technology to achieve extremely high throughput with extremely low power, energy, and memory requirements.

4.1 A Configurable BNN ASIC Using a Network of Programmable Threshold Logic Standard Cells

This section presents TULIP [99], a binary neural network ASIC accelerator that was constructed with the goal of maximizing energy efficiency per classification. TULIP is composed of an array of processing elements called (TULIP-PEs) which process the operations of a BNN in parallel. Each TULIP-PE contains interconnected binary neurons, with a small local register per neuron. The binary neurons are *mixed-signal* circuits that evaluate threshold functions. The unique aspect of the binary neuron is that its functionality can be reconfigured with a change in a single parameter. Using this property, we present algorithms for mapping arbitrary nodes of a BNN onto the TULIP-PEs. To provide a fair comparison, we also implement a MAC-based BNN accelerator that was recently reported and show that TULIP is 3X more energy-efficient than the conventional design, consistently, without any penalty in performance, area, or accuracy. Furthermore, the reported gains in energy efficiency do not rely on standard low-power techniques such as voltage scaling and approximate computing.

TULIP is a scalable SIMD machine, consisting of a collection of concurrently executing TULIP-PEs. Each TULIP-PE is constructed using a collection of neurons (FTL cells). TULIP-PEs are capable of computing a node of a BNN. The node of a BNN is first decomposed into a network of threshold functions, which will be explained in further detail in Sect. 4.1.1. Then, these threshold functions are scheduled on the neurons (perceptrons) of the TULIP-PE.

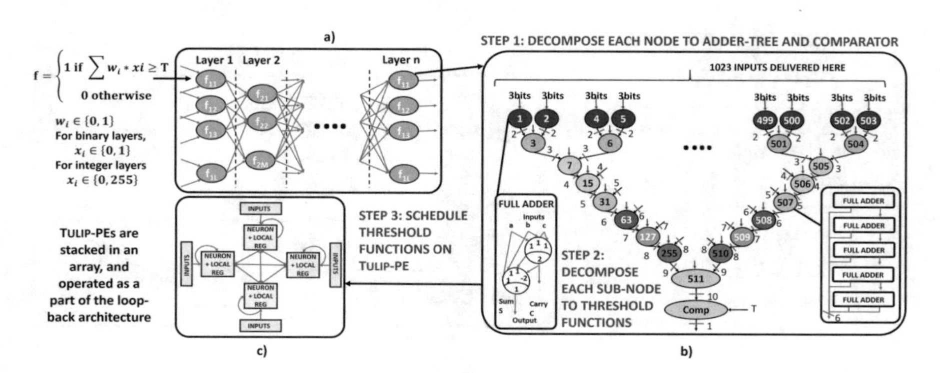

Fig. 18 TULIP flow: Each node of a BNN is decomposed into an adder tree. Each sub-node of an adder tree is decomposed into a network of two-level threshold functions. The decomposed network is scheduled using reverse post-order (RPO) scheduling (indicated using node numbers with unmarked red arrows indicate 1-bit input), on a TULIP-PE built using a cluster of four hardware neurons. (**a**) Binary neural network. (**b**) Threshold logic adder tree. (**c**) TULIP-PE

4.1.1 Binary Neural Network Using Binary Neurons

Figure 18 depicts the design flow and the main components of TULIP. Each node of a BNN is first expressed as a network of threshold functions f_{ij} (Fig. 18a). For each threshold function, the weighted sum is decomposed into a tree of adders (Fig. 18b), such that each adder can be further decomposed as a linear iterative array that uses binary neurons of bounded fain-in size (Fig. 18b; see insets). Note that the use of an adder tree delivers better energy efficiency than conventional accumulators. This is because unlike accumulators [100, 100–107] that use operators of max width for each addition operation, each adder in the adder tree only requires a bit width that is equal to the bit width of its operands. In Fig. 18b the labels inside the node show the order in which that node is executed on a TULIP-PE for a threshold function with 1023 inputs. Note that although accumulation can be implemented by using conventional adders of varying sizes, the key difference with TULIP is that *all* the operations that arise in a BNN (addition, accumulation, comparison, and max pooling) are implemented by the same, single configurable binary neuron in TULIP.

The main processing element (TULIP-PE) of TULIP consists of a complete network of four configurable binary neurons (see Fig. 18c). The operations of the adder tree and all other operations of the BNN are scheduled to be performed in TULIP-PE. Each full adder is implemented as a cascade of two binary neurons (Fig. 18b, left inset). Larger adders are implemented using a cascade of full adders (Fig. 18b, right inset).

The TULIP top-level structure consists of several PEs, as well as image and kernel buffers (Fig. 22). TULIP is a scalable architecture. This means that throughput can be scaled by using larger image and kernel buffers and adding PEs, without changing the scheduling algorithm.

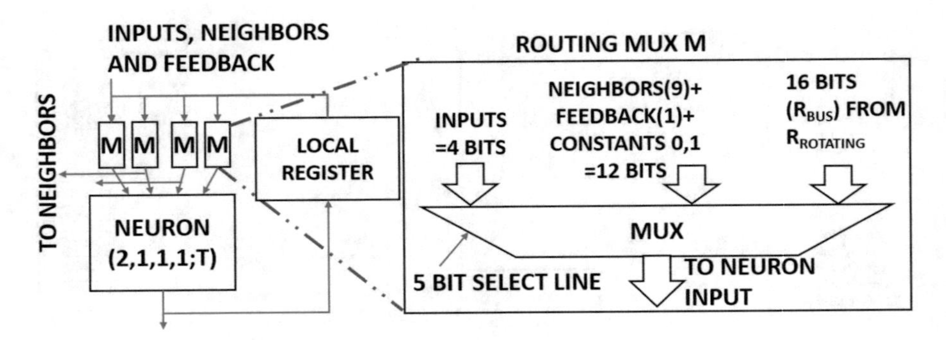

Fig. 19 The hardware neuron and its connections

4.1.2 Hardware Architecture of TULIP-PE

As shown in Fig. 18c, a TULIP-PE has four fully connected neurons $N1, \cdots, N4$. Each neuron has a 16-bit local register associated with it. These neurons and their local registers are interconnected using multiplexers, as shown in Fig. 19. Each neuron has four inputs a, b, c, and d with weights 2, 1, 1, and 1, respectively, and a threshold T, which can be modified at runtime using control signals. The choice of using four neurons is determined by the computation requirements. Since four is the minimum number of neurons needed to perform all the basic operations needed for the operations of a BNN (addition, comparison, max pooling, and RELU), it was chosen for this architecture. For the implementation of local registers, latches were used, as they provide low-power access to temporary data.

The following subsection provides details on how the operations are scheduled and executed on a TULIP-PE.

4.1.3 Addition and Accumulation Operation

Consider a node p in the adder tree that adds two operands stored in R_1 and R_4, using the threshold function shown in Fig. 20a bottom-right inset. The sum and carry bits of p are generated by $N2$ and $N3$, respectively, over multiple cycles, using the threshold function shown in Fig. 20a top-right inset. Since the bits of the sum operation are computed using $N2$, the final result of p is stored in the local register $N2$, i.e., R_2. Figure 20a shows two 4-bit operands x and y, i.e., $\{x_3, x_2, x_1, x_0\}$ and $\{y_3, y_2, y_1, y_0\}$. The final result $x + y$ is stored in R_2.

Next, consider the adder tree nodes p, q, and r, as shown in the in Fig. 20b. r sums the results of p and q.

The results of p and q are added together in r. Because the result of p is placed in R_2, the result of q is saved in R_3 to allow operands to be read simultaneously while computing r. r generates its sum bits on $N1$ and carry on $N4$ by reading R_2

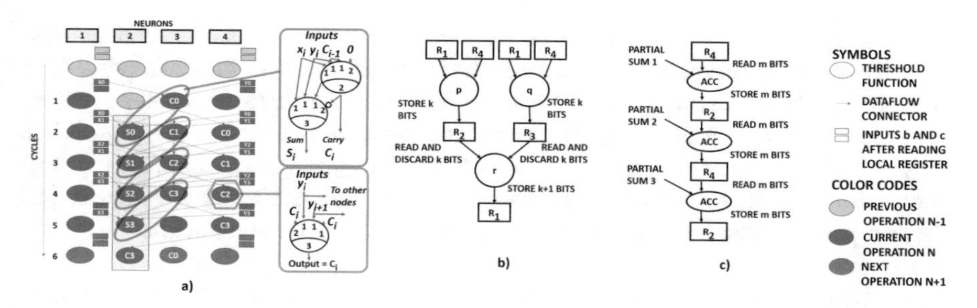

Fig. 20 Adder, adder tree, and accumulator schedule. (**a**) Addition operation. (**b**) Adder-tree memory management. (**c**) Accumulation operation to add partial sums

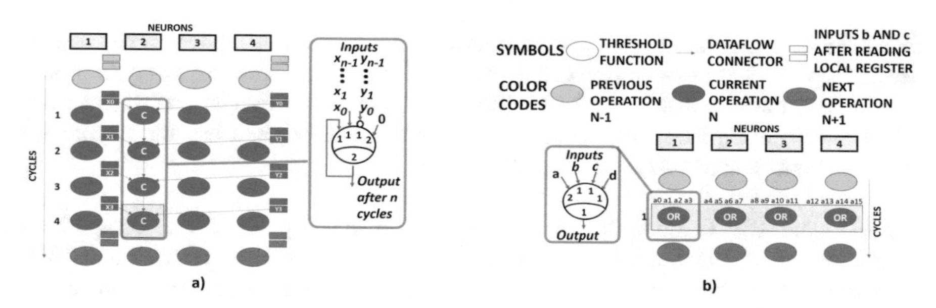

Fig. 21 Comparator and max pooling schedule. (**a**) Comparison operation. (**b**) Maxpooling operation

and R_3. The memory utilized by the p and q results can now be released once r is processed. Each addition operation stores its result to a specific memory location in the local registers so that the data in the memory is not prematurely overwritten during RPO scheduling.

The adder tree used in this work handles additions of up to 10 bits on the TULIP-PE. However, this range can be further extended by configuring the TULIP-PE for accumulation. A multi-cycle addition operation can be performed to an accumulated term stored in the local registers. Figure 20c shows the addition of an input number p with the accumulated term q. The storage of q is alternated between the R_2 and R_4, as local registers that provide the operands cannot store the results simultaneously.

4.1.4 Comparison, Batch Normalization, Max Pooling, and RELU Operation

Comparison A multi-cycle sequential comparator can be implemented using three-input threshold functions, as shown in Fig. 21a. Two n-bit numbers x and y are serially delivered from LSB to MSB to the comparator that returns the result of the comparison $x > y$. Starting with the comparison of the LSBs of both operands, the ith cycle of the comparison compares the ith position bits of the operands. If the ith bit of the first operand is greater than the second, it overrides the result generated in cycle $i - 1$. The inset in Fig. 21a shows the logic for bitwise comparison. After n cycles, if the output is 1, then $x > y$; otherwise $x \leq y$. As an example, Fig. 21a shows the schedule for a 4-bit comparison. The 4-bit inputs x and y are streamed to the comparator either through the local registers or through the input channels.

Batch Normalization This operation performs biasing of an input value in BNNs. For BNNs, it is realized by subtracting the value of bias from the threshold T of the binary neuron, as described in [108]. Therefore, batch normalization in TULIP is implemented using the comparison operation.

Max Pooling In a BNN, this operation is an OR operation on a pooling window of layer outputs. An OR operation can be implemented using the threshold gate shown in Fig. 21b. Each neuron in the TULIP-PEs implements one four-input OR function. Note that local registers are not needed for this operation, because the results are streamed out as soon as they are computed.

RELU The RELU operation in TULIP is realized by using the comparison operation. First, the operand is compared against 0. An AND operation is then performed using the operand and the result of the comparison operation, using a two-input threshold function [1,1;2]. Note that although the implementation of RELU is not as efficient as using a dedicated circuit built using standard cells, implementing this operation on a TULIP-PE removes the need of integrating custom hardware for special operations such as RELU.

4.1.5 Top-Level View of the Architecture

The top-level TULIP architecture is shown in Fig. 22. It is designed to achieve high energy efficiency per operation while achieving the throughput of state-of-the-art implementations. This architecture consists of four main types of components: image buffers, kernel buffers, one or more processing units, and controllers. The kernel buffer stores the BNN weights. The image buffer is a two-level standard cell memory (SCM) named L2 and L1 that is used to send input pixels to all processing units present in the design. The processing unit contains TULIP-PEs which is mainly responsible for performing the convolution operation. These units also get the appropriate weights from the kernel buffer.

Each TULIP-PE in a processing unit is used to handle one output feature map (OFM) of the binary layers, while the MAC units are used for integer layers. Note that although TULIP-PEs can also be used to compute integer layers, doing so results in reduced throughput.

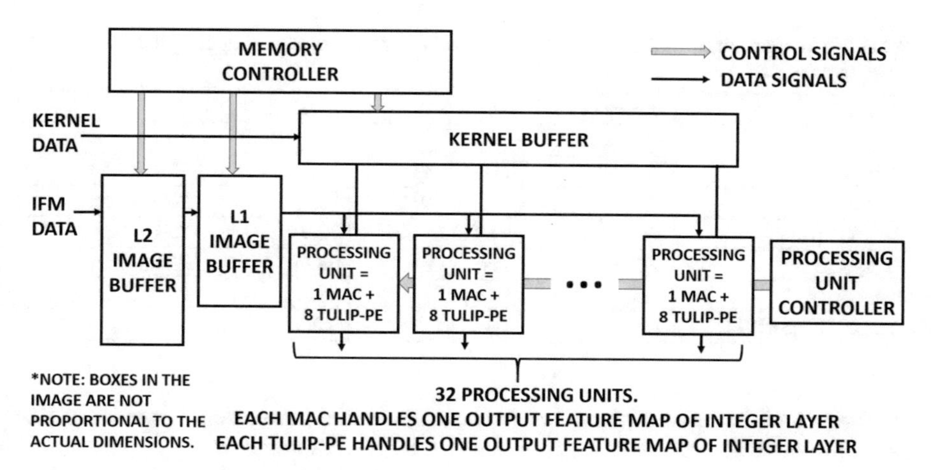

Fig. 22 TULIP top-level architecture

Table 6 Reconfigurable MAC unit [106] vs TULIP-PE, for 288-input neuron evaluation

Single PE metrics	YodaNN MAC (B)	TULIP-PE (T)	Ratio (X) (B/T)
Area (μm^2)	3.54E+04	1.53E+03	23.18
Power (mW)	7.17	0.12	59.75
Cycles	17	441	0.038
Time period (ns)	2300	2300	1
Time (ns)	39	1014	0.038

4.1.6 Experimental Results

Table 6 compares a TULIP-PE against the 15-bit reconfigurable MAC unit used in YodaNN[106], a recent ASIC-based BNN accelerator. The MAC unit used in YodaNN can be reconfigured to support kernel sizes 3×3, 5×5, and 7×7. Note that both the MAC unit and TULIP-PE support integer inputs and binary weights. In large BNN architectures such as AlexNet[109], the initial layers are integer layers, which means that the inputs are integers but the weights are binary. The remaining layers are binary layers which means that both inputs and weights are binary. YodaNN uses MAC units for all layers, while TULIP uses TULIP-PEs for binary layers and simplified MACs (which support only 5×5 and 7×7 kernel windows) for integer layers.

Since the calculation method of YodaNN and TULIP is different only in the binary layer, the comparison of MAC and TULIP is performed in the binary layer. That is, both modules perform a weighted sum using binary activations and binary weights of 288 inputs, i.e., 3×3 kernel for 32 IFMs. Based on Table 6, we can see that the TULIP-PE is 23 times smaller than a MAC unit and consumes 1/60th of the power. However, a TULIP-PE takes 27 times longer than the MAC unit because it

Table 7 Comparison of YodaNN with TULIP architecture for accelerating convolution layers of standard datasets

Conv only	BinaryNet		AlexNet	
	CIFAR10		Imagenet	
Dataset	YodaNN	TULIP (X)	YodaNN	TULIP (X)
Op.(MOp)	1017	1017 (1.0)	2050	2050 (1.0)
Perf.(GOp/s)	47.6	49.5 (1.0)	72.9	79.1 (1.1)
Energy(uJ)	472.6	159.1 (3.0)	678.8	224.5 (3.0)
Time (ms)	21.4	20.6 (1.0)	28.1	25.9 (1.1)
En.Eff. (TOp/s/W)	2.2	6.4 (3.0)	3.0	9.1 (3.0)

performs addition on a bit-by-bit basis. As a result, the power delay product of PE is 2.27X lower than the MAC unit, while also being 23X smaller.

Using an adder tree-based schedule helps TULIP deliver a better power delay product than a traditional MAC unit. Also, since the MAC unit cannot perform operations such as compare and maximum pooling, the data for these operations need to be sent to other parts of the chip for further processing [106]. On the other hand, the TULIP-PE maintains the locality of the data and can perform internal comparison and maximum pooling operations without moving the data to other parts of the chip, thereby saving additional energy.

To ensure that the chip area of TULIP matches that of YodaNN, TULIP was designed with 32 simplified MAC units and 256 TULIP-PEs. As a result, convolution in TULIP is performed in batches of 32 OFMs for integer layers and 256 OFMs for binary layers. Table 7 compares the characteristics of YodaNN with TULIP for the convolution layers. For the convolution layers, the TULIP architecture surpasses YodaNN by around 3X in terms of energy efficiency. This is due to the combined use of an adder tree-based schedule, coupled with clock gating. Increased reuse of input pixels also improves energy efficiency and increases throughput. Note that the results show that the gains are consistent across different neural networks.

4.1.7 Conclusion

This section demonstrates how a BNN accelerator TULIP has 3X better energy efficiency than another current state-of-the-art architecture with the help of threshold logic gates/neurons. Gains were achieved without using standard low-power techniques such as voltage scaling and approximate computing.

4.2 A Flash-Based Current-Mode IC to Realize Quantized Neural Networks

This section presents a mixed-signal architecture for implementing quantized neural networks (QNNs) using flash transistors to achieve extremely high throughput with very low power, energy, and memory requirements [110]. Its low resource consumption makes our design especially suited for use in edge devices. The network weights are stored in-memory using flash transistors, and nodes perform operations in the analog current domain. Our design can be programmed with any QNN whose hyperparameters (the number of layers, filters, or filter size, etc.) do not exceed the maximum provisioned in the hardware. Once the flash devices are programmed with a trained model and the IC is given an input, our architecture performs inference with zero access to off-chip memory. We demonstrate the robustness of our design under current-mode nonlinearities arising from process and voltage variations. We test validation accuracy on the ImageNet dataset, and show that our IC suffers only 0.6% and 1.0% reduction in classification accuracy for top-one and top-five outputs, respectively. Our implementation results in a $\sim$50$\times$ reduction in latency and energy when compared to a recently published mixed-signal ASIC implementation, with similar power characteristics. Our approach provides layer partitioning and node sharing possibilities, which allow us to trade off latency, power, and area among each other.

In our IC, we implement all aspects of the QNN needed for inference. Convolutional (CONV), max pooling (MAXPOOL), and fully connected (FC) layers, batch normalization, intermediate data storage, and control flow are all implemented on-chip. At programming time, all network hyperparameters can be chosen up to the maximum allowed. The maximum values of these hyperparameters are summarized in Fig. 23. Note that not all layer types need be used, resulting in a highly flexible architecture.

Each layer is comprised of circuits implementing the nodes of that layer. The nodes in the first CONV layer accept 8-bit inputs. Nodes in all other layers accept 1-bit inputs whose values are in $\{-1, 1\}$. The nodes in all layers can be programmed with any weight programming scheme from binary up to 9-valued, at zero additional cost to delay, area, or power. All nodes have 1-bit outputs whose values are in $\{-1, 1\}$.

In the rest of this section, we describe our implementation of every aspect of a QNN. First, we discuss the designs of the FC node (Sect. 4.2.1), CONV node (Sect. 4.2.2), and MAXPOOL node (Sect. 4.2.3). Then, we describe how batch normalization is implemented (Sect. 4.2.4), and we illustrate our dataflow architecture (Sect. 4.2.5). Finally, we present our experiments and results (Sect. 4.2.6) and our conclusion (Sect. 4.2.7).

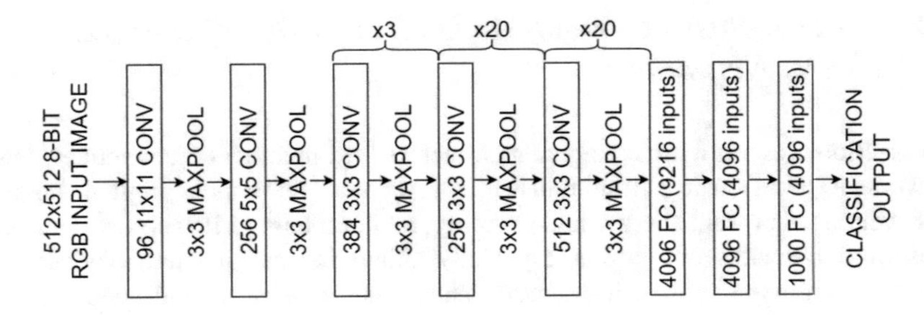

Fig. 23 Maximum architecture provisioned on-chip

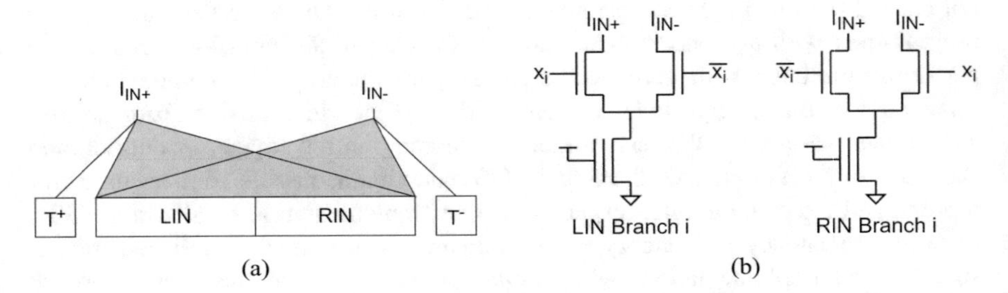

Fig. 24 FC and CONV node input network (IN). (**a**) IN and threshold blocks. (**b**) IN branch design and connectivity

4.2.1 Fully Connected (FC) Node Design

Input Network (IN) The FC node accepts inputs through an input network, consisting of two halves. We refer to these halves as the left input network (LIN) and the right input network (RIN), as shown in Fig. 24a. Each branch in the IN is made from one flash FET and two MOSFETs, as shown in Fig. 24b. For IN branch i, the flash FET V_{th} is programmed to either 0.862 V, 0.908 V, 0.962 V, 1.037 V, or 2 V according to the magnitude of weight parameter w_i. These threshold voltages are chosen such that the $I_{DS} \propto |w_i|$ when the FET is conducting. The LIN (RIN) stores positive (negative) weight parameters. For a given weight w_i, $sign(w_i)$ will determine whether it is programmed to branch i of the LIN or RIN, yielding 9 weights in all. The programming of these branches is mutually exclusive; if branch i of the LIN (RIN) is programmed with a weight w_i, branch i of the RIN (LIN) will be programmed with a weight of zero.

The FC node accepts a digital voltage input x_i and its logical complement $\overline{x_i}$. If the corresponding input is $1(-1)$, then $x_i = 1(0)$ and $\overline{x_i} = 0(1)$. These signals are connected as shown in Fig. 24b, which can encode an input in $\{-1, 1\}$ as follows. Recall that if weight $w_i > 0$, it will be programmed to the LIN. Now, if $x_i = 1$ $(x_i = 0)$, the branch current will flow to the node I_{IN+} (I_{IN-}). Similarly, recall that if weight $w_i < 0$, it will be programmed to the RIN. Then, if $x_i = 1$ $(x_i = 0)$, the branch current will flow to I_{IN-} (I_{IN+}). This effectively computes $w_i x_i$ in the current domain, for a given branch i, with 9 possible weight values.

At I_{IN+} and I_{IN-}, all branch currents are summed by Kirchhoff's current law. The current flowing to I_{IN+} (I_{IN-}) will be $\sum w_i x_i$ over all i for which $w_i x_i$ is positive (negative). Both the LIN and the RIN have a number of input branches equal to the maximum number of node inputs. To realize a node with a smaller number of inputs than the maximum, the unused branches will simply be programmed with zero weight.

Each LIN/RIN branch pair can be programmed with any nine-valued weight parameter. Our experiments prove reliable node performance for weights with up to and including nine values. However, it should be noted that we can use a larger number of weights as well. It is known that one can choose flash V_{th} with very fine precision [11, 111], and hence we could have hundreds of unique weight values within a typical V_{th} range (a few hundred mV). Importantly, this choice of weights comes at no memory cost, because the weight value is stored in the flash FET alone. That said, our work currently uses nine weight values as described above.

Current Mirrors and Comparator Each node uses two current mirrors to convert the differential currents from the nodes I_{IN+} and I_{IN-} to a common node V_{CMP}, as shown in Fig. 25. On the positive (left) side of the figure, the current I_{IN+} is transferred to V_{CMP} through a current mirror connected to V_{DD}, so that an increase in this current causes V_{CMP} to increase. On the negative (right) side of the figure, we use a two-stage current mirror, with the second stage connected to GND, so that an increase in current I_{IN-} causes V_{CMP} to decrease. The left and right current mirrors have identical scaling factors, and hence $V_{CMP} > V_{DD}/2$ when $I_{IN+} > I_{IN-}$, and $V_{CMP} < V_{DD}/2$ otherwise. Thus, $V_{CMP} > V_{DD}/2$ if $\sum w_i x_i > 0$ and $V_{CMP} < V_{DD}/2$ otherwise. The comparator (CMP) compares V_{CMP} against $V_{REF} = V_{DD}/2$, to produce the node output bit. This results in an output bit of 1 when $\sum w_i x_i > 0$, and zero otherwise. The output bit encodes a value in $\{-1, 1\}$.

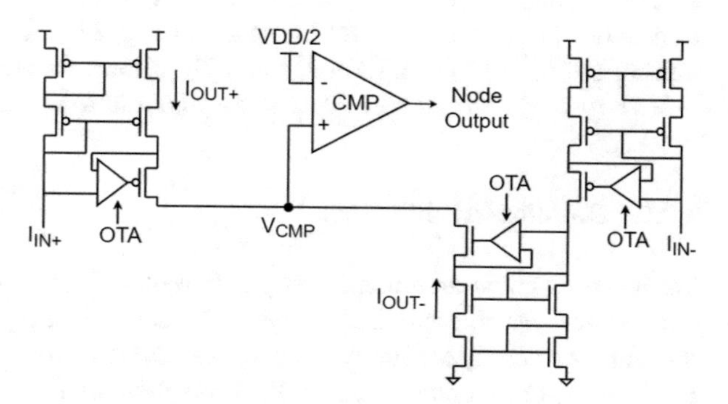

Fig. 25 Neuron output computation hardware

4.2.2 Convolution (CONV) Node Design

Input Network The input network (IN) of CONV nodes is exactly the same as in FC nodes. This is because the convolution operation is also a summation $\sum w_i x_i$, over a portion of a 3D feature map. To realize a CONV node with a $k \times k$ kernel and c channels, the node will have k^2c inputs. In our IC, each CONV node will implement a different filter. Our design can implement different numbers of filters, with different filter sizes as shown in Fig. 23.

The IN design is different for the first CONV layer however, because this CONV layer accepts 8-bit image inputs with three (RGB) channels. We accomplish this by duplicating the IN block into 8 blocks $\mathrm{IN_0}$-$\mathrm{IN_7}$.

Current Mirrors and Comparator In a CONV node with 1-bit inputs, the current mirror design is identical to that in the FC nodes. For a CONV node with 8-bit inputs, the current through each block $\mathrm{IN_j}$ must have twice the magnitude of the current of block $\mathrm{IN_{j-1}}$, so that the bits of x_i are given their proper significance. To accomplish this, we duplicate current mirrors and design them to have binary-weighted scaling factors.

4.2.3 MAXPOOL

We implement MAXPOOL as an OR operation on binary values in a feature map. We use 9-input OR gates to accomplish MAXPOOL of a 3×3 kernel or smaller.

4.2.4 Batch Normalization

We implement batch normalization as described in [112], by changing the threshold T used in the node output computation $\sum w_i x_i > T$. We do this with threshold blocks T^+ and T^-, as depicted in Fig. 24a. T^+ and T^- are implemented with flash transistors (as in the node IN branches in Fig. 24b), but without any MOSFETs. Setting the V_{th} of the FETs in T^+ and T^- allows for setting the node threshold to $-Max \leq T \leq Max$, where Max is the maximum $\sum w_i x_i$ value of that node.

4.2.5 Dataflow Architecture

We use a dataflow architecture to pipeline data through our chip. Our architecture allows for a variable number of layers, for each layer type, to be programmed onto the chip. We supply the input image to the CNN in a row-by-row manner. Each layer produces its output feature map (OFM) pixel-by-pixel. This allows all CONV layers to compute in parallel, because each layer can begin computing once it has enough inputs for a single output pixel to be produced. The user can disable some number of layers as desired. We route data around unused layers, giving the user a flexible choice for the number of CONV, MAXPOOL, and FC layers, up to the maximum for each (see Fig. 23). Because layers can be disabled, some layers can receive their data from different source layers. We use multiplexors to determine the origin of the input data for each layer. We list below the options for input data sources for each layer:

- The first CONV layer can only receive the input image.
- MAXPOOL layers may only receive inputs from the preceding CONV layer.
- All CONV layers after the first may receive inputs from the preceding MAX-POOL layer *or* the preceding CONV layer. This allows for any and all MAX-POOL layers to be disabled.
- All FC layers may receive inputs from any preceding layer. This allows any and all CONV, FC, and MAXPOOL layers to be disabled, and skipped.

4.2.6 Experiment and Results

We implement our designs in a 45 nm process technology. We simulate circuit designs using Synopsys HSPICE [113], to test node correctness and measure performance metrics like latency, area, and power. All the CNN components described are accounted for in our computations. For CMOS devices, we use a 45 nm PTM model card [114]. For flash devices, we use the model card derived in [115]. Our nominal $V_{DD} = 1.2$ V for all HSPICE experiments. We use the Python library TensorFlow [116] to measure the ImageNet classification accuracy of Binary AlexNet implemented with our approach.

We first measure node output bit error rates and node latency. The simulations for these sections are conducted in HSPICE. We then export node error rates to a Python simulation to measure classification accuracy on ImageNet. Finally, we quantify and compare metrics like layout area and inference latency, power, and energy, and we also discuss variations of our design that can allow us to trade off these metrics.

Bit Error This experiment tests node performance in the presence of process and voltage variation. We implemented our largest node (a 1-bit 9216-input FC node) in HSPICE, and performed 18,000 Monte Carlo simulations. For each simulation, we choose the weights and inputs of the node, such that the mean weight w_{mean} and the mean input x_{mean} follow a uniform distribution. We test the entire range of node weights and inputs by sampling the ranges of w_{mean} and x_{mean} equally, with increments of 0.01. In this experiment, we select node weights from $\{-1, 1\}$, i.e., 1-bit weights, so we can compare our ImageNet classification accuracy against that of a baseline BNN that also uses 1-bit weights. We note that performing this experiment with 9-valued weights resulted in *lower* error rates, a drop of about 3.7% on average per bin.

For the Monte Carlo simulations, we model process variation by randomizing the length (L), width (W), and threshold voltage (V_{th}) of all MOSFETs. For L and W, we take the absolute value of the variation from [117], which was reported for a 65 nm process. We use the V_{th} variation presented in [118]. We derive our MOSFET

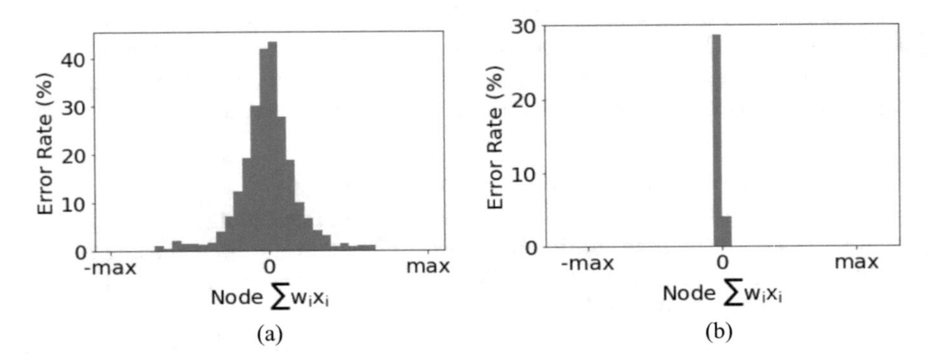

Fig. 26 FC node bit error rates. (**a**) 9216-input FC bit error. (**b**) 1200-input FC latency error

average channel doping N_a from the average value of recent papers ([119–122]). For each simulation we also vary V_{DD}, with a 10% (120 mV) standard deviation. Our CMP reference voltage V_{REF} is assumed to be constant, being generated by a precise bandgap voltage reference [123]. For each simulation, if the node output is incorrect, we record this as a bit error for that $\sum w_i x_i$ over all node inputs and weights. We split the range of values for $\sum w_i x_i$ into bins of width 512. In Fig. 26a, we plot the node output error rates for each bin.

Latency Error In our design, we bound the latency of a node, thereby incurring some error. We use our smallest, slowest nodes to measure latency and its associated error. For Binary AlexNet this is a 1200-input 1-bit FC node. For BinaryNet, this is a 27-input 8-bit CONV node. For both nodes, we perform 50,000 HSPICE transient simulations. Node weights and inputs are selected as in the bit error experiment. For each simulation, we record the $\sum w_i x_i$ of the node, and the latency from the time the node inputs arrive to the time the node output bit reaches its final value. Just before the node inputs are applied, we precharge V_{CMP} to $V_{DD}/2$, using the CMP V_{REF}. We do not take the maximum latency as our clock period, but rather, we select the clock period < maximum latency, and accept node output errors for situations when node latency exceeds the selected clock period. This gives additional error rates per bin, as shown in Fig. 26b. Our chosen clock period is 4.2 ns for Binary AlexNet and 7.2 ns for BinaryNet. The latency error rates for Binary AlexNet are shown in Fig. 26b. In general, a user can choose the clock period based on their tolerance to reduced classification accuracy, and this can be done in real time.

CNN Classification Accuracy on ImageNet This experiment will test how much the bit error and latency error rates impact classification accuracy of the CNN as a whole. We use the Binary AlexNet architecture, implemented in the Larq Python library [124]. We measure top-one and top-five validation accuracy on ImageNet. We model total node error behavior by summing the two histograms depicted in Fig. 26, thus adding the bit error and latency error rates. We apply error rates on a per-bin basis, to all nodes in the CNN. It should be noted that the two error histograms in Fig. 26 were measured for our most error-prone node and our slowest node respectively, and therefore give a conservative estimate for error rates of all nodes.

Compared to the digital implementation of Binary AlexNet, our mixed-signal implementation results in a minimal decrease in CNN classification accuracy. We measure a 0.6% dtop in top-one accuracy (from 36.3% to 35.7%) and a 1.0% drop in top-five accuracy (from 61.5% to 60.5%).

Latency, Power, and Layout Area In this section we present results on performance metrics of latency, area, power, and energy of a single inference for the Binary AlexNet and BinaryNet architectures implemented with our approach. We compare the performance metrics of our approach against TULIP [125], a recent BNN implementation, which had the best performance at the time it was published. TULIP is implemented in a TSMC 40 nm LP process. In our results, we include the contribution to latency, layout area, power, and energy of all the component circuits described (including digital and memory blocks). We note that TULIP does not report measurements for Binary AlexNet, but for a slightly different architecture, proposed in [96]. These two architectures are both binarized versions of AlexNet and are very similar, and therefore provide a fair comparison. We list TULIP entries under the Binary AlexNet heading in Table 8 for simplicity. We implement BinaryNet as implemented in [125], for a direct comparison.

As described in Sect. 4.2.5, for each CONV filter, we pass different values over the node inputs to emulate the sliding of the CONV kernel over the IFM over time. We call this modification *node sharing*, and use it to achieve lower power and area requirements. In order to further reduce our power consumption, we only allow one node per layer to fire in any given cycle. We call this modification *layer partitioning*. For both CONV and FC layers, this means that only a single OFM pixel per layer will be computed in any given clock cycle. We present three combinations of these modifications while presenting performance metrics, in Table 8. Variation 3 uses only CONV node sharing. Variation 2, in addition, uses FC layer partitioning. Variation 1, in addition to the modifications of Variation 2, uses CONV layer partitioning.

We perform HSPICE simulations to measure the latency and maximum active power of all circuits described. To compute inference latency, power, and energy, we perform an architecture-level simulation using Python. For Binary AlexNet and BinaryNet, we use the clock periods determined in Sect. 4.2.6. The inference latency, power, and energy are shown in Table 8.

To estimate chip layout area, we first generate a layout design for all IN blocks. For the digital blocks, we estimate their layout area using Synopsys Design Compiler [113]. The chip area used by our architectures adds the layout area of the component circuits and is shown in Table 8. If we provision the "maximum" network of Fig. 23, our area is 27.1 mm^2, with a maximum power of 91 mW per classification.

Table 8 Chip design variations

Setting/metric	Binary AlexNet on ImageNet			
	Variation 1	Variation 2	Variation 3	TULIP [125]
Max power	6.47 mW	1.6 W	48.5 W	Unknown
Average power	2.72 mW	614 mW	24.8 W	2.59 mW
Latency	3.07 ms	13.6 μs	336 ns	165 ms
Energy	8.35 μJ	8.35 μJ	8.35μJ	428 μJ
Area	17.7 mm^2	17.7 mm^2	64.7 mm^2	1.80 mm^2
	BinaryNet on Cifar-10			
Setting/metric	Variation 1	Variation 2	Variation 3	TULIP [125]
Max power	4.59 mW	1.2 W	20.8 W	Unknown
Average power	1.86 mW	433 mW	3.46 W	6.36 mW
Latency	1.99 ms	8.50 μs	1.07 μs	28.9 ms
Energy	3.71 μJ	3.71 μJ	3.71 μJ	184 μJ
Area	4.36 mm^2	4.36 mm^2	41.8 mm^2	1.80 mm^2

4.2.7 Conclusion

In this section, we present a flash-based mixed-signal architecture for accelerating QNNs, with the goal of minimizing power, energy, and latency. We utilize flash technology for network weight storage and node computation, which enables us to achieve extremely low power, energy, and latency per inference. Flash also allows for storage of many weight bits in a single transistor. We test up to nine-valued weights, but the number of weight values used in our architecture could be increased past the nine-valued weights we test in this paper, at no additional cost to latency, area, power, or memory.

We implement all the circuits needed for QNN inference, to quantify classification error as well as performance metrics like latency, area, power, and energy. As a part of these experiments, we perform a Monte Carlo analysis to demonstrate the robustness of our designs in the presence of process and voltage variation. We model the performance of our chip as applied to the Binary AlexNet architecture and demonstrate that ImageNet classification accuracy is minimally degraded (by $\sim$1%) across circuit variations. We show that our chip design has $\sim$50$\times$ lower latency and energy than a recent ASIC approach, for ImageNet classification. We present three variants of our architecture to enable the trade-off between latency, area, and power.

5 CIDAN: Computing in DRAM with Artificial Neurons

This section describes a processing in-memory (PIM) architecture called CIDAN [126]. It is composed of a new processing element called TLPE, which is built using the FTL gate described in Sect. 3.2. A TLPE contains digital logic in addition to the

FTL gate which enables it to compute a few non-threshold functions by a sequence on threshold gate evaluations. CIDAN is designed on a DRAM platform with an array of TLPE(s) integrated into its memory arrays. An evaluation on a represented set of workloads shows that CIDAN improves upon the state of the art by 3X in performance and 2X in energy consumption.

5.1 Introduction

The widening performance gap between the CPU and the memory has severely constrained the throughput of the traditional von Neumann architectures. This well-known issue is referred to as the *memory wall* problem in the literature. It has been a principal reason for degraded throughput of high bandwidth applications consisting of bulk bitwise operations such as machine learning[127], database management[128], encryption[129], etc. Over the years different proposals have been investigated to bridge the performance gap between CPU and memory. Several efforts have been made to improve the data bandwidth of the off-chip memory which resulted in memory designs such as double data rate (DDR) memory and 3D high bandwidth memory (HBM). Even with these improved bandwidth memory architectures, a greater increment in the CPU performance hasn't solved the memory wall problem. To reduce the memory transfers over the channel connecting the CPU and memory, some proposals have advocated bringing computation and memory closer. The work presented in [130, 131] introduced larger on-chip cache memory and added computation to the cache units. This method has limitations arising from the limited size of the cache. Other solutions broadly classified as processing in-memory (PIM) architectures have been proposed as well. Examples of PIM architectures include [132–134]. These architectures are based on the fundamental idea of using the memory to perform certain computations on the data stored in the memory and eliminate the use of the CPU for such computations. The PIM architectures use the high parallelism of memory to process large amounts of data simultaneously and can improve the throughput of high bandwidth applications. The number of data transactions over the bandwidth-limited memory channel is reduced in PIM architectures, and this alleviates the communication bottleneck between the memory and the processor.

This section describes a new PIM architecture called CIDAN, which is based on DRAM with threshold logic processing elements (TLPEs) embedded in it. CIDAN adds computation capability to the DRAM without sacrificing its area, changing its access protocol, or violating any timing constraints.

5.2 Threshold Logic Processing Element (TLPE)

This section will describe the architecture of CIDAN. The reader is encouraged to review [135] to get a better understanding of how a DRAM works and [136] to understand what threshold functions are, and how threshold logic gates are constructed.

The design of a TLPE is shown in Fig. 27a. It is composed of a threshold gate, two latches (L1, L2), and four XOR gates. The threshold gate performs computations and implements a threshold function [−2, 1, 1, 1, 1, 1; T], where T can switch between 1 and 2. The two latches are used to store the output temporarily. The four XOR gates at the input of the threshold gate are used to invert the signals from the memory bank. The inverted/non-inverted inputs to the threshold gate are controlled by the control signals *C0–C3*. The control signals en_{ri} and en_{li} are used to select the value of T and other two inputs to the threshold gate.

The threshold gate performs logic operations which are threshold functions such as (N)AND, (N)OR in a single cycle when the required inputs (I1–I3) and the appropriate threshold T are selected in the TLPE. The enabled inputs and the threshold value for the logic operations are shown in Table 9. The X(N)OR operations are non-threshold functions and are implemented in a two-cycle schedule on the TLPE.

The implementation of the addition operation is carried as a schedule on the TLPE and is shown in Fig. 27b. An addition operation on the ith significant bits of the operands A and B and the previous carry bit $C[i]$ stored in L1 is carried as follows. The threshold gate is configured to perform a majority operation on the input bits $A[i]$ and $B[i]$ and the previous carry $C[i]$ in the first cycle and produce output carry $C[i+1]$. The carry bit $C[i+1]$ is stored in the latch L2 and also fed back to the input of the threshold gate. In the second cycle, the threshold gate is configured to perform threshold function [−2, 1, 1, 1; 1] to generate the sum bit

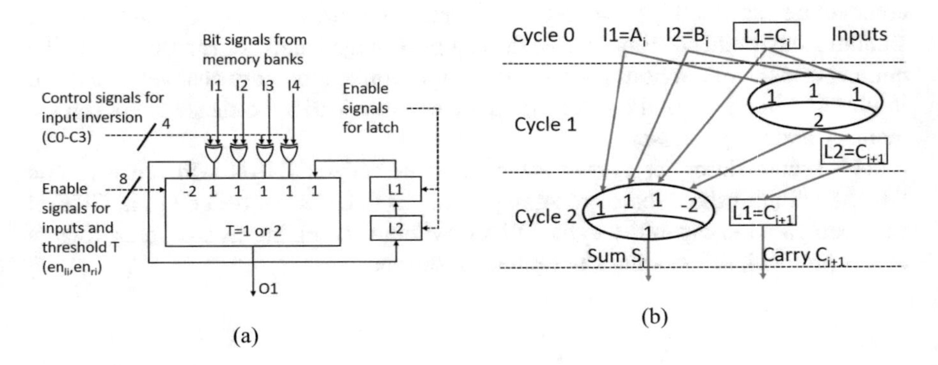

Fig. 27 TLPE and its schedule for basic addition operation. (**a**) Architecture of threshold logic processing element (TLPE). (**b**) Schedule for addition operation on TLPE for 3-bits A_i, B_i, and C_i. Outputs are S_i and C_{i+1}

Table 9 Basic logic operations using threshold logic processing element. For demonstration, operands are I1 and I2

Func	Cycle number	Weights			T
		−2	1	1	
NOT	1	X	~I1	X	1
AND	1	X	I1	I2	2
OR	1	X	I1	I2	1
NAND	1	X	~I1	~I2	1
NOR	1	X	~I1	~I2	2
XOR	1	X	I1	~I2	2
	2	OP1	~I1	I2	2
XNOR	1	X	I1	I2	2
	2	OP1	~I1	~I2	2

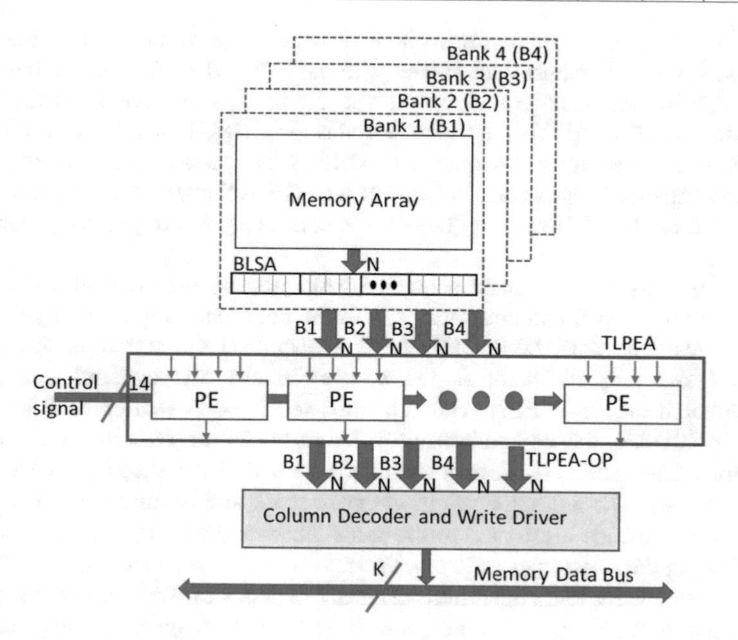

Fig. 28 Threshold logic processing element array (TLPEA) connected to banks in a DRAM device

$S[i+1]$ using the inputs $C[i+1]$, $A[i]$, $B[i]$, and $C[i]$. Simultaneously, the L1 latch is written with $C[i+1]$ stored in L2 to serve as the input carry bit for the addition of the next significant bits of operands $A[i+1]$ and $B[i+1]$.

5.3 Top-Level Architecture of CIDAN

CIDAN is designed using the existing architecture of the DRAM chips used in DIMMs. The TLPEs are integrated with the DRAM memory array as shown in Fig. 28. The TLPEs form an array TLPEA interfaced with the outputs of the bit line sense amplifiers (BLSA). A single TLPEA is connected to four banks with each TLPE taking four inputs from bit lines of four different banks $B1$, $B2$, $B3$, and $B4$. Consequently, the number of TLPEs is equal to the size of the row in the bank (N). The output of the TLPEA is connected to the column decoder and write driver block as shown in Fig. 28. The control signals are used to write data from the TLPEA output instead of the other bank outputs (B1, B2, B3, and B4) when computation is being carried out on the TLPEA.

5.4 System-Level Integration and the Controller Design

CIDAN is a DRAM-based platform and is designed to serve the dual purpose of an accelerator and as a memory interfaced with the CPU. The computation operations on CIDAN are encoded as assembly-level instructions and added to the CPU's instruction set. The CIDAN controller decodes the CIDAN-specific instructions for computations. It generates a sequence of DRAM instructions and activates control signals to implement operations in CIDAN. The CIDAN platform requires extra bits to be added on the CPU-memory bus for the control signals required to operate the TLPEA.

Table 10 lists the command sequence generated by the controller for all the operations on CIDAN and compares it with the command sequence of prior work architectures. The data D_i and D_j for the operation are stored in the rows at address i and j in the bank m and n, respectively. The operand data is read and computed on the TLPEA. The generated result D_r is written back to bank o row r. In CIDAN, the operands from different banks are read using consecutive activation commands on two banks separated by a period specified by DRAM timing parameter t_{RRD} (7.5 ns), whereas, in the prior work architectures, the consecutive activation commands are issued to the same bank separated by a period specified by the timing parameter t_{RAS} (35 ns). Prior work uses a sequence of AAP (82.5 ns each) commands for logic operations. The large latency of AAP commands and the requirement of multiple such commands degrade the throughput of the prior work. As the complexity of the operations increases, the number of AAP commands in the prior work increases. In contrast, the command sequence in CIDAN remains short and nearly the same for all the operations. The PIM architectures such as GraphiDe [137] and SIMDRAM [138] build upon ReDRAM and Ambit, respectively, perform addition and report (7 AAP) and (6 AAP + 2 AP) commands for 1-bit addition, respectively. Hence, the advantages of using CIDAN over the prior work increase for complex instructions.

Table 10 Basic functions and command sequence for CIDAN and other PIM platforms. D_i = Data in row i, A_{mi} = Activate bank m row i, W_{or} = Write bank o row r, $PREA$ = Precharge all open banks, AAP = ACT ACT PRE a bank, AP = Activate Precharge a bank

		Command sequence			
Func	Operation	CIDAN	ReDRAM [139]	Ambit [132]	DRISA[140]
Copy	$D_r \leftarrow D_i$	A_{mi} A_{nr} 1 clk cycle W_{nr} PREA	AAP	AAP	AP AP
NOT	$D_r \leftarrow \overline{D_i}$	A_{mi} A_{nr} 1 clk cycle W_{nr} PREA	AAP	AAP AAP	AAP AAP
AND	$D_r \leftarrow D_i \wedge D_j$	A_{mi} A_{nj} A_{or} 1 clk cycle W_{nr} PREA	AAP AAP AAP	AAP AAP AAP AAP	AP AAP AAP
OR	$D_r \leftarrow D_i \vee D_j$	A_{mi} A_{nj} A_{or} 1 clk cycle W_{nr} PREA	AAP AAP AAP	AAP AAP AAP AAP	N/A
XOR	$D_r \leftarrow D_i \oplus D_j$	A_{mi} A_{nj} A_{or} 2 clk cycles W_{nr} PREA	AAP AAP AAP	AAP AAP AAP AP AP AAP AAP	N/A
ADD	$D_r \leftarrow D_i \oplus D_j \oplus C_{in}$ $C_{out} \leftarrow Maj(D_i, D_j, C_{in})$	A_{mi} A_{nj} A_{or} 2 clk cycles W_{nr} PREA	N/A	N/A	N/A

5.5 Experimental Results

CIDAN is evaluated and compared against the prior PIM architectures, such as ReDRAM [139] and Ambit [132] for raw performance and energy. The TLPE is functionally verified using SPICE, and its delay, energy, and area are extracted and scaled to the 45 nm DRAM technology using [141]. Gem5 [142] is used for system-level simulation using these extracted values. Gem5 is integrated with Ramulator [143]—a DRAM simulator—to run the applications and to obtain the performance statistics for CIDAN as well as the other benchmark architectures. The simulator DRAMPower [144] is used to evaluate the energy consumption.

For the evaluation of PIM architectures, custom benchmarks are prepared which are composed of bulk bitwise operations NOT, AND, OR, and XOR on large bit vectors of size 1 Mb, 2 Mb, and 4 Mb. A memory array of size $16{,}384 \times 1024 \times 8$ bits and 8 banks is used for all the platforms to have a fair comparison. The performance and energy of all the evaluated platforms are shown in Table 11.

Table 11 shows that ReDRAM requires about 3X more DRAM cycles than CIDAN to compute bitwise AND, OR, and XOR for different operand sizes. These improvements stem from the fact that CIDAN required far less internal DRAM operations than ReDRAM and any other PIM platform. Table 11 also shows that CIDAN's energy consumption is nearly half the energy of ReDRAM for bulk

Table 11 Average latency, energy, and throughput for basic operations on PIM platforms computed for input vectors of size 1 MB, 2 MB, and 4 MB. Latency and energy are normalized to CIDAN

	Latency (CIDAN=1)		Energy (CIDAN=1)		Throughput (GOps/s)		
	Ambit	ReDRAM	Ambit	ReDRAM	Ambit	ReDRAM	CIDAN
NOT	2.4	1.2	1.64	0.82	94.7	189.6	227.5
AND	4.32	3.24	2.61	1.96	47.3	63.1	205.03
OR	4.32	3.24	2.61	1.96	47.3	63.1	205.03
XOR	6.54	3.19	4.12	1.94	30.7	63.1	201.8

Table 12 Latency and energy comparison for executing AES on different platforms normalized to CIDAN

	Latency (CIDAN = 1)	Energy (CIDAN = 1)
ReDRAM	1.15	1.10
CPU	4.04	3.74

Table 13 Latency and energy comparison for executing Graph Matching Index problem and DNA sequence mapping algorithm[129] on different platforms normalized to CIDAN. Graph Matching Index problem is carried out on three datasets; facebook, amazon, dblp [145]

	Latency (CIDAN =1)		Energy (CIDAN = 1)	
Workload	Graph matching index	DNA sequence mapping	Graph matching index	DNA sequence mapping
ReDRAM	3.24	3.14	1.96	2.12
Ambit	4.32	4.35	2.61	2.88

bitwise operations, and is significantly better than Ambit due to the same reason. It must be noted that the TLPEs only consume 1% of the overall DRAM chip area.

The evaluation of TLPE is also extended to several other practical applications such as AES, Graph Matching Index problem, and DNA sequence mapping algorithm [129]. Their latency and energy comparison against the benchmark architectures are shown in Tables 12 and 13.

5.6 Conclusion

In this section, integration of highly reconfigurable and low-power threshold logic processing elements (TLPEs) with DRAM is presented as PIM architecture CIDAN. It performs binary bitwise operations such as NOT, (N)AND, (N)OR, X(N)OR, etc., and full adder on large bit vectors. CIDAN improves upon the equivalent prior architectures by a factor of 3X in performance and 2X in energy consumption. An evaluation and comparison of CIDAN with prior PIM architectures on real-world applications such as AES encryption, graph processing operation, and DNA sequence mapping algorithm is also presented.

6 Flash Devices in Analog Circuits

So far, we have seen the benefits of using flash transistors in digital logic circuits and neural network accelerators and for processing in-memory. In this section, we discuss the use of flash devices for the design of analog and mixed-signal circuits. The tunability of flash transistors makes them remarkably useful in analog circuits. Designs may be tuned after fabrication, to achieve the accuracy, power consumption, or latency that is desired by the user. Further, this tuning ability empowers the user to overcome non-idealities including process and voltage variations. In this section, we present two recent works that demonstrate the use of flash transistors in analog designs. In Sect. 6.1, we describe a digital to analog converter (DAC) design which uses flash transistors as tunable current sources, to create a current-steering DAC that is suitable for use in edge devices. In Sect. 6.2, we present a digital low-dropout (digital LDO) regulator that uses flash transistor subarrays for voltage regulation, and we demonstrate the high degree of tunability that this approach achieves.

6.1 Flash-Based Digital to Analog Conversion

The applications of Internet of Things (IoT) circuits are expected to grow significantly between 2025 and 2035. Low-power and energy-efficient chips are necessary for IoT applications. Because IoT devices are mainly stand-alone, battery-operated sensors and microprocessors, they benefit greatly from ultralow power chips. Traditional digital architectures and designs may struggle to deliver the required energy efficiency [146], limiting their utility in IoT devices. Additionally, IoT devices typically do not benefit from the high processing speeds offered by traditional digital designs, and instead only require moderate processing speeds, from sub-kHz up to 100 MHz [147].

Digital to analog converters (DACs) are one crucial element of sensor-processor interfaces that are common in IoT nodes. Traditional analog DACs struggle to meet the ultralow power requirements of IoT devices, and thus recent works have developed novel designs suitable for IoT chips. For example, recent contributions in [148, 149] propose all-digital low-cost DAC designs based on dyadic digital pulse modulation (DDPM). These DACs meet some of the requirements of IoT chips, like low voltage and area, but because they drive a voltage output to a high-impedance node, they would be unsuitable for driving a resistive load.

Current-steering DACs [150] offer a reliable method for driving resistive loads. The design naturally provides a differential current output, which can be easily converted to a voltage output using a load resistor. Perhaps the biggest problem faced by current-steering DACs is the error caused by mismatch-induced nonlinearities. Designers must make use of costly calibration hardware and large devices in order to minimize error. The extra calibration hardware comes at a cost to power, area, and energy, making such designs unsuitable for IoT applications. We solve this problem

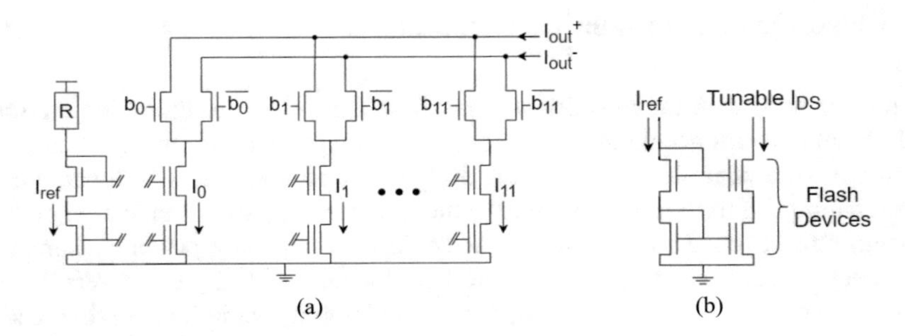

Fig. 29 (**a**) The proposed 12-bit DAC. (**b**) Flash current source

by using flash transistors which provide in-memory calibration to eliminate the need for calibration hardware and large devices.

We present (Fig. 29a) a current-steering digital to analog converter (DAC) which utilizes flash transistors to achieve low latency, area, and power, making it suitable for the internet of things (IoT) and other applications with a small resource budget. We use flash transistors to create a tunable design that is robust to non-idealities such as process and voltage variation, and chip aging. The use of flash devices allows our DAC to give precise and accurate output without the need for costly calibration hardware. We demonstrate the robustness of our design under non-idealities arising from process and voltage variations. Compared to other state-of-the-art DACs intended for IoT, we achieve extremely high throughput and extremely low energy per conversion while reporting competitive error and power metrics. The proposed DAC achieves a maximum INL (DNL) of 0.96 LSB (0.71 LSB), an average INL (DNL) of 0.25 LSB (0.08 LSB), and an ENOB of 11.90 bits

6.1.1 Design

Current Sources As depicted in Fig. 29b, our binary-weighted current sources are implemented using flash transistors. A reference current I_{ref} is provided as input to a current mirror. Multiple current mirror output branches are constructed from flash devices. By programming the flash devices and setting their V_{th}, the drain to source current (I_{ds}) can be tuned with fine precision. In our design, these current sources can be programmed to deliver current ranging from the pA-level to the μA-level. However, we select an LSB current of 0.5nA, and an MSB current of 1024 nA, because we found this choice to result in the lowest error. We choose $I_{ref} = 1uA$.

6.1.2 Features

Flash Devices Unlike other DAC implementations, our design utilizes flash transistors as programmable, tunable current sources. Historically, flash transistors have only been used for non-volatile memory (NVM) technologies, but recent results demonstrate their promise for use in digital and analog design, due to their programmable V_{th} [115]. Flash is a well-understood, high-yielding technology, making the use of flash devices practical and reliable. The use of flash devices naturally provides an in-memory calibration mechanism, eliminating the need for additional calibration hardware which is a burden to other designs.

Current Steering for IoT To the best of our knowledge, this is the first current-steering DAC implementation that is suited for IoT. We reap the benefits of IoT DACs (extremely low power, area, and energy), as well as current-steering DACs (differential output, high throughput, and the ability to drive a resistive load). As we will demonstrate, we achieve extremely high energy efficiency relative to other IoT DACs.

Resilience to Aging and Variation The tunability of flash devices enables the mitigation of non-idealities arising from process and voltage variation. These are flash current source mismatch, flash current source variation caused by chip aging, and I_{ref} variation caused by VDD variations or chip aging.

6.1.3 Performance Metrics

The proposed DAC was implemented in a 45 nm process technology and simulated using Synopsys HSPICE [113]. For CMOS devices, we use a 45 nm PTM model card [114]. For flash devices, we use the model card derived in [115]. Our nominal $VDD = 1.2$ V for all HSPICE experiments.

Error The dynamic performance of the DAC was tested to measure the Integral Nonlinearity (INL) and Differential Nonlinearity (DNL). We tune each bit of our DAC using a write-verify scheme, such that the current produced for bit i approaches $I_{LSB} * 2^i$. The two flash transistors for each bit are programmed in unison, so they will always have the same V_{th}. Once the DAC is programmed, we perform a transient analysis to measure the dynamic performance. In Fig. 30a we plot the Integral Nonlinearity (INL) over all input codes, and in Fig. 30b we plot the Differential Nonlinearity (DNL) over all input codes. Our maximum INL (DNL) over all inputs is 0.96 LSB (0.71 LSB). Our average INL (DNL) over all inputs is 0.25 LSB (0.08 LSB). The output voltage swing was measured to be 600 mV.

We conduct a frequency analysis of the DAC output to measure the effective number of bits (ENOB) and spurious-free dynamic range (SFDR). The proposed DAC is analyzed in response to a sine input with 1Hz frequency. We achieve an ENOB of 11.90 bits and an SFDR of 84.97 dB.

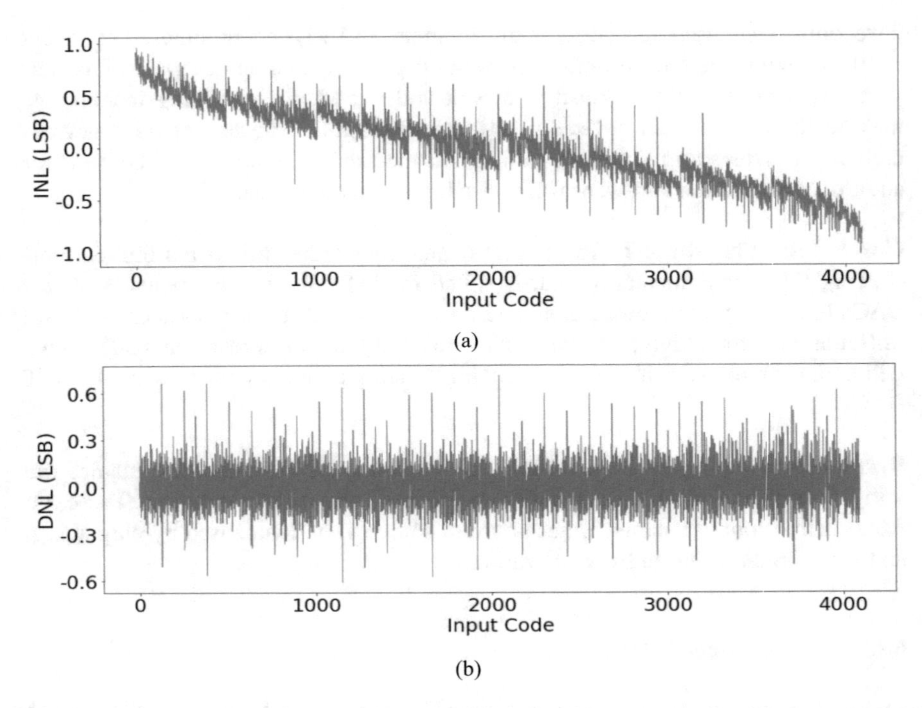

Fig. 30 INL and DNL over all input codes. (**a**) INL over all input codes. (**b**) DNL over all input codes

Throughput, Power, Area, and Energy We estimate the chip area of our design using Synopsys Design Compiler, and report an area of 406 μm^2. The total current through our DAC, for any input code, is $I_{ref} + I_{out}^{+} + I_{out}^{-} = 3.05\,\mu A$. Our nominal $VDD = 1.2\,V$, so our design therefore has a constant power draw of $3.05\,\mu A * 1.2\,V = 3.66\,\mu W$. We perform transient HSPICE simulations to measure the latency of our DAC. We report a latency of $10\,ns$, which gives a throughput of 100 MSamples/s (MS/s). The total energy consumed by our DAC per conversion is therefore $3.66\,\mu W * 10\,ns = 36.6\,fJ$.

Table 14 compares the performance metrics of our DAC to recent DACs suitable for use in IoT devices. We leave cells in Table 14 blank if the value is not reported by the author, or if we are unable to compute it. With comparable error metrics and chip area, our design beats the fastest approach by 50× in throughput, and the most energy efficient approach by ∼30× in energy per conversion. Although our power consumption of our DAC cannot compete with micro-resonator-based designs, our power is still at a level that is suitable for IoT applications.

Table 14 Comparison of our flash-based DAC to other IoT DACs

Design	Resolution (bits)	Throughput (kS/s)	INL/DNL (LSB)	ENOB (bits)	Power (μW)	Energy (pJ/S)	Area (μm^2)
Flash-based DAC (this work)	12	100,000	0.96 / 0.71	11.90	3.66	0.037	406
Relaxation [151]	10	2000	1/1	9.06	–	–	–
Micro-resonator [152]	4	0.64	0.6 / 1	–	0.001	150	–
Micro-resonator [153]	4	100	–	–	0.27	2.7	–
Relaxation [154]	10	400	0.33/0.2	9.9	0.44	1.1	1000
Relaxation [155]	10	0.3	2.4/3.3	7.1	–	–	–
DDPM [149]	12	27	2/1	11.3	55	2000	500
DDPM [148]	16	–	2/1	15.4	–	–	2000

6.1.4 Summary

In this section, we presented a digital to analog converter (DAC) which utilizes flash transistors, resulting in a tunable design that is suitable for IoT and robust to non-idealities such as process and voltage variation and chip aging. We implement our design in a 45 nm process technology, and perform simulations to measure performance metrics like INL, DNL, ENOB, throughput, power, and energy per conversion. We report competitive error metrics and chip area, extremely high throughput, and extremely low energy per conversion.

6.2 *Pseudo-Flash-Based Digital Low-Dropout Regulator*

With the rapid growth in demand for low-power Internet of Things (IoT) devices, and the increased use of advanced fabrication processes, the power supply voltage of VLSI ICs is decreasing. Traditional analog low-dropout (analog LDO) regulators are not able to maintain good regulation due to the poor gain of error amplifiers operating at low supply voltages. To address this problem, digital low-dropout (digital LDO) regulators [161] are increasingly gaining traction. In contrast with analog LDOs, digital LDOs can be operated at a lower supply voltage since they do not need error amplifiers. In addition, digital LDOs don't require large passive components (like capacitors) for stability, and are therefore more compact than analog LDOs. Lastly, digital LDOs are much easier to realize because most of the circuit components in a digital LDO are designed using standard cells, while custom circuitry is needed in analog LDOs.

Figure 31 shows a conventional digital LDO which consists of a comparator, a digital controller, and an array of N identical PMOS transistors. The comparator compares the difference between the output voltage (V_{out}) and the reference voltage (V_{ref}), and the digital controller implements the control algorithm which determines the number of PMOS transistors that need to be turned on based on the output of the comparator and the control algorithm used, so that the output load current requirement is met.

However, a traditional digital LDO has some limitations. Since the number and the size of the transistors in the PMOS array are fixed, a traditional digital LDO has a predetermined current and voltage range. If it is used for different current or voltage ranges, the performance of the digital LDO can degrade, and exhibit unstable behavior due to the impedance mismatch of the PMOS transistors. Another concern is that the impedance of each PMOS transistor in the array varies due to device length, width, and threshold voltage (V_{th}) variations. This causes the current provided by each PMOS transistor to be different, negatively impacting output ripple (V_{ripple}), output voltage overshoot/undershoot (V_{shoot}), output voltage (V_{out}), and recovery time (t_{rec}).

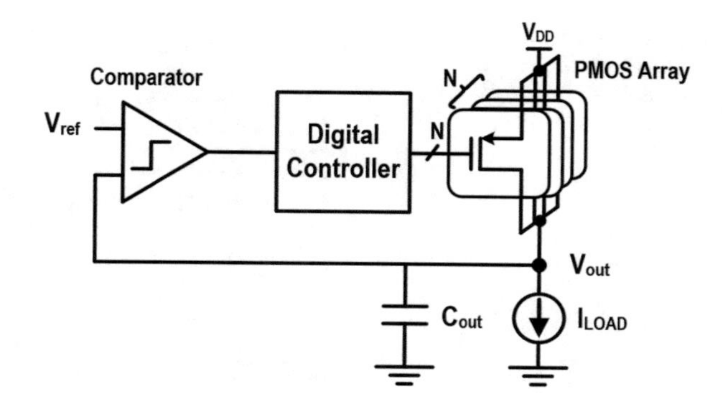

Fig. 31 A conventional digital LDO

In this section, we propose a pseudo-flash[1]-based digital LDO. We use a traditional CMOS process as described in [42–53], and hence call it a "pseudo-flash transistor." Using pseudo-flash PMOS transistor arrays for coarse and fine regulation, we can achieve many significant improvements over traditional digital LDOs, such as:

- The current range of the regulator can be adjusted (at the factory or in-field) by adjusting the V_{th} of the pseudo-flash transistors.
- The resistance of each pseudo-flash transistor can be precisely tuned by adjusting its V_{th}, effectively cancelling variations due to device length, width, and V_{th} variations that are faced by traditional digital LDOs.
- It allows manufacturers to change the V_{th} of the pseudo-flash subarrays, and hence use the same design to obtain different regulators with different electrical specifications such as V_{DD}, V_{ripple}, V_{out}, t_{rec}, V_{shoot}, maximum output current (I_{max}), and minimum output current (I_{min}). This reduces manufacturing costs significantly.
- The threshold voltage can even be changed in the field by the user, to compensate for temperature or aging effects.
- Although we present our results using pseudo-flash technology, one could use a traditional flash technology as well.

6.2.1 Proposed Pseudo-Flash-Based Digital LDO

Figure 32 shows the proposed pseudo-flash-based digital LDO. It consists of two pseudo-flash arrays, a 3-bit comparator bank, and a coarse-fine controller. The pseudo-flash array is divided into two parts: a coarse subarray with $N = 100$ pseudo-flash transistors and a fine subarray with $M = 5$ pseudo-flash transistors. V_{fg} (V_{fg_c} for the coarse subarray and V_{fg_f} for the fine subarray) is programmed to different voltage levels to change the effective threshold voltage V_{th} of the

[1] Although we use pseudo-flash transistors, one could use flash transistors as well.

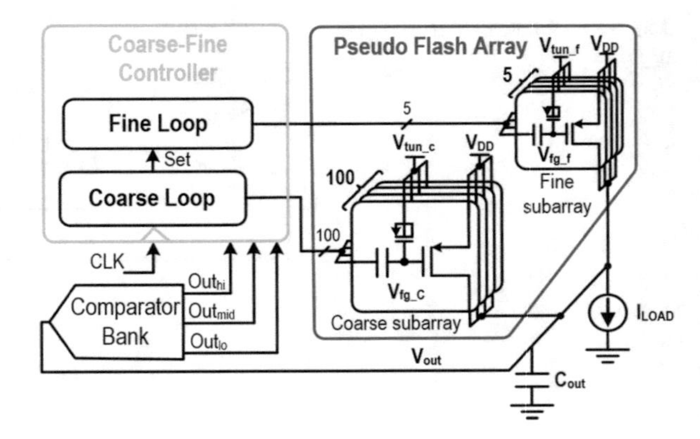

Fig. 32 Schematic of proposed pseudo-flash-based digital LDO

subarray devices, and therefore modify the resistance of the pseudo-flash transistors. This enables the same digital LDO circuit to be used to obtain different I_{max}, I_{min}, V_{ripple}, and V_{shoot}. The comparator bank consists of three comparators to produce outputs Out_{hi}, Out_{mid}, and Out_{lo}. The signals Out_{hi} and Out_{lo} of the comparator bank form a *dead zone* in which we want to maintain V_{out} in steady state. If the load current changes, and V_{out} is no longer in the *dead zone*, the coarse-fine controller is enabled immediately. The remaining bit (Out_{mid}) of the comparator bank detects the difference between V_{out} and V_{ref} and is used as a reference to regulate V_{out}. The coarse-fine controller contains an inner coarse loop and a outer fine loop. The coarse loop regulates V_{out} by determining the number n of transistors to turn on such that $(V_{DD} - I_{load} \cdot \frac{R_C}{n}) < V_{ref} < (V_{DD} - I_{load} \cdot \frac{R_C}{n+1})$, where I_{load} is the load current and R_C is the resistance of each coarse pseudo-flash transistor. Once n is found, the coarse loop sends a *set* signal to activate the fine loop, and we proceed to further reduce the output voltage ripple (V_{ripple}) by adjusting the fine part of the pseudo-flash array. The fine controller now finds m, the number of fine transistors to turn on such that $(V_{DD} - I_{load} \cdot (\frac{R_C}{n} || \frac{R_F}{m})) < V_{ref} < (V_{DD} - I_{load} \cdot (\frac{R_C}{n} || \frac{R_F}{m+1}))$, where R_F is the resistance of each fine pseudo-flash transistor. At this point, the coarse-fine controller has converged, with a ripple of $I_{load} \cdot ((\frac{R_C}{n} || \frac{R_F}{m}) - (\frac{R_C}{n} || \frac{R_F}{m+1}))$.

6.2.2 Coarse-Fine Controller

Figure 33a shows the state transition diagram of the finite-state machine (FSM) of the coarse-fine controller, while Fig. 33b shows an example timing of the operation of the FSM. When the output load current increases sufficiently, an undershoot occurs, causing V_{out} to be pulled out of the dead zone. The controller now enters the *transient* state of the FSM and turns on more coarse transistors to provide the increased load current immediately. When V_{out} crosses V_{ref}, the number of coarse transistors which are turned on is recorded in the *Max* register. Since an incorrect (excess) number of coarse transistors are turned on, a small overshoot occurs. Thus, the controller turns off coarse transistors, to stabilize the V_{out}. Once the V_{out} crosses

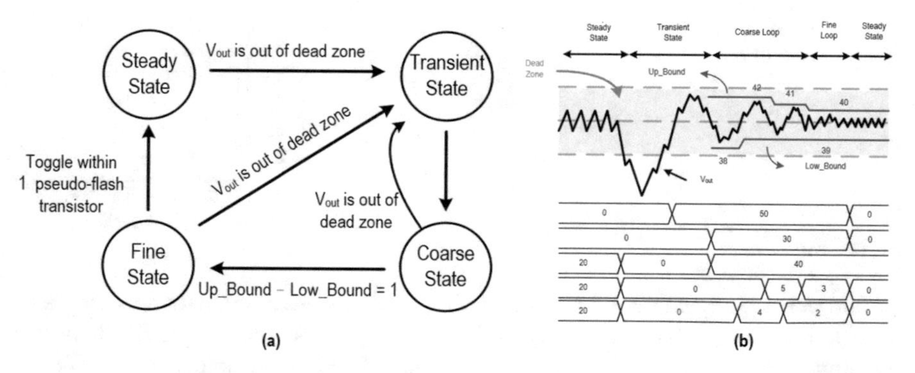

Fig. 33 (**a**) The FSM of the coarse-fine controller and (**b**) operation of the coarse-fine controller

V_{ref} again, the number of transistors which are turned on is recorded in the *Min* register, and the controller switches to the *coarse* state of the FSM. The controller averages the value of *Max* and *Min* to compute *Mean*. The controller adds/subtracts 2 to *Mean* to create Up_Bound/Low_Bound to limit V_{out} within these bounds by adjusting the number of pseudo-flash transistors. Then, the controller measures the duration t_{Low} for which V_{out} is below the V_{ref}, and the duration t_{Up} for which V_{out} is above the V_{ref}. If t_{Up} is longer than t_{Low}, it means the controller has turned on too many transistors, and so the value of Up_Bound is decreased by 1. Otherwise, the value of Low_Bound is increased by 1. By comparing t_{Up} and t_{Low}, the controller narrows the bounds. When the difference between Up_Bound and Low_Bound is 1, it means that the controller has found the correct value for the number of coarse transistors that need to be turned on to provide a sufficient output load current. The controller now fixes the number of coarse transistors to the value of Low_Bound and switches to the *fine* state of the FSM to further reduce V_{ripple}. The same method is used in the fine loop, in which the controller regulates five fine pseudo-flash transistors, by again comparing the durations of t_{Up} and t_{Low}. Once the fine loop converges as well, the controller enters steady state to keep V_{out} stable, and V_{ripple} is minimized.

6.2.3 Simulation Results

We performed experiments to show that the same pseudo-flash-based digital LDO can be reprogrammed to achieve different I_{max} and V_{out} for different V_{DD}. This illustrates how the change in V_{fg} can be used to change the performance characteristics of the LDO based on application requirements, *using the same LDO circuit*. By changing V_{fg_c}, I_{max} and I_{min} can be altered. The fine subarray determines V_{ripple} by adjusting V_{fg_f}. Therefore, the proposed pseudo-flash-based digital LDO can be operated for different V_{DD}, V_{out}, I_{max}, and I_{min} by simply modifying V_{fg_c} and V_{fg_f}. We conceive of a situation where the manufacturer

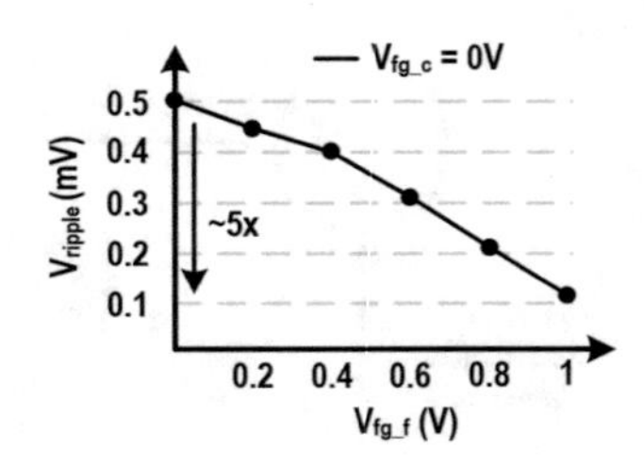

Fig. 34 Measured results of V_{ripple} under different V_{fg_f}

would precompute values of V_{fg_c} and V_{fg_f}, for different regulator specifications (V_{DD}, V_{out}, I_{max}, I_{min}, and V_{ripple}). Based on the customer's specifications, the manufacturer would program the proposed digital LDO in the factory (or provide the ability to the customer to do this in the field). Hence, the same design could be used for several LDO applications, reducing design cost. Table 15a and b illustrates how the same digital LDO design is used for several different LDO specifications with different V_{DD}, V_{out}, I_{max}, I_{min}, and V_{ripple}. Each of the rows in these tables uses different V_{fg_c} and V_{fg_f} values. Table 15a shows the simulation results when the output load current changes from $20\% I_{max}$ to $40\% I_{max}$, while Table 15b shows the simulation results when the output load current changes from $40\% I_{max}$ to $80\% I_{max}$. On average, V_{shoot}, t_{rec}, and V_{ripple} are 41 mV, 0.21 μs and 0.22 mV respectively when the output load current changes from $20\% I_{max}$ to $40\% I_{max}$. When the output load current changes from $40\% I_{max}$ to $80\% I_{max}$, V_{shoot}, t_{rec}, and V_{ripple} are 72 mV, 0.31 μs, and 0.13 mV, respectively. Figure 34 shows that our proposed pseudo-flash digital LDO achieves an improvement of up to 5× in output voltage ripple by programming V_{fg_f} to different values.

Table 16 shows the comparison of the proposed pseudo-flash-based digital LDO with prior works. The proposed pseudo-flash-based digital LDO is designed to regulate V_{out} which ranges from 1.1 V to 1.7 V with the supply voltage operated at 1.2–1.8 V, as described in Table 15a and b. The output capacitance is set to $0.1nF$ and the operating frequency is $100MHz$.

Unlike other state-of-the-art approaches [156–160], our digital LDO can operate over a large range of V_{DD}, I_{max}, I_{min}, and V_{ripple}. Our V_{shoot} values are competitive, and we achieve a very low ripple while using a much smaller C_{out} compared to the other approaches.

6.3 Summary

In this section, we presented a pseudo-flash-based digital LDO in which we replace PMOS transistors by pseudo-flash transistors. Our digital LDO can be tuned to alter a variety of specifications such as V_{DD}, V_{out}, I_{max}, I_{min}, V_{shoot}, and V_{ripple} by changing the V_{th} of the pseudo-flash transistors. Process variations can be effectively eliminated as well. By changing the threshold voltage (and thus the resistance) of the flash transistors, a manufacturer can use the *same* design for

Table 15 Simulation results with a load current changes

(a) From $20\% I_{max}$ to $40\% I_{max}$							(b) From $40\% I_{max}$ to $80\% I_{max}$						
V_{DD} (V)	V_{out} (V)	I_{max} (mA)	I_{min} (mA)	V_{shoot} (mV)	t_{rec} (μs)	V_{ripple} (mV)	V_{DD} (V)	V_{out} (V)	I_{max} (mA)	I_{min} (mA)	V_{shoot} (mV)	t_{rec} (μs)	V_{ripple} (mV)
1.8	1.7	100	1	36	0.14	0.23	1.8	1.7	100	1	58	0.28	0.13
1.8	1.7	1	0.01	63	0.58	0.19	1.8	1.7	1	0.01	123	0.46	0.18
1.5	1.4	90	0.9	38	0.15	0.14	1.5	1.4	90	0.9	62	0.3	0.04
1.5	1.4	60	0.6	37	0.16	0.15	1.5	1.4	60	0.6	67	0.31	0.07
1.2	1.1	58	0.58	35	0.13	0.23	1.2	1.1	58	0.58	57	0.26	0.12
1.2	1.1	45	0.45	35	0.13	0.31	1.2	1.1	45	0.45	65	0.27	0.21
Average				41	0.21	0.22	Average				72	0.31	0.13

Table 16 Comparison of the pseudo-flash-based digital LDO with prior works

	This work			[156] ISSCC'20	[157] JSSC'17	[158] EDSSC'19	[159] ICTA'20	[160] TCAS-I'19
Type	Digital			Digital	Digital	Digital	Digital	Digital
Process (nm)	65 nm			28 nm	28 nm	65 nm	65 nm	65 nm
Operating frequency	100 MHz			168 MHz–2.0 GHz	N.A.	10 MHz	10 MHz	100 MHz
V_{DD} (V)	1.2	1.5	1.8	0.5–1.0	1.1	0.6–0.8	N.A.	0.6
V_{out} (V)	1.1	1.4	1.7	0.45–0.95	0.9	0.5–0.75	0.5–0.95	0.5
V_{ripple} (mV)	0.14	0.15	0.13	N.A.	N.A.	N.A.	N.A.	N.A
V_{shoot} @ ΔI_{load}	57 mV@ 23 mA	67 mV@ 36 mA	58 mV@ 40 mA	112 mV@ 430 mA	120 mV@ 180 mA	90 mV@ 15 mA	16.4 mV@ 5 mA	53 mV@ 100 mA
Max current (mA)	58	90	100	160–480	200	N.A.	N.A.	100
C_{out} (nF)	0.1			Cap-free	23.5	0.13	0.052	1

different specifications like maximum output current (I_{max}), output voltage (V_{out}), ripple (V_{ripple}), and transient recovery time (t_{rec}). We design a coarse pseudo-flash subarray and a fine pseudo-flash subarray for voltage regulation. The coarse pseudo-flash subarray is used to generate sufficient output current immediately, and to stabilize the V_{out} when the output current load changes. The fine pseudo-flash subarray minimizes the V_{ripple} once the coarse pseudo-flash subarray has stabilized. By tuning V_{fg_f}, our digital LDO achieves up to 5× output voltage ripple reduction. Over a variety of LDO specifications, the V_{shoot}, t_{rec}, and V_{ripple} are on average 41 mV, 0.21 μs, and 0.22 mV (72 mV, 0.31 μs, and 0.13 mV), respectively, when a load current changes from $20\% I_{max}$ to $40\% I_{max}$ ($40\% I_{max}$ to $80\% I_{max}$) under different V_{DD} and V_{out}. We compare our work with several recent state-of-the-art LDOs, and show that our approach has very competitive metrics for several characteristics (like V_{DD}, V_{out}, I_{max}, and V_{ripple}), which we can implement with the same circuit.

7 Conclusions and Future Outlook

Floating gate (flash) technology has become the ubiquitous technology for many non-volatile memory (NVM) applications today. Flash is widely used in NVM applications like SSDs, USB drives, SD cards, and boot ROMs, and is a high-volume, mature, and high-yielding technology.

With the slowing of Moore's law, CMOS technology scaling has slowed significantly, and processing variation issues have continued to plague the predictability of CMOS designs. In the search of alternate post-CMOS technologies, research has directed attention to exotic approaches (which include nanotube and nanoribbon devices, molecular electronics, spintronics, photonic devices, etc.) that are highly experimental and not mature.

In this landscape, our position is that there is an existing technology which could be a solution, but which has been largely neglected in the design of a variety of general-purpose circuits, both digital and analog. This overlooked technology is flash. Not only does flash have the maturity and high yields from years of development in the NVM space, but also provides several unique benefits over CMOS. These benefits include in-factory and in-field performance tunability, the ability to counteract circuit aging in the field, the control of speed binning, and the ability to mitigate process variations by fine-grained threshold voltage control. These advantages can improve on the performance of CMOS designs, and can also result in completely novel applications that have not been conceived with CMOS technologies to date.

The goal of this chapter is to demonstrate these benefits of flash, and also to illustrate several digital and analog applications where flash can be used. We demonstrate this via case studies that demonstrate two styles of flash-based ASIC design (including a secure variant), flash-based convolutional neural network accelerators (both analog and digital variants), flash-based in-memory computing designs, as well as flash-based analog circuits like DACs and LDOs.

Based on our findings, we posit that the programmability, robustness, stability, and maturity of flash give it a significant edge over the class of "emerging" post-CMOS technologies, making flash a viable technology to eventually replace CMOS. Doubtless, there exist fabrication challenges that need to be addressed to allow flash to scale to smaller process node geometries, but by shining a light on flash as a means to implement general-purpose digital and analog circuits, we hope to generate more interest in the fabrication as well as the design communities to conduct more research into flash—both in terms of scaling to smaller technology nodes and also alternative design approaches that use flash technology. We hope that this chapter will encourage further interest in this arena, allowing flash to become a key technology for digital and analog circuits in the future.

References

1. Fowler, R.H., Nordheim, L.: Electron emission in intense electric fields. Proc. R. Soc. Lond. Series A **119**(781):173–181 (1928). https://doi.org/10.1098/rspa.1928.0091
2. Heremans, P., Bellens, R., Groeseneken, G., Maes, H.E.: Consistent model for the hot-carrier degradation in n-channel and p-channel MOSFETs. IEEE Trans. Electron Devices **35**(12), 2194–2209 (1988). https://doi.org/10.1109/16.8794
3. An, H., Kim, K., Jung, S., Yang, H., Kim, K. Song, Y.: The threshold voltage fluctuation of one memory cell for the scaling-down NOR flash. In: 2010 2nd IEEE InternationalConference on Network Infrastructure and Digital Content, 2010, pp. 433–436. https://doi.org/10.1109/ICNIDC.2010.5657806
4. Hwang, J.-R., et al.: 20 nm gate bulk-finFET SONOS flash. In: IEEE International Electron Devices Meeting, 2005. IEDM Technical Digest, pp. 154–157 (2005) https://doi.org/10.1109/IEDM.2005.1609293
5. Kamigaichi, T., et al.: Floating gate super multi level NAND flash memory technology for 30 nm and beyond. In: 2008 IEEE International Electron Devices Meeting, pp. 1–4 (2008). https://doi.org/10.1109/IEDM.2008.4796825
6. Lue, H.-T., et al.: Scaling feasibility study of planar thin floating gate (FG) NAND flash devices and size effect challenges beyond 20 nm. In: 2011 International Electron Devices Meeting, pp. 9.2.1–9.2.4 (2011). https://doi.org/10.1109/IEDM.2011.6131519
7. Sakamoto, W., et al.: Reliability improvement in planar MONOS cell for 20 nm-node multi-level NAND Flash memory and beyond. In: 2009 IEEE International Electron Devices Meeting (IEDM), pp. 1–4 (2009). https://doi.org/10.1109/IEDM.2009.5424211
8. Seol, K.S., et al.: A new floating gate cell structure with a silicon-nitride cap layer for sub-20 nm NAND flash memory. In: 2010 Symposium on VLSI Technology, pp. 127–128 (2010). https://doi.org/10.1109/VLSIT.2010.5556197
9. Bez, R., Camerlenghi, E., Modelli, A., Visconti, A.: Introduction to flash memory. Proc. IEEE **91**(4), 489–502 (2003). https://doi.org/10.1109/JPROC.2003.811702
10. Suh, K.-D., et al.: A 3.3 V 32 Mb NAND flash memory with incremental step pulse programming scheme. In: Proceedings ISSCC '95—International Solid-State Circuits Conference, pp. 128–129 (1995). https://doi.org/10.1109/ISSCC.1995.535460
11. Jung, T.-S., et al.: A 117-mm/sup 2/ 3.3-v only 128-MB multilevel NAND flash memory for mass storage applications. IEEE J. Solid-State Circuits **31**(11), 1575–1583 (1996)

12. Boboila, S., Desnoyers, P.: Write endurance in flash drives: measurements and analysis. In: Proceedings of the 8th USENIX Conference on File and Storage Technologies, FAST'10, p. 9. USENIX Association (2010)
13. Jung, D., Chae, Y.-H., Jo, H., Kim, J.-S., Lee, J.: A group-based wear-leveling algorithm for large-capacity flash memory storage systems. In: Proceedings of the 2007 International Conference on Compilers, Architecture, and Synthesis for Embedded Systems - CASES '07, p. 160, Salzburg, Austria. ACM Press (2007)
14. Cuppens, R., Hartgring, C., Verwey, J., Peek, H.: An EEPROM for microprocessors and custom logic. In: 1984 IEEE International Solid-State Circuits Conference. Digest of Technical Papers, volume XXVII, pp. 268–269 (1984)
15. Gogl, D., Burbach, G., Fiedler, H.L., Verbeck, M., Zimmermann, C.: A single-poly EEPROM cell in SIMOX technology for high-temperature applications up to 250/spl deg/C. IEEE Electron Device Lett. **18**(11), 541–543 (1997)
16. Dagan, H., Teman, A., Fish, A., Pikhay, E., Dayan, V., Roizin, Y.: A low-cost low-power non-volatile memory for RFID applications. In: 2012 IEEE International Symposium on Circuits and Systems (ISCAS), pp. 1827–1830 (2012). ISSN: 2158-1525
17. Roizin, Y., Aloni, E., Birman, A., Dayan, V., Fenigstein, A., Nahmad, D., Pikhay, E., Zfira, D.: C-flash: an ultra-low power single poly logic NVM. In: 2008 Joint Non-Volatile Semiconductor Memory Workshop and International Conference on Memory Technology and Design, pp. 90–92 (2008). ISSN: 2159-4864
18. Cui, Z.-Y., Choi, M.-H., Kim, Y.-S., Lee, H.-G., Kim, K.-W., Kim, N.-S.: Single poly-EEPROM with stacked MIM and n-well capacitor. Electron. Lett. **45**(3), 185 (2009)
19. Ogura, T., Ogura, N., Kirihara, M., Park, K.T., Baba, Y., Sekine, M., Shimeno, K.: Embedded twin MONOS flash memories with 4 ns and 15 ns fast access times. In: 2003 Symposium on VLSI Circuits. Digest of Technical Papers (IEEE Cat. No.03CH37408), pp. 207–210, Kyoto, Japan. Japan Soc. Appl. Phys. (2003)
20. Torricelli, F., Milani, L., Richelli, A., Colalongo, L., Pasotti, M., Kovacs-Vajna, Z.M.: Half-MOS single-poly EEPROM cell in standard CMOS process. IEEE Trans. Electron Devices **60**(6), 1892–1897 (2013)
21. Chung, C.-P., Chang-Liao, K.-S.: A highly scalable single poly-silicon embedded electrically erasable programmable read-only memory with tungsten control gate by full CMOS process. IEEE Electron Device Lett. **36**(4), 336–338 (2015)
22. Raszka, J., Advani, M., Tiwari, V., Varisco, L., Hacobian, N.D., Mittal, A., Han, M., Shirdel, A., Shubat, A: Embedded flash memory for security applications in a 0.13 μm CMOS logic process. In: 2004 IEEE International Solid-State Circuits Conference (IEEE Cat. No.04CH37519), pp. 46–512, San Francisco, CA, USA. IEEE, Piscataway (2004)
23. Cho, I.W., Lim, B.R., Kim, J., Kim, S.S., Kim, K.C., Lee, B.J., Bae, G.J., Lee, N.I., Kim, S.H., Koh, K.W., Kang, H., Seo, M.K., Kim, S.W., Hwang, S.H., Lee, D.Y., Kim, M.C., Chae, S.D., Seo, S.A., Kim, C.W.: Full integration and characterization of localized ONO memory (LONOM) for embedded flash technology. In: Digest of Technical Papers. Symposium on VLSI Technology, pp. 240–241 (2004).
24. Fang, L., Gu, J., Zhang, B., Kong, W.R., Zou, S.C.: A highly reliable 2-bits/cell split-gate flash memory cell with a new program-disturbs immune array configuration. IEEE Trans. Electron Devices **61**(7), 2350–2356 (2014)
25. Mitani, H., Matsubara, K., Yoshida, H., Hashimoto, T., Yamakoshi, H., Abe, S., Kono, T., Taito, Y., Ito, T., Krafuji, T., Noguchi, K, Hidaka, H., Yamauchi, T.: 7.6 A 90 nm embedded 1T-MONOS flash macro for automotive applications with 0.07mJ/8kB rewrite energy and endurance over 100M cycles under Tj of 175°C. In: 2016 IEEE International Solid-State Circuits Conference (ISSCC), pp. 140–141, San Francisco, CA, USA. IEEE, Piscataway (2016)
26. Park, S., Kim, S., Lee, B.: Development of 2T-SONOS cell using a contamination-free process integration for a highly reliable code storage eNVM. IEEE Trans. Electron Devices **67**(3), 922–928 (2020)

27. Bartoli, J., Della Marca, V., Delalleau, J., Regnier, A., Niel, S., La Rosa, F., Postel-Pellerin, J., Lalande, F.: A new non-volatile memory cell based on the flash architecture for embedded low energy applications: ATW (asymmetrical tunnel window). In: 2014 International Semiconductor Conference (CAS), pp. 117–120, Sinaia. IEEE, Piscataway (2014)
28. Park, S.-K., Song, H.-M., Kim, N.-Y., Cho, I.-W., Yoo, K.-D.: Novel select gate lateral coupling single poly ENVM for an HVCMOS process. IEEE Electron Device Lett. **35**(3), 351–353 (2014)
29. Song, S., Chun, K.C., Kim, C.H.: A logic-compatible embedded flash memory featuring a multi-story high voltage switch and a selective refresh scheme. In: 2012 Symposium on VLSI Circuits (VLSIC), pp. 130–131 (2012). ISSN: 2158-5636
30. Chu, Y.S., Wang, Y.H., Wang, C.Y., Lee, Y.H., Kang, A.C., Ranjan, R., Chu, W.T., Ong, T.C., Chin, H.W., Wu, K.: Split-gate flash memory for automotive embedded applications. In: 2011 International Reliability Physics Symposium, pp. 6B.1.1–6B.1.5 (2011) ISSN: 1938-1891
31. Shukuri, S., Ajika, N., Mihara, M., Kobayashi, K., Endoh, T., Nakashima, M.: A 60 nm NOR flash memory cell technology utilizing back bias assisted band-to-band tunneling induced hot-electron injection (B4-flash). In: 2006 Symposium on VLSI Technology, 2006. Digest of Technical Papers., pp. 15–16, Honolulu, HI, USA. IEEE, Piscataway (2006)
32. Lee, Y.K., Seo, B., Yu, T.K., Lee, B., Kim, E., Jeon, C., Park, W., Kim, Y., Lee, D., Lee, H., Cho, S.: A 45-nm logic compatible 4Mb-split-gate embedded flash with 1M-cycling-endurance. In: 2014 IEEE 6th International Memory Workshop (IMW), pp. 1–4, Taipei, Taiwan. IEEE, Piscataway (2014)
33. Kono, T., Ito, T., Tsuruda, T., Nishiyama, T., Nagasawa, T., Ogawa, T., Kawashima, Y., Hidaka, H., Yamauchi, T.: 40-nm embedded split-gate MONOS (SG-MONOS) flash macros for automotive with 160-MHz random access for code and endurance over 10 M cycles for data at the junction temperature of 170 °C. IEEE J. Solid-State Circuits **49**(1), 154–166 (2014)
34. Agrawal, V., Prabhakar, V., Ramkumar, K., Hinh, L., Saha, S., Samanta, S., Kapre, R.: In-memory computing array using 40 nm multibit SONOS achieving 100 TOPS/W energy efficiency for deep neural network edge inference accelerators. In: 2020 IEEE International Memory Workshop (IMW), pp. 1–4 (2020). ISSN: 2573-7503
35. Luo, L.Q., Teo, Z.Q., Kong, Y.J., Deng, F.X., Liu, J.Q., Zhang, F., Cai, X.S., Tan, K.M., Lim, K.Y., Khoo, P., Jung, S.M., Siah, S.Y., Shum, D., Wang, C.M., Xing, J.C., Liu, G.Y., Diao, Y., Lin, G.M., Tee, L., Lemke, S.M., Ghazavi, P., Liu, X., Do, N., Pey, K.L., Shubhakar, K.: Functionality demonstration of a high-density 2.5 V self-aligned split-gate NVM cell embedded into 40 nm CMOS logic process for automotive microcontrollers. In: 2016 IEEE 8th International Memory Workshop (IMW), pp. 1–4, Paris, France. IEEE, Piscataway (2016)
36. Tehrani, S., Pak, J., Randolph, M., Sun, Y., Haddad, S., Maayan, E., Betser, Y.: Advancement in charge-trap flash memory technology. In: 2013 5th IEEE International Memory Workshop, pp. 9–12 (2013). ISSN: 2159-4864
37. Viraraghavan, J., Leu, D., Jayaraman, B., Cestero, A., Kilker, R., Yin, M., Golz, J., Tummuru, R.R., Raghavan, R., Moy, D., Kempanna, T, Khan, F., Kirihata, T., Iyer, S.: 80 Kb 10 ns read cycle logic embedded high-K charge trap multi-time-programmable memory scalable to 14 nm FIN with no added process complexity. In: 2016 IEEE Symposium on VLSI Circuits (VLSI-Circuits), pp. 1–2, Honolulu, HI, USA. IEEE, Piscataway (2016)
38. Taito, Y., Nakano, M., Okimoto, H., Okada, D., Ito, T., Kono, T., Noguchi, K., Hidaka, H., Yamauchi, T.: 7.3 A 28 nm embedded SG-MONOS flash macro for automotive achieving 200MHz read operation and 2.0 MB/S write throughput at Ti, of 170 oC. In: 2015 IEEE International Solid-State Circuits Conference - (ISSCC) Digest of Technical Papers, pp. 1–3, San Francisco, CA, USA. IEEE (2015)
39. Ma, S., Donato, M., Lee, S.K., Brooks, D., Wei, G.Y.: Fully-CMOS multi-level embedded non-volatile memory devices with reliable long-term retention for efficient storage of neural network weights. IEEE Electron Device Lett. **40**(9), 1403–1406 (2019)
40. Khan, F., Han, M.S., Moy, D., Katz, R., Jiang, L., Banghart, E., Robson, N., Kirihata, T., Woo, J.C., Iyer, S.S.: Design optimization and modeling of charge trap transistors (CTTs) in 14 nm FinFET technologies. IEEE Electron Device Lett. **40**(7), 1100–1103 (2019)

41. Tsuda, S., Kawashima, Y., Sonoda, K., Yoshitomi, A., Mihara, T., Narumi, S., Inoue, M., Muranaka, S., Maruyama, T., Yamashita, T., Yamaguchi, Y., Hisamoto, D.: First demonstration of FinFET split-gate MONOS for high-speed and highly-reliable embedded flash in 16/14 nm-node and beyond. In: 2016 IEEE International Electron Devices Meeting (IEDM), pp. 11.1.1–11.1.4, San Francisco, CA, USA. IEEE, Piscataway (2016)
42. Rahimi, K., Diorio, C., Hernandez, C., Brockhausen, M.D.: A simulation model for floating-gate MOS synapse transistors. In: 2002 IEEE International Symposium on Circuits and Systems (ISCAS), pp. II–II (2002). https://doi.org/10.1109/ISCAS.2002.1011042
43. Hasler, P., Minch, B.A., Diorio, C.: Adaptive circuits using pFET floating-gate devices. In: Proceedings 20th Anniversary Conference on Advanced Research in VLSI, pp. 215–229 (1999). https://doi.org/10.1109/ARVLSI.1999.756050
44. Basu, A., Hasler, P.E.: A fully integrated architecture for fast and accurate programming of floating gates over six decades of current. IEEE Trans. Very Large Scale Integr. Syst. **19**(6), 953–962 (2011). https://doi.org/10.1109/TVLSI.2010.2042626
45. Srinivasan, V., Serrano, G.J., Gray, J., Hasler, P.: A precision CMOS amplifier using floating-gates for offset cancellation. In: Proceedings of the IEEE 2005 Custom Integrated Circuits Conference, 2005, pp. 739–742 (2005). https://doi.org/10.1109/CICC.2005.1568774
46. Hasler, P., Dugger, J.: Correlation learning rule in floating-gate pFET synapses. In: 1999 IEEE International Symposium on Circuits and Systems (ISCAS), vol. 5, pp. 387–390 (1999). https://doi.org/10.1109/ISCAS.1999.777590
47. Bandyopadhyay, A., Serrano, G.J., Hasler, P.: Adaptive algorithm using hot-electron injection for programming analog computational memory elements within 0.2% of accuracy Over 3.5 decades. IEEE J. Solid-State Circuits **41**(9), 2107–2114 (2006). https://doi.org/10.1109/JSSC.2006.880621
48. Hasler, P.E., et al.: Impact ionization and hot-electron injection derived consistently from boltzmann transport. VLSI Design **1998**, 454–461 (1998)
49. Serrano, G., et al.: Automatic rapid programming of large arrays of floating-gate elements. In: 2004 IEEE International Symposium on Circuits and Systems (IEEE Cat. No.04CH37512), pp. I-I (2004). https://doi.org/10.1109/ISCAS.2004.1328209
50. Hasler, P., Basu, A., Koziol, S.: Above threshold pFET injection modeling intended for programming floating-gate systems. In: 2007 IEEE International Symposium on Circuits and Systems, pp. 1557–1560 (2007). https://doi.org/10.1109/ISCAS.2007.378709
51. Hasler, P., Dugger, J.: Correlation learning rule in floating-gate pFET synapses. IEEE Trans. Circuits Syst. II: Analog Digit. Signal Process. **48**(1), 65–73 (2001). https://doi.org/10.1109/82.913188
52. Srinivasan, V., Serrano, G., Twigg, C.M., Hasler, P.: A floating-gate-based programmable CMOS reference. IEEE Trans. Circuits Syst. I: Regul. Papers **55**(11), 3448–3456 (2008). https://doi.org/10.1109/TCSI.2008.925351
53. Hasler, J., Kim, S., Adil, F.: Scaling floating-gate devices predicting behavior for programmable and configurable circuits and systems. J. Low Power Electron. Appl. **6**(13), 1–19 (2016)
54. Song, S., Kim, J., Kim, C.H.: A comparative study of single-poly embedded flash memory disturbance, program/erase speed, endurance, and retention characteristic. IEEE Trans. Electron Devices **61**(11), 3737–3743 (2014). https://doi.org/10.1109/TED.2014.2359388
55. Raszka, J., et al.: Embedded flash memory for security applications in a 0.13 μm CMOS logic process. In: 2004 IEEE International Solid-State Circuits Conference (IEEE Cat. No.04CH37519), vol. 1, pp. 46–512 (2004). https://doi.org/10.1109/ISSCC.2004.1332586
56. Wang, B., Nguyen, H., Ma, Y., Paulsen, R.: Highly reliable 90-nm logic multitime programmable NVM cells using novel work-function-engineered tunneling devices. IEEE Trans. Electron Devices **54**(9), 2526–2530 (2007). https://doi.org/10.1109/TED.2007.903199
57. Kim, M., et al.: A 68 parallel row access neuromorphic core with 22 K multi-level synapses based on logic-compatible embedded flash memory technology. In: 2018 IEEE International Electron Devices Meeting (IEDM), pp. 15.4.1–15.4.4 (2018). https://doi.org/10.1109/IEDM.2018.8614599

58. Song, S., Chun, K.C., Kim, C.H.: A logic-compatible embedded flash memory for zero-standby power system-on-chips featuring a multi-story high voltage switch and a selective refresh scheme. IEEE J. Solid-State Circuits **48**(5), 1302–1314 (2013). https://doi.org/10.1109/JSSC.2013.2247691
59. Song, S., Kim, J., Kim, C.H.: Program/erase speed, endurance, retention, and disturbance characteristics of single-poly embedded flash cells. In: 2013 IEEE International Reliability Physics Symposium (IRPS), pp. MY.4.1–MY.4.6 (2013). https://doi.org/10.1109/IRPS.2013.6532095
60. Abusultan, M., Khatri, S.P.: A flash-based digital circuit design flow. In: 2016 IEEE/ACM International Conference on Computer-Aided Design (ICCAD), pp. 1–6 (2016). https://doi.org/10.1145/2966986.2966990
61. Brayton, R.K., Sangiovanni-Vincentelli, A.L., McMullen, C.T., Hatchel, G.D.: Logic Minimization Algorithms for VLSI Synthesis. Kluwer Academic Publishers, Norwell (1984)
62. Brglez, F., Bryan, D., Kozminksi, K.: Combinational profiles of sequential benchmark circuits. In: IEEE International Symposium on Circuits and Systems, 1989, vol. 3, pp. 1929–1934 (1989)
63. Corno, F., Reorda, M.S., Squillero, G.: Rt-level ITC'99 benchmarks and first ATPG results. In: IEEE Design Test of Computers, vol 17, pp. 44–53 (2000)
64. Amarù, L., Gaillardon, P.-E., De Micheli, G.: The EPFL combinational benchmark suite. In: Proceedings of the 24th International Workshop on Logic & Synthesis (IWLS) (2015)
65. Wagle, A., Singh, G., Yang, J., Khatri, S., Vrudhula, S.: Threshold logic in a flash. In: 2019 IEEE 37th International Conference on Computer Design (ICCD), pp. 550–558 (2019). https://doi.org/10.1109/ICCD46524.2019.00081
66. Kulkarni, N., Yang, J., Seo, J.S., Vrudhula, S.: Reducing power, leakage, and area of standard-cell ASICs using threshold logic flip-flops. IEEE Trans. Very Large Scale Integr. Syst. **24**(9), 2873–2886 (2016)
67. Yang, J., Davis, J., Kulkarni, N., Seo, J.-S., Vrudhula, S.: Dynamic and leakage power reduction of ASICs using configurable threshold logic gates. In: 2015 IEEE Custom Integrated Circuits Conference (CICC), pp. 1–4, San Jose, CA, USA. IEEE, Piscataway (2015)
68. Muroga, S.: Threshold Logic and Its Applications. Wiley-Interscience, New York (1971)
69. Commission on the Theft of American Intellectual Property, The National Bureau of Asian Research (NBR) (2021). https://www.nbr.org/program/commission-on-the-theft-of-intellectual-property/
70. Cormen, T.H., Leiserson, C.E., Rivest, R.L.. Stein, C.: Introduction to Algorithms, 3rd edn., pp. 253–280. Massachusetts Institute of Technology, Cambridge (2009). ISBN 978-0-262-03384-8
71. Brayton, R., Mishchenko, A.: ABC: an academic industrial-strength verification tool. In: Proc. CAV'10. LNCS 6174, pp. 24–40. Springer, Berlin (2010)
72. Roy, J.A., Koushanfar, F., Markov, I.L.: Ending piracy of integrated circuits. Computer **43**(10), 30–38 (2010). https://doi.org/10.1109/MC.2010.284
73. Yasin, M., Rajendran, J.J., Sinanoglu, O., Karri, R.: On improving the security of logic locking. IEEE Trans. Comput.-Aided Design Integr. Circuits Syst. **35**(9), 1411–1424 (2016). https://doi.org/10.1109/TCAD.2015.2511144
74. Subramanyan, P., Ray, S., Malik, S.: Evaluating the security of logic encryption algorithms. In: 2015 IEEE International Symposium on Hardware Oriented Security and Trust (HOST), pp. 137–143 (2015). https://doi.org/10.1109/HST.2015.7140252
75. Dupuis, S., Flottes, M.L.: Logic locking: a survey of proposed methods and evaluation metrics. J Electron Test **35**, 273–291 (2019). https://doi.org/10.1007/s10836-019-05800-4
76. Krishnan, S., M.K.N., N.D.M.: Weighted logic locking to increase hamming distance against key sensitization attack. In: 2019 3rd International conference on Electronics, Communication and Aerospace Technology (ICECA), pp. 29–33 (2019). https://doi.org/10.1109/ICECA.2019.8821880
77. Yasin, M., Mazumdar, B., Rajendran, J.J.V.,Sinanoglu, O.: SARLock: SAT attack resistant logic locking. In: 2016 IEEE International Symposium on Hardware Oriented Security and Trust (HOST), pp. 236–241 (2016). https://doi.org/10.1109/HST.2016.7495588

78. Yasin, M., Sengupta, A., Nabeel, M.T., Ashraf, M., Rajendran, J., Sinanoglu, O.: Provably-secure logic locking: from theory to practice. In: Proceedings of the 2017 ACM SIGSAC Conference on Computer and Communications Security (CCS '17). Association for Computing Machinery, New York, NY, USA, pp. 1601–1618 (2017). https://doi.org/10.1145/3133956.3133985
79. Yasin, M., Mazumdar, B., Sinanoglu, O., Rajendran, J.: Removal attacks on logic locking and camouflaging techniques. IEEE Trans. Emer. Topics Comput. **8**(2), 517–532 (2020). https://doi.org/10.1109/TETC.2017.2740364
80. Cook, S.A.: The complexity of theorem-proving procedures. In: IN STOC, pp 151–158. ACM, New York (1971)
81. Tseytin, G.S.: On the complexity of derivation in propositional calculus. In: Slisenko, A.O. (ed.) Studies in Constructive Mathematics and Mathematical Logic, Part II, Seminars in Mathematics, pp. 115–125. Steklov Mathematical Institute (1970). Translated from Russian: Zapiski Nauchnykh Seminarov LOMI **8**, 234–259 (1968)
82. Sörensson, N., Eén, N.: MiniSat v1.13–a SAT solver with conflict-clause minimization (2005)
83. Vemuri, R., Chen, S.: Split Manufacturing of Integrated Circuits for Hardware Security and Trust: Methods, Attacks and Defenses, 1st edn. Springer, Springer (2021)
84. Cai, Y., Haratsch, E.F., Mutlu, O., Mai, K.: Threshold voltage distribution in MLC NAND flash memory: characterization, analysis and modeling. In: Design, Automation & Test in Europe Conference & Exhibition (DATE), 2013, pp. 1285–1290, Grenoble, France. IEEE Conference Publications (2013)
85. Perricone, R., Ahmed, I., Liang, Z., Mankalale, M.G., Hu, X.S., Kim, C.H., Niemier, M., Sapatnekar, S.S., Wang, J.-P.: Advanced spintronic memory and logic for non-volatile processors. In: Design, Automation & Test in Europe Conference & Exhibition (DATE), 2017, pp. 972–977, Lausanne, Switzerland. IEEE, Piscataway (2017)
86. Yang, J., Kulkarni, N., Yu, S., Vrudhula, S.: Integration of threshold logic gates with RRAM devices for energy efficient and robust operation. In: 2014 IEEE/ACM International Symposium on Nanoscale Architectures (NANOARCH), pp. 39–44, Paris, France. IEEE, Piscataway (2014)
87. Gupta, P., Jha, N.K.: An algorithm for nanopipelining of RTD-based circuits and architectures. IEEE Trans. Nanotechnol. **4**(2), 159–167 (2005)
88. Berezowski, K.S., Vrudhula, S.B.K.: Automatic design of binary and multiple-valued logic gates on RTD series. In: 8th Euromicro Conference on Digital System Design (DSD'05), pp. 139–143, Porto, Portugal. IEEE, Piscataway (2005)
89. Krizhevsky, A., Sutskever, I., Hinton, G.: ImageNet classification with deep convolutional neural networks. In: NIPS'2012 (2012).
90. Girshick, R., Donahue, J., Darrell, T., Malik, J.: Rich feature hierarchies for accurate object detection and semantic segmentation. In: Proceedings of the IEEE Conference on Computer Vision and Pattern Recognition (2014)
91. Jaderberg, M., Vedaldi, A., Zisserman, A.: Deep features for text spotting. In: Computer Vision—ECCV 2014. Springer, Cham (2014)
92. Han, S., Pool, J., Tran, J., Dally, W.J.: Learning both weights and connections for efficient neural networks. In: NIPS (2015)
93. Han, S., Mao, H., Dally, W.J.: Deep compression: compressing deep neural networks with pruning, trained quantization and Huffman coding. In: International Conference on Learning Representations (2016)
94. Howard, A., et al.: Searching for MobileNetV3. In: The IEEE International Conference on Computer Vision (ICCV) (2019)
95. Sandler, M., et al.: MobileNetV2: inverted residuals and linear bottlenecks. In: Proceedings of the IEEE Conference on Computer Vision and Pattern Recognition (2018)
96. Rastegari, M., Ordonez, V., Redmon, J., Farhadi, A.: XNOR-Net: imagenet classifcation using binary convolutional neural networks. In: European Conference on Computer Vision, pp. 525–542. Springer, Berlin (2016)

97. Zhou, S., et al.: DoReFa-net: training low bitwidth convolutional neural networks with low bitwidth gradients (2016). arXiv preprint arXiv:1606.06160
98. Hubara, I., et al.: Binarized neural networks. In: Advances in Neural Information Processing Systems (2016)
99. Wagle, A., Khatri, S., Vrudhula, S.: A configurable BNN ASIC using a network of programmable threshold logic standard cells. In: 2020 IEEE 38th International Conference on Computer Design (ICCD), pp. 433–440 (2020). https://doi.org/10.1109/ICCD50377.2020.00079
100. Umuroglu, Y., Fraser, N.J., Gambardella, G., Blott, M., Leong, P., Jahre, M., Vissers, K.: FINN: a framework for fast, scalable binarized neural network inference. In: Proceedings of the 2017 ACM/SIGDA International Symposium on Field-Programmable Gate Arrays, pp. 65–74, Monterey California USA. ACM, New York (2017)
101. Anderson, A.G., Berg, C.P.: The high-dimensional geometry of binary neural networks (2017). CoRR, abs/1705.07199
102. Li, Y., Liu, Z., Liu, W., Jiang, Y., Wang, Y., Goh, W.L., Yu, H., Ren, F.: A 34-FPS 698-GOP/s/W binarized deep neural network-based natural scene text interpretation accelerator for mobile edge computing. IEEE Trans. Ind. Electron. **66**(9), 7407–7416 (2019)
103. Sun, X., Yin, S., Peng, X., Liu, R., Seo, J.S., Yu, S.: XNOR-RRAM: a scalable and parallel resistive synaptic architecture for binary neural networks. In: 2018 Design, Automation & Test in Europe Conference & Exhibition (DATE), pp. 1423–1428, Dresden, Germany. IEEE, Piscataway (2018)
104. Geng, T., Wang, T., Wu, C., Yang, C., Song, S.L., Li, A., Herbordt, M.: LP-BNN: ultra-low-latency BNN inference with layer parallelism. In: 2019 IEEE 30th International Conference on Application-Specific Systems, Architectures and Processors (ASAP), pp. 9–16, New York, NY, USA. IEEE (2019)
105. Al Bahou, A., Karunaratne, G., Andri, R., Cavigelli, L., Benini, L.: XNORBIN: a 95 TOp/s/W hardware accelerator for binary convolutional neural networks. In: 2018 IEEE Symposium in Low-Power and High-Speed Chips (COOL CHIPS), pp. 1–3, Yokohama. IEEE, Piscataway (2018)
106. Andri, R., Cavigelli, L., Rossi, D., Benini, L.: YodaNN: an architecture for ultralow power binary-weight CNN acceleration. IEEE Trans. Comput.-Aided Design Integr. Circuits Syst. **37**(1), 48–60 (2018)
107. Nakahara, H., Yonekawa, H., Sasao, T., Iwamoto, H., Motomura, M.: A memory-based realization of a binarized deep convolutional neural network. In: 2016 International Conference on Field-Programmable Technology (FPT), pp. 277–280, Xi'an, China. IEEE, Piscataway (2016)
108. Simons, T., Lee, D.J.: A review of binarized neural networks. Electronics **8**(6), 661 (2019)
109. Rastegari, M., Ordonez, V., Redmon, J., Farhadi, A.: XNOR-net: imagenet classification using binary convolutional neural networks. In: Bastian Leibe, Jiri Matas, Nicu Sebe, and Max Welling, editors, Computer Vision—ECCV 2016, Lecture Notes in Computer Science, pp. 525–542. Springer, Cham (2016)
110. Scott, K.R., Lee, C.-Y., Khatri, S.P., Vrudhula, S.: A flash-based current-mode IC to realize quantized neural networks. In: Design, Automation & Test in Europe Conference & Exhibition (DATE) (2022)
111. Bez, R., Camerlenghi, E., Modelli, A., Visconti, A.: Introduction to flash memory. Proc. IEEE **91**(4), 489–502 (2003)
112. Simmons, T., Lee, D.: A review of binarized neural networks. Electronics **8**(6), 661 (2019)
113. Synopsys website. http://www.synopsys.com/
114. PTM website. http://ptm.asu.edu/
115. Abusultan, M., Khatri, S.P.: Implementing low power digital circuits using flash devices. In: 2016 IEEE 34th International Conference on Computer Design (ICCD), pp. 109–116 (2016). https://doi.org/10.1109/ICCD.2016.7753268
116. TensorFlow website. https://www.tensorflow.org/

117. Zhao, W., et al.: Rigorous extraction of process variations for 65 nm CMOS design. In: Proc. of the European Solid State Device Research Conf. (2007)
118. Bernstein, K., Pearson, D.J., Rohrer, N.J., et al.: High-performance CMOS variability in the 65-nm regime and beyond. In: IBM J. Res. Dev. **50**(4/5), 433–449 (2006)
119. Rezali, F.A.M., et al.: Scaling impact on design performance metric of sub-micron CMOS devices incorporated with halo. 2015 IEEE Regional Symposium on Micro and Nanoelectronics (RSM) pp. 1–4 (2015)
120. Ali, N., et al.: TCAD analysis of variation in channel doping concentration on 45 nm double-gate MOSFET parameters. In: 2015 Annual IEEE India Conference (INDICON), New Delhi, pp. 1–6 (2015)
121. Lemoigne, P., Quenette, V., Juge, A., Rideau, D.: Monitoring variability of channel doping profile in the 45 nm node MOSFET through reverse engineering of electrical back-bias effect. In: 2009 Proceedings of the European Solid State Device Research Conference, pp. 383–386 (2009)
122. Nayfeh, H.M., et al.: Impact of lateral asymmetric channel doping on 45-nm-technology N-type SOI MOSFETs. IEEE Trans. Electron. Devices **56**, 3097–3105 (2009)
123. Kok, C.-W., Tam, W.-S.: Bandgap voltage reference. In: CMOS Voltage References: An Analytical and Practical Perspective, pp. 71–101. IEEE, Piscataway (2013)
124. Geiger, L., Team, P.: Larq: an open-source library for training binarized neural networks. J. Open Source Softw. **5**(45), 1746 (2020)
125. Wagle, A., Khatri, S., Vrudhula, S.: A configurable BNN ASIC using a network of programmable threshold logic standard cells. In: 2020 IEEE 38th International Conference on Computer Design (ICCD), pp. 433–440 (2020)
126. Singh, G., Wagle, A., Vrudhula, S., Khatri, S.: CIDAN: computing in DRAM with artificial neurons. In: 2021 IEEE 39th International Conference on Computer Design (ICCD), pp. 349–356 (2021). https://doi.org/10.1109/ICCD53106.2021.00062
127. Angizi, S., et al.: Accelerating deep neural networks in processing-in-memory platforms: analog or digital approach? In: ISVLSI'19
128. Li, Y., et al.: BitWeaving: fast scans for main memory data processing. In: SIGMOD'13
129. Myers, G.: A fast bit-vector algorithm for approximate string matching based on dynamic programming. In: JACM'99
130. Fujiki, D., et al.: Duality cache for data parallel acceleration. In: ISCA'19
131. Eckert, C., et al.: Neural cache: bit-serial in-cache acceleration of deep neural networks. In: ISCA'18
132. Seshadri, V., et al.: Ambit: in-memory accelerator for bulk bitwise operations using commodity DRAM technology. In: MICRO'17
133. He, M., et al.: Newton: a DRAM-maker's accelerator-in-memory (AiM) architecture for machine learning. In: MICRO'20
134. Yin, S., et al.: Vesti: energy-efficient in-memory computing accelerator for deep neural networks. In: TVLSI'20
135. Jacob, B., et al.: Memory Systems: Cache, DRAM, Disk. MK Publishers (2008)
136. Wagle, A., et al.: Threshold logic in a flash. In: ICCD'19
137. Angizi, S., et al.: GraphiDe: a graph processing accelerator leveraging in-DRAM-computing. In: GLSVLSI'19 (2019)
138. Hajinazar, N., et al.: SIMDRAM: a framework for bit-serial SIMD processing using DRAM. In: ASPLOS'21
139. Angizi, S., et al.: ReDRAM: a reconfigurable processing-in-DRAM platform for accelerating bulk bit-wise operations. In: ICCAD'19
140. Li, S., et al.: DRISA: a DRAM-based reconfigurable in-situ accelerator. In: IEEE/ACM MICRO'17
141. Kim, Y., et al.: Assessing merged DRAM/logic technology. In: ISCAS'96
142. Binkert, N., et al.: The gem5 simulator. In: SIGARCH'11
143. Kim, Y., et al.: Ramulator: a fast and extensible DRAM simulator. IEEE Comp. Arch. Lett. **15**(1), 45–49 (2015)

144. Chandrasekar, K., et al.: DRAMPower: open-source DRAM power and energy estimation tool. http://www.drampower.info/
145. Stanford Large Network Datasets (2021). https://snap.stanford.edu/data/
146. Frank, D.J.: Power-constrained CMOS scaling limits. IBM J. Res. Dev. **46**(2–3), 235–244 (2002). https://doi.org/10.1147/rd.462.0235
147. Blaauw, D., et al.: IoT design space challenges: circuits and systems. In: 2014 Symposium on VLSI Technology (VLSI-Technology): Digest of Technical Papers, pp. 1–2 (2014). https://doi.org/10.1109/VLSIT.2014.6894411
148. Crovetti, P.S.: All-digital high resolution D/A conversion by dyadic digital pulse modulation. IEEE Trans. Circuits Syst. I: Regul. Pap. **64**(3), 573–584 (2017). https://doi.org/10.1109/TCSI.2016.2614231
149. Aiello, O., Crovetti, P.S., Alioto, M.: Fully synthesizable low-area digital-to-analog converter with graceful degradation and dynamic power-resolution scaling. IEEE Trans. Circuits Syst. I: Regul. Pap. **66**(8), 2865–2875 (2019). https://doi.org/10.1109/TCSI.2019.2903464
150. Razavi, B.: The current-steering DAC [a circuit for all seasons]. IEEE Solid-State Circuits Mag. **10**(1), 11–15 (2018). https://doi.org/10.1109/MSSC.2017.2771102
151. Crovetti, P.S., Rubino, R., Musolino, F.: Relaxation digital-to-analog converter with foreground digital self-calibration. In: 2020 IEEE International Symposium on Circuits and Systems (ISCAS), pp. 1–5 (2020). https://doi.org/10.1109/ISCAS45731.2020.9180696
152. Ahmed, S., Zou, X., Jaber, N., Younis, M.I., Fariborzi, H.: A low power microelectromechanical resonator-based digital to analog converter. J. Microelectromech. Syst. **29**(3), 320–328 (2020). https://doi.org/10.1109/JMEMS.2020.2988790
153. Ahmed, S., Zou, X., Fariborzi, H.: A micro-resonator based digital to analog converter for ultralow power applications. In: 2019 20th International Conference on Solid-State Sensors, Actuators and Microsystems & Eurosensors XXXIII (TRANSDUCERS & EUROSENSORS XXXIII), pp. 821–824 (2019). https://doi.org/10.1109/TRANSDUCERS.2019.8808759
154. Rubino, R., Crovetti, P.S., Aiello, O.: Design of relaxation digital-to-analog converters for internet of things applications in 40 nm CMOS. 2019 IEEE Asia Pacific Conference on Circuits and Systems (APCCAS), pp. 13–16 (2019). https://doi.org/10.1109/APCCAS47518.2019.8953168
155. Crovetti, P.S., Rubino, R., Musolino, F.: Relaxation digital–to–analogue converter. Electron. Lett. **55**(12), 685–688 (2019)
156. Oh, J., Park, J.-E., Hwang, Y.-H., Jeong, D.-K.: 25.2 A 480 mA output-capacitor-free synthesizable digital LDO using CMP-triggered oscillator and droop detector with 99.99% current efficiency, 1.3 ns response time, and 9.8 A/mm^2 current density. In: 2020 IEEE International Solid-State Circuits Conference—(ISSCC), pp. 382–384 (2020). https://doi.org/10.1109/ISSCC19947.2020.9063018
157. Lee, Y.-J., et al.: A 200-mA digital low drop-out regulator with coarse-fine dual loop in mobile application processor. IEEE J. Solid-State Circuits **52**(1), 64–76 (2017). https://doi.org/10.1109/JSSC.2016.2614308
158. Shi, J., Zhao, B., Wang, B.: A coarse-fine dual loop digital low dropout regulator with fast transient response. In: 2019 IEEE International Conference on Electron Devices and Solid-State Circuits (EDSSC), pp. 1–3 (2019). https://doi.org/10.1109/EDSSC.2019.8754116
159. Chen, R., Zhou, S., Wu, Z., Li, B., Huang, M.: A fast response digital low-dropout regulator based on enhanced analog assisted loop. In: 2020 IEEE International Conference on Integrated Circuits, Technologies and Applications (ICTA), pp. 55–56 (2020). https://doi.org/10.1109/ICTA50426.2020.9332011
160. Cai, G., Zhan, C., Lu, Y.: A fast-transient-response fully-integrated digital LDO with adaptive current step size control. IEEE Trans. Circuits Syst. I: Regul. Pap. **66**(9), 3610–3619 (2019). https://doi.org/10.1109/TCSI.2019.2917558
161. Akram, M.A., Hwang, I.-C., Ha, S.: Architectural advancement of digital low-dropout regulators. IEEE Access **8**, 137838–137855 (2020). https://doi.org/10.1109/ACCESS.2020.3012467

Nonvolatile Memory Technologies: Characteristics, Deployment, and Research Challenges

Sadhana Rai and Basavaraj Talawar

1 Introduction

Memory plays a prominent role in any computing system, be it a handheld device or a super computing system; it is essential for storing the data and information needed by the central processing unit. Memory plays a significant role in power consumption, reliability, and deciding the application performance [46]. Since its inception, dynamic random access memory (DRAM) has been used as memory in almost all computing devices. However, new technologies like artificial intelligence (AI), deep learning (DL), and cloud computing demand large memory footprints with low power consumption and good performance. It is expected that data generation rate doubles every 3 years [17]. There is a necessity to build large memories with excellent performance and low power consumption in order to satisfy the demands of the emerging applications. Though DRAM has excellent properties as a memory device, it fails to provide good density and has high static power consumption. It is estimated that DRAM consumes nearly 30–50 percent of the total power consumption of the system [68]. Building large memories requires more memory cells, but technologies like DRAM and SRAM are facing issues when they are scaled down 10 nm or below [77]. These limitations of existing memory devices have led to exploring alternative technologies. New memory technologies should integrate high-performance features similar to DRAM/SRAM: scalability, persistence, and cost-effectiveness similar to existing flash. Emerging nonvolatile memory (NVM) devices are excellent candidates as they provide persistence similar to secondary storage devices and access latencies comparable to DRAM. Another advantage of these NVM devices is that they exhibit low leakage power as they

S. Rai (✉) · B. Talawar
SPARK Lab, Department of Computer Science and Engineering, National Institute of Technology Karnataka (NITK), Surathkal, India
e-mail: sadhana.197cs002@nitk.edu.in; basavaraj@nitk.edu.in

A. Iranmanesh (ed.), *Frontiers of Quality Electronic Design (QED)*,
https://doi.org/10.1007/978-3-031-16344-9_4

do not require a constant refresh. NVMs such as NAND flash are used these days extensively along with other nonvolatile memories, phase change memory (PCM), spin-transfer torque random access memory (STT-RAM), resistive random access memory (ReRAM), and carbon nanotube random access memory (N-RAM). NVM devices are not a new invention, and they have been used even in early supercomputers such as Cray EL92. Still, the only difference now and then is that they were not energy efficient, targeted only specific workloads, and were limited [68]. Emerging NVM devices are often known as storage class memory devices as they exhibit characteristics of both memory and storage. There is a possibility of a paradigm shift in the memory hierarchy with the advent of these devices. A two-tier hierarchy that currently persists between the slow nonvolatile devices and fast memory devices may be merged as a single hierarchy [8]. However, there are several challenges to using these NVM devices at different levels in the memory hierarchy. This chapter discusses emerging NVM technologies and their characteristics, advantages, pitfalls, deployment, and other issues.

2 Characteristics of NVM Devices

Devices such as PCM, ReRAM, STT-MRAM, FeRAM, and NRAM are emerging; while some of them are still in prototype, few are commercially available. These devices have features in common such as byte addressability; nonvolatility, low leakage power, and access latency much lower than existing NVM devices (flash). Some of the devices exhibit multilevel cell (MLC), which means that a single cell can store multiple bits, which helps to improve the density of the device. In the subsequent subsections, we discuss the characteristics of these devices.

2.1 Flash Storage Devices

Before we discuss the modern NVM technologies in detail, we will throw light on its predecessor, i.e., flash memory technology. Flash memory was invented in the 1900s and it uses floating gates (FG); they can be NAND-based or NOR-based [3]. They do not have mechanical moving parts and allow random access. In addition to this, they also have excellent density and are cost-effective. Pertaining to these advantages, they soon replaced the existing hard disk drive (HDD)-based storage devices. Flash-based storage devices reduced the speed gap between memory and storage by three orders of magnitude compared to HDD [5]. Three types of operations are performed on flash devices (read, write, and erase), while most other memory devices have only read and write operations. Erase operations are required because writes can only change the bit from one to zero, but it is impossible to perform reverse functions; hence, it is necessary to have erase operations. However, these operations are slower than write operations [46]. Another problem of flash-based storage devices is that they have limited endurance, which means that after a certain number of write operations, cells become physically unfit for use. Though flash devices had advantages over HDD, the access latency is still slower when compared to DRAM. Hence, if a better technology is available, it can replace flash; that is where NVM devices show up – while providing all the features of flash base storage they have added advantage of access latencies close to DRAM [27].

2.2 *Phase Change Memory (PCM)*

This type of memory device has been studied since the 1960s, ever since the inception of the ovonic threshold switching (OTS) phenomenon that was measured by Ovshinsky [23]. PCM devices are constructed using phase change materials that can exist in two states amorphous or crystalline. The switching between the states is achieved by heating the materials either by applying electrical current/voltage pulse (T. [35]). Chalcogenide alloys such as $Ge_2Sb_2Te_5$ (GST) are widely used phase change materials to construct PCM devices. The SET operation stores 1 in the cell; this is achieved by heating GST materials above the heating temperature (300 °C) and below the melting temperature (600°C) over some time period; this changes the state to crystalline, thereby storing the information 1. GST is heated above the melting point and quenched quickly; this leads to high resistance amorphous states, thereby storing the value 0. There is a huge difference between crystalline and amorphous state; this can be utilized to store multiple bits in one cell; hence, they are known as multilevel cell (MLC), while many other devices can store only a single bit per cell and are known as single bit cell (SLC) [20, 76]. The data stored in the cell can be read by applying an electrical bias without altering the data. PCM has several advantages like good scalability, reliability, and low device-to-device variation; all these features promote PCM to be an excellent candidate for memory devices [23]. PCM is studied extensively in most of the research works carried out recently. One reason for this could be the commercial availability of devices based on PCM technology. 3D XPoint, a product of Intel and Micron, is one of the commercially available products [24]. Figure 1 depicts the cell structure of PCM.

2.3 *Resistive Random Access Memory (ReRAM/RRAM)*

Resistive RAMs have been studied and researched since around the 1960s. However, these devices started gaining prominence in the early 2000s and were studied extensively between 2005 and 2015 [13]. The unique property of this device is that it uses the change in the material's resistance to store the information. The memory cell of RRAM consists of an insulator sandwiched between two metal electrodes. Data is stored by applying an external electrical voltage across the cells, which changes the state of the cell from a high resistance state to a low resistance

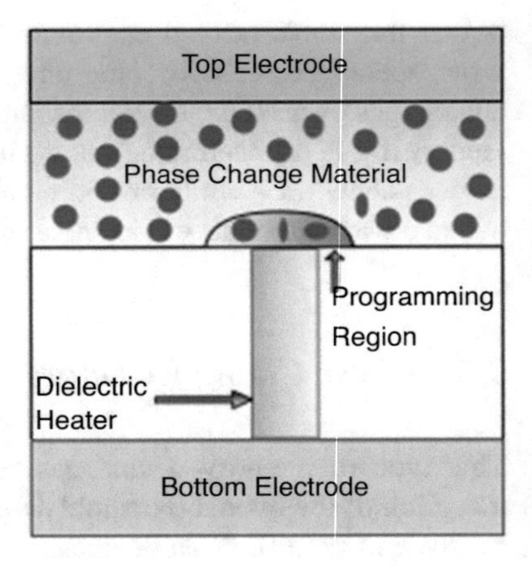

Fig. 1 Structure of PCM cell

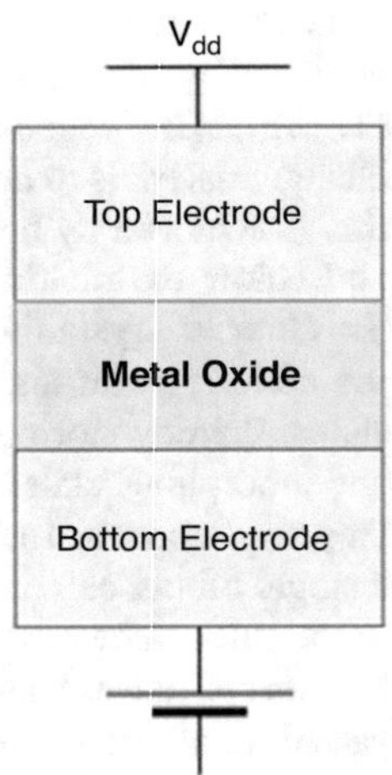

Fig. 2 Structure of ReRAM cell [77]

state and vice versa. The transition from high resistance to low resistance results in storing the value 0, and vice versa results in storing 1. The biggest challenge in the design of these devices is choosing the electrode. Metal oxide metal structures are widely used as electrodes because of the ease of fabrication [77]. Figure 2 shows the cell structure of ReRAM. Significant advantages of ReRAM are ease of fabrication, simple structure, data retention, and compatibility with existing CMOS technology which make it one of the strong candidates for digital memory. Limitation of uniformity on wide device characteristics is a major hindrance for large-scale manufacture [77].

2.4 Ferro-Electric Random Access Memory (FeRAM)

FeRAM comprises one transistor-one capacitor structure (1T1C), more like a DRAM cell structure. The major difference is in the capacitor structure, which is made of ferro-electric-based layers, making it nonvolatile. The most commonly used ferroelectric material is PZT (lead zirconate titanate) [3]. The polarization of the ferroelectric capacitor achieves data storage. In FeRAM devices the gate dielectric is substituted with ferroelectric polarization [37]. Unlike flash devices they do not require high voltages for performing write operations [3]. Conventional 1T1C-based FeRAM devices provided high performance but they suffered from destructive reads due to device-to-device interference. To solve this issue, there was a necessity to design separate read-write paths. 1T1T cell-based FeRAM devices are developed to avoid destructive reads. 1T1T-based FeRAM cell is fabricated on a plastic substrate using a ferroelectric memory transistor (MT) and a control transistor (CT). A single-walled carbon nanotube (SWNT) was inkjet printed and used as a semiconducting channel between two transistors. Omega-shaped ferroelectric gate was fabricated by incorporating organic poly(vinylidene fluoride-co-trifluoroethylene) (P(VDF-TrFE)) layer on top of SWNT. The 1T1T cell structure separated read-write paths, thus avoiding destructive reads [37]. Apart from changing the cell structure, some techniques change the methods of reading such as an electro-optic method, acoustic reading, photovoltaic reading, and pyroelectric reading [30]. Usage of hafnium oxide (HfO2)-based ferroelectrics has brought tremendous improvement in the FeRAM-based structures, and being one of the mature technologies, FeRAM is in the race to be in the future NVM-based memory device. Figure 3 depicts the basic cell structure of FeRAM.

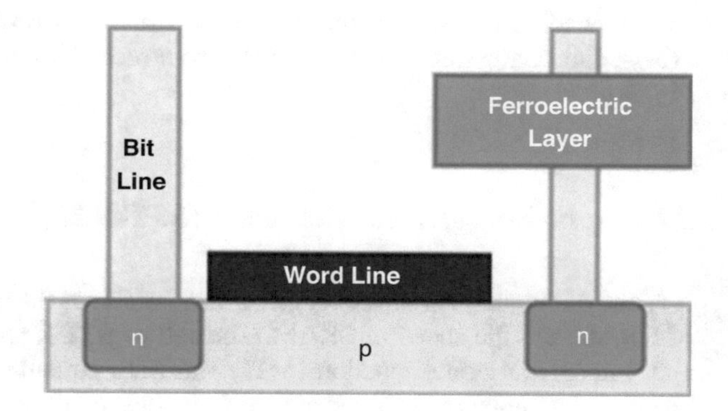

Fig. 3 Structure of FeRAM cell [44]

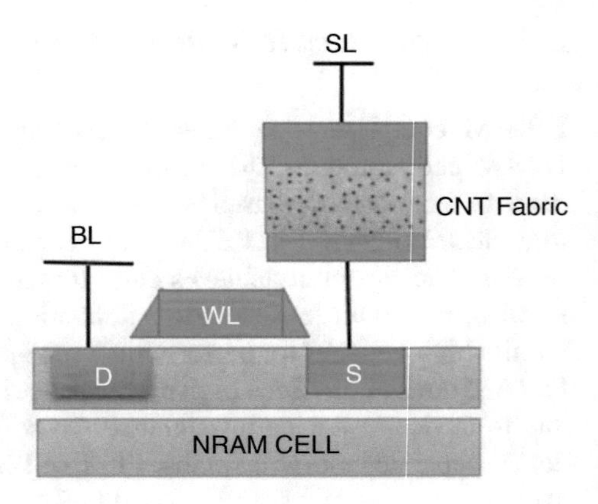

Fig. 4 Structure of NRAM cell [26]

2.5 Carbon Nanotube Random Access Memory (NRAM)

The memory cell of a NRAM consists of carbon nanotubes (CNT) interconnected in the form of a matrix. These CNTs are sandwiched between a pair of electrodes that are bound to be semiconductor fab friendly. Application of positive bias voltage on the bottom electrode while keeping the top grounded, the CNTs are in contact with each other; this reduces the resistance of the cell and represents 1. Similarly, by reversing the direction of applied voltage, the CNTs are forced to be separated, and the resistance of the cell increases, representing 0. The Van der Waals force effect keeps the CNTs in contact to be connected and those apart to remain separated until an external voltage is applied; this makes the device nonvolatile. The storage element is integrated into the back end of the line (BEOL) and, depending on applications, can be in a one transistor-one resistor (1T1R) or cross-point configuration [25, 26, 54]. The structure of NRAM cell is depicted in Fig. 4.

2.6 Spin-Transfer Torque RAM (STT-RAM)

It is one of the unique kinds of NVM which combines the cost and capacity benefits of DRAM and the speed of SRAM. The cell of STT-RAM consists of a transistor and a magnetic tunnel junction (MTJ). An MTJ contains a pair of ferromagnetic layers and a tunnel oxide layer sandwiched between them. Out of the two magnetic layers, one has a fixed magnetic orientation (known as the reference layer). At the same time, the other has a free magnetic orientation (also known as the free layer). The free layer is used to store the information and is relatively thin so that it can be switched quickly. When both the ferromagnetic layers are parallel, they follow

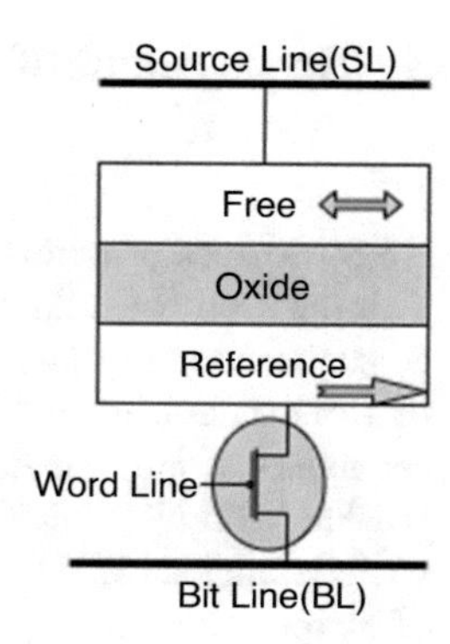

Fig. 5 Structure of 1 T-1MTJ STT-RAM cell

Table 1 Comparison of device properties on NVM technologies [19]

Device	PCM	STT-RAM	ReRAM	FeRAM	NRAM
Cell element	1 T(D) 1R	1(2)T1R	1 T(D), 1R	1T1C/2T1C	1T1R/1 T-1CNT
Cell size (F^2)	4–20	6–20	<4(3D)	30	<6(2D)
Endurance	10^7	10^5	10^5	10^{15}	>10^{12}
Retention	>10 Y	>10 Y	>10 Y	> 10 Y	>10 Y
Multilevel cell (MLC)	Yes	Yes	Yes	Yes	Yes

low resistance, and it is used to store zero. When anti-parallel, they exhibit high resistance, thereby storing value one [6, 40]. The data is read by applying voltage and detecting the current flow. Figure 5 shows the cell structure of the STT-RAM cell. Several free layers are connected to the bitline, while access transistors are connected to the source line. Due to the high switching current of magnetic tunnel junction (MTJ), write operations consume high energy. In addition, since the MOS device determines this switching current, this plays a role in determining the density of memory [78]. STT-RAM is mostly used as a substitute for SRAM-based cache.

Table 1 displays the comparison of the device properties of different technologies. All these devices have good retention time and simple cell structure. In the subsequent sections, we will discuss about the possible interactions of these NVM devices at different levels of memory organization. In addition, we highlight the various issues faced by them while integrating them and also the possible solutions. However, some of the NVM devices are not explored much and there are very few evidences of their usage in memory. Hence, most of our discussions are limited to usage of widely used NVM devices such as PCM, ReRAM, and STT-MRAM, and in few sections we have discussed about FeRAM based on availability of resources.

3 Deployment of Nonvolatile Memories in Computing System

NVM devices provide potential benefits both in terms of performance and power. This makes them suitable to be used at different levels of the memory hierarchy, such as cache, memory subsystem, or secondary storage devices. It was also believed that these devices could emerge as universal memory, replacing both DRAM and secondary storage devices [23]. However, it is not ideal for replacing the existing DRAM with NVM devices because of the limited endurance and differential read-write latencies [23]. It is not easy to fit all the NVM devices in the common level of memory hierarchy [64]. This section will discuss the possibilities and challenges of using NVM devices at different levels of the memory hierarchy. NVM devices are still not able to completely replace either of these because they cannot satisfy the cost benefits of NAND and performance benefits of DRAM [35]. It has been an active area of research for several years, and several researchers have explored the suitable position for these emerging devices in the memory hierarchy; as a result, several outcomes have been observed. Another challenge in integrating new technology is its adaptability to existing hardware and software. When a new memory device is added, it often requires changing the operating system and the file system interfaces, thereby changing the behavior of existing applications; hence, it is necessary to consider all these factors before bringing the change in the hierarchy.

3.1 Deploying NVM Devices as Processor Cache

SRAM is used as cache memory because of the performance benefits it provides. The challenge SRAM is facing is its high static energy consumption, and it has low density because its cell is made up of six transistors. Another major drawback of SRAM is that it has very high leakage power. Energy consumption is a significant issue in the modern era, and an alternative device is required. As we know, cache memories are used to improve performance; it is highly recommended that a device that stands as an alternative to SRAM should have excellent performance. Nonvolatile memory devices can be fabricated using CMOS technology, and this makes it possible to integrate directly with the processors [71]. Some of the NVM devices can be fabricated without per cell access transistors. A recent work integrated ReRAM into an architecture, which supports many-core CPU and a large number of in-order multi-threaded cores. They have incorporated a 3D ReRAM into the last-level cache of all tiles of the core; this creates a monolithic architecture. The advantage of such a design is that the area is utilized [71] efficiently. STT-RAM is another device that has emerged as an alternative to SRAM, thereby changing cache memory. The qualities such as high density, low leakage power, and access performance similar to SRAM have attracted STT-RAM to be used as a candidate in cache memory. The improved density of STT-RAM can help store more data in the

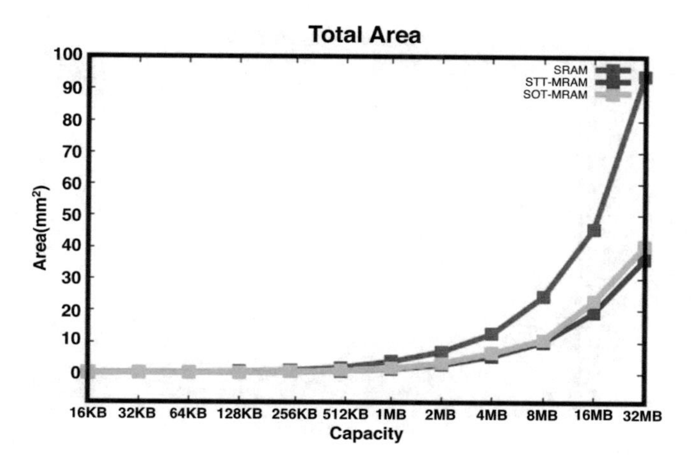

Fig. 6 Comparison of area efficiency of SRAM and NVM devices [32]

same area of SRAM; this helps to improve the cache hit ratio. Another advantage of this substitution is that STT-RAM has a low leakage current. Replacing SRAM will help improve energy consumption [64]. Though STT-RAM supports multiple-bit storage, write disturbance between two bits reduces performance and energy efficiency [40]. STT-RAM is studied widely as an alternative to existing SRAM-based caches among all the existing NVM devices. Spin-orbit torque magnetic random access memory (SOT-MRAM) is also a MRAM-based memory device, which is being experimented recently. It is expected to overcome the drawback of high write latency and energy of STT-MRAM. However, we will not discuss SOT-MRAM in detail in this chapter. Figures 6 and 7 depict the area efficiency and leakage power of SRAM with STT-RAM- and SOT-MRAM-based devices; these results are obtained by experiments which were conducted in our lab by our colleagues. In both the cases, NVM-based caches prove to be more efficient as the capacity increases. This clearly indicates that NVM devices are better option for large capacity caches. Figures 6 and 7 shows the comparison of SRAM,STT-MRAM and SOT-MRAM.

3.2 Integrating NVM as Main Memory

As mentioned earlier, a memory device should be capable of delivering data and instructions with speed comparable to that of processors. Since most of the NVM devices have access latencies similar to that of DRAM, it was believed that conventional memory devices could be replaced with NVMs. NVM devices like PCM, STT-RAM, ReRAM, and NRAM have been studied as alternatives to DRAM. Features such as low power consumption, byte address ability, and low access

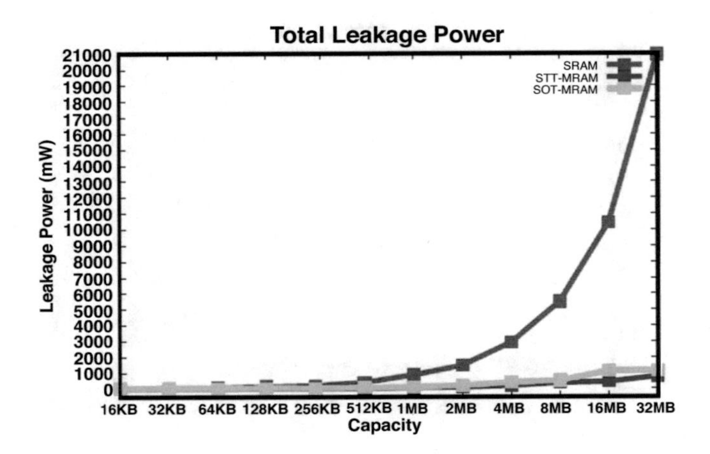

Fig. 7 Comparison of leakage power of SRAM and NVM devices [32]

latencies (compared to HDD) attract them to be excellent memory candidates. However, to replace DRAM, a memory device in almost all computing devices for several decades, the new devices should provide comparable features in terms of speed, cost, energy, and density. NVM devices have been integrated as memory components in different ways. This subsection discusses the possibilities, problems, and challenges faced when using NVMs as memory devices.

3.2.1 Replacing Conventional Memory Devices with NVM

Latest applications like big data, cloud computing, and deep learning require huge memory footprints with optimized energy consumption. DRAM is facing scalability issues, and the static energy consumption of DRAM increases with the increase in the size of the device. To compromise these issues, NVMs can be treated as alternative devices. Experiments were conducted by Gamati'e et al. [24] to study the impact of using ReRAM and PCM at different levels of the memory hierarchy. This study revealed that ReRAM was ideal for being used as a memory device as it consumes less energy with little compromise in performance.

At the same time, PCM was considered more suitable to be a storage device rather than a memory. Recent PCM-based devices are modified to improvise write latency and endurance issues, making them suitable for use as memory devices. Figure 8 shows the conventional DRAM-based memory and NVM-based memory organization. Replacing DRAM with NVM-based memories may help for low-frequency and high energy-efficient processors, but it may not be a good idea to be used in high-performance computing systems because of the prolonged write operations [15]. DRAM can provide excellent performance, while NVM devices can provide excellent capacity to combine the performance benefits of DRAM and

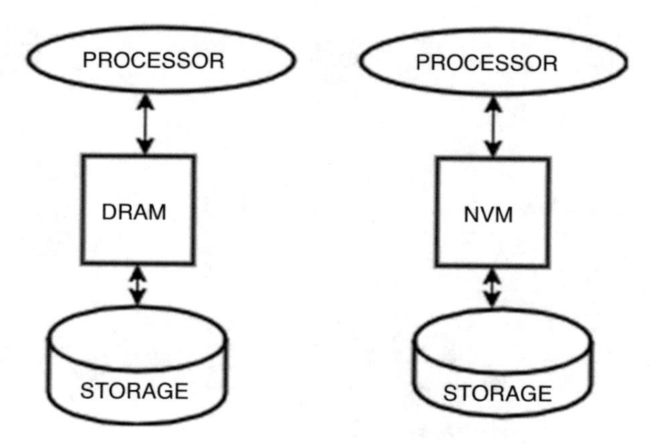

Fig. 8 Conventional DRAM-based memory and NVM-based memory architectures [15]

capacity benefits of NVM; a new architecture was proposed by integrating both the technologies, that is, hybrid memory. This kind of architecture was initially proposed by Qureshi et al. [52]. Thereafter, this architecture has been widely studied and researched, and various modifications to this are being developed.

3.2.2 Integrating NVM and DRAM as Hybrid Memories

The major setbacks for replacing conventional DRAM-based memory with NVMs were:

1. *Disparity in read-write latency*: The reason for this is that devices like PCM store the information by changing the resistance of the cell, especially by heating; this makes writes slow. The content of the cell can be read by checking the resistance, and this generally takes less time [5].
2. *High dynamic energy consumption*: NVM devices do not require refresh and have low leakage current compared to DRAM. They have high dynamic energy consumption; this is because write operations in most NVM are usually performed by changing the resistance, which consumes more energy.
3. *Limited endurance*: Most of the NVM devices have comparatively low endurance compared to DRAM. The reason for this could be due to the materials used to build the memory device and the way how read and write operations are performed.

The benefit of hybrid memory is that we can increase the capacity of memory (NVM) without compromising performance (DRAM). NVMs are integrated with DRAM to form memory hierarchy in cache/hierarchical organization or flat/parallel organization. Figure 9 depicts two different ways of organizing hybrid memories.

Cache/hierarchical organization Here conventional DRAM acts as a cache and NVM devices are used as memory components. Similar to the cache memory here, DRAM will hold the data that is frequently in use (hot data), while NVM contains

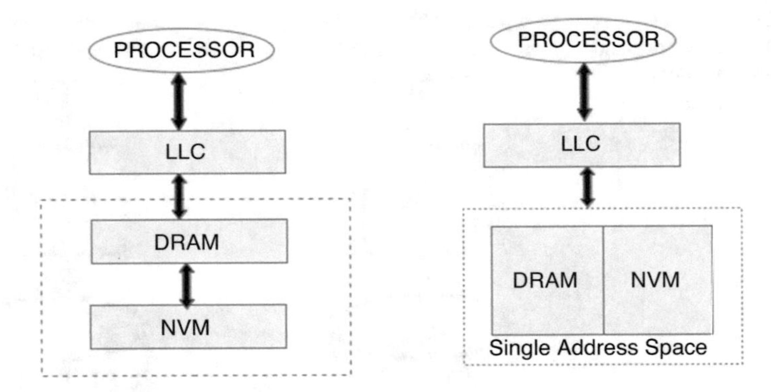

Fig. 9 Hybrid memory architectures

the data that is used rarely often, which is termed cold data. The challenge in implementing this architecture is that we need some hardware to manage tags; DRAM memory is hidden from the operating system's perspective, which reduces the overall size of the memory. The advantage of this design is that it does not require any modification to the existing software.

Flat/parallel organization Both DRAM and NVM share a common address space, unlike a hierarchical organization. Flat/parallel organization increases the overall capacity, as both memory spaces are visible to the operating system. If capacity is the primary requirement, it is better to utilize this organization. However, we should be careful about data placement because of the slow access latencies of the NVM devices. Generally, capacity-intensive applications benefit from flat organization [38].

Hierarchical architectures can be more beneficial for applications that exhibit good data locality [31]. A study of different heterogeneous memory architectures [10] showed that cache misses do consume the bandwidth and have an impact on bandwidth utilization. Hence, cache-based architectures are ideal for those systems with a good cache hit ratio. The study also revealed that though cache hit rates are high, maintaining sufficiently high bandwidth between the chosen memories is essential for effective resource utilization. Hence, cache-based architectures are more suitable if the cache hit rates are high and energy and bandwidth ratios between stacked and on-chip DRAM are at least four times [10].

3.3 Deploying NVM as Storage

This subsection discusses the characteristics that make NVMs suitable as storage devices and the challenges in deploying them. Currently, NAND flash devices are being used as secondary storage devices in place of traditional HDD due to their significant I/O performance [68]. An ideal storage media is expected to have good density and low cost and be nonvolatile. NVM devices like PCM aim to provide good density compared to NAND flash. It is observed that NAND flash is not cost-effective with a density below 20 nm; in contrast, PCM devices are forecasted to be stable below the 5 nm node, which makes PCM an ideal storage media. Another feature that helps to increase the density of the NVM device is its ability to store multi-bits [5]. Subsequently, access times of NVMs are far better than hard disk drives. It is also possible that NVMs can be exclusively used as storage devices in large-scale HPC systems to store global data that all the nodes in the network can access; this can help manage unstructured datasets [68]. Experiments were carried out by Bahn and Cho [5] to analyze the impact of adapting NVM devices like PCM and STT-RAM as storage media. These experiments revealed that due to the challenges faced with increasing memory size, it is ideal for building systems with limited memory and huge storage with NVM devices without compromising performance. Such designs would be suitable for memory-intensive applications. Another advantage of using NVM in storage media is their endurance compared to existing flash-based storage media; devices like PCM can sustain 1000x times more writing than existing storage devices. NRAM also has a density that is much higher than DRAM and consumes less power when compared to flash; these characteristics make it suitable for storage [26]. Although ReRAM also emerged as an alternative to existing storage media, it could not cope with NAND flash storage, and now it is kept away from storage. However, it has found good implications in machine learning-based applications [13]. All NVM devices possess the quality for replacing flash-based storage devices. Most of the research work is carried out by utilizing them as memory device alternatives to DRAM rather than as storage devices.

4 Challenges in Adopting NVM Devices at Different Levels of Memory

In the previous sections, we discussed the use of NVM devices at different levels of memory hierarchy. To replace or to integrate NVM with conventional devices, several modifications are required. In this section we discuss the different problems and possible solutions which we come across at different levels of memory hierarchy.

4.1 Design Issues in Utilizing NVM as Cache

SRAM is facing challenges with power consumption and areal density; hence, it is necessary to find an alternative to this technology. NVM devices have good density and low leakage current as mentioned earlier, and they can be an alternative to SRAM-based memories. Among all NVMs, STT-RAM has emerged alternative to SRAM due to its excellent density, which is near four times more enormous than SRAM, and low leakage current [56]. Besides these advantages, the costly NVM writes in terms of energy and time are major setbacks in adopting them directly as caches. Hence, several modifications are suggested in terms of write intensity reduction techniques and write latency reduction techniques to overcome costly write operations in NVMs. Modifications to the replacement policies in last-level caches are often adopted to solve the issues related to costly writes. However, the straightforward approach is to place read-intensive blocks in NVM part of the cache; they should be handled intelligently such that the hit ratio of upper-level caches is well maintained without affecting the performance [64]. Previous works have shown that rather than replacing existing caches with NVM-based caches, it is ideal to use both in the form of a hybrid cache.

4.1.1 Management of Hybrid Caches

Apart from good density and leakage, current STT-RAM also has good endurance and good read performance [42]. Nevertheless, it is not ideal to replace SRAM with NVMs because they consume high write energy and have long write latencies. To mitigate the drawbacks of individual devices, they are integrated and used as hybrid caches. These designs have several challenges; we discuss the issues and possible solutions in the subsequent sections.

4.1.2 Challenges in Adopting Hybrid Caches

1. *Handling Writes in Hybrid Cache*: Unlike a monolithic cache, a hybrid cache comprising SRAM and NVM has differential write latencies. Since writes are costly in NVM, we need to handle them with care. Migration is often applied by monitoring the read-write operations, directing read-intensive blocks to NVM, and retaining write-intensive blocks in SRAM. Some counters are used to keep track of these operations, and generally, based on threshold values, migrations are triggered. These counters should be designed to be optimal in terms of space and time. We should also take care of demand writes, i.e., when a new block is loaded from memory to cache, write-intensive blocks should be loaded to SRAM and read-intensive to NVM-based cache. Next, care should be taken while handling thrashing blocks; these blocks keep hopping between the two layers and may degrade the performance [42].
2. *Modified cache replacement policies*: Existing cache replacement policies are designed for monolithic caches. However, these policies may not be efficient when adapted directly to hybrid caches because of the disparity in read-write latencies of the two technologies. All the replacement policies concentrate on reducing writes being performed on NVM devices and directing the reads to

NVM as reads are cheaper and nondestructive in those devices. Care should be taken when performing demand writes as well, such that a new block brought to cache from memory should be loaded to hybrid cache and that write-intensive blocks are placed in SRAM.

3. *Data placement and migration*: Data placing is challenging in a hybrid cache because there is a difference in read and write latencies in SRAM and NVM. When a new block is loaded, rather than just placing an analysis about past usage in some part of the cache, type of requests on the block are often considered to make intelligent placement decisions. Placement and migration decisions based on thrashing information also help in minimizing energy consumption and improving performance by reducing the number of LLC misses [42].
4. *Modification in eviction policies*: In monolithic caches, when a block from the last-level cache is evicted, it will be placed in the next level of the memory hierarchy. Sometimes there is a possibility that the same block may be needed shortly; in such cases again, the block has to be loaded from memory. One advantage of a hybrid cache is that the capacity of NVM is much larger than SRAM. Hence, rather than evicting the blocks directly to the next level, some of the blocks which are used frequently can be placed in an NVM-based cache if they are read-intensive because writes are costly in NVM. When evicting blocks, preferences will usually be given for clean blocks in the case of hybrid caches; preferences should be shown for NVM caches because writes in NVM are more costly in terms of energy and time. Several modifications are suggested for the existing cache eviction policies so that both energy and performance are not compromised.

4.2 Challenges in Adapting NVM as a Candidate for Hybrid Memories

The most common architecture discussed and researched in the past years is hybrid memory architecture. Despite the advantages the NVM devices provide, they are not good candidates for replacing DRAM-based memories, at least in the near future, unless and until the write energy and endurance of the NVMs are improvised. Hybrid memories are pretty famous, and they are studied heavily around the globe to be utilized in full-fledged forms in the upcoming computers. However, before adopting these in real-world systems, modifications are required in several areas of the current memories.

1. *Data placement and migration*: Since there is inequality in writing energy consumption and time compared to reading, data placement in hybrid memories is challenging. Hence, most hybrid memory architectures observe static data placement and migration. Access patterns may observe tremendous change during application execution, resulting in poor performance. The reason for this could be more writes being performed in NVM devices. Migration is applied

in hybrid memories to balance the number of writes being performed and improve energy and performance. Migration is performed by monitoring the access patterns of applications through profiling and memory tracing activity. During this process, data objects which are used extensively, termed hot objects, will be directed toward fast memory (DRAM). In turn, objects experiencing less access, termed as cold objects, will be brought to slow but large NVM. The challenge in data migration is to identify the most appropriate candidates for the memory device. A good migration technique should place data such that energy and performance are well optimized. Migrations can be either active or passive in the case of dynamic; migration is performed by scrutinizing the access pattern of applications, whereas passive migrations are triggered when DRAM is full [80]. Apart from identifying the candidates for migration, other challenges are selecting the right proportion for migration, known as the granularity of migration. In most cases, if the migration is performed with the involvement of the operating system, then the granularity of migration will be a page. However, sometimes, it may not be necessary to migrate an entire page because the only portion of the page may be accessed frequently. It may not be ideal for fixing to a common granularity because different applications would benefit from different sizes. An application that exhibits random memory access behavior would benefit from small fragments, while stream-based applications would incur more benefit from large sizes. Hence, rather than fixing static granularity, it is always beneficial to choose dynamic granularity, which changes based on the application behavior [55]. Once the candidates for migration are chosen, and the migration granularity is fixed, the next challenge is when to trigger the migration. The most common approach is to trigger migration based on threshold values; proper analysis should be performed before fixing the threshold. In addition, the good idea is to modify the threshold value based on the application behavior dynamically. Several types of research are conducted to perform migration in a DRAM-NVM-based hybrid; however, there is a scope for improving these techniques.

2. *Handling page replacement in hybrid memory*: A page fault occurs if the page requested by the processor is not available in the memory devices. In such a case, pages should be loaded from secondary storage media. If the memory device does not have sufficient space, then a page from memory should be migrated to secondary storage, creating space for the new page. Identifying the candidate for replacement is a challenging task. Well-known techniques like least recently used (LRU), least frequently used (LFU), first in-first out (FIFO), and round-robin are often used to identify the victim page for a replacement. However, these techniques are ideal for conventional monolithic-based memory designs and cannot be used directly on hybrid memories. In this regard, several modifications are suggested making the replacement techniques suitable for hybrid memories considering the property of the devices. Page replacement techniques can be broadly classified as LRU-based techniques and clock-based techniques. LRU-based techniques use the LRU queue and classify the pages as most recently used and least recently used. Pages that are not used recently are

considered victims for replacement; this technique is widely accepted, but it has some computational drawbacks, such as shifting in the LRU queue to mark the recency of access on every access page. Another major problem with the LRU queue is that it does not keep track of the type of access, but this information is essential in the case of hybrid memories. Rather than just keeping track of recency of access, other information such as the number of writes and reads performed on a page are also essential when using page replacement in hybrid memories. Several modifications are applied to LRU-based techniques to make them suitable for hybrid memories. Application pattern prediction aware LRU (APP-LRU), maintain high hit ratio LRU (MHR-LRU), and double LRU are a few LRU-based techniques that are modified suitably to work with DRAM-NVM-based hybrid memories. Another category of page replacement algorithms is clock based; in these techniques, pointers are used similar to the hands of the clock, hence the name. If a page is accessed recently, but the pointer points to it, indicating it has a victim for replacement, a bit called reference bit is checked; if the value is one, then it will survive the replacement making some other page a victim. Comparatively, these techniques take less time for computation. Once again, even existing clock-based techniques cannot be used directly on hybrid memories. Hence, several modifications are suggested for the primary clock algorithm making it suitable for hybrid memories. Some of the prominent clock-based algorithms which were developed to satisfy the demands of hybrid memory are clock with dirty bits and write frequency (Clock-DWF), clock for page cache in hybrid memory architecture (Clock-HM), migration optimized page replacement algorithm (M-Clock), adaptive classification clock (AC-Clock), and tendency-aware page replacement policy (TA-Clock). These techniques show that both LRU-based and clock-based techniques require some modifications before applying them in hybrid memories with differential read-write properties.

3. *Impact of hybrid memory on cache design*: In Sect. 4.1, we discussed how to use nonvolatile devices as hybrid caches and the challenges faced when using them. If our design choice is to adopt NVM devices in the memory hierarchy keeping the cache intact, certain modifications are required in the cache for performance reasons. One of the areas of cache designs that have several modifications observed due to the adaptation of hybrid memory is the cache replacement policy. In monolithic DRAM-based memories, eviction of the block was straightforward. However, in the case of hybrid memories, a cache block belonging to NVM should be given more priority, and it should be made to stay for a longer duration because fetching a block from NVM is costly compared to DRAM. Retaining blocks from NVM may affect the DRAM blocks because if most of the space is occupied by NVM, there may be significantly less space for DRAM blocks, and giving priority recklessly to NVM blocks may increase thrashing DRAM cache blocks and poor utilization of cache as well. Hence, it is necessary to redesign cache replacement policies for hybrid memories, considering the impact on performance and power consumption. Several modifications are suggested making them suitable for DRAM-NVM-based hybrid memories. Some of the prominent cache replacement policies

designed exclusively for hybrid memories consisting of DRAM and NVM are hybrid memory-aware partition in shared last-level cache (HAP), dynamic adaptive replacement policy in shared last-level cache of DRAM/PCM hybrid memory for big data storage (DARP), miss penalty-aware cache replacement policy (MALRU), victim-aware cache replacement policy for improving NVM lifetime (VAIL), and reuse distance-based victim cache; many more techniques might exist as well. All these techniques ensure that the hit ratios of both the devices are well maintained, and the overall system's performance remains unaffected to a major extent.

4. *Power and performance trade-offs in hybrid memory*: Power plays a vital role in modern systems, which work with huge memory footprints. NVM devices are introduced as a substitute to DRAM because of their low power consumption and ability to support the huge size. However, the setback for these devices is their dynamic energy consumption, primarily because of the high energy consumption for performing write operations. Hence, care should be taken such that power and performance are well balanced. The only way to reduce the dynamic energy consumption of NVM devices is by controlling the number of writes being performed on the device. Some of the techniques which can reduce energy consumption are migration, as already discussed in the previous section, which can help in reducing dynamic memory consumption. Another approach is following intelligent data placement; these techniques ensure that write-intensive data gets placed in DRAM and read-intensive data in NVM. Even following good wear-leveling techniques can enhance the lifetime and dynamic energy consumption. Despite all these techniques, there is still scope for improvement as different applications exhibit different access patterns. Most of the research is directed to one common research that can be carried out to increase the performance while not trading much with the energy.

4.3 Challenges in Adopting NVM as Storage Media

The invention of NVM devices is a boon to memory technology; nevertheless, replacing traditional devices with these new design choices is not easy and often requires modification in several aspects of a memory hierarchy. One major challenge is a modification to the operating system. Adapting NVMs as storage requires some alterations to the existing software, which was developed for HDDs. Some modifications in memory management are also required as the access time of storage becomes closer to that of DRAM [5]. Most designs use it as persistent memory instead of separate storage media.

4.3.1 Changes to the Operating System

Unlike traditional HDD and SDD, current NVM technologies are byte-addressable, indicating that it is possible to target the actual byte rather than accessing data blocks. However, to exploit this advantage, it is necessary to modify the operating system so that we can explore all the benefits of these NVM devices. The challenge is to build the new OS without affecting the existing applications. Generally, operating systems provide data abstractions depending on the type of hardware for which they have been designed. The data storage and access behavior would change with NVM designs; this attracts some modifications to the operating system. Some of the significant changes that are required to work with the new NVM devices are:

1. NVM devices have access latencies comparable with DRAM, while the existing APIs may have latencies longer than access time.
2. Processors can directly access NVM devices with load and store instructions. There is a necessity to redesign system calls to persistent data by allowing programs to work directly on persistent data, and also pointers to persistent data should be designed such that they have a long lifetime similar to the objects they are pointing [8]. Generally, operating system working with conventional hard disk-based storage media makes use of a buffer to accumulate the data and then perform writes to the HDD after certain time limits; in addition, while performing reads, look-ahead read was performed by reading multiple pages ahead of time, but this had a poor impact when NVM devices were used as storage. Even though there is a massive difference in the performance gaps of HDDs and NVM, there was no significant improvement because of the presence of buffers. Hence, when using NVM devices as storage, buffers must be redesigned. Another feature that was adopted by most operating systems is the read-ahead feature, but this can enhance performance only when there is sequential access. However, this did not improve the performance when the STT-RAM and PCM were used as storage media because they can perform faster reads than traditional HDDs. Also, colossal software stack overhead was observed while using NVMs as storage devices. Several kinds of research are currently being carried out such that the operating system supports the use of NVM devices with ease. Also, there were a few suggestions to use NVM devices as a universal memory replacing both memory and storage; however, this could not be achieved proficiently due to a few drawbacks of NVM devices.

4.3.2 Modifications in the File System

Several file systems exist that have been used and are comfortable with underlying storage media, such as hard disk drives. However, when HDDs or SSDs are replaced with emerging NVM devices such as PCM, ReRAM, etc., the existing file systems may not provide good performance. Some designs have adopted NVMs as a cache between memory and secondary storage media to bridge the performance gap. In traditional storage media, maintaining consistency of data synchronization was essential, but this adds additional overhead; hence, NVM caches are used to hold the dirty pages. Another advantage of using these caches is holding the journalling information. HasFS file system was proposed recently to handle file system consistency problems. In this case, metadata is stored in DRAM while actual data is stored in NVM. This file system achieved good performance as it avoided costly access to storage media. Some modifications are suggested to the existing file system to adapt them suitably to the NVM devices [41]. Other than this, file systems such as BPFS, PMFS, and NOVA were also designed and developed to support NVM-based storage media. These systems avoid page cache and block-based I/O software stack. All these file systems follow direct memory access via load-store accesses. When used on NVM devices, conventional file systems need an extra block transition layer that adds additional cost in terms of performance and space. Researches were carried out to study the impact of file systems on commercially available 3D XPoint-based NVMs; this revealed that read performance was good when using either traditional file systems or NVM-aware file systems. However, write performances were optimal with NVM-based file systems when compared to traditional file systems [83]. Some of the factors that may help improve the performance of the file systems are as follows: If the file systems use journaling, some may experience degradation in performance if the directory width is increased. Increasing the number of threads may lead to performance degradation; hence, there should be control over the number of threads. Access granularity has a significant impact on performance; the smaller the granularity, the larger the latency [58]. All these facts must be considered while designing new file systems for NVM-based devices.

5 Current Research Challenges

Though storage class memories have emerged as alternatives for existing memory/storage devices, they have certain pitfalls such as high write energy consumption, limited endurance, and long write latency, and some of them also suffer from security issues. Hence, various studies have been conducted to overcome these drawbacks; this section gives insights into the possible enhancements and solutions which are provided to solve these issues.

5.1 Lifetime Improvement

NVM devices have limited endurance when compared to existing memory devices. This subsection will discuss the different methodologies proposed for improving the lifetime of NVM devices. Nonvolatile memory devices are being used in deep neural networks because of the capacity it provides them. However, there are challenges to adopting them; weights in the neural networks need to be updated periodically, which may affect the endurance of a few cells of the NVM. The lifetime of the NVM cells can be improved by embracing techniques such as wear-leveling, fault-tolerant, and write reduction techniques. Wear-leveling techniques focus on distributing writes evenly among all rows so that few rows do not suffer from endurance limits. To prolong the lifetime of the NVM devices, techniques such as data comparison write, flip-and-write, and dead write bypassing techniques are applied [15].

5.1.1 Wear-Leveling Techniques

Due to the varying behaviors of applications, it is possible that the NVM may be written several times in certain cells, which might affect the endurance. Wear-leveling techniques help balance the writes performed on the rows so that all of them get a balanced number of writes. These techniques can be either proactive or reactive; in the former approach, the wear leveling is always active, while in the latter, they are activated when the number of writes reaches some threshold [60]. Most of the techniques try to balance the writes by keeping track of the number of writes and swapping more frequently written rows with less used ones so that all rows have a uniform distribution of writes. The challenging task here is to optimize the number of counters used to keep track of the number of writes performed.

Wear leveling in NVMs at memory hierarchy Age-aware row swapping is one of the techniques which aim to balance the writes in deep learning applications. Rows that are written several times during training have the possibility of being written in the future as well; such rows are identified and swapped with less frequently written rows. Registers are used to keep track of writes since the number of writes performed on cells is common to a row; the number of counters depends on rows, significantly reducing the number of counters [12].

Wear-leveling in NVMs at cache hierarchy Write variations in the cache can be either inter-set or intra-set. If there is write variation among sets, it is known as inter-set variation, while variations across ways are known as intra-set variations. Since data caches are written heavily than instruction caches, logically reorganizing can be helpful to balance the writes when NVMs are used as caches [60]. When NVMs are used as caches, static window write restriction (SWWR) and dynamic window write restriction techniques are applied to evenly distribute writes by partitioning the cache into windows and restricting the writes to certain windows after they have reached the threshold [1].

5.1.2 Write Reduction Techniques

Wear-leveling techniques try to balance the writes by distributing them evenly among all locations, whereas write reduction techniques concentrate on reducing the write operations themselves. Writes are more costly in NVM than reading operations in terms of energy and latency. Reducing write operations will significantly impact the lifetime of NVMs due to the limitations in endurance. Several approaches have been proposed to reduce writes in NVM, thereby enhancing performance. One of the popular write reduction techniques is flip-and-write: in this case, before performing a write operation, existing data and data to be written are compared. If more than N/2 bits are different, the flip operation is performed, and the flip bit is set. After this operation, only the bits that need to be updated are written. This operation was able to reduce write, thereby increasing both performance and endurance [18]. Another method that was adopted was two-stage write: in this case, all 0 s are written together, and all 1 s are written together, performing the writes in two stages. Flip-and-write was also adopted if more than N/2 bits had the value 1 to be written. The advantage of this approach is that 0 s are written faster. Flip-mirror rotate is yet another approach that concentrates on reducing the writes in nonvolatile memories. Though the flip-and-write approach reduced the writes, it may not be efficient if less than N/2 bits require updates. Adaptive flip-write combines flip and write with compression techniques by dynamically adapting data width. The second method applied in this technique is performing a mirroring operation before writing. This will further reduce the number of bits to be written [48]. In devices like PCM, standard write units followed were 4, 8, or 16 bits; hence, to complete one cache block write, several series of writes were necessary; this increases the latency for write operations and impacts the performance. In some applications generally, for data structure alignment, multiple 0 s might be appended; such data is known as zero-extended values. Min-WU (minimize write units) tries to reduce the number of write units by applying data coding techniques. Another approach this method follows is by encapsulating more bits in write operations such that the number of serial writes required to complete one cache write is lowered. Employing this method helps in reducing the time consumed for writing operations.

5.1.3 Error Correction

Another method to improve the lifetime of NVM devices is the development of error correction techniques. These techniques focus on improving reliability as well as limited endurance. Nonvolatile memories suffer from permanent and transient failures due to manufacturing defects, decreased feature size, or multi-bit storage. Sometimes they may suffer from hard error, which means the cell is damaged permanently and cannot be used in the future. At the same time, transient errors can be rectified by reprogramming the cell. If these errors are not corrected, they may affect the reliability of the applications. Irrespective of the type of NVM technology, they may become victims of these errors. Conventional error correction codes were not effective when applied to NVMs because of the memory overhead; hence, there was a necessity to build robust error correction techniques. This was an active area of research for several years. Hard error correction techniques can be generalized to different technologies; however, soft error correction techniques need to be modified as per the underlying NVM. In conventional technologies, hard errors were mainly caused due to manufacturing defects; however, in NVM devices, excessive writes may cause hard errors [66]. The most common methods that are followed to reduce hard errors are dynamically replicated memory (DRM), error-correcting pointers (ECP), SAFER, FREE-p, Pay-As-you-go, and Zombie. The major challenge in these techniques is to implement these with minimal overhead. ECP emerged as an excellent alternative to ECC; this technique used pointers to point to the failed cell and save the correct values. The advantage of this approach is that it uses a fixed number of pointers to address the problem in a given block. The major drawback of this approach is that it uses a fixed number of ECPs for the entire memory; however, some of these may remain unutilized. If a page contains weak rows (i.e., rows that experience many failures), then using 6 ECP may not be sufficient, and this may result in failure of the page even though several rows exist on the page that is still viable for use. Pay-As-you-go and zombie are ECP-based techniques that are improvised with added latency cost [67]. To overcome the drawback of limited pointer availability of ECP, another major technique was proposed for handling hard errors in single-bit, multi-bit, and triple-bit cells, which is error-correcting strings. In this approach, variable-length offsets are used instead of fixed-length pointers; this enhances the ability to handle more errors. Compared to ECP, this method can correct more errors with less overhead because it uses offsets in error correction.

5.2 Multilevel Cell

Most of the NVM devices support multi-bit storage per cell called multilevel cell (MLC); this subsection details this property. NVM devices exhibit multiple resistive states; this can be exploited to store multiple bits. The advantage of MLC is that it provides high density. Several types of research are carried out to reduce the physical dimensions of NVM devices so that density can be improved. However, this requires modifications to the fabrication methodology; an alternative to this is MLC, i.e., store multiple bits in an individual cell without reducing its physical dimensions. Devices like PCM, ReRAM, STT-RAM, and NRAM have the potential to store multiple bits per cell, but this is possible by varying multiple high resistance and low resistance instead of single high resistance and low resistance. However, to achieve this, a proper control over different resistance levels is essential; otherwise, the device may suffer from resistance variability as well as reliability issues [77]. Multilevel storage in NVM devices can be achieved in many ways and will be discussed in the subsequent subsections.

5.2.1 Multilevel Cell Property in PCM

PCM supports multiple-bit storage in a single cell. The huge disparity in resistance between the crystalline and amorphous states helps in intermediate resistance states. However, it is challenging to store multiple bits because of the reliability issues. The major hindrance to this problem is resistance drift and variability. The resistance drift problem also decreases the retention power of the cell. It is observed that if the write latency is longer, then such writes have a longer retention time, whereas smaller latency writes have less retention time. Hence, a small trade-off between performance and retention time is essential while dealing with MLC, because MLC is more vulnerable to resistance drifts than SLC. Hence, before adopting this property, there is a necessity to make suitable adaptations so that reliability will not affect the performance. Hot data blocks can survive low retention as they get accessed and updated frequently, whereas, cold data blocks should experience long latency writes. Techniques like region retention monitor (RRM) can be used to monitor the number of access being performed on the data blocks and set appropriate write latency [79]. Another approach that is followed for improving the retention time of short-latency writes is quick and dirty write (QnD). This technique can be applied if the applications experience heavy write operations only during certain periods. Short-latency writes are performed when the memory controller has too many writes to handle. In the duration where the memory controller is idle or performing no write operations, long latency writes are issued, once again thereby refreshing the short-latency writes [81]. There is always a trade-off between latency and retention time in MLC of PCM; hence, there should be intelligent techniques to deal with these issues. Multi-level PCM cells are more suitable to be used in storage media [53]. MLC support for PCM is proved in 1T1R array, but there are limitations due to the difficulty in scaling transistors and other problems regarding read-write verification. To solve this issue, MLC in PCM is supported by using ovonic threshold switch (OTS)-based cross-point array. OTS-based PCM cross-point array does not use transistor and is free from read-write verification problem, but the challenge is to manage unselected cells while selected cells are programmed, because there are no transistors to select a particular group of cells. Tight distribution of memory window can help in solving this issue. Methods like open loop programming are used to avoid read-write verification problem. Another major issue even in OTS-based MLC is threshold voltage (vt) drift. To solve this issue, optimized programming methods can be used [27]. Tremendous research is underway to solve the issues in MLC which can provide promising results in the future.

5.2.2 Multilevel Cell Property in ReRAM

Multiple bits can be stored in ReRAM either by controlling the reset voltage, changing the compliance current, or altering the pulse width of the program/erase operation. From the experiments, it was evident that ReRAM can support up to eight resistance states in a single cell [77]. ReRAMs constructed using oxides can also support MLC property. In most cases, MLC operations are validated at device, while operations at circuit and system level were less explored. When using oxide-based ReRAM, a huge variation is observed in MLCs concerning HRS and LRS; to handle this multiple-step programming can be helpful. By varying the voltage during a ReSET operation, multiple bits can be stored, and a recent work tried to terminate the reset operation by itself to obtain different high resistance states. The advantage of this approach is that it can achieve MLC without read-verify [4]. Another issue faced by MLC in ReRAM is retention failure; sudden transitions may change the state of the cell from low resistance to high resistance, known as LRS retention failure, and vice versa, known as HRS retention failure. Reprogramming the cell may resume the normal state. A proper analysis of materials, resistance allocation schemes, and programming strategies can help in making decisions regarding retention failures [74].

5.2.3 Multilevel Cell Property of STT-RAM

Multi-level cell support is possible in STT-RAM if the magnetic tunnel junctions support four or more resistance levels. The major problem for support of MLC is reliability and performance. These can be affected by process variations in the design of MTJs and MOS and due to the thermal fluctuations during the switching process [78]. Though MLCs support good density compared to single-level cells (SLC), they suffer from read and write disturbance issues. The challenge is to explore the advantage of MLC by solving these issues. Two popular designs which support MLC in STT-RAM are serial and parallel designs. In serial design, a small and a large MTJ are placed vertically above the other. Small MTJ is referred to as soft bit and large MTJ as hard bit. Writing hard bits of the MLC consumes more energy. It may also disturb the value stored in the corresponding soft bit line; hence, some modification is required to mitigate this problem before utilizing MLC. Techniques like word-splitting schemes are used to reduce the costly writes to the hard region [15], and one-step writes are proposed by Zhao et al. [82] instead of two-step writes to improve the lifetime of MLCs.

5.2.4 Multilevel Cell of FeRAM

Compared to other NVM devices, multi-bit storage in FeRAM is not explored much yet; however, there are few shreds of evidence supporting multi-bit storage capacity. Many shreds of evidence show multi-bit storage capacity in ferro electric-based field-effect transistors (FeFET) [22]. Multi-bit cells in ferro electric-based devices can be realized by using very thin films of ferroelectric oxides [7]. In the recent work carried out by Kim et al. [37], a FeRAM cell structure was developed by substituting 1T1C by 1T1T structure; this structure could perform nondestructive reads as well supported data storage up to 5 bits per cell. In the future, we can expect more innovation in this regard.

5.3 Accelerators

Some previous works have used NVMs to perform computations rather than using them only for storage. This subsection will provide details about the use of NVMs as accelerators. Research has shown that PCM [11], STT-RAM [69, 70], and ReRAM [28, 39] are capable of performing operations and computations in addition to their capability of storing data. When devices are capable of performing some kind of computing, then they can be used as accelerators in memory which is sometimes referred as in-memory computing. ReRAM is said to exhibit crossbar array structure which makes it an excellent candidate for processing matrix-vector multiplication. The crossbar structure of ReRAM cell is shown in Fig. 10. Owing to this nature of ReRAM, it is used as accelerator in many researches [14, 36, 51]. There are also evidences where ReRAM is being used to perform computations in neural network as well as convolution neural networks (CNN); more details about this can be found in studies by Chi et al. [16] and Shafiee et al. [59]. There were problems associated with training and weight updating when ReRAM was used for CNN computation; to overcome this drawback, Song et al. [61] developed networks utilizing the advantage of intra-level parallelism, thereby creating architectures which could enhance performance of CNN at both the stages of training and inference. Apart from neural network, ReRAMs are also used as accelerators in Boltzmann machine [9]. In addition, ReRAM-based accelerators are used in graph algorithms especially for performing sparse matrix-vector multiplication [62]. A more detailed survey of using ReRAM in neural networks as accelerators can be found in a study by Mittal [45]. Other than ReRAM, recently PCM-based devices are being explored as well. There are few evidences of PCM usage in accelerators that can be explored [34, 57, 63]. Conductance drift is a major hindrance for usage of PCM as an accelerator

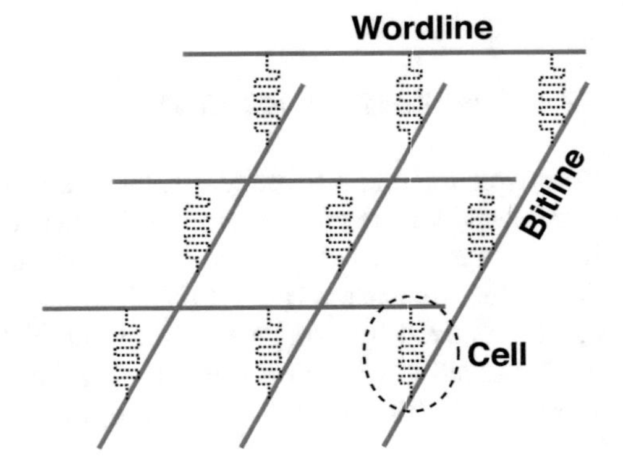

Fig. 10 Crossbar structure of ReRAM cells [33]

time-dependent amplification can help to mitigate this issue. Despite this challenge PCM is a good candidate as an accelerator.

5.4 Security

This subsection will give details about how to handle the issues related to security in NVM devices. The major drawback of the NVM device is the limited endurance; attackers take this as an advantage and write the NVM cell several times, thus making them weak. To avoid this problem, there is a necessity to develop techniques to enhance the security of NVM devices. Another property of NVM-based systems is their long retention time; an attacker with physical access to the system can readily scan the main memory content and extract all valuable information from the main memory. This vulnerability can be addressed by memory encryption using a dedicated AES engine, but it comes with extra concerns. First is that it should be fast enough during encryption and provide an instant response when unlocked. Second, it should be energy-efficient considering the limited battery life. Hence, a new alternative of AES, AIM, has been introduced by Xie et al. [73] for fast and energy-efficient NVM encryption that will encrypt the whole/part memory only when it is required. AIM leverages the nondestructive read in NVMs for performing efficient XOR operations, which dominate AES. AIM utilizes the intrinsic logic operation capability of NVMs to implement the AES task and provide a bandwidth-intensive encryption application by doing the AES procedure in place. Mao et al. [43] explored the wear-leveling attack in NVMs such as PCM and ReRAM. The attacker can wear out NVMs using the row buffer hit latency. Latency can reveal LAs (logical addresses) mapped to the same physical row. Thus, an attacker can discover a group of LAs mapped to a specific physical row and keep detecting it using latency. The attacker can then figure out the new LA mapped (for wear leveling) to the same physical address written last, therefore cracking the protection of wear leveling. To solve this attack, intra-row swap (IRS) was introduced, which would hide the wear-leveling details. The basic idea is to enable an additional intra-row block swap when a new logical address is remapped to the memory row. It is observed that the cells of NVM devices have varying endurance. Some of the attackers use this; they try to write all the rows of the device uniformly several times, which may make a few rows weak and finally affect the device. These attacks are known as uniform address attacks (UAA). These kinds of attacks will impact the wear-leveling schemes as well. To avoid such attacks, techniques like MAX-WE have been employed. This technique tries to maximize the endurance of weak cells and protects UAA [75]. In deep learning applications, security can be breached by the optimizer as it repeatedly updates the common location, thereby affecting the cell's endurance. As the training phase incurs more writes it can be victim of such attacks. Generally, attacks are carried out in two steps: localization and targeting. In the first step, rows that are to be targeted are identified, and in the second step, these rows are written several times, hampering the endurance and thereby affecting the security. Techniques like random age-aware row swapping are used to confuse the attackers by randomizing the rows to be swapped because following a typical pattern for swapping may attract attackers to perform easy attacks [12]. Security is also an essential factor to consider in applications. Though several software-based approaches exist to ensure application security, hardware-based security can also be used. The random switching mechanism used in ReRAM devices is less vulnerable to data attacks compared to conventional devices. ReRAM devices are used in key authentication and generation. However, the use of NVM devices to enhance security is still under research, and much progress is expected in the near future [77].

6 Application of NVM Devices in IoT and AI

In the previous sections, we have discussed the different NVM devices and how they can be integrated to different levels of memory hierarchy. In this subsection we will shed some light on how NVM devices have changed the way of processing in emerging fields like artificial intelligence (AI), big data, and Internet of Things (IoT). Initially, IoT was not able to process complex information; however, with the improvement of processors in IoT devices, they can now process complex information, including AI. When such complex information can be processed in IoT devices, they are referred to as AIoT. However, these IoT devices are generally battery-based devices where power consumption is an essential factor. Most of the time, IoT devices operate on standby mode; in such cases, it is hard to satisfy the needs only with DRAM-based devices due to their power consumption.

Nevertheless, it is not a good idea to completely replace DRAM with NVM devices because of the issues discussed previously; hence, hybrid architectures are adopted. By adopting specific changes to exploit the best of power and performance, NVM-based hybrid memories can satisfy the need for enormous memory with low power consumption [65]. In the previous sections, we discussed how NVM could be used as cache devices. In the processors meant for processing AI applications, NVM devices like STT-RAM are used as caches. The advantage of using NVM as cache in AIoT-based devices is that they enhance the performance with reduced power consumption, which is essential for those systems. Despite these advantages, long write latencies of STT-RAM may hamper the performance; to avoid this, certain modifications are essential for the existing SRAM-based policies. Techniques like adaptive nonvolatile cache prefetch (ANCP) are developed to handle the issues faced by STTRAM devices. Prefetching techniques generally increase the writes performed on the cache, which may adversely affect performance. To avoid this, stream-based prefetching was adopted in ANCP, which improved the performance slightly [47]. In-memory computing is often used in edge computing to reduce the time spent in exchange of information between processing elements and storage. Most of the AI edge devices perform multiply and accumulate (MAC) operation. NVM devices like PCM-, ReRAM-, and MRAM-based memories like STT-MRAM

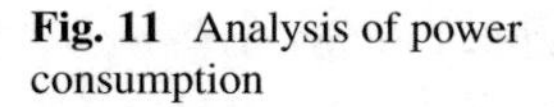

Fig. 11 Analysis of power consumption

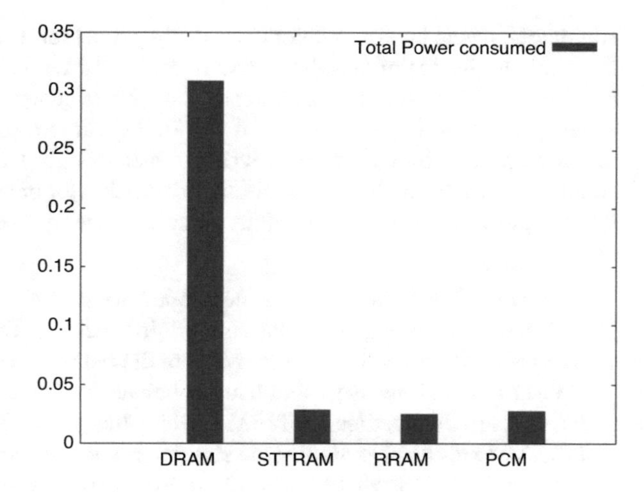

have been used as they can provide precise results with low energy consumption and improved performance. Figure 11 depicts the total power consumed by various memory devices for a common application program. From the figure it is evident that NVM devices have very low power consumption compared to DRAM, giving the proof that they can be right candidates for devices which have constraints on power consumption. However, this is an active area of research, and many solutions can be explored.

7 Simulators

Generally, computer architecture research uses various simulation tools to analyze the performance and power consumption and check the system's behavioral changes with new modifications and adoptions. Rather than making changes directly in the hardware device, it is advisable to use simulators. This section details about the different simulators used and how one can use them in their research. Since most of the NVM devices are still prototype devices, simulators are used to analyze the behavior of NVM devices. A plethora of simulation tools are currently available, and we should choose the tool that satisfies our requirements accordingly. Simulators can be broadly classified as functional simulators, timing simulators, and integrated timing and functional simulators based on the simulation detail. Functional simulators are more like an emulator; they try to imitate the behavior of ISAs, but they do not work at the micro-architectural level. However, they will give us the details regarding the program's behavior. Timing simulators give the details at the micro-architectural level. These simulators will be able to give details performance of memory systems, time taken by a program for execution, etc. Functional simulators can be either cycle level simulators or event-driven simulators. Cycle level simulators imitate the processor's behavior in each cycle, but they cannot give the minute-level details associated with the hardware. On the other hand, event-driven simulators try to simulate the behavior of the system based on the events [2]. Depending on the type of research we are involved in, we should choose the appropriate simulators. However, another category of simulators is memory simulators, which can be used to simulate the behavior of access to memory devices. Given below is the list of simulators that are widely used in the study of NVM devices.

1. *NVMain*: This simulator was developed to study the behavior of emerging NVM devices along with the support for existing DRAM. While simulators like DRAMSim can be used to evaluate the behavior of DRAM-based memory, NVMain will allow us to simulate the behavior of emerging NVM devices. This has built-in config files for NVM devices like PCM, STT-RAM, ReRAM, and DRAM. One can also modify the simulator; add new config files to support other types of NVM provided we have well-defined data sheet. It uses the memory traces of the applications as input and provides details with respect how the reads and writes are distributed in various levels of memory such as channels, ranks, and banks giving performance details as well as power details. It also has support for simulating the behavior of hybrid memories that combine DRAM and NVM. If we are interested in simulation of benchmark programs, then it can be integrated with other simulators. Though it is integrated with many simulators, most of the researches have used it by integrating gem5. More details about the simulator and its usage can be found in a study by Poremba and Xie [49].
2. *Gem5*: A modular platform for computer -system architecture research, encompassing system-level architecture as well as processor micro-architecture. Though it was initially developed for academic purpose, now it is being used by both academic and industry experts for computer design research. It is an open-source architectural-level simulator which supports simulation in full system mode and system emulation mode. It has well-supported documentation and tutorial that help the beginners to gain insights into the simulator. More details of the simulator can be found in https://www.gem5.org. Though early versions of gem5 did not support NVM simulation, latest versions do have.
3. *NVSim*: This a circuit-level simulator which can be used to analyze the behavior of emerging nonvolatile memory devices. It supports devices such as PCM, ReRAM, STT-RAM, and legacy NAND Flash. It can be used to optimize the design metrics or to evaluate the performance, area, or energy of the NVM devices [21]. Unlike gem5 and NVMain, this simulator can be used only to analyze the changes in the circuit level. Suppose we want to analyze the impact of changing the cell size on area efficiency, power, and other factors; then it would be the ideal choice.
4. *DESTINY*: This enables to study the behavior of both conventional and emerging nonvolatile memory devices in 2D as well as 3D. 3D stacking helps to improve the density as well as flexibility in power of the devices. This simulator is an

ideal tool for those researchers who want to make a detailed study of the devices in 3D [50].

5. *NVM Streaker*: This simulator can be used for simulation of big data applications especially using Spark-based applications. To compensate the time consumption observed in simulation of big data applications while using NVM, this simulator utilizes DRAM-based accesses and modifies it suitably to adapt to the NVM. This simulator provides good performance for big data applications in NVM devices [29].
6. *VANS*: Validated cycle accurate NVRAM simulator (VANS) helps to simulate the nonvolatile memory systems. It can integrate with other full system simulators like gem5 as well. It is an architectural-level simulator which can be used to simulate the performance and experiment with different architecture designs associated with NVM devices. Though it is experimented with optane, it is said to support other NVM devices as well [72]. This is just a list of simulators that are widely used by researches; however, there are many more simulators which are not listed here. Readers need not restrict yourself to use only these simulators; you can always explore the available list and choose as per your requirement. We have given the list by integrating the simulators which are widely used in our reference materials.

8 Conclusions and Future Scope

Throughout the chapter, we have discussed how NVM devices are bringing the paradigm shift in memory organization. However, specific issues need to be addressed before putting these NVM devices to full-fledged use. ReRAM devices are facing issues with reliability; hence, there is a necessity to develop techniques to detect the operation failure of the device. Though ReRAM has good endurance, it is not able to reach up to the level of DRAM [77]. Multi-bit storage capacity is an added advantage for all nonvolatile memory devices, but the challenge is handling writes and making them more reliable. Hence, there is a scope for developing techniques that can improvise this facility of NVM devices. Nevertheless, another challenge is that most enhancements and progress are limited to individuals or a few nonvolatile memory devices. Devices like NRAM are still not explored much; hence, there is scope for improving these devices. Most of the NVM devices support multi-level storage capacity, which helps to enhance the density without changing the physical dimensions. However, this adds complications; many types of research are carried out in this regard, yet it remains an active area of research. The major drawback of the FeRAM device is the destructive reads. To enhance this property, recently, a cell structure with 1T1T was proposed by Kim et al. [37], and this has opened up several kinds of research with FeRAM-based devices. Once again, MLC in ReRAM has much scope for improvising as it is in the early stages of development [4]. Ultimately, there is no doubt that future computing devices will all have NVM-based devices either in the form of cache, memory, or storage.

References

1. Agarwal, S., Kapoor, H.K.: Improving the lifetime of non-volatile cache by write restriction. IEEE Trans. Comput. **68**(9), 1297–1312 (2019). https://doi.org/10.1109/TC.2019.2892424
2. Akram, A., Sawalha, L.: A survey of computer architecture simulation techniques and tools. IEEE Access. **7**, 78120–78145 (2019). https://doi.org/10.1109/ACCESS.2019.2917698
3. Aswathy, N., Sivamangai, N.: Future nonvolatile memory technologies: challenges and applications. In: 2021 2nd International Conference on Advances in Computing, Communication, Embedded and Secure Systems (ACCESS), pp. 308–312 (2021). https://doi.org/10.1109/ACCESS51619.2021.9563288
4. Aziza, H., Hamdioui, S., Fieback, M., Taouil, M., Moreau, M.: Density enhancement of RRAMS using a reset write termination for MLC operation. In: 2021 Design, Automation Test in Europe Conference Exhibition (DATE), pp. 1877–1880 (2021). https://doi.org/10.23919/DATE51398.2021.9473967
5. Bahn, H., Cho, K.: Implications of NVM based storage on memory subsystem management. Appl. Sci. **10**(3) (2020). Retrieved from https://www.mdpi.com/2076-3417/10/3/999, https://doi.org/10.3390/app10030999
6. Banerjee, W.: Challenges and applications of emerging nonvolatile memory devices. Electronics **9**(6) (2020). Retrieved from https://www.mdpi.com/2079-9292/9/6/1029, https://doi.org/10.3390/electronics9061029
7. Baudry, L., Lukyanchuk, I., Vinokur, V.M.: Ferroelectric symmetry-protected multibit memory cell. Sci. Rep. **7**(1), 42196 (2017, February 08). Retrieved from https://doi.org/10.1038/srep42196
8. Bittman, D., Alvaro, P., Mehra, P., Long, D.D.E., Miller, E.L.: Twizzler: a data-centric OS for non-volatile memory. In: 2020 USENIX Annual Technical Conference (USENIX ATC 20) (2020, July)
9. Bojnordi, M.N., Ipek, E.: Memristive Boltzmann machine: a hardware accelerator for combinatorial optimization and deep learning. In: 2016 IEEE International Symposium on High Performance Computer Architecture (HPCA), pp. 1–13 (2016, March). https://doi.org/10.1109/HPCA.2016.7446049
10. Bolotin, E., Nellans, D., Villa, O., O'Connor, M., Ramirez, A., Keckler, S.W.: Designing efficient heterogeneous memory architectures. IEEE Micro. **35**(4), 60–68 (2015)
11. Burr, G.W., Shelby, R.M., Sidler, S., di Nolfo, C., Jang, J., Boybat, I., et al.: Experimental demonstration and tolerancing of a large-scale neural network (165 000 synapses) using phase-change memory as the synaptic weight element. IEEE Trans. Electron Devices. **62**(11), 3498–3507 (2015, Nov). https://doi.org/10.1109/TED.2015.2439635
12. Cai, Y., Lin, Y., Xia, L., Chen, X., Han, S., Wang, Y., Yang, H.: Long live time: improving lifetime and security for NVM-based training-in-memory systems. IEEE Trans. Comput. Aided Design Integr. Circuits Syst. **39**(12), 4707–4720 (2020). https://doi.org/10.1109/TCAD.2020.2977079
13. Chen, Y.: RERAM: history, status, and future. IEEE Trans. Electron Devices. **67**(4), 1420–1433 (2020). https://doi.org/10.1109/TED.2019.2961505
14. Chen, Z., Gao, B., Zhou, Z., Huang, P., Li, H., Ma, W., … Chen, H.: Optimized learning scheme for grayscale image recognition in a rram based analog neuromorphic system. In: 2015 IEEE International Electron Devices Meeting (IEDM), pp. 17.7.1–17.7.4 (2015, December) https://doi.org/10.1109/IEDM.2015.7409722
15. Chen, X., Wang, J., Zhou, J.: Promoting MLC STT-ram for the future persistent memory system. In: 2017 IEEE 15th International Conference on Dependable, Autonomic and Secure Computing, 15th International Conference on Pervasive Intelligence and Computing, 3rd International Conference on Big Data Intelligence and Computing and Cyber Science and Technology Congress (DASC/PICOM/ Datacom/ Cyberscitech), pp. 1180–1185 (2017). https://doi.org/10.1109/DASC-PICom-DataCom-CyberSciTec.2017.189

16. Chi, P., Li, S., Xu, C., Zhang, T., Zhao, J., Liu, Y., ... Xie, Y.: Prime: a novel processing-in-memory architecture for neural network computation in RERAM-based main memory. In: Proceedings of the 43rd International Symposium on Computer Architecture, pp. 27–39. IEEE Press, Piscataway (2016). Retrieved from https://doi.org/10.1109/ISCA.2016.13
17. Chiu, C.-H., Huang, C.-W., Hsieh, Y.-H., Chen, J.-Y., Chang, C.-F., Chu, Y.-H., Wu, W.-W.: In-situ tem observation of multilevel storage behavior in low power FERAM device. Nano Energy **34**, 103–110 (2017). Retrieved from https://www.sciencedirect.com/science/article/pii/S2211285517300794, https://doi.org/10.1016/j.nanoen.2017.02.008
18. Cho, S., Lee, H.: Flip-n-write: a simple deterministic technique to improve pram write performance, energy and endurance. In: 2009 42nd Annual IEEE/ACM International Symposium on Microarchitecture (Micro), pp. 347–357 (2009).
19. Daulby, T., Savanth, A., Weddell, A.S., Merrett, G.V.: Comparing NVM technologies through the lens of intermittent computation. In: Proceedings of the 8th International Workshop on Energy Harvesting and Energy-Neutral Sensing Systems, pp. 77–78. Association for Computing Machinery, New York (2020). Retrieved from https://doi.org/10.1145/3417308.3430268
20. Ding, K., Chen, B., Chen, Y., Wang, J., Shen, X., Rao, F.: Recipe for ultrafast and persistent phase-change memory materials. NPG Asia Mater. **12**(1), 63 (2020, September 25). Retrieved from https://doi.org/10.1038/s41427-020-00246-z
21. Dong, X., Xu, C., Xie, Y., Jouppi, N.P.: NVSIM: a circuit-level performance, energy, and area model for emerging nonvolatile memory. IEEE Trans. Comp-Aid. Des. Integr. Circuits Syst. **31**(7), 994–1007 (2012). https://doi.org/10.1109/TCAD.2012.2185930
22. Fey, D., Reuben, J., Slesazeck, S.: Comparative study of usefulness of FEFET, FTJ and RERAM technology for ternary arithmetic. In: 2021 28th IEEE International Conference on Electronics, Circuits, and Systems (ICECS), pp. 1–6 (2021). https://doi.org/10.1109/ICECS53924.2021.9665635
23. Fong, S.W., Neumann, C.M., Wong, H.-S.P.: Phase-change memory—towards a storage-class memory. IEEE Trans. Electron Devices. **64**(11), 4374–4385 (2017). https://doi.org/10.1109/TED.2017.2746342
24. Gamatié, A., Nocua, A., Weloli, J.W., Sassatelli, G., Torres, L., Novo, D., Robert, M.: Emerging NVM Technologies in Main Memory for Energy-Efficient HPC: an Empirical Study (2019, May). Retrieved from https://hal-lirmm.ccsd.cnrs.fr/lirmm-02135043 (working paper or preprint)
25. Gilmer, D.C., Rueckes, T., Cleveland, L., Viviani, D.: Nram status and prospects. In: 2017 IEEE International Conference on IC Design and Technology (ICICDT), pp. 1–4 (2017, May). https://doi.org/10.1109/ICICDT.2017.7993504
26. Gilmer, D. C., Rueckes, T., Cleveland, L.: NRAM: a disruptive carbon-nanotube resistance-change memory. Nanotechnology **29**(13), 134003 (2018, February). Retrieved from https://doi.org/10.1088/1361-6528/aaaacb
27. Gong, N.: Multi level cell (MLC) in 3D crosspoint phase change memory array. Sci. China Inf. Sci. **64**, 166401 (2021). https://doi.org/10.1007/s11432-021-3184-5
28. Hu, M., Li, H., Wu, Q., Rose, G. S.: Hardware realization of BSB recall function using memristor crossbar arrays. In: Proceedings of the 49th Annual Design Automation Conference, pp. 498–503. ACM, New York (2012). Retrieved from https://doi.org/10.1145/2228360.2228448
29. Hu, D., Lv, F., Wang, C., Cui, H.-M., Wang, L., Liu, Y., Feng, X.-B.: NVM streaker: a fast and reconfigurable performance simulator for non-volatile memory-based memory architecture. J. Supercomput. **74**(8), 3875–3903 (2018, August 01). Retrieved from https://doi.org/10.1007/s11227-018-2438-y
30. Iuga, A., Lindfors-Vrejoiu, I., Boni, G.: Ultrafast nondestructive pyroelectric reading of feram memories. Infr. Phys. Technol. **116**, 103766 (2021). Retrieved from https://www.sciencedirect.com/science/article/pii/S1350449521001389, https://doi.org/10.1016/j.infrared.2021.103766
31. Jin, H., Chen, D., Liu, H., Liao, X., Guo, R., Zhang, Y.: Miss penalty aware cache replacement for hybrid memory systems. IEEE Trans. Comp-Aid. Des. Integr. Circuits Syst. **39**(12), 4669–4682 (2020). https://doi.org/10.1109/TCAD.2020.2966482

32. Kallinatha, H.D., Talawar, B.: Comparative analysis of non-volatile memory on-chip caches (2022)
33. Kamath, A.K., Monis, L., Karthik, A.T., Talawar, B.: Storage class memory: principles, problems, and possibilities. arXiv (2019). Retrieved from https://arxiv.org/abs/1909.12221, https://doi.org/10.48550/ARXIV.1909.12221
34. Kariyappa, S., Tsai, H., Spoon, K., Ambrogio, S., Narayanan, P., Mackin, C., et al.: Noise-resilient DNN: tolerating noise in PCM-based AI accelerators via noise-aware training. IEEE Trans. Electron Devices. **68**(9), 4356–4362 (2021). https://doi.org/10.1109/TED.2021.3089987
35. Kim, T., Lee, S.: Evolution of phase-change memory for the storage class memory and beyond. IEEE Trans. Electron Devices. **67**(4), 1394–1406 (2020). https://doi.org/10.1109/TED.2020.2964640
36. Kim, Y., Zhang, Y., Li, P.: A reconfigurable digital neuromorphic processor with memristive synaptic crossbar for cognitive computing. J. Emerg. Technol. Comput. Syst. **11**(4), 38:1–38 (2015, April). Retrieved from http://doi.acm.org/10.1145/2700234
37. Kim, S., Sun, J., Choi, Y., Lim, D.U., Kang, J., Cho, J.H.: Carbon nanotube ferroelectric random access memory cell based on omega-shaped ferroelectric gate. Carbon **162**, 195–200 (2020). Retrieved from https://www.sciencedirect.com/science/article/pii/S0008622320301901, https://doi.org/10.1016/j.carbon.2020.02.044
38. Kokolis, A., Skarlatos, D., Torrellas, J.: PageSeer: using page walks to trigger page swaps in hybrid memory systems. In: 2019 IEEE International Symposium on High Performance Computer Architecture (HPCA), pp. 596–608 (2019)
39. Li, B., Shan, Y., Hu, M., Wang, Y., Chen, Y., Yang, H.: Memristor-based approximated computation. In: Proceedings of the 2013 International Symposium on Low Power Electronics and Design, pp. 242–247. IEEE Press, Piscataway (2013). Retrieved from http://dl.acm.org/citation.cfm?id=2648668.2648729
40. Liang, Y.-P., Chen, S.-H., Chang, Y.-H., Liu, Y.-F., Wei, H.-W., Shih, W.-K.: A cache consolidation design of MLC STT-ram for energy efficiency enhancement on cyber-physical systems. SIGAPP Appl. Comput. Rev. **21**(1), 37–49 (2021). Retrieved from https://doi.org/10.1145/3477133.3477136
41. Liu, Y., Li, H., Lu, Y., Chen, Z., Xiao, N., Zhao, M.: HASFS: optimizing file system consistency mechanism on NVM-based hybrid storage architecture. Clust. Comput. **23**(4), 2501–2515 (2020, December 01). Retrieved from https://doi.org/10.1007/s10586-019-03023-y
42. Luo, J.-Y., Cheng, H.-Y., Lin, I.-C., Chang, D.-W.: Tap: reducing the energy of asymmetric hybrid last-level cache via thrashing aware placement and migration. IEEE Trans. Comput. **68**(12), 1704–1719 (2019). https://doi.org/10.1109/TC.2019.2917208
43. Mao, H., Zhang, X., Sun, G., Shu, J.: Protect nonvolatile memory from wear-out attack based on timing difference of row buffer hit/miss. In: Design, Automation Test in Europe Conference Exhibition (DATE), 2017, pp. 1623–1626 (2017, March). https://doi.org/10.23919/DATE.2017.7927251
44. Meena, J.S., Sze, S.M., Chand, U., Tseng, T.-Y.: Overview of emerging nonvolatile memory technologies. Nanoscale Res. Lett. **9**(1), 526 (2014, September 25). Retrieved from https://doi.org/10.1186/1556-276X-9-526
45. Mittal, S.: A survey of ReRam-based architectures for processing in-memory and neural networks. Mach. Learn. Knowl. Extr. **1**(1), 75–114 (2018). Retrieved from http://www.mdpi.com/2504-4990/1/1/5, https://doi.org/10.3390/make1010005
46. Mittal, S., Vetter, J.S.: A survey of software techniques for using non-volatile memories for storage and main memory systems. IEEE Trans. Parallel Distrib. Syst. **27**(5), 1537–1550 (2016). https://doi.org/10.1109/TPDS.2015.2442980
47. Ni, M., Chen, L., Hao, X., Sun, H., Liu, C., Zhang, Z., ... Pan, L.: A novel prefetching scheme for non-volatile cache in the AIOT processor. In: 2020 5th International Conference on Universal Village (UV), pp. 1–7 (2020). https://doi.org/10.1109/UV50937.2020.9426214

48. Palangappa, P.M., Mohanram, K.: Flip-mirror-rotate: an architecture for bit-write reduction and wear leveling in non-volatile memories. In: Proceedings of the 25th Edition on Great Lakes Symposium on VLSI, pp. 221–224. Association for Computing Machinery, New York (2015). Retrieved from https://doi.org/10.1145/2742060.2742110
49. Poremba, M., Xie, Y.: NV Main: an architectural-level main memory simulator for emerging non-volatile memories. In: 2012 IEEE Computer Society Annual Symposium on VLSI, pp. 392–397 (2012). https://doi.org/10.1109/ISVLSI.2012.82
50. Poremba, M., Mittal, S., Li, D., Vetter, J.S., Xie, Y.: Destiny: a tool for modeling emerging 3D NVM and EDRAM caches. In: 2015 Design, Automation Test in Europe Conference Exhibition (DATE), pp. 1543–1546 (2015). https://doi.org/10.7873/DATE.2015.0733
51. Prezioso, M., Merrikh-Bayat, F., Hoskins, B.D., Adam, G.C., Likharev, K.K., Strukov, D.B.: Training and operation of an integrated neuromorphic network based on metal-oxide memristors. Nature **521**(61) (2015). Retrieved from https://doi.org/10.1038/nature14441
52. Qureshi, M.K., Srinivasan, V., Rivers, J.A.: Scalable high performance main memory system using phase-change memory technology. SIGARCH Comput. Archit. News **37**(3), 24–33 (2009, June). Retrieved from https://doi.org/10.1145/1555815.1555760
53. Rashidi, S., Jalili, M., Sarbazi-Azad, H.: A survey on PCM lifetime enhancement schemes. ACM Comput. Surv. **52**(4) (2019, August). Retrieved from https://doi.org/10.1145/3332257
54. Rosendale, G., Viviani, D., Manning, M., Henry Huang, X.M., Rueckes, T., Wen, S.J., Wong, R.: Storage element scaling impact on CNT memory retention and on/off window. In: 2014 IEEE 6th International Memory Workshop (IMW), pp. 1–3 (2014, May). https://doi.org/10.1109/IMW.2014.6849391
55. Ryoo, J.H., John, L.K., Basu, A.: A case for granularity aware page migration. In: Proceedings of the 2018 International Conference on Supercomputing, pp. 352–362 (2018). Association for Computing Machinery, New York. Retrieved from https://doi.org/10.1145/3205289.3208064
56. Samavatian, M.H., Arjomand, M., Bashizade, R., Sarbazi-Azad, H.: Architecting the last-level cache for Gpus using STT-ram technology. ACM Trans. Des. Autom. Electron. Syst. **20**(4) (2015, September). Retrieved from https://doi.org/10.1145/2764905
57. Sebastian, A., Boybat, I., Dazzi, M., Giannopoulos, I., Jonnalagadda, V., Joshi, V., ... Eleftheriou, E.: Computational memory-based inference and training of deep neural networks. In: 2019 Symposium on VLSI Technology, pp. T168–T169 (2019). https://doi.org/10.23919/VLSIT.2019.8776518
58. Sehgal, P., Basu, S., Srinivasan, K., Voruganti, K.: An empirical study of file systems on NVM. In: 2015 31st Symposium on Mass Storage Systems and Technologies (MSST), pp. 1–14 (2015). https://doi.org/10.1109/MSST.2015.7208283
59. Shafiee, A., Nag, A., Muralimanohar, N., Balasubramonian, R., Strachan, J.P., Hu, M., ... Srikumar, V.: Isaac: a convolutional neural network accelerator with in-situ analog arithmetic in crossbars. In: Proceedings of the 43rd International Symposium on Computer Architecture, pp. 14–26 (2016). IEEE Press, Piscataway. Retrieved from https://doi.org/10.1109/ISCA.2016.12
60. Sivakumar, S., Abdul Khader, T., Jose, J.: Improving lifetime of non-volatile memory caches by logical partitioning. In Proceedings of the 2021 on Great Lakes Symposium on VLSI, pp. 123–128 (2021). Association for Computing Machinery, New York. Retrieved from https://doi.org/10.1145/3453688.3461488
61. Song, L., Qian, X., Li, H., Chen, Y.: Pipelayer: a pipelined reram-based accelerator for deep learning. In: 2017 IEEE International Symposium on High Performance Computer Architecture (HPCA), pp. 541–552 (2017, February). https://doi.org/10.1109/HPCA.2017.55
62. Song, L., Zhuo, Y., Qian, X., Li, H., Chen, Y.: Graphr: accelerating graph processing using ReRam. In: 2018 IEEE International Symposium on High Performance Computer Architecture (HPCA), pp. 531–543 (2018, February). https://doi.org/10.1109/HPCA.2018.00052
63. Spoon, K., Ambrogio, S., Narayanan, P., Tsai, H., Mackin, C., Chen, A., ... Burr, G.W.: Accelerating deep neural networks with analog memory devices. In: 2020 IEEE International Memory Workshop (IMW), pp. 1–4 (2020). https://doi.org/10.1109/IMW48823.2020.9108149

64. Sun, G., Zhao, J., Poremba, M., Xu, C., Xie, Y.: Memory that never forgets: emerging nonvolatile memory and the implication for architecture design. Natl. Sci. Rev. **5**(4), 577–592 (2017, August). Retrieved from https://doi.org/10.1093/nsr/nwx082
65. Sun, H., Chen, L., Hao, X., Liu, C., Ni, M.: An energy-efficient and fast scheme for hybrid storage class memory in an AIoT terminal system. Electronics **9**(6) (2020). Retrieved from https://www.mdpi.com/2079-9292/9/6/1013, https://doi.org/10.3390/electronics9061013
66. Swami, S., Mohanram, K.: Reliable nonvolatile memories: techniques and measures. IEEE Design Test. **34**(3), 31–41 (2017). https://doi.org/10.1109/MDAT.2017.2682252
67. Swami, S., Palangappa, P.M., Mohanram, K.: ECS: error-correcting strings for lifetime improvements in nonvolatile memories. ACM Trans. Archit. Code Optim. **14**(4) (2017, December). Retrieved from https://doi.org/10.1145/3151083
68. Vetter, J.S., Mittal, S.: Opportunities for nonvolatile memory systems in extreme-scale high-performance computing. Comput. Sci. Eng. **17**(2), 73–82 (2015). https://doi.org/10.1109/MCSE.2015.4
69. Vincent, A.F., Larroque, J., Zhao, W.S., Romdhane, N.B., Bichler, O., Gamrat, C., … Querlioz, D.: Spin-transfer torque magnetic memory as a stochastic Memristive synapse. In: 2014 IEEE International Symposium on Circuits and Systems (ISCAS), pp. 1074–1077 (2014, June). https://doi.org/10.1109/ISCAS.2014.6865325
70. Vincent, A.F., Larroque, J., Locatelli, N., Ben Romdhane, N., Bichler, O., Gamrat, C., et al.: Spin-transfer torque magnetic memory as a stochastic memristive synapse for neuromorphic systems. IEEE Trans. Biomed. Circ. Syst. **9**(2), 166–174 (2015, April). https://doi.org/10.1109/TBCAS.2015.2414423
71. Walden, C., Singh, D., Jagasivamani, M., Li, S., Kang, L., Asnaashari, M., … Yeung, D.: Monolithically integrating non-volatile main memory over the last-level cache. ACM Trans. Archit. Code Optim. **18**(4) (2021, July). Retrieved from https://doi.org/10.1145/3462632
72. Wang, Z., Liu, X., Yang, J., Michailidis, T., Swanson, S., Zhao, J.: Characterizing and modeling non-volatile memory systems. In: 2020 53rd Annual IEEE/ACM International Symposium on Microarchitecture (Micro), pp. 496–508 (2020). https://doi.org/10.1109/MICRO50266.2020.00049
73. Xie, M., Li, S., Glova, A.O., Hu, J., Wang, Y., Xie, Y.: Aim: fast and energy-efficient AES in-memory implementation for emerging nonvolatile main memory. In: 2018 Design, Automation Test in Europe Conference Exhibition (DATE), pp. 625–628 (2018, March). https://doi.org/10.23919/DATE.2018.8342085
74. Xu, C., Niu, D., Muralimanohar, N., Jouppi, N.P., Xie, Y.: Understanding the trade-offs in multi-level cell ReRam memory design. In: 2013 50th ACM/EDAC/IEEE Design Automation Conference (DAC), pp. 1–6 (2013)
75. Xu, J., Feng, D., Hua, Y., Huang, F., Zhou, W., Tong, W., Liu, J.: An efficient spare-line replacement scheme to enhance nvm security. In: 2019 56th ACM/IEEE Design Automation Conference (DAC), pp. 1–6 (2019)
76. Xue, C.J., Sun, G., Zhang, Y., Yang, J.J., Chen, Y., Li, H.: Emerging non-volatile memories: opportunities and challenges. In: 2011 Proceedings of the Ninth IEEE/ACM/IFIP International Conference on Hardware/Software Codesign and System Synthesis (Codes+ISSS), pp. 325–334 (2011). https://doi.org/10.1145/2039370.2039420
77. Zahoor, F., Azni Zulkifli, T.Z., Khanday, F.A.: Resistive random access memory (RRAM): an overview of materials, switching mechanism, performance, multilevel cell (MLC) storage, modeling, and applications. Nanoscale Res. Lett. **15**(1), 90 (2020, April 22). Retrieved from https://doi.org/10.1186/s11671-020-03299-9
78. Zhang, Y., Zhang, L., Wen, W., Sun, G., Chen, Y.: Multi-level cell STT-ram: is it realistic or just a dream? In: 2012 IEEE/ACM International Conference on Computer-Aided Design (ICCAD), pp. 526–532 (2012)
79. Zhang, M., Zhang, L., Jiang, L., Liu, Z., & Chong, F.T. Balancing performance and lifetime of MLC PCM by using a region retention monitor. In: 2017 IEEE International Symposium on High Performance Computer Architecture (HPCA), pp. 385–396 (2017a). https://doi.org/10.1109/HPCA.2017.45

80. Zhang, Z., Fu, Y., Hu, G.: Dualstack: a high efficient dynamic page scheduling scheme in hybrid main memory. In: 2017 International Conference on Networking, Architecture, and Storage (NAS), pp. 1–6 (2017b)
81. Zhang, M., Zhang, L., Jiang, L., Chong, F.T., Liu, Z.: Quickand-dirty: an architecture for high-performance temporary short writes in MLC PCM. IEEE Trans. Comput. **68**(9), 1365–1375 (2019). https://doi.org/10.1109/TC.2019.2900036
82. Zhao, W., Tong, W., Feng, D., Liu, J., Xu, J., Wei, X., ... Liu, B.: OSwrite: improving the lifetime of MLC STT-ram with one-step write (2020)
83. Zhu, G., Han, J., Lee, S., Son, Y.: An empirical evaluation of NVM-aware file systems on intel Optane DC persistent memory modules. In: 2021 International Conference on Information Networking (ICOIN), pp. 559–564 (2021). https://doi.org/10.1109/ICOIN50884.2021.9333911

Data Analytics and Machine Learning for Coverage Closure

Raviv Gal, Wesam Ibraheem, Ziv Nevo, Bilal Saleh, and Avi Ziv

1 Introduction

Verification in general and specifically functional verification are, without a doubt, some of the most important and labor-intensive parts of a hardware design cycle. Some market estimations claim that verification costs reach 50%–70% of the overall design development effort [14]. Simply stated, the goal of functional verification is to answer the question: "Does the proposed design meet its specification?"

Many methodologies, technologies, and tools have been developed to address the ever-increasing complexity of modern hardware systems and their stringent requirements for reliability [46]. Today, functional verification is a highly automated process that comprises random stimuli generators and sophisticated checkers. The random stimuli generators feed the *Design Under Verification* (DUV) with high-quality stimuli. Then, sophisticated checkers verify that the DUV behaves as expected. Verification teams utilize large compute farms to simulate, or emulate, large numbers of tests. The use of advanced random test generators can increase the quality of generated tests, but it cannot detect cases in which some areas of the design are not exercised, while others are verified repeatedly.

Test coverage analysis is the main technique for checking and showing that the verification has been thorough [35]. The idea is to create, in a systematic fashion, a large and comprehensive list of coverage events and check that each such event occurred (i.e., was hit or was covered) during the verification process. Coverage analysis can help monitor the quality of verification and can direct the stimuli generators, whether manually or automatically, to generate test that cover parts of the design that have not been adequately verified. Coverage metrics, and specifically

R. Gal · W. Ibraheem · Z. Nevo · B. Saleh · A. Ziv (✉)
IBM Research, Haifa, Haifa, Israel
e-mail: ravivg@il.ibm.com; wesam@il.ibm.com; nevo@il.ibm.com; bilal@il.ibm.com; aziv@il.ibm.com

A. Iranmanesh (ed.), *Frontiers of Quality Electronic Design (QED)*,
https://doi.org/10.1007/978-3-031-16344-9_5

functional coverage, have evolved to become a standard for monitoring the state of the verification process and its progress. Specifically, coverage is used to ensure the completeness of the verification process by directing the verification efforts toward unexplored areas of the design. Coverage is also used to improve the utilization of simulation resources [37].

For coverage to fulfill its role in the verification process and produce the needed information to monitor its state and progress, three main aspects need to be handled: coverage events, coverage data, and coverage analysis. First, the "right" coverage events need to be defined and implemented. Coverage events should cover all the blocks and functions in the DUV. They should focus on risky aspects of the design, such as new functions or complex blocks, and ensure that every corner-case or near corner-case scenario occurs often enough during verification. That said, if there are too many coverage events, it is not always feasible to ensure that almost all the events are covered during the verification process. Building a coverage plan and defining the right coverage events is beyond the scope of this chapter. These topics are part of the *coverage-driven verification* (CDV) methodology, a well-established verification methodology. The interested reader can find more on CDV in textbooks, such as [7, 37, 46] and research papers, such as [11, 28].

The second aspect is creating, transferring, and processing the coverage data. A complex DUV can contain tens or hundreds of thousands of coverage events. During the verification of such a DUV, thousands to millions of simulations are executed daily. This means that a massive amount of coverage data is continuously being produced. Verification environments need to effectively process and store this data to allow the required analyses. This aspect, which is mostly an engineering challenge, is also not in the scope of this chapter. We discuss this issue in Sect. 3 when we introduce *Template Aware Coverage* (TAC).

This chapter focuses on the third aspect of coverage, namely, coverage analysis and closure automation. The goal of coverage analysis is to extract concise and useful information out of the vast amount of coverage data produced during the verification process. Coverage analysis covers the three main facets of data analysis: descriptive, predictive, and prescriptive analysis [38]. Coverage analysis information can be used by the verification team to answer questions ranging from specific corner-case scenarios, such as "Did we fill the overflow buffer in the DUV?", to generic questions on the state of verification, such as "Have we stressed the cache coherency mechanism enough?". The analysis can also help the verification team see the progress of verification and predict whether they will meet their goals for the next milestone. The outcome of the coverage analysis can also be fed back directly to the verification process. For example, coverage analysis can identify the test-cases or test-templates (i.e., the specification to the stimuli generator) that contribute most to the verification of a risky function and increase their execution frequency.

This chapter summarizes more than two decades of research on coverage analysis conducted at IBM Research. It showcases state-of-the-art coverage analysis techniques that are used in IBM and elsewhere in the electronic design industry. We illustrate how coverage analysis can simplify the work of the verification team and improve the efficiency and quality of the coverage process, and through it the entire verification process. Special attention is given to the coverage closure process. Coverage closure is the process of advancing coverage goals in general, and specifically coverage levels [37]. The importance of coverage in determining whether verification goals have been reached means that coverage status is an important criterion for many project milestones, such as tape-outs. The verification team can spend significant time and effort on coverage closure. To achieve coverage closure, the verification team needs to analyze the uncovered events and understand whether these events can be hit and if so what is needed to hit them. Only then can they write or modify tests or test-templates (the input to the random generator) that hit the uncovered events or improve the probability of hitting them.

The chapter proceeds with the increasing complexity of the analysis method used, from descriptive to prescriptive analysis. Section 2 focuses on the descriptive analysis of coverage data, that is, the process of extracting useful information out of the coverage data collected during the verification process and presenting it to the users. The section begins with a quick overview of basic types of coverage reports. It then moves to techniques that utilize the structure of coverage models, allow navigation in the coverage space, can zoom in to focus on specific areas, and zoom out to see the entire picture in that space [3].

The navigation and zooming capabilities provide users with the means to identify areas in the coverage space, and thus in the DUV, which require attention. But finding such areas requires time and expertise from the verification team. In the second part of Sect. 2, we deal with advanced coverage analysis techniques that are used to identify large areas of uncovered events or lightly covered events. These hole analysis techniques [4] exploit the structure of cross-product coverage models to identify such areas. The last part of Sect. 2 describes a machine learning-based algorithm that is used to identify structures among the coverage events when the coverage events are individually defined [4].

Most coverage tools maintain a single database that summarizes the coverage data of all the passing test-cases. Such coverage repositories can provide information about what is covered and using the analysis techniques presented in Sect. 2 identify areas that need attention. But they cannot answer queries regarding why certain events are covered, that is, which test-cases or test-templates cover them. In Sect. 3, we present *Template Aware Coverage* (TAC) [16]. TAC maintains a matrix with the test-templates as its rows and the coverage events as its columns. Each entry in the matrix contains the probability that a test-case generated from a given test-template hits a given event. With such a matrix, queries that relate events and test-templates, such as "Does a given test-template hit the events it is supposed to hit?", can be answered. The section describes many potential use-cases for TAC. Due to the challenges in maintaining the very large TAC matrix, we provide some details of its actual implementation.

Section 4 moves from descriptive to prescriptive analysis. It deals with the holy grail of coverage closure, namely, automatic closing of the coverage feedback loop from coverage data to stimuli generation. Even when the best coverage information is available, the verification team still needs to invest a significant amount of time and effort to translate this information into actions that improve the coverage state. Section 4 describes several techniques that can (partially) close this loop automatically.

The first part of the section shows how the TAC data can be used to create regression suites that address various needs of the verification team; these needs range from a small suite ensuring that deposited versions of the design are alive for *continuous integration* and *continuous delivery* (CI/CD) to regression suites that improve the hit rate for hard-to-hit events that are not hit often enough. The section provides the information about the coverage criteria for each use-case and describes the various algorithms used to create the regression suites [16].

The second part of the section deals with coverage-directed generation (CDG), where the goal is more ambitious. Namely, CDG is focused on creating new test-cases or test-templates that hit previously uncovered coverage events. Over the years, CDG has received significant attention in the hardware verification research community. We provide a short overview of the main techniques and methods that have been proposed in the past. We focus on two newer techniques for CDG. The first uses optimization techniques for noisy functions to achieve the CDG goal [21]. The second technique uses machine learning to improve the quality of data-driven CDG solutions [18].

2 Descriptive Coverage Analysis

The goal of descriptive coverage analysis, as its name suggests, is to extract simple, concise, and useful information out of the coverage data and present it to users. In many cases, descriptive coverage analysis exploits the structure of coverage models to improve the quality of the analysis and help users adapt the analysis and resulting reports to their needs. In this section, we focus on cross-product coverage [24, 37].

Any cross-product functional coverage model is based on a schema that comprises the following: a semantic description (story) of what needs to be covered, a list of the attributes mentioned in the story, a set of all the possible values for each attribute, and an optional set of partitions for each attribute. A coverage event is a point in the multidimensional space defined by the cross-product of the attributes.

Table 1 Floating-point schema attributes

Attribute	Values	Size
Instr	fadd, fadds, fsub, fmul, fdiv, fmadd, fmsub, fres, frsqrte, fabs, fneg, fsel, ...	54
Result	± 0, $\pm\infty$, $\pm$Norm, $\pm$MinNorm, $\pm$MaxNorm, $\pm$DeNorm,$\pm$MinDeNorm, $\pm$MaxDeNorm, SNaN, QNaN	18
Round mode	ToNearest, To0, To$+\infty$, To$-\infty$	4
Round occur	True, False	2

Usually, not all the points in the coverage space are legal or interesting. The coverage model is the intersection of the legal and interesting subspaces.

The following is an example of a simple cross-product functional coverage model taken from the floating-point domain. We use this model throughout this section to demonstrate the various analysis techniques presented. This model is part of a generic floating-point verification plan, and it has been used in the verification of many floating-point units in IBM processors [1]. The semantic description of the schema is: test that all *instructions* produce all possible *target results* in the various *rounding modes* supported by the processor, both when *rounding* did and did not occur. Table 1 shows the attributes and their values of the coverage schema. The schema consists of four attributes—Instr, Result, Round Mode, and Round Occur—each with the possible values shown. The Size column indicates the number of possible values for each attribute.

Each attribute of a schema may be partitioned into one or more disjoint sets of semantically similar values. This provides a convenient way for users to conceptualize their model and for analysis tools to report on coverage data. For example, the instruction attribute is partitioned into arithmetic and non-arithmetic instructions, and the resulting attribute is partitioned according to its sign.

In the coverage process, it is critical to define a *coverage model* that contains only the legal and interesting events of the schema. This is because not all events in the coverage space of the schema are legal. For example, the results of executing a floating-point absolute value instruction can never be negative. It is imperative that illegal events be disregarded when doing coverage analysis; otherwise, the coverage information will be skewed.

Generally speaking, descriptive coverage reports can be divided into two types: status reports and progress reports. Status reports present the state or status of coverage at a given point in time (usually, the present). In most cases, these reports present information about coverage events or groups of coverage events in a tabular form. The information can include items such as the hit count of the events and the first and last time they were covered. Table 2 shows a simple coverage status report for a few events in the floating-point schema.

Table 2 Coverage data

Attributes				Coverage data		
Instr	Res	RM	RO	Count	First	Last
fadd	+0	To0	False	4	08/04	08/30
fadd	+0	To0	True	0	–	–
fadd	+0	To$+\infty$	False	1	08/04	08/04
fadd	+0	To$+\infty$	True	0	–	–
fadd	+0	To$-\infty$	False	3	07/28	08/30
fadd	+0	To$-\infty$	True	0	–	–

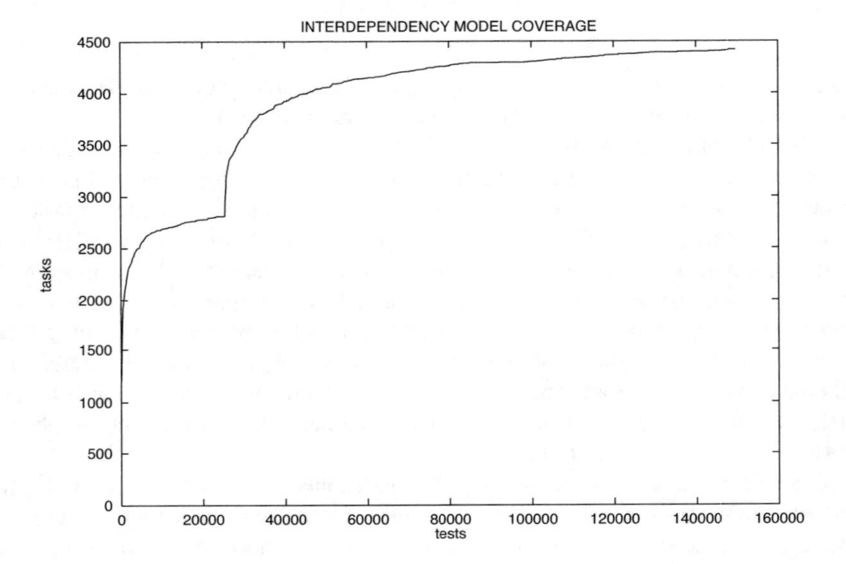

Fig. 1 Coverage progress

Progress reports play an important role in presenting the coverage picture because they show changes in a given coverage measure over time.[1] They can reveal trends in the coverage progress, and thus in the progress of the entire verification process. This can help identify when the progress slows down and the verification process runs out of steam. Figure 1 shows the coverage progress of a model for the execution pipelines of a superscalar processor. The figure shows the number of events covered as a function of the number of tests executed. We can see that after about 20,000 tests, the coverage progress started to slow down, but corrective actions taken by the verification team helped renew the progress and almost doubled the number of covered events.

In addition to monitoring the coverage progress, progress reports can help the verification team in several other ways. Looking at the coverage levels backward

[1] Time can be measured by clock time, number of tests, number of cycles, etc.

from the current time can reveal how long it takes to cover all the covered events and detect aged-out events, namely, events that have been covered in the past but have not been covered for a long time [5]. Moreover, analysis of the time series data used in progress reports can be used as the source for predictive analysis and help provide predictions such as when coverage will converge and at what level [25].

This section and the rest of the chapter focus on status reports.

2.1 Coverage Views

The basic status report of Table 2 is somewhat misleading. Our simple floating-point coverage model contains thousands of coverage events. Viewing such a large number of events and extracting useful information out of it is almost impossible. While most coverage tools provide filtering capabilities that are similar to the capabilities of spreadsheets, these capabilities are not sufficient to provide a brief summary of the coverage status or help the verification team focus on the arithmetic instructions.

A coverage tool should allow users to focus on certain aspects of the coverage data, such as certain attributes, specific values of attributes, and coverage during a given time interval. Moreover, the tool should allow the users to rapidly shift their focus. To do this, the coverage tool must provide a simple way to define views into the coverage data that allow users to concentrate on the specific aspects of the model in which they are interested while ignoring those areas they are not. These views provide useful coverage reports that address the needs of the users. Next, we describe three basic operations that take advantage of the semantics and structure of cross-product coverage models. These operations enable users to fit the coverage status reports to their current needs and perform efficient analysis of the coverage data. These operations are:

- *Selecting* out a subset of the coverage events that satisfy some criteria based on specific attribute values or value combinations, or on the coverage data itself.
- *Projecting* the coverage model onto a subset of its attributes so that some attributes are essentially ignored.
- *Grouping* together the coverage data of related attribute values that belong to an attribute partition.

Applying these operations to a coverage model, either individually or in combinations, produces a new coverage model that is, in some sense, a subset of the model. We call the resulting coverage model a *coverage view* on the model. It is both smaller and more focused than the original coverage model, and therefore simpler to analyze. The coverage view does not in any way alter the coverage data that has been accumulated for each event in the schema. It simply acts as an overlay or filter that is applied to the coverage data when performing analysis.

2.1.1 Selections

The most straightforward way to reduce the number of events that need to be analyzed is to focus on a selected set of events or, conversely, to filter out unwanted ones. This is not only a practical approach to take, but a logical one as well, since the set of events that are of interest changes according to the verification effort that is being performed. For example, when the focus of the verification team is on simple arithmetic instructions, they select fadd, fsub, etc. But when the focus shifts to instructions that require complex algorithms, they change the selection to fsqrt, fres, etc.

Selections are expressed as queries over the coverage model. Formally, we write this as $\sigma(pred)$, where `pred` is a logical expression over the attributes of the schema. For example, the selection $\sigma(Instr \in Arith \bigwedge RoundOccur = true)$ defines a coverage view that contains only arithmetic instructions with inexact results. Coverage reports using this view will display only those events that have this combination of attribute values. No other events will appear in the report.

The selection noted above selects events that have some desired attribute values. Equally important, and often more so, are selections based on the coverage data itself. Such selections filter out events based upon their hit count or when they were first or last covered. For example, the query $\sigma(count = 0 \bigvee last < 08/01)$ defines a coverage view that contains only those events that have never been covered or have not been covered recently. We call selections involving coverage data *dynamic selections*, since the set of events they define changes as coverage progresses.

From a methodology perspective, dynamic selections are crucial not just for defining coverage views but in the coverage analysis itself. All selections, whether dynamic or static, reduce the size of the coverage space, thereby making it easier to analyze. Dynamic selections have the additional benefit that the information they extract is directly relevant to coverage analysis. Knowing, for example, which events have not been covered, or which events have been covered a disproportionate number of times, is the essence of coverage analysis. This information is also what is explicitly obtained from dynamic selections.

Dynamic selections are often combined with selections of attribute values. This enables us to look for holes or other such coverage-related constraints, within a restricted set of events. The selection, $\sigma(count < 5 \bigwedge (Instr \in \{fadd, fsub\} \bigwedge Result \in Positive)$ for example, finds all lightly covered combinations of `fadd` or `fsub` with positive results.

2.1.2 Projections

The purpose of projection is to answer such questions as "Have all instructions been covered?" and "What is the coverage data for all combinations of results and rounding modes in the coverage model?" Looking at the coverage data of individual events is far too cumbersome and error-prone to effectively answer such questions. A better approach is to project the coverage schema onto a subset of its attributes.

Table 3 Projected coverage data

Attributes		Coverage data			
Instr	Result	Count	First	Last	Density
fabs	+0	3	08/02	08/11	1/4
fabs	$+\infty$	0	–	–	0/4
fadd	+0	18	07/14	08/30	3/4
fadd	−0	5	08/11	08/11	2/4
fadd	$+\infty$	0	–	–	0/8

Formally, this is expressed as $\pi(A_1, ..., A_k)$, where A_i is the name of some attribute in the schema.

In Table 3, we show the partial results of projecting the floating-point coverage model defined earlier in the section onto the attributes `Instr` and `Result`. Every event in this projected coverage view corresponds to a number of reflected events in the original model. For example, the event `<fadd,+0>` corresponds to all reflected events whose `Instr` is `fadd` and whose `Result` is `+0`, that is, to `<fadd,+0,To0,false>`, `<fadd,+0,To+∞,false>`, and so on. The coverage count of each projected event is equal to the sum of the counts of all its reflected events. The first and last times a projected event was covered is the earliest and latest times of any of these reflected events, respectively.

Projections essentially control the granularity of the coverage information presented. Clearly there is some loss of information when we look at events in a projected model. The fact that `<fabs,+0>` has been covered three times tells us nothing about which events were actually covered and which were not. But looking at the data at this scale gives a better overall picture of the coverage. With projection, we can zoom in or zoom out to whatever level of detail is needed.

The last column in Table 3 shows the *density* of a projected event; this is equal to the ratio of reflected events covered to the total number of reflected events of the projected event. Density gives an indication of the distribution of coverage for the projected event. A low density, even for events with high coverage counts, means that coverage is not evenly distributed among the reflected events and many reflected events are not covered. If we look at the event `<fadd,+0>`, we see that its count is 18 and its density is 3/4. This means that of the four legal events with `fadd` and `+0` in the pre-projected coverage model, 3 were covered a total of 18 times. The dynamic selection, $\sigma(count = 0 \bigwedge Instr = fadd \bigwedge Result = +0)$ applied to the original coverage model, can be used to find the missing non-covered event.

More expressive coverage views can be built by combining projections with selections. We use selection to focus on particular events, and then projection to coalesce their coverage data. For example, the coverage view $\pi(Instr) \circ \sigma(Instr \in Arith \bigwedge Result = +0)$ projects the model onto the single attribute dimension `Instr`, but only considers events with `Arith` instructions and `+0` results when computing the coverage data of each instruction (we use $\circ$ to show the composition of two operations). Similarly, we can combine projections with dynamic selections to look at coverage data in a more compact form. This enables us to define coverage views like $\sigma(count = 0) \circ \pi(Instr)$, which computes the set of instructions that have not been covered at all, and the coverage view $\sigma(density < 1.0) \circ \pi(Instr)$, which computes the set of instructions that are not fully covered.

As long as the selection criteria involve only projected attributes, the order of the selection and projection operations does not matter. This situation changes, however, for dynamic selections. Consider the two coverage views, $\sigma(count < 5) \circ \pi(Instr)$ and $\pi(Instr) \circ \sigma(count < 5)$. Both views select events that were covered less than five times, and both views project onto the single attribute `Instr`. They differ only in the order in which the operations are invoked. This difference, however, has a critical effect on the semantics of the views defined. The first view generates a list of all lightly covered instructions. In contrast, the second view generates a list of all instructions that have at least one reflected event that was lightly covered. If the `fadd` instruction, for example, was covered many times, but one of its reflected events, say `<fadd,+0,To0,false>`, was only covered twice, then the projected event `fadd` will appear in the second view, but not in the first. The ability to express queries based upon the coverage data of events, either before or after they are projected, is yet another powerful feature of coverage analysis.

2.1.3 Groupings

Another technique for defining a smaller functional coverage view is to group the coverage data of semantically related events, such as the instruction type (arithmetic and non-arithmetic) and the sign of the result. We can use these partitions to create a coverage view that looks collectively at related events in the model. This is formally expressed as $\lambda(A_1.P_1, A_2.P_2, ..., A_n.P_n)$ where P_i is a partition on attribute A_i, defined in the coverage schema. Note that not all attributes in the schema need to be partitioned. Like projections, using partitions to analyze coverage information allows us to analyze clusters of events rather than individual ones, thereby raising the level of abstraction at which we look at the data. Another benefit is that attribute values that are partitioned together are frequently tested together, so that patterns of coverage activity are likely to be present throughout the entire group.

In Table 4, we show the partial results obtained for the coverage view $\pi(Instr, Result) \circ \lambda(Instr.Type, Result.Sign)$, which groups and projects events according to the `Instr` and `Result` attributes. For the grouping operation, the partitions `Type` and `Sign` are used for the `instr` and `result` attributes, respectively. As we can see, individual events are grouped according to their instruction type and according to the sign of the result.

The algorithm for computing the coverage data for a grouped event is the same as for a projected event. The count, first, and last time an event was covered is equal to the sum, earliest, and latest values, respectively, of all of its reflected events. Its density is the ratio of covered events to the total number of reflected events. As seen in this example, grouping can be combined with other operations to create a coverage view. In particular, it can be used with projections to focus on specific

Table 4 Partitioned coverage data

Attributes		Coverage data			
Instr	Res	Count	First	Last	Density
Arith	Pos	1922	07/11	08/30	621/712
Arith	Neg	1337	07/11	09/04	473/684
Arith	NaN	194	08/04	08/17	88/144
NonArith	Pos	542	07/11	08/30	168/192

attributes, and with selections to focus on certain attribute values or certain coverage data statistics.

2.2 *Hole Analysis: Automatic Descriptive Coverage Analysis*

One of the main activities of the verification team during coverage closure is to identify large and meaningful coverage holes that indicate weaknesses in the definition and execution of the verification plan. The coverage views presented earlier provide a means for finding such holes, but using coverage views to find such holes often requires time and expertise. Hole analysis [4, 32] is a technique to automatically detect and report such large holes. The main idea of the hole analysis technique is to group together sets of uncovered events that share some commonality, thus allowing the coverage tool to provide shorter, more meaningful, and actionable coverage reports to the user.

To illustrate the importance of hole analysis, consider a model with just two integer attributes, X and Y, each capable of taking on values between 0 and 9. Figure 2a shows the individual uncovered events in the model after 70% of the events are covered. The two meaningful holes that exist in the coverage data, indicated by arrows in the figure, are not immediately obvious. One hole occurs whenever Y equals 2, and a second hole exists when both attributes have values 6, 7, and 8. These holes are, however, readily seen in Fig. 2b, which shows the coverage of the model in a 2D plot, with covered events marked as black squares. Such plots provide a convenient way to present holes that are clustered along ordered values in models with a small number of attributes. The challenge for automatic holes analysis is to discover more complex holes in arbitrarily large models and to present these holes in such a way that their root cause can be more easily discerned. Note that the diagonal lines are explained later in Sect. 2.2.2.

Meaningful holes can be automatically discovered between uncovered events that have some similarity. Next, in Sects. 2.2.1 and 2.2.2, we describe two algorithms for finding such holes. The first looks for holes where all the possible values of one or more attributes have not been covered for some specific values of the other attributes. The second algorithm aggregates close uncovered events and holes to form larger, more meaningful holes.

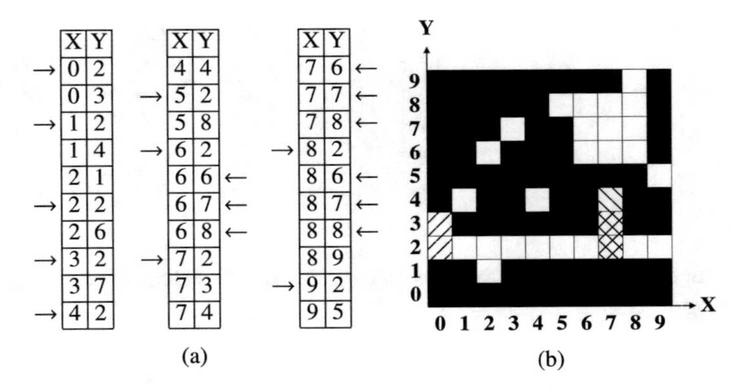

Fig. 2 Coverage hole example. (**a**) List of uncovered events. (**b**) Hole visualization

2.2.1 Algorithm for Projected Holes

Of special interest are those cases where all the possible values of one (or more) attributes were not covered for some specific values of the other attributes. An example of this is the value 2 for the Y attribute in Fig. 2. We call this a projected hole and mark it $< *, 2 >$, where the $*$ sign indicates a wild card.

A coverage model of n attributes can be viewed as a set of coverage events or points in an n-dimensional space. A projected subspace of dimension 1 contains events that lie on a line parallel to the axis in the hyperspace. In other words, a projected hole of dimension 1 corresponds to a set of events whose attributes all have fixed values, except a single attribute that has a wild card. A projected subspace of dimension 2 describes events that lie on a plane parallel to the axis and so on. A projected subspace of a higher dimension subsumes all contained subspaces of a lower dimension. For example, the subspace $p =< *, *, x_3, \ldots x_n >$ of dimension 2 contains all the events described by the subspace $q =< *, x_2, x_3, \ldots x_n >$. In such a case, we say that p is the ancestor of q and that q is the descendant of p in dimension 2. We denote by S_v the set of dimensions of subspace v that are not fixed (i.e., their value is $*$). s_v is the attribute with the smallest index in S_v. For example, for the subspace p above, $S_p = \{1, 2\}$ and $s_p = 1$. In Algorithm 1, which describes the projected holes algorithm, descendants (v, s) denote all the direct descendants of subspace v in dimension s. That is, it refers to all the subspaces that are created by replacing the $*$ in the sth dimension of v with a specific value for that dimension. Similarly, descendants (v, S_v) denote all direct descendants of v in all possible dimensions. Holes of higher dimension are larger and, in general, more meaningful than the holes of lower dimension they subsume. Therefore, it is desirable to report holes of the highest possible dimension and leave the subsumed holes out of the report.

The skeleton of the algorithm for finding projected holes is shown in Algorithm 1. The algorithm is based on the double traversal of all the possible subspaces in the coverage space, first in increasing and then in decreasing order of dimension. In the

Algorithm 1 Algorithm for projected holes

```
procedure HOLEPROJECTION
  // Mark covered subspaces
  for i = 0 to n − 1 do
    for all subspace v with dimension i do
      if v is marked covered then
        Mark all direct ancestors of v as covered
  //Mark and report holes
  for i = n downto 0 do
    for all subspace v with dimension i do
      if v is not marked (as subsumed hole or covered) then
        Report v as hole
        recursively mark all descendants of v as subsumed holes
```

first phase, the algorithm marks all the subspaces that are covered; that is, at least one event in the subspace is covered. This is done by marking all the ancestors of a covered space, starting with the covered events themselves (subspaces of dimension 0). During the second phase, the algorithm traverses all the subspaces in decreasing order of dimensions. For each subspace, if it is not marked as covered or as a subsumed hole, the algorithm reports it as a hole and recursively marks all its descendants as subsumed holes because they are subsumed by this hole.

The complexity of the algorithm described here is exponential to the number of dimensions (attributes) in the coverage model. The performance of the algorithm can be significantly improved by applying pruning techniques, both during the construction of the projected subspaces graph and while traversing the graph to report the holes.

2.2.2 Algorithm for Aggregated Holes

A simple metric for similarity is the rectilinear distance between two holes. This corresponds to the number of attributes on which the two differ. The distance will be one for holes that differ in only one attribute, two for holes that differ in two attributes, and so on, up to the number of attributes in the coverage space. We aggregate together any two holes whose rectilinear distance is one. Thus, the two uncovered events <0,2> and <0,3> from Fig. 2a can be aggregated into the single hole <0,{2,3}>, and similarly for other such events.

Rectilinear distances can be computed on aggregated holes as well. Again, the distance between any two holes is equal to the number of differing attributes, but now the comparison is done for aggregated sets as well as atomic values. The process can be applied iteratively until no more new aggregated holes are discovered. Figure 3 shows how the five events <0,2>, <0,3>, <7,2>, <7,3>, and <7,4> can be aggregated together until only the holes <{0,7},{2,3}> of size 4 and <7,{2,3,4}> of size 3 remain. This technique is similar to Karnaugh binary mapping [30] and is useful for much the same reasons. As in Karnaugh binary

$$
\begin{array}{l} <0,2> \\ <0,3> \\ <7,2> \\ <7,3> \\ <7,4> \end{array} \Longrightarrow \begin{array}{l} <\{0,7\},2> \\ <\{0,7\},3> \\ <0,\{2,3\}> \\ <7,\{2,3,4\}> \end{array} \Longrightarrow \begin{array}{l} <\{0,7\},\{2,3\}> \\ \\ <7,\{2,3,4\}> \end{array}
$$

Fig. 3 Aggregated hole calculation

Table 5 Coverage hole report

Instr	Result	Round mode	Round occur	Hole size
fadd	*	*	*	152
Square Roots	+DeNorm +MaxDeNorm +MinDeNorm	*	*	72
Estimates	*	*	True	144

mapping, holes may intersect. This is fine because every hole provides a different view of the uncovered events, each with different functional meaning and potential action. The aggregated holes of Fig. 3 are marked by left and right diagonal lines in Fig. 2b. When the values of a given attribute are ordinal, the aggregation rule of rectilinear distance equals one, which can be replaced with a stricter rule of rectilinear distance equals one, and then the distance in the changed attribute is also one. In this case, the events $< 0, 2 >$ and $< 7, 2 >$ are not aggregated, as in Fig. 3. Using the stricter rule increases the chances that the aggregated hole has semantic meaning.

The projection holes and aggregated holes algorithms can be combined. The order in which the aggregations are applied with respect to the marking phase of the projection algorithm affects the reported holes. If we do not mark subsumed holes immediately, but defer that marking until after aggregation, those holes may be aggregated to form different and more meaningful holes, at least from the size aspect.

Table 5 shows some of the holes found in the floating-point model when the model was implemented as part of the floating-point verification plan in an IBM project. The first hole indicates that none of the events related to the fadd instruction were covered. The hole was caused by a bug in the specification to the test generator that omitted the fadd instruction from the list of instructions that needed to be generated. The next two holes show two groups of events that are impossible to hit. The first points to square root instructions (fsqrt, fsqrts, frsqrte) with denormalized results, while the second points to estimation instructions (fres, frsqrte) with inexact results. Both holes are covered by restrictions that eluded the developer of the model. After their detection, the two holes were converted to restrictions.

2.3 *Machine Learning-Based Discovery of Coverage Model Structure*

Earlier, we showed how the inherent structure of coverage models can be used to improve the analysis of the coverage space. Specifically, we showed how hole analysis can find large uncovered holes in cross-product coverage models. In many cases, coverage models are defined as a set of individual coverage events without an explicit structure. In such cases, a manual search for large coverage holes is needed. This search involves scanning a list of uncovered events to find sets of closely related events. This is a tedious and time-consuming task.

Machine learning techniques have proven effective in many classification and clustering problems. Therefore, applying such techniques can improve the efficiency and quality of finding coverage holes, and reduce the manual effort involved in this task. Next, we describe a clustering-based technique that finds large coverage holes in verification environments in which coverage models do not have an explicit structure, such as cross-product. The proposed technique exploits the relations between the names of the coverage events. For example, event `reg_msr_data_read` is close to event `reg_pcr_data_write`, but not to event `instr_add_flush`.

To find large coverage holes, this technique combines a classic clustering technique with domain-specific optimizations and uses it to map individually defined coverage events to cross-product spaces. Then, it uses cross-product hole analysis techniques, such as the ones described earlier in the section, to find large coverage holes in these spaces. The analysis comprises three main steps. In the first step, it clusters all the coverage events and maps the clusters into cross-product spaces. The second step applies domain-specific optimizations to improve the cross-products created, by cleaning imperfections left by the clustering algorithm. Finally, in the third step, a standard cross-product hole analysis is performed and the holes are reported to the user. The first two steps of the analysis need to be repeated only when major changes in the coverage definition occur, which is infrequent. Therefore, complex and time-consuming computations can be used. The following is a brief description of the first two steps in the analysis. A more detailed description can be found in [20].

2.3.1 Clustering Events into Cross-Products

The first step in the analysis is to cluster all the coverage events based on their names. This clustering begins by breaking each coverage event name into words. In the example here, the names of coverage events are in the form of `w1_w2_..._wn`, where each `wi` is a word. However, breaking the names of events into words can work with any other naming convention, such as a capital first letter (camel case). The words of all the events are used as the features for the clustering algorithm. This technique is common to many document clustering techniques [39], and indeed a classic clustering technique is used here. The one major difference is that we add the index of each word to the word. The reason is that events sharing the same word in the same location are much more likely to be related than events that share the same word in different locations. That is, the event `reg_msr_read` is more likely to be related to the event `reg_pcr_write` than to the event `set_data_reg_to_0`. To address this, the index of the word in the event's name is added to every word feature. Practically, each word in the event is used with two indices: a positive index from the start of the name and a negative index from the end. This helps to relate events that share the last word, second to last word, and so on.

Once features are extracted, a clustering algorithm is used. There are many available clustering algorithms. Some, like the *Latent Dirichlet Allocation* (LDA) [6] algorithm, are designed for the task of document clustering, while others, like K-means [29], are general-purpose clustering algorithms. The performance of the clustering algorithm, both in terms of the quality of its results and its runtime, strongly depends on the actual algorithm used and the values of its hyperparameters. Specifically, the two most important hyperparameters are the number of clusters and the features used.

For the discovery of coverage model problem, it turns out that these factors have a negligent effect on the quality of the overall results of the algorithm [20]. The reason for this is the domain-specific optimizations that follow the clustering phase and are explained later. These optimizations allow the algorithm to fix imperfections in the initial clustering and identify the correct cross-product models. In fact, the results in [20] show that a combination of the simple K-means algorithm [29] with nonnegative matrix factorization (NMF) [34] that significantly reduces the number of features from thousands to 30 dramatically speeds up the clustering execution time while having no impact on the quality of the cross-product models found.

After the clusters are formed, the next step extracts the cross-product structure out of each cluster. Algorithm 2 describes this step and Fig. 4 provides an example of its application. The first step in the algorithm is to create, for each possible location, the set of all words that appear in that location with the number of times each word appears in the location. For example, the cluster of the events in Fig. 4a yields the ten location sets in Fig. 4b. Note that because we mark locations both from the start and from the end, the number of sets is twice the length of the longest events.

The location sets are used to find *anchor locations*. Anchors are locations that have a single word in them, and that word appears in all the events in the cluster. In the example, location 1 with the set {`reg (6)`} and location −2 with the set {`data (6)`} are anchors. Location 5 with the set {`rmw (2)`} is not an anchor because the word `rmw` does not appear in all the events. The underlined locations in Fig. 4b mark the anchors in the example cluster.

After the anchors are identified, all the locations between the anchors are considered the dimensions or attributes of the cross-product space. This includes the locations before the first anchor, if it is not in location 1, and the locations after the last anchor, if it is not in the last location. The dimensions of the cross-product in the example are shown in Fig. 4c. In the general case, dimensions may include several words and have a different number of words in each event. For example, the first dimension in the example includes two words for the second and last events and

```
reg_msr_data_read             reg_msr_data_write
reg_msr_atomic_data_rmw       reg_pcr_data_read
reg_pcr_data_write            reg_ir_atomic_data_rmw
```

(a)

```
1: {reg(6)}
2: {msr(3), pcr(2), ir(1)}
3: {atomic(2), data(4)}
4: {data(2), read(2), write(2)}
5: {rmw(2)}
-1: {read(2), rmw(2), write(2)}
-2: {data(6)}
-3: {atomic(2), msr(2), pcr(2)}
-4: {ir(1), msr(1), reg(4)}
-5: {reg(2)}
```

(b)

```
1: loc. 2 – -3 {msr, msr_atomic, pcr, ir_atomic}
2: loc. -1     {read, rmw, write}
```

(c)

```
1: loc. 2      {msr, pcr, ir}
2: loc. 3 – -3 {atomic, ϕ}
3: loc. -1     {read, rmw, write}
```

(d)

```
1: loc. 2  {msr, pcr, ir}
2: loc. -1 {read, rmw, write}
```

(e)

Fig. 4 Extracting cross-product example. (**a**) Events in the cluster. (**b**) Location sets. (**c**) Initial cross-product dimensions. (**d**) Cross-product dimensions after removing redundant second dimension. (**e**) Cross-product dimensions after removing redundant second dimension

one word for the other events. This dimension semantically corresponds to two real dimensions. In the next subsection, we show how to reveal these real dimensions.

2.3.2 Improving the Cross-Product Quality

After forming the initial cross-products, the algorithm improves their quality by increasing their size, and increasing the number of dimensions in the clusters. The technique also improves their density, which is defined as the ratio between the number of events in the cross-product and the size of the cross-product space. The

Algorithm 2 Extract cross-product from a cluster

```
procedure EXTRACTCROSSPRODUCT(cluster)
    // Build location sets
    Clear all location sets L_i
    for all event e in the cluster do
        for all location i in e with word w_i do
            Add w_i to L_i and increase its count
    // Identify anchors
    for all locations L_i do
        if L_i has one word with count equals to the cluster size then
            Mark L_i as an anchor
    // Extract dimensions
    Sort anchor locations in increasing order, first positives then negatives
    Make gaps between anchors attributes
```

improvement process can be divided into two main steps. The first step adds events to the cross-products and combines similar cross-products to overcome deficiencies of the clustering algorithm described in the previous section. The second step utilizes domain knowledge on cross-product coverage to apply heuristics that increase the number of dimensions and improve the density of the models.

The clustering and mapping to cross-product algorithm leaves behind some dirt in the form of: outlier events, orphan events that should belong to a cross-product but are not clustered with it, and several clusters that should belong to the same cross-product. The reason for this dirt is that the distance measure used by the clustering algorithm does not exactly match the distance needed for the cross-products.

The cleaning step consists of two substeps: adding orphan events to cross-products and combining similar cross-products. The cleaning process continues until convergence. To find orphan events, the algorithm compares all the events in the unit to the pattern of each cross-product. If an event matches the pattern exactly or almost exactly, it is added to that cross-product. An event almost matches a pattern if it matches all the anchors and differs in at most one dimension. In this case, the dimension value of the event is added to the cross-product dimension. For example, when event `reg_ir_data_read` that is not part of the initial cluster in Fig. 4a is compared against its cross-product pattern, it matches the two anchors and the second dimension, but not the first dimension. In this case, the event is added to the cross-product and the first dimension is updated to include the value `ir`.

To combine close clusters, the algorithm compares the patterns of each pair of clusters. If the two patterns differ in just one location, either an anchor or a dimension, it combines the two clusters into a larger cluster. In this comparison, patterns are equal in a dimension if they have the same values in the dimension or if the values in one pattern are a subset of the values in the second dimension. Note that if two patterns that differ in a single anchor are combined, the anchor becomes an attribute in the combined pattern.

The second part of the improvements is specific to cross-product coverage models. These heuristics were developed to improve the quality of the cross-products after examining the results of many real-life models. The heuristics improve the quality of the cross-product by adding more dimensions to the cross-products on the one hand and by removing redundant dimensions on the other hand. The resulting cross-product can have more dimensions, may be denser, or both. The process repeats the two steps of breaking dimensions and removing redundant dimensions until convergence is reached for each cross-product.

Dimensions in the cross-products found in the previous steps can span multiple words. For example, the first dimension in Fig. 4c is from location 2 to location −3 in the events, and it spans one to two words. In such cases, breaking the long attribute into two or more smaller attributes can improve the quality of the cross-product. In the example, the first attribute comprises two separate semantic meanings. The first word in the attribute (location 2) is the name of a register (`ir, msr, pcr`), and the second word is an indicator of whether or not the data access is atomic. Understanding the semantic of each word is hard. However, if breaking a dimension into two or more dimensions results in a dense cross-product, then, in most cases, this break is semantically correct and improves the quality of the cross-product.

For each single breaking point, the heuristics check the ratio between the number of values in the dimension and the size of the cross-product of the broken dimension. If the ratio is high enough ($\geq$ 50% in the current implementation), the algorithm performs the break. In the example, the number of values in the first dimension is 3. Breaking it into two attributes (location 2 with values $\{$`ir, msr, pcr`$\}$ and locations 3-−3 with values $\{$`atomic,` $\phi\}$). The size of the original dimension equals 50% of the product of the sizes of the new attributes ($3 \cdot 2 = 6$), so the algorithm performs the break. The resulting dimensions appear in Fig. 4d.

Removing redundant dimensions from the cross-product reduces the size of the cross-product space, thus improving its density. A dimension d is redundant if a projection of the n dimensional space into the $n - 1$ dimension space that does not include d leaves any point in the new space with at most one event associated with it. The implementation uses a simpler form of a pair-wise redundancy: dimension d_1 is redundant with respect to dimension d_2 if considering the value pairs for all the events, the values in dimension d_1 partition the values in dimension d_2.

In the cross-product in Fig. 4d and the events in Fig. 4a, the pair of values for dimensions 2 and 3 are `(atomic, rmw), (`ϕ`, read), (`ϕ`, write)`. Therefore, the second dimension partitions the third dimension into `(atomic,` $\{$`rmw`$\}$`), (`ϕ`,` $\{$`read, write`$\}$`)`, which makes it redundant. Figure 4e shows the dimensions after removing the redundant dimension.

After completing all the improvements, a final cleanup removes cross-products that are either too large (more than one million events) or too sparse (less than 10% density). Any cross-products that have just one dimension are also removed. Therefore, the final set of cross-products does not cover all the events in the unit the algorithm started with.

2.3.3 Usage Results

The clustering analysis technique is incorporated into the coverage reports that are produced by the Verification Cockpit (VC) [2], a platform for a holistic centralized data model for the arsenal of verification tools used in a modern verification process. The clustering analysis results are part of the coverage status reports, and they are used regularly by all major hardware design and verification projects in IBM. The units on which the clustering technique is used contain anywhere from several thousands to tens of thousands of events. The analysis is able to map between 25% and 60% of the events in each unit to clusters. The analysis found many large clusters (30%–60% of the clusters) that contain more than 100 events as well as several (about 5%) clusters with more than 1000 events. Hole analysis and compact hole reports are extremely important for these clusters because of the difficulty in looking at all the events and manually identifying the holes in these clusters.

Figures 5 and 6 show the clustering analysis results for the load-store unit (LSU) of a high-end processor core. Figure 5 shows a summary of the clusters found in that unit and the status of each cluster in terms of covered and uncovered events. For example, the first cluster, `lX_xi_threadX_syn_matchX_syn_mismatchX_X`, has a space of 1024 events, out of which 640 are defined. Of these 640 events, 320 (or 50%) are covered.

Figure 6 shows snippets of a report for the LSU cluster `lX_bias_lX_req_id_dX`. The cluster contains events for requests from two sources to two caches. Figure 6a shows the definition of this four-dimensional cross-product space. The first dimension (`V1`) contains the sources, the second dimension (`V2`) is the caches, the third dimension (`V4`) is the response, and the last dimension (`VM1`) contains the request id. Figure 6b shows the events defined in this cross-product. All the events with source `ls` or with response `roi` are defined, or, in other words, requests from `ic` cannot have response `norm` or `rej`. This leaves 424 events in the model out of the 636 events in the cross-product space.

Figure 6c shows the holes found in that cluster. The cluster contains one simple and large hole: there are no requests from `ic`. This representation of the hole is much shorter and easier to understand than the partial list of events that appears in Fig. 6d. Figure 6d shows that all the events in the hole are not covered at the unit level (as expected), but they are covered at the core level.

142	Space	Events	Covered		Uncovered	
Cluster	Size	#	#	%	#	%
lX_xi_threadX_syn_matchX_syn_mismatchX_X	1024	640	320	50	320	50
lX_bias_lX_req_id_dX	636	424	318	75	106	25
thX_X	2112	384	277	72.1	107	27.9
trace_array_X_bit_X_is_1	260	260	245	94.2	15	5.8

Fig. 5 List of clusters for the LSU

Name	Type	Size	Values
V1	string	2	{ic,ls}
V2	string	2	{l30,l31}
V4	string	3	{norm,rej,roi}
VM1	string	53	{d127,d16,d17,d18,d19,d20,d21,d22,d23,d

(a)

V1	V2	V4	VM1
{ls}	*	*	*
*	*	{roi}	*

(b)

Id	Events/Size	V1	V2	V4	VM1
1	106/318	{ic}	*	*	*

(c)

Entity	Varclass	Varname ↑	Unit	Core	Total
ex_ztop	za_intf_l3_xi	ic_l30_bias_roi_l3_req_id_d17	0	185 K	185 K
ex_ztop	za_intf_l3_xi	ic_l30_bias_roi_l3_req_id_d18	0	185 K	185 K
ex_ztop	za_intf_l3_xi	ic_l30_bias_roi_l3_req_id_d19	0	185 K	185 K
ex_ztop	za_intf_l3_xi	ic_l30_bias_roi_l3_req_id_d20	0	184 K	184 K
ex_ztop	za_intf_l3_xi	ic_l30_bias_roi_l3_req_id_d21	0	184 K	184 K

(d)

Fig. 6 Details of `lX_bias_lX_req_id_dX` cluster. (**a**) Space definition. (**b**) Defined events. (**c**) Hole description. (**d**) Events in the hole

Since the introduction of the clustering analysis to the users, the analysis and the holes it identified have helped to simplify and improve the coverage closure work. In addition, some of the holes identified by the analysis led to the discovery of a number of problems in the verification environment and in the test templates of several units. For example, they helped identify missing events that should have been defined.

3 Template Aware Coverage

The analysis described in the previous section does not include information about the relations between test-templates and coverage, and thus cannot provide answers to questions such as "What is the best test-template to hit a given coverage event?" and "Does a given test-template achieve its coverage goals?" The common practice in verification environments is that when the simulation of a test-case completes successfully, its coverage results are submitted to the coverage engine. The coverage engine accumulates the results into a unified database that provides the data for the coverage status and progress reports. Due to the large number of test-cases, which can reach millions of tests a day, the coverage engine aggregates the coverage of all test-cases together, regardless of the test-template from which they originated. We call this approach *template blind*, because the information about the test-template is forgotten or lost.

Template Aware Coverage (TAC) [16] maintains a hit matrix H, with the test-templates as its rows and the coverage events as its columns. Each entry in the matrix contains the number of times a given test-template hit a given event. Given the hit matrix and the total number of test-cases successfully simulated for each test-template $W = (w_i)$, we define P, the hit-probability matrix, as $p_{i,j} = h_{i,j}/w_i$. With such a matrix, the questions above and other queries that relate events and test-templates can be answered.

There are two main approaches for accumulating the coverage data. The simplest and most common approach is to accumulate all data from the beginning of the project (or the last reset). The second approach looks only at fresh coverage data. To accomplish this, it accumulates coverage within a rolling window of, say, the last 2 weeks. TAC is designed to support the latter approach, which fits the IBM coverage methodology [40].

Based on the rolling window approach, the cumulative TAC matrix is defined as the matrix resulting from summing up the last K daily TAC matrices (e.g., 28 days). Figure 7 illustrates the daily update process, where the newest daily matrix is added to the current cumulative matrix, and the oldest daily matrix (from 28 days ago) is, first, subtracted from the current cumulative matrix, and then permanently deleted from the disk to save storage.

The next section describes how TAC data can be used in some potential use-cases. Due to the unique challenges in handling the big data involved, we briefly describe the implementation.

3.1 TAC Use-Cases and Queries

Understanding the relationship between coverage and test templates is an important layer in understanding the state and progress of the verification process. These relations can be used, for example, to show descriptive information such as "does

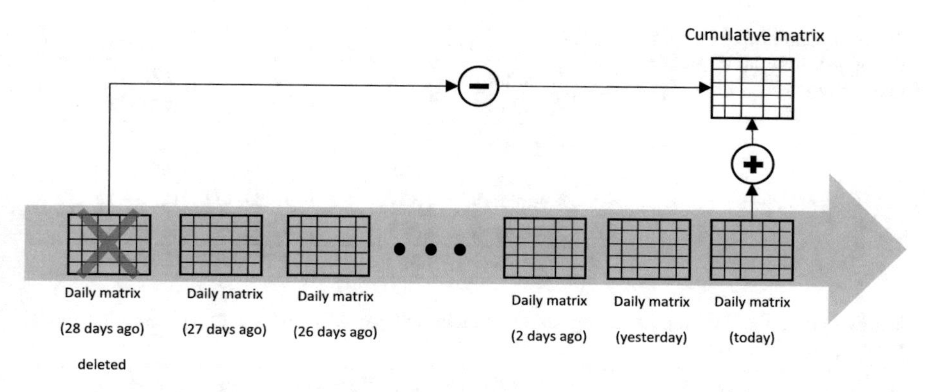

Fig. 7 The incremental update of the cumulative TAC matrix

a test template achieve its target verification goal?" To this end, the first and most basic type of TAC usage is the generation of descriptive reports that explore the relation between coverage and test templates. Some of these queries and reports are described below. In Sect. 4, we show how TAC data can be used to prescribe test policies that help achieve predefined coverage goals. These include items such as which test-templates to run and how many times to run each template.

3.1.1 Best Test-Templates to Hit an Event

The most basic TAC query is about the best test-template to hit an event. This query is used when a verification engineer decides to improve the coverage of a specific event. To achieve this goal, the engineer can run a TAC query to get a full list of test-templates that hit the event of interest.

Table 6 shows an example of such a report. By default, the table is sorted by the hit probability, from best to worst. With this information in hand, the verification engineer first tries to better understand the conditions for hitting the event, as given in the test-templates. Based on this information, the engineer can decide, for example, to allocate more simulation resources to the top three test-templates to increase the coverage of the event. Note that the test count and the cycle columns presented in the table refer to the number of times the test-template was run and its total number of simulation cycles. These statistics can give some confidence about the reliability of the data. Because the probability presented is empirical, the number of runs is a good indicator for the confidence in the data.

Table 6 Sample TAC report listing test-templates that hit a given coverage event

#	Test template	Test count	Cycles	Hit probability
1	Template1	384	24,823,452	0.828125
2	Template2	355	22,521,283	0.811268
3	Template3	398	39,382,138	0.809045
4	Template4	386	22,052,294	0.803109
5	Template5	357	21,831,271	0.792717
6	Template6	388	24,735,940	0.786082
7	Template7	389	24,705,637	0.766067

Table 7 Sample TAC report listing test-templates that partially cover a given coverage model consisting of 156 events

#	Test	Test count	Cycles	Hit events %	Hit events	Mean hit probability
1	Template1	2400	117,241,582	80.77	126	0.114391
2	Template2	2403	118,006,467	76.92	120	0.11444
3	Template3	390	23,579,020	75.64	118	0.131114
4	Template4	363	21,658,257	75	117	0.112012
5	Template5	386	23,653,018	74.36	116	0.129866
6	Template6	367	21,477,811	73.72	115	0.065867
7	Template7	333	20,296,061	73.08	114	0.10882
8	Template8	343	16,684,543	71.15	111	0.127177
9	Template9	335	19,523,881	71.15	111	0.063261
10	Template10	380	19,410,258	70.51	110	0.11164

3.1.2 Best Test-Templates to Hit a Coverage Model

A straightforward extension of the previous use-case is to improve the coverage of a set of events, or a coverage model. The TAC data offers the verification team additional information they can use. They can start with a listing of the test-templates that cover (fully or partially) this coverage model. Table 7 shows, as an example, the per test-template coverage report of an arbitrary coverage model that consists of 156 events. This report tells us that $Template1$ hits the maximum number of events in this coverage model (126 out of 156). It also shows the per test-template average hit probability in this coverage model (excluding the non-hit events). This report provides a lot of information to its users. For example, drilling down to each test-template can show the detailed list of the events hit by each test-template. Nevertheless, deeper analysis is required to decide which test-templates should be run to improve the coverage of this coverage model. This topic is discussed in Sect. 4.1.

Table 8 An example of a TAC report displaying a list of events hit by a given test-template. This test-template was run 20 times

#	Event	Hit probability	Test hits
1	Event1	1	20
2	Event2	0.9	18
3	Event3	0.9	18
4	Event4	0.9	18
5	Event5	0.8	16
6	Event6	0.7	14

Table 9 An example of a TAC report displaying a list of coverage-models partially covered by a given test-template

#	Coverage model	Number of events	Hit events	Hit events %
1	CovModel1	190	95	50
2	CovModel2	26	7	26.92
3	CovModel3	1535	280	18.24
4	CovModel4	159	27	16.98
5	CovModel5	156	17	10.9
6	CovModel6	1960	140	7.14
7	CovModel7	16	1	6.25

3.1.3 Coverage of a Test-Template

In this use-case, we shift our focus to the test-template. Specifically, we would like to explore the performance of a given test-template, to answer questions such as "Is it achieving its coverage goals?" and "What is its contribution to the overall coverage state?" To answer these questions, the verification engineer starts by examining the list of events that were hit by the test-template. Table 8 shows an example report including 6 events that were hit by a given test-template that was run 20 times. From such a report, the verification engineer can read the probability of her test-template hitting each of the events, as well as the actual number of hits. From this information, she can learn whether the test-template is hitting the events it is supposed to hit, hence whether it is doing its job or not.

An important extension of this report is the report showing all the coverage models that were covered fully, or partially, by a given test-template. Table 9 exemplifies such a report. Each row in this report presents the name of the coverage model, the number of events belonging to it, how many of these events were hit by the test-template, and the percentage of hit events.

3.1.4 Uniquely Hit Events

A *uniquely hit event* is an event hit by test-cases generated from a single test-template. That is, no other test-template hits it. Test-templates responsible for uniquely hit events offer a unique contribution to the coverage, reflecting the fact that they exercise singular scenarios or areas in the code.

Using TAC data, we can create a report of these uniquely hit events along with the test-templates that hit them. Table 10 exemplifies such a report. Note that a single test-template may hit many uniquely hit events, like $Template3$ in Table 10. This observation takes us to the next report, which lists all test-templates that hit uniquely hit events. An example of such a report is presented in Table 11. It is worth mentioning that this set of test-templates is the minimum necessary set to ensure that all the uniquely hit events are covered in the DUV.

Table 10 An example report of TAC listing the uniquely-hit events along with the test-templates that hit them

#	Event	Test template	Test hits	Test runs	Hit probability
1	Event1	Template4	2	36,802	0.000054
2	Event2	Template65	2	79,916	0.000025
3	Event3	Template3	1	116,599	0.000009
4	Event4	Template12	2	36,802	0.000054
5	Event5	Template23	1	36,802	0.000027
6	Event6	Template27	3	79,916	0.000038
7	Event7	Template9	9	71,264	0.000126
8	Event8	Template3	2	35,652	0.000056
9	Event9	Template11	2	131,517	0.000015

Table 11 An example report of TAC listing the test-templates that hit uniquely-hit events

#	Test template	Test count	Cycles	Uniquely hit events
1	Template1	14,820	623,538,730	42
2	Template2	3253	39,943,270	38
3	Template3	14,770	382,130,326	32
4	Template4	68,520	1,284,360,421	17
5	Template5	7737	110,414,380	10
6	Template6	79,916	2,007,702,100	10
7	Template7	322	6,717,904	10
8	Template8	36,802	1,783,010,147	8
9	Template9	2618	105,486,836	5
10	Template10	460	3,353,615	3
11	Template11	128,384	2,620,074,221	3
12	Template12	73,897	3,821,274,928	2

The query for uniquely hit events can be easily extended to determine whether a set of test-templates has a unique contribution to coverage. This is done by summing the rows corresponding to the test-templates in the hit matrix into one row, and checking whether it has a unique contribution.

3.1.5 Aged-Out Events

Since coverage is tracked on a daily basis, the coverage monitoring system can detect changes in the coverage trends and alert the verification team about these changes. One important trend is related to *aged-out events* [5]. These are events that were previously hit, but have not been hit in the last L (e.g., 14) days. Aged-out events are important because they indicate that something changed in the model or that the relevant test-templates are no longer running. Aged-out events or events that are about to age-out can be detected and reported using standard coverage reports. With TAC, the verification team can take the next step and find out which test-template hit (or best hit) these events last and take corrective actions. For example, if a test-template that previously hit an aged-out event is no longer running, it can be scheduled with increased priority. If the test-templates that last hit the event are still running but not hitting the event, a further investigation is needed to determine which changes in the DUV, the drivers, or the test-template caused the test-templates to stop hitting the event and why.

3.2 *Implementation*

3.2.1 Data Structures

At the heart of TAC lies the coverage hit matrix, which contains the hit counts of each test-template for each coverage event defined in the verification model. In large verification projects, there may be thousands of test-templates, and hundreds of thousands of coverage events. Keeping such matrices in memory is impractical, and storing them on disk would yield long processing and querying time. Most of the test-templates are written to target specific areas or features in the DUV. Therefore, it is reasonable to think of using some kind of sparse matrix representation (i.e., a matrix that holds the non-zero elements only). Empirically, the sparsity of these matrices, as measured in the verification environments of IBM's high-end processors, ranges from 65% to 85%. Here, sparsity is defined as the number of zero-valued elements in the matrix divided by the total number of elements in the matrix.

A suitable sparse matrix representation that fits our requirements is the *Coordinate List* (a.k.a. COO) representation [10]. The COO stores an ordered list of <*row, column, value*> tuples for each non-zero element of the matrix (see Fig. 8). The list is kept sorted according to the **row-major order** format. That is, consecutive elements of a row reside next to each other. As we shall see next, this representation is efficient for matrix construction, addition, and subtraction. The row-major order is

$$\begin{pmatrix} 2 & 0 & 0 & 0 \\ 5 & 0 & 7 & 0 \\ 0 & 0 & 0 & 0 \\ 0 & 1 & 0 & 0 \end{pmatrix} \rightarrow \begin{bmatrix} (0,0,2) \\ (1,0,5) \\ (1,2,7) \\ (3,1,1) \end{bmatrix}$$

Fig. 8 An example of a regular matrix representation and a COO sparse matrix representation

also efficient for template-based queries that are based on rows of the matrix, such as the events covered by the test-template in Sect. 3.1.3. For event-based queries, such as the best template to hit an event as in Sect. 3.1.1, the transposed matrix is also kept in COO representation.

To allow quick access to the rows of the hit matrix, an index file that points to the beginning and end of each row in the matrix is kept alongside the matrix. This index allows an $O(1)$ random-access operation to rows of the matrix as well as an efficient row-iterator interface. Creating the index requires a single pass over the matrix cells.

3.2.2 Sparse Matrix Operations

Big matrices pose a technical challenge because it is impractical to hold the entire matrix in memory. One way to overcome this hurdle is to keep the matrices on disk and only load a small portion of it to memory. This, in turn, creates the need to implement matrix operations and queries in such a way that minimizes disk access. Efficiently implementing the operations needed to create and maintain the cumulative hit matrix of TAC (i.e., construction, transposition, addition, subtraction) is based on merging two or more COO matrices residing on disk into a single COO matrix. The merge is performed using a minimum priority queue to prioritize reading from the input matrices. Specifically, the queue is filled with at least one cell from each input matrix, and then, the following is repeated until the queue is empty: remove the cell with the minimal key from the queue, add the removed cell to the output matrix, and refill the queue with cells from the input matrix that the last removed cell came from. To improve efficiency, the reading of input matrices and the writing of the output matrix are done in chunks.

Given the merge operation, the daily matrix is constructed from a stream of unordered cells by collecting a manageable number of cells, converting them into a COO in memory, and storing the COO in a temporary file. Once all the incoming data is handled, all the temporary files are merged into the final COO matrix.

Transposing a matrix is done by breaking the matrix into chunks that can be handled in memory, transposing each chunk, converting the transposed chunk into a COO, and storing the transposed chunk in a temporary file. Once the entire input matrix is handled, all the temporary files are merged, similar to the construction operation.

Adding and subtracting two matrices is simply a merge of the two matrices. This merge must handle the case when the minimal entries in the matrices have the same index. In this case, the two entries are consumed and their values are added

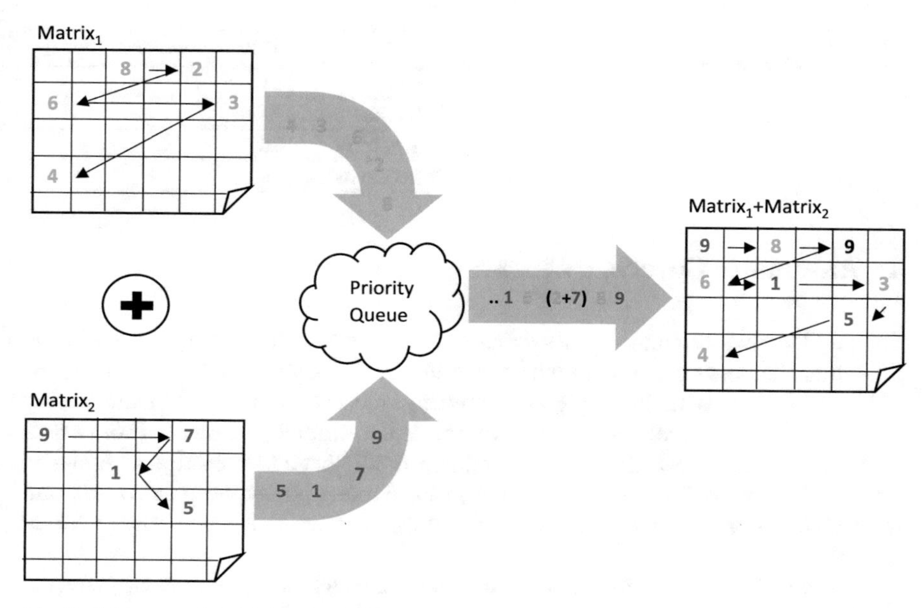

Fig. 9 Sparse matrix addition operation

or subtracted. Figure 9 shows an example of a sparse matrix addition. The input matrices are $Matrix_1$ and $Matrix_2$, with only the non-zero elements showing. The figure shows how the entries for index $(1, 5)$ in the two matrices are combined in the resulting matrix.

3.2.3 Performance

In a verification environment with thousands of test-templates and hundreds of thousands of coverage events, the cumulative sparse TAC matrix produced can still be very large (e.g., 10GB on disk in binary format). But, with the help of the index file, querying such a matrix takes only few seconds or a fraction of a second, instead of several minutes without the index file.

It is noteworthy that the matrix operations and the index creation are performed offline. This is done during the phase of preparing the cumulative TAC matrix. The random-access operation is used during the phase of serving the descriptive and prescriptive coverage reports. Table 12 shows the approximate time measurements for the offline operations when performed on a cumulative TAC matrix with 17k test-templates and 135k events, 82% sparsity, and 11 GB size on disk.

Table 12 Time measurements of matrix operations

Operation	Time measurement (mins)
Add	21
Subtract	25
Transpose	16
Create Index	6

4 Automatic Coverage Closure

The previous two sections discuss extracting information out of coverage data, and providing the verification team with a handful of descriptive analytics techniques. This section deals with the holy grail of coverage closure, namely, automatic closing of the coverage feedback loop from coverage data to stimuli generation. From a data analytics point of view, this means providing predictive and prescriptive analytics that predict how coverage will improve given the test policy we plan to run, and furthermore suggesting new policies and test-templates that can improve future coverage.

It is worthwhile looking at how coverage closure is done manually by the verification team. There can be multiple reasons for a coverage hole, or a set of events that are never hit: the events may be impossible to hit under the design and environment limitations, hence should be removed; a driver that is too restricted may need to be fixed; a test-template may be missing to exercise the underlying functionality related to these coverage events; and last, it may be a bug in the implementation of the coverage. We now focus our discussion on the third problem, and suggest ways for dealing with it automatically.

Consider a verification expert with the task of thoroughly covering a set of coverage events; some of them are lightly hit, while the rest are never hit. There is a known hidden assumption in the way a task is defined, which is that the events are related and hence can be handled together. This property of the likelihood of events is well known. A coverage model for the different fill level of a buffer is an example of this likelihood. A test that hits a *buffer-almost-full* event is clearly a good starting point for hitting a *buffer-full* event. A cross-product coverage model also defines a useful metric for measuring the distance between events. Events that differ in a single attribute can be considered close to each other; this is referred to as the Manhattan distance.

Using the likelihood property, our verification expert starts by looking for a test-template that hits the lightly hit events. If this test was not used in the regression to generate enough test instances, she will run it more times, and see if this is enough to hit all the target events. If not, she will look for a way to improve it, or write a new test-template that will increase the probability of hitting the target events.

We describe automatic coverage closure using two approaches. These approaches mimic and automate similar actions taken by experts. The first approach for improving coverage is to optimize the utilization of the existing inventory of test-templates. Given the limited resources for simulation, the verification team cannot run each test-template an infinite number of times, hence exhausting its overall potential. We can use TAC data to suggest an optimized policy that minimizes simulation runs while maximizing coverage. This approach is discussed in Sect. 4.1.

A second, and in a sense more ambitious approach, is to use analytics to suggest new test-templates that will improve the coverage by hitting events that are not hit (or very rarely hit) by the current test-templates. This approach is discussed in Sects. 4.2–4.3.

4.1 Coverage-Based Regression

We describe here the approach, techniques, and use-cases for optimizing the use of the existing inventory of test-templates to improve the coverage state. In addition to providing a deeper analysis of the coverage state, TAC data, presented in Sect. 3 can also be used to create efficient coverage-based regression suites [9]. There are many different uses for regression suites, each with its own requirements. The most trivial prescriptive policy is the answer to the query about the best test-template t to hit an event. This is an example of prescriptive analytics, with straightforward action following: running t more times may improve the coverage of the event.

A *test policy* $TP = ((t, w)_i)$, or regression suite, is defined by the list of test-templates t from which we will generate test instances, the weights w, which indicate how many times to run each test-template. For simplicity, we will use $TP = (w_i)$.

To validate a new model under CI/CD methodology, an ideal regression suite would be a light one (given limited budget) that targets a wide coverage, but with limited guarantee that each event is hit. To minimize the use of resources while achieving a similar coverage goal, again we target wide coverage, but this time we would like to ensure a high probability for hitting each event, even at the cost of large regression suite. We can also target a subset of the events, whether representing a feature in the design on which the verification team is focusing, or events that are lightly hit. We define lightly hit events as events that were hit less than a given threshold, for example, 100.

Next, we formally define the problem of finding an optimized policy as two optimization problems and discuss the methods to solve it.

4.1.1 Finding an Optimized Test Policy

We define two optimization problems that suit different verification goals and yield different policies. We follow [16] and [9] and use the TAC hit matrix. We denote the probability that a test instance, generated from test template i, will hit event j by $p_{i,j}$ (computed by TAC matrix); for a given test policy, we denote the probability of hitting event j by P_j. P_j can be computed by looking at the probability that the event j was never hit, giving $P_j = 1 - \prod_i (1 - p_{i,j})^{w_i}$. For the following two definitions, we assume that events that were not hit by any test are omitted, as the probability for hitting event j, $P_j = 0$ for them.

Definition 1 (*Optimized Test Policy–Strict*) Find a test policy $TP = (w_i)$ that minimizes the number of simulation runs, with probability Ψ of covering each event:

$$\min_{TP} \sum w_i \tag{1}$$

$$\texttt{s.t.} \forall j \, P_j = 1 - \prod_i (1 - p_{i,j})^{w_i} \geq \Psi \tag{2}$$

$$\forall i \; w_i \geq 0, w_i \in \mathbb{N} \tag{3}$$

This is an integer programming problem. Because the resulting $\{w_i\}$ are large, it can be relaxed to the real domain with rounding of the final solution. We can further use common techniques [9] to transform it into a linear programming (*LP*) problem. Applying the log operation to both sides of Eq. 2, we get

$$\sum_i w_i \cdot log(1 - p_{i,j}) \leq log(1 - \Psi) \tag{4}$$

Definition 2 (*Optimized Test Policy–Average*) Find a test policy $TP = (w_i)$ that minimizes the number of simulation runs, with an **average** probability Ψ of covering all events:

$$\min_{TP} \sum w_i \tag{5}$$

$$\texttt{s.t.} \frac{1}{|E|} \sum_j P_j \geq \Psi \tag{6}$$

$$\forall i \; w_i \geq 0, w_i \in \mathbb{N} \tag{7}$$

This is a nonlinear problem. The "log trick" used to transform the strict problem to linear cannot work here, due to the summation over the events.

The two problems model different verification goals. The strict problem models the goal of hitting every event with a high probability. Here we are ready to invest many test runs to cover the most hard-to-hit events (i.e., events with a low hit probability). This results in large test policy solutions due to the many runs needed to target these hard-to-hit events. The average problem, on the other hand, allows us to leave some of these hard events not covered, leading to smaller solutions.

To solve the strict policy problem, we can use linear programming algorithms and commercial solvers like [27]. For the average algorithm, we suggest the greedy algorithm described in Algorithm 3, which can be thought of as a probabilistic version of the set-cover problem greedy algorithm. At each step, the algorithm computes the potential contribution of each test-template to the average hit probability of all the events, and chooses the test-template that gives the maximal added value. It is important to note that the marginal potential contribution of a test-template decreases as its weight in the policy increases. Furthermore, the potential contribution of any test-template needs to be computed at each iteration, since the current policy (w_i) is updated on each iteration. The complexity of the greedy algorithm is $|w| \cdot |E| \cdot |T|$. Practically, for big TAC matrices, finding a policy for a strict solution is efficient (less than 1 min), while for the average problem, very large policy computations can take several hours.

We suggest two more practical improvements to the greedy algorithm. First, we add the $size$ parameter, which allows us to limit the size of the returned policy. Second, we saw that in many cases, the same test-template is selected for a few consecutive times. Hence, after we find the best test to improve our policy average probability, we add it K times instead of one. Although this may impact the output policy, it can dramatically improve performance. In our solution, we use $K = 10$. This factor reduces the runtime of the algorithm by almost a factor of 10, and while it increases the number of times most tests are used in the policy, for most tests and for the entire policy size, the change is relatively small.

We are now ready to map verification goals, which are described by coverage goals, as optimization problems, to find optimized test policies.

4.1.2 Mapping Verification Goals to TAC-Optimized Test Policies

Light Regression for a New Model Building a new model is a task frequently done during the verification cycle. A new model includes bug fixes, alongside new features in the design and in the test bench. As CI/CD (continuous integration and continuous deployment) methodology becomes common practice, the frequency of releasing a new model can reach several times a day. When building a new model, it is important to quickly confirm that it is not broken and can be used by the entire team. This can be achieved by covering events from all features and coverage models; no specific event is important. The proper regression suite for this task includes the entire coverage space, with a relatively low value for average hit probability (e.g., 50%), and is based on the average algorithm. The resulting regression suite should contain a small number of test instances (this can also be a parameter for the average algorithm), so it can be executed quickly and provide immediate feedback on broken models. An improvement to this policy can look at the areas of the design that were modified from the last stable model, and allocate part of the CI/CD testing budget to it. Using TAC for this goal requires mapping the design changes into the coverage space. Once this mapping is provided, we may again create another light regression, using the average problem with relatively low value for average hit probability (e.g., 50%).

Wide Regression When approaching a major milestone, it is important to ensure that everything that was verified still works. A common way to achieve this is by re-

Algorithm 3 Greedy algorithm for optimized test policy (average problem, Definition 2)

procedure FINDOPTAVGPOLICY(T, E, Ψ)
  // Inputs: A test templates set for selection T, target events E, target average probability ψ
  // Output: A test policy $w = (w_1, ..., w_{|T|})$*, average probability* $avgP$ *and policy size* $tpSize$

  // Initialization
  $w = (0)$
  for each event j **do**
    $P_j = 0$
  $avgP = 0$
  $tpSize = 0$

  // Execution
  while ($avgP < \Psi$) **do**
    for each $testi$ **do**
      $d_i = \sum_j p_{i,j}(1 - P_j)$
    $k = argmax(d_i)$
    // Update policy
    $w_k = w_k + 1$
    for each $eventj$ **do**
      // update P_j *assuming we run test k*
      $P_j = 1 - (1 - P_j)(1 - p_{k,j})$

    // Update average probability and policy size
    $avgP = \frac{1}{|E|} \sum_j P_j$
    $tpSize+ = 1$

  Return W, avgP, tpSize

hitting all coverage events that were hit in the past. For this goal, we use a regression suite based on the entire coverage space, along with all the test-templates. To ensure high coverage, we use the strict optimization policy of Definition 1 with a high hit probability threshold Ψ (e.g., 90%).

This policy raises questions regarding the test-templates that are left aside; this may be more than half of the test-templates. Are these test-templates needed or are they redundant given the selected policy? We claim that the answer is complex since not all the information is given by the coverage. For example, test-templates that found bugs, or are easy to debug, may still be valuable. Nevertheless, the data provided by the optimization engine is a valuable input for the decision to remove redundancy from the regression.

Optimized Policy to Verify a Feature Following a bug fix, it is important to run a thorough regression on the feature related to the bug. The proper regression suite is based on the coverage models related to the feature, uses the strict optimization policy, and has a high hit probability (e.g., 90%).

Covering Hard-to-Hit Events Hard-to-hit events are events that were rarely hit. Our definition for hard-to-hit events is events that have been hit less than 100 times in the last 2 weeks. This means that not enough exploration was done surrounding these events and there is a higher probability of unexposed bugs there. Therefore, it is important to hit hard-to-hit events more often. To do that, one needs to create a regression suite that focuses specifically on these hard-to-hit events. This can be achieved by selecting these events as the coverage goal and using the strict optimization policy.

Note that there can be several reasons for events to be lightly hit. We may have a good test-template for the event (i.e., hitting it with high probability) that was not run enough times. However, it may also be that the best test-template to hit these events did run many times, but hit the event with very low probability. Many events of the second type will yield a very big optimized policy that is sometimes not feasible for the team to run. In this case, the verification team has to improve the test suite, either by improving an existing test-template or by writing a new one. From our experience, the latter is almost always the case.

A TAC-optimized policy can still help the team save valuable time. Instead of the strict algorithm, we use the average algorithm to create a policy with a given budget of test runs. This policy will be biased toward the easier hard-to-hit events. Covering these events in a sufficient way, within a limited budget, means that the verification team can focus on the events that are hardest to hit. Figure 10 shows the impact of using this policy for the core-level verification of a high-end processor. It can be seen that within 12 days of running this policy, the number of lightly hit events was reduced by over 25%.

Usage results show that using the hard-to-hit policy not only improves the coverage of the target events, it usually hit events that were never hit before. This is a result of the likelihood property of events described in Sect. 4. Running more test-templates that hit lightly hit events can also increase the probability of hitting their close neighbor events. This property lies at the heart of our CDG solution described in Sect. 4.2. Furthermore, this policy has already found many bugs; some declared by the team as *quality bugs*. This is an indication of the quality of the coverage, since this policy steers the generation to exercise areas in the design that were lightly tested.

4.2 Coverage-Directed Generation

So far, we discussed methods to optimize test-template selection and scheduling to advance coverage goals. None of these methods modified the test-templates themselves, and their impact is thus inherently limited. If all existing test-templates have zero probability of hitting a given coverage event, changing test-template scheduling alone will not help. *Coverage-directed generation* (CDG) [44] is a generic name used for a multitude of techniques that **create** tests or test-templates for hitting uncovered events.

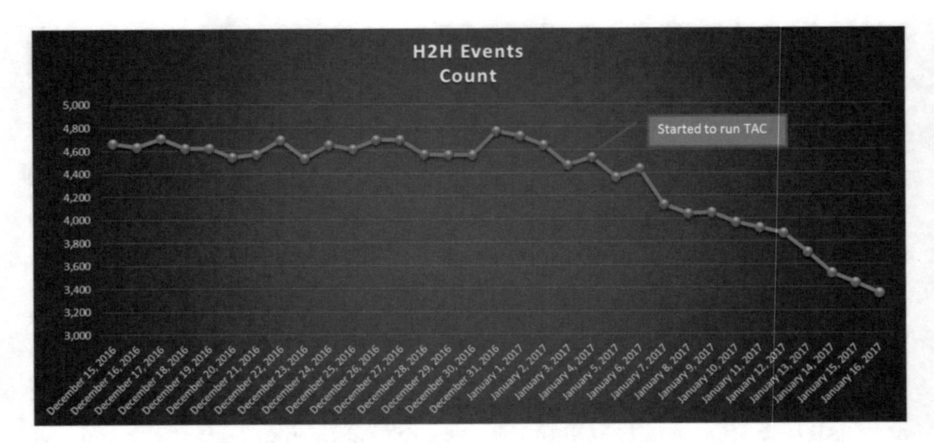

Fig. 10 The impact of running TAC hard-to-hit policy, on the core-level verification of high-end processor

Over the last decades, CDG has received a lot of attention in both academia and industry, due to its huge potential to preserve verification resources. Early CDG papers showed encouraging results; however, none of them matured into an industrial solution. This is due to scalability issues, usage complexity, and having DUV-specific components that limit the generality of the solution. In general, approaches for CDG can be classified into two main categories: model-based CDG [36, 44] and data-driven CDG [12, 42, 43, 45].

In model-based CDG, a model of the DUV is used to generate test instances or test-templates. Reference [36] uses a formal architecture model, while [44] uses a micro-architecture model. Both techniques failed to scale due to the limitations of formal methods. In addition, these techniques require the model definition to be relatively accurate in order to hit hard events, making its definition a complex task that requires high maintenance during a project's lifetime.

In data-driven CDG, the system discovers and learns the complex relations between the test-template parameters and the coverage events. Reference [45] uses Markov chains to capture the relationship, where the chain is design and domain specific. Reference [42] uses a genetic algorithm approach to generate new test instances. One major problem with this method is how it handles the validity of the evolutionary tests. Reference [12] uses Bayesian networks to guide the input generation. While the network weights are learned automatically, the network topology is DUV-specific and requires domain knowledge.

Fuzzing techniques are another option that originated in the software testing realm. Reference [33] suggests using this to mutate existing test instances. This technique requires changes in the DUV and has limitations regarding the validity of the tests created.

In the rest of this section, we describe a recent field-proven approach first described in [21], called AS-CDG, which casts the CDG problem as an optimization

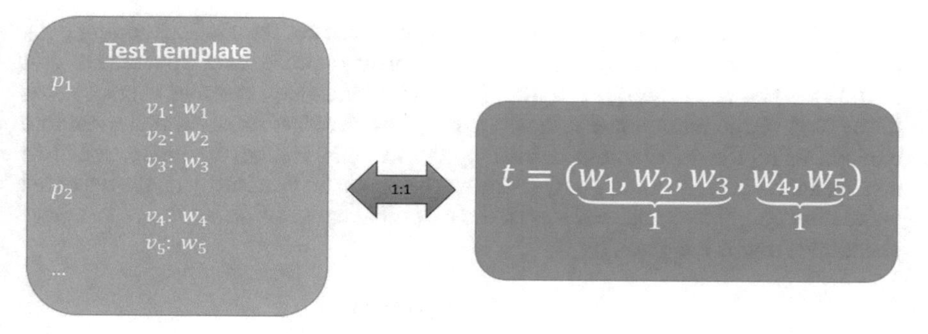

Fig. 11 Representing a test-template as a test-template vector

problem. This approach assumes that for a given group of coverage events, a candidate test template was already chosen, e.g., using the methods described in Sect. 4.1. Furthermore, it assumes the candidate test-template has a set of parameters or directives that may affect its probability of hitting the target coverage events. Finally, it assumes that each parameter is a set of value-weight pairs. When the random stimuli generator needs to make a random decision related to a given parameter, it uses the weights as a distribution function for selecting random values. AS-CDG attempts to find the weights that maximize the probability of hitting the target coverage events. By modifying only weights, the test validity is easily ensured.

We first formally define the problem of tuning test-template parameters to maximize their probability of hitting target events. We then describe two techniques, namely, random sampling and optimization, which AS-CDG uses to solve this parameter-tuning problem. Finally, Sect. 4.3 presents utilization of machine learning algorithms to further improve the overall flow.

4.2.1 Problem Definition

We assume each test-template t comprises a set of parameters $\{p_1, p_2, \ldots, p_k\}$. Each parameter p_i can be assigned with a set of values $\{v_{i,1}, \ldots, v_{i,m_i}\}$, and each value $v_{i,j}$ is associated with a weight $w_{i,j}$ that corresponds to the probability of choosing this value. The weights of all values of a given parameter define a distribution space; they are all nonnegative and their sum is 1.

The optimization process in AS-CDG starts from a given test-template t_{orig}, and modifies only its weights to produce mutant test-templates. We can therefore think of any test-template generated in this process as a vector of weights, as demonstrated in Fig. 11. Note that the sum of each parameter's weights is 1. Thus, the objective of the optimization process is to find the weight vector that maximizes the probability of hitting target events.

Let $\mathcal{C} = \{c_1, \ldots, c_m\}$ be the set of coverage events for the DUV. Running $\theta(t)$, a *test-instance* generated from some test-template t, produces a coverage vector $s(\theta(t)) = \{s_1, s_2, \ldots, s_m\}, s_j \in \{0, 1\}$, where s_j indicates whether event c_j was hit by $\theta(t)$. Generating a test instance from a test-template involves many random choices, using the distributions defined by the weights. Hence, the outcome of two test instances, $s(\theta_1(t))$ and $s(\theta_2(t))$, may not be equal. To this end, $e(t)$, the expected value of hitting the events, and $e_N(t)$, the empirical expectation of N test instances generated from t, are given by

$$e(t) = \mathbb{E}[s(\theta(t))]$$

$$e^N(t) = \frac{1}{N}\sum_{i=1}^{N} s(\theta_i(t))$$

While $\forall j, s_j \in \{0, 1\}$, the vectors $e(t)$ and $e^N(t)$ contain real values in the interval $[0, 1]$, where the j-th values are the probability and estimated probability of the event c_j being hit by a test instance generated from template t. Clearly, $e(t)$ cannot be directly observed or calculated and so we must rely on its sampled estimation, $e^N(t)$. This estimation is subject to *sampling noise*, which may depend on the number of samples N or the specific template t, and may even change from one coverage event to another.

Consider a single target event c_j for which we would like to maximize the probability of it being hit. The goal of AS-CDG would be to find the test template t_j^{max} that is a variant of t_{orig} and maximizes the probability of hitting c_j. That is,

$$t_j^{max} = \underset{t}{\operatorname{argmax}}\, e_j(t) \tag{8}$$

As mentioned before, $e_j(t)$ is unknown. We therefore try to maximize its sampled estimation $e_j^N(t)$. Of course, we must ensure that sampling noise does not disturb the optimization process and does not distort our results.

4.2.2 Approximated Target Function

Noise is not the only hurdle the optimization procedure will have to overcome. Because our target events are exactly those events that were not yet hit (or were very rarely hit), it is likely that most variants of our original test template t_{orig} will also have nearly zero probability of hitting these events. In other words, for most variants, $e_j^N(t)$ is likely to be very small or even zero.

This means that any search or optimization technique is doomed to aimlessly wander in a mostly "flat" landscape. The presence of sampling noise makes this hurdle even worse. Weak signals suffer from an unfavorable signal-to-noise ratio, which can divert search algorithms from the right direction.

To overcome this problem, AS-CDG defines an approximated target that provides a stronger signal. This approximated target is based on the coverage of events that are in strong correlation with the target events (a.k.a. neighbor events) and that are easier to hit. The idea, which mimics the work of verification experts, is that by improving the probability of hitting these neighbors, we exercise the relevant area in the DUV. This, in turn, increases the probability of hitting the target event itself. Once again, this is a result of the likelihood property of events described in Sect. 4.

There are many possible ways to automatically find the neighbors of a coverage event. For example, in [45], the natural order of buffer utilization is used to learn how to fill a buffer. Reference [12] exploits the structure of a cross-product coverage. In [19], formal methods are used to find a set of neighboring events with positive and negative information regarding the probability of a test hitting the target event.

The above methods provide a set of relevant coverage events $\{c_{j_1}, \ldots, c_{j_n}\}$ that include both target events and neighboring events. The methods may also provide coefficients $\{a_1, \ldots, a_n\}$ for the set of events, indicating the importance of each event in improving the probability of hitting the target events. For example, when trying to hit a buffer-full event, the events describing the buffer's fill level may have larger coefficients assigned to events that indicate the buffer is fuller. The goal of the optimization process is to find the test template t^{max} that is a variant of t_{orig} and maximizes the approximated target function. That is,

$$t^{max} = \underset{t}{\operatorname{argmax}} \left[\sum_{k=1}^{n} a_k \cdot e_{j_k}^{N}(t) \right] \tag{9}$$

Note that using a larger set of more-easily hit coverage events can also help us in selecting t_{orig}, as the available hit statistics are now much more significant. This makes the techniques described in Sect. 4.1 more reliable.

4.2.3 Random Sample

A lightweight search technique performs a random sample of the parameter settings and measures the statistical estimate for the approximated target on each sampling point. In the random sampling process, we create J random test-templates that uniformly span the weights defined in t_{orig}. We then generate and simulate N test instances from each template. The coverage obtained from the simulations is then used to estimate the probabilities of hitting each event c_{j_i} by each template t, and to calculate the approximated target function according to Eq. 9.

The random sample requires $J \times N$ simulations, and while its probability of hitting the actual target events is low, it serves two purposes. First, the results of this step can be used to perform sensitivity analysis on the relevant parameters, to check if there are parameters that do not affect the relevant coverage events or have a minimal effect. Then, such parameters can be removed if the number of relevant parameters is too high.

A second purpose of the random sample is to find a good starting point for the subsequent optimization step. Specifically, the optimization step can begin with the test-template that reaches the highest target-function value. This good starting point can save the optimization algorithm many iterations of wandering in an almost flat area reached by a random start. This makes the investment of $J \times N$ simulations worthy.

Moreover, as shown in [21], the random sampling itself, which uses weights that were not considered by the verification team, often improves the hits of lightly hit neighboring events. In some cases, it even hits some never-hit target events. Figure 14 shows such a case.

4.2.4 Optimization

At the heart of AS-CDG lies an optimization algorithm that searches for test-templates to maximize the probability of hitting each of the target events. While casting CDG as an optimization problem is natural, there are two main challenges that complicate the use of optimization techniques for CDG. The first challenge is the nature of the objective function. As explained earlier, we do not have direct access to the objective function. Instead, we must rely on an estimate of the function obtained from simulations. Therefore, commonly used optimization methods that rely on first-order derivatives (gradient methods) and second-order derivatives (Hessian methods) of the objective function cannot be applied. To address this challenge, we rely on *derivative-free optimization* (DFO) methods [8]. These methods require only samples of the objective function itself, without the need to calculate its derivatives. However, this leads to the second challenge, which is the unknown dynamic noise in the observed objective value. To solve both challenges, the *implicit filtering* algorithm can be used. This is a DFO algorithm, known to work well on noisy target functions. Indeed, it was proven to be efficient in CDG settings [17].

Algorithm 4 describes the implicit filtering algorithm. The algorithm starts with t_0, the best random sample from the random sampling step. At each iteration of the algorithm, it selects n random directions and samples the objective function at points with distance h (a.k.a. step-size or stencil) from the current center in each of the selected directions. If the best value of these samples is better than the value at the center, the center is moved to that point and the process repeats. Otherwise, when the best result is at the center, the distance h is halved and the process repeats. This is done to reduce the possibility of overshooting the maximum. The algorithm stops when a stopping criterion is met. The stopping criterion is usually a combination of the number of iterations, the current stencil value, and the hit probability of the target event.

The implicit filtering algorithm has several hyperparameters: n, the number of directions used in each iteration; h, the initial stencil; and the stopping criteria. Each of these hyperparameters can affect the convergence rate of the algorithm in terms of iterations and number of samples.

Algorithm 4 Implicit filtering algorithm

```
procedure IF(n, N, h, t_0, stopping criteria)
    repeat
        best ← T_N(t_0)
        next_center ← t_0
        D ← vector of n random directions
        for each direction d in D do
            t ← (t_0 + d × h)
            if T_N(t) > best then
                best ← T_N(t)
                next_center ← t
        if next_center == t_0 then
            h ← h/2
        t_0 ← next_center
    until stopping criteria is met
    return t_0
```

When dealing with dynamic noise, we make two small modifications to the base algorithm. First, we use another hyperparameter N, the number of test instances to create from each test template. Increasing N reduces the effective noise and thus can lead to a faster convergence. On the other hand, increasing N increases the number of simulations needed per iteration. It is also a common practice to resample the center point in each iteration, even though it was sampled in the previous iteration. This resampling is used to reduce the effect of extremely high noise.

At each iteration of the implicit filtering algorithm, it creates $n+1$ test-templates, one for each of the n random directions and one for the center. Each of these templates is simulated N times, and the empirical expectation $e_N(t)$ for each of the events is calculated. This result is used to calculate an estimation for the approximated target, which is in turn used to calculate the starting point for the next iteration. The output of the algorithm is the best template found in the last iteration. Figure 12 shows how the optimization algorithm interacts with an existing simulation environment.

Figure 13 shows the maximal value of the target function per optimization iteration. The data refers to an optimization process that was applied to a group of coverage events in an L3 cache unit of a high-end processor. It can be seen that the optimization process makes gradual progress toward a local maximum value. The peak at iteration 10 is the result of sampling noise. As desired, the optimization algorithm was able to absorb this disturbance and get back on its track.

4.2.5 Combining Random Sampling and Optimization

As mentioned before, random sampling is a relatively low-cost, low-precision technique. Nevertheless, AS-CDG applies random sampling as a preliminary stage to the high-cost, high-precision optimization process, as this brings many advantages.

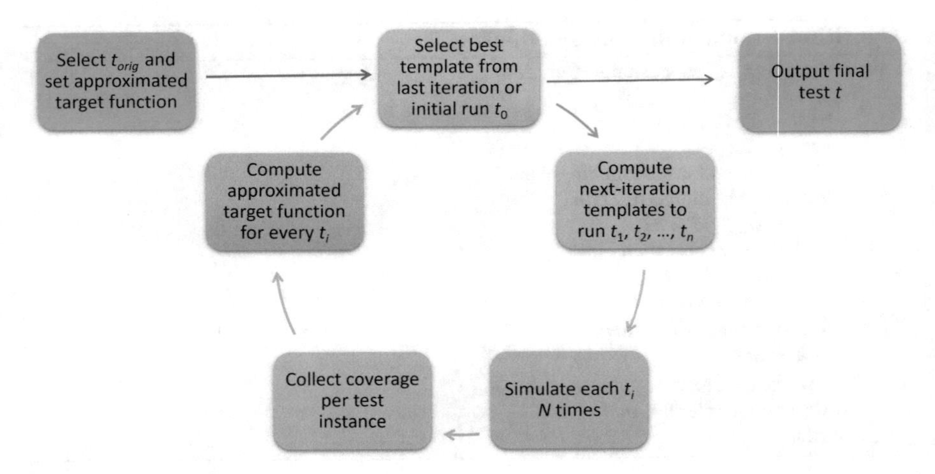

Fig. 12 Optimization flow – the two bottom stages are part of the existing simulation environment

Fig. 13 Optimization progress

The main advantage is providing a good starting point for the optimization process, to significantly increase its chances of converging to a higher local maximum.

Figure 14 shows the results of applying the overall AS-CDG flow to an L3 cache unit of a high-end processor. The table shows hit counts and hit rates at the various phases for a given family of coverage events. The approximated target function has set identical coefficients for all events. The color coding follows a convention where an event with a hit count smaller than 100 is considered lightly hit and is colored in orange. In addition, events with a hit rate smaller than 1% are also considered lightly

Event name	Before CDG (1,000,000 sims)		Sampling phase (210 tests x 100 sims each)		Optimization phase (25 iteration x 12 tests x 100 sims)		Running best test (15000 sims)	
	# hits	hit rate	# hits	hit rate	# hits	hit rate	# hits	hit rate
byp_reqs01	865,000	86.500%	18,109	86.233%	24,957	83.190%	14,478	96.520%
byp_reqs02	284,800	28.480%	10,015	47.690%	19,569	65.230%	13,038	86.920%
byp_reqs03	25,600	2.560%	4,425	21.071%	13,251	44.170%	9,608	64.053%
byp_reqs04	2,000	0.200%	3,109	14.805%	11,232	37.440%	8,409	56.060%
byp_reqs05	300	0.030%	2,558	12.181%	10,339	34.463%	7,989	53.260%
byp_reqs06	0	0.000%	2,052	9.771%	9,492	31.640%	7,536	50.240%
byp_reqs07	0	0.000%	1,598	7.610%	8,485	28.283%	7,026	46.840%
byp_reqs08	0	0.000%	1,242	5.914%	7,424	24.747%	6,387	42.580%
byp_reqs09	0	0.000%	903	4.300%	6,268	20.893%	5,559	37.060%
byp_reqs10	0	0.000%	640	3.048%	4,957	16.523%	4,537	30.247%
byp_reqs11	0	0.000%	389	1.852%	3,494	11.647%	3,401	22.673%
byp_reqs12	0	0.000%	193	0.919%	2,148	7.160%	2,245	14.967%
byp_reqs13	0	0.000%	90	0.429%	1,096	3.653%	1,180	7.867%
byp_reqs14	0	0.000%	33	0.157%	450	1.500%	462	3.080%
byp_reqs15	0	0.000%	6	0.029%	125	0.417%	127	0.847%
byp_reqs16	0	0.000%	0	0.000%	24	0.080%	15	0.100%

Fig. 14 Hit statistics for a family of events in an L3 cache unit of a high-end processor

hit. Never-hit results are colored in red. All other results (i.e., well-hit events) are colored in green.

The first two columns after the event name column show hit counts and hit rates before AS-CDG was applied. These are the results of applying mainstream unit simulation for several weeks, utilizing multiple test-templates. Using simulation statistics from TAC and expert advice, a test-template t_{orig} was selected, with parameters that should best hit the family of events at hand. Then, random sampling was applied, producing 210 test templates; each was simulated 100 times. This gives the numbers in the third and fourth columns. The next two columns show hit counts and hit rates after running the *optimization* phase. The starting point for this phase is the best test-template from the sampling phase (i.e., the one with the highest score for the target function). Finally, the last two columns show the results of running the best test-template from the optimization phase multiple times. Adding such best test-templates to the daily regression should be very beneficial.

These results show that the suggested flow improves the hit counts and hit rates for both families of events. Moreover, each phase improves upon its predecessor. Starting with 5 well-hit events and 11 never-hit events (looking at hit counts), sampling phase alone, using just 21,000 simulations, was able to turn 7 uncovered events into well-hit events and 3 uncovered events into lightly hit events. The optimization phase was able to then turn the 3 lightly hit events into well-hit events and the remaining uncovered event into a lightly hit event, using just 30,000 simulations. Finally, the best test-template shows significantly better hit rates.

Algorithm 5 Event after event implicit filtering (EE-IF)

1: Use random sampling, $T = [t_1^\top, \ldots, t_M^\top]$ and compute, $E = [e_1^\top, \ldots, e_M^\top]$
2: Set $\mathcal{I}$ to the events that are not hit above the threshold τ
3: **while** $\mathcal{I} \neq \emptyset$ **do**
4: Find e, the event in $\mathcal{I}$ with the highest probability in E.
5: Use implicit filtering to improve the hit probability of e to above τ
6: Augment the test-templates used in the implicit filtering to the matrices T and E
7: Remove from $\mathcal{I}$ all the events that are hit above the threshold τ

4.3 CDG for Large Sets of Events

In the previous section, we showed how DFO algorithms can be used to hit previously uncovered or lightly covered events. We also showed that the implicit filtering algorithm (Algorithm 4) nicely covers single events or small groups of related events. Often, during coverage closure, the verification team is required to improve the coverage of larger sets of events that are not necessarily related. In this section, we extend the implicit filtering CDG algorithm to handle large sets of coverage events and show how machine learning can be used to improve the performance of the implicit filtering algorithm specifically, and DFO search algorithms in general. Note that we slightly change the basic search problem here from finding the optimal test-template that maximizes the probability of hitting the target event to finding a test-template that hits the target event with a probability greater than a given threshold τ. This modification can be easily achieved by changing the stopping conditions in line 14 of Algorithm 4.

4.3.1 Event After Event Implicit Filtering for Multiple Targets

The simplest method for covering multiple unrelated events is to apply the implicit filtering algorithm to each of these events serially. Algorithm 5 provides some small improvement to this naive approach. Specifically, the algorithm keeps all the coverage results of all test-templates T used throughout its iterations in a matrix E. The algorithm uses these results to remove any event covered above the threshold τ from the list of remaining target events $\mathcal{I}$, even if they are not targeted by the implicit filtering algorithm.

At each iteration of the algorithm, it selects from $\mathcal{I}$, the set of remaining targets the event with the maximal probability and applies the implicit filtering algorithm to it. The implicit filtering algorithm starts its search from the test-template that best hits this event so far and searches for a test-template that hits the event with probability greater than τ. After the implicit filtering algorithm finishes, all events that are hit with probability greater than τ are removed from $\mathcal{I}$, and the algorithm continues until $\mathcal{I}$ is empty.

In line 4, the algorithm selects the remaining target with the highest hit probability as the next event to improve. This is a greedy approach that attempts to eliminate the easy targets first. This approach can be replaced with other approaches, such as randomly selecting the next event or selecting the events from the most difficult to the easiest. Our experimental results show that the greedy approach works best.

Within the optimization process, the algorithm collects a relatively large amount of simulation data that is added to T and E, but it uses this data very lightly. In the next section, we discuss how to use this data to learn more about the search space and speed up the implicit filtering algorithm.

4.3.2 Machine Learning Accelerated Implicit Filtering

To benefit from the collected data and reduce the number of simulations needed, we can replace the DUV simulation with a machine learning model that captures the relation between test-templates and the coverage. To this end, machine learning techniques, such as *deep neural networks* (DNNs) [15], can be used to build and train DNN models on the collected data to help cover new target events.

Building and training such a model presents many challenges. First, training an accurate model requires many data samples. Therefore, such a solution may not be more effective than optimizing the DUV directly. In addition, the flow of the verification information is from test-templates to coverage, while the flow of a generative CDG is in the other direction from coverage to test-templates. This challenge can be handled using machine learning techniques that work in both directions such as Bayesian networks [12], and using generative methods such as generative adversary networks (GAN) [23]. Another alternative is to construct the ML model in the CDG flow from coverage to test-templates, instead of the verification flow direction. The final challenge is the need to handle previously uncovered events. This can be handled using approximated targets (see Sect. 4.2.2) or by exploiting the structure of the coverage model [13]. The machine learning literature describes several techniques for one-class classification [31], but these techniques cannot handle the large number of missing labels in CDG. It is due to these challenges that generative (or direct) CDG never became a common practice, although it received considerable attention in the research community (e.g., [12, 26]).

We have a different approach for using machine learning models with CDG. Instead of using the machine learning model as the main ingredient of the CDG system, the machine learning model is used as a helper for the DFO search algorithm. Specifically, we show how a DNN trained on the T and E matrices collected during the execution of Algorithm 5 can help the implicit filtering algorithm progress faster and use fewer simulations. A detailed description of the method can be found in [18].

The DNN can help the search algorithm in two ways. First, if the DNN model is highly accurate, it can directly provide a test-template that hits the target event well enough by finding the test-template that maximizes the probability of hitting the event in the model. This can be done, for example, using gradient decent methods.

Algorithm 6 DNN accelerated search algorithm (EE-DNN-IF)

1: Use random sampling, $T = [t_1^\top, \ldots, t_M^\top]$ and compute, $E = [e_1^\top, \ldots, e_M^\top]$
2: Set $\mathcal{I}$ to the events that are not hit above the threshold τ
3: **while** $\mathcal{I} \neq \emptyset$ **do**
4: *Train the DNN using T and E*
5: Find e, the event in $\mathcal{I}$ with the highest probability in E.
6: *Using the DNN, find the test-template $t_{\text{best}}^{\text{DNN}}$ that maximizes the probability of hitting e*
7: *Simulate $t_{\text{best}}^{\text{DNN}}$*
8: **if** *The hit probability of e by $t_{\text{best}}^{\text{DNN}} \leq \tau$* **then**
9: Use *the modified* implicit filtering to improve the hit probability of e until it is above τ
10: Augment the test-templates used for the matrices T and E
11: Remove from $\mathcal{I}$ all the events that are hit above the threshold τ

Because of the small amount of data available for training the model, we don't expect this to work very well. Still, the cost of adding this predicted test-template is small compared to the number of simulations required by the search algorithm, so using this test-template can be beneficial.

The second method in which the DNN model can help the implicit filtering algorithm is by helping it select its search candidate in each iteration. Here, the DNN model replaces the random selection of the k direction used by the implicit filtering algorithm with a set of the best directions predicted by the model, out of a much larger set of random directions. It is easy to show that if the DNN model is even slightly better than random, this method for selecting the direction speeds up the implicit filtering and causes it to require fewer simulations to reach its target. To address the case when the DNN model is worse than random, some of the k directions (say half) are selected randomly. This guarantees that the implicit filtering algorithm converges, although at a slower pace, even with a bad DNN.

Algorithm 6 shows the accelerated algorithm. The emphasized lines in the algorithm are the changes from Algorithm 5 that use the DNN model. Specifically, in line 4 the DNN model is trained in each iteration of the algorithm. In lines 6–7, the optimal test-template is obtained from the model and simulated. If the simulation results are good, the call to the implicit filtering algorithm is skipped. Otherwise, in line 8, the modified implicit filtering that uses the DNN model to select directions is used.

In many aspects, the proposed method resembles *active learning* [41] techniques. In both cases, the model is used to generate the next data points that will be used to improve its quality. The main difference between the methods is that in active learning, the main goal of the generated data points is to explore the input space to improve the quality of the model. In our method, the goal of the generated test-templates is to exploit the model for our coverage goals, and the exploration aspect is merely an important side effect.

4.3.3 Experimental Results

To test the ability of the accelerated search algorithm in Algorithm 6 to improve the performance of the EE-IF algorithm (Algorithm 5), we deployed an experimental environment that used an abstract high-level model of a simple in-order processor called NorthStar. The coverage model used in the experiment was a cross-product of the state of several pipe stages in the two arithmetic units of the processor. The coverage model contained 185 legal events. The goal of CDG was to hit each of these events with a probability of at least 0.25.

The experiment measured the number of events left uncovered as a function of the number of test-templates used. It compared between four different CDG algorithms:

1. Random—Generate random tests to cover events
2. EE-IF—Use the event after events implicit filtering (Algorithm 5)
3. EE-DNN—Use the DNN to eliminate events one after the other without implicit filtering, that is, Algorithm 6 without the if block in line 8
4. EE-IF-DNN—Use the DNN accelerated search algorithm (Algorithm 6)

Figure 15 shows the experiment results. The results are the average of five runs of each algorithm. The results show that the EE-IF-DNN algorithm outperforms the other algorithms. It covers all the events with about half the number of test-templates needed by the EE-IF algorithm, while the other two algorithms leave many events uncovered. This indicates that the DNN model is able to assist the implicit filtering algorithm in its search. The results also show that when a small number of test-templates are used, the DNN alone in the EE-DNN algorithm is better than EE-IF. But when more test-templates are used, the progress of the DNN slows significantly while EE-IF continues to progress and starts to perform better than the DNN. This is caused by the limited ability of the DNN to find good test-templates for hard-to-hit events. Note that the addition of the DNN have a negligent affect on the runtime of the algorithm because the simulation time is much longer than the time it takes to train the DNN.

Reference [22] shows that a surrogate DNN can be used as an accelerator. This is the case not just for the implicit filtering algorithm but also when using other stateless DFO algorithms, such as simulated annealing (SA), genetic algorithms (GA), and particle swarm (PS).

5 Conclusions

Coverage is one of the main measures for the quality of the verification process because it points the design and verification teams to areas in the design that are not verified thoroughly. Therefore, coverage closure, or the process of advancing coverage goals, plays a major role in the verification process. To achieve coverage closure, the verification team must first analyze the coverage data to understand

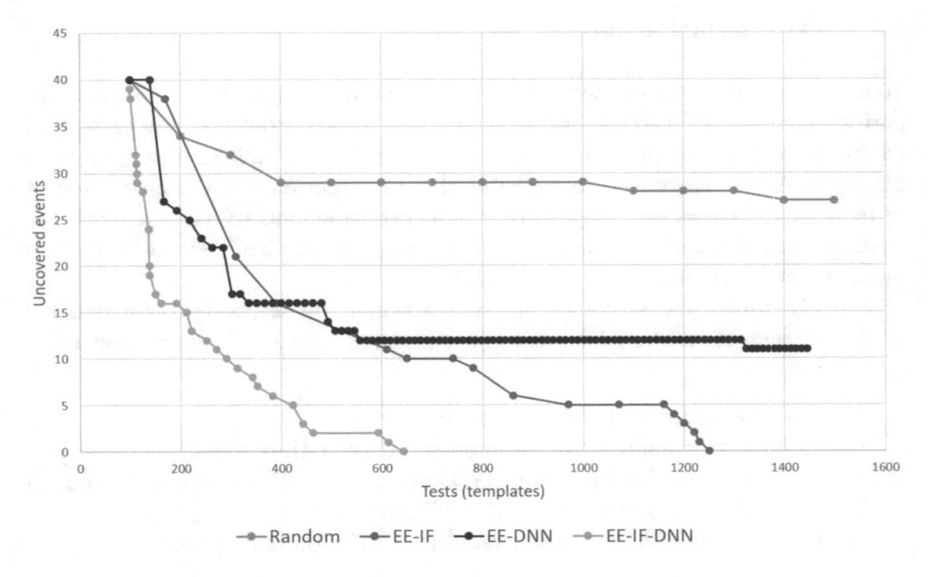

Fig. 15 Comparing the different algorithms, starting from the same initial 100 random test-templates

the coverage picture and identify weaknesses in it. Then, it needs to close holes identified by the coverage analysis.

The vast amount of coverage data produced by the verification process calls for the use of data science techniques to help the verification team in achieving its coverage closure goals. In this chapter, we presented several techniques and tools that utilize data science methods for coverage closure goals. In the first part of the chapter, we focused on descriptive analysis that helps extracting concise information out of the coverage data when the coverage space is structured. We also showed how machine learning clustering technique combined with domain-specific optimizations can be used to find structure in coverage models when this structure is not explicitly defined. In the second part of the chapter, we used statistics, optimization techniques, and machine learning to exploit the relation between test-templates and coverage to help closing coverage holes.

All the tools and techniques described in this chapter are integrated in the IBM verification process and are used in the verification of all IBM processor systems. Many of these techniques are also used in many other places. For example, the cross-product analysis techniques described in Sect. 2 are integral part of most coverage analysis and visualization tools and tools that exploit the relations between test-templates and coverage start to emerge.

While data science methods already play an essential role in the coverage closure process, they can still do a lot more. Coverage-directed generation (CDG) is one area where more data analytics and machine learning can help improving the overall results. Despite the recent advancements in CDG techniques, they are still far from reaching their potential. Several aspects of the CDG process can be improved to increase its capabilities. These include enhancing the capabilities of CDG engine by introducing new analysis and learning techniques that better explore the relations between test-templates and coverage and helping identifying whether event is not hit due to missing test or is impossible to hit, adapting verification environments to better utilize CDG, and making the use of CDG user friendlier. In addition, data analytics and machine learning techniques can be used to improve the definition of coverage plans and coverage models.

References

1. Aharoni, M., Asaf, S., Fournier, L., Koyfman, A., Nagel, R.: FPgen - a deep-knowledge test generator for floating point verification. In: Proceedings of the 8th High-Level Design Validation and Test Workshop, pp. 17–22 (2003)
2. Arar, M., et al.: The verification cockpit - creating the dream playground for data analytics over the verification process. In: Proceedings of the 11th Haifa Verification Conference, pp. 104–119 (2015)
3. Asaf, S., Marcus, E., Ziv, A.: Defining coverage views to improve functional coverage analysis. In: Proceedings of the 41st Design Automation Conference, pp. 41–44 (2004)
4. Azatchi, H., Fournier, L., Marcus, E., Ur, S., Ziv, A., Zohar, K.: Advanced analysis techniques for cross-product coverage. IEEE Trans. Comput. **55**(11), 1367–1379 (2006)
5. Birnbaum, A., Fournier, L., Mittermaier, S., Ziv, A.: Reverse coverage analysis. In: K. Eder, J. Lourenço, O. Shehory (eds.) Hardware and Software: Verification and Testing, pp. 190–202. Springer, Berlin, Heidelberg (2012)
6. Blei, D., Ng, A., Jordan, M.: Latent dirichlet allocation. J. Mach. Learn. Res. **3**, 993–1022 (2003)
7. Carter, H.B., Hemmady, S.G.: Metric Driven Design Verification: An Engineer's and Executive's Guide to First Pass Success. Springer (2007)
8. Conn, A., Scheinberg, K., Vicente, L.: Introduction to Derivative-Free Optimization. SIAM, Philadelphia (2009)
9. Copty, S., Fine, S., Ur, S., Yom-Tov, E., Ziv, A.: A probabilistic alternative to regression suites. Theor. Comput. Sci. **404**(3), 219–234 (2008)
10. Duff, I.S., Erisman, A.M., Reid, J.K.: Direct Methods for Sparse Matrices. Oxford University Press, USA (1986)
11. Elakkiya, C., Murty, N., Babu, C., Jalan, G.: Functional coverage - driven uvm based jtag verification. In: 2017 IEEE International Conference on Computational Intelligence and Computing Research (ICCIC), pp. 1–7 (2017). https://doi.org/10.1109/ICCIC.2017.8524556
12. Fine, S., Ziv, A.: Coverage directed test generation for functional verification using Bayesian networks. In: Proceedings of the 40th Design Automation Conference, pp. 286–291 (2003)
13. Fine, S., Fournier, L., Ziv, A.: Using bayesian networks and virtual coverage to hit hard-to-reach events. Int. J. Softw. Tools Tech. Trans. **11**(4), 291–305 (2009)
14. Foster, H.: The 2020 wilson research group functional verification study part 8 IC/ASIC resource trends. https://blogs.sw.siemens.com/verificationhorizons/2021/01/06/part-8-the-2020-wilson-research-group-functional-verification-study
15. Friedman, J., Hastie, T., Tibshirani, R.: The Elements of Statistical Learning, vol. 1. Springer Series in Statistics. Springer, Berlin (2001)
16. Gal, R., Kermany, E., Saleh, B., A.Ziv, Behm, M.L., Hickerson, B.G.: Template aware coverage: Taking coverage analysis to the next level. In: Proceedings of the 54th Design Automation Conference, pp. 36:1–36:6 (2017)

17. Gal, R., Haber, E., Irwin, B., Saleh, B., Ziv, A.: How to catch a lion in the desert: on the solution of the coverage directed generation (CDG) problem. Optim. Eng. (2020)
18. Gal, R., Haber, E., Ziv, A.: Using dnns and smart sampling for coverage closure acceleration. In: Proceedings of the 2020 ACM/IEEE Workshop on Machine Learning for CAD, pp. 15–20 (2020). https://doi.org/10.1145/3380446.3430627
19. Gal, R., Kermany, H., Ivrii, A., Nevo, Z., Ziv, A.: Late breaking results: Friends - finding related interesting events via neighbor detection. In: Proceedings of the 57th Design Automation Conference (2020)
20. Gal, R., Simchoni, G., Ziv, A.: Using machine learning clustering to find large coverage holes. In: 2020 ACM/IEEE 2nd Workshop on Machine Learning for CAD (MLCAD), pp. 139–144 (2020). https://doi.org/10.1145/3380446.3430621
21. Gal, R., Haber, E., Ibraheem, W., Irwin, B., Nevo, Z., Ziv, A.: Automatic scalable system for the coverage-directed generation (CDG) problem. In: Proceedings of the Design, Automation and Test in Europe Conference (2021)
22. Gal, R., Haber, E., Irwin, B., Mouallem, M., Saleh, B., Ziv, A.: Using deep neural networks and derivative free optimization to accelerate coverage closure. In: Proceedings of the 2021 ACM/IEEE Workshop on Machine Learning for CAD (2021)
23. Goodfellow, I., Bengio, Y., Courville, A.: Deep Learning. MIT Press (2016)
24. Grinwald, R., Harel, E., Orgad, M., Ur, S., Ziv, A.: User defined coverage - a tool supported methodology for design verification. In: Proceedings of the 35th Design Automation Conference, pp. 158–165 (1998)
25. Hajjar, A., Chen, T., Munn, I., Andrews, A., Bjorkman, M.: High quality behavioral verification using statistical stopping criteria. In: Proceedings of the 2001 Design, Automation and Test in Europe Conference, pp. 411–418 (2001)
26. Hsiou-Wen, H., Eder, K.: Test directive generation for functional coverage closure using inductive logic programming. In: Proceedings of the High-Level Design Validation and Test Workshop, pp. 11–18 (2006)
27. IBM—ILOG: https://www.ibm.com/products/ilog-cplex-optimization-studio. [Online; accessed 5-August-2015]
28. Imková, M., Kotásek, Z.: Automation and optimization of coverage-driven verification. In: 2015 Euromicro Conference on Digital System Design, pp. 87–94 (2015). https://doi.org/10.1109/DSD.2015.34
29. James, G., Witten, D., Hastie, T., Tibshirani, R.: An Introduction to Statistical Learning with Applications in R. Springer Text in Statistics. Springer (2013)
30. Karnaugh, M.: The map method for synthesis of combinational logic circuits. Trans. Am. Inst. Electr. Eng. **72**(9), 593–599 (1953)
31. Khan, S.S., Madden, M.G.: A survey of recent trends in one class classification. In: Coyle, L., Freyne, J. (eds.) Artificial Intelligence and Cognitive Science, pp. 188–197. Springer, Berlin, Heidelberg (2010)
32. Lachish, O., Marcus, E., Ur, S., Ziv, A.: Hole analysis for functional coverage data. In: Proceedings of the 39th Design Automation Conference, pp. 807–812 (2002)
33. Laeufer, K., Koenig, J., Kim, D., Bachrach, J., Sen, K.: RFUZZ: coverage-directed fuzz testing of RTL on fpgas. In: Proceedings of the International Conference on Computer-Aided Design, pp. 1–8 (2018)
34. Lee, D.D., Seung, H.S.: Algorithms for non-negative matrix factorization. In: Leen, T.K., Dietterich, T.G., Tresp, V. (eds.) Advances in Neural Information Processing Systems, vol. 13, pp. 556–562. MIT Press (2001)
35. Marick, B.: The Craft of Software Testing, Subsystem Testing Including Object-Based and Object-Oriented Testing. Prentice-Hall (1985)
36. Mishra, P., Dutt, N.: Automatic functional test program generation for pipelined processors using model checking. In: Seventh Annual IEEE International Workshop on High-Level Design Validation and Test, pp. 99–103 (2002)
37. Piziali, A.: Functional Verification Coverage Measurement and Analysis. Springer (2004)
38. Pyne, S., Rao, B.P., Rao, S.: Big Data Analytics Methods and Applications. Springer (2016)

39. Reddy, C.K., Aggarwal, C.C.: Data Clustering. Chapman and Hall/CRC (2016)
40. Schubert, K.D., Roesner, W., Ludden, J.M., Jackson, J., Buchert, J., Paruthi, V., Behm, M., Ziv, A., Schumann, J., Meissner, C., Koesters, J., Hsu, J., Brock, B.: Functional verification of the IBM POWER7 microprocessor and POWER7 multiprocessor systems. IBM J. Res. Dev. **55**(3), 308–324 (2011)
41. Settles, B.: Active learning literature survey. Computer sciences technical report 1648, University of Wisconsin–Madison (2009)
42. Smith, J., Bartley, M., Fogarty, T.: Microprocessor design verification by two-phase evolution of variable length tests. In: Proceedings of the 1997 IEEE Conference on Evolutionary Computation, pp. 453–458 (1997)
43. Tasiran, S., Fallah, F., Chinnery, D.G., Weber, S.J., Keutzer, K.: A functional validation technique: biased-random simulation guided by observability-based coverage. In: Proceedings of the 2001 International Conference on Computer Design, pp. 82–88 (2001)
44. Ur, S., Yadin, Y.: Micro-architecture coverage directed generation of test programs. In: Proceedings of the 36th Design Automation Conference, pp. 175–180 (1999)
45. Wagner, I., Bertacco, V., Austin, T.: Microprocessor verification via feedback-adjusted Markov models. IEEE Trans. Comput. Aided Des. Integr. Circuits Syst. **26**(6), 1126–1138 (2007)
46. Wile, B., Goss, J.C., Roesner, W.: Comprehensive Functional Verification - The Complete Industry Cycle. Elsevier (2005)

Cell-Aware Model Generation Using Machine Learning

Pierre d'Hondt, Aymen Ladhar, Patrick Girard, and Arnaud Virazel

1 Introduction

Digital integrated circuits (ICs) are commonly synthesized with predefined libraries of standard cells of various nature and complexity. As the semiconductor industry moves to increasingly smaller geometries, new types of manufacturing defects appear and need to be targeted by industrial test flows. Conventional fault models like stuck-at, transition, and layout-aware (e.g., bridging) fault models are becoming less effective for ensuring desired test and diagnosis quality levels. Indeed, these fault models only consider faults at the boundary of library cells. However, an increasing number of defects in circuits fabricated with the most recent manufacturing technologies occur within the logic cell structures. They are called *intra-cell* or *cell internal defects* [1–3]. These defects are only covered fortuitously with conventional fault models, and hence not surprisingly, these defects are found to be the root cause of a significant fraction of test escape [4].

Cell-aware (CAs) test and diagnosis have been proposed recently to target those subtle defects in ICs requiring highest product quality [5–9]. The realistic assumption under this concept is that the excitation of a defect inside a cell is

P. d'Hondt (✉)
STMicroelectronics, LIRMM, University of Montpellier / CNRS, Montpellier, France
e-mail: pierre.dhondt@st.com

A. Ladhar
STMicroelectronics, Crolles, France
e-mail: aymen.ladhar@st.com

P. Girard · A. Virazel
LIRMM, University of Montpellier / CNRS, Montpellier, France
e-mail: patrick.girard@lirmm.fr; arnaud.virazel@lirmm.fr

A. Iranmanesh (ed.), *Frontiers of Quality Electronic Design (QED)*,
https://doi.org/10.1007/978-3-031-16344-9_6

highly correlated with the logic values at the input pins of the cell [10, 11]. A preliminary step when performing CA test and diagnosis is to characterize each standard cell of a given library with respect to all possible cell internal defects. Analog (SPICE) simulations are performed to identify which cell internal defects are detected by which cell patterns. The simulation results are encoded in a cell internal-fault dictionary or *CA model* (also referred to as *CA fault model* or *CA test model* in the literature) [12, 13].

One bottleneck of CA model generation is that it requires extensive computational efforts to characterize all standard cells of a library [14, 15]. Typically, the generation time of cell-aware models for few hundreds of cells may reach up to several months considering a single SPICE license. Reducing the generation runtime of CA models and easing the characterization process are therefore mandatory to swiftly deploy the CA methodology on industrial ICs and make it a standard in the qualification process of silicon products [16]. To this end, machine learning (ML) can be used to drastically accelerate the CA model generation flow.

This chapter presents a comprehensive flow experimented on industrial cell libraries and preliminary introduced in [17]. The flow is based on a learning method that uses existing CA models of various standard cells developed using different technologies to predict CA models for new standard cells independently of the technology. This is the first work to address this problem since previous works on ML focused on cell library characterization without defect injection [18–20]. Experiments performed on a standard cell population of reasonable size (about two thousand cells from different technology nodes and transistor sizes) show that the generation time of CA models can be reduced by more than 99% (a few hours instead of almost 3 months when CA models are generated using a single SPICE license). Part of these results are extracted from [17] in which the proposed flow has been experimented on combinational cells of industrial libraries.

The remainder of this chapter is organized as follows. Section 2 gives some background on standard cell characterization, first for design purpose and then for test and diagnosis purposes. The last part of the section explains why using ML for cell characterization can help in reducing the generation time of CA models. Section 3 presents the ML-based CA model generation flow and details the two main steps of the flow, namely, the generation of training data and the generation of new data. Section 4 shows how cell transistor netlists and cell internal defects are represented and manipulated by the proposed methodology. Section 5 presents experimental results gathered on industrial cell libraries and proposes a performance comparison with a simulation-based approach. Section 6 presents the hybrid CA model generation flow developed for an industrial usage of the ML-based methodology. Section 7 summarizes the contribution and concludes the chapter.

2 Background on Standard Cell Characterization

2.1 Standard Cell Characterization for Design Purpose

Digital circuit designers use predefined standard cells to synthesize circuits with various sizes and complexities [21–27]. As the simulation of a full circuit design can take a huge amount of time, designers rely on standard cell characterization, a process that produces simple models of functionality, timing, and power consumption at the cell level. The (simplified) design and characterization flow for a standard cell is summarized in Fig. 1. It starts with the functional specification, which describes the logical function of the cell (AND, flip-flop, etc.) using a hardware description language (HDL). The next step defines the cell's transistors and their connections in a SPICE netlist. This netlist is known as the cell's schematic or structure. The layout describes the physical implementation of the cell on silicon, using several layers and materials (metal, polysilicon, etc.) [28], and is designed from the SPICE netlist. A parasitic extraction is then performed on the obtained layout, in order to specify the parasitic resistors and capacitors introduced in the physical implementation. The parasitic components are appended to the SPICE transistors netlist in the detailed standard parasitic format (DSPF).

Cell characterization for design purpose uses the generated cell descriptions (also called cell views) to perform electrical simulations of standard cells and extract the power and delay information, as well as the identification of timing constraints (setup and hold times). Typically, cell characterization requires the definition of global parameters such as process, voltage, and temperature, known as PVT corners, and global constraints such as wire loads and time limits for transitions. The cell schematic and layout are iteratively modified until quality

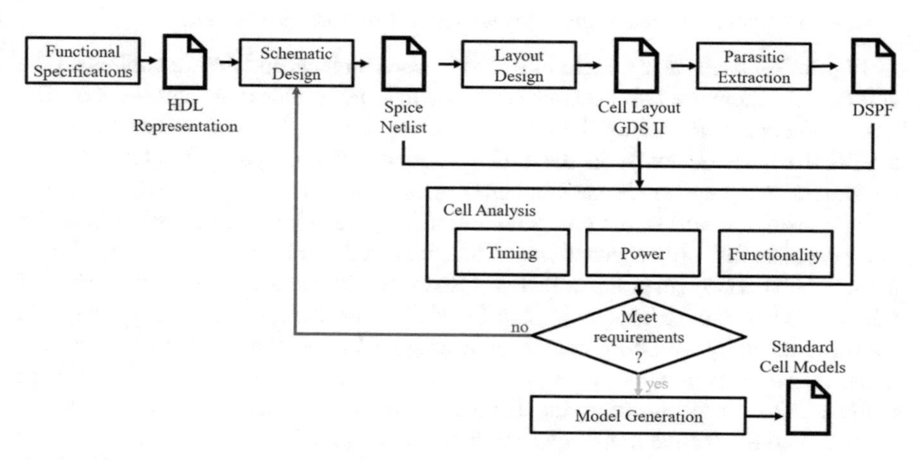

Fig. 1 Schematized process of standard cell creation and characterization

and constraint requirements are met in terms of functionality, timing, and power consumption.

Once done, data describing every aspect (transition time, internal power, capacitance, sequential cells constraints, etc.) of the cell are written to dedicated files, known as cell models. Using electrical simulations considering different values of the global parameters, cell models are created to determine the behavior of standard cells in every condition that may occur during the lifetime of the circuit.

2.2 Cell Internal Defect Universe

The first step of a standard cell characterization process for test and diagnosis purpose (i.e., CA model generation) is to extract all potential and realistic defects within each cell to be able to simulate their effect in a defective cell [29–33].

Figure 2 presents an example of internal defects that may occur at the cell level. These defects can be classified into two main categories:

- *Transistor defects*, which are defects occurring at the transistor ports (source, drain, gate, and bulk). These defects can be modeled as short or open defects at the transistor ports. As illustrated in Fig. 2a, for a CMOS transistor, six short (gate-drain, source-drain, gate-source, and each port to bulk) and three open defects (gate, source, and drain) can be identified. These nine defects are added to the potential defects list for every transistor in the standard cell.
- *Inter-transistor defects*, which are defects occurring at the interconnexion between two different transistors. These defects can also be modeled as short or open defects between two internal nodes. Their existence is bound to the actual layout of the cell (e.g., two close polygons may be defectively shorted), so inter-transistors defects require layout extraction to be identified.

Figure 3 presents an example of inter-transistor defects and their locations on the cell layout. There are two possible solutions to extract inter-transistor defects. The first one consists in reading the layout database of each standard cell and creating a SPICE transistor netlist in the DSPF format including parasitic elements like resistors and capacitors. These elements represent the list of inter-transistor defects to be considered during the characterization. A parasitic capacitor exists between two polygons that are supposed not to be connected. Consequently, the location of a potential short defect and a defective resistor can become an open defect. Even if this method is easy to apply, its main drawbacks are the huge number of parasitic elements listed by the DSPF netlist (on average, 61 times the number of transistors in the cell) and the fact that some of these parasitic elements cannot be considered as realistic defect locations (e.g., the distance between two nets may be large enough to ensure non-defective manufacturing but still described by a small value parasitic capacitor, some layers are not sensitive to open defects but still described with their own resistors, etc.). In addition, several parasitic elements are equivalent, and there is no solution to recognize them without characterization (e.g., a single physical net

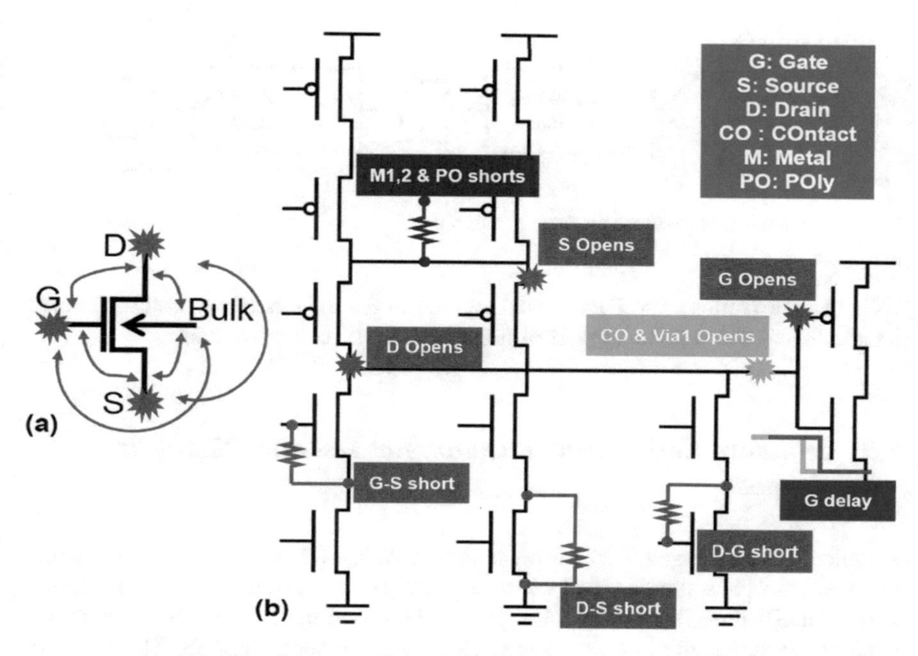

Fig. 2 (**a**) Illustration of the six short defects and three open defects that can affect a CMOS transistor's ports, (**b**) example of cell internal defects in a simple structure made of various transistors

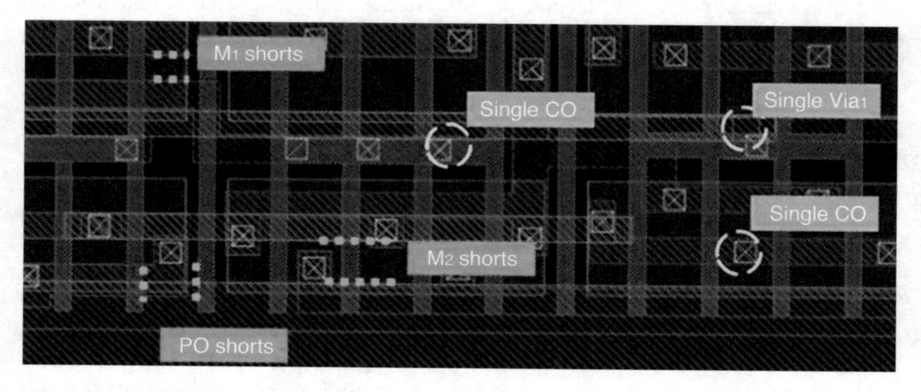

Fig. 3 Example of inter-transistor defects

is described by several serial resistors and any defect on one of these resistors is equivalent to a defect on the whole net).

To address these limitations, a second method based on design rule checking (DRC) can be used. This solution allows the localization of neighbored internal nets as well as the localization of potential open defects that can be identified for the

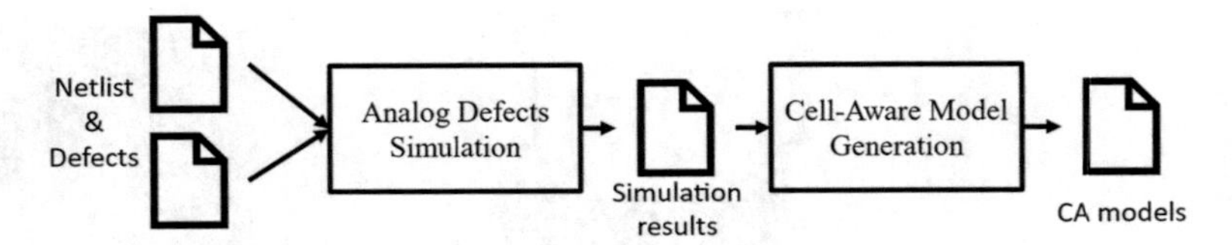

Fig. 4 Conventional cell-aware model generation flow

cell characterization. The DRC-based method limits the number of potential defect locations to 4.3 times the number of transistors in the cell, on average.

2.3 *Standard Cell Characterization for Test and Diagnosis Purpose*

A typical CA model generation flow, as shown in Fig. 4, has as input a SPICE netlist representation of a standard cell which is usually derived from a layout description, e.g., a GDSII file. This DSPF cell netlist is then used by an electrical simulator to simulate each potential defect against an exhaustive set of stimuli. Those stimuli include static (one vector) and dynamic (two vectors) input patterns of the cell (called cell patterns in the sequel). Once the simulation is completed, all cell internal defects are classified into defect equivalence classes with their detection information (required input values for each defect within each cell) and are synthetized into a CA model. As standard cells may have more than ten inputs, and thousands of cells with different complexities are usually used for a given technology, the generation time of CA models for complete standard cell libraries of a given technology may reach up to several months, thus drastically increasing the library characterization process cost.

Once the CA model of a given standard cell is generated, it can be used either for automatic test pattern generation (ATPG) or for fault diagnosis:

- *ATPG usage*. Using the CA models, which is a dictionary mapping cell patterns to the cell internal faults they detect, an ATPG tool identifies for each cell in the CUT the minimum set of stimuli detecting all cell internal defects. Then, it generates test patterns exercising this test stimuli at the input pins of the cell under test and ensures the fault propagation to an observation point.
- *Fault diagnosis*. A diagnostic tool extracts the failing and passing logic values at the input pins of the defective cell. This information is then matched with the CA model of the defective cell in order to identify the suspect internal defect.

2.4 Cell-Aware Model Generation: A Machine-Learning Friendly Process

Machine learning can be used to significantly accelerate the CA model generation process. The motivation behind the use of ML is the result of several observations made while performing comparisons between several CA models coming from different standard cell libraries and technologies:

- Several cell internal defects, such as stuck-open defects, are independent of the technology and transistor size [34, 35].
- For the same function, two cell-internal structures are usually quite similar for two different technologies.
- Detection tables for static and dynamic defects, in the form of binary matrices describing the detection patterns for each cell internal defect, are ML friendly.
- CA models may change with respect to test conditions and PVT corners. In fact, CA model generation for the same cell with different test conditions may exhibit slight differences. Few defects can be of different types (i.e., static or dynamic) or may have different detection patterns. Since CA models are generated for specific test conditions and can be used with different ones, it may lead to inaccurate characterization. This inaccuracy is usually allowed in the industry since it is marginal. This indicates that we can also tolerate few error percentages in the ML-based prediction.
- Very simple CA models are used to emulate short and open defects, for which resistance values are often identical for all technologies.
- A large database of CA models is usually available and can be used to train a ML algorithm.

All these observations intuitively indicate that CA model generation through ML is possible. However, the first challenging task is to be able to describe cell transistor netlist as well as corresponding cell internal defects in a uniform (standardized) manner, so that a ML algorithm can learn and infer from data irrespective of their incoming library and technology. Indeed, similar cells (e.g., cells with same logic function, same number of inputs, and same number of transistors) may be described differently in transistor-level (SPICE) netlists of various libraries (e.g., a transistor label does not always correspond to the same transistor in two similar cells coming from two different libraries). It is therefore mandatory to standardize the description of cells and corresponding defects for the ML-based defect characterization methodology. Heuristic solutions developed to this purpose are described in Sect. 4. The second challenging task is to find a way to represent all these information/input data so that they can be ML friendly. A matrix description of cells and corresponding defects is used to this purpose.

3 Learning-Based Cell-Aware Model Generation Flow

The learning-based CA model generation flow initially introduced in [17] is used to predict the behavior of a cell (combinational or sequential) when affected by intra-cell defects. The flow is presented in Fig. 5. It is based on supervised learning that takes a set of input data and known responses (*labeled data*) used as training data, trains a model to classify those data, and then uses this model to predict (*infer*) the class of new data.

Figure 5 depicts the *two main steps* of the *supervised learning process* used for ML-based CA model generation. A random forest classifier is used for predicting the class of each new data instance. This choice comes from the results obtained after experimenting several learning algorithms (k-NN, support vector machine, random forest, linear, ridge, etc.) and observing their inference accuracies.

The *first main step* of the CA model generation flow consists in generating a random forest model and to train it by using the training dataset. A random forest classifier is composed of several decision tree classifiers, which are models predicting class of samples by applying simple decision rules. During training, a decision tree tries to classify data samples, and its decision rules are modified until it reaches a given quality criterion. Then, the forest averages the responses of all trees and outputs the class of the data sample.

The *second main step* consists in using the random forest classifier to make prediction (or inference) when a new data instance has to be evaluated. Prediction for a new data instance amounts to answer the question: "Does this stimulus detects this defect affecting this cell?" Answering this question allows obtaining a new CA model for a given standard cell.

3.1 Generation of Training Data

Training data are made of various and numerous CA models formerly generated by relying to brute-force electrical defect simulations. For each cell (combinational

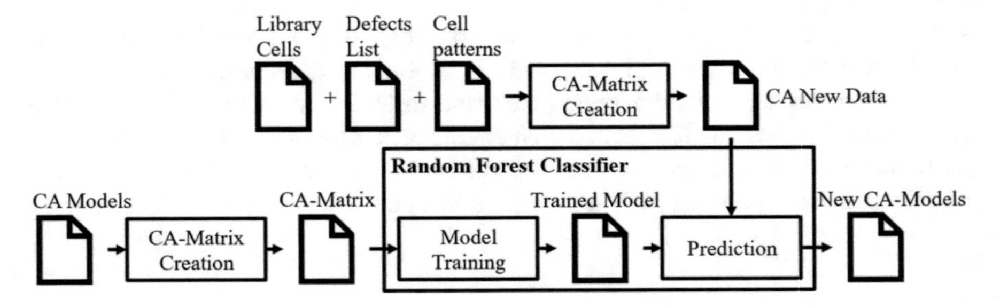

Fig. 5 Generic view of the ML-based CA model generation flow

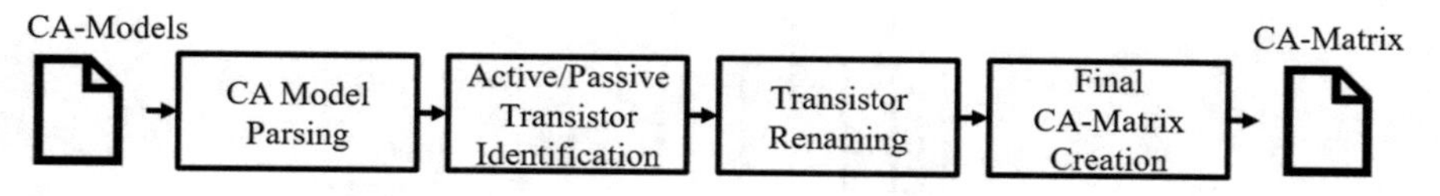

Fig. 6 CA matrix creation flow

or sequential) in a library, the CA model is transformed into a so-called CA matrix and filled in with meaningful information. Cells with the same number of inputs and having the same number of transistors are grouped together to form the training dataset.

The CA matrix creation flow is depicted in Fig. 6. The flow starts by rewriting the CA model so that it can be ML friendly. To this end, the CA model file is parsed and its content is organized into a matrix which contains numbers and categories of certain values (more details are given later on). Then, it identifies the activation conditions of each transistor inside the cell with respect to input stimuli. Once the activation conditions for each transistor have been identified, transistor renaming is done. This is a critical step in this flow since it allows the usage of the training data across different libraries and technologies. Finally, the CA matrix is created with the above information.

Table 1 shows an example of a training dataset for a combinational NAND2 cell. It is composed of four types of information:

- *Cell patterns and responses*. This gives the values applied on inputs (A, B) of the cell as well as the cell response on output Z. As can be seen, the test pattern sequence provides all the possible input stimuli that can be applied to the cell. These stimuli must also be efficient to detect sequence-dependent defects like stuck-open defects. For this reason, a four-valued logic algebra made of 0, 1, R, and F is used to represent input stimuli in the CA matrix. R (resp. F) represents a rising (resp. falling) transition from 0 to 1 (resp. from 1 to 0).
- *Transistor switching activity*. This indicates the activation conditions of each transistor in the cell schematic. Each transistor can be in the following state: active (1), passive (0), switching to active state (R), or switching to passive state (F).
- *Defect description*. This gives information about defect locations inside the cell transistor schematic. This part contains a column for each transistor's ports. In Table 1, "N1_D" stands for the drain port of the NMOS transistor named N1 and "N1_S" for its source port. In these columns, a "1" (resp. "0") indicates that the port is concerned (resp. non-concerned) by the described defect. For example, D15 is a short between the drain and the source of transistor N1, so columns "N1_D" and "N1_S" contain a one, while other columns are filled with zeros. The name and type of each defect are also given in this description. The matrix also includes rows describing the cell with no defects ("free"). This is presented in more detail in Sect. 4.4.

Table 1 Example of training dataset for a NAND2 cell

Cell inputs and responses			Transistor switching activity				Defect description				About defect		Defect detection
A	B	Z	N0	N1	P0	...	N1_D	N1_G	N1_S	...	Name	Type	fZ
0	0	1	0	0	1	...	0	0	0	...	Free	Free	0
0	1	1	0	1	1	...	0	0	0	...	Free	Free	0
0	F	1	0	F	1	...	0	0	0	...	Free	Free	0
⋮	⋮	⋮	⋮	⋮	⋮	⋮	⋮	⋮	⋮	⋮	⋮	⋮	⋮
0	1	1	0	1	1	...	1	0	1	...	D15	Short	1
1	1	0	1	1	0	...	1	0	1	...	D15	Short	0
⋮	⋮	⋮	⋮	⋮	⋮	⋮	⋮	⋮	⋮	⋮	⋮	⋮	⋮

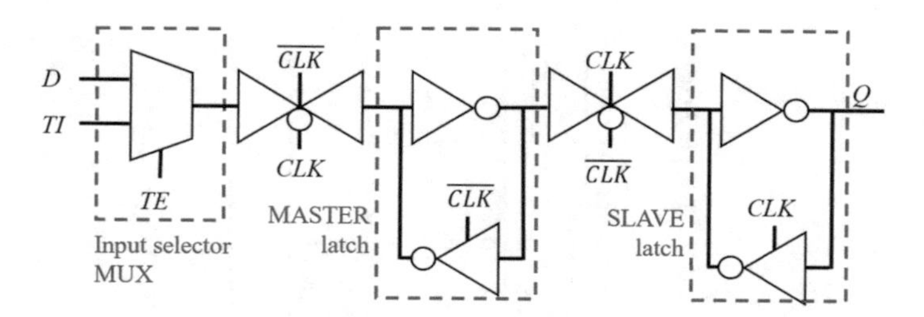

Fig. 7 Block-level representation of a scan flip-flop example

- *Defect detection.* This is the class of the data sample (the output of the ML classifier). A value "1" ("0") means that the defect is detected (undetected) by the cell pattern.

The first three types of information constitute the inputs of the ML algorithm.

In order to illustrate the various steps of the CA matrix creation flow in the case of sequential cells, let us consider the block-level representation of a scan flip-flop as depicted in Fig. 7. It consists of three main blocks (MUX, MASTER latch, and SLAVE latch) plus two transmission gates. It has four inputs (D, TI, TE, CLK), one virtual input (Q-), and one output (Q). The virtual input represents the value loaded in the flip-flop before applying the test stimulus.

Table 2 shows an example of a training dataset for the scan flip-flop shown in Fig. 7. It is composed of four types of information:

- *Cell inputs and outputs.* This gives the values applied on inputs (D, CLK, TE, TI, Q-) of the cell as well as the cell response on output Q. The test pattern sequence provides all the possible input stimuli that can be applied to the cell. For the sake of readability, they are represented partially in Table 2. These stimuli must also be efficient to detect sequence depending defects like stuck-open defects. For this reason, a six-valued logic algebra made of 0, 1, R, F, P, and A is used to represent input stimuli in the CA matrix. R (resp. F) represents a rising (resp. falling) transition from 0 to 1 (resp. from 1 to 0). P (resp. A) represents a pulse 010 (resp. anti-pulse 101) and is used for the input clock signal of the cell.
- *Transistor switching activity.* This indicates the activation conditions of each transistor (e.g., N0, N1, etc. for NMOS transistors, and P0, P1, etc. for PMOS transistors) in the cell schematic. Each transistor can be in one of the following states: active (1), passive (0), switching to active state (R), switching to passive state (F), pulsing (P), or anti-pulsing (A).
- *Defect description.* This gives information about all defect locations in the cell transistor schematic. In Table 2, "N1_D" stands for "defect on the drain of transistor N1," "N1_G" stands for "defect on the gate of transistor N1," and so on. The name and type of each defect are also given in this description.
- *Defect detection.* This is the class of the data sample (the output of the ML classifier). A value "1" ("0") means that the defect is detected (undetected) by the input pattern at the corresponding output of the cell.

Table 2 Example of training dataset for a scan flip-flop with a non-inverting output. The cell has four physical input pins: data (D), clock (CLK), test enable (TE), test input (TI), and a virtual input (Q-) which correspond to the previous state of the output pin (Q)

Cell inputs and outputs						Transistor switching activity						Defect description							About defect		Defect detection
D	CLK	TE	TI	Q-	Q	N0	N1	…	P0	P1	…	N1_D	N1_G	N1_S	…	P3_D	P3_G	P3_S	name	type	fZ
0	P	0	0	0	0	P	0	…	P	1	…	0	0	0	…	0	0	0	Free	Free	0
R	P	0	1	0	R	P	R	…	P	F	…	0	0	0	…	0	0	0	Free	Free	0
0	P	0	F	1	0	P	0	…	P	1	…	0	0	0	…	0	0	0	Free	Free	0
⋮	⋮	⋮	⋮	⋮	⋮	⋮	⋮	⋮	⋮	⋮	⋮	⋮	⋮	⋮	⋮	⋮	⋮	⋮	⋮	⋮	⋮
0	P	0	1	0	0	P	0	…	P	1	…	1	0	1	…	0	0	0	D15	Short	1
F	P	0	1	1	F	P	F	…	P	R	…	0	0	0	…	0	0	1	D47	Open	1
⋮	⋮	⋮	⋮	⋮	⋮	⋮	⋮	⋮	⋮	⋮	⋮	⋮	⋮	⋮	⋮	⋮	⋮	⋮	⋮	⋮	⋮

As for combinational cells, the first three types of information are used as inputs for the ML algorithm.

3.2 Generation of New Data

New data represent the cells to be characterized and are obtained for each standard cell from the cell description, corresponding list of defects, and cell patterns. The format of a new data instance is similar to that of the training data, except that the class (label) of the new data instance is missing. The ML classifier is used to predict that class. As for training data, new data are grouped together according to their number of cell inputs and transistors, so that inference can be done at the same time for cells with the same number of inputs and transistors.

4 Cell and Defect Representation in the Cell-Aware Matrix

This section details the various steps required to represent a standard cell in a CA matrix. The starting point of this process is a transistor-level (SPICE) netlist of the standard cell. The CA matrix must be accurate enough to clearly identify each transistor and each net of the cell transistor schematic. This description also associates each transistor with its sensitization patterns and reports the output response for each cell pattern. For this reason, the cell description process requires several successive operations that are detailed below. Note that this process is applied to all cells in a library to be characterized.

4.1 Identification of Active, Passive, and Pulsing Transistors

The first step consists in identifying active and passive transistors in the cell netlist with respect to an input stimulus. To this purpose, a single golden (defect-free) electrical simulation of the cell to be characterized is first performed. By monitoring the voltage of cell's transistors gates, active and passive transistors are identified for each input stimulus (cell pattern). An active NMOS (resp. PMOS) transistor is a transistor with a logic-1 (resp. logic-0) value measured on its gate port. A passive NMOS (resp. PMOS) transistor is a transistor with a logic-0 (resp. logic-1) value measured on its gate port. Note that for sequential cells, an active NMOS (resp. PMOS) transistor is a transistor with a logic-1 (resp. logic-0) value appearing on its gate port during application of the test pattern whose duration is one clock cycle. A passive NMOS (resp. PMOS) transistor is a transistor with a logic-0 (resp. logic-1) value appearing on its gate port during application of the test pattern. Clock signal-controlled transistors can be pulsing (resp. anti-pulsing), which means a 0–

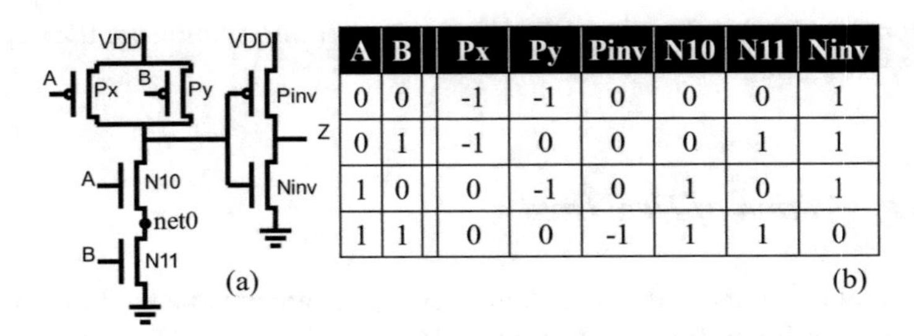

A	B	Px	Py	Pinv	N10	N11	Ninv
0	0	-1	-1	0	0	0	1
0	1	-1	0	0	0	1	1
1	0	0	-1	0	1	0	1
1	1	0	0	-1	1	1	0

Fig. 8 Example of a AND2 cell: (**a**) cell transistor schematic and (**b**) partial CA matrix representation

1-0 (resp. 1–0-1) sequence appears on the transistor gate port during application of the test pattern. *Note also that a Verilog simulation, with a CDL (circuit description language) netlist that should be written using NMOS and PMOS primitives, can replace the single defect-free electrical simulation.* This simulation also provides the cell output value. With this information, each cell pattern can be associated with the list of active transistors in the cell. After this step, the CA matrix contains the following columns:

- *Cell input and response columns*. They contain all input stimuli (cell patterns) that can be applied to the cell and the corresponding responses.
- *Transistor switching activity columns*. They contain six possible values indicating if the transistor is active (1), passive (0), switching from an active state to a passive one (F), switching from a passive state to an active one (R), pulsing (P), and anti-pulsing (A). Note that "P" and "A" are only used for sequential cells. Since PMOS and NMOS transistors are activated in opposite way, the "-" character is used before the PMOS values.

Figure 8 shows (a) the transistor schematic of a six-transistor AND2 cell and (b) a partial representation of the CA matrix of the cell. Columns A and B list all the possible input stimuli for this cell. For each stimulus, active and passive information about each transistor of the cell is entered in the CA matrix. For example, AB = 00 leads to two active PMOS transistors and two passive NMOS transistors in the NAND2 block and one passive PMOS transistor and one active NMOS transistor in the output inverter.

Figure 9a presents the transistor schematic of a LATCH such as the ones used inside the scan flip-flop depicted in Fig. 7. In the partial representation of the CA matrix of the latch (Fig. 9b), columns D and CLK list all the possible input stimuli for this structure.

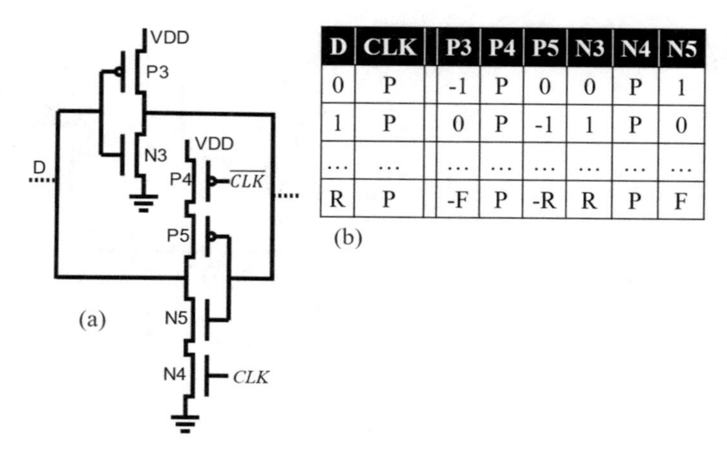

D	CLK	P3	P4	P5	N3	N4	N5
0	P	-1	P	0	0	P	1
1	P	0	P	-1	1	P	0
…	…	…	…	…	…	…	…
R	P	-F	P	-R	R	P	F

Fig. 9 Example of a LATCH structure: (**a**) cell transistor schematic and (**b**) partial CA matrix representation

4.2 Renaming of Transistors

In the CA model generation flow, the goal is to train a ML algorithm using this representation of standard cells coming from different libraries and technologies. However, this matrix representation is dependent on the transistor names and the order they are defined in the SPICE netlist. Two standard cells having the same schematic may have different transistor naming, and the order of transistors in the SPICE netlist may differ as well. This is because standard cell libraries are created several months or years apart, by different teams, with sometimes new guidelines in terms of best practices. Without an accurate naming convention of each cell transistor in the CA matrix, any ML algorithm will fail to predict the behavior of the cell in presence of a defect. To mitigate this issue, a second step consisting in renaming all cell transistors independently of their initial names and order in the input SPICE netlist is required. The algorithm developed to this purpose is detailed in the following.

In order to ensure that the CA matrix is unique for a given cell and that the CA matrices of two cells having the same structure have identical transistor switching activity columns (i.e., they have the same transistor names irrespective of their incoming library and technology), a transistor renaming procedure is required. The first step consists in sorting the transistors of a standard cell in an algorithmic way that only depends on the cell's *transistor structure*. A transistor structure is a virtual SPICE netlist without specification of the connections between transistor gates, i.e., only source and drain connections between transistors are listed. Once the transistors are sorted, they are consistently and unambiguously renamed. The transistor-renaming algorithm consists of the following two steps: determination of branch equations and sorting of branch equations.

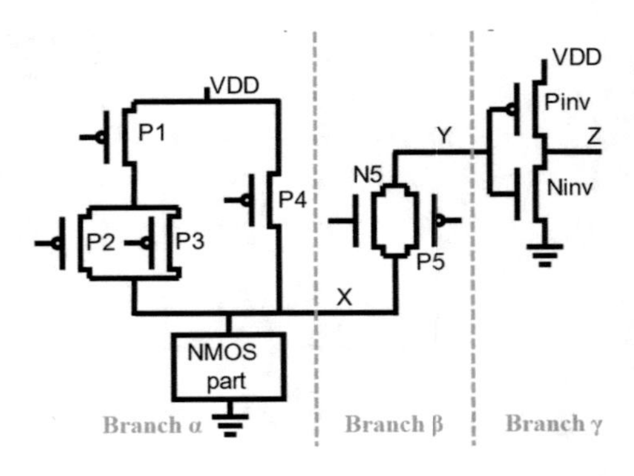

Fig. 10 Example schematic

4.2.1 Determination of Branch Equations

The transistor structure of a standard cell is composed of one or more branches. A branch is a group of transistors connected by their drain and source ports. The entry (or gate) of each branch is the set of transistor gates and its exit (or drain) is the connection net between the NMOS and PMOS transistors, which drives the gate of the next branch. A branch's source is connected to a power and/or a ground net. A branch equation is a Boolean-like equation describing how the transistors of the branch are connected, using Boolean and (symbolized by "&") for serial transistors or serial groups of transistors and Boolean or (symbolized by "|') for parallel transistors or parallel groups of transistors.

Sequential cells and complex combinational cells tend to integrate transmission gates in their structures. A transmission gate is a transistor configuration acting as a relay that can conduct or block depending on the control signal. It is composed of one PMOS and one NMOS transistors in parallel (i.e., sharing drain and source), and the control signal applied to the gate of the NMOS transistor is the opposite (i.e., NOT-ed) of the signal applied to the gate of the PMOS transistor. A transmission gate directly connects the exit of a branch to the entry of another branch. As such, a transmission gate is considered as an autonomous branch of the transistor structure. The entry of such a branch is the set of transistor gates plus the exit of the previous branch, and the exit is the entry of the following branch.

Figure 10 shows an example of transistors structure. It is composed of three branches, namely, α, β, and γ. The two-transistor output inverter is the simplest branch whose input is net Y and output is net Z (branch γ). The inverter creates two paths between the branch output and the power nets, so its branch equation is (N_{inv}!P_{inv}). The equation of the left-most branch α (PMOS branch driving net X) is (P4|(P1&(P2|P3))). In order not to rely on any name present in the SPICE netlist, the branch equations are anonymized, i.e., a NMOS is described by "1n" and a PMOS by "1p." The anonymized equation of the PMOS branch driving net X in Fig. 10 is therefore (1p|(1p&(1p|1p))). These two branches are separated by the β branch, i.e., a transmission gate composed of N5 and P5, which is anonymized as "1t."

Table 3 Branch equations for the schematic of Fig. 10

Level	Number of transistors	Anonymized equation	Comment
1	2	(1n|1p)	Branch α inverter
2	2	1 t	Branch β transmission gate
3	4	(1p|(1p&(1p|1p)))	Branch γ PMOS structure

4.2.2 Sorting of Branch Equations

Once all the branch equations for the considered cell have been determined, they are sorted by using the following deterministic criteria:

- *Level of each branch.* It is defined in ascending order with respect to the cell output (level 1 branches drive the cell output, level 2 branches drive the gates of transistors in level 1 branches, and so on and so forth).
- *Number of transistors in each branch* – in ascending order.
- *Anonymized branch equation* – in alphabetical order.

Table 3 reports all the branch equations of the schematic in Fig. 10 sorted according to the above criteria.

4.3 Identification of Parallel Transistors

Because of parallel transistors, the identification of branch equations is not enough to unambiguously rename all transistors. Specifically, two or more parallel transistors in a branch share the same drain and source, making their identification quite difficult. For example, transistors P2 and P3 in Fig. 10 can be either represented as "P2|P3" or "P3|P2," thus leading to a confusing situation. A solution to solve this problem consists in sorting transistors inside their branch according to their activity with respect to the input stimuli. The algorithm developed to this purpose proceeds as follows. For each transistor, an activity value is computed. This value summarizes the states of the transistor (active, passive, pulsing) for all possible stimuli applied to the cell. The input stimuli range from (0...,0) to (1...,1) for combinational cells and from (P,0...,0) to (P,1...,1) for sequential cells. For each of these stimuli, the transistor is either active (1), passive (0), pulsing (P), or anti-pulsing (A). The activity value is defined as a word made of 0, 1, P, and A, in which the first symbol corresponds to the state of the transistor when the first stimulus is applied, second symbol for second stimulus, and so on for the whole stimuli range.

To compute the activity values, one needs to know whether the transistor is active or passive for each input stimulus. This information is already available in the CA matrix as described in Sect. 3. To illustrate this process, activity values for the transistors of the AND2 cell given in Fig. 8 are listed in Table 4.

Table 4 Activity values for the AND2 cell in Fig. 8

			Old names					
A	B	Comments	Px	Py	N10	N11	Pinv	Ninv
0	0	First stimulus	1	1	0	0	0	1
0	1		1	0	0	1	0	1
1	0		0	1	1	0	0	1
1	1	Last stimulus	0	0	1	1	1	0
Activity value			1100	1010	0011	0101	0001	1110
			↓ *Renaming* ↓					
			P2	P1	N0	N1	P0	N2

Table 5 Activity values for the LATCH structure in Fig. 9

<table>
<tr><td></td><td colspan="2">Level</td><td colspan="2">1</td><td colspan="4">2</td></tr>
<tr><td></td><td colspan="2">Number of transistors</td><td colspan="2">2</td><td colspan="4">4</td></tr>
<tr><td>Branch extraction and sorting (left to right)</td><td colspan="2">Anonymized equation</td><td colspan="2">(1n|1p)</td><td colspan="4">(1n&1n|1p&1p)</td></tr>
<tr><td rowspan="4">Activity value within branches</td><td>CLK</td><td>D</td><td>N3</td><td>P3</td><td>N4</td><td>N5</td><td>P4</td><td>P5</td></tr>
<tr><td>P</td><td>0</td><td>0</td><td>1</td><td>P</td><td>1</td><td>P</td><td>0</td></tr>
<tr><td>P</td><td>1</td><td>1</td><td>0</td><td>P</td><td>0</td><td>P</td><td>1</td></tr>
<tr><td colspan="2">Activity value</td><td>01</td><td>10</td><td>PP</td><td>01</td><td>PP</td><td>01</td></tr>
<tr><td rowspan="2">Renaming of transistors</td><td colspan="2" rowspan="2"></td><td colspan="6">↓ Renaming ↓</td></tr>
<tr><td>N1</td><td>P1</td><td>N3</td><td>N2</td><td>P3</td><td>P2</td></tr>
</table>

Finally, transistors of each branch are sorted by their activity values (*alphabetical order*) to give the final description of the cell in the CA matrix. For the AND2 cell in Fig. 8, the renaming process is illustrated in Table 4.

The whole transistor renaming process for the transistors of the LATCH structure presented in Fig. 9 is summarized in Table 5, starting with the structure branch extraction and sorting, the computation of activity values, and then the renaming process itself.

4.4 Defect Representation in the Cell-Aware Matrix

To describe cell internal defects in a standardized and ML-friendly manner, the CA matrix contains a set of categorical columns representing the cell transistors' ports. Cell internal defects are classified into:

- *Intra-transistor defects*. These defects affect transistor ports (source, drain, gate, and bulk) and can be either an open defect or a short. In order to describe these defects, all transistor ports are listed as a column in the CA matrix (cf. Table 1). For an open defect, a value "1" indicates that this transistor port is affected by the defect, "0" otherwise. For a short, a value "1" on two-transistor ports indicates that a short exists between these two ports, "0" otherwise.

Table 6 Example of defect columns for the AND2 presented in Fig. 8

Py_S	Py_D	Px_S	Px_D	...	N10_S	N10_D	N11_S	N11_D	Comment
0	0	0	0	...	1	1	0	0	Source-drain short on N10
1	0	1	0	...	1	0	0	1	net0 and VDD short

- *Inter-transistor defects*. These defects affect a connection(s) between at least two different transistors. Though these defects are not considered in this work, the matrix representation is flexible enough to represent them. For these defects, the same representation mechanism as for intra-transistor defects is used.

Table 6 is an example of defect description in the CA matrix of the AND2 cell in Fig. 8. The row with red cells describes the intra-transistor short defect between drain and source ports of transistor N10 (newly N0). The row with purple cells describes the inter-transistor short defect between VDD at PMOS sources and "net0" (net0 connects N10-source and N11-drain).

5 Validation on Industrial Cell Libraries

The ML-based CA model generation flow has been implemented in a Python program. The ML algorithms were taken from the publicly available Python module called scikit-learn [35]. A dataset composed of 1712 combinational standard cells coming from standard cell libraries developed using three technologies – C40 (446 cells), 28SOI (825 cells), and C28 (441 cells) – was assembled. Another dataset composed of 219 sequential cells coming from the same libraries and technologies – C40 (27 cells), 28SOI (108 cells), and C28 (84 cells) – was also used for validation. All these cells already had a CA model generated by a commercial tool. The CA matrix was generated for each cell. The flow was experimented in two different ways. First, the ML model was trained and evaluated using cells belonging to the same technology. Second, the model was trained on one technology and evaluated on another one. Combinational and sequential cells were considered separately. Part of these results are extracted from [17].

5.1 *Predicting Defect Behavior on the Same Technology*

5.1.1 Combinational Standard Cells

The ML model was first trained on cells of the 28SOI standard cell library. Cells were grouped according to their number of transistors and inputs. For m cells

Table 7 Average prediction accuracy for combinational cells in the same technology

Predicion accuracy (%)	Number of inputs				
Number of transistors	2	3	4	5	6
6	99.98	99.99			
8	99.91	99.96	99.91		
9		100.0			
10	99.98	99.81	99.96		
12	99.72	99.73	100.0	99.91	99.93
14	99.7	99.56	99.83	99.92	99.96
16	99.99	100.0	99.94		99.98
18	99.99	99.94			
20	100.0	99.98	100.0	99.73	
22		99.84	99.98	99.62	
24	100.0	99.84	99.97		99.85
26	100.0	99.7	100.0		99.89
28	99.49	99.98	100.0	99.88	99.81
30	99.75	100.0	100.0		
32	100.0	100.0			99.98
42		100.0			
44		100.0			
46		99.81			
47		99.98	99.95		

available in a given group, the ML model was trained over *m-1* cells and its *prediction accuracy* was evaluated on the *m*-th cell. A loop ensured that each cell is used as the *m*-th cell. On average, a group contains 8.6 cells. All possible open and short defects (static and dynamic) were considered for each cell. Results presented in Table 7 report the prediction accuracy for open defects. Results achieved for short defects are similar.

Table 7 presents the prediction accuracy achieved for open defects. For the sake of conciseness, only results for cells with less than 7 inputs and 48 transistors are reported, although experiments have been done on cells with up to 8 inputs and 112 transistors. Non-empty boxes report the *average* prediction accuracy obtained *for a group of cells*. Empty boxes mean that there is zero or one cell available and that the group cannot be evaluated. A green background indicates that the maximum prediction accuracy in this group is 100%, i.e., the ML model can perfectly predict the defective behavior of at least one cell. In contrast, white background indicates that no cell was perfectly predicted in that group (all prediction accuracies are less than 100%). For example, let us consider the circled box in Table 7 that corresponds to 24 cells having 4 inputs and 24 transistors: (i) 15 cells are perfectly predicted (100% accuracy), which leads to a green background, (ii) the prediction accuracy for the 9 remaining cells ranges from 99.82% to 99.99%, and (iii) the average prediction accuracy over all 24 cells is 99.97%.

Table 8 Average prediction accuracy for sequential cells in the same technology

Prediction accuracy (%)		Number of inputs 4	5	6	7
Number of transistors	32	100			
	34				
	36				
	38				
	40			100	
	42			100	
	44			100	
	46			100	
	48			100	100
	50				100
	52				100

5.1.2 Sequential Standard Cells

For these experiments, the ML model was trained on a group of sequential standard cells coming from C40 standard cell libraries. Cells were grouped according to their number of transistors and inputs. As for combinational cells, for *m* cells available in a given group, the ML model was trained over *m-1* cells and the *prediction accuracy* was evaluated on the *m*-th cell. A loop ensured that each cell is used as the *m*-th cell. On average, a group contains 4.5 cells. All possible open and short defects (static and dynamic) in each cell were considered. Results in Table 8 report the prediction accuracy for short defects. Results achieved for open defects are similar.

Results are reported according to the number of transistors and number of inputs of each cell. Non-empty boxes report the *average* prediction accuracy obtained *for a group of sequential cells*. Empty boxes mean that there is zero or one cell available and that the group cannot be evaluated. As can be seen in Table 8, the maximum prediction accuracy (100%) was always obtained for each group, i.e., the ML model can perfectly predict the defective behavior of all cells in each group. This means that *the CA model generated by ML fit the real behavior achieved with electrical simulations*.

The above cell category with good prediction score has been analyzed manually to identify why it led to good results. The analysis showed that all these cells have at least one cell in the training dataset with the same transistors structure or a very similar one.

These results show that the ML model can accurately predict the behavior of a sequential cell affected by a given defect. The goal of the next subsection is to leverage on existing CA models to generate CA models for a new technology.

Table 9 Average prediction accuracy for combinational cells in different technologies

Predicion accuracy (%)		Number of inputs				
		2	3	4	5	6
Number of transistors	6	98.21	99.47			
	8	94.56	96.86	99		
	9					
	10	94.69	96.01	99.27		
	12	87.73	98.05	99.1		99.76
	14	85.69	97.35	98.75		
	16	91.74		99.2		
	18	88.18	96.28			
	20	90.29	94.37			
	22	78.73		98.37		
	24	87.91	96.88	99.37		99.79
	26	87.24	98.92			
	28	88.18	98.68			
	30			97.52		
	32	88.73	95.6			
	42					
	44					
	46					
	47					

5.2 *Predicting Defect Behavior on Another Technology*

5.2.1 Combinational Standard Cells

Another set of experiments was conducted on combinational standard cells belonging to two different technologies. Evaluation was slightly different compared to the previous one. Here, the ML model was trained over all available cells of a given technology and the evaluation was done on one cell of another technology. A loop was used to allow all cells of the second technology to be evaluated. Cells were grouped according to their number of inputs and transistors. Table 9 shows the prediction accuracy achieved on open defects of the C28 cells after training on the 28SOI cells. Results are averaged over all cells in each group (same number of inputs and number of transistors). The average prediction accuracies are globally lower compared to those of Table 7. After investigation, it appears that the behavior of most of the cells (68% of cells) is accurately predicted (accuracy >97%), while accuracy for few cells is quite low. This phenomenon is discussed in Sect. 5.2.2.

To verify the efficiency of the ML-based CA model generation method when different transistor sizes are considered, the ML model was trained over the 28SOI standard cells and used to predict the behavior of C40 cells. Table 10 shows the

Table 10 Average prediction accuracy for combinational cells using different transistor sizes

Predicion accuracy (%)	Number of inputs				
Number of transistors	2	3	4	5	6
6	100.0	99.8			
8	87.39	99.14	99.03		
9		97.19			
10	92.07	95.49	99.32	98.46	
12	91.71	98.07	99.24	98.47	99.46
14	90.1	95.84	98.63	98.79	99.52
16	91.17	93.59	99.23		99.59
18	88.5	97.15	97.14	97.74	
20	83.87	97.73	97.15	98.94	
22	87.26		98.98	98.44	
24	93.96	99.34	99.58	98.84	99.63
26	87.52	97.55	99.04	99.02	99.92
28	98.19		98.79	99.31	99.44
30			99.13	99.37	99.58
32	92.91			98.92	99.78
42					
44	92.03	98.82			
46		99.23			
47		98.29	99.76		

prediction accuracy achieved on open defects of the C40 cells after training on the 28SOI cells. Results are averaged over all cells in each group (same number of inputs and transistors). This time, 80% of cells are accurately predicted (accuracy >97%), proving that the ML-based characterization methodology could be used to generate CA models for a (large) part of combinational cells of a new technology.

5.2.2 Analysis and Discussion

A first analysis was done on cells for which the defect characterization methodology gives excellent prediction accuracy as well as those for which the prediction accuracy was quite low. Then, the limitations of the CA model generation method were investigated. After running several experiments on different configurations using one fault model at a time, the following behaviors were noticed:

- Accuracy for most of the cells is excellent, i.e., more than 97% prediction accuracy for 70% of cells. In this case, *the CA model generated by ML fit the real behavior achieved with electrical simulations*.
- Accuracy for few cells (30%) is quite low and the ML prediction is not accurate.

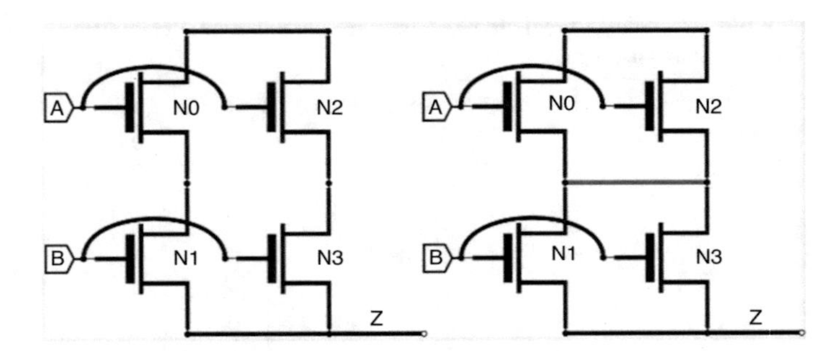

Fig. 11 Typical transistor configurations leading to good prediction

For the first cell category with good prediction score, cells have been analyzed manually to identify why they led to good results. The analysis showed that all these cells had at least one cell in the training dataset with the same transistor structure or a very similar one. The difference between very similar cells is always the same and is presented in Fig. 11. More precisely, cells giving good results are always composed of one of the configurations presented in Fig. 11, and at least one cell of the training dataset contains the other configuration. The difference between these two transistor configurations is the presence or absence of the red net. The logic function of these configurations is the same. These configurations are mostly found in high-drive cells.

For the second cell category – cells leading to poor prediction accuracy – the manual analysis showed that they have (i) new logic functions that do not appear in the cells of the training dataset or (ii) a transistor configuration which is completely new when compared to cells in the training dataset.

5.2.3 Sequential Standard Cells

We also conducted experiments on sequential cells belonging to two different technologies. As for combinational cells, the ML model was trained over all available sequential standard cells of a given technology and the evaluation was done on one cell of another technology. Cells were grouped according to their number of inputs and transistors. Table 11 shows the prediction accuracy achieved on short defects of the 28SOI cells after training on the C28 cells. Results are averaged over all cells in each group.

Similarly, we conducted experiments to analyze the efficiency of our method when different transistor sizes are considered. This time, we trained the ML model over the C40 standard cells and used it to predict the behavior of more technologically advanced 28SOI cells. Table 12 shows the prediction accuracy achieved on short defects.

In the above two scenarios, the average prediction accuracies are globally very low (around 50%), indicating that our method needs good training dataset

Table 11 Average prediction accuracy for sequential cells different technologies

Prediction accuracy (%)		Number of inputs			
		4	5	6	7
Number of transistors	32		53		
	34		50		
	36		50	54	
	38		50	51	
	40			50	
	42			50	
	44				
	46				
	48				
	50				
	52				

Table 12 Average prediction accuracy for sequential cells with different transistor sizes

Prediction accuracy (%)		Number of inputs			
		4	5	6	7
Number of transistors	32				
	34				
	36				
	38		50		
	40			50	
	42		48	50	
	44			50	
	46			48	
	48			48	57
	50				
	52				58

representative of every type of standard cells and transistor structures. Indeed, investigations showed that in most cases, functionally equivalent sequential cells in libraries from different technologies were designed differently and hence do not have neither the same transistor structure nor a very similar one (as discussed in Sect. 5.2.2). A manual analysis showed that they have (i) new logic functions that do not appear in the cells of the training dataset or (ii) a transistor configuration which is completely new when compared to cells in the training dataset. Considering the main property of our learning method for CA model generation, which is based on the recognition and use of identical structures, it is not surprising to get such low-quality results. Adding more cells and thus more known structures to the training database should help to increase the prediction accuracy.

5.2.4 Controlled Experiments

In an attempt to check the above hypothesis, "controlled experiments" were performed by considering three scan flip-flops (SDFPQ cells) coming from three different technologies (C40, 28SOI, C28). The SPICE description of each cell was manually modified so as to get *the same schematic* for all of them. This was done by removing some buffers and duplicate transistors, which were initially inserted in the cell descriptions for driving strength purpose. After modification, each flip-flop contained 5 inputs, was made of 32 transistors, and can be affected by the same intra-transistor defects. The physical layouts of the modified cells have not been made identical and carefully modified to the minimum in an attempt to keep the technological specificities of each cell. Therefore, the list of potential inter-transistor defect locations is different for each cell. A CA model has been generated for each modified flip-flop, using the simulation-based flow implemented by a commercial tool.

Three types of experiments were performed. First, the ML model was trained by considering all short defects of the C28 SDPFQ cell, and its prediction accuracy was successively evaluated over all short defects (576) of the C40 SDPFQ cell and over all short defects (928) of the 28SOI SDPFQ cell. The same procedure was done for open defects (the C40 SDPFQ cell and the 28SOI SDPFQ cell each contain 387 open defects). Next, the ML model was trained by considering all short defects of the 28SOI SDPFQ cell, and its prediction accuracy was successively evaluated over all short defects of the C40 SDPFQ cell and over all short defects (1016) of the C28 SDPFQ cell. Again, the same procedure was done for open defects (the C28 SDPFQ cell contains 394 open defects). Finally, the ML model was trained by considering all short defects of the C40 SDPFQ cell, and its prediction accuracy was successively evaluated over all short defects of the 28SOI SDPFQ cell and over all short defects of the C28 SDPFQ cell. The same procedure was done for open defects.

To visualize the efficiency of the ML-based characterization method, a confusion matrix was generated in which each row of the matrix represents the instances in a predicted class (defects that are predicted by the ML algorithm to be detected/not detected by a given cell pattern), while each column represents the instances in an actual class (defects that are actually detected/not detected by a given input pattern). By this way, the confusion matrix reports the number of true positives, false positives, false negatives, and true negatives.

Results are reported in Table 13 for the three types of experiments. The confusion matrix can be found at the center of the table (green- and red-headed columns), this time represented using only a horizontal axis. For example, let us consider the first experiment, when the ML model is trained by considering all defects of the C28 SDPFQ cell, and the prediction accuracy is evaluated over all defects of the 28SOI SDPFQ cell (third row in Table 13). The number of true positives, false positives, false negatives, and true negatives is 442, 23, 40, and 423, respectively, thus leading to a prediction accuracy of 93%. From the overall results reported in Table 13, this time it appears that the prediction accuracy achieved with the ML-

Table 13 Results of the controlled experiments

Train	Predict	Defect type	True P	False P	False N	True N	Accuracy
C28	C40	short	248	40	52	236	84%
		open	215	8	2	162	97%
	28SOI	short	442	23	40	423	93%
		open	215	2	2	168	98%
28SOI	C40	short	252	36	59	229	83%
		open	215	8	2	162	97%
	C28	short	483	27	77	429	89%
		open	221	2	3	168	98%
C40	28SOI	short	416	49	153	310	78%
		open	215	2	8	162	97%
	C28	short	446	64	182	324	75%
		open	221	2	9	162	97%

based method ranges from 75 to 98%, thus clearly demonstrating its efficiency. As for combinational cells, one or more structural patterns have been identified in functionally equivalent cells from various libraries, so that the ML algorithm can exploit them efficiently for training and inference purpose.

6 Hybrid Flow for CA Model Generation

Considering the above analysis, it appears that the ML-based CA model generation flow cannot be used for all cells in a standard cell library to be characterized. A mixed solution, which consists in combining ML-based CA model generation and conventional (simulation-based) CA model generation, should be preferably used. This is illustrated in the following.

The hybrid flow for accelerating the CA model generation is sketched in Fig. 12. Typically, when the CA model for a new cell is needed, the first step consists in checking whether the ML-based generation will lead to high-quality CA models. This is done by analyzing the structure of the new cell and check whether the training dataset contains a cell with identical or similar structure (as discussed in Sect. 5.2.2). If the ML algorithm is expected to give good results, the new cell is prepared (representation in a CA matrix) and submitted to the trained ML algorithm. The output information is then parsed to the desired file format. Conversely, if the ML algorithm is expected to give poor prediction results, the standard generation flow presented in Fig. 4 is used to obtain the CA model. A feedback loop uses this new simulated CA model to supplement the training datasets and improve the ML algorithm for further prediction.

The experiments performed to estimate the improvement in CA model generation time achieved with the hybrid flow in Fig. 12 are described in the following.

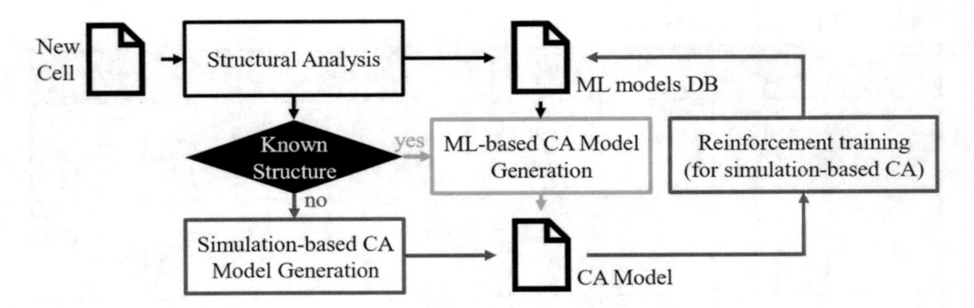

Fig. 12 Hybrid flow for CA model generation

6.1 Runtime Saving for Combinational Cells

For these experiments, the random forest model was first trained on 28SOI combinational standard cells, and CA models were then generated for a subgroup of the C40 combinational standard cell libraries. A subgroup is composed of cells representing all the cell functions available in C40 libraries. In these experiments, this subgroup contained 409 cells: 118 (29%) have a cell with an identical structure in the training dataset, 87 (21%) have a cell with an equivalent structure (as explained in Sect. 5.2.2) in the training dataset, and 204 (50%) have no identical or equivalent structure in the training dataset (a simulation-based generation is thus needed). For these 204 cells, the generation time was calculated and found to be equal to ~172 days (~ 5.7 months) considering a single SPICE license. Using the ML-based CA model generation for the 118 + 87 = 205 (50%) remaining cells requires 21,947 seconds (~ 6 h), again considering a single SPICE license. Considering that a simulation-based generation for these 205 cells would require ~78 days, we can estimate the reduction in generation time to *99.7%*. Now, when considering the whole C40 subgroup composed of 409 cells, the hybrid generation flow would require ~172 days + ~6 h, compared with ~172 days + ~78 days = ~250 days using only the simulation-based generation. This represents a reduction in generation time of about *38%*. After investigating results of these experiments, it appears that the ML-based CA model generation works well for about 80% of cells of the C40 subgroup. Surprisingly, the structural analysis revealed that only 50% (205 cells) could be evaluated using the ML-based generation part of the flow. This shows that there is still room for further improvement of the structural analysis in the flow and hence better performance of the ML-based CA model generation process.

6.2 Runtime Saving for Sequential Cells

In these experiments, the goal was to (re-)generate CA models for the 27 C40 sequential standard cells in an efficient manner. To achieve this goal, the number of CA models obtained by the simulation-based flow had to be minimized. The lowest achievable number of simulation-based CA models is given by one cell per training group (same number of inputs and transistors), plus the number of cells that are alone in their training group (and thus cannot go through the ML-based flow). In these experiments, 13 cells had to go through the simulation-based flow (9 groups + 4 individual cells). The simulation-based generation flow for those 13 cells took ~*4.1 h*. The ML-based generation flow for the remaining 14 cells took ~*35 s*. By comparison, the simulation-based model generation for these 14 cells would take ~*4.2 h*. The ML-based flow thus provides a runtime reduction of *99.8%* for the cells it can handle. Now, if we consider the whole C40 group composed of 27 cells, the hybrid generation flow would require ~4.1 h + ~35 s, compared with ~4.1 h + ~4.2 h = 8.3 h by using only the simulation-based generation flow. This represents a reduction in generation time of about *51%*.

It is worth mentioning that most of the runtime in the hybrid flow is taken by the simulation-based flow. Therefore, as long as new CA models generated by simulation are added to the database, the ML-based flow can use them and then handle more and more cells, further reducing the generation runtime.

7 Discussion and Conclusion

A novel approach based on machine learning was presented in this chapter to generate CA models. The main goal is to speed up the characterization process of standard cell libraries for test and diagnosis purpose, which usually resort to SPICE simulations and hence is very time-consuming. The methodology is based on the recognition of identical structural patterns between cells already characterized by simulation and those to be characterized by machine learning.

Experiments done on both combinational and sequential cells from industrial libraries demonstrate the accuracy and performance of the method when predicting defect behavior has to be done on the same technology. In this case, the generation runtime of CA models can be significantly reduced for experimented cells having other cells with similar structure in the training dataset.

In order to deal with functionally equivalent cells having different internal structures, a hybrid flow combining learning-based and simulation-based CA model generation can be used. Experiments carried out on a subset of cells from an industrial library have shown that the generation time of CA models can be reduced by more than 50%.

Experiments reported in this chapter have been carried out on a small size of standard cell population. Considering that more than 10,000 cells have usually to be characterized for a given technology, the hybrid flow described in this chapter is expected to provide even better results, especially owing to the reinforcement training that uses simulation generated models for supplementing the training datasets and hence reduce the number of electrical simulations.

References

1. Ladhar, A., Masmoudi, M., Bouzaida, L.: Efficient and accurate method for intra-gate defect diagnoses in nanometer technology. In: Proceedings of IEEE/ACM Design Automation and Test in Europe (2009)
2. Sun, Z., Bosio, A., Dilillo, L., Girard, P., Virazel, A., Auvray, E.: Effect-cause intra-cell diagnosis at transistor level. In: Proceedings of IEEE International Symposium on Quality Electronic Design (2013)
3. Hapke, F., Reese, M., Rivers, J., Over, A., et al.: Cell-aware production test results from a 32-nm notebook processor. In: Proceedings of International Test Conference (2012, November)
4. Gao, Z., Hu, M.-C., Swenton, J., Magali, S., Huisken, J., Goosens, K., Marinissen, E.J.: Optimization of cell-aware ATPG results by manipulating library cells' defect detection matrices. In: Proceedings of IEEE International Test Conference in Asia (ITC-Asia) (2019)
5. Amyeen, E., Nayak, D., Venkataraman, S.: Improving precision using mixed-level fault diagnosis. In: Proceedings of International Test Conference (2006, October)
6. Tang, H., Jain, A., Pillai, S.K.: Using cell aware diagnostic patterns to improve diagnosis resolution for cell internal defects. In: Proceedings of Asian Test Symposium, pp. 231–236 (2017, November)
7. Fan, X., Sharma, M., Cheng, W.-T., Reddy, S.M.: Diagnosis of cell internal defects with multi-cycle test patterns. In: Proceedings of Asian Test Symposium (2012, November)
8. Archer, B., Schuermyer, C.: Cell-aware test for lower DPPM and faster silicon diagnosis. In: Proceedings of Synopsys User Group (SNUG) (2017, March)
9. Feldman, N.: Accelerating silicon diagnosis using a cell-aware flow. In: Proceedings of Synopsys User Group (SNUG) (2017, March)
10. Hapke, F., et al.: Cell-aware test. IEEE Trans. Comput-Aid. Des. **33**(9), 13–16 (2014)
11. Maxwell, P., Hapke, F., Tang, H.: Cell-aware diagnosis: defective inmates exposed in their cells. In: IEEE European Test Symposium (2016)
12. Hapke, F., Krenz-Baath, R., Glowatz, A., Schloeffel, J., Weseloh, P., Wittke, M., Kassab, M., Schuermyer, C.W.: Cell-aware fault model creation and pattern generation. US Patent 12/718,799 (2010)
13. Mhamdi, S., Girard, P., Virazel, A., Bosio, A., Ladhar, A.: A learning-based cell-aware diagnosis flow for industrial customer returns. In: Proceedings of IEEE International Test Conference (2020)
14. Lorenzelli, F., Gao, Z., Swenton, J., Magali, S., Marinissen, E.J.: Speeding up cell-aware library characterization by preceding simulation with structural analysis. In: Proceedings of IEEE European Test Symposium (2021)
15. Gao, Z., Malagi, S., Chun Hu, M., Swenton, J., Baert, R., Huisken, J., Chehab, B., Goossens, K., Marinissen, E.J., Application of cell-aware test on an advanced 3nm CMOS technology library. In: Proceedings of IEEE International Test Conference (2019)
16. Guo, R., Archer, B., Chau, K., Cai, X.: Efficient cell-aware defect characterization for multi-bit cells. In: Proceedings of IEEE International Test Conference in Asia (2018)
17. d'Hondt, P., Ladhar, A., Girard, P., Virazel, A.: A learning-based methodology for accelerating cell-aware model generation. In: Proceedings of IEEE/ACM Design Automation and Test in Europe (2021)
18. Nanometer Library Characterization: Challenges and Solutions, Webinar. Silvaco (2019, March)
19. Improving Library Characterization with Machine Learning, White Paper. Mentor, A Siemens Business (2018)
20. Unified Library Characterization Tool Leverages Machine Learning in the Cloud, White Paper. Cadence (2018)
21. Poornima, H.S., Chethana, K.S.: Standard Cell Library design and characterization using 45nm technology. IOSR J. VLSI Signal Process. (IOSR-JVSP). **4**, 29–33 (2014)

22. Rabaey, J., et al.: Digital Integrated Circuit – A Design Perspective. 2nd ed., Prentice Hall (2003)
23. bin Bahari Tambek, A., bin Mohd Noor Beg, A.R., Rais Ahmad, M.: Standard Cell Library development. In: Proceedings of the 11th International Conference on Microelectronics (1999), pp. 22–24
24. Weste, N.H.E., Harris, D., Banerjee, A.: CMOS VLSI Design: A Circuits and Systems Perspective. Wesley (1993)
25. Shoji, M.: CMOS Digital Circuit Technology. Prentice Hall (1988). ISBN 978-0131388505
26. Naga Lavanya, M., Pradeep, M.: Design and characterization of an ASIC standard cell library industry–academia chip collaborative project. In: Microelectronics, Electromagnetics and Telecommunications (2018)
27. H. Tang et al., Diagnosing cell internal defects using analog simulation-based fault models. In: Proceedings of Asian Test Symposium, pp. 318–323 (2014)
28. Clein, D.: CMOS IC LAYOUT concepts, methodologies, and tools. ISBN 978-0750671941
29. Hapke, F., Krenz-Baath, R., Glowatz, A., Schloeffel, J., Hashempour, H., Eichenberger, S., et al.: Defect-oriented cell-aware ATPG and fault simulation for industrial cell libraries and designs. In: Proceedings of IEEE International Test Conference (2009, November)
30. Goncalves, F.M., Teixeira, I.C., Teixeira, J.P.: Integrated approach for circuit and fault extraction of VLSI circuits. In: Proceedings of EEE International Symposium on Defect and Fault Tolerance in VLSI Systems (1996, November)
31. Goncalves, F.M., Teixeira, I.C., Teixeira, J.P.: Realistic fault extraction for high-quality design and test of VLSI systems. In: Proceedings of IEEE International Symposium on Defect and Fault Tolerance in VLSI Systems (1997, October)
32. Stanojevic, Z., Walker, D.M.: Fed–x - a fast bridging fault extractor. In: Proceedings of IEEE International Test Conference (2001, November)
33. Venkataraman, S., Drummonds, S.D.: A Technique for logic fault diagnosis of interconnect open defect. In: Proceedings of IEEE VLSI Test Symposium (2000)
34. Li, C.-M., McCluskey, E.J.: Diagnosis of resistive-open and struck-open defects in digital CMOs ICs. IEEE Trans. CAD Integr. Circuits Syst. **24**(11), 1748–1759 (2005)
35. Pedregosa, F., et al.: Scikit-learn: machine learning in python. J. Mach. Learn. Res. **12**(Oct), 2825–2830 (2011)

Neuromorphic Computing: A Path to Artificial Intelligence Through Emulating Human Brains

Noah Zins, Yan Zhang , Chunxiu Yu, and Hongyu An

1 Introduction

The human brain is regarded as the most intelligent and computationally efficient machine [1]. The human brain is constructed with billions of neurons that are connected through trillions of synapses, also known as neural junctions transmitting electric nerve impulses between two neurons. Each neuron can communicate with an average of more than ten thousand other neurons forming a complicated network, normally referred to as a biological neural network. The topology of the biological neural networks is dynamic and adjustable in accordance with different exterior stimuli which are captured by different sensory organs. The change in the neural networks is caused by modifying the connecting strength among neurons. The connecting strength is designated as synaptic plasticity. The intricate and dynamic biological neural network is widely believed to be the source of intelligence [2, 3].

The extraordinary intelligence of the biological neural systems motivates scientists to think about whether the intelligence can be recreated artificially. This thought leads to a birth of a new discipline: *artificial intelligence (AI)* [4]. AI aims to build a machine or a system that has the cognition and intelligence of humans or even outperforms them. These intelligent machines/systems are expected

N. Zins · H. An (✉)
Department of Electrical and Computer Engineering, Michigan Technological University, Houghton, MI, USA
e-mail: nwzins@mtu.edu; hongyua@mtu.edu

Y. Zhang
Department of Biological Sciences, Michigan Technological University, Houghton, MI, USA
e-mail: yzhang49@mtu.edu

C. Yu
Department of Biomedical Engineering, Michigan Technological University, Houghton, MI, USA
e-mail: chunxiuy@mtu.edu

A. Iranmanesh (ed.), *Frontiers of Quality Electronic Design (QED)*,
https://doi.org/10.1007/978-3-031-16344-9_7

to perceive the surroundings, make decisions, and conduct actions to accomplish the goals in complicated environments. The essential purposes of developing artificial intelligence can be summarized as liberating humans from tedious and dangerous work, understanding the functions of human brains and explaining why and how we have intelligence, and even creating advanced intelligence to handle the challenges beyond our capabilities. Nowadays, AI has been widely used in social media, web searching, and online store and services and is part of our daily life. AI systems will cause a fundamental change in society.

Therefore, numerous approaches have been studied for AI implementation since the last century. Alan Turing originally introduced the idea of creating and evaluating intelligent machines in the article: *Computing Machinery and Intelligence* [5], where he first introduced the idea of creating intelligent machines and how to evaluate whether this machine has intelligence. Later, this method is named as *Turing test*. In Turing test, the intelligence of a machine is measured in a scenario in which a human interacts with another entity. During the interaction, the entity will not show at the sight of the human. If the human in Turing test is unable to distinguish this entity as a machine or human, then it says that the machine has the intelligence. The underlying rationale of the Turing test is that the machine must have the substantial intelligence to handle all difficult questions being asked so that it can camouflage itself as a human. Passing the Turing test requires the intelligent machine to have an incredibly deep understanding and insight into human society and the natural world so that it is able to answer the challenging, even very trick, questions appropriately.

One of the essential purposes of developing artificial intelligence is to liberate humans from laborious or dangerous work. Whether the AI systems can accomplish the tasks that originally only can be handled by humans is an important indicator of evaluating the capabilities of AI systems. These tasks include natural language processing, games, etc. ELIZA, developed by Joseph Weizenbaum, was one of the famous natural language processing programs that attempted to pass the Turing test [6]. In the game field, which is traditionally dominated by human players, the supercomputer *Deep Blue* developed by IBM defeated the chess world champion, Gary Kasparov, in 1997 [7]. The defeat of the machines to human players in chess is widely considered a big achievement of AI, demonstrating their computational capability far more performing than human brains.

Nowadays, the most successful approach to AI is deep learning (DL) [8–10]. The idea of DL stems from artificial neural networks (ANNs) proposed by Warren in 1943. ANNs are initially used as a computational model to emulate biological neural networks [11]. In McCulloch and Walter Pitts's model, the threshold function was used for modeling biological neurons. The most important improvement of DL is a replacement of these threshold functions with the differential nonlinear functions, such as the sigmoid function. In this way, the gradient descent algorithms become applicable for training ANNs. The training process is generally to adjust the weights of ANNs with massive high-dimensional data being used for inference and classification. This is the basic mechanism of these training processes. The underlying mechanism of these training processes is to map the data into the high-

Table 1 Comparison of deep learning and neuromorphic systems

Features	Deep learning/von Neumann computational platform	Neuromorphic system
Information representation	Binary signals in square waves	Spikes
Operating frequency	High frequency	Low frequency
Architecture	Centralized von Neumann architecture	Distributed non-von Neumann architecture
Basic modules	Arithmetical modules and memory	Electronic neurons and synapses
Neural network models	CNN, RNN, etc.	Spiking neural networks

dimensional space for inference and classification. The study exhibits that deeper layers of the ANNs lead to a much higher training/inference accuracy. This is the reason why the phrase *deep learning* is named. Recently, big data and powerful computational machines, e.g., GPU, boost the deployment of DL on a variety of tasks. For example, the recommendation systems are used by YouTube and Amazon. The voice assistance systems are on mobile devices, such as Siri [12] and Alexa [13]. There are also intelligent recognition and navigation algorithms on autonomous vehicles in Tesla [14, 15]. In addition is the overwhelming victory against human players in strategic games, such as the Go game [16]. However, accompanying the countless achievements, shortcomings emerge, such as overfitting, long training time, dependence on massive data, extremely high-power consumption, and low immunity to *adversarial attack*. In order to resolve these issues, scientists again have begun to seek possible solutions and inspirations in biological neural systems.

Neuromorphic computing is the approach that stays on track of realizing artificial intelligence through the emulation of biological neural systems [17]. One of the advantages of neuromorphic computing against deep learning is it aims to build the neural structure physically. The physical building of a biological neural system involves designing electronic neurons and synapses at the microscopic level and non-centralized architecture at the macroscopic level. This unique attempt means neuromorphic computing essentially aims to reinvent computing machines that are fundamentally different from the current digital computer. In digital computers, the algorithms and calculations are performed by arithmetic circuits. Moreover, the data is encoded into binary numbers by analog-to-digital converters. In neuromorphic systems, the computations are performed through electronic neurons and synapses. In an addition, neuromorphic systems convert the exterior stimuli into discrete spike trains. These spiking signals are generated with the voltage potential between the membrane of the neurons [1]. The reinventions from basic computation units to decentralized architecture are inherently more suitable for the demands of ANNs. Table 1 particularizes the differences between deep learning associated with conventional computational platforms and neuromorphic systems.

This chapter introduces the fundamentals of neuromorphic computing including biological neural networks, spiking neural networks, electronic neurons and synapses, and state-of-the-art neuromorphic chips. This chapter is organized as follows: Section 2 introduces the basics of biological neural systems; Sect. 3 presents the models of neural systems; Sect. 4 expresses the designs of electronic neurons and synapses; Sect. 5 discusses the state-of-the-art neuromorphic chips; and Sect. 6 summarizes the challenges and opportunities of neuromorphic computing.

2 Biological Neural System

The basic organs forming biological neural systems will be introduced in this section, covering neurons and synapses. In addition, associative memory learning, which widely exists in animals, will be discussed at both cellular and behavior levels in this section.

2.1 Neurons and Synapses

The studies of the neural system can trace back hundreds of years ago. In 1899, Santiago Cajal first discovered and determined the basic signal processing cell in brains: neurons [1, 18]. His drawings of neuron structure are shown in Fig. 1a. He observed that the neurons are tree-like structures with a number of branches connecting with each other to form a complicated network, as shown in Fig. 1b. There are more than 100 billion neurons within an individual human brain. A typical neuron consists of four central parts: dendrites, soma, axon, and synapses, as depicted in Fig. 1c [1, 18].

There are more than ten thousand dendrites within a single neuron and their function is to connect and receive signals from other neurons [1]. Due to a large number of dendrites, each neuron is able to communicate with thousands of other neurons at the same time. The communication signals among the neurons are electrical spikes, as illustrated in Fig. 1d. The spiking signals received by the dendrites are integrated/summed into the soma, the body of the neuron (see Fig. 1c). If the integrated value exceeds a specific voltage, the soma generates a sequence of electrical spiking signals to the axon. This behavior is referred to as neuron firing.

The mechanisms of electrical signal generation/propagation within the neurons are fundamentally different from the current flowing in conductors. The spiking signals are generated due to the opening of Na ions (Na^+) channel, resulting in Na^+ ion transport from the outside to the inside of the neuron membrane. In 1939, Hodgkin and Huxley recorded the spiking signal propagating along the axon of the cat [1], which is depicted in Fig. 1d. The magnitude of the spiking signals is at millivolt-level with up to 10-millisecond duration [1, 18]. The spikes propagating on the axons of neurons are named *membrane potential*, as illustrated in Fig. 2. The membrane potential (V_m) can be described by the following equation:

$$V_m = V_{\text{in}} - V_{\text{out}}, \tag{1}$$

where V_{in} is the voltage on the inside of the neuron and the V_{out} is the voltage on its outside [1].

When these spiking signals travel along the axon and reach the end of an axon, they convey to other neurons through the synapse. A synapse acts as a connecting organ between neurons. The neuron sending or generating spikes is referred to as a presynaptic neuron, whereas the neuron receiving spikes from the synapse is called a postsynaptic neuron. The presynaptic and postsynaptic neurons are not physically connected, and there is a space between the presynaptic neuron and postsynaptic called the synaptic cleft. The transfer of spikes is achieved through a biochemical reaction. When the presynaptic neuron fires, the neuron releases chemical messengers, known as *neurotransmitters*, into the synapse cleft, as illustrated in Fig. 3 [1]. The neurotransmitters diffuse from a presynaptic end to a postsynaptic end. After that, they bind to their receptors expressed on the postsynaptic cell. Neurotransmitters cross the synaptic cleft (4–40 nm) to interact with their receptor at the postsynaptic neuron, resulting in a time delay between the membrane potentials (spikes) of the presynaptic neuron and the postsynaptic neuron. The signals propagated through the synapses can be either attenuated or amplified, which is referred to as the plasticity of synapses [1]. Synaptic plasticity has been widely believed as a critical feature for memory and other functions of human brains [1]. The synaptic plasticity which represents the connection strength between neurons determines the magnitude of the membrane potential (spikes) stimulated at the postsynaptic neurons. An excitatory synapse will excite a larger membrane potential at the postsynaptic neuron; in contrast, the inhibitory synapses

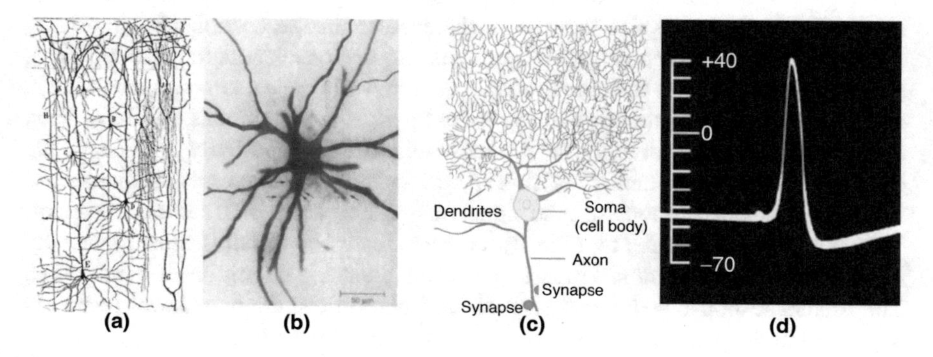

Fig. 1 The neuron structure and spiking membrane potential in neural systems: (**a**) the hand drawing of the neural network by Santiago Cajal [19, 20]; (**b**) image of a motor neuron [1]; (**c**) the illustration of a typical neuron includes four critical parts: soma, dendrites, axon, and synapse [1]; (**d**) the membrane potential measured [1]

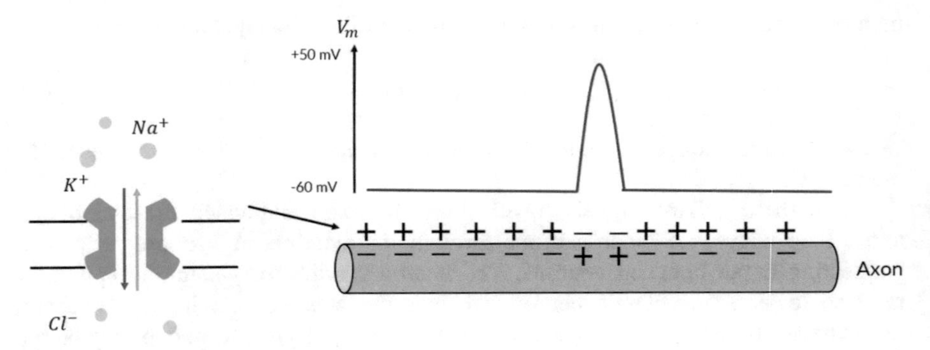

Fig. 2 Illustration of membrane potential propagation on the axon

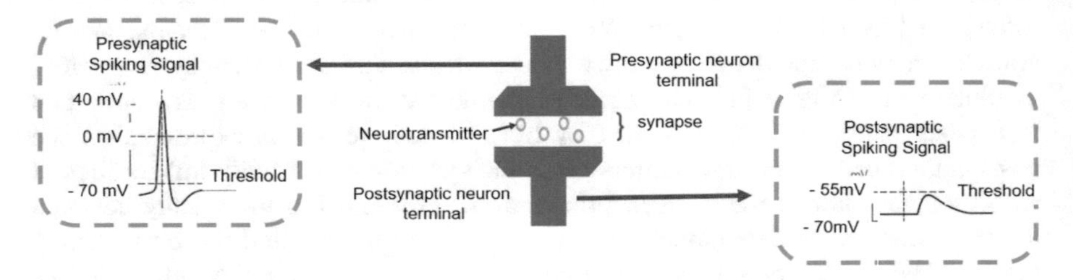

Fig. 3 Spiking signal propagation across the synapse

reduce the probability of generating a membrane potential at postsynaptic neurons [1], as illustrated in Fig. 3.

The change of the synaptic connection is widely believed to be responsible for the memory and learning of neural systems. For instance, habituation is one of the implicit memories that can be interpreted and explained using the plasticity of synapses. Habituation is a phenomenon wherein an animal is capable of memorizing unharmful stimuli and consequently presents less responsiveness to these repetitive exterior stimuli, such as the noisy sound that repeatedly appears in its surroundings. The phenomena of habituation are prevalently existing in animals and have been experimentally tested in *Aplysia*. A tactile stimulus is repeatedly applied to the siphon of the *Aplysia*, and the shrink of gill indicates the neuronal response. As the number of stimuli increases, the signal magnitude of neuronal response reduces. This experiment indicates that the *Aplysia* habituate the stimulus after repeatedly receiving exterior stimulus. These experiments yield the definition of habituation. The smaller response is due to fewer released neurotransmitters or fewer receptors expressed in the postsynaptic neurons under the repetitive stimulus.

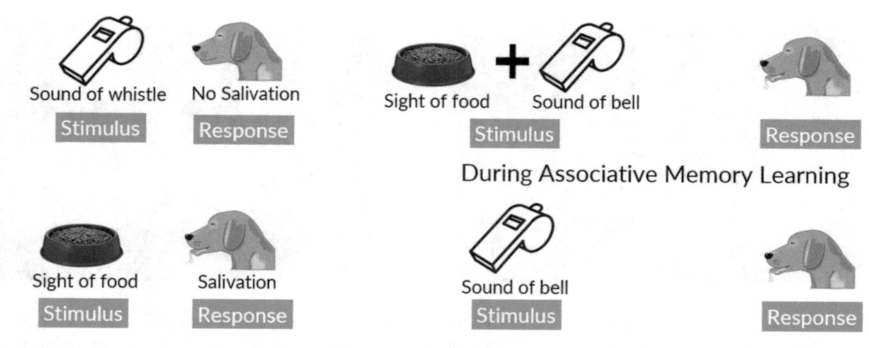

Fig. 4 Behavior-level associative memory learning experiment on the physiology of digestion of dogs

2.2 *Associative Memory Learning*

Associative memory is a ubiquitous self-learning method that animals have. In associative memory learning, the neural systems of the animals remember the causality of two concurrent events. Unlike the prevailing deep learning that relies on a large amount of labeled data. Associative memory enables the animals to correlate two events that occur at the same time, thereby memorizing the relationship between them [1]. As the classic Pavlov's experiment presented, dogs can learn the sound of whistles as a sign of food; people can remember a word representing an object [1, 21]. The neuromorphic system which emulates associative memory learning will learn based on its own experience and perception. Moreover, the rebuilding of associative memory is not only to reveal a way of designing a self-learning neuromorphic system but also to help with understanding the learning/memory mechanism of a nervous system [22].

Associative learning is first studied by Ivan Pavlov through a series of experiments on dogs [1]. Originally, dogs salivated only when presented with food and show no salivation response to whistle sound. Then, the food was offered accompanied by whistle sound to the dogs [1]. After this process was repeated multiple times, the dogs started to exhibit a salivation reaction when they heard a whistle sound alone, as illustrated in Fig. 4. These experiments demonstrate a learning phenomenon that salivation from the sight of food can also be stimulated by other different stimuli, e.g., the sound of whistles. This learning phenomenon is named associative memory. Typically, two types of signal pathways are involved in associative memory. In Pavlov's experiment, the visual perception (seeing food) which unconditionally stimulates the salivation response is an unconditional stimulus (US) signal pathway, whereas the auditory perception (sound from the whistles) is referred to as a conditional stimulus (CS) because it cannot evoke salivation reaction without learning.

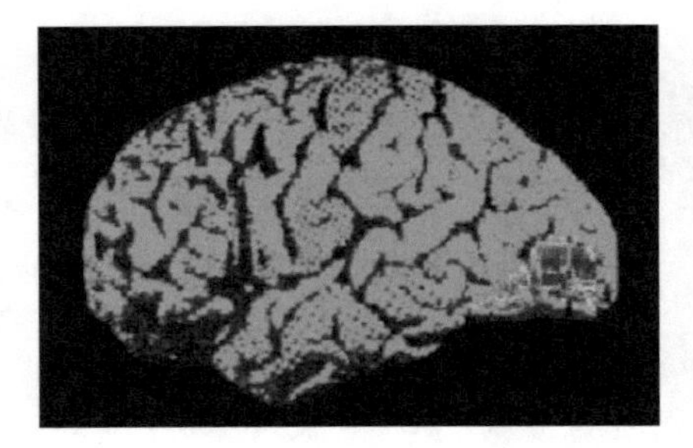
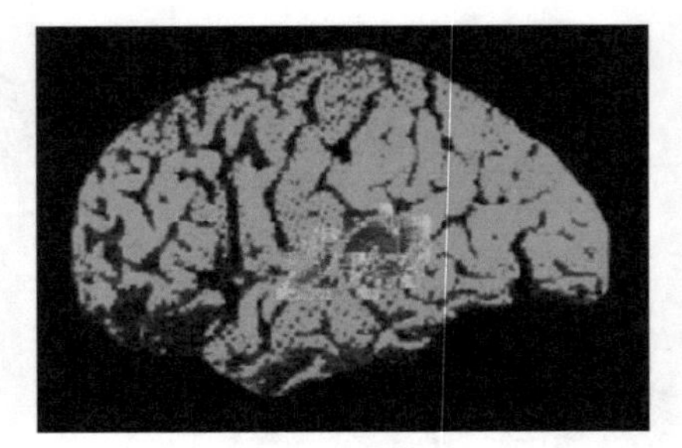

Fig. 5 Different parts of the brain process different stimuli [1]. (**a**) Positron Emission Tomography (PET) image of looking at words. (**b**) Positron Emission Tomography (PET) image of listening at words

Aforementioned, there are two signal pathways involved in associative memory learning for processing different signals. In the neural system, e.g., human brains, the function of the signal is not determined by the signal itself but by its pathway and the processing regions in the brain [3]. In a nervous system, the shapes and durations of spiking signals share the same shape and duration independent of sensory stimuli such as light or sound [1].

This fact leads to a reasonable question: if the spiking signals are stereotyped reflecting limited properties of the stimulus, how do the neural signals carry and convey particular information? Studies on the nervous system reveal that distinct sensation signals are routed and processed in different regions of the brain and that signals are distinguished by their pathway rather than their particular magnitudes or shapes [1]. Positron emission tomography (PET) is a powerful tool to reveal brain activities, as shown in Fig. 5 [1]. Visual pathways which are activated by cells in the retina in response to light are completely different from the auditory sensory pathways activated by ear sensory cells stimulated by sound.

After preprocessing at different regions, output signals from different regions will be integrated during associative memory learning [1]. For instance, Fig. 6 depicts how signals are converged in the brain of rats. The captured somatosensory and auditory signals are processed in the auditory and somatosensory thalamus, respectively, which are different regions in rat brains. After that, the output spiking signals from these two regions converge together at lateral nucleus, as illustrated in Fig. 6. The auditory somatosensory cortex is at the conditional signal pathway.

Originally, the rat ignores a neutral tone since the signal pathway from the lateral nucleus is at the conditional signal pathway and cannot stimulate a fear response. Nevertheless, when the tone is presented with an electric shock, the rat learns to correlate the neural stimulus of tone with the noxious stimulus of the electric shock. After the concurrent events are repeated multiple times, the tone alone will stimulate a fear response as well. This experiment indicates some changes happen when the tone and shock are converged at the lateral nucleus region of the rat.

The studies on *Aplysia* reveal the cellular mechanism of how the blocked unconditional signal pathway becomes unblocked during associative memory learning:

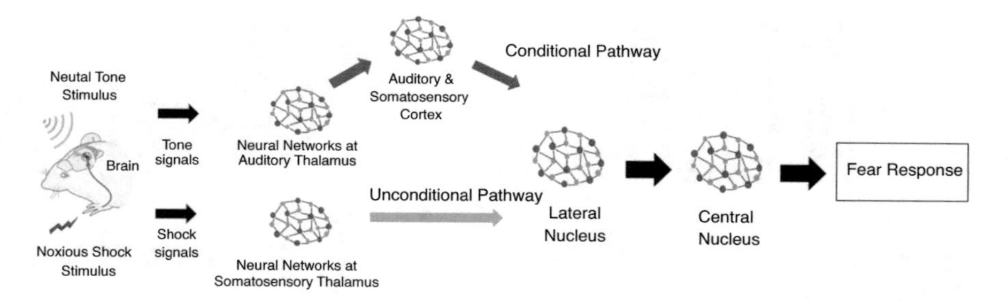

Fig. 6 Associative memory learning in mouse brain

the signal pathway is modified through the change of synaptic transmission. Dr. Kandel studied associative memory learning at the cellular level using *Aplysia* and discovered synaptic transmission change. Dr. Kandel won the Nobel Prize in 2000 because of this work [1]. During the learning process, the synaptic connection between the sensory and response neurons is enhanced, increasing the received signals of response neurons [1]. In the *Aplysia* experiment, the gill motor is typically unresponsive to the siphon stimulation. However, by applying a shock to the tail, an unconditional stimulus, and touching the siphon (CS) simultaneously and repeatedly, the gill motor neuron became more responsive to inputs from the siphon sensory neuron (CS). As depicted in Fig. 7, the stimulus from the US and CS are paired and overlapped with each other in the time that is considered a trigger condition of associative memory learning at the cellular level [1]. The increased magnitude of the gill motor response results from a stronger synaptic connection induced between the sensory neuron of the siphon and the motor neuron of the gill during the associative learning process. This cellular association learning behavior comes from the increment of connection strength between the sensory neuron and response neuron. The strong connection strength comes from higher synaptic transmission [1].

The observation of associative memory learning raises the question: *How does the neural system correlate two initially unrelated events*? Investigations on associative memory at the cellular level reveal that the changes in synaptic weight play a critical role in associative memory [1]. The weight of a synapse, which refers to the amount of the chemical neurotransmitters, reflects the connection strength between two neurons. With the increase of the connecting strength between neurons, the relationship between two concurrent stimuli is memorized [1]. The studies on *Aplysia* reveal the mechanism of associative memory at the cell level results from the synaptic plasticity and signal pathway modification [1]. The associative memory learning in Aplysia involves two signal pathways from the sensory and response neurons. These two signal pathways are marked in blue and red, respectively, in Fig. 7 [1]. Normally, the gill motor is unresponsive to the siphon stimulation of the siphon before learning. However, by applying a shock to the tail and touching the siphon simultaneously and repeatedly, the gill motor neuron became more

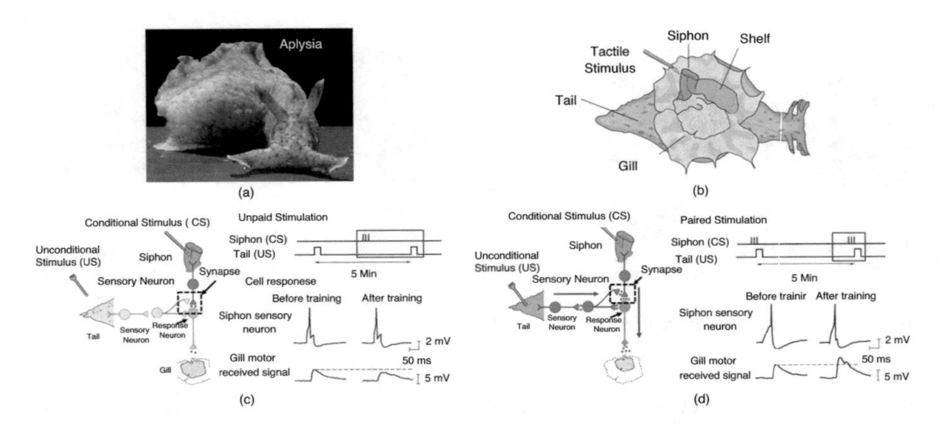

Fig. 7 (**a**) Image of *Aplysia* (**b**), the experimental setup, (**c**) touching and shocking stimulate the *Aplysia* siphon and tail, respectively. Under unpaired stimulation, the received signal of the response neuron (gill motor neuron) is nearly identical before and after training (**d**) and a greater magnitude of the received signal at the gill motor neuron is measured [1]

responsive to the stimulus coming from the siphon. The stimulus from the US and CS are paired, as illustrated in Fig. 7d [1]. The larger magnitude of the gill motor response results from a stronger synaptic connection induced between the sensory neuron of the siphon and the motor neuron of the gill. This cellular association learning behavior comes from the strengthening synaptic connection between the sensory and response neurons. The strong connection strength comes from higher synaptic transmission [1].

3 Modeling Neural System

In order to build a neuromorphic system, the biological neural system should be first accurately modeled, including the neurons and synapses. The neuron models are the sets of mathematical equations that are able to describe the biochemistry behaviors of the biological neurons. These behaviors should mainly express the threshold-based firing behavior and integration feature of the neurons. These neuron models need to balance biological plausibility and computational complexity.

The basic function of a biological neuron can be simplified into two steps, integrating received spikes from other neurons and generating other spike streams once the integrated voltage crosses the specific threshold voltage. When a neuron is generating a spike scream, it is referred to as the neuron firing. The accumulated charge at the body (soma) of a neuron is named excitatory postsynaptic potential (EPSP). Figure 8 illustrates the phenomena of EPSP. If the accumulated charges reach a specific threshold, the soma of the neuron launches the spiking signals on the axon, as shown in Fig. 8. These spikes will be conveyed to other neurons through the

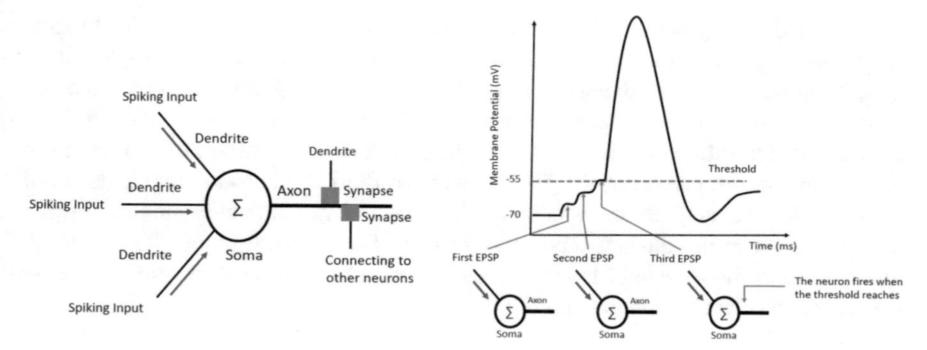

Fig. 8 Illustration of the neuron firing

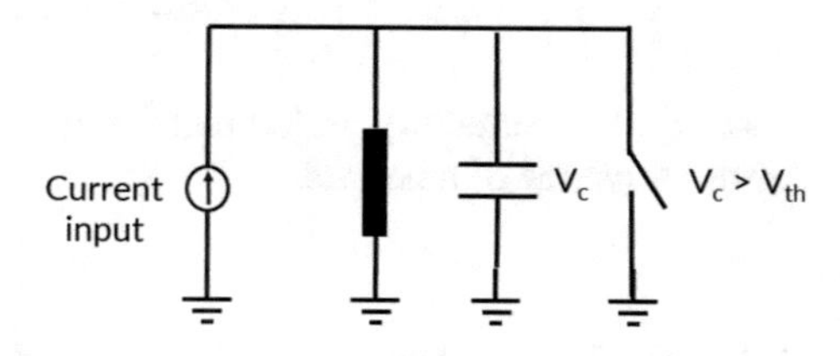

Fig. 9 Leaky integrate and fire model

synapses forming the biological neural networks. In this section, we will introduce four typical neuron models: Hodgkin-Huxley neuron model, leaky integrate and fire model, Izhikevich model, and McCulloch-Pitts model.

3.1 *Leaky Integrate and Fire Neuron Model*

The leaky integrate and fire (LIF) neuron model, proposed by Louis Lapicque in 1907, can be used to represent charge accumulation and fire behavior [23, 24]. As illustrated in Fig. 8, a typical LIF model consists of an RC circuit and a switch. Figure 9 illustrates the circuitry of a LIF model. The current input stimulates the received spikes from other neurons. The charges will be accumulated within the capacitor resulting in an increment of voltage (V_c) across the capacitor. Once the voltage Vc exceeds the threshold value (V_{th}), the switch will be closed and the capacitor discharges to the ground generating a spike signal.

Thus, the discharging of the capacitor can be described by the following equation:

$$\mathrm{RC}\frac{\mathrm{d}V_m}{\mathrm{d}t} = -V_m + \mathrm{RI}(t) \tag{2}$$

where V_m is membrane potential, R is the membrane resistance, C is the membrane capacitance, and I is input current. The model equation is valid for input current until a threshold V_{th} is reached, where the membrane potential is V_{reset}. The biggest advantage of the LIF model is its simplicity and feasibility for circuit implementation. Meanwhile, the disadvantage is it omits neuronal adaptation features. In order to resolve this limitation, several variants of the LIF models are developed, i.e., the exponential integrate and fire model, the adaptive exponential integrate and fire model, and so on. However, the disadvantage of the LIF neuron is that it does not contain neuronal adaptation, so it cannot describe an experimentally measured spike train in response to constant input current [25]. For constant input, the minimum input to reach the threshold is $I_{\text{th}} = V_{\text{th}} / R_m$. Assuming a reset to zero, the firing frequency thus can be calculated with:

$$f(I) = \begin{cases} 0, & I \leq I_{\text{th}} \\ \left[t_{\text{ref}} - R_m C_m \log\left(1 - \frac{V_{\text{th}}}{I R_m}\right)\right]^{-1}, & I > I_{\text{th}} \end{cases} \tag{3}$$

where t_{ref} is a refractory period that the neuron will not fire resulting in limiting the firing frequency of a neuron.

3.2 *Hodgkin-Huxley Neuron Model*

The Hodgkin-Huxley neuron model was first introduced by Alan Hodgkin and Andrew Huxley in 1952 [26]. Hodgkin-Huxley neurons precisely describe the ionic mechanisms of the initiation and propagation of action potentials in the axons of squid giants. The flow of charges, including K^+, Na^+, etc., that forms membrane potential in the axons is modeled using the circuits in Hodgkin-Huxley neuron model, as illustrated in Fig. 10.

The relationship between the flow of ionic currents across the cell membrane and the membrane voltage can be described using the following set of equations:

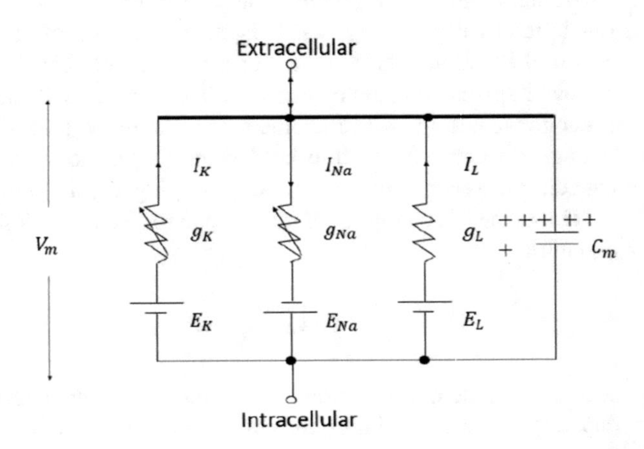

Fig. 10 Equivalent circuit diagram of the Hodgkin–Huxley model

$$C_m \frac{\mathrm{d}V}{\mathrm{d}t} = \overline{g}_K n^4 (V_k - V) + \overline{g}_{Na} m^3 h (V_{Na} - V) + g_l (V_l - V) + I(t), \tag{4}$$

$$\frac{\mathrm{d}n}{\mathrm{d}t} = \alpha_n (1 - n) - \beta_n n, \tag{5}$$

$$\frac{\mathrm{d}m}{\mathrm{d}t} = \alpha_m (1 - m) - \beta_m m, \tag{6}$$

$$\frac{\mathrm{d}h}{\mathrm{d}t} = \alpha_h (1 - h) - \beta_h h, \tag{7}$$

where C_m is the membrane capacitance ($C_m = \tau_m/R$, where R is the membrane resistance, τ_m is the timescale); V is the membrane potential; $\overline{g}_K$ and $\overline{g}_{\mathrm{Na}}$ are maximal values of the K^+ and Na^+ channel conductance, respectively; g_l is the leak conductance; n, m, and h are functions to describe K^+ channel activation, Na^+ channel activation, and Na^+ channel inactivation, respectively; V_k, V_{Na}, and V_l are K^+, Na^+, and leaky channel reversal potential; I is the time-dependent input current; and α_i and β_i ($i = n, m, h$) are rate constants for associate channel depend on voltage value, which can be determined by:

$$\alpha_n(V) = \frac{0.01\,(10 + V_{\mathrm{reset}} - V)}{\exp\left(\frac{10 + V_{\mathrm{reset}} - V}{10}\right) - 1}, \tag{8}$$

$$\beta_n(V) = 0.125 \exp\left(\frac{V_{\mathrm{rest}} - V}{80}\right), \tag{9}$$

$$\alpha_m(V) = \frac{0.1\,(25 + V_{\mathrm{rest}} - V)}{\exp\left(\frac{25 + V_{\mathrm{reset}} - V}{10}\right) - 1}, \tag{10}$$

$$\beta_m(V) = 4 \exp\left(\frac{V_{\mathrm{rest}} - V}{18}\right), \tag{11}$$

$$\alpha_h(V) = 0.07 \exp\left(\frac{V_{\mathrm{rest}} - V}{20}\right), \tag{12}$$

$$\beta_h(V) = \frac{1}{\exp\left(\frac{30 + V_{\mathrm{rest}} - V}{10}\right) + 1}, \tag{13}$$

where V_{rest} is the resting potential and $V_{\mathrm{rest}} - V$ denotes the negative depolarization.

Although the Hodgkin-Huxley (H–H) model accurately describes the membrane potential on the axon of neurons, it fails to accurately respond to the stochastic current injection that comes from the isolated nature of ion pathways. Moreover, the Hodgkin-Huxley model specifies the ion channel activities in detail, but it exhibits an advanced computational cost than the other neuron models such as the LIF model.

3.3 *Izhikevich Neuron Model*

Izhikevich neuron model balances the complexity and biological plausibility [27]. It also has the capability of implementing a variety of firing patterns of distinct neurons. With adjustable parameters, the Izhikevich neuron model is capable of producing a wide range of firing patterns, such as intrinsically bursting, chattering, resonator, and thalamocortical firing. In Izhikevich neurons describe the membrane potential of neurons with equations:

$$\frac{\mathrm{d}V}{\mathrm{d}t} = 0.04V^2 + 5V + 140 - u + I(t), \tag{14}$$

$$\frac{\mathrm{d}u}{\mathrm{d}t} = a\,(bV - u)\,, \tag{15}$$

$$\text{if } V \geq 30 \text{ mV, then } \begin{Bmatrix} V \leftarrow c \\ u \leftarrow u + d' \end{Bmatrix} \tag{16}$$

where V is the membrane potential, u is the membrane recovery variable, and a, b, c, and d are scalar parameters. With different combinations of these parameters, the Izhikevich neuron model can represent various neuron firing patterns [27]. Specifically, the parameter a expresses the timescale of the recovery variable u. The parameter b influences the sensitivity of the recovery variable u to the subthreshold fluctuations of the membrane potential v. The parameter c describes the after-spike reset value of the membrane potential v. The parameter d defines the after-spike reset of the recovery variable u.

3.4 *McCulloch-Pitts Neuron Model*

Another simple neuron model is the McCulloch-Pitts model which simply simulates the neurons as threshold functions in 1943 [11]. The equations expression is simply as:

$$y_j = f\left(\sum_{i=0}^{N} w_{i,j}x_i\right), \tag{17}$$

where y_j and x_i are the output and input of the neuron j, f represents a threshold function, N is the total inputs of neuron j, and $w_{i,j}$ is the synaptic weight connecting neuron i and neuron j. Equation (17) indicates that the outputs of McCulloch-Pitts models are either one or zeros. Therefore, the McCulloch-Pitts neuron model typically is used to implement the Boolean logic. Theoretically, the McCulloch-Pitts models can be used for implementing arbitrary logic with stacking layers. The McCulloch-Pitts model omits other considerable features of biological neural systems, such as the relationship between integrated input and the firing rate of the output. In realistic neural systems, the firing rate of a neuron generally is proportional to the input stimulus. For example, the duration of a muscle neuron depends on the intensity of the muscle stretch which means a more intensive stretch stimulates a spiking signal with a higher firing rate [1, 28].

3.5 *Neural Coding*

As previously introduced, the communication among neurons is in a form of spikes that are the voltage potentials between the inner and outer membranes of the neurons. Thus, a reasonable hypothesis is that the information from one neuron to another neuron is encoded into these spikes. Several neural coding paradigms have been proposed and studied, including rate coding, temporal coding, etc. Neural coding characterizes pristine exterior analog data into neural responses.

One of the most prevalent neural coding schemes is rate coding, which is built upon the observation that the neural firing rate is proportional to the input intensity. This relationship dates back to the work of Adrian that revealed the firing rate of stretch receptor neurons in the muscles is higher when the applied force on muscle is larger [29]. The rate coding is typically defined by a temporal average of spike count in a specific period, as shown in Fig. 11a, governed by the equation:

$$v = \frac{n_{\mathrm{sp}}(T)}{T}, \tag{18}$$

where n_{sp} is the number of spikes and T is the time window of measuring time and v is the average firing rate of a single neuron in units of Hertz. Besides the relationship between firing rate and input intensity, a number of studies exhibit that precise spike timing also plays an important role in modifying synaptic plasticity. For instance, the modifications of synapse plasticity depend not only on spike rates (rate encoding) but also on the precise arrival timing of presynaptic and postsynaptic neurons, expressed with spike timing-dependent plasticity (STDP). In order to accurately model the timing characteristic of the neurons, temporal encoding is

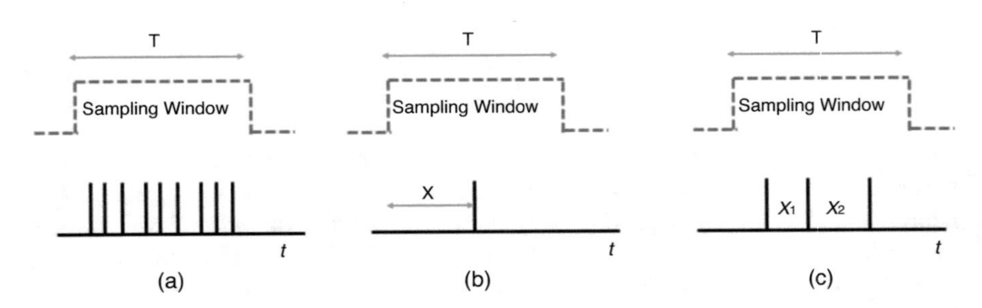

Fig. 11 (**a**) Rate coding, (**b**) time-to-first-spike latency coding, and (**c**) inter-spike-interval temporal coding

proposed [30]. Two types of temporal encoding are prominent, as illustrated in Fig. 11, namely, time-to-first-spike (TTFS) latency encoding [31] and inter-spike-interval (ISI) encoding [32]. In TTFS, the input signal is encoded in the measure of the timing interval between the sampling start point and the timing of the spike, as illustrated in Fig. 11. Thereby, in each sampling period, only one input can be encoded. On the contrary, the ISI encoding encodes the input signal into the internal time between spikes. The ISI code is capable of carrying more information within a sampling period compared to the TTFS latency code, as illustrated Fig. 11.

The rate and temporal encoding schemes only exhibit the mechanism of communication among neurons in the format of spikes but do not describe the activity of a group of neurons. In biological neural systems, a group of neurons can simultaneously respond to the same input stimulus with same the firing pattern group [33–35]. This implies that the exterior stimulus can be encoded and represented by the firing pattern of a group of neurons with a particular topology. This encoding scheme is referred to as population encoding. Population encoding represents stimuli with the joint activities of a number of neurons. The population coding can reach a much faster response time than other encoding paradigms, e.g., rate and temporal encoding. The group of neurons can almost instantaneously reflect the change of the stimulus [36]. The different firing patterns are influenced by the threshold voltage, connection topology, synaptic strength, etc. When the external input exceeds the threshold voltage, the group of neurons, marked as red and solid cycles, will fire resulting in different output responses. The population activity of N neurons in a small-time interval Δt is defined by the number of spikes $n_{\text{act}}(t; t + \Delta t)$, which is described by the following equation:

$$A(t) = \lim_{\Delta t \to 0} \frac{1}{\Delta t} \frac{n_{\text{act}}(t; t + \Delta t)}{N}, \tag{19}$$

where activity $A(t)$ describes the population average activity (firing) of the neurons [37].

4 Silicon Brain

As aforementioned, neuromorphic computing aims to recreate the neural systems physically, which naturally involves the implementation of neurons and synapses. In this section, the implementations of electronic neurons and synapses using CMOS (complementary metal-oxide-semiconductor) technology and memristors will be highlighted.

4.1 Electronic Neurons

Hardware implementations of electronic neurons typically contain several approaches such as CMOS technology [38–43], emerging devices [44–49], etc. Typically, the CMOS circuits are used for implementing complementary and symmetrical pairs of p-type and n-type MOSFETs for logic functions and memories in digital systems. But in neuromorphic systems, computing units are electronic neurons that have different functions rather than Boolean algebra. Therefore, the electronic neurons mainly realize the generation of spiking signals, threshold function, and other biological plausible features, such as different firing patterns and refractory periods of realistic biologically neurons. These implementations typically are built upon the aforementioned mathematic neuron models. One of the typical examples is a large family of integrate and fire neuron models.

The pioneer in neuromorphic computing, Carver Mead, first proposed a simple circuit implementing the firing of neurons with CMOS, namely, Axon-Hillock circuits [17, 39, 50]. The Axon-Hillock circuit [17, 39, 50] is a simple example of implementing the behaviors of integration, threshold firing, and refractory periods. The typical design of the Axon-Hillock circuit is illustrated in Fig. 12. The Axon-Hillock circuit generates a sequence of spikes when the membrane voltage (V_{mem})

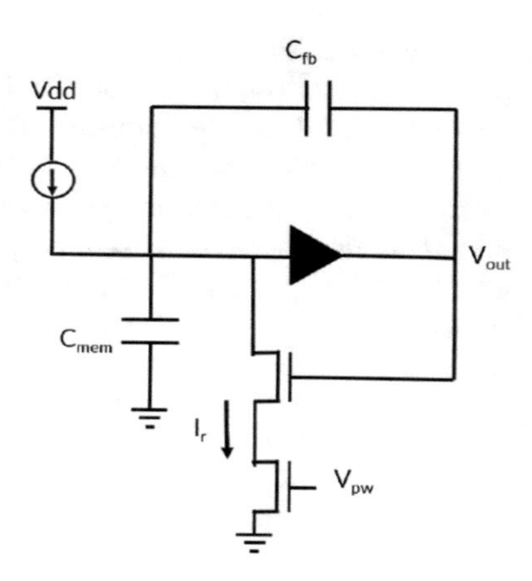

Fig. 12 The schematic of the Axon-Hillock circuit [17, 39, 50]

Table 2 Neuron models and designs

Neuron model	Description	Implementation
Hodgkin-Huxley Model [26]	Computationally complex but accurately models the ion channels of neurons and the propagation mechanism of membrane potential	[59, 60]
Leaky integrate and fire model [61, 62]	Simple for implementation	[63–65]
Izhikevich model [27]	Mathematical neuron model with high flexibility for implementing different firing behaviors. Balance the realistic and the computational efficiency	[58, 66]

crosses a voltage threshold. The specific value of the voltage threshold can be design by transistor sizes. The capacitance C_{mem} in the circuit simulates the membrane of a biological neuron. With no inputs, the membrane voltage will be drawn to the ground. While if the input is provided (I_{in}), the charges will be accumulated within the membrane capacitance (C_{mem}) resulting in an increment of the volage of (V_{mem}) accordingly. A threshold voltage (V_{th}) is set for measuring whether the membrane potential is larger than the expected value using a simple transconductance amplifier. A spike (action potential) is produced with if V_{mem} exceeds V_{th}, which is determined by the amplifier switching threshold [17, 39, 50]. After that, V_{out} changes from 0 to V_{dd} turning on the reset transistor. Consequently, the membrane capacitor (C_{mem}) is discharged repeating another firing cycle (charging and discharging of C_{mem}).

Although the typical Integrate and Fire (I&F) neuron circuits require far fewer transistors and parameters than the biophysically realistic models, they omit a large number of behaviors of biological neurons that exhibit the computational properties of neural systems [27, 51–53].

An acceptable compromise design between the computational complexity and hardware feasibility is the *generalized* I&F models [54, 55]. It has the capability of emulating numerous classic behaviors of biological neurons, meanwhile achievable due to relatively simple designs compared to H–H-based models. The generalized I&F models balance the hardware simplicity and biological plausibility [56–58]. The design generalized I&F models significantly reduce the transistor count, power consumption, and design area. It implements refractory period and spike frequency adaptation, which are essential properties of realistic neurons for producing resonances and oscillatory behaviors often emphasized in more complex models [56–58]. Table 2 summarizes the typical implementations of the classic neuron models.

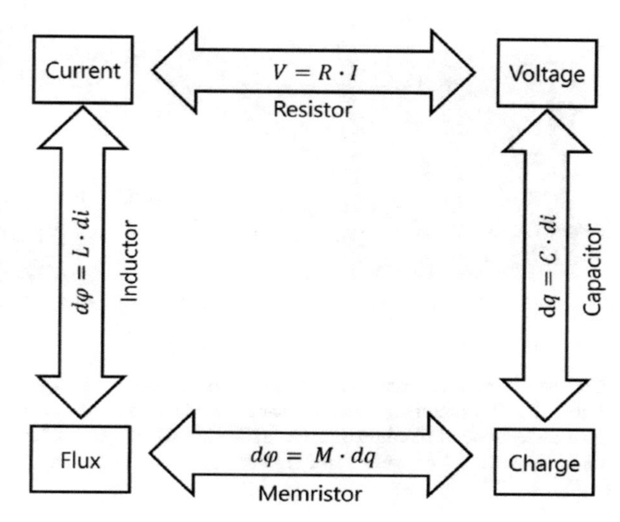

Fig. 13 Relationships among the four basic circuit variables

4.2 *Memristive Synapses*

The synaptic plasticity can be implemented with an emerging device *memristor*. In 1971, Leon Chua first introduced the mathematical theory of memristor [67]. The mathematical model of the memristor is initially an additional circuit component equivalent to the status of resistors, capacitors, and inductors. In the circuit theory, four basic circuit variables exist, namely, current i, voltage v, charge q, and flux φ. The mathematical relationship among them is illustrated in Fig. 13. For instance, resistances demonstrate the positive and linear relationship between the voltage and the current. Similarly, the relationships between *voltage and charge* and *current and flux* are expressed with the capacitance and the inductance, respectively. But the relationship between the flux and electric charge was missing originally in the chart (Fig. 13). Thus, Leon Chua predicted that there should be another basic circuital element expressing the relationship between the flux and electric charge. He named this hypothetical element the *memristor*.

After almost four decades, HP labs found the memristor in 2008 [68]. The scientists in HP Labs fabricate a nano device with a metal/dielectric/metal configuration (Pt/TaO_x/Pt). Figure 14 exhibits a crossbar structure of the nano device that can be accessed through the nano wires. The resistance of this device can be changed by applying voltages at its two terminals, which perfectly match the most important characteristics of memristors that show a nonlinear and butterfly-shaped current-voltage switching [67–70]. Another name for this metal/dielectric/metal device is resistive RAM (RRAM) in the memory field [71–73]. In a memristor or RRAM, the insulator layer is usually fabricated with a resistive switching material [74, 75]. The adjustable resistance of memristors is caused by the construction/deconstruction

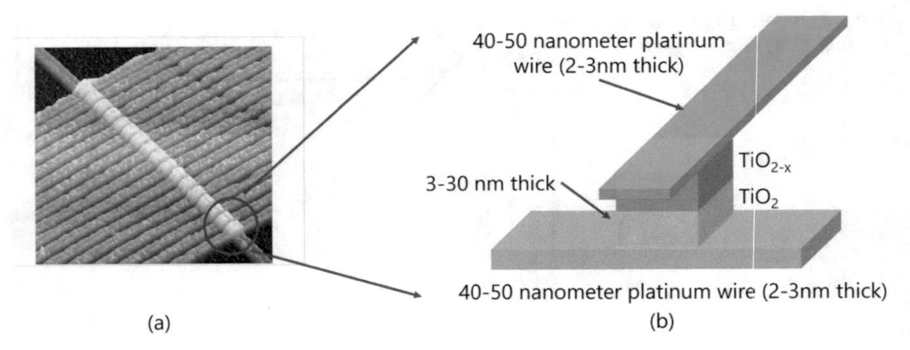

Fig. 14 The memristors from HP labs: (**a**) microscope image of a memristor fabricated by HP Labs. (**b**) The memristor cell is located at the cross-point of the crossbar structure with a 40-nanometer cube of titanium dioxide (TiO_2) in two layers. The lower layer is traditional of titanium dioxide with a 2:1 oxygen-titanium ratio

processes of the conductive filaments in their oxide layer. Figure 15 illustrates four typical stages of the resistive changes of a memristor. With the exterior voltage/current applied to the terminals of memristors, their resistances gradually change between their low resistance state (LRS) and high resistance state (HRS). The decrease of resistance is caused by the conductive filament (CF) formed within their oxide layers. At the initial state shown in Fig. 15, the atomic structure of the oxide layers is intact. The bonding among oxygen ions and metal atoms in the metal dioxides is stable. However, this bonding is breakable with external high electrical fields. Some oxygen ions in the metal oxide escape from the constraint of the bonding force under applied electrical fields, as shown in Fig. 15 [71]. Consequently, the oxygen vacancies or metal precipitates form the conductive filaments. The conductive filaments reduce the resistance of memristors by some alternative current pathways. The TEM (transmission electron microscopy) images demonstrate these conductive filaments (Fig. 15). Moreover, these oxygen ions can migrate back into the oxide to refill the oxygen vacancy and re-oxidize the metal precipitates if an opposite voltage is applied. Then, the resistances of memristors restore to their initial high value.

The recoverable resistive switching of the MIM materials has been observed and studied for many years; no one connects this particular phenomenon of resistors to the concept of memristor until HP Labs successfully found a connection [69]. A short history of memristor exploration is summarized in Table 3.

Another important benefit of memristive synapses is that they can be produced vertically forming a three-dimensional integrated circuit (3D-IC). The current neuromorphic systems are limited to two-dimensional (2D) structures suffering a long signal transportation distance, low energy efficiency, and high design area [43, 81–83]. The novel neuromorphic systems fabricated with 3D-IC offer vertical signal propagation paths, resulting in a significant reduction of the design area and power

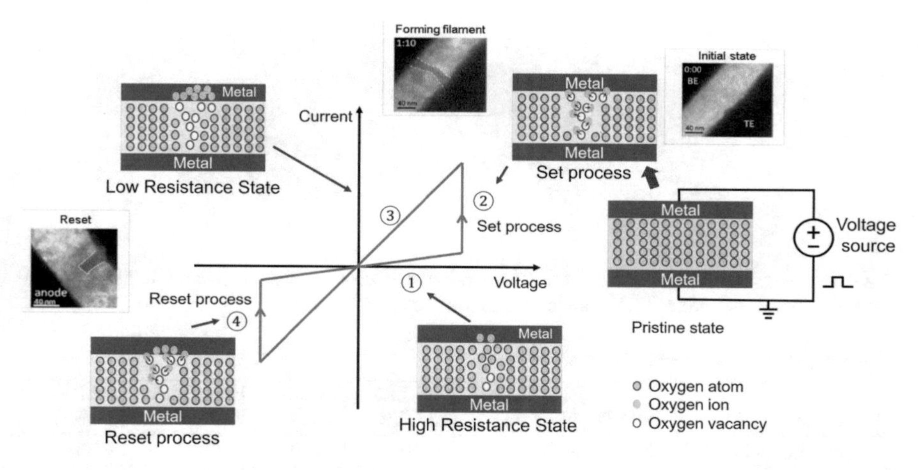

Fig. 15 Illustration of the switching mechanism of a memristor. The memristor has two states (HRS and LRS) marked as ① and ③ and two transition states (set and reset processes) marked as ② and ④, respectively [76]

Table 3 Historical introduction to memristors

Year	Memristor discovery milestones
1967	In a silicon oxide thin film with gold ions injected, J. G. Simmons and R. R. Verderber discovered a stress-strain resistance shifting phenomenon [77]
1968	In a metal oxide thin film, F. Argall observed a resistance shifting behavior [78]
1971	The notion of memristor was foreseen by the Leon Chua, which is a similar mathematical attempt by Constantine A. Balanis [79]. In his paper, his assumption was that there might be another relationship between charge and flux [67]
1998	Bhagwat Swaroop, William West, Gregory Martinez, Michael Kozicki, and Lex Akers showed how to improve the robustness of an artificial synapse by using a programmable resistance device [80]
2008	Dmitri et al. HP Labs published an article in Nature introducing a relationship between the two-terminal resistance switching characteristic of TaOx [68]
2008	Leon Chua, Stan Williams, Greg Snider, Wolfgang Porod, Massimiliano Di Ventra, Rainer Waser, and Blaise Mouttet talked about the theoretical foundations of utilizing memristor for RRAM and neuromorphic architectures on memristive systems [70]

consumption. Two fabrication technologies stack memristors (RRAM) into three-dimensional space: horizontal RRAM (H-RRAM) and vertical RRAM (V-RRAM) depicted in Fig. 16.

Both design paradigms demonstrate a $4F^2/n$ device size, where n is the number of vertical layers and F is the minimal fabrication feather [84]. Two types of 3D integration technologies can be used for integrating other transistor-based neurons and supportive circuitry into the 3D memristor-based synapse array. One

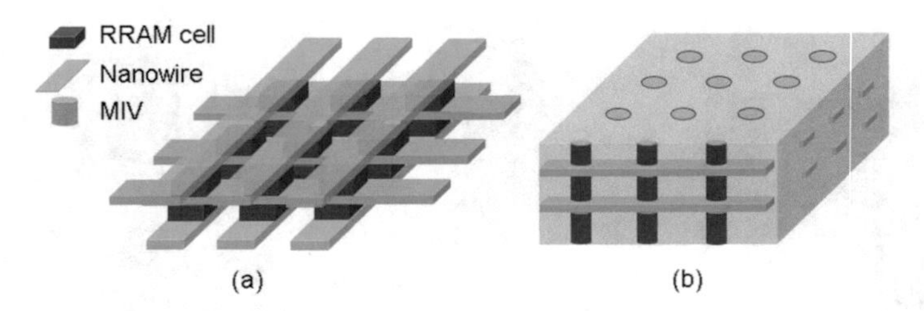

Fig. 16 Two types of state-of-the-art 3D memristor structures: (**a**) horizontal integration; (**b**) vertical integration. (Note: MIV stands for monolithic inter-tier via)

Table 4 The low-temperature transistors [87]

3D device	FinFET	Epi-like Si NWFET	Epi-like Si UTB	SOI-Si UTB	Poly-Si/Ge FinFET	IGZO OSFET
Thermal budget (°**C**)	<400	<400	<400	<650	<400	<500
I_on/I_off	$>10^7$	$>5 \times 10^5$	$>5 \times 10^5$	$>10^7$	$>10^7$	$>10^{21}$

is more traditional, mature, and closer to the commercialization level: TSV-based 3D integration technology [85, 86]. The second one is a more emerging monolithic 3D technology that sequentially fabricates the transistors in a single wafer at a low process temperature.

For the TSV-based 3D integration technology, power delivery is one of the challenges. As multiple dies are stacked together with small footprints, delivering current to all circuitry located at different vertical layers while meeting the power noise and thermal constraints becomes more and more challenging. This is mainly because the number of TSVs available for power distribution networks is limited. Currently, another state-of-the-art 3D-IC technology with no TSVs delivers a transformative impact on the silicon industry, which is referred to as monolithic 3D integration. With no TSVs, the monolithic 3D integration directly fabricates the memristors on the top of silicon wafers with nanoscale monolithic inter-tier vias (MIVs). MIVs are much smaller than the traditional TSVs. The main challenge for the monolithic 3D integration technology is the low-temperature fabrication constraint for the upper layers. In order to protect the former fabricated devices in lower layers, the higher layers need to be fabricated at a lower temperature. Thus, the conventional CMOS transistor is not applicable for monolithic 3D technology since they are fabricated more than 1000 °C.

Several low-temperature transistors have been investigated as the candidates to meet lower temperature fabrication processes, including fin field-effect transistors (FinFETs) [87], carbon nanotube FETs [88], etc. Table 4 summarizes the parameters of these transistors [87]. Furthermore, the functional chip combining monolithic 3D integration technology, memristor, and CNTFETs has been fabricated recently by Stanford, which is demonstrated in Fig. 17 [89].

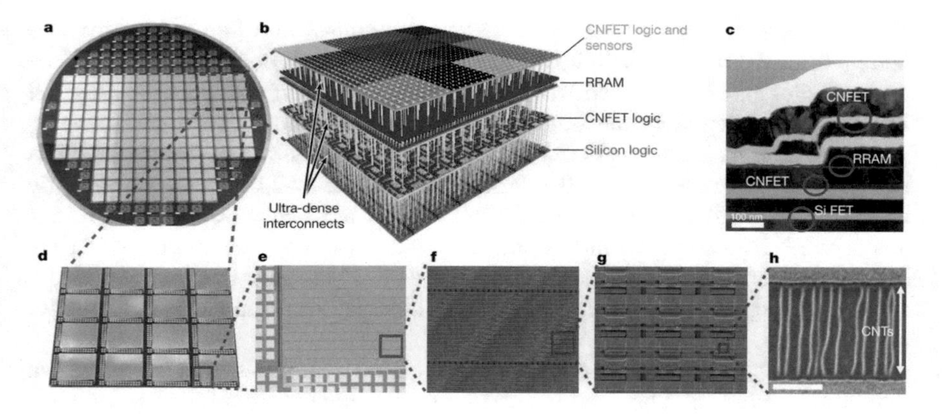

Fig. 17 3D chip integrated with RRAM and CNFET logics fabricated by Stanford [89]

5 Neuromorphic Chips

Nowadays, neuromorphic systems are not just a concept living in the articles but the actual intelligent chips. In this section, several neuromorphic chips and their applications will be introduced, including Loihi chips, TrueNorth, etc. Due to the incomparable ultra-high energy efficiency and fast response, these neuromorphic chips have been applied to many applications, including robotics, speech and image recognition, edge computing, etc.

5.1 Loihi Chips

Loihi chips developed by Intel Labs are one of the top-tier neuromorphic chips for research purposes [90, 91]. Unlike conventional GPUs and CPUs under von Neumann architecture operating digital data, Loihi chips are specifically designed for neuromorphic computing and asynchronous SNNs. To date, two generations of Loihi chips have been released. The first generation of Loihi chip was revealed in 2017 [90, 91], namely, Loihi-1. Loihi-1 chips consist of 130,000 electronic neurons and 130 million synapses that are located in 128 neuromorphic cores. The advanced 14 nm process of Intel renders the area of the Loihi-1 chip as small as 60 mm^2. Loihi-1 chips implement the digital leaky and fire neurons, which are partitioned on 128 cores. At each core, the communication among neurons is organized in a mesh configuration. The synapses in Loihi-1 chips are fully configurable and further support weight sharing and compression features. The plasticity of synapses can be manipulated with various biologically plausible learning rules, such as Hebbian rules, STDP, and reward-modulated rules. The firing behavior of neurons in Loihi chips is implemented when received spikes accumulate to a threshold value in a certain time; the neurons will fire off their own spikes to its connected neurons.

Loihi-1 chips are offered with several neuromorphic platforms providing distinct interfaces for integrating the Loihi-1 chip with other computer systems or field-programmable gate array (FPGA) devices. Kapoho bay includes 1–2 Loihi chips with a USB interface. The USB interface enables the Loihi chip to conveniently communicate with computer systems. Nahuku is a 32-chip Loihi board with a standard FPGA mezzanine card (FMC) connector. The FMC connector allows the Nahuku system to communicate with the Arria FPGA development board. Pohoiki Spring is a large-scale Loihi chip with 100 million neurons equipped as a server for remote access.

The second generation of the Loihi chips, namely, Loihi-2, was introduced in late 2021. Loihi-2 is fabricated in Intel 4 process, previously referred to as 7 nm technology. Powered by this advanced technology, the area of the Loihi-2 reduces to 31 mm^2 from 60 mm^2 of the first-generation Loihi chips. Unlike the rigid neuron models in the last generation of Loihi chips, Loihi-2 realizes fully programmable neuron models. In Loihi-2, the specific behavior of the neurons can be programmed with microcode instructions. The microcode instructions support basic bitwise and math operations. Loihi-2 chip is dedicatedly designed for neuromorphic computing and edge devices with parallel computations achieving high computational and energy efficiency. Two developing platforms, which are Oheo Gulch and Kapoho Point, scale Loihi-2 to a large number of neurons and synapses. Kapoho Point is a 4 by 4 inch development board equipped with eight Loihi-2 chips. These eight Loihi-2 chips include 8.4 million neurons and 960 million synapses. The comparison between the two generations of Loihi chips is summarized in Table 5.

Recently, Intel Labs also develop a neuromorphic software framework, namely, Lava. Lava supports state-of-the-art neuromorphic algorithms, such as SLAYER [92]. Lava is extensible and user-friendly to third-party development frameworks, including robotic operating system (ROS) [93], TensorFlow, PyTorch [94], Nengo [95], Brain [96], etc. These third-party frameworks enable the developers to apply Loihi-2 chips to a range of applications.

The most prominent application of Loihi chips is the realization of olfactory function [97]. In [97], an online training algorithm for identifying the smell of samples is deployed in the first generation of Loihi neuromorphic system. Each inference cycle only requires 2.75 ms and 0.43 mJ energy [97]. Additionally, the

Table 5 Introduction to Loihi-1 and Loihi-2 chips

Features	Loihi-1	Loihi-2
Technology	Intel 14 nm	Intel 4 (7 nm)
Die area	60 mm^2	31 mm^2
Max # neurons/chip	128,000	1 million
Max # synapses/chip	128 million	120 million
Neuron model	Generalized digital LIF	Fully programmable

inference/training time is not affected by the scale of the problem and data due to the computational parallelism of Loihi chips. The colocalization of memory and computing units further minimizes the demand for energy for data transfer.

5.2 *Dynamic Neuromorphic Asynchronous Processors*

Dynamic neuromorphic asynchronous processors (DYNAPs) are developed by SynSense and ETH Zurich. The family of DYNAP includes several chip models: Dynap-CNN, DYNAP-SE1, and DYNAP-SE2, as illustrated in Fig. 18.

Dynap-CNN chip is a configurable and digital neuromorphic chip specifically designed for spiking convolutional neural networks (SCNN). Dynap-CNN chip can constitute more than one million ReLU (rectified linear unit) spiking neurons. It includes four cores, which comprise 256 neurons. It has hierarchical asynchronous routers and implements SRAM (static random access memory) and CAM (content addressable memory) connected with routers across the cores. The asynchronous communication router design allows a single neuron being able to communicate with more than 230 K neurons. Each Dynap-CNN chip has one through adaptive exponential integrate-and-fire neurons (AdExp-I&F) and 65 K configurable synapses. Dynap-CNN chips support a large range of CNN layers including ReLU, pooling, padding, etc. Additionally, several network models, e.g., ResNet, LeNet, Inception, etc., have been incorporated into Dynap-CNN chips. Dynap-CNN chips are fabricated in the 22 nm advanced process occupying a 12mm^2 area. The low-amplitude spiking signals significantly reduce the power consumption of the Dynap-CNN chip. The low latency and high energy efficiency render the Dynap-CNN chips ideal candidates for the low power and low latency event-driven applications, such as surveillance, medical applications, etc. For instance, The Dynap-CNN development kit offers the interface of dynamic vision sensors (DVS) that can process event streams from DVS in real time.

The Dynap-SE chips consist of four cores, as illustrated in Fig. 18b. Each neural processor core has 16×16 analog neurons and 64 programmable synapses. The analog neurons (AdExp-I&F) in DYNAP-SE directly emulate the firing biological

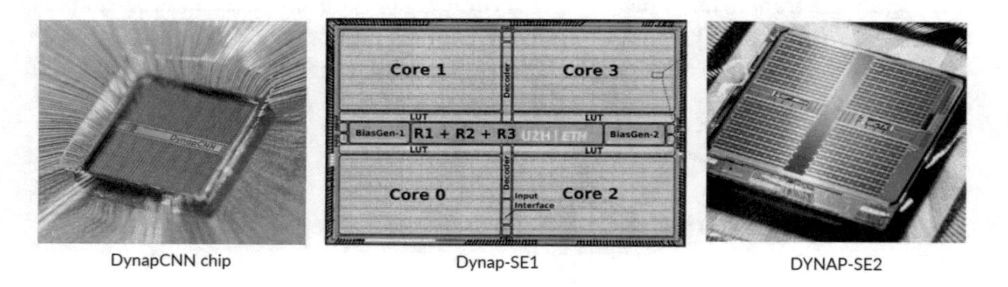

Fig. 18 Die photos of Dynap-CNN, DYNAP-SE1, and DYNAP-SE2

behavior of neurons. The DYNAP-SE chips use traditional SRAM and CAM [98, 99]. The CAM and SRAM of the DYNAP-SE chips are organized into small blocks placed on the vicinity of the neuron and synapse arrays. This computing-in-memory architecture extremely minimizes power consumption and memory bandwidth requirement compared to the traditional von Neumann architecture. Additionally, DYNAP-SE utilizes the subthreshold technology that reduces the core supply voltage of DYNAPs as small as 1.3 V leading to an outperformed energy efficiency.

The communications among neurons yield classic asynchronous address-event representation (AER) protocol. Through AER protocol, the generated spikes can be routed one core within one core, further among cores, and even multiple chips. The distinct levels of routing are implemented with mesh and hierarchical routing methods. The mesh routing has low bandwidth usage but high latency, while the hierarchical routing has low latency but high bandwidth. Therefore, the DYNAP trade off these two routing methods. The routing levels are characterized into three categories: Levels 1, 2, and 3. The generated spikes in Level 1 are only routed within the single core. In Level 2, the spikes can be sent to other cores, but still within the same chip. In Level 3, the spikes can be transferred to other chips leading to a large-scale neuromorphic system. This capability significantly enhances the scalability of DYNAP-SE chips. Moreover, the AER protocol also enables the communication between the DYNAP-SE chip and other neuromorphic sensors such as DVS.

DYNAP-SE2 development kit also integrates several supportive circuitries that preprocess the external analog signals, such as amplifying, filtering, neural encoding, etc. The recurrent and feedback connections supported by DYNAP-SE chips meet the requirements of the rich dynamics and memory characteristics. DYNAP-SE2 integrates numerous advanced features of brain dynamics, such as spike frequency adaptation, synaptic delay, homeostasis, short-term plasticity, etc. Due to the intrinsic similarity of semiconductor physics and biological dynamics, many complex models, e.g., the AdExp-I&F neuron, ion channel conductance could be implemented simply with several transistors and capacitors. The wearable healthcare devices need 24/7 monitoring of physiological signals that drained the battery shortly. The ultra-low latency and power consumption of DYNAP-SE2 can provide the desired energy efficiency and real-time processing capability. In [100], an ECG anomaly detection algorithm is proposed and prototyped on DYNAP-SE1. The algorithm transformed the ECG recordings into an event stream. Additionally, the study of iEEG [101] signal of epilepsy patients demonstrates that the DYNAP-SE2 chip can identify particular features in intracranial human data. Table 6 summarizes the specifications of two generations of NYNAP chips [98].

5.3 *TrueNorth Chips*

In 2014, IBM released a neuromorphic chip, TrueNorth, which has 4096 neurosynaptic cores. These neurosynaptic cores consist of 256 million synapses and

Table 6 DYNAP chips

Features	DYNAP-SE1	DYNAP-SE2
Technology	180 nm	180 nm
Die area	44 mm^2	99 mm^2
Number of cores	4	4
Number of analog parameters	25	70
Number of neurons	256	256/64
Type of synapses	AMPA, NMDA, GABAA, GABAB	AMPA, NMDA, GABAA, GABAB
Number of synapses	64	64/256
Neuron model	AdExp-I&F	AdExp-I&F

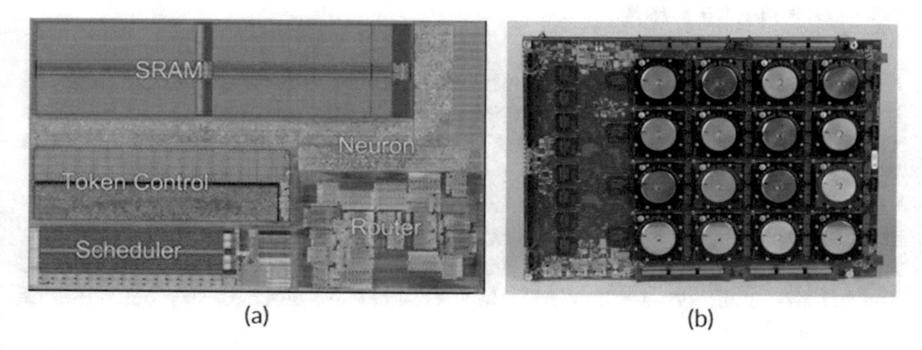

Fig. 19 (**a**) The neurosynaptic core of TrueNorth. (**b**) TrueNorth neuromorphic system with 16 chips, photo courtesy of IBM [81]

1 million electronic neurons. TrueNorth utilizes 28 nm process technology and occupies an area of 4.3 cm^2. The mesh communication configuration renders the TrueNorth chips a significant capability of fan-in and fan-out.

Figure 19a illustrates the layout of an individual neuromorphic core of TrueNorth chips marked by several modules, such as the scheduler, controller, SRAM, and router. The SRAM stores the data of each neuron. The scheduler buffers incoming spike signals mimicking the delay of axons in biological neurons. The controller ordinates and manages the overall operations of the core. With the digital neurons and synapses, TrueNorth achieves a 20 mW/cm^2 power density, whereas a typical CPU is 50–100 W/cm^2. In the a real-time object detection tasks, TrueNorth chips consume merely 65 mW [102]. TrueNorth systems also utilize an asynchronous event-driven communication protocol for maximizing energy efficiency.

Fig. 20 Neurogrid development kit with 16 neurocores [103]

5.4 *Neurogrid Chips*

Neurogrid is a digital and analog mix multichip system developed by Stanford University in 2014 [103] that integrates 16 neurocores, as illustrated in Fig. 20a. These neurocores are fabricated using 180 nm technology on a 168 mm^2 die. The 16 neurocores are integrated on a 6.5 $\times$ 7.5 in^2 board in a tree routing network. With 16 neurocores, Neurogrid system is capable to simulate up to 65,536 silicon neurons [103], up to 1 M neurons with 16 neurocores in total. Neurogrid is a system specifically designed for simulating large-scale biological neural networks. The electron neurons in Neurogrid chips emulate the biochemistry behaviors of biological neurons.

Specifically, it models the soma, dendrites, synapses, and axons [103]. The individual neurocore consumes $\sim$150 mW resulting in the whole Neurogrid with 16 neurocores merely consuming an average of ~3 W of power.

5.5 *BrainScaleS Project*

A multichip design methodology is inevitable for achieving an excessively large scale of neurons. However, the high-degree network structure requires an unusually high communication bandwidth among the neuromorphic chips. BrainScaleS addresses this challenge by using wafer-scale integration.

The BrainScaleS project is developed collaboratively by the University of Heidelberg. The hardware of the BrainScaleS project utilizes wafer-scale analog silicon circuits [104]. A total of 384 neuromorphic cores are placed on a 20 cm diameter silicon wafer [105], as shown in Fig. 21. Each chip fabricates 512 AdExp-

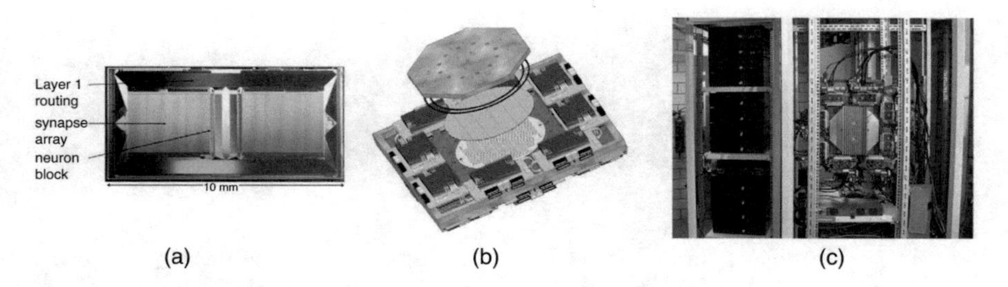

Fig. 21 (**a**) Photograph of the HICANN die [106]; (**b**) silicon wafer and aluminum back panel and a silicon wafer with 48 reticles; (**c**) assembled BrainScaleS module [106]

I&F neurons and 128,000 synapses. The wafers are assembled by integrating an uncut silicon chip, as depicted in Fig. 21c. Neuromorphic chips are fabricated with a 180 nm process. A neuromorphic chip is 5×10 mm^2 in size. Therefore, each silicon wafer consists of 49 million synapses and 200 K neurons. The single wafer module consumes roughly 1 kW power. Due to the large size of the wafer modules, BrainScaleS project is typically only remotely accessible [107].

5.6 *Human Brain Project*

The neuromorphic chips from the BrainScaleS project are applied to Human Brain Project (HBP) as a large-scale server [108–110]. *The computational platforms in HBP are built upon* EBRAINS research infrastructure [111], which consists of *SpiNNaker* and BrainScaleS. The *SpiNNake***r** machine developed by Manchester and the *BrainScaleS* machine in the University of Heidelberg implements analog electronic models of more than 1 billion synapses and 4 million neurons on multiple wafers chips. With the outperformed computational resources, HPB empowers the studies on neuroscience, cognitive computing, and other real-world applications worldwide. Several tasks are addressed particularly by HPB. One of the outstanding achievements of HBP is training the dexterous robot hands to perform vision-based object reorientation [112]. The anthropomorphic dexterous hand of a shadow robot is equipped with 129 sensors and 24 joints and 20 degrees of freedom for mimicking the movements of human hands. Additionally, over 100 proprioception sensors in the dexterous hand will collect substantial data for reinforcement training. With the cutting-edge computational infrastructure of HBP, the humanoid hand can be trained for manipulating a block from an initial configuration to a goal configuration using vision alone, as illustrated in Fig. 22. The dexterous hand is integrated with the HBP neurorobotics simulation platform.

Fig. 22 The rolling of a block from an initial configuration (A) to a goal configuration (E) with only vision input using neuromorphic computing

6 Challenges and Opportunities

Nowadays, the most promising AI approach is deep learning which is built upon massive data and deep ANNs. Although deep learning has demonstrated its excellent competencies in cognition tasks, the excessive demands on large-scale data and computational resources limit its feasibility and practicality [113]. The present studies reveal that the larger datasets allow a higher learning accuracy [114, 115] resulting in an inevasible demand for excessively large datasets. The scale of datasets is nearly linearly increasing over the years [115, 116], whereas the neural networks are scaling accordingly, at a double rate every year [115, 116]. On the contrary, the increment of computational capabilities of the traditional computational platforms, such as GPU and CPU, are far behind these fierce climbing demands [114]. Thus, a more efficient computational platform is vital for future AI development.

Neuromorphic computing emulates the physical structure of neural systems using an approach of software and hardware co-design that potentially can offer a more efficient and reliable AI [117, 118]. In the software, neuromorphic systems emulate the spiking-based information representation methods forming an SNN system. The operating frequency, which is a firing rate in neural systems, is significantly lower than the modern computer, which leads to an enhancement of energy efficiency [90, 91, 102, 119]. Digital computers are designed for Boolean algebra and are organized in von Neumann architecture. In a von Neumann computer system, the CPUs and memory are at different locations and are connected by a high-speed bus, as illustrated in Fig. 23b [114–116, 120]. This separation between CPUs and memory results in a back-and-forth transfer of data. The frequent data transfer ultimately becomes too costly and infeasible in terms of energy consumption and latency. The issue becomes more severe for ANNs because of their excessive amounts of data.

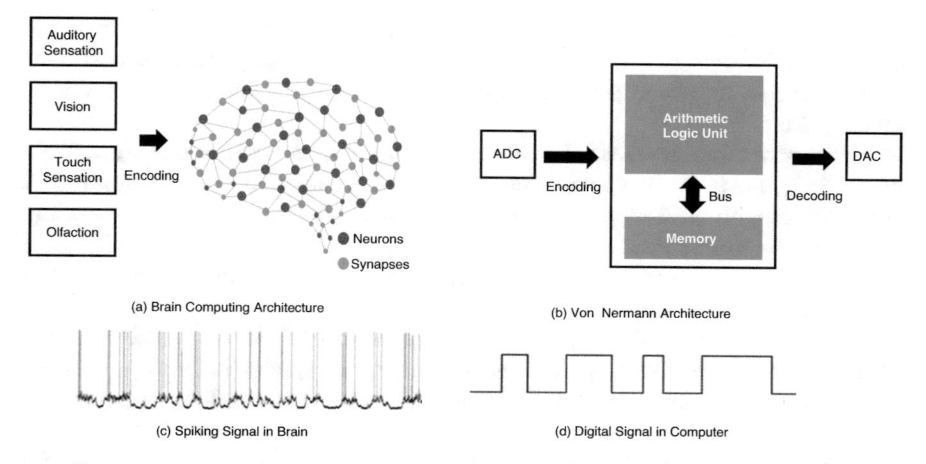

Fig. 23 Comparison between the computational paradigm of the brain and von Neumann computers

On the contrary, neural systems utilize an essentially distinct computing paradigm from digital computers. The computing units (neurons) and memory units (synapses) in human brains are placed adjacently (see Fig. 23a) forming a distributed network structure. The proximity of neurons and synapses leverages the data transferring efficiency. The data in neural systems is represented in spikes rather than square waveforms. The sensory organs, e.g., eyes, ears, etc., are responsible for the transformation between analog signals and spiking signals, which is referred to as neural coding. The firing rate of these spiking signals is as low as the level of kilohertz, which further minimizes the power consumption of neural systems. Thus, the current studies of the neuromorphic system aim to build a neuromorphic chip with brain-like architecture [49, 121], which operates spiking neural networks, expecting an extra-power efficiency for edge computing and real-time systems [92, 122].

Another fascinating capability of the neural system in animals is that they can memorize the events occurring at the same time. This capability of memorization is referred to as associative memory learning. Associative memory learning dissevers the dependency on the massive datasets. Building numerous large scales of datasets is a time-consuming and financially expensive task. In some scenarios, the data is even unobtainable, such as underwater robots. The co-design with novel hardware (silicon neurons and non-von Neumann architecture) and software (SNN and associative memory learning) will offer a promising design strategy for efficient artificial intelligence implementation. More and more applications prefer local data processing due to security and energy concerns. Other applications may have a life and death issue that extremely relies on fast response, e.g., autonomous cars. However, the current autonomous navigation of self-driving vehicles requires sending data to a remote server to process. This approach excessively relies on stable communication that is usually interfered. In addition, some specific applications have no access to powerful servers, for instance, the spaceships operating on the moon or Mars.

Furthermore, besides these advantages of neuromorphic systems that are actively under study, the brains have much more remarkable capabilities that currently still are not thoroughly studied yet, such as the mechanism of memory, decision-making, robustness to noise scenarios, etc. Thus, to further draw the benefits of biological neural systems and thrust the potential of neuromorphic systems, several aspects can be explored in the future.

Since the information is represented in spikes instead of square waves in the neural systems, the first step for any neuromorphic system is to transform the exterior signal into spike screams. In biological neural systems, the transformation is realized by sensory organs, e.g., eyes, ears, etc. Currently, many specifically designed neuromorphic sensors can implement similar tasks for these organs. For example, dynamic vision sensors (DVS) [123–126] can convert visual data into spike screams. The spike screams of DVS are triggered by the motion of the object. Each pixel inside a DVS generates spiking signals independently in response to the motion of objects and keeps rest otherwise. The resting mode of DVS leads to high energy efficiency compared to the traditional cameras, which are always in an active state. In addition, the DVS has an excessive capability of dealing with the motion blur than traditional frame cameras since it can reconstruct images from the spike screams [127, 128]. Another biological sensor is the silicon cochlea *dynamic audio sensor (DAS) which emulates the function of the hearing* [129–132]. In a conventional audio system, the sound is first transformed into digital data with ADCs and then processed using Fourier transform, bandpass filtering, etc. This transformation process converts continuous analog signals (audio data) into discrete binary data. However, in *DAS*, the analog audio signals are encoded, in parallel, into trains of spikes events. Analogously, besides audio and visual signaling, more sensations, such as skin and posture sensations, should be explored and encoded into spikes for neuromorphic systems. These aspects are uncultivated research fields.

Moreover, the current studies of deep learning and ANNs merely process different signals with distinct neural networks independently without a fusion/associative learning like what happens in the neural systems. In human brains, the captured signals are processed at different regions in parallel significantly enhancing computation efficiency and cognitive capabilities.

In addition, what we can learn from emulating the human brain is not only how to build an efficient artificial system but also how our brains work including the possible explanation for neurological diseases, optical illusions, visual agnosia, and Parkinson's disease [1]. For instance, recent studies indicate that synapse loss highly causes Alzheimer's disease [133]. Thus, the emulation of human brains by neuromorphic systems offers a platform for exploring the mechanism of diseases.

References

1. Kandel, E.R., Schwartz, J.H., Jessell, T.M., Siegelbaum, S.A., Hudspeth, A.: Principles of Neural Science. McGraw-Hill, New York (2000)
2. Baird, E., Srinivasan, M.V., Zhang, S., Cowling, A.: Visual control of flight speed in honeybees. J. Exp. Biol. **208**(20), 3895–3905 (2005)
3. Kern, R., Boeddeker, N., Dittmar, L., Egelhaaf, M.: Blowfly flight characteristics are shaped by environmental features and controlled by optic flow information. J. Exp. Biol. **215**(14), 2501–2514 (2012)
4. Haenlein, M., Kaplan, A.: A brief history of artificial intelligence: on the past, present, and future of artificial intelligence. Calif. Manag. Rev. **61**(4), 5–14 (2019)
5. A. M. Turing, Computing machinery and intelligence," in Parsing the Turing Test: Springer, Dordrecht 2009, pp. 23–65
6. Weizenbaum, J.: ELIZA—a computer program for the study of natural language communication between man and machine. Commun. ACM. **9**(1), 36–45 (1966)
7. Campbell, M., Hoane Jr., A.J., Hsu, F.-H.: Deep blue. Artif. Intell. **134**(1–2), 57–83 (2002)
8. Goodfellow, I., Yoshua, B., Aaron, C.: Deep Learning, p. 785 (2016). https://doi.org/10.1016/B978-0-12-391420-0.09987-X
9. Bengio, Y., Goodfellow, I., Courville, A.: Deep Learning. MIT Press, Cambridge, MA (2017)
10. Goodfellow, I., et al.: Generative adversarial nets. In: Advances in Neural Information Processing Systems, pp. 2672–2680 (2014)
11. McCulloch, W.S., Pitts, W.: A logical calculus of the ideas immanent in nervous activity. Bull. Math. Biophys. **5**(4), 115–133 (1943)
12. Aron, J.: How innovative is Apple's new voice assistant, Siri? ed: Elsevier (2011)
13. Hoy, M.B.: Alexa, Siri, Cortana, and more: an introduction to voice assistants. Med. Ref. Serv. Q. **37**(1), 81–88 (2018)
14. Greenblatt, N.A.: Self-driving cars and the law. IEEE Spectr. **53**(2), 46–51 (2016)
15. Pedrycz, W., Chen, S.-M.: Deep Learning : Algorithms and Applications. Springer, Cham (2020)
16. Gibney, E.: Google AI algorithm masters ancient game of Go. Nat. News. **529**(7587), 445 (2016)
17. Mead, C.: Neuromorphic electronic systems. Proc. IEEE. **78**(10), 1629–1636 (1990)
18. Soediono, B.: The handbook of brain theory and neural networks. J. Chem. Inf. Model. **53**, 719–725 (1989). https://doi.org/10.1017/CBO9781107415324.004
19. y Cajal, S.R.: Comparative Study of the Sensory Areas of the Human Cortex, Clark University, Worcester (1899)
20. Bear, M.F., Connors, B.W., Paradiso, M.A.: Neuroscience. Lippincott Williams & Wilkins, Philadelphia (2007)
21. P. I. Pavlov, "Conditioned reflexes: an investigation of the physiological activity of the cerebral cortex," Ann. Neurosci., vol. 17, no. 3, p. 136, Jul 2010, doi: https://doi.org/10.5214/ans.0972-7531.1017309
22. H. An, An, Q., Yi, Y.: Realizing behavior level associative memory learning through three-dimensional Memristor-based neuromorphic circuits. In: IEEE Transactions on Emerging Topics in Computational Intelligence (2019)
23. Brunel, N., Van Rossum, M.C.: Lapicque's 1907 paper: from frogs to integrate-and-fire. Biol. Cybern. **97**(5), 337–339 (2007)
24. Orhan, E.: The Leaky Integrate-and-Fire Neuron Model, pp. 1–6 (2012)
25. Fuortes, M., Mantegazzini, F.: Interpretation of the repetitive firing of nerve cells. J. Gen. Physiol. **45**(6), 1163–1179 (1962)
26. Hodgkin, A.L., Huxley, A.F.: A quantitative description of membrane current and its application to conduction and excitation in nerve. Bull. Math. Biol. **52**, 25–71 (1990). https://doi.org/10.1007/BF02459568

27. Izhikevich, E.M.: Simple model of spiking neurons. IEEE Trans. Neural Netw. **14**, 1569–1572 (2003). https://doi.org/10.1109/TNN.2003.820440
28. Darian-Smith, I., Johnson, K., Dykes, R.: "Cold" fiber population innervating palmar and digital skin of the monkey: responses to cooling pulses. J. Neurophysiol. **36**(2), 325–346 (1973)
29. Adrian, E.D.: The impulses produced by sensory nerve endings: part I. J. Physiol. **61**(1), 49–72 (1926)
30. Panzeri, S., Brunel, N., Logothetis, N.K., Kayser, C.: Sensory neural codes using multiplexed temporal scales. Trends Neurosci. **33**, 111–120 (2010). https://doi.org/10.1016/j.tins.2009.12.001
31. Zhao, C., Yi, Y., Li, J., Fu, X., Liu, L.: Interspike-interval-based analog spike-time-dependent encoder for neuromorphic processors. IEEE Trans. Very Large Scale Integr. (VLSI) Syst. **25**, 2193–2205 (2017). https://doi.org/10.1109/TVLSI.2017.2683260
32. Zhao, C., et al.: Energy efficient temporal spatial information processing circuits based on STDP and spike iteration. IEEE Trans. Circuits Syst. II. **67**(10), 1715–1719 (2019)
33. Averbeck, B.B., Latham, P.E., Pouget, A.: Neural correlations, population coding and computation. Nat. Rev. Neurosci. **7**(5), 358–366 (2006)
34. Pasupathy, A., Connor, C.E.: Population coding of shape in area V4. Nat. Neurosci. **5**(12), 1332–1338 (2002)
35. Panzeri, S., Macke, J.H., Gross, J., Kayser, C.: Neural population coding: combining insights from microscopic and mass signals. Trends Cogn. Sci. **19**(3), 162–172 (2015)
36. Pouget, A., Dayan, P., Zemel, R.: Information processing with population codes. Nat. Rev. Neurosci. **1**(2), 125–132 (2000)
37. Gerstner, W., Kistler, W.M.: Spiking Neuron Models: Single Neurons, Populations, Plasticity. Cambridge University Press, Cambridge (2002)
38. Liu, J.-H., Wang, C.-Y., An, Y.-Y.: A Survey of neuromorphic vision system: biological nervous systems realized on silicon. In: 2009 International Conference on Industrial Mechatronics and Automation, IEEE, pp. 154–157 (2009)
39. Indiveri, G., et al.: Neuromorphic silicon neuron circuits (in English). Front. Neurosci., Review **5** (2011, May 31) https://doi.org/10.3389/fnins.2011.00073
40. Poon, C.S., Zhou, K.: Neuromorphic silicon neurons and large-scale neural networks: challenges and opportunities. Front. Neurosci. **5**, 2009–2011 (2011). https://doi.org/10.3389/fnins.2011.00108
41. Ahmed, M.R., Sujatha, B.K.: A review on methods, issues and challenges in neuromorphic engineering. In: 2015 International Conference on Communications and Signal Processing (ICCSP), pp. 899–903 (2015). https://doi.org/10.1109/ICCSP.2015.7322626
42. Schuman, C.D., Ridge, O., Disney, A.: Dynamic adaptive neural network arrays: a neuromorphic architecture. In: Proceedings of the Workshop on Machine Learning in High-Performance Computing Environments – MLHPC'15, pp. 1–4 (2015). https://doi.org/10.1145/2834892.2834895
43. Yi, Y., et al.: FPGA based spike-time dependent encoder and reservoir design in neuromorphic computing processors. Microprocess. Microsyst. **46**, 175–183 (2016). https://doi.org/10.1016/j.micpro.2016.03.009
44. Sun, J.: CMOS and Memristor Technologies for Neuromorphic Computing Applications (2015)
45. Babacan, Y., Kaçar, F., Gürkan, K.: A spiking and bursting neuron circuit based on memristor. Neurocomputing. **203**, 86–91 (2016). https://doi.org/10.1016/j.neucom.2016.03.060
46. An, H., Ehsan, M.A., Zhou, Z., Shen, F., Yi, Y.: Monolithic 3D neuromorphic computing system with hybrid CMOS and memristor-based synapses and neurons. Integr. VLSI J. (2017)
47. An, H., Al-Mamun, M.S., Orlowski, M., Yi, Y.: A three-dimensional (3D) Memristive Spiking Neural Network (M-SNN) system. In: International Symposium on Quality Electronic Design (2021)

48. An, H., Ha, D.S., Yi, Y.: Powering next-generation industry 4.0 by a self-learning and low-power neuromorphic system. In: Proceedings of the 7th ACM International Conference on Nanoscale Computing and Communication, pp. 1–6 (2020)
49. An, H.: Powering Next-Generation Artificial Intelligence by Designing Three-Dimensional High-Performance Neuromorphic Computing System with Memristors. Virginia Tech (2020)
50. Mead, C.: How we created neuromorphic engineering. Nat. Electron. **3**(7), 434–435 (2020)
51. Izhikevich, E.M.: Dynamical systems in neuroscience computational neuroscience. Dyn. Syst. **25**, 227–256 (2007). https://doi.org/10.1017/S0143385704000173
52. Izhikevich, E.M.: Which model to use for cortical spiking neurons? IEEE Trans. Neural Netw. **15**(5), 1063–1070 (2004)
53. Brette, R., Gerstner, W.: Adaptive exponential integrate-and-fire model as an effective description of neuronal activity. J. Neurophysiol. **94**(5), 3637–3642 (2005)
54. Jolivet, R., Lewis, T.J., Gerstner, W.: Generalized integrate-and-fire models of neuronal activity approximate spike trains of a detailed model to a high degree of accuracy. J. Neurophysiol. **92**(2), 959–976 (2004)
55. Livi, P., Indiveri, G.: A current-mode conductance-based silicon neuron for address-event neuromorphic systems. In: 2009 IEEE international symposium on circuits and systems, IEEE, pp. 2898–2901 (2009)
56. Wijekoon, J.H.B., Dudek, P.: Compact silicon neuron circuit with spiking and bursting behaviour. Neural Netw. **21**, 524–534 (2008). https://doi.org/10.1016/j.neunet.2007.12.037
57. Van Schaik, A., Jin, C., McEwan, A., Hamilton, T.J., Mihalas, S., Niebur, E.: A log-domain implementation of the Mihalas-Niebur neuron model. In: Proceedings of 2010 IEEE International Symposium on Circuits and Systems, IEEE, pp. 4249–4252 (2010)
58. Schaik, V., Jin, C., McEwan, A., Hamilton, T.J.: A log-domain implementation of the Izhikevich neuron model. In: ISCAS 2010–2010 IEEE International Symposium on Circuits and Systems: Nano-Bio Circuit Fabrics and Systems, pp. 4253–4256 (2010). doi: https://doi.org/10.1109/ISCAS.2010.5537564
59. Ma, Q., Haider, M.R., Shrestha, V.L., Massoud, Y.: Bursting Hodgkin–Huxley model-based ultra-low-power neuromimetic silicon neuron. Analog Integr. Circ. Sig. Process. **73**(1), 329–337 (2012)
60. Yu, T., Sejnowski, T.J., Cauwenberghs, G.: Biophysical neural spiking, bursting, and excitability dynamics in reconfigurable analog VLSI. IEEE Trans. Biomed. Circuits Syst. **5**(5), 420–429 (2011)
61. Abbott, L.F.: Lapicque's introduction of the integrate-and-fire model neuron (1907). Brain Res. Bull. **50**, 303–304 (1999). https://doi.org/10.1016/S0361-9230(99)00161-6
62. Stein, R.B.: A theoretical analysis of neuronal variability. Biophys. J. **5**(2), 173 (1965)
63. Rozenberg, M., Schneegans, O., Stoliar, P.: An ultra-compact leaky-integrate-and-fire model for building spiking neural networks. Sci. Rep. **9**(1), 1–7 (2019)
64. Chatterjee, D., Kottantharayil, A.: A CMOS compatible bulk FinFET-based ultra low energy leaky integrate and fire neuron for spiking neural networks. IEEE Electron Device Lett. **40**(8), 1301–1304 (2019)
65. Dutta, S., Kumar, V., Shukla, A., Mohapatra, N.R., Ganguly, U.: Leaky integrate and fire neuron by charge-discharge dynamics in floating-body MOSFET. Sci. Rep. **7**(1), 1–7 (2017)
66. Demirkol, A.Ş., Özoğuz, S.: A low power real time izhikevich neuron with synchronous network behavior. İstanbul Ticaret Üniversitesi Fen Bilimleri Dergisi. **12**(24), 39–52 (2013)
67. Chua, L.: Memristor-the missing circuit element. IEEE Trans. Circuit Theory. **18**(5), 507–519 (1971)
68. Strukov, D.B., Snider, G.S., Stewart, D.R., Williams, R.S.: The missing memristor found (in English). Nature. **453**(7191), 80–83 (2008). https://doi.org/10.1038/nature06932
69. Williams, S.R.: How we found the missing memristor. Spectrum IEEE. **45**(12), 28–35 (2008)
70. Keshmiri, V.: A Study of the Memristor Models and Applications (2014)
71. Wong, H.S.P., et al.: Metal-oxide RRAM. Proc. IEEE. **100**, 1951–1970 (2012). https://doi.org/10.1109/JPROC.2012.2190369

72. Strukov, D.B., Borghetti, J.L., Williams, R.S.: Coupled ionic and electronic transport model of thin-film semiconductor memristive behavior (in English). Small. **5**(9), 1058–1063 (2009). https://doi.org/10.1002/smll.200801323
73. Jo, S.H., Chang, T., Ebong, I., Bhadviya, B.B., Mazumder, P., Lu, W.: Nanoscale memristor device as synapse in neuromorphic systems. Nano Lett. **10**(4), 1297–1301 (2010)
74. Stefanovich, G., Pergament, A., Stefanovich, D.: Electrical switching and Mott transition in VO2. J. Phys. Condens. Matter. **12**(41), 8837 (2000)
75. Honig, J., Reed, T.: Electrical properties of Ti 2 O 3 single crystals. Phys. Rev. **174**(3), 1020 (1968)
76. Chen, J.Y., et al.: Dynamic evolution of conducting nanofilament in resistive switching memories. Nano Lett. **13**(8), 3671–3677 (2013). https://doi.org/10.1021/nl4015638
77. Simmons, J., Verderber, R.: New conduction and reversible memory phenomena in thin insulating films. Proc. R. Soc. London, Ser. A. **301**(1464), 77–102 (1967)
78. Argall, F.: Switching phenomena in titanium oxide thin films. Solid State Electron. **11**, 535–541 (1968). https://doi.org/10.1016/0038-1101(68)90092-0
79. Balanis, C.A.: Advanced Engineering Electromagnetics. John Wiley & Sons, New York (2012)
80. Swaroop, B., West, W., Martinez, G., Kozicki, M., Akers, L.: Programmable current mode Hebbian learning neural network using programmable metallization cell. In: ISCAS'98. Proceedings of the 1998 IEEE International Symposium on Circuits and Systems (Cat. No. 98CH36187), vol. 3, IEEE, pp. 33–36 (1998)
81. Akopyan, F., et al.: True north: design and tool flow of a 65 mW 1 million neuron programmable neurosynaptic chip (in English). IEEE Trans. Comput-Aided Des. Integr. Circuits Syst. **34**(10), 1537–1557 (2015). https://doi.org/10.1109/tcad.2015.2474396
82. An, H., Ehsan, M.A., Zhou, Z., Yi, Y.: Electrical Modeling and Analysis of 3D Neuromorphic IC with Monolithic Inter-tier Vias.
83. Yi, Y., Li, P., Sarin, V., Shi, W.: Impedance extraction for 3-D structures with multiple dielectrics using preconditioned boundary element method. In: 2007 IEEE/ACM International Conference on Computer-Aided Design, IEEE, pp. 7–10 (2007)
84. Xu, C., Niu, D., Yu, S., Xie, Y.: Modeling and design analysis of 3D vertical resistive memory—a low cost cross-point architecture. In: 2014 19th Asia and South Pacific Design Automation Conference (ASP-DAC), IEEE, pp. 825–830 (2014)
85. Yi, Y., Li, P., Sarin, V., Shi, W.: A preconditioned hierarchical algorithm for impedance extraction of three-dimensional structures with multiple dielectrics. IEEE Trans. Comput-Aided Des. Integr. Circuits Syst. **27**(11), 1918–1927 (2008)
86. Yi, Y., Zhou, Y., Fu, X., Shen, F.: Modeling differential through-silicon-vias (TSVs) with voltage dependent and nonlinear capacitance. Cyber J. **3**(6), 234–241 (2013)
87. Yang, C.-C., et al.: Footprint-efficient and power-saving monolithic IoT 3D+ IC constructed by BEOL-compatible sub-10nm high aspect ratio (AR>7) single-grained Si FinFETs with record high Ion of 0.38 mA/μm and steep-swing of 65 mV/dec. and I<inf>on</inf>/I<inf>off</inf> ratio of 8," pp. 9.1.1–9.1.4 (2016). https://doi.org/10.1109/iedm.2016.7838379
88. Shulaker, M.M., et al.: Monolithic 3D integration of logic and memory: Carbon nanotube FETs, resistive RAM, and silicon FETs. In: Electron Devices Meeting (IEDM), 2014 IEEE International, IEEE, pp. 27.4.1–27.4.4 (2014). https://doi.org/10.1109/IEDM.2014.7047120
89. Shulaker, M.M., et al.: Three-dimensional integration of nanotechnologies for computing and data storage on a single chip. Nature. **547**(7661), 74–78 (2017). https://doi.org/10.1038/nature22994
90. Davies, M., et al.: Advancing neuromorphic computing with Loihi: a survey of results and outlook. Proc. IEEE. **109**, 911–934 (2021)
91. Davies, M., et al.: Loihi: a neuromorphic manycore processor with on-chip learning. IEEE Micro. **38**(1), 82–99 (2018)
92. Shrestha, S.B., Orchard, G.: Slayer: Spike layer error reassignment in time, arXiv preprint arXiv:1810.08646 (2018)

93. DiLuoffo, V., Michalson, W.R., Sunar, B.: Robot operating system 2: the need for a holistic security approach to robotic architectures. Int. J. Adv. Robot. Syst. **15**(3), 1729881418770011 (2018)
94. Rao, D., McMahan, B.: Natural Language Processing with PyTorch: Build Intelligent Language Applications Using Deep Learning, 1st ed. O'Reilly Media, Beijing, p. 1 online resource [Online] (2019). Available: http://proquest.safaribooksonline.com/9781491978221
95. Bekolay, T., et al.: Nengo: a python tool for building large-scale functional brain models. Front. Neuroinform. **7**, 48 (2014)
96. Goodman, D.F., Brette, R.: The brian simulator. Front. Neurosci. **3**, 26 (2009)
97. Imam, N., Cleland, T.A.: Rapid online learning and robust recall in a neuromorphic olfactory circuit. Nat. Mach. Intell. **2**(3), 181–191 (2020)
98. Moradi, S., Qiao, N., Stefanini, F., Indiveri, G.: A scalable multicore architecture with heterogeneous memory structures for dynamic neuromorphic asynchronous processors (DYNAPs). IEEE Trans. Biomed. Circuits Syst. **12**(1), 106–122 (2017)
99. Thakur, C.S., et al.: Large-scale neuromorphic spiking Array processors: a quest to mimic the brain (in English). Front. Neurosci., Review. **12**(891) (2018). https://doi.org/10.3389/fnins.2018.00891
100. Bauer, F.C., Muir, D.R., Indiveri, G.: Real-time ultra-low power ECG anomaly detection using an event-driven neuromorphic processor. IEEE Trans. Biomed. Circuits Syst. **13**(6), 1575–1582 (2019)
101. Sharifshazileh, M., Burelo, K., Sarnthein, J., Indiveri, G.: An electronic neuromorphic system for real-time detection of high frequency oscillations (HFO) in intracranial EEG. Nat. Commun. **12**(1), 1–14 (2021)
102. Akopyan, F., et al.: TrueNorth: design and tool flow of a 65 mW 1 million neuron programmable neurosynaptic chip. IEEE Trans. Comput-Aided Des. Integr. Circuits Syst. **34**(10), 1537–1557 (2015). https://doi.org/10.1109/TCAD.2015.2474396
103. Benjamin, B., et al.: Neurogrid: a mixed-analog-digital multichip system for large-scale neural simulations (in English). Proc. IEEE. **102**(5), 699–716 (2014). https://doi.org/10.1109/Jproc.2014.2313565
104. Models, P., Circuits, N., Project, H.B.: Physical Models of Neural Circuits in BrainScaleS and the Human Brain Project Status and Plans
105. Meier, K.: A mixed-signal universal neuromorphic computing system. In: 2015 IEEE International Electron Devices Meeting (IEDM), IEEE, pp. 4.6.1–4.6.4 (2015)
106. Schemmel, J., Bruderle, D., Grubl, A., Hock, M., Meier, K., Millner, S.: A wafer-scale neuromorphic hardware system for large-scale neural modeling. In: Circuits and Systems (ISCAS), Proceedings of 2010 IEEE International Symposium on, IEEE, pp. 1947–1950 (2010)
107. Appukuttan, S., Bologna, L., Migliore, M., Schürmann, F., Davison, A.: EBRAINS Live Papers-Interactive resource sheets for computational studies in neuroscience (2021)
108. Markram, H.: The human brain project. Sci. Am. **306**(6), 50–55 (2012)
109. Calimera, A., Macii, E., Poncino, M.: The human brain project and neuromorphic computing. Funct. Neurol. **28**, 191–196 (2013). https://doi.org/10.11138/FNeur/2013.28.3.191
110. Peppicelli, D., et al.: Human Brain Project. Neurorobotics Platform Specification, pp. 1–79 (2015)
111. Schirner, M., et al.: Brain Modelling as a Service: The Virtual Brain on EBRAINS, arXiv preprint arXiv:2102.05888 (2021)
112. Andrychowicz, O.M., et al.: Learning dexterous in-hand manipulation. Int. J. Robot. Res. **39**(1), 3–20 (2020)
113. Bahdanau, D., Cho, K., Bengio, Y.: Neural machine translation by jointly learning to align and translate, arXiv preprint arXiv:1409.0473 (2014)
114. Devlin, J., Chang, M.-W., Lee, K., Toutanova, K.: Bert: Pre-training of deep bidirectional transformers for language understanding, arXiv preprint arXiv:1810.04805 (2018)
115. Goodfellow, I., Bengio, Y., Courville, A., Bengio, Y.: Deep Learning. MIT Press, Cambridge, MA (2016)

116. Sun, C., Shrivastava, A., Singh, S., Gupta, A.: Revisiting unreasonable effectiveness of data in deep learning era. In: Proceedings of the IEEE International Conference on Computer Vision, vol. 2017-Octob, pp. 843–852 (2017). https://doi.org/10.1109/ICCV.2017.97
117. Deng, L., Tang, H., Roy, K.: Understanding and bridging the gap between neuromorphic computing and machine learning. Front. Comput. Neurosci. **15** (2021)
118. Roy, K., Jaiswal, A., Panda, P.: Towards spike-based machine intelligence with neuromorphic computing. Nature. **575**(7784), 607–617 (2019)
119. Arthur, I.J., Dada, P.: Algorithm Prototyping, Development, and Deployment for TrueNorth: The Caffe Tea Case Study (2015)
120. Sze, V., Chen, Y.-H., Yang, T.-J., Emer, J.S.: Efficient processing of deep neural networks: a tutorial and survey. Proc. IEEE. **105**(12), 2295–2329 (2017)
121. An, H., Zhou, Z., Yi, Y.: Opportunities and challenges on nanoscale 3D neuromorphic computing system. In: Electromagnetic Compatibility & Signal/Power Integrity (EMCSI), 2017 IEEE International Symposium on, IEEE, pp. 416–421 (2017)
122. Severa, W., Vineyard, C.M., Dellana, R., Verzi, S.J., Aimone, J.B.: Training deep neural networks for binary communication with the Whetstone method. Nat. Mach. Intell. **1**(2), 86 (2019)
123. Drazen, D., Lichtsteiner, P., Häfliger, P., Delbrück, T., Jensen, A.: Toward real-time particle tracking using an event-based dynamic vision sensor. Exp. Fluids. **51**(5), 1465 (2011)
124. Delbruck, T., Lang, M.: Robotic goalie with 3 ms reaction time at 4% CPU load using event-based dynamic vision sensor. Front. Neurosci. **7**, 223 (2013)
125. Blum, H., Dietmüller, A., Milde, M., Conradt, J., Indiveri, G., Sandamirskaya, Y.: A neuromorphic controller for a robotic vehicle equipped with a dynamic vision sensor. Robot. Sci. Syst. **2017** (2017)
126. Dominguez-Morales, M.J., Jimenez-Fernandez, A., Jiménez-Moreno, G., Conde, C., Cabello, E., Linares-Barranco, A.: Bio-inspired stereo vision calibration for dynamic vision sensors. IEEE Access. **7**, 138415–138425 (2019)
127. Choi, S.-Y., Kim, J.-S., Seo, J.-H.: A study on the reduction of power consumption and the improvement of motion blur for OLED displays. J. Korean Inst. IIlum. Electr. Install. Eng. **30**(3), 1–8 (2016)
128. Chen, G., et al.: Neuromorphic vision based multivehicle detection and tracking for intelligent transportation system. J. Adv. Transport. **2018**, 4815383 (2018)
129. Anumula, J., Neil, D., Delbruck, T., Liu, S.-C.: Feature representations for neuromorphic audio spike streams. Front. Neurosci. **12**, 23 (2018)
130. Liu, S.C., Delbruck, T.: Neuromorphic sensory systems. Curr. Opin. Neurobiol. **20**, 288–295 (2010). https://doi.org/10.1016/j.conb.2010.03.007
131. Richter, C., et al.: Musculoskeletal robots: scalability in neural control. IEEE Robot. Autom. Mag. **23**, 128–137 (2016). https://doi.org/10.1109/MRA.2016.2535081
132. Vanarse, A., Osseiran, A., Rassau, A.: A review of current neuromorphic approaches for vision, auditory, and olfactory sensors. Front. Neurosci. **10**, 115 (2016)
133. Sheng, M., Sabatini, B.L., Südhof, T.C.: Synapses and Alzheimer's disease. Cold Spring Harb. Perspect. Biol. **4**(5), a005777 (2012)

AI for Cybersecurity in Distributed Automotive IoT Systems

Vipin Kumar Kukkala, Sooryaa Vignesh Thiruloga, and Sudeep Pasricha

1 Introduction

Modern vehicles consist of several distributed processing elements called electronic control units (ECUs) that communicate using an in-vehicle network. Each ECU runs various mixed-criticality real-time applications that range from advanced vehicle control to entertainment. Each ECU takes input from different sensors or information from other ECUs to control or actuate different components in the vehicle. Additionally, some of the ECUs in the cars connect to various external systems such as OEM servers to receive over-the-air (OTA) updates via the Internet, other vehicles to communicate traffic information, etc. These unique characteristics of automotive systems make them one of the best examples of a complex distributed time-critical cyber-physical IoT system.

The number of ECUs along with the complexity of software running on these ECUs has been steadily increasing in emerging vehicles. This is mainly driven by the need to support state-of-the-art advanced driver assistance system (ADAS) features such as collision warning, lane keep assist, parking assist, blind spot warning, etc. These advancements have resulted in an increase in the complexity of the in-vehicle network, which is the backbone over which huge volumes of heterogeneous sensor data and safety-critical real-time decisions and control commands are communicated. Moreover, the state-of-the-art ADAS solutions are increasingly communicating with various external systems using advanced communication standards such as 5G technology and Vehicle-to-X (V2X) [1]. This increased interaction with external systems makes modern vehicles highly vulnerable to various cybersecurity attacks that can have catastrophic consequences. Several

V. K. Kukkala · S. V. Thiruloga · S. Pasricha (✉)
Department of Electrical and Computer Engineering, Colorado State University,
Fort Collins, CO, USA
e-mail: vipin.kukkala@colostate.edu; sooryaa@colostate.edu; sudeep.pasricha@colostate.edu

A. Iranmanesh (ed.), *Frontiers of Quality Electronic Design (QED)*,
https://doi.org/10.1007/978-3-031-16344-9_8

cyberattacks on multiple vehicles have been demonstrated in [2–4] showing various approaches to gain access to the in-vehicle network and take control of the vehicle via malicious messages. As connected and autonomous vehicles are becoming increasingly ubiquitous, the problem of security in automotive systems will be further aggravated. Thus, it is highly essential to prevent unauthorized access of vehicular networks from external attackers to ensure the security of automotive systems.

Traditional computer networks use firewalls as a defense mechanism to protect the network from various unauthorized accesses. However, no firewall is flawless and no network can be impenetrable. Therefore, there is a need for an active monitoring system that scans the network to detect cyberattacks manifesting in the system. An intrusion detection system (IDS) actively monitors network traffic and triggers alerts when malicious behavior or known attack signatures are detected. The IDS acts as the last line of defense in distributed automotive IoT systems.

General IDSs can be classified into two categories: (i) signature-based and (ii) anomaly-based. The signature-based IDSs observe for traces of any known attack signatures, while the anomaly-based IDSs observe for any deviation from the known normal system behavior to indicate the presence of an attacker. Signature-based IDS typically have fewer false alarms (false positives) and faster detection times but can only detect pre-modeled attack patterns that were observed previously. On the other hand, anomaly-based IDS can detect both previously observed and novel attack patterns, while they can suffer from high false alarms and relatively longer detection times when designed sub-optimally. An efficient IDS needs to be robust, lightweight, and scalable with diverse system sizes. In addition, a practical IDS needs to be able to detect a large spectrum of attacks with high confidence in detection. A low false-positive rate is also important because in time-critical systems such as automotive systems, recovery from a false positive can be very expensive.

With the increasing adoption of deep learning and artificial intelligence (AI) in emerging vehicles in an attempt to move toward achieving complete autonomy, their power can be leveraged to develop an effective anomaly-based IDS to detect cyberattacks. The large availability of data and the increasing computational power of ECUs further bolsters the case for an AI-based IDS to detect cyberattacks that are active over the in-vehicle networks. The ability of AI to learn the highly complex features in the data that are hard to capture with traditional techniques gives AI-based IDS a unique edge over other techniques. Moreover, the ability of AI to operate on heterogeneous data can provide an AI-based IDS the ability to detect both known and unknown cyberattacks. Thus, AI-based IDS can be a promising solution for the problem of automotive cybersecurity.

In this chapter, we provide an overview of a novel AI-based vehicle IDS cybersecurity framework called *INDRA* [33] that actively monitors messages in the controller area network (CAN) (a popular in-vehicle network protocol) bus to detect cyberattacks. During the *offline* phase, *INDRA* uses advanced deep learning models to learn the normal system behavior in an unsupervised fashion. At *runtime*, the *INDRA* framework leverages the knowledge of previously learned normal system behavior, to monitor and detect various cyberattacks. *INDRA* aims to maximize detection accuracy and minimize false-positive rate while incurring a very low overhead on the ECUs. The key contributions of the *INDRA* framework are as follows:

- A gated recurrent unit (GRU)-based recurrent autoencoder network to learn the latent representation of normal system behavior during the offline phase.
- A metric called intrusion score (IS), to quantify the deviation from normal system behavior.
- A thorough analysis toward the selection of thresholds for this intrusion score metric.
- A comprehensive analysis that demonstrates the effectiveness of *INDRA* for vehicle cybersecurity, with superior results compared to the best known state-of-the-art prior works in the area.

2 Related Work

Various techniques have been proposed to design IDS for securing time-critical distributed automotive IoT systems. These works attempt to detect various types of cyberattacks by monitoring the network traffic.

Signature-based IDS reckon on detecting known and pre-modeled attack signatures. In [5], the authors used a language theory-based model to derive attack signatures. However, this technique fails to detect intrusions when it misses the packets transmitted during the early stages of an attack. The authors in [6] used transition matrices to detect intrusions in a CAN bus-based system. This technique achieves a low false-positive rate for trivial attacks but failed to detect more realistic attacks such as replay attacks. In [7], the authors identify prominent attack patterns such as a sudden increase in the message frequency and missing messages to detect intrusions. The authors in [8] proposed a specification-based approach that analyzes the behavior of the system and compare it with the predefined attack patterns to detect intrusions. However, their system can only detect predefined attack patterns and fails to detect unknown attacks. The authors in [9] proposed an IDS technique using the Myers algorithm [10] under the map-reduce framework. In [11], the authors use a time-frequency analysis of CAN messages to detect multiple intrusions. A rule-based regular operating mode region is derived in [12] by analyzing the message frequency at design time. This region is observed for deviations at runtime to detect intrusions. The authors in [13] proposed a technique that uses the fingerprints of the sender ECU's clock skew and the messages to detect intrusions by observing for variations in the clock-skew at runtime. A formal analysis for clock-skew-based IDS is presented in [14] and evaluated on a real vehicle. In [15], a memory heat map is used to characterize the memory behavior of the operating system to detect intrusions. An entropy-based IDS is proposed in [16] that observes for change in system entropy to detect intrusions. However, this technique fails to detect small scale attacks where the change in entropy is minimal.

In summary, signature-based techniques offer a quick solution to the intrusion detection problem with low false-positive rates but cannot detect more complex and novel cyberattacks. Moreover, modeling signatures of every possible attack is impractical.

On the other hand, an anomaly-based IDS aims to learn the normal system behavior in an offline phase and observe for any deviation from the learned normal behavior to detect intrusions (as anomalies) at runtime. In [17], a sensor-based IDS was proposed, where the attack detection sensors are used to monitor various system events to observe for any deviations from the normal behavior. This approach is not only expensive but also suffers from poor detection rates. A one-class support vector machine (OCSVM)-based IDS was proposed in [18]. However, this technique suffers from poor detection latency and has high tuning overhead. The authors in [19] used four different nearest neighbor classifiers to distinguish between a normal and an attack-induced payloads in CAN bus. A decision tree-based detection model is proposed in [20] that monitors the physical features of the vehicle to detect intrusions. However, this model is not realistic and suffers from high detection latencies. A hidden Markov model (HMM)-based technique was proposed in [21] that monitors the temporal relationships between messages to detect intrusions. In [22], a deep neural network-based approach was proposed to monitor the message payloads in the in-vehicle network. This approach is tuned for a low priority tire pressure monitoring system (TPMS), which makes it hard to adapt to high priority safety-critical powertrain applications. The authors in [23] proposed a long short-term memory (LSTM)-based IDS for multi-message ID detection. Due to the high complexity of the model architecture, this technique incurs high overhead on the ECUs. The authors in [24] use an LSTM-based IDS to detect insertion and dropping attacks (explained later in Sect. 4.3). An LSTM-based predictor model is proposed in [25] that predicts the next time step message value at a bit level granularity and examines for large variations in loss to detect intrusions. A recurrent neural network (RNN)-based IDS was proposed in [26] that learns the normal patterns in CAN messages in the in-vehicle network. A hybrid IDS to detect anomalies in time-series data was proposed in [27], which utilizes a specification-based system in the first stage and an RNN-based model in the second stage to detect anomalies. *However, none of these techniques provides a holistic system-level cybersecurity solution that is lightweight, scalable, and reliable to detect multiple types of cyberattacks for in-vehicle networks.*

This chapter describes a novel lightweight recurrent autoencoder-based IDS framework called *INDRA* [33] that utilizes gated recurrent units (GRUs) to monitor messages at a signal level granularity to detect various types of attacks more effectively and successfully than the state of the art. Table 1 summarizes some of the state-of-the-art IDS works' performance under different metrics and shows how *INDRA* fills the existing research gap. The *INDRA* framework aims at improving multiple performance metrics compared to the state-of-the art IDS works that target a subset of performance metrics. A detailed analysis of each metric and evaluation results are presented later in Sect. 6.

Table 1 Comparison between *INDRA*[33] framework and state-of-the-art IDS works

Technique	IDS performance			
	Lightweight	Low false -positive rate	High accuracy	Fast inference
PLSTM [25]	X	✓	X	X
RepNet [26]	✓	X	X	✓
CANet [23]	X	✓	✓	X
INDRA	✓	✓	✓	✓

3 Background on Sequence Learning

The availability of increased computing power from GPUs and custom accelerators led to training neural networks with many hidden layers (known as deep neural networks) that resulted in the creation of powerful models for solving difficult problems in many domains. One such problem is detecting intrusions in the distributed automotive IoT systems, specifically in the in-vehicle network that connects them. In an in-vehicle network, the communication between ECUs happens in a timely manner. Hence, there exist temporal relationships between the messages, which is crucial to exploit, in order to detect intrusions. However, this cannot be achieved using traditional feedforward neural networks as the output of any input is independent of the other inputs. One of the solutions is to use sequence models as they are more appropriate and are designed to handle sequences and time-series data.

3.1 Sequence Models

A sequence model can be thought of as a function which ensures that the current output is dependent not only on the current input but also on the previous inputs. Recurrent neural network (RNN) is one of the first sequence models which was introduced in [28]. In recent years, improved sequence models such as long short-term memory (LSTM) and gated recurrent unit (GRU) have also been developed.

3.1.1 Recurrent Neural Network (RNN)

An RNN is a type of artificial neural network that takes sequential data (such as sequence or time-series data) as the input and learns the relationship in the data. RNNs achieve this by using the hidden states, which allows learned information to persist over time steps. Moreover, the hidden states also enable the RNN to connect previous information to current inputs. An RNN cell with feedback is shown in Fig. 1a, and an RNN unrolled in time is shown in Fig. 1b.

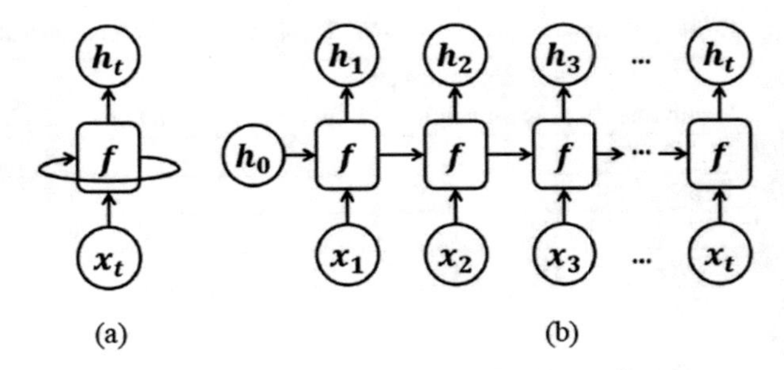

Fig. 1 (**a**) A single RNN cell and (**b**) RNN unit unrolled in time, where f is the RNN cell, x is the input, and h represents hidden states [33]

The output of an RNN cell at a time step t (h_t) is a function of both the input at time step t (x_t) and the previous time step output (h_{t-1}):

$$h_t = f\,(Wx_t + Uh_{t-1} + b) \tag{1}$$

where W, U represent the weight matrices, b is a bias term, and f is a nonlinear activation function (such as a sigmoid or tanh). One of the limitations of RNNs is that they are very hard to train. As RNNs and other sequence models deal with sequence or time-series inputs, backpropagation happens through various time samples (known as backpropagation through time (BPTT)). During the BPTT process, the feedback loop in RNNs causes the errors to shrink or grow rapidly (resulting in vanishing or exploding gradients respectively), destroying the information in backpropagation. This problem of vanishing gradients hampers the RNNs from learning *long-term dependencies*. This problem was solved with the introduction of additional states and gates in the RNN cell to remember long-term dependencies, which led to the introduction of long short-term memory networks [29].

3.1.2 Long Short-Term Memory (LSTM) Networks

LSTMs are improved RNNs that use cell state and the hidden state information along with several gates to remember long-term dependencies in the input sequence. The cell state can be visualized as a transport highway that carries relevant information throughout the processing of a sequence. The cell state accommodates the information from earlier time steps, which can be used in the later time steps, thereby reducing the effects of short-term memory. The information in the cell state is modified using various gates, which helps the network decide which information needs to be retained and which information to forget.

An LSTM cell consists of three gates: (i) forget gate (f_t), (ii) input gate (i_t), and (iii) output gate (o_t), as shown in Fig. 2a. The forget gate is a binary gate that controls which information to retain from the previous cell state (c_{t-1}). The input gate is responsible for adding relevant information to the current cell state (c_t). Lastly, the output gate controls the output layer, which uses information from the forget and input gates to produce an appropriate output. An unrolled LSTM unit is shown in Fig. 2b.

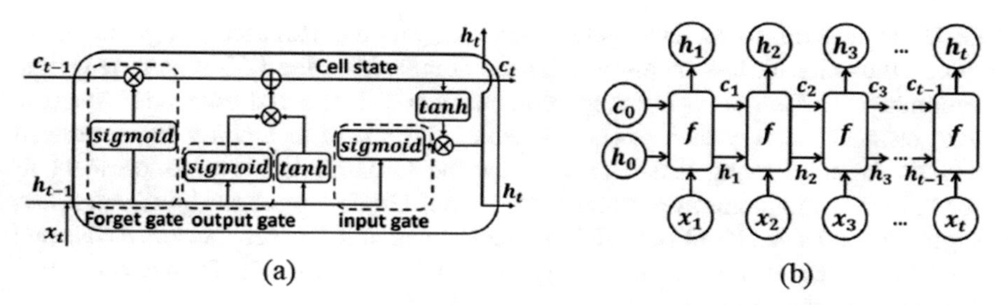

Fig. 2 (**a**) A single LSTM cell with different gates and (**b**) LSTM unit unrolled in time, where f is an LSTM cell, x is input, c is cell state, and h is the hidden state [33]

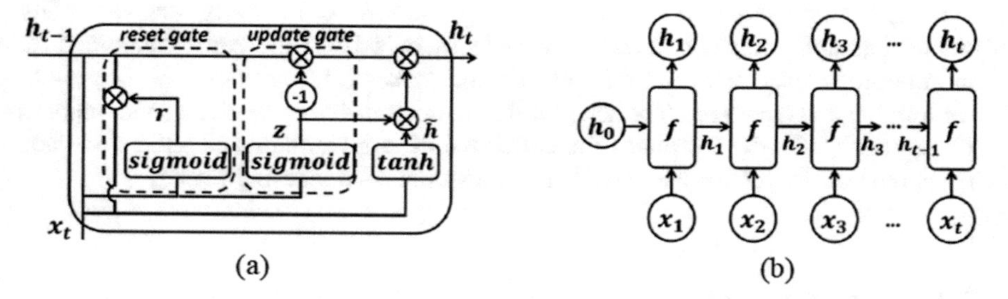

Fig. 3 (**a**) A single GRU cell with different gates and (**b**) GRU unit unrolled in time, where f is a GRU cell, x is input, and h represents hidden states [33]

The combination of the abovementioned different gates, along with the cell and hidden states, enables LSTMs to learn long-term dependencies in sequences. However, they are not computationally efficient as the addition of multiple gates increased the complexity of the sequence path (more than in RNNs) and also require more memory at runtime. Additionally, training LSTMs is compute-intensive even with advanced training methods such as truncated backpropagation. To overcome these limitations, a simpler recurrent neural network called gated recurrent unit (GRU) network was introduced in [30] that can be trained faster than LSTMs and also remembers dependencies in long sequences with relatively low memory overhead while solving the vanishing gradient problem.

3.1.3 Gated Recurrent Unit (GRU)

A GRU cell uses an alternate route for gating information by combining the input and forget gate of the LSTM into a solitary *update* gate. GRUs furthermore combine the hidden and cell states, as shown in Fig. 3a, b.

A GRU cell consists of two gates: (i) reset gate and (ii) update gate. The reset gate combines new input with past memory, while the update gate selects the amount of relevant data that should be held. This enables the GRU cell to control the data stream like an LSTM by uncovering its hidden layer contents. Moreover,

GRUs achieve this using fewer gates and states, which makes them computationally more efficient with low memory overhead compared to the LSTMs. As real-time automotive ECUs are highly resource-constrained distributed embedded systems with stringent energy and power budgets, it is crucial to employ low overhead models for inferencing tasks. This makes the GRU-based networks an ideal fit for inference in automotive systems. Moreover, GRUs are relatively new and less explored and have a lot of potential to offer compared to its predecessors RNNs and LSTMs. Hence, in this chapter, a lightweight GRU-based IDS framework called *INDRA* is presented (explained in detail in Sect. 5).

The sequence models can be trained using both supervised and unsupervised learning approaches. Due to the large volume of automotive network data in a vehicle, labeling the data can become very tedious. Additionally, the variability in the messages between different vehicle models from the same manufacturer and the proprietary nature of this information makes it furthermore challenging to accurately label messages. The accessibility to automotive network data via onboard diagnostics (OBD-II) facilitates the collection of large amounts of unlabeled data. Thus, the IDS in *INDRA* uses GRUs in an unsupervised learning setting.

3.2 Autoencoders

An autoencoder is an unsupervised learning algorithm that tries to reconstruct the input by learning the latent input features. Autoencoders achieve this by encoding the input data (x) toward a hidden layer and finally decoding it to produce a reconstruction $\tilde{x}$ (as shown in Fig. 4). The encoding produced at the hidden layer is called an embedding. The layers that create this embedding are called the encoder, and the layers that utilize the embedding and reconstruct the original input are called the decoder. When training the autoencoders, the encoder tries to learn a nonlinear mapping of the inputs, while the decoder tries to learn the nonlinear mapping of the embedding to the inputs. Both encoder and decoder achieve this with the help of nonlinear activation functions, such as tanh and rectified linear unit (ReLU). Moreover, the autoencoder network tries to recreate the input as accurately as possible by selectively extracting the key features from the inputs with a goal of minimizing reconstruction error. The most commonly used loss functions in autoencoders are mean squared error (MSE) and Kullback-Leibler (KL) divergence.

Since the autoencoders aim to reconstruct the input by learning the underlying distribution of the input data, it makes them an excellent choice to learn and reconstruct highly correlated time-series data efficiently by learning the temporal relations between signals. *Thus, the INDRA framework uses lightweight GRUs in an autoencoder to learn latent representations of CAN messages in an unsupervised learning setting.*

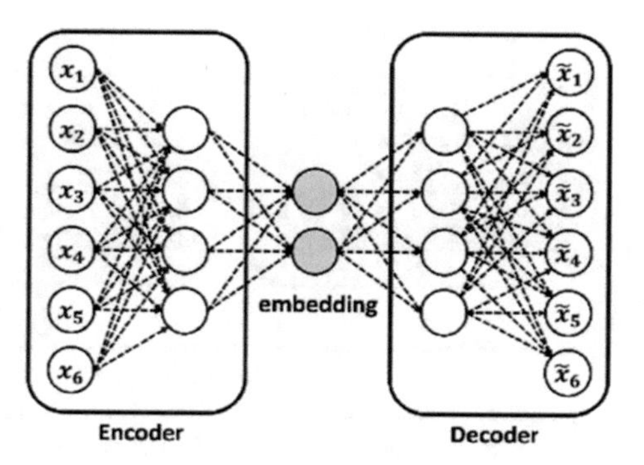

Fig. 4 Autoencoders [33]

4 Definitions and Problem Formulation

4.1 System Model

The *INDRA* framework considers a generic distributed automotive *system* consisting of multiple ECUs connected using a CAN-based in-vehicle network, as shown in Fig. 5. Each ECU runs a set of hard real-time automotive applications that have strict timing and deadline constraints. Additionally, each ECU also executes intrusion detection applications that monitors and detects intrusions in the in-vehicle network. *INDRA* employs a distributed IDS approach, where the intrusion applications are collocated with real-time automotive applications as opposed to a centralized IDS approach where a single central ECU handles all intrusion detection tasks. This design decision is driven by the following reasons:

- A centralized IDS approach is particularly prone to single-point failures, which can fully open up the system to the attacker.
- In some extreme scenarios such as during a distributed denial-of-service (DDoS) or flooding attack (explained in Sect. 4.3), the in-vehicle network can get highly congested and the centralized IDS might not be able to communicate with the victim ECUs.
- If an attacker succeeds in fooling the centralized IDS ECU, attacks can go undetected by the other ECUs, compromising the entire system. However, with a distributed IDS approach, fooling multiple ECUs is required which is much harder. Even if one of the ECUs is compromised, the attack can still be detected by the decentralized intelligence.
- In a distributed IDS approach, ECUs can stop accepting suspicious messages as soon as an intrusion is detected without waiting for a centralized system to notify them, resulting in faster detection times.
- In a distributed IDS approach, the computation load of intrusion detection is split among the ECUs, and the monitoring can be limited to only the required messages. This facilitates multiple ECUs to monitor a subset of messages independently, with very lower overhead.

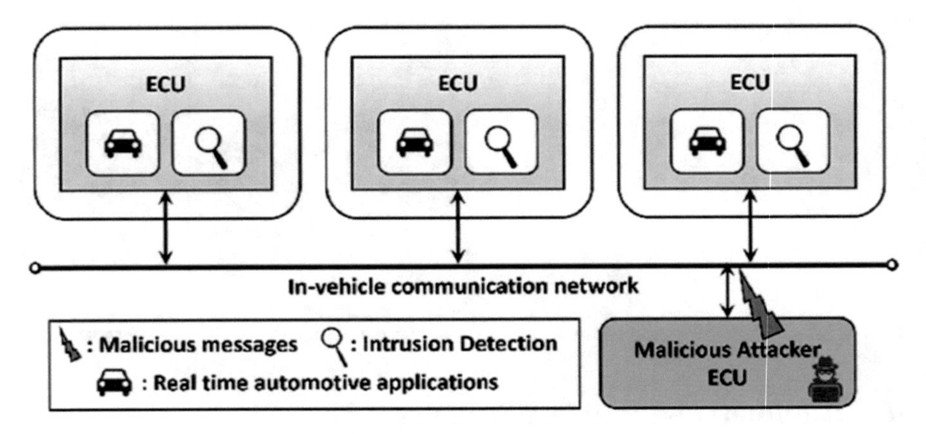

Fig. 5 Overview of the system model considered in *INDRA* [33]

Many prior works, such as in [5, 12], consider a distributed IDS approach for these reasons. Moreover, with automotive ECUs becoming increasingly powerful, the collocation of IDS applications with real-time automotive applications in a distributed manner is feasible, provided the overhead from the IDS is minimal. The *INDRA* framework is not only lightweight but also highly scalable and achieves superior intrusion detection performance, as discussed in Sect. 6.

An efficient IDS design should have low susceptibility to noise, low cost, and a low power/energy footprint. Additionally, *INDRA* considers the following design objectives in the development process of the IDS:

- *Lightweight*: Intrusion detection tasks can incur additional overhead on the ECUs, which could result in poor application performance or missed deadlines for real-time applications. This can have catastrophic consequences in some cases. Thus, it is important to have a lightweight IDS that incurs very minimal overhead on the system.
- *Few false positives*: This is a highly desired quality in any kind of IDS (even outside of the automotive domain), as handling false positives can become expensive very quickly. An efficient IDS needs to have few false positives or false alarms.
- *High attack coverage*: Attack coverage is the range of attacks an IDS can detect. A good IDS needs to be able to detect more than one type of attack. A high attack coverage for IDS will make the system resilient to multiple attack surfaces.
- *Scalability*: This is a crucial requirement as the numbers of ECUs, software, and network complexity have been increasing in the emerging vehicles. A practical IDS should be highly scalable and be able to support various system sizes.

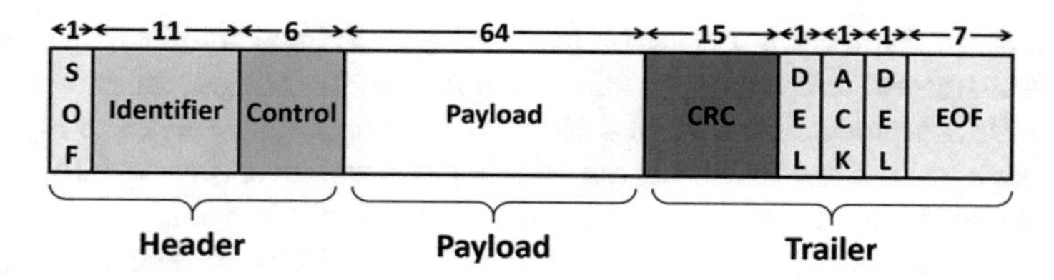

Fig. 6 Standard CAN frame format [33]

Signal Name	Message	Start bit	Length	Byte Order	Value Type
Battery_Current	Status	0	16	Intel	Signed
Battery_Voltage	Status	16	16	Intel	Unsigned
Motor_Current	Status	32	16	Intel	Signed
Motor_Speed	Status	48	8	Intel	Signed
Motor_Direction	Status	56	8	Intel	Unsigned

Fig. 7 Real-world CAN message with signal information [33]

4.2 *Communication Model*

A brief overview of the vehicle communication model that was considered in *INDRA* is presented in this subsection. The *INDRA* framework mainly focuses on detecting intrusions in a CAN bus-based automotive system. Controller area network (CAN) is the de facto industry standard in-vehicle network protocol for modern-day automotive systems. CAN is a low-cost, lightweight event-triggered communication protocol that transmits messages in the form of CAN frames. The structure of a standard CAN frame is shown in Fig. 6, and the length of each field (in bits) is shown on the top. The standard CAN frame consists of header, payload, and trailer segments. The header segment consists of information such as the message identifier (ID) and the length of the message. The actual data that needs to be transmitted is in the payload segment. Lastly, the information in the trailer segment is mainly used for error checking at the receiver. A variation of the standard CAN, called CAN-extended or CAN 2.0B, is also becoming increasingly common in modern vehicles. CAN extended consists of a 29-bit identifier compared to 11-bit identifier in the CAN standard, allowing for more number of unique message IDs.

The IDS design in *INDRA* focuses on monitoring the payload segment of the message and observe for anomalies to detect intrusions. This is mainly because an attacker needs to modify the message payload to accomplish a malicious activity. An attacker could also target the header or trailer segments, but it would result in the message getting rejected at the receiver. The payload segment consists of multiple data entities called signals. An example real-world CAN message with multiple signals is shown in Fig. 7 [31]. Each signal has a fixed length (in bits), an associated data type, and a start bit that specifies its starting location in the 64-bit payload segment of the CAN message.

The *INDRA* framework focuses on monitoring individual signals within message payloads to observe for anomalies and detect intrusions. The neural network model in the *INDRA* framework is trained to learn the temporal dependencies between the messages at a signal level during training and observes for deviations during the deployment (at runtime) to detect intrusions in the in-vehicle network. This signal level monitoring would not only give the capability to detect the presence of an intruder but also helps in identifying the signal within the message that is being targeted during an attack. This information can be crucial in understanding the intentions of the attacker, which can be used for developing countermeasures. The signal level monitoring mechanism in *INDRA* is discussed in detail in Sect. 5.2. *Note*: Even though the *INDRA* framework mainly focuses on detecting intrusions by monitoring CAN messages, this approach can be extended to be used with other in-vehicle network protocols as the framework is agnostic to the underlying communication protocol.

4.3 Attack Model

The *INDRA* framework aims to protect the vehicle from various types of cyberattacks that are listed below. These are some of the most commonly seen and hard to detect automotive attack patterns that have been widely considered in literature to evaluate IDS models.

1. *Flooding attack*: This is the most common and easy to launch attack and requires little to no knowledge about the system. In this attack, the attacker floods the in-vehicle network with a random or specific message and prevents the other ECUs from communicating. This is also known as the denial-of-service (DoS) attack. These attacks are generally detected and prevented by the bridges and gateways in the in-vehicle network and often do not reach the last line of defense (the IDS). However, it is important to consider these attacks in the IDS evaluation as they can have a severe impact on the system when handled incorrectly.
2. *Plateau attack*: In this attack, an attacker overwrites a signal value with a constant value over a period of time. The attack severity depends on the magnitude of the change in signal value and the attack duration. Large changes in magnitude of the signal values are easier to detect compared to shorter changes.
3. *Continuous attack*: In this attack, an attacker slowly overwrites the signal value until some target value is achieved and tries to avoid the triggering of IDS. This attack is hard to detect and can be sensitive to the IDS parameters (discussed in Sect. 5.2).
4. *Suppress attack*: In this attack, the attacker suppresses the signal value(s) by either disabling the communication controller or by powering off the target ECU. These attacks can be easily detected, when the message transmissions are shut down for long durations, but are harder to detect for shorter durations.

5. *Playback attack*: In this attack, the attacker replays a valid series of message transmissions from the past trying to trick the IDS. This attack is hard to detect if the IDS does not have the ability to capture the temporal relationships between messages.

Moreover, the *INDRA* framework assumes that the attacker can gain access to the vehicle using the most common attack vectors, which include connecting to V2X systems that communicate with the outside world (e.g., infotainment and connected ADAS systems), connecting to the OBD-II port, probing into the in-vehicle bus, and replacing an existing ECU. Furthermore, the *INDRA* framework assumes that the attacker has access to the bus parameters (such as BAUD rate, parity, flow control, etc.) that can help in gaining access to the in-vehicle network.

Problem objective The goal of *INDRA* is to implement a lightweight IDS that can detect various types of cyberattacks (mentioned earlier) in a CAN bus-based distributed automotive system, with a high detection accuracy and low false-positive rate, and while maintaining a large attack coverage.

5 *INDRA* Framework Overview

The *INDRA* framework aims to achieve a signal level anomaly-based IDS for monitoring CAN messages in automotive embedded systems. An overview of the *INDRA* framework is shown in Fig. 8. At a high level, the *INDRA* framework consists of design-time and runtime phases. At design time, *INDRA* uses trusted CAN message data to train a recurrent autoencoder-based model to learn the normal operating behavior of the system. At runtime, the trained recurrent autoencoder model is used for observing deviations from normal behavior (inference) and detect intrusions based on the deviation computed using the proposed intrusion score metric (detection). The following subsections describe these steps in more detail.

5.1 *Recurrent Autoencoder*

Recurrent autoencoders are powerful neural networks that are designed to behave like an autoencoder network but handle time-series or sequence data inputs. They can be visualized similar to the regular feed-forward neural network-based autoencoders, except with the neurons being either RNN, LSTM, or GRU cells (discussed in Sect. 3). Similar to regular autoencoders, the recurrent autoencoders consist of an encoder and a decoder stage. The encoder is responsible for generating a latent representation of the input sequence data in an n-dimensional space. The decoder uses the generated latent representation from the encoder and tries to reconstruct the input with minimal reconstruction error. In this section, a lightweight recurrent autoencoder model that is customized for the design of IDS to detect

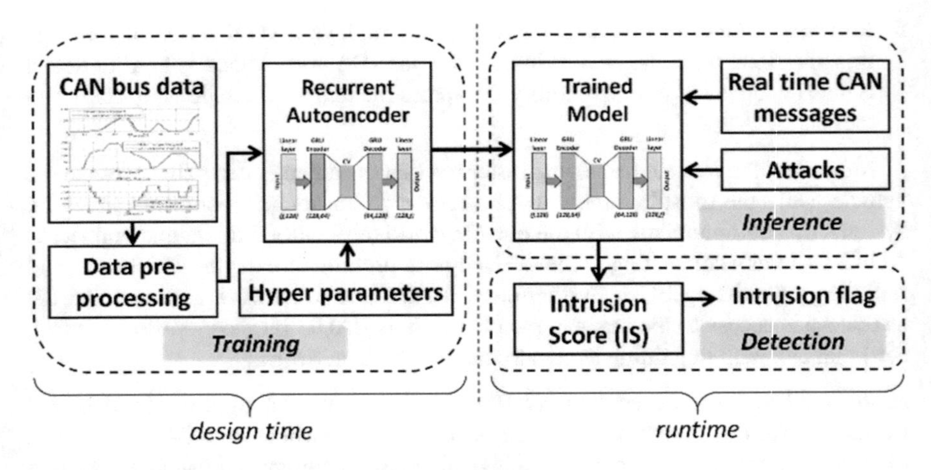

Fig. 8 Overview of *INDRA* framework [33]

intrusions in the in-vehicle network data is presented. The details related to the recurrent autoencoder model architecture and the different stages involved in its training are discussed in the subsequent subsections.

5.1.1 Model Architecture

The proposed recurrent autoencoder model architecture in *INDRA* with the dimensions (input, output) of each layer is illustrated in Fig. 9. The model consists of a linear layer at the input, GRU-based encoder, GRU-based decoder, and a final linear layer before the output. The time-series CAN message data with signal level values with f features (where f is the number of signals in that particular message) is given as the input to the first linear layer. The output of the first linear layer is passed to the GRU-based encoder to generate the latent representation of the time-series signal inputs. This latent representation is referred to as a message context vector (MCV). The MCV captures the context of different signals in the input message data in the form of a vector, hence the name. Each MCV can be thought of as a point in an n-dimensional space that contains the context of the series of signal values given as input. The MCV is given as the input to a GRU-based decoder, which feeds its output as an input to the final linear layer. The linear layer at the end produces the reconstructed input time-series that represents the CAN message data with individual signal level values. Mean square error (MSE) loss function is used to compute the loss between the input and the reconstructed input. The model weights are updated using backpropagation through time (BPTT). The *INDRA* framework builds a recurrent autoencoder model for each message ID.

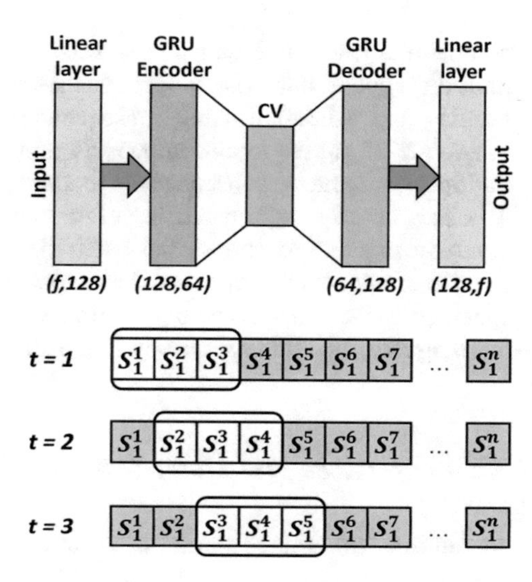

Fig. 9 Recurrent autoencoder network architecture in *INDRA* (*f* is number of features, i.e., number of signals in the input CAN message, MCV is message context vector) [33]

Fig. 10 Illustration of rolling window-based approach [33]

5.1.2 Training Procedure

The first step of the training process is preprocessing the input CAN message data. Each sample in the dataset consists of a message ID and corresponding values of the signals within that message ID. The signal values are scaled between 0 and 1 for each signal type, as the range of signal values can be very large in some cases. Using unscaled signal values as inputs can result in an extremely slow or very unstable training process. Moreover, scaling the signal values also helps in avoiding the problem of exploding gradients.

The final preprocessed data is split into training data (85%) and validation data (15%) and is prepared for training using a rolling window-based approach. This involves selecting a window of fixed size and rolling it to the right by one-time sample every time step. A rolling window size of three samples for three time steps ($t = 1, 2, 3$) is illustrated in Fig. 10, where the term S_i^j represents the ith signal value at j^{th} sample. The elements in the rolling window are collectively called as a subsequence and the subsequence length is equal to the size of the rolling window. As each subsequence consists of a set of signal values over time, the recurrent autoencoder model in *INDRA* tries to learn the existing temporal relationships between the series of signal values. These signal level temporal relationships play a crucial role in identifying more complex cyberattacks such as *continuous* and *playback* (as discussed in Sect. 4.3). The process of training using subsequences is carried out iteratively until the end of the sequence in training data.

Each iteration in the training step consists of a forward pass and a backward pass using BPTT to update the weights and biases of the neurons (discussed in Sect. 3) based on the error value. At the end of the training, the model's learning is evaluated (forward pass only) using the validation data, which was not seen by the model during the training. By the end of validation, the model has seen the complete dataset once and this is known as an *epoch*. The model is trained for multiple epochs until the model reaches convergence. Moreover, the process of training and validation using subsequences is sped up by training the input data in groups of subsequences known as mini-batches. Each mini-batch consists of multiple consecutive subsequences that are given as input to the model in parallel. The size of each mini-batch is commonly referred to as *batch size*, and it is a common practice to choose the batch size as a power of two. Lastly, to control the rate of update of parameters during backpropagation, a learning rate needs to be specified to the model. The hyperparameters such as subsequence size, batch size, learning rate, etc., that are chosen in the *INDRA* are presented later in Sect. 6.1.

5.2 *Inference and Detection*

At runtime, the trained model is set to evaluation mode, where only the forward passes occur and the weights and biases are not updated. In this phase, the trained model is tested under multiple attack scenarios (mentioned in Sect. 4.3), by simulating appropriate attack conditions in the CAN message dataset.

Every data sample that is given as the input to the model gets reconstructed at the output, and the reconstruction loss is fed to the detection module to compute a metric called *intrusion score* (IS). The IS helps in identifying whether a signal is normal or malicious. The IS metric is computed at a signal level to predict the signal that is under attack. The IS metric is computed at every iteration during inference, as a *squared error* to estimate the prediction deviation from the input signal value, as shown below:

$$\mathrm{IS}_i = \left(\left(S_i^j - \hat{S}_i^j\right)^2\right) \quad \forall i \in [1, m] \tag{2}$$

where, S_i^j represents i^{th} signal value of the jth sample, $\hat{S}_i^j$ denotes its reconstruction, and m is the number of signals in the message. The predicted value would have a large deviation from the input signal value (i.e., large IS value), when the input signal pattern is not seen during the training phase, and a minimal IS value otherwise. This is the basis for the detection step in *INDRA*.

Additionally, the *INDRA* framework combines the signal level IS information into a message-level IS, by taking the maximum IS of the signals in that message as shown in Eq. (3). This is mainly to facilitate the lack of signal level intrusion label information in the dataset.

$$\mathrm{MIS} = \max\left(IS_1, IS_2 \ldots, IS_m\right) \tag{3}$$

To get adequate detection accuracy, an *intrusion threshold* (IT) needs to be selected for flagging messages appropriately. *INDRA* explores multiple choices for IT, using the best model from the training process. The best model is the model with

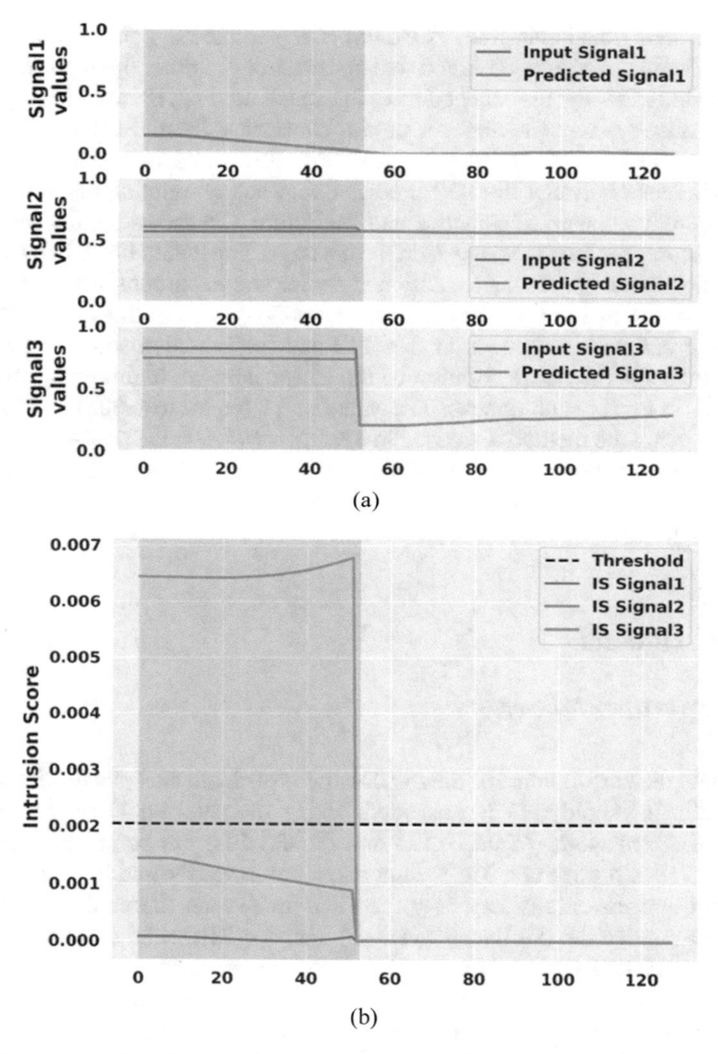

Fig. 11 Snapshot of *INDRA* IDS checking a message with three signals under a plateau attack, where (**a**) shows the signal comparisons and (**b**) shows IS for signals and IS for the message and intrusion flag [33]

the lowest validation running loss during the training process. Using the best model, multiple metrics such as maximum, mean, median, and 99.99%, 99.9%, 99%, and 90% validation loss across all iterations are logged as the choices for the IT. The detailed analysis on selection of IT is presented in Sect. 6.2.

A snapshot of *INDRA* IDS working in an environment with cyberattacks is illustrated in Fig. 11a, b, with a plateau attack on a message with three signals, between time 0 and 50. The highlighted area in red represents the attack interval.

Figure 11a shows the input (true) vs *INDRA* IDS predicted signal value comparisons for three signals. It can be observed that for most of the time, the reconstruction is close for almost all signals except during the attack interval. Signal 3 is subjected to a plateau attack where the attacker held a constant value until the end of attack interval as shown in the third subplot of Fig. 11a (note that this resulted in a larger difference between the IDS predicted and actual input signal values in the third subplot, compared to signals 1 and 2). Figure 11b shows the different signal intrusion scores for the three signals in the message. The dotted black line represents the intrusion threshold (IT). As mentioned earlier, the maximum of signal intrusion scores is chosen as message intrusion score (MIS), which in this case is the IS of signal 3. It can be clearly seen in Fig. 11b that the intrusion score of signal 3 is above the IT, for the entire duration of the attack interval, highlighting the ability of *INDRA* to detect such attacks. The value of IT (equal to 0.002) in Fig. 11b is computed using the method discussed in Sect. 6.2. *Note*: This IT value is specific to the example case shown in Fig. 11 and is not a representation of the IT value used for the remaining experiments. Section 6.2 describes the selection of IT value in the *INDRA* framework.

6 Experiments

6.1 Experimental Setup

To evaluate the performance of the *INDRA* framework, an analysis for the selection of intrusion threshold (IT) is presented. Using the selected IT, two variants of the *INDRA* framework (*INDRA*-LED and *INDRA*-LD) are compared against the baseline *INDRA* framework. The former variant removes the end linear layer before the output and essentially has only the GRU to decode the context vector. The term LED implies the (L) linear layer, (E) encoder GRU, and (D) decoder GRU. The latter variation replaces the GRU and the linear layer at the decoder with a series of linear layers (LD implies linear decoder). These variants were studied mainly to understand the importance of different layers in the network. However, the encoder portion of the network remained unchanged in the variants as a sequence model is needed to generate an encoding of the time-series data. The study in *INDRA* explored other variants, but they are not included in the discussion as their performance was inferior compared to the LED and LD variants.

Subsequently, the best variant of the *INDRA* framework is compared with three prior works: predictor LSTM (PLSTM [25]), replicator neural network (RepNet [26]), and CANet [23]. The first comparison work (PLSTM) employs an LSTM-based network that is trained to predict the signal values in the next message transmission. PLSTM achieves this by taking the 64-bit CAN message payload as the input and learns to predict the next signal values in the message at a bit-level granularity by minimizing the prediction error. A log loss or binary cross-entropy loss function is used to measure the bit level deviations between the real next signal values and the predicted next signal values. At runtime, PLSTM uses the calculated prediction loss value to decide whether a particular message is malicious or not. The second comparison work (RepNet) uses a series of RNN layers to increase the dimensionality of the input data and reconstruct the signal values by reducing back to the original dimensionality. RepNet achieves this by minimizing the mean squared error (MSE) between the input and the reconstructed signal values. During runtime, large deviations between the input signal and the reconstructed signal values are used to detect intrusions. Lastly, the third comparison work (CANet) unifies multiple LSTMs and linear layers in an autoencoder architecture and adapts a quadratic loss function to minimize the signal reconstruction error. Details related to all experiments conducted with the *INDRA* variants and comparison works are discussed in further subsections.

To evaluate the *INDRA* framework with its variants and against prior works, an open-source dataset called SynCAN, developed by ETAS and Robert Bosch GmbH [23] is used. The dataset consists of CAN message data for ten different message IDs that were modeled based on the real-world CAN message data. The dataset comes with both training and test data with multiple attacks, as discussed in Sect. 4.3. Each row in the dataset consists of a timestamp, message ID, and individual signal values. Additionally, the test data consists of a label column with either 0 or 1 values indicating normal or malicious messages, respectively. The label information is available on a per message basis and does not indicate which signal within the message is subjected to the cyberattack. This label information is used to evaluate the *INDRA* IDS over several metrics such as detection accuracy and false-positive rate (discussed in detail in the next subsections). Moreover, to simulate a more realistic attack scenario in the in-vehicle networks, the test data has normal CAN traffic between the attack injections. *Note*: The training phase does not use any label information, as *INDRA* learns the patterns in the input data in an unsupervised manner.

All the machine learning-based frameworks including *INDRA* and its variants, and comparison works are implemented using PyTorch 1.4. Additionally, several experiments were conducted to select the best performing model hyperparameters (including number of layers, hidden unit sizes, and activation functions). The final recurrent autoencoder model presented in Sect. 5.1 was trained using the SynCAN dataset by splitting 85% of train data for training and the remaining for validation. The validation data is primarily used to evaluate the performance of the trained model at the end of every epoch. The model was trained for 500 epochs, using a rolling window approach (as discussed in Sect. 5.1.2) with the subsequence size of 20 messages and the batch size of 128. An early stopping mechanism was employed during the training phase that monitors the validation loss across epochs and stops the training process if there is no improvement after ten (patience) epochs. A learning rate of 1e-4 is chosen, and tanh activations are applied after each linear and GRU layers. Lastly, an ADAM optimizer with the mean squared error (MSE) loss criterion is used for back propagation. During testing, the trained model is evaluated using multiple test data inputs to simulate various attack scenarios. The intrusion threshold is computed based on the intrusion score metric (as described in Sect. 5.2), which was used in determining a message as malicious or normal. Various performance metrics such as detection accuracy, false positives, etc. are computed to quantify the performance of *INDRA*. All the simulations are run on an AMD Ryzen 9 3900X server with an Nvidia GeForce RTX 2080Ti GPU.

Before looking at the experimental results, the following terminologies are defined in the context of IDS:

- True positive (TP): when an IDS detects an actual malicious message as malicious.
- False negative (FN): when an IDS detects an actual malicious message as normal.
- False positive (FP): when an IDS detects a normal message as malicious (aka false alarm).
- True negative (TN): when an IDS detects an actual normal message as normal.

The INDRA framework focuses on two key performance metrics: (i) *detection accuracy*, which is the measure of an IDS ability to detect intrusions correctly, and (ii) *false-positive rate*, also known as false alarm rate. These metrics are calculated using Eqs. (4) and (5):

$$\text{Detection Accuracy} = \frac{\text{TP} + \text{TN}}{\text{TP} + \text{FN} + \text{FP} + \text{TN}} \tag{4}$$

$$\text{False Positive Rate} = \frac{\text{FP}}{\text{FP} + \text{TN}} \tag{5}$$

6.2 Intrusion Threshold Selection

A comprehensive analysis for the selection of intrusion threshold (IT) by considering various options such as max, median, mean, and different quantile bins of validation loss of the final model is presented in this subsection. The reconstruction error of the model for the normal messages should be much smaller than the error for malicious messages. Hence, several candidate options for the IT are explored to achieve this goal that would work across multiple attack and no-attack scenarios. In some scenarios, having a large IT value can make it harder for the model to detect the attacks that change the input pattern minimally (e.g., continuous attack). In contrast, having a small threshold value can potentially trigger multiple false alarms, which is highly undesirable in time-critical systems. Thus, it is crucial to select an appropriate IT value to optimize the performance of the model.

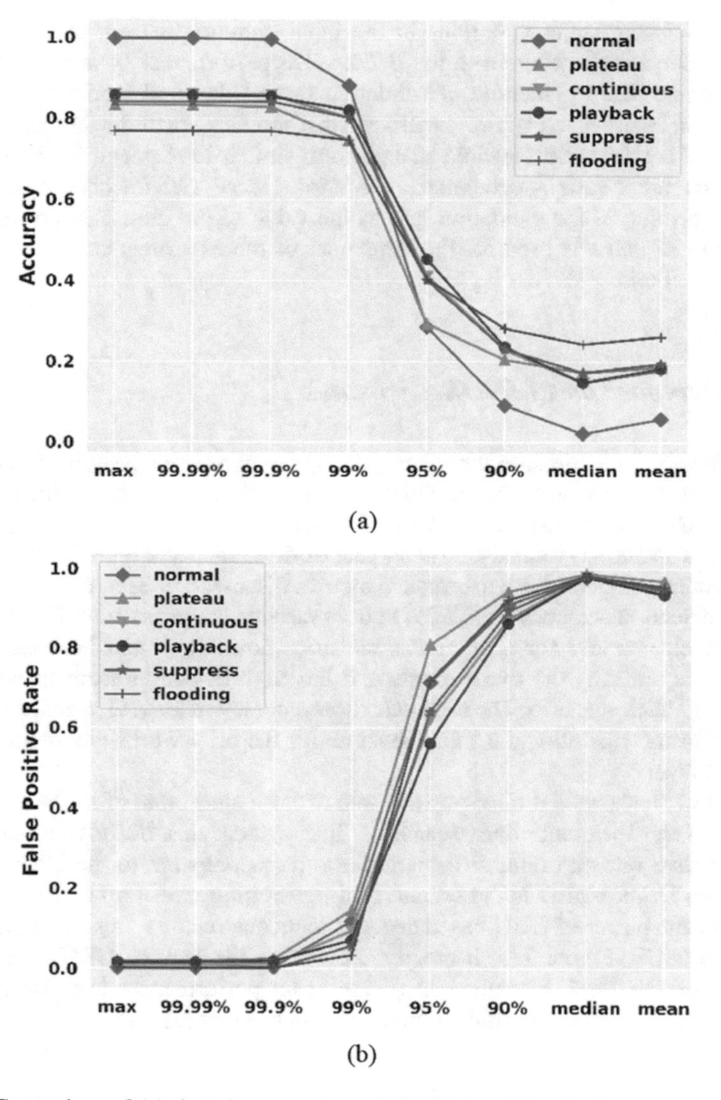

Fig. 12 Comparison of (**a**) detection accuracy and (**b**) false-positive rate for various candidate options of intrusion threshold (IT) as a function of validation loss under different attack scenarios. (% refers to percentile not percentage) [33]

Figure 12a, b illustrates the detection accuracy and false-positive rate, respectively, for various candidate options to calculate IT, under different attack scenarios. From the results in Fig. 12a, b, it can be seen that selecting higher validation loss as the IT can result in a high detection accuracy and low false alarm rate. However, choosing a very high value (such as “max” or “99.99 percentile”) can sometimes result in missing small variations in the input patterns that are caused by more sophisticated attacks. Moreover, the *INDRA* IDS performance is very similar when maximum or 99.99 percentile of validation loss of the final model is selected as the IT. But, in order to capture the attacks that produce small deviations, a slightly smaller IT is selected that would still perform similar to max and 99.99 percentile thresholds under various cyberattack scenarios. Hence, *INDRA* chooses the 99.9th percentile value of the validation loss as the value of the intrusion threshold (IT). The same IT value is used for the remainder of the experiments discussed in the next subsections.

6.3 Comparison of INDRA Variants

After selecting the intrusion threshold using the methodology presented in previous subsection, the performance of *INDRA* framework is evaluated with two other variants: *INDRA*-LED and *INDRA*-LD. The motivation behind evaluating different variants of *INDRA* is to analyze the impact of different layer types in the recurrent autoencoder model on the performance metrics discussed in Sect. 6.1.

The detection accuracy of *INDRA* and its variants is illustrated in Fig. 13a under different attacks and for a no-attack scenario (normal). It can be observed that *INDRA* outperforms the two variants and has high detection accuracy in normal and every attack scenario. The high detection accuracy of *INDRA* is achieved due to its monitoring capability at a signal level unlike the prior works that monitor at the message level.

Figure 13b shows the false-positive rate or false alarm rate of *INDRA* and other variants under different attack scenarios. It is evident that *INDRA* has the lowest false-positive rate and highest detection accuracy compared to the other variants. Moreover, *INDRA*-LED is the second best-performing model after *INDRA*, which leverages the power of GRU-based decoder to reconstruct the original signal values from the MCV. Figure 13a, b clearly shows that the lack of GRU layers in the decoder of *INDRA*-LD resulted in a significant performance degradation. Thus, *INDRA* is chosen as the candidate model for subsequent experiments.

6.4 Comparison with Prior Works

In this subsection, a comparison of the *INDRA* framework with PLSTM [25], RepNet [26], and CANet [23], which are some of the best known prior works in the IDS area, is presented. Figure 14a, b shows the detection accuracy and false-positive rate, respectively, for the various techniques under different attack scenarios.

From Fig. 14a, b, it is evident that *INDRA* achieves a high detection accuracy for each attack scenario and also has low positive rates for most scenarios. The ability to monitor signal level variations along with the more cautious selection of intrusion

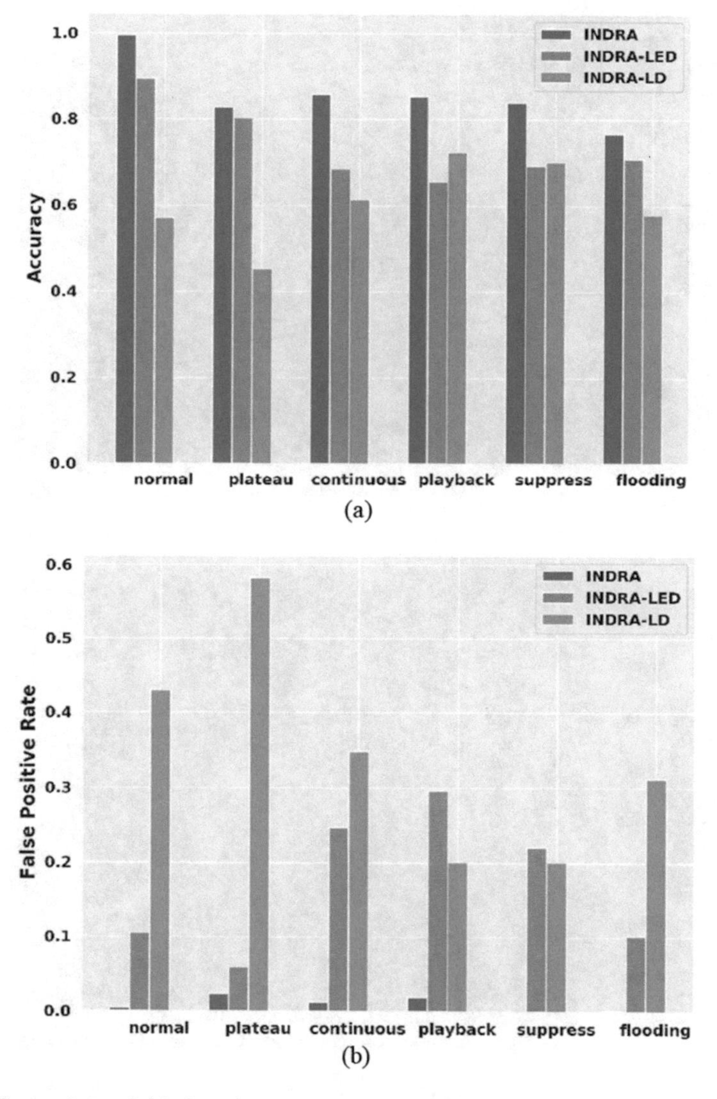

Fig. 13 Comparison of (**a**) detection accuracy and (**b**) false-positive rate for *INDRA* and its variants *INDRA*-LED and *INDRA*-LD under different attack scenarios [33]

threshold gives *INDRA* an advantage over comparison works. Both PLSTM and RepNet use the maximum validation loss in the final model as the threshold to detect intrusions in the system, while CANet uses an interval-based monitoring to detect cyberattacks. The larger threshold value helped PLSTM to achieve slightly lower false-positive rates for few scenarios, but it hurt the ability of both PLSTM and

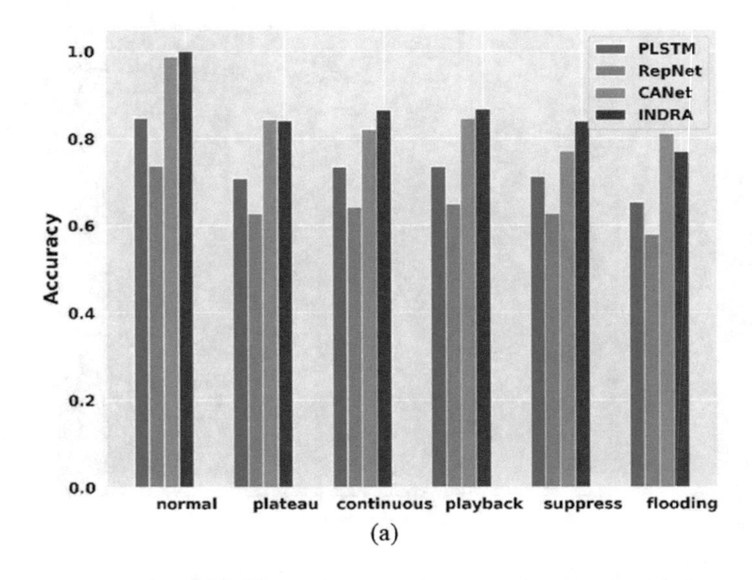

(a)

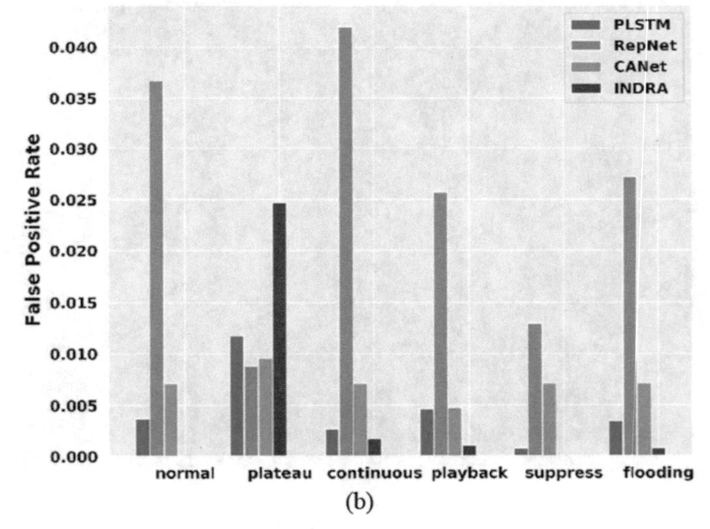

(b)

Fig. 14 Comparison of (**a**) detection accuracy and (**b**) false-positive rate of *INDRA* [33] and the prior works PLSTM [25], RepNet [26], and CANet [23]

RepNet to detect cyberattacks that produce small variations in the input data. This is because the deviations produced by some of the complex attacks are small and the attacks go undetected due to the large thresholds. Moreover, the interval-based monitoring in CANet struggles with finding an optimal threshold value. Lastly, the false-positive rates of *INDRA* are still significantly low with the maximum of 2.5%

Table 2 Memory footprint comparison between *INDRA* framework and the prior works PLSTM [25], REPNET [26], and CANET [23]

Framework	Memory footprint (KB)
PLSTM [25]	13,417
RepNet [26]	55
CANet [23]	8718
INDRA	443

Note: Data in this table is adapted from [33]

for plateau attacks. It is important to note that the y-axis in Fig. 14b has a much smaller scale than in Fig. 14a and the magnitude of the false positive rate is very small.

6.5 IDS Overhead Analysis

In this subsection, a detailed analysis of the *INDRA* IDS overhead is presented. The overhead is quantified in terms of both memory footprint and time taken to process an incoming message, i.e., inference time. The former metric is important as the resource-constrained automotive ECUs have limited available memory, and it is crucial to have a low memory overhead to avoid interference with real-time automotive applications. The inference time not only provides important information about the time taken to detect the attacks but also can be used to compute the utilization overhead on the ECU. Thus, the abovementioned two metrics are used to analyze the overhead and quantify the lightweight nature of *INDRA* IDS.

To accurately capture the overhead of the *INDRA* framework and the prior works, they are implemented on an ARM Cortex-A57 CPU on a Jetson TX2 board, which has similar specifications to the state-of-the-art multi-core ECUs. The memory footprint of the *INDRA* framework and the comparison works mentioned in the previous subsections are shown in Table 2. It is clear that the *INDRA* framework has a low memory footprint compared to the comparison works, except for the RepNet [26]. However, it is important to observe that even though the *INDRA* framework has slightly higher memory footprint compared to RepNet [26], *INDRA* outperforms all prior works including RepNet [26] in every performance metric under different cyberattack scenarios, as shown in Fig. 14. Even though the heavier (high memory footprint) models can provide the ability to capture a large variety of details about the system behavior, they are not an ideal choice for resource-constrained automotive embedded systems. On the other hand, a much lighter model such as RepNet cannot capture crucial details about the system behavior due to limited parameters and therefore suffers from performance issues.

In order to understand the inference overhead, different IDS frameworks are benchmarked on an ARM Cortex-A57 CPU. In this experiment, different system configurations are considered to encompass a wide variety of state-of-the-art ECU

Table 3 Inference time comparisons between *INDRA* framework and the prior works PLSTM [25], REPNET [26], and CANET [23] using single-core and dual-core configurations

Framework	Average inference time (μs)	
	Single-core ARM Cortex A57 CPU	Dual-core ARM Cortex A57 CPU
PLSTM [25]	681.18	644.76
RepNet [26]	19.46	21.46
CANet [23]	395.63	378.72
INDRA	80.35	72.91

Note: Data in this table is adapted from [33]

hardware in vehicles. Based on the available hardware resources on the Jetson TX2, two different system configurations are selected. The first configuration utilizes only one CPU core (single core), while the second configuration uses two CPU cores (dual core).

Each framework is run ten times for two different CPU configurations, and the average inference time (in μs) is computed, as shown in Table 3. From the results in Table 3, it can be seen that *INDRA* has significantly faster inference times compared to the prior works (excluding RepNet) under all configurations. This is partly associated with the lower memory footprint of the *INDRA* IDS. As mentioned earlier, even though RepNet has a lower inference time, it has the worst performance out of all frameworks, as shown in Fig. 14. The large inference times for the better performing frameworks can impact the real-time performance of the control systems in the vehicle and can result in missing of critical deadlines, which can be catastrophic. These inference times can be further improved by employing a dedicated deep learning accelerator (DLA) compared to the above presented configurations.

Thus, from Fig. 14 and Tables 2 and 3, it is evident that *INDRA* achieves a clear balance of having superior intrusion detection performance while maintaining low memory footprint and fast inference times, making it a powerful and lightweight IDS solution.

6.6 Scalability Results

In this subsection, a scalability analysis of the *INDRA* IDS is presented by studying the system performance using the ECU utilization metric as a function of increasing system complexity (number of ECUs and messages).

Each ECU has a real-time utilization (U_{RT}) and an IDS utilization (U_{IDS}) from running real-time and IDS applications, respectively. The IDS overhead (U_{IDS}) is analyzed as a measure of the compute efficiency of the IDS. Since the safety-critical messages monitored by the IDS are periodic in nature, the intrusion detection task can be modeled as a periodic application with a period that is same as the message period [32]. Thus, monitoring an ith message m_i results in an induced IDS utilization

(U_{IDS,m_i}) at an ECU and can be computed as:

$$U_{\mathrm{IDS},m_i} = \left(\frac{T_{\mathrm{IDS}}}{P_{m_i}}\right) \tag{6}$$

where T_{IDS} and P_{mi} indicate the time taken by the IDS to process one message (inference time) and the period of the monitored message, respectively. Moreover, the sum of all IDS utilizations as a result of monitoring different messages is the overall IDS utilization at that ECU (U_{IDS}) and is given by:

$$U_{\mathrm{IDS}} = \sum_{i=1}^{n} U_{\mathrm{IDS},m_i} \tag{7}$$

To evaluate the scalability of *INDRA*, six different system sizes are studied. Moreover, a pool of commonly used message periods {1, 5, 10, 15, 20, 25, 30, 45, 50, 100} (all periods in ms) in automotive systems are uniformly assigned to various messages in the system. These messages are evenly distributed among different ECUs in each system configuration and the IDS utilization is computed using Eqs. (6) and (7). To analyze the worst case overhead, a pessimistic scenario consisting of only a single core per each ECU in the system is considered in this experiment.

The average ECU utilization under various system sizes is illustrated in Fig. 15. The system size is denoted by $\{p, q\}$, where p is the number of ECUs and q is the number of messages in the system. Additionally, a very pessimistic estimate of 50% real-time ECU utilization for real-time automotive applications is assumed ("RT Util," as shown in the dotted bars) for all system configurations. The overhead incurred by the IDS executing on the ECUs is represented by the solid bars on top of the dotted bars, and the red horizontal dotted line represents the 100% ECU utilization mark. It is important to avoid exceeding the 100% ECU utilization under any scenario, as it could induce undesired latencies that could result in missing deadlines for time-critical automotive applications, which can be catastrophic. From the results in Fig. 15, it is evident that the prior works PLSTM and CANet incur heavy overhead on the ECUs while RepNet and *INDRA* have a very minimal overhead that scales favorably with increasing system sizes. Thus, from the results in this section (Figs. 14 and 15; Tables 2 and 3), it is apparent that not only does *INDRA* achieve superior performance in terms of both detection accuracy and low false-positive rate for intrusion detection than state-of-the-art prior works, but it is also lightweight and scalable.

7 Conclusion

In this chapter, a novel recurrent autoencoder-based lightweight intrusion detection system framework called *INDRA* for distributed automotive embedded systems was presented. The *INDRA* framework uses the proposed metric called the intrusion score (IS) to quantify the deviation of the prediction signal from the actual input signal. Moreover, a thorough analysis of the intrusion threshold selection process

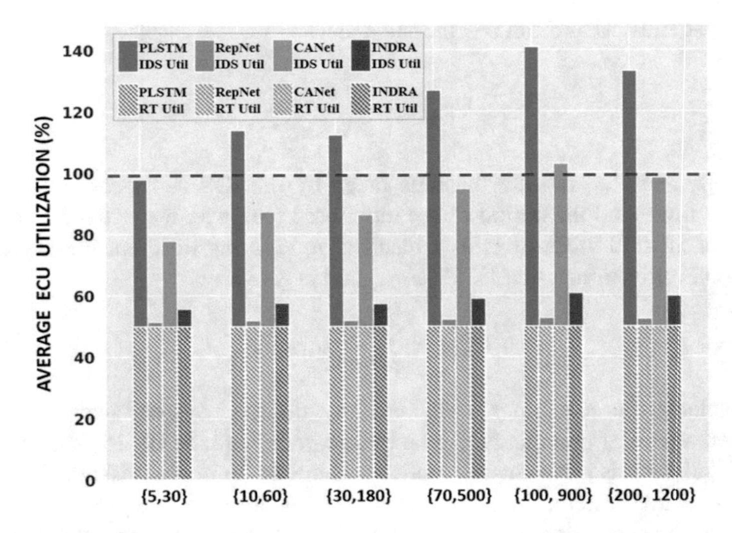

Fig. 15 Scalability results of the *INDRA* [33] IDS for different system sizes compared to the prior works PLSTM [25], RepNet [26], and CANet [23]

and the comparison of *INDRA* with the best known prior works in this area is presented in this chapter. The promising results of *INDRA* indicate a compelling potential for being adapted to enhance cybersecurity in emerging automotive platforms. Our ongoing work is exploring newer and more powerful algorithms [34–36] for intrusion detection in automotive embedded systems.

Acknowledgments This research is supported by a grant from NSF (CNS-2132385).

References

1. Kukkala, V., Tunnell, J., Pasricha, S.: Advanced driver assistance systems: a path toward autonomous vehicles. IEEE Consum. Electron. **7**(5), 18–25 (2018)
2. Koscher, K., et al.: Experimental security analysis of a modern automobile. In: IEEE SP (2010)
3. Valasek, C., et al.: Remote exploitation of an unaltered passenger vehicle. https://ioactive.com/pdfs/IOActive_Remote_Car_Hacking.pdf (2015)
4. Izosimov, V., et al.: Security-aware development of cyber-physical systems illustrated with automotive case study. In: IEEE/ACM DATE (2016)
5. Studnia, I., et al.: A language-based intrusion detection approach for automotive embedded network. In: IEEE PRISDC (2015)
6. Marchetti, M., et al.: Anomaly detection of CAN bus messages through analysis of ID sequences. In: IEEE IV (2017)
7. Hoppe, T., et al.: Security threats to automotive CAN networks- practical examples and selected short-term countermeasures. In: RESS (2011)
8. Larson, U.E., et al.: An approach to specification-based attack detection for in-vehicle networks. In: IEEE IV (2008)

9. Aldwairi, M., et al.: Pattern matching of signature-based IDS using Myers algorithm under MapReduce framework. In: EURASIP (2017)
10. Myers, E.W.: An O(ND) difference algorithm and its variations. In: Algorithmica (1986)
11. Hoppe, T., et al.: Applying intrusion detection to automotive IT-early insights and remaining challenges. In: JIAS (2009)
12. Waszecki, P. et al.: Automotive electrical and electronic architecture security via distributed in-vehicle traffic monitoring. In: IEEE TCAD (2017)
13. Cho, K.T., et al.: Fingerprinting electronic control units for vehicle intrusion detection. In: USENIX (2016)
14. Ying, X., et al.: Shape of the Cloak: formal analysis of clock skew-based intrusion detection system in controller area networks. In: IEEE TIFS (2019)
15. Yoon, M.K., et al.: Memory heat map: anomaly detection in real-time embedded systems using memory behavior. In: IEEE/ACM/EDAC DAC (2015)
16. Müter, M., et al.: Entropy-based anomaly detection for in-vehicle networks. In: IEEE IV (2011)
17. Müter, M., et al.: A structured approach to anomaly detection for in-vehicle networks. In: ICIAS (2010)
18. Taylor, A., Japkowicz, N., Leblanc, S.: Frequency-based anomaly detection for the automotive CAN bus. In: Proceedings of WCICSS (2015)
19. Martinelli, F., et al.: Car hacking identification through fuzzy logic algorithms. In: FUZZ-IEEE (2017)
20. Vuong, T.P., et al.: Performance evaluation of cyber-physical intrusion detection on a robotic vehicle. In: IEEE CIT/IUCC/DASC/PICOM (2015)
21. Levi, M., Allouche, Y., et al.: Advanced analytics for connected cars cyber security [Online]. Available: https://arxiv.org/abs/1711.01939 (2017)
22. Kang, M.-J., et al.: A novel intrusion detection method using deep neural network for in-vehicle network security. In: IEEE, VTC Spring (2016)
23. Hanselmann, M., et al.: CANet: an unsupervised intrusion detection system for high dimensional CAN bus data: In: IEEE Access (2020)
24. Loukas, G., et al.: Cloud-based cyber-physical intrusion detection for vehicles using deep learning. In: IEEE Access (2018)
25. Taylor, A., et al.: Anomaly detection in automobile control network data with long short-term memory networks. In: IEEE DSAA (2016)
26. Weber, M., et al.: Online detection of anomalies in vehicle signals using replicator neural networks. In: ESCAR (2018)
27. Weber, M., et al.: Embedded hybrid anomaly detection for automotive can communication. In: Embedded Real Time Software and Systems (2018) [Online]. Available: https://hal.archives-ouvertes.fr/hal-01716805
28. Schmidhuber, J.: Habilitation thesis: system modeling and optimization (1993)
29. Hochreiter, S., et al.: Gradient flow in recurrent nets: the difficulty of learning long-term dependencies. In: IEEE Press (2001)
30. Cho, K., et al.: Learning phrase representations using RNN encoder-decoder for statistical machine translation. In: EMNLP (2014)
31. DiDomenico, G., et al.: Colorado State University EcoCAR 3 final technical report. In: SAE, WCX (2019)
32. Kukkala, V., Pasricha, S., Bradley, T.H.: SEDAN: Security-aware design of time-critical automotive networks. IEEE Trans. Veh. Technol. (TVT) **69**(8), 9017–9030 (2020, August)
33. Kukkala, V.K., Thiruloga, S.V., Pasricha, S.: INDRA: intrusion detection using recurrent autoencoders in automotive embedded systems. IEEE Trans. Comput-Aid. Des. Integr. Circuits Syst. (TCAD) **39**(11) (2020, November)
34. Thiruloga, S.V., Kukkala, V.K., Pasricha, S.: TENET: temporal CNN with attention for anomaly detection in automotive cyber-physical systems. In: IEEE/ACM Asia & South Pacific Design Automation Conference (ASPDAC) (2022, January)

35. Kukkala, V.K, Thiruloga, S.V., Pasricha, S.: LATTE: LSTM self-attention based anomaly detection in embedded automotive platforms. In: ACM Transactions on Embedded Computing Systems (TECS) (2021)
36. Kukkala, V.K., Thiruloga, S.V., Pasricha, S.: Roadmap for cybersecurity in autonomous vehicles. In: IEEE Consumer Electronics (2022)

Ultralow-Power Implementation of Neural Networks Using Inverter-Based Memristive Crossbars

Shaghayegh Vahdat, Mehdi Kamal, Ali Afzali-Kusha, and Massoud Pedram

1 Introduction

Data-intensive application domains of neural networks (NNs), such as image classification, have highlighted the inefficiency of traditional Von Neumann-based computing architectures due to their memory wall bottleneck [1]. Furthermore, performing matrix-vector multiplication (MVM) operations, as the main computation part of NNs, using digital platforms such as CPUs (GPUs) may not enjoy fully from efficient computation parallelism (suffering from high-energy consumption) [2]. To overcome the memory wall bottleneck, one may perform, for example, computations in memory [3] or maximize data reuse [4]. In the case of the MVM operations, approximate computing paradigm may be used as another approach for reducing the energy consumption of the required computations [5, 6].

Analog computing is an emerging approach utilized for the implementation of NNs offering significant reductions in the energy consumption as well as large improvements in the computation speed compared to those of the digital ones. As an example, the fabricated memristive-based NN of reference [7] led to 30× and 110× better speed and energy consumption, respectively, compared to those of the GPU platform for the MNIST benchmark [8]. Switched-capacitor matrix multipliers [9], multiple input floating gate MOS (MIFGMOS) transistors [10], and memristive crossbars [11] are examples of elements utilized in the analog implementation of

S. Vahdat · M. Kamal (✉) · A. Afzali-Kusha
School of Electrical and Computer Engineering, College of Engineering, University of Tehran, Tehran, Iran
e-mail: vahdat_s@ut.ac.ir; mehdikamal@ut.ac.ir; afzali@ut.ac.ir

M. Pedram
Department of Electrical and Computer Engineering, University of Southern California, Los Angeles, CA, USA
e-mail: pedram@usc.edu

A. Iranmanesh (ed.), *Frontiers of Quality Electronic Design (QED)*,
https://doi.org/10.1007/978-3-031-16344-9_9

NNs. These analog implementations, however, suffer from a number of limitations. More specifically, when utilizing MIFGMOS transistors, the number of inputs of neurons is limited by the number of the floating gates (e.g., nine in [10]). In the structure of a switched-capacitor matrix multiplier [9], inputs should be applied to the multiplier serially which degrades its speed efficacy. In the memristive crossbars, weights of the neurons are determined based on the resistance of the memristors. It is, however, possible that the resistance of the actual memristor differs from the design value, which can have a destructive effect on the output of the MVM operation. Compared to the other analog realizations of NNs, memristive structures can perform MVM operations with a higher speed since the MVM operation of each layer is performed in parallel using memristors. In addition, they can be fabricated more densely which leads to a lower area occupation [26]. In this chapter, we deal with the hardware implementation of memristive NNs.

Operational amplifiers (Op-Amps) and current buffers [11–16] on the one hand and inverters [17–25] on the other hand are two available choices for hardware implementation of memristive NNs. Between these two options, employing inverters in the structure of the neurons results in a considerable reduction in the power consumption compared to the Op-Amp option. As another advantage, although the required activation function of the neurons (i.e., tangent hyperbolic function) can be easily generated using the VTC of the inverters, generating the same function (the tangent hyperbolic or sigmoid) requires additional circuit components in other memristive structures (see, e.g., [14–16, 27]). Efficient implementation of the activation functions further improves the energy efficacy of the NNs providing the ability to realize ultralow-power implementations of NNs using inverter-based memristive (*IM*) neurons. In other words, employing inverters and memristors in the structure of artificial neurons leads to an ultralow-power implementation of NNs compared to the digital ASIC counterparts (i.e., 5000× smaller power consumption [11]) or Op-Amp-based implementations (i.e., 800× smaller power consumption [11]). The implementation is highly desirable for applications where low-power consumption and high speed are the key design drivers (i.e., internet of things (IoT) smart sensors [19]).

To connect the *IM*-NNs to the digital realm, input and output interfaces are required to produce digital inputs for the NN from the analog input signals and digital signals from the analog outputs of the NN. The existing analog-to-digital and digital-to-analog converters (ADCs and DACs) consume considerably larger power compared to the memristive crossbars. As an example, the employed DACs (ADCs) in the RENO structure [28], which is a memristor-based neuromorphic accelerator, consume 7200× (5300×) higher power compared to those of the memristor crossbar. To fully benefit from the low-power consumption of memristive implementations of NNs, especially, the *IM* implementations, it is important to employ low power ADCs and DACs.

Regarding the modeling of these networks, it should be noted that the mathematical equations describing the functionality of an *IM* neuron are very different from those of the other memristive implementations in which Op-Amps or current buffers are utilized. This implies that the design (proper network training) and analysis of

IM-NNs require the use of mathematical model with high accuracy. Finally, it should be noted that the design of *IM*-NNs faces some challenges which if not tackled properly, the network would suffer from non-negligible accuracy degradation. The sources of the degradations include the non-idealities of memristors, the non-idealities of the transistors, and the loading effect of the memristive crossbars causing inverter/buffer VTC (voltage transfer characteristic) variations. They cause the output voltage of the *IM*-NNs become different than the expected value, degrading the accuracy of the NNs. To take full advantage of the *IM*-NN benefits, the impact of these undesired yet real-life non-idealities should be minimized. Additionally, the loading effect of the memristive crossbars should be considered in the training phase of the NNs.

Based on the above explanations, the topics covered in this chapter are outlined in the following:

- Prior work about implementation and reliability enhancement of memristive NNs
- Mathematical model of *IM* neurons and state-of-the-art training methods for *IM*-NNs
- Low-power memristive DACs and ADCs which can be employed as the input and output interfaces of *IM*-NNs
- Variation mitigation approaches for reducing the impacts of non-idealities (i.e., memristors conductance variations as well as inverters VTC variations) on the accuracy of the *IM*-NNs
- Analysis of the strengths and weaknesses of the considered training methods and the variation mitigation approaches
- Discussion of the open problems in the utilization of *IM*-NNs

The rest of the chapter includes a review of the general structure of artificial NNs (ANNs), their implementation using memristive crossbars, and their reliability enhancement (Sect. 2). Next, the mathematical model of *IM*-NNs as well as the effects of circuit elements non-idealities on the NN outputs are presented (Sect. 3). After this part, the input and output interfaces for connecting *IM*-NNs to the digital realm are described (Sect. 4). In addition, different training methods as well as variation mitigation approaches for *IM*-NNs are explained as the subsequent topics (Sects. 5 and 6) where their efficacies are dealt with afterward (Sect. 7).

2 Literature Review

2.1 Memristor

Due to the emergence of nonvolatile memories such as memristors, new implementations of NNs have been proposed. The memristor was theoretically proposed in 1971 [29] and fabricated in 2008 [30]. Resistive random-access memories (RRAMs) or memristors consist of a metal oxide layer sandwiched between two metal

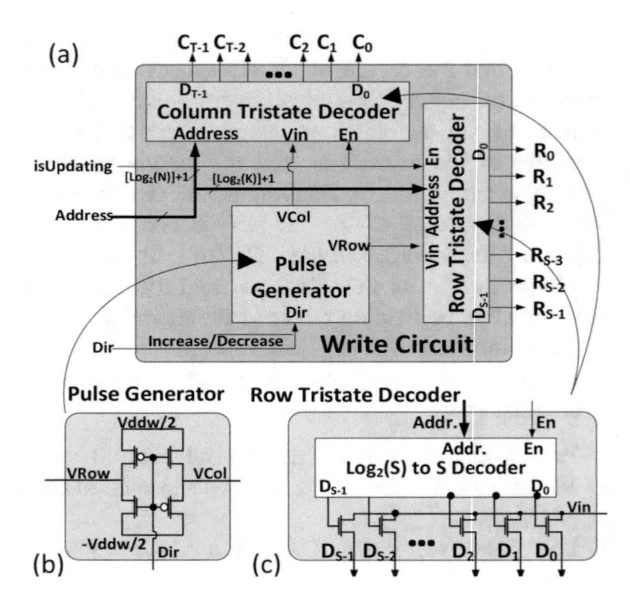

Fig. 1 The structure of the (**a**) write circuit, (**b**) pulse generator, and (**c**) tristate decoder [21]

electrodes called top and bottom electrodes [31]. By applying voltage or current pulses to the memristor, the oxygen ions move in the lattice, and oxygen vacancies appear which lead to the generation of conductive filaments (CFs) [32]. The applied voltage or current pulses may result in the formation or disruption of conductive filaments which give rise to the memristor conductance increase or decrease.

Programming the memristors can be performed using open-loop or closed-loop processes. In the open-loop process, the characteristic of the memristor is mathematically modeled, and based on the desired conductance change, the number of voltage pulses is estimated. In the closed-loop process, voltage pulses are applied to the memristor, and the resistance of the memristor is measured. This procedure is repeated until achieving the desired resistance value with an acceptable error. Although the programming accuracy of closed-loop approach is higher than the open-loop one, its power consumption is also higher. In addition, the lifetime of the memristors is degraded due to more writings.

As an example, the hardware implementation of an open-loop programming process [21], which consists of row/column tristate decoder and pulse generator units, is depicted in Fig. 1a. The circuit implementation of the pulse generator unit having one input (two outputs) called Dir (VRow and VCol) is shown in Fig. 1b. Based on the value of Dir signal, the difference between VRow and VCol which is applied across the considered memristor becomes almost Vddw or –Vddw increasing or decreasing the memristor conductance. Here, Vddw should be larger than the write threshold voltage of the memristors.

The circuit implementation of the row/column tristate decoder unit, which exerts the write voltage across the corresponding memristor, is depicted in Fig. 1c. This unit consists of an address decoder which converters the row/column address bits to the controlling signals of the pass transistors. Based on the address bits, one of the pass transistors connected to the corresponding row or column becomes activated and passes VRow or VCol voltage to the considered row or column.

2.2 *Memristive Neuron Circuit*

A feed forward ANN is composed of some neurons in several layers. In each layer, the sum of weighted inputs of each neuron is added by the bias to which the activation function is applied to generate the neuron output. Let us assume an ANN with L layers where the number of inputs of the lth layer is N^l and x_i^l (y_j^l) represents the *ith* input (*jth* output) of the lth layer. Therefore, y_j^l can be obtained from

$$y_j^l = f\left(net_j^l\right) = f\left(\sum_{i=1}^{N^l} w_{j,i}^l \times x_i^l + wb_j^l\right), \tag{1}$$

where the weights (biases) are denoted by $w_{j,i}^l$ (wb_j^l) and f represents the activation function. In addition, $x_j^l = y_j^{l-1}$ because the outputs of the $(l-1)^{st}$ layer are the inputs of the lth layer. To clarify the notations better, the structure of a fully connected two-layer ANN is depicted in Fig. 2a. In this figure, n, p, and q denote the number of inputs, hidden neurons, and outputs, respectively.

The nonvolatility, high access speed, low power consumption, and high layout density make memristors a very good candidate for realizing analog NNs [26]. More specifically, the MVM operation of NNs can be implemented using a memristor crossbar. As an example, a current-based memristive crossbar composed of memristors and Op-Amps is depicted in Fig. 2b. In this structure, the negative input terminal of each Op-Amp becomes a virtual ground, and its input current is almost zero. This means that the current that passes through each memristor (i.e., $R_{j,i}^+$) is equal to the weighted input (i.e., $x_i^+ \times \sigma_{j,i}^+$ where $\sigma_{j,i}^+ = 1/R_{j,i}^+$) which should finally pass through R_1. Consequently, the negative output voltage of the neuron can be obtained from

$$y_j^- = f_1\left(-R_1 \times \sum_{i=1}^{n}\left(\frac{x_i{}^+}{R_{j,i}{}^+} + \frac{x_i{}^-}{R_{j,i}{}^-}\right)\right), \tag{2}$$

where

$$f_1(Z) = \begin{cases} V_{DD} & \text{if} \quad Z > V_{DD} \\ Z & \text{if} \quad |Z| \le V_{DD} \\ -V_{DD} & \text{if} \quad Z < -V_{DD} \end{cases}, \tag{3}$$

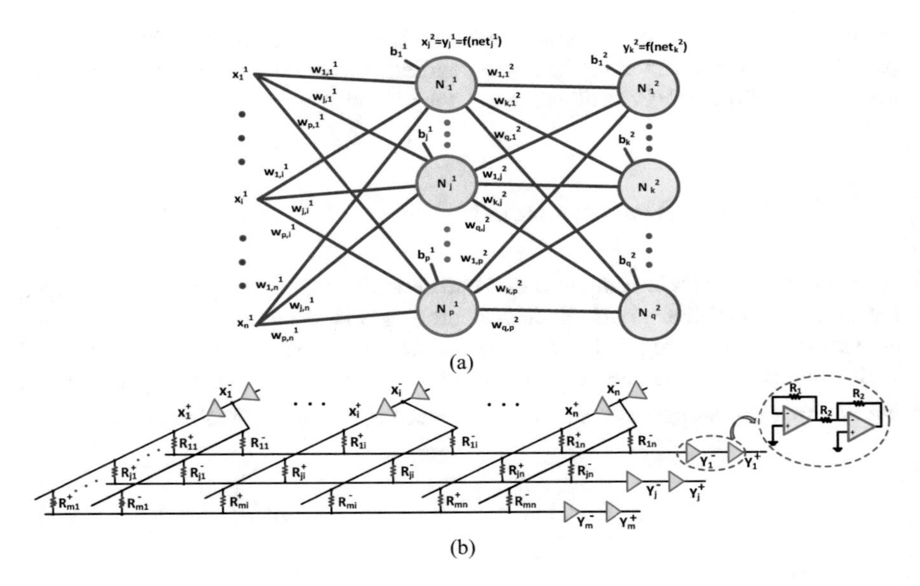

Fig. 2 (**a**) A general two-layer ANN. (**b**) Implementation of MVM operation using current-based memristor crossbars [33]

and V_{DD} ($-V_{DD}$) represents the positive (negative) voltage sources of the Op-Amp. Furthermore, the positive output voltage can be calculated as $y_j^+ = -y_j^-$. Therefore, the output voltage of the neuron is equal to the sum of weighted inputs which is passed through a symmetric saturating linear transfer function. Furthermore, other activation functions such as sigmoid, tangent hyperbolic, and rectified linear unit (ReLU) may be implemented using dedicated hardware such as the ones presented in [16, 27, 34]. To implement tangent hyperbolic activation functions without requiring Op-Amps in the structure of the neuron, one may employ inverters which lead to a voltage-based neuron design. In this case, the VTC of the inverter implements the required activation function where the input voltage of the inverter is equal to the sum of weighted inputs. The details of the hardware implementation, mathematical model, and training procedure of *IM*-NNs will be explained in the rest of this chapter.

2.3 Memristor Non-idealities

After introducing the hardware implementation of memristive NNs, it is required to study the effect of memristors non-idealities on the accuracy of the NN. More specifically, since the actual conductance values of the memristors may differ from the desired values, considerable degradation in the NN accuracy may occur. As examples, the results of [34, 35] revealed that the accuracy of two-layer current-based memristive NNs for the MNIST dataset may degrade from 98.3% and 96% in the ideal condition to about 39.7% and 56% when considering 30% variations for the conductance of the memristors.

To deal with the non-ideality issues of memristors, in the first step, it is required to study the sources of non-idealities. First, note that the reliability issues of memristors may be divided into static and dynamic categories [31]. Static issues may be caused by defects (fabrication yield) or process variations [31]. As examples, the resistance of a memristor may stick to a minimum or maximum value which is called stuck at faults (SAFs). Based on the reported values in [31], it may occur for 16% of the memristors in a crossbar fabricated with a yield of 84%. Also the thickness of a memristor may differ from its desired value which may cause larger resistance variations on the lower resistance levels of the memristors [36] (e.g., 6% variations for the lowest resistance state versus 0.1% variations for the highest resistance state [39]). To handle SAFs, one may train the network assuming constant values for the weights mapped on the defected cells [12] or make use of redundant memristors to implement important weights having strong impacts on the NN accuracy [37]. The improvement of the fabrication yield (i.e., 99.8% device yield of [38]) would considerably reduce the concern on SAFs. To mitigate thickness variation issues, one may parallelize memristors with high resistances to implement lower resistance values [36] or retrain the network and adjust higher resistance values to compensate the variation effects of lower resistance memristors (after fabricating the device) [39].

When it comes to dynamic reliability issues of memristive crossbars, examples include the conductance drift of the memristors [13], endurance limitations [40], and the IR-drop [41]. As a specific example for the conductance drift, the work of [43] reports a 2% variation on the memristor resistance value compared to its initial state resistance caused by a voltage of 0.1 V applied to the memristor for 1 s. To overcome the conductance drift issues, one may employ error correction approaches [42] or encode (decode) the weights (MVM outputs) where the error due to the memristor non-idealities can be mitigated [43]. Writing memristors many times would degrade the lifetime and endurance of memristors. To manage this issue, one may only update the weights of the neurons which have higher impacts on the outputs of the NN [40]. This approach has the advantage of smaller number of writings on the memristors which increases the lifetime of the memristors while also reducing the power consumption of the weight updating phase. As the next issue, voltage drops caused by the resistance of the wires may have impact on the voltage of the crossbar nodes during the inference phase or may affect the resistance of the programmed memristors during the weight-updating phase owing to not receiving the desired programming voltage. As an example, a 0.2 V voltage drop in the programming phase caused by the wire resistance can change the resistance of the programmed memristor from 900 $K\Omega$ (the desired value) to 200 $K\Omega$ (the actual value) [41]. To overcome this issue, one may consider the voltage drop of the wires during the weight update and training phases [41]. Furthermore, programming procedure of the memristors may also be erroneous (both of open-loop and closed-loop programming approaches) with respect to the desired conductance value. The results reveal that the writing error induced by the programming process follows Gaussian distributions which can be reduced by increasing the number of writing iterations at the price of increased power consumption [38].

3 Mathematical Analysis

In this section, first, the output voltage of an *IM* neuron based on the circuit characteristics is mathematically modeled. Next, the effect of circuit elements non-idealities on the output voltage of an *IM* neuron is mathematically modeled. Finally, the presented mathematical analysis is used to determine the factors which have the highest impact on the output voltage of the neuron. This helps us to improve the resiliency of *IM* neurons with respect to the non-idealities of circuit elements.

3.1 Circuit Model of an IM Neuron

In *IM* neurons, the multiplication and addition operations can be implemented using memristive crossbars, and the activation functions can be generated using inverters. More specifically, the weights and biases of the neurons can be mapped to the conductance values of the memristors, while the VTCs of inverters generate the activation functions. In the memristive-based neurons, the positive and negative weights should be implemented using memristors noting that the conductance of the memristors is always positive. To have the capability of implementing positive and negative weights in *IM* neurons, the authors of [11] proposed to apply the inputs of each neuron in differential form (see Fig. 3a). As shown in Fig. 3a, the neuron has two outputs, called negative and positive outputs, denoted by y_j^- and y_j^+, which are applied to the neurons of the next layer as their negative and positive inputs. In this neuron, the net_j voltage is equal to the weighted sum of the inputs, while the VTC of the inverter and buffer (two consecutive inverters) generate the activation functions for the negative and positive outputs, respectively. To further clarify it, one may find the value of net_j by writing the Kirchhoff's current law (KCL) equation for the input node of the negative inverter (the left-side inverter of Fig. 3a) as

$$net_j\left(\sum_{i=1}^{n}\left(\frac{1}{R_{j,i}{}^+}+\frac{1}{R_{j,i}{}^-}\right)+\frac{1}{Rb_j{}^+}+\frac{1}{Rb_j{}^-}\right)=\sum_{i=1}^{n}\left(\frac{x_i{}^+}{R_{j,i}{}^+}+\frac{x_i{}^-}{R_{j,i}{}^-}\right)+\frac{V_{DD}}{Rb_j{}^+}-\frac{V_{DD}}{Rb_j{}^-}, \tag{4}$$

which can be rewritten as

$$net_j=\sum_{i=1}^{n}\left(x_i{}^+\times\frac{\sigma_{j,i}{}^+}{\gamma_j}+x_i{}^-\times\frac{\sigma_{j,i}{}^-}{\gamma_j}\right)+V_{DD}\times\left(\frac{\sigma b_j{}^+-\sigma b_j{}^-}{\gamma_j}\right), \tag{5}$$

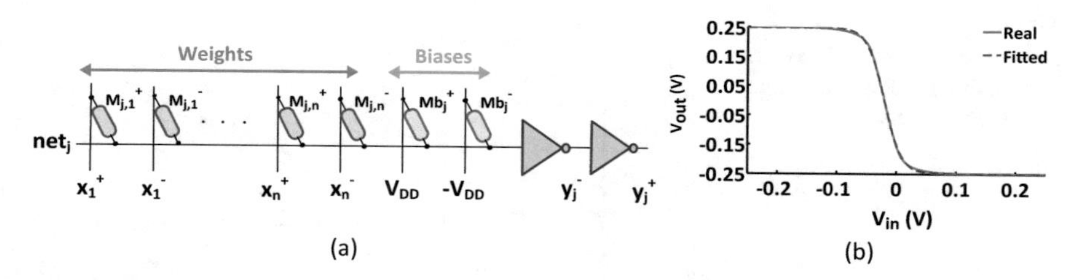

Fig. 3 (**a**) An inverter-based memristive neuron [22]. (**b**) The VTC curve of an inverter and the tangent hyperbolic function fitted to this VTC

where $\sigma_{j,i}^{+/-}$ represents the conductance of the memristors and $\gamma_j = \sum_{i=1}^{n} \left(\sigma_{j,i}{}^{+} + \sigma_{j,i}{}^{-}\right) + \sigma b_j{}^{+} + \sigma b_j{}^{-}$. Considering positive and negative weights (biases) for each neuron denoted by $w_{j,\,i}{}^{+/-}$ ($wb_j{}^{+/-}$), one may find the net_j value similar to Eq. (1) as

$$net_j = \sum_{i=1}^{n} \left(w_{j,i}{}^{+} \times x_i{}^{+} + w_{j,i}{}^{-} \times x_i{}^{-}\right) + V_{DD} \times \left(wb_j^{+} - wb_j^{-}\right). \tag{6}$$

By comparing Eqs. (5) and (6), it can be inferred that the positive and negative weights and biases of an *IM* neuron can be found as the following:

$$w_{j,i}{}^{+/-} = \frac{\sigma_{j,i}^{+/-}}{\gamma_j}, \tag{7}$$

$$wb_j{}^{+/-} = \frac{\sigma b_j^{+/-}}{\gamma_j}. \tag{8}$$

To model the activation function of the neurons, the VTC curve of an ideal inverter can be extracted using HSPICE simulations and then fitted to a tangent hyperbolic function (i.e., $a + b \times \tanh\,(d \times (x - c))$ [11]). Therefore, the positive and negative outputs of an *IM* neuron can be calculated as

$$y_j^{+/-} = f_j^{+/-}\left(net_j\right) = a_j^{+/-} + b_j^{+/-} \times \tanh\left(d_j^{+/-} \times \left(net_j - c_j^{+/-}\right)\right), \tag{9}$$

where $a_j^{+/-}$, $b_j^{+/-}$, $c_j^{+/-}$, and $d_j^{+/-}$ are the VTC-fitting coefficients of the inverter and buffer of the jth neuron. To find the VTC coefficients of an inverter, one may extract the VTC of the inverter in the considered technology using HSPICE simulations and then fit it to a tangent hyperbolic function (i.e., using MATLAB simulations) by minimizing the fitting error across the entire range of the inputs. As

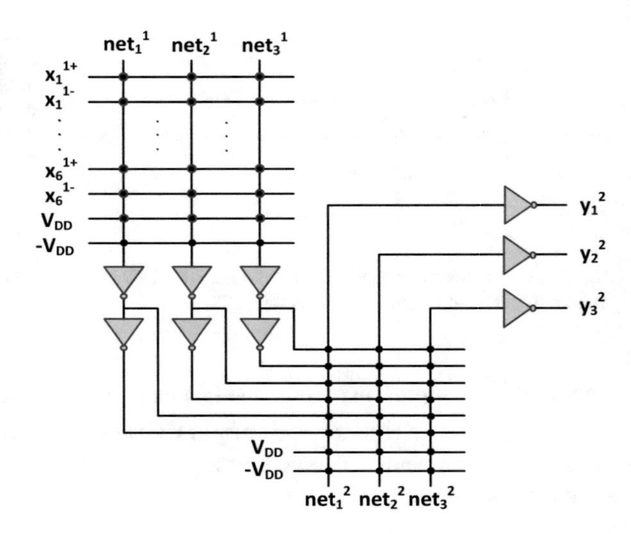

Fig. 4 Circuit diagram of a two-layer *IM*-NN with six inputs, three hidden neurons, and three outputs

shown in Fig. 3b, the fitted curve can model the behavior of the inverter VTC with a low error in both high and low slope regions of the VTC.

In an *IM*-NN, the negative (positive) output of each neuron is applied as the negative (positive) input of the next layer. As an example, the structure of a two-layer *IM*-NN with six inputs, three hidden neurons, and three outputs is depicted in Fig. 4 where the black circles represent the memristors. For this network, two inverters are utilized in the structure of the hidden neurons, while the output neurons have only one inverter.

3.2 Effects of Non-idealities on the Outputs of IM Neurons

The outputs of *IM* neurons may change due to the conductance variations of the memristors as well as the characteristic variations of the inverters. More specifically, the difference between the desired and actual conductance values of the memristors leads to variations in the weights and biases of the neuron giving rise to variations in the *net* and output voltage of the neuron (see Eqs. (5), (6), (7), (8), and (9)). In addition, the VTC coefficients of the inverters may change due to the parameter variations of the transistors including width, length, and threshold voltage or the variations in the supply voltages leading to the variations in the neuron output voltage (see Eq. (9)). As an example, the distribution of the VTC coefficients of an inverter in a 90-nm technology is depicted in Fig. 5. The results are based on assuming Gaussian random variations for the transistor parameters and the voltage sources. As the figure shows, the standard deviations of coefficients a and b are larger when variations are applied to the voltage sources compared to the parameter

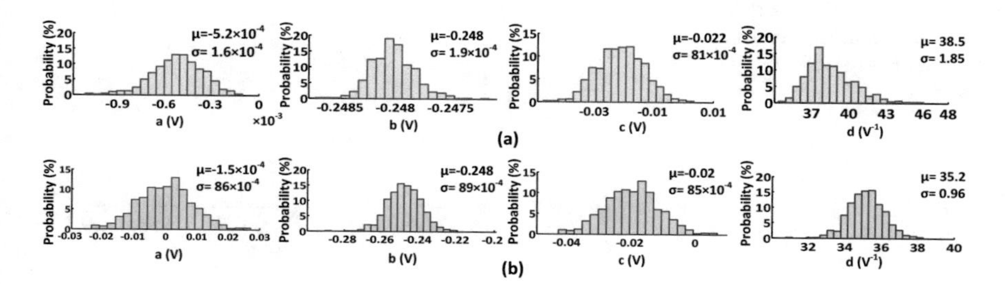

Fig. 5 The distribution of VTC coefficients of an inverter considering 5% variations for the (**a**) transistors characteristics and (**b**) voltage sources [22]

variations. In other words, these coefficients are highly dependent on the value of voltage sources. The dependency of the coefficient d on the characteristic variations of the transistors, however, is higher than its dependency to the variations of voltage sources.

To evaluate the non-ideality effects of each circuit element, the sensitivity of the neuron output voltage to the conductance variations of the memristors, and the variations of VTC coefficients are expressed analytically in the rest of this section.

3.2.1 Variations of the Activation Functions Coefficients

In this section, the sensitivity of the negative output of an *IM* neuron with respect to the variations of the inverter VTC coefficients is analyzed. The formulation can be simply extended to the positive output of the neuron and the VTC coefficients of the buffers. As the negative output of the neuron can be found using Eq. (9), its sensitivity with respect to the variations of activation functions coefficients may be written as [22]

$$\frac{\partial\left(y^{-}\right)}{\partial\left(a^{-}\right)}=1, \tag{10}$$

$$\frac{\partial\left(y^{-}\right)}{\partial\left(b^{-}\right)}=\tanh\left(d^{-}\times\left(net-c^{-}\right)\right), \tag{11}$$

$$\frac{\partial\left(y^{-}\right)}{\partial\left(c^{-}\right)}=-b^{-}d^{-}\times\left(1-\tanh^{2}\left(d^{-}\times\left(net-c^{-}\right)\right)\right)=-b^{-}\times d^{-}\times v^{-}, \tag{12}$$

$$\frac{\partial\left(y^{-}\right)}{\partial\left(d^{-}\right)}=b^{-}\times\left(net-c^{-}\right)\times v^{-}, \tag{13}$$

Fig. 6 The output voltage (y^-) and parameter v^- of an inverter based on Net^-

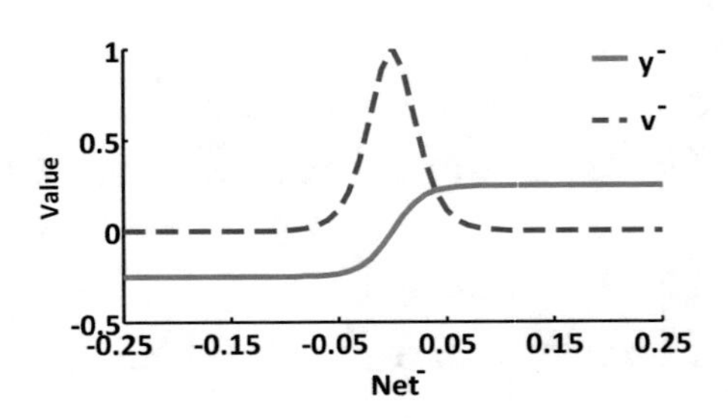

where $v^- = (1 - \tanh^2(d^- \times (net - c^-)))$. The parameter v^- determines whether the inverter is operating in its high- or low-gain regions. The output voltage (y^-) and parameter v^- of an inverter based on the Net^- (=$net - c^-$) voltage of that inverter are plotted in Fig. 6 where the VTC of the inverter was assumed to be modeled by $b^- \times \tanh(d^- \times Net^-)$. As is observed from the figure, if v^- is around 0 (1), it means that the inverter operates in its low (high)-gain region. Assume that, however, the VTC of the inverter is horizontally shifted by c_0^- making its VTC curve as $b^- \times \tanh\left(d^- \times \left(Net^- - c_0^-\right)\right)$ instead of $b^- \times \tanh(d^- \times Net^-)$. In this case, the Net^- values that are near c_0^- exist in the high-gain region of the inverter, while in the case where the VTC function is $b^- \times \tanh(d^- \times Net^-)$, the Net^- values that are around "0" exist in the high-gain region of the inverter. Thus, the parameter v^- shows the operating region of the inverter independent of the VTC shift.

Based on Eqs. (12) and (13), one may reduce the sensitivity of the neuron output to the variations of coefficients c and d by decreasing v^- or forcing the inverter to operate in the low-gain regions of its VTC. The parameter v^- depends on Net^-(=$net - c^-$) and d^-. The diagram of the parameter v^- versus Net^- is plotted in Fig. 7a considering different values for the coefficient d^-. As shown in this figure, the value of v^- is small for large $|Net^-|$ values, and the slope of the v^- curve depends on the value of d^-. Therefore, similar to Fig. 7b, the value of v^- can be approximated as

$$v_{app}^- \approx f_{app}\left(Net^-\right) = \begin{cases} 1 - \frac{|Net^-|}{\varepsilon^-} & \text{if} \quad \left|Net^-\right| \leq \varepsilon^- \\ 0 & \text{Otherwise} \end{cases}, \tag{14}$$

where f_{app} is the mathematical representation of a triangular function with the slope of $1/\varepsilon^-$ and crosses the curve of v^- at $v^- = 0.5$ (see Fig. 7b). It should be noted that $v^- = 0.5$ occurs when $Net^- = 0.88/d^-$. Therefore, the value of v_{app}^- should be equal to 0.5 when $Net^- = 0.88/d^-$, which leads to

$$0.5 = 1 - \frac{\left|0.88/d^-\right|}{\varepsilon^-}. \tag{15}$$

Consequently, one may find ε^- as

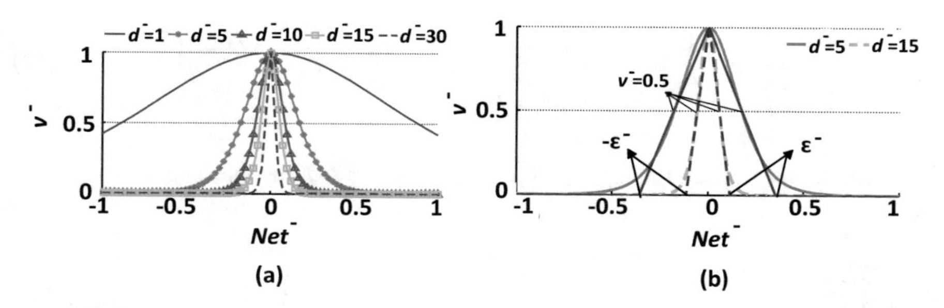

Fig. 7 (**a**) The value of v^- versus Net^- (d^- as the running parameter). (**b**) The exact and approximate values of v^- for the cases where $d^- = 5$ or $d^- = 15$ [22]

$$\varepsilon^- = \left|\frac{1.76}{d^-}\right|, \tag{16}$$

which shows that the slope of v^- is proportional to d^- value (see Eqs. (14) and (16)). In other words, v^- becomes almost zero for larger ranges of Net^- values when d^- is large.

To have a comparison between the sensitivity of the neuron outputs to different VTC coefficients, recall that $\partial(y^-)/\partial(b^-) = \tanh(d^- \times (net - c^-))$ which is never greater than 1. The sensitivity of the neuron output to the variations of coefficient a^-, therefore, is greater than that of the coefficient b^- (see Eqs. (10) and (11)). Additionally, the sensitivity of the neuron output to the variations of coefficient d^- depends on $Net^- \times v^-$. It is obvious that for small $|Net^-|$ values, this sensitivity becomes small. When $|Net^-|$ is large (i.e., $|Net^-| > \varepsilon^-$), however, the value of v^- becomes small (see Eq. (14)). Therefore, the amount of $\partial(y^-)/\partial(d^-)$ becomes small for small or large $|Net^-|$ values. Furthermore, $\partial(y^-)/\partial(c^-)$ depends on d^- which is large due to the large VTC slope of the inverters ($d^- \approx 30$ in [11] and $d^- \approx 15$ in [25]). It is expected that the negative output of an *IM* neuron has the highest (lowest) sensitivity to the variations of the coefficient c^- (d^-). The abovementioned equations and explanations are valid for the positive output of the neuron (y^+) and its corresponding activation function coefficients (a^+, b^+, c^+, d^+).

3.2.2 IM Neuron Output Sensitivity to the Conductance Variation of Memristors

To calculate the sensitivity of the neuron output (i.e., y_j^-) to the conductance variations of its memristors (i.e., the variations of $\sigma_{j,k}^+$), one may use Eq. (9) to obtain

$$\frac{\partial\left(y_j^-\right)}{\partial\left(\sigma_{j,k}^+\right)} = \frac{\partial\left(y_j^-\right)}{\partial\left(net_j\right)} \times \frac{\partial\left(net_j\right)}{\partial\left(\sigma_{j,k}^+\right)} = b_j^- \times d_j^- \times v_j^- \times \frac{\partial\left(net_j\right)}{\partial\left(\sigma_{j,k}^+\right)}. \tag{17}$$

Additionally, based on Eq. (5), $\partial\left(net_j\right)/\partial\left(\sigma_{j,k}^+\right)$ can be obtained as

$$\frac{\partial\left(net_j\right)}{\partial\left(\sigma_{j,k}^+\right)} = \sum_{i=1}^{n}\left(x_i^+ \times \frac{\partial\left(w_{j,i}^+\right)}{\partial\left(\sigma_{j,k}^+\right)} + x_i^- \times \frac{\partial\left(w_{j,i}^-\right)}{\partial\left(\sigma_{j,k}^+\right)}\right) + V_{DD} \times \left(\frac{\partial\left(wb_j^+\right)}{\partial\left(\sigma_{j,k}^+\right)} - \frac{\partial\left(wb_j^-\right)}{\partial\left(\sigma_{j,k}^+\right)}\right), \tag{18}$$

where

$$\frac{\partial\left(w_{j,k}^+\right)}{\partial\left(\sigma_{j,k}^+\right)} = \frac{\gamma_j - \sigma_{j,k}^+}{\left(\gamma_j\right)^2} = \frac{1 - w_{j,k}^+}{\gamma_j}, \tag{19}$$

and for the other weights or biases of the neuron ($i \neq k$)

$$\frac{\partial\left(w_{j,i}^{+/-}\right)}{\partial\left(\sigma_{j,k}^+\right)} = \frac{-\sigma_{j,i}^{+/-}}{\left(\gamma_j\right)^2} = \frac{-w_{j,i}^{+/-}}{\gamma_j}, \frac{\partial\left(wb_j^{+/-}\right)}{\partial\left(\sigma_{j,k}^+\right)} = \frac{-\sigma b_j^{+/-}}{\left(\gamma_j\right)^2} = \frac{-wb_j^{+/-}}{\gamma_j}. \tag{20}$$

Substituting Eqs. (19) and (20) in Eq. (18) leads to

$$\frac{\partial\left(net_j\right)}{\partial\left(\sigma_{j,k}^+\right)} = \frac{x_k^+}{\gamma_j} - \frac{1}{\gamma_j} \times \begin{pmatrix} \sum_{i=1}^{n}\left(x_i^+ \times w_{j,i}^+ + x_i^- \times w_{j,i}^-\right) + \\ V_{DD} \times \left(wb_j^+ - wb_j^-\right) \end{pmatrix} = \frac{x_k^+ - net_j}{\gamma_j}, \tag{21}$$

which can be used to rewrite Eq. (17) as

$$\frac{\partial\left(y_j^-\right)}{\partial\left(\sigma_{j,k}^+\right)} = b_j^- \times d_j^- \times v_j^- \times \frac{x_k^+ - net_j}{\gamma_j}. \tag{22}$$

Using a similar approach, the sensitivity of the positive output with respect to the memristor conductance variations can be obtained.

As mentioned previously, the output voltage of an inverter has the highest sensitivity to the variations of the coefficient c^- compared to the other VTC coefficients. Comparing Eqs. (12) and (22) shows that both sensitivities are proportional to $b_j^- \times d_j^- \times v_j^-$. However, $\partial\left(y_j^-\right)/\partial\left(\sigma_{j,k}^+\right)$ inversely depends on γ_j which is equal to the sum of the conductance values of the memristors connected to the corresponding neuron. It is, therefore, expected that γ_j becomes large which leads to a smaller sensitivity with respect to $\sigma_{j,k}^{+}$ compared to c_j^{-}. This means that the output of an *IM* neuron is frequently more sensitive to the VTC variations of inverters or buffers compared to the conductance variations of memristors.

3.2.3 IM-NN Primary Output Sensitivity to Characteristic Variations of Circuit Elements

The sensitivity of the NN output (e.g., the jth output of the NN denoted by y_j^L) to the variations of the VTC coefficients (e.g., the VTC coefficients of the tth buffer of the lth layer denoted by $coef_t^{l+}$ where *coef* can be each of coefficients a, b, c, d) may be obtained from

$$\frac{\partial\left(y_j^L\right)}{\partial\left(coef_t^{l+}\right)}=\frac{\partial\left(y_j^L\right)}{\partial\left(net_j^L\right)}\times\frac{\partial\left(net_j^L\right)}{\partial\left(coef_t^{l+}\right)}, \tag{23}$$

where

$$\frac{\partial\left(net_j^L\right)}{\partial\left(coef_t^{l+}\right)}=\sum_{i=1}^{N^L}\left(w_{j,i}^{L\ +}\times\frac{\partial\left(y_i^{L-1^+}\right)}{\partial\left(coef_t^{l+}\right)}+w_{j,i}^{L\ -}\times\frac{\partial\left(y_i^{L-1^-}\right)}{\partial\left(coef_t^{l+}\right)}\right). \tag{24}$$

In the next step, $\partial\left(y_i^{L-1^+}\right)/\partial\left(coef_t^{l+}\right)$ and $\partial\left(y_i^{L-1^-}\right)/\partial\left(coef_t^{l+}\right)$ are found. Similar to Eq. (24), these may be calculated recursively by substituting L with $L-1$ until reaching $\partial\left(y_t^{l+}\right)/\partial\left(coef_t^{l+}\right)$, which is calculated based on Eqs. (10), (11), (12), and (13). Note that $\partial\left(y_t^{l-}\right)/\partial\left(coef_t^{l+}\right)$ is zero. As an example, for an *IM*-NN with two hidden layers ($L=3$), $\partial\left(y_j^3\right)/\partial\left(coef_t^{1+}\right)$ may be found from

$$\frac{\partial\left(y_j^3\right)}{\partial\left(coef_t^{1+}\right)}=\frac{\partial\left(y_j^3\right)}{\partial\left(net_j^3\right)}\times\sum_{i=1}^{N^3}\left(w_{j,i}^{3\ +}\times\frac{\partial\left(y_i^{2^+}\right)}{\partial\left(coef_t^{1+}\right)}+w_{j,i}^{3\ -}\times\frac{\partial\left(y_i^{2^-}\right)}{\partial\left(coef_t^{1+}\right)}\right), \tag{25}$$

where

$$\frac{\partial\left(y_i^{2+/-}\right)}{\partial\left(coef_t^{1+}\right)}=\frac{\partial\left(y_i^{2+/-}\right)}{\partial\left(net_i^2\right)}\times w_{i,t}^{2\ +}\times\frac{\partial\left(y_t^{1+}\right)}{\partial\left(coef_t^{1+}\right)} \tag{26}$$

and $\partial\left(y_t^{1+}\right)/\partial\left(coef_t^{1+}\right)$ may be calculated using Eqs. (10), (11), (12), and (13).

Similarly, the sensitivity of the NN outputs (i.e., y_j^L) to the conductance variation of the memristors (i.e., $\sigma_{k,t}^{l\ +}$) is found from

$$\frac{\partial\left(y_j^L\right)}{\partial\left(\sigma_{k,t}^{l\ +}\right)}=\frac{\partial\left(y_j^L\right)}{\partial\left(net_j^L\right)}\times\sum_{i=1}^{N^L}\left(w_{j,i}^{L\ +}\times\frac{\partial\left(y_i^{L-1^+}\right)}{\partial\left(\sigma_{k,t}^{l\ +}\right)}+w_{j,i}^{L\ -}\times\frac{\partial\left(y_i^{L-1^-}\right)}{\partial\left(\sigma_{k,t}^{l\ +}\right)}\right), \tag{27}$$

where $\partial\left(y_i^{L-1^{+/-}}\right)/\partial\left(\sigma_{k,t}^{l\ +}\right)$ may be recursively calculated using Eq. (27) by substituting L with $L-1$ until achieving $\partial\left(y_k^{l^{+/-}}\right)/\partial\left(\sigma_{k,t}^{l\ +}\right)$ which is calculated using Eq. (22).

3.3 Inductions Drawn from the Above Analysis

One may draw the following inductions from the above analysis regarding the *IM*-NN output sensitivity to the parameter variations:

1. The output voltage of an *IM* neuron has the highest sensitivity to the horizontal shift of the inverter and buffer VTCs (i.e., variations of coefficients $c^{+/-}$).
2. The output voltage of an *IM* neuron with large γ value has low sensitivity to the conductance variations of the memristors.
3. The output voltage of an *IM* neuron has lower sensitivity to the conductance variations of memristors compared to the variations of VTC coefficients.
4. The sensitivity of an *IM* neuron output to the non-idealities of circuit elements can be mitigated by lowering $d^{+/-}$, which occurs when the VTC slopes of the inverters and buffers are reduced, or by decreasing $v^{+/-}$ values which occurs when the inverters and buffers are forced to operate in the low gain regions of their VTCs.

4 Input/Output Interfaces

To connect *IM*-NNs to the digital part of the system, digital-to-analog and analog-to-digital converters (DACs and ADCs) are required which consume considerably large power consumption compared to that of the memristive crossbars. The authors of [19] proposed memristive implementations for DAC and ADC which can be utilized as the input and output interfaces for *IM*-NNs. The structure of the mentioned ADC and DAC will be explained next.

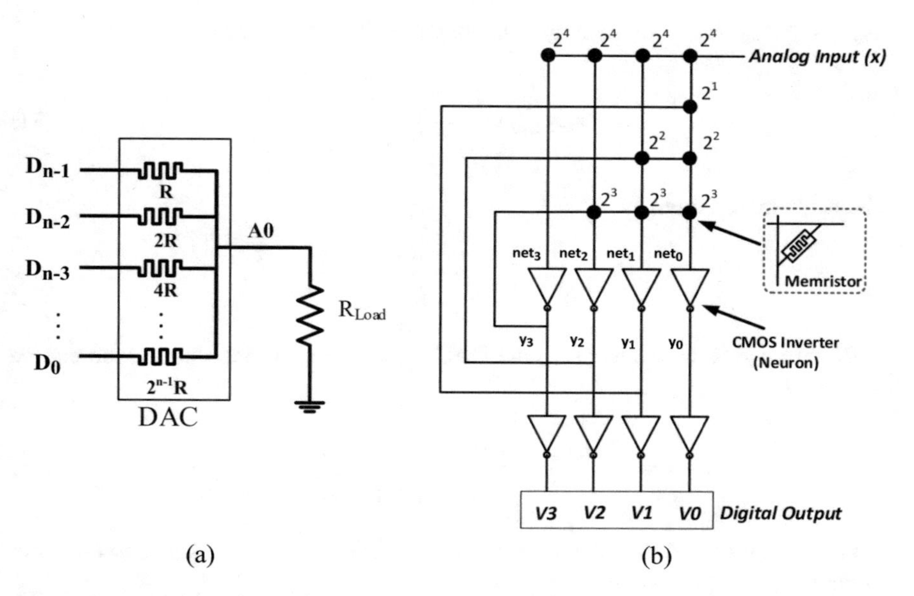

Fig. 8 The circuit implementation of a memristive (**a**) DAC and (**b**) ADC modified from [19]

4.1 Memristive DAC

The circuit implementation of a memristive DAC is depicted in Fig. 8a where R_{Load} represents the equivalent loading resistance connected to the output node of the DAC. Since the DAC is utilized as the input interface of *IM*-NNs, R_{Load} may be determined by the equivalent resistance of the memristive crossbar. Writing the KCL equation for the A0 node (in Fig. 8a) leads to

$$\sum_{i=0}^{n-1} \frac{V_{D_i} - V_{A0}}{2^{n-1-i}R} = \frac{V_{A0}}{R_{\text{Load}}}, \tag{28}$$

which can be rewritten as

$$\sum_{i=0}^{n-1} \frac{V_{D_i}}{2^{n-1-i}R} - \sum_{i=0}^{n-1} \frac{V_{A0}}{2^{n-1-i}R} = \frac{V_{A0}}{R_{\text{Load}}}. \tag{29}$$

Accordingly, V_{A0} can be calculated as

$$V_{A0} = \frac{\sum_{i=0}^{n-1} \frac{V_{D_i}}{2^{n-1-i}R}}{\frac{1}{R_{\text{Load}}} + \sum_{i=0}^{n-1} \frac{1}{2^{n-1-i}R}}. \tag{30}$$

Assuming that R_{Load} is very large, one may find the ideal value of V_{A0} as

$$V_{A0\text{ideal}} = \frac{\sum_{i=0}^{n-1} \frac{V_{D_i}}{2^{n-1-i} R}}{\sum_{i=0}^{n-1} \frac{1}{2^{n-1-i} R}}, \tag{31}$$

which can be rewritten as

$$V_{A0\text{ideal}} = \frac{\sum_{i=0}^{n-1} \left(2^i V_{D_i}\right)}{2^n - 1}. \tag{32}$$

Based on Eqs. (30) and (31) , the DAC error caused by the output load can be calculated as

$$\text{Error} = \frac{V_{A0\text{ideal}} - V_{A0}}{V_{A0\text{ideal}}} = \frac{\frac{1}{R_{\text{Load}}}}{\frac{1}{R_{\text{Load}}} + \sum_{i=0}^{n-1} \frac{1}{2^{n-1-i} R}}. \tag{33}$$

Let us assume that the input error tolerance of the NN is α. This puts a constraint on the value of R as shown below

$$\frac{\frac{1}{R_{\text{Load}}}}{\frac{1}{R_{\text{Load}}} + \sum_{i=0}^{n-1} \frac{1}{2^{n-1-i} R}} < \alpha. \tag{34}$$

which may be rewritten as

$$R < \frac{\alpha \left(2^n - 1\right)}{(1-\alpha)\, 2^{n-1}} R_{\text{load}}, \tag{35}$$

which shows that the maximum acceptable value for R is a function of the tolerable error (i.e., α), the input bit width of the DAC (i.e., n), and its output load (i.e., R_{load}).

4.2 Memristive ADC

In a memristive ADC, first, the most significant bit (MSB) of the output is generated where the other bits are produced one after another from the most significant to the least significant bit (LSB). This makes the value of the ith bit dependent on other output bits with higher significances. Next, we explain the functionality of this structure. Assume that the voltage sources are V_{dd} and $-V_{dd}$ where each output with the voltage of $-V_{dd}$ (V_{dd}) represents logical 0 (1). In addition, the analog input is in the range of $[-V_{dd}, V_{dd}]$. In an n-bit memristive ADC, the input with the value of $-V_{dd}$ (V_{dd}) should be converted to 0 ($2^n - 1$) implying that the MSB of ADC output for the inputs with negative (positive) values is 0 (1). Assume that an inverter can be employed as a comparator whose output becomes logically 1 (0) if its input is negative (positive). The MSB of the ADC output, thus, can be generated by applying the analog input to two consecutive inverters as shown in Fig. 8b which shows the circuit implementation of a 4-bit memristive ADC. In this figure, the black circles show the memristors, and the numbers written beside them represent their corresponding weights. In other words, the resistance of the memristor with the weight of 2^k is equal to $2^{(n-k)}R$ where n represents the resolution of the ADC. The other bits of the output can be generated by subtracting the impact of the outputs with higher positions from the analog input. As an example, assume that the input voltage is equal to V_{dd} which should be converted to $(1111)_2$. In this case, y_3 and V_3 (see Fig. 8b) become $-V_{dd}$ and V_{dd} as the analog input is greater than 0. In addition, net_2 can be found from

$$net_2 = \frac{x}{R} + \frac{y_3}{2R} = \frac{V_{dd}}{R} - \frac{V_{dd}}{2R} = \frac{V_{dd}}{2R}, \tag{36}$$

which is greater than 0 and leads to $y_2 = -V_{dd}$ and $V_2 = V_{dd}$. Similarly, net_1 can be obtained from

$$net_1 = \frac{x}{R} + \frac{y_3}{2R} + \frac{y_2}{4R} = \frac{V_{dd}}{R} - \frac{V_{dd}}{2R} - \frac{V_{dd}}{4R} = \frac{V_{dd}}{4R}, \tag{37}$$

which leads to $V_1 = -y_1 = V_{dd}$ and

$$net_0 = \frac{x}{R} + \frac{y_3}{2R} + \frac{y_2}{4R} + \frac{y_1}{8R} = \frac{V_{dd}}{R} - \frac{V_{dd}}{2R} - \frac{V_{dd}}{4R} - \frac{V_{dd}}{8R} = \frac{V_{dd}}{8R}, \tag{38}$$

which results in $V_0 = -y_0 = V_{dd}$.

5 Training of IM-NNs

In this section, PHAX [11], RIM [25], LATIM [23], and ERIM [18] methods, which are recent work focusing on offline training of *IM*-NNs using back-propagation method, and OCTAN [21], which is an on-chip training method, are discussed.

5.1 *PHAX*[1]

The conductance range of the memristors is limited, a fact that should be considered during the training phase. To address this limitation, in the PHAX method, a continuously differentiable weight mapping function (denoted by g_1) is employed

[1] Physical characteristics aware ex situ training.

to determine the conductance of the memristors (here as an example, the ith positive conductance of the j^{th}neuron) as [11]

$$\sigma_{j,i}{}^{+} = g_1\left(\theta_{j,i}{}^{+}\right) = \sigma_{\min} + \frac{\sigma_{\max} - \sigma_{\min}}{1 + e^{-\theta_{j,i}{}^{+}}}, \tag{39}$$

where $\sigma_{\max}$ and $\sigma_{\min}$ represent the maximum and the minimum conductance values of the memristors and $\theta_{j,\,i}{}^{+}$ is the training variable. Based on Eqs. (7) and (39), the positive weight of the neuron is calculated from

$$w_{j,i}{}^{+} = g\left(\theta_{j,i}{}^{+}\right) = g_2\left(g_1\left(\theta_{j,i}{}^{+}\right)\right), \tag{40}$$

where

$$g_2\left(\sigma_{j,i}{}^{+}\right) = \frac{\sigma_{j,i}{}^{+}}{\sum_{p=1}^{n}\left(\sigma_{j,p}^{+} + \sigma_{j,p}^{-}\right) + \sigma b_j^{+} + \sigma b_j^{-}}. \tag{41}$$

The parameters θ are updated after each training iteration where the conductance of the memristors is determined based on Eq. (39) and the weights are updated based on Eq. (40). Assuming that J represents the cost function of the training phase, the positive and negative θ variables of the output layer (L^{th} layer) are updated as

$$\Delta\theta_{k,j}{}^{L} = -\alpha\left(\frac{\partial J}{\partial y_k^L}\right) \times \left(\frac{\partial y_k^L}{\partial net_k^L}\right) \times \left(\frac{\partial\left(\sum_{q\in(L-1)} g\left(\theta_{k,q}{}^{L}\right) y_q{}^{L-1}\right)}{\partial\theta_{k,j}{}^{L}}\right), \tag{42}$$

and for the hidden layers as

$$\Delta\theta_{j,i}{}^{l} = -\alpha\left(\frac{\partial y_j^l}{\partial net_j^l}\right)\left[\sum_{k\in(l+1)} \delta_k g\left(\theta_{k,j}{}^{l+1}\right)\right] \times \left(\frac{\partial\left(\sum_{p\in(l-1)} g\left(\theta_{j,p}{}^{l}\right) y_p{}^{l-1}\right)}{\partial\theta_{j,i}{}^{l}}\right), \tag{43}$$

where α is the learning rate, L and l represent the last and the current layers, and δ_k is the portion of the error of the kth output in the next layer [11].

In PHAX, the VTC curves of the inverter and buffer, extracted by HSPICE simulations, are fitted to tangent hyperbolic functions (see Eq. (9)) where the fitting coefficients are used to model the activation functions. The VTC coefficients of an inverter, however, depend on the size and the output load of that inverter (caused by the loading effect of the memristive crossbar). To demonstrate it better, the VTC coefficients of an inverter based on its size and its output load (R_M) are shown in Fig. 9. Note that, in [11], the size of transistors for an inverter of size s were considered as $(W/L)_p = 2s$ and $(W/L)_n = 1.2s$. To reduce the sensitivity of the VTC coefficients

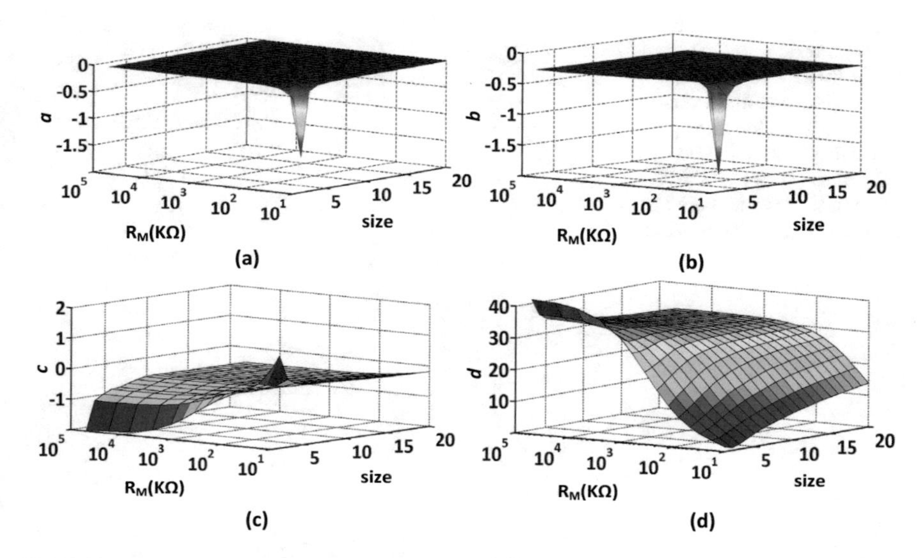

Fig. 9 The fitted coefficients (**a**) a, (**b**) b, (**c**) c, and (**d**) d of the inverter VTC versus the output loading resistance (R_M) as well as the size of the inverter [23]

with respect to the output load of the inverter, the size of all the inverters of the NN was chosen as a constant value (i.e., 5) in the PHAX method.

5.2 *RIM*[2]

As mentioned previously, the loading effect of the memristive crossbar may change the VTC curve of the inverters which was not considered in the PHAX method. The VTC changes may cause large errors in the modeling of the NN outputs trained by the method. In addition, according to Eqs. (12) and (22), the sensitivity of the output voltage of an *IM* neuron to the non-idealities of circuit elements is proportional to coefficients $d^{+/-}$. The VTC slope of an inverter (S) can be approximated by the slope of $a + b \times \tanh((x - c) \times d)$ when $x = c$ [25]. In other words, S can be approximated by $b \times d$. Consequently, to reduce the sensitivity of the VTC coefficients to the loading effect of the memristors as well as decreasing the effects of the circuit elements non-idealities on the output voltage of *IM* neurons, adding a grounded resistor to the output node of each inverter was proposed in [25]. To explain this further, assume that the output node of an inverter is connected to a memristive crossbar with an equivalent resistance of R_M and a grounded resistor with the resistance of R. The VTC slope of this inverter may be defined as

[2] Resistor-inverter-based memristive NN training

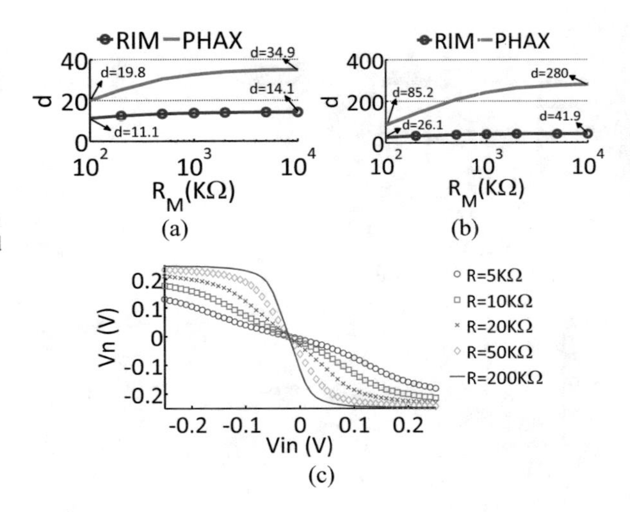

Fig. 10 The coefficient *d* of the (**a**) inverter- and (**b**) buffer-fitted VTC curves in the nominal condition versus the memristors loading resistance for PHAX and RIM methods. (**c**) The VTC curve of the inverter based on the resistance of the grounded resistor (*R*) [25]

$$S = \frac{V_{out}\left(V_{in} = V_{IH}\right) - V_{out}\left(V_{in} = V_{IL}\right)}{V_{IH} - V_{IL}}. \tag{44}$$

In addition, *S*, approximated by the small signal gain of the inverter when the NMOS and PMOS transistors are in the saturation mode, may be calculated from

$$S \approx \left(g_{mn} + g_{mp}\right) \times \left(r_{o,n} \left\| r_{o,p} \right\| R_M \middle\| R\right), \tag{45}$$

where g_{mn} (g_{mp}) and $r_{o,n}$ ($r_{o,p}$) are the small signal parameters of the NMOS (PMOS) transistor [25]. Based on Eq. (45), it can be inferred that by adding the grounded resistor; the sensitivity of the VTC slope to the loading effect of the memristors is reduced as this slope is proportional to $r_{o,n} \| r_{o,p} \| R_M \| R$ instead of $r_{o,n} \| r_{o,p} \| R_M$. It is, therefore, expected that when R_M decreases, the value of the coefficient *d* decreases with a lower rate compared to the case in which the grounded resistor is not utilized. In Fig. 10, the parameter *d* versus R_M is depicted for an inverter and a buffer. In the RIM method, the resistance of the grounded resistor (*R*) is equal to 50 $K\Omega$. As shown in this figure, the sensitivity of *d* to the loading effect of the memristive crossbar is reduced in the RIM method compared to PHAX. In addition, the VTC curve of an inverter versus *R* is plotted in the Fig. 10c demonstrating the dependency between the characteristic of the inverters and their equivalent output resistance. As shown in this figure, the VTC slope of the inverter decreases when *R* reduces. Thus, by adding the grounded resistor (*R*), the coefficient *d* of the inverters and buffers can be reduced, leading to a lower sensitivity for the neuron outputs to the non-idealities of circuit elements.

Furthermore, the results of [25] revealed that by adding the grounded resistor, the effects of process variations on the VTC of the inverters decreased. The VTC curves of 1000 inverters are plotted in Fig. 11a. For this figure, random variations with a

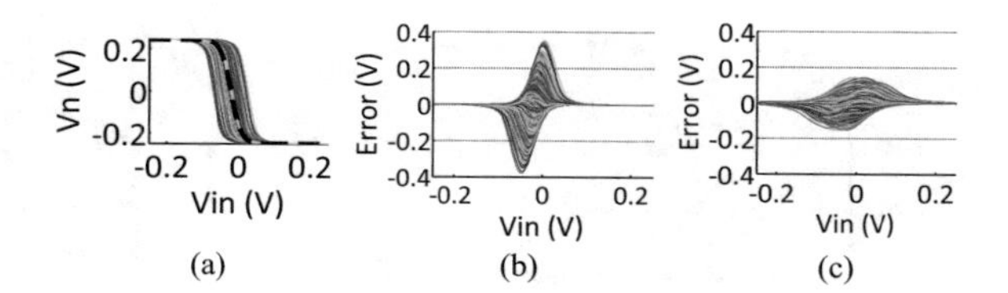

Fig. 11 (**a**) VTC curves of 1000 inverters under 10% variation condition and their modeling error for the (**b**) PHAX and (**c**) RIM methods [25]

Gaussian distribution and a variability (σ/μ) of 10% were applied to the threshold voltage, width, and length of the transistors. The dotted black curve shows the fitted tangent hyperbolic function of the nominal condition. In the PHAX method, this curve is used in the training phase to model the behavior of all inverters of the NN. The modeling error (the difference between the fitted curve of the nominal condition and the real output values of 1000 inverters) is shown in Fig. 11b and Fig. 11c for the PHAX and RIM methods (with $R = 50\ K\Omega$). As is evident from the figures, the modeling error of the VTCs due to the transistors non-idealities reduces when the VTC slope of the inverter is reduced by the added grounded resistor.

To find proper values for the resistance of the grounded resistors in the RIM method, it is required to study the functionality of the inverters in an *IM* neuron. Since the conductance of the memristors is always positive, to have the capability of implementing negative and positive weights, two inverters are employed in the structure of each *IM* neuron. In other words, the second inverter is employed to generate an output with the opposite sign compared to the first one which makes it possible to produce negative weights. As the VTC of an inverter is not the same as a negating function, it is expected that the positive and negative outputs of an *IM* neuron have opposite signs with different magnitudes. Furthermore, the slope of the activation functions is determined by the VTC slope of the negative and positive inverters. Based on the above discussion, however, one may employ the positive inverter as a negating function with the slope of -1, enabling us to generate negative weights while controlling the slope of the activation function based on the VTC slope of the negative inverter. This is the main idea behind the RIM method which is explained in the following.

As mentioned in Sect. 3.2, lower values for coefficients $d^{+/-}$ lead to smaller sensitivity of the inverter output to the variations of circuit elements which is desirable. Based on Fig. 10c, however, using lower-resistance grounded resistors leads to undesirable smaller output swings for the inverters. Hence, in the RIM method, a resistor (R_2) was added to the output node of the positive inverter to make its VTC curve similar to a negating function without a significant reduction in the inverter output swing. Since the VTC slope of an inverter may be approximated by $b \times d$, to implement a negating function using an inverter, the resistance of the grounded resistor (i.e., R_2) should be chosen such that $b \times d = -1$. In addition, a resistor (R_1) was added to the output node of the negative inverter to reduce its VTC slope as well as control the slope of the tangent hyperbolic activation function.

Even though reducing the VTC slope of the inverters leads to lower output variations in the case of non-idealities (see Fig. 11), it decreases the accuracy

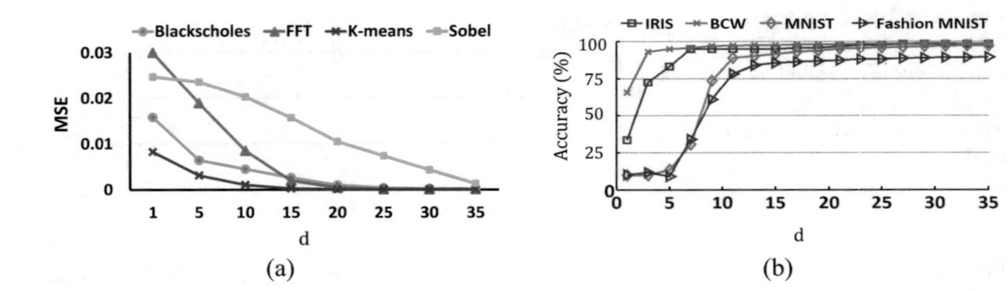

Fig. 12 The (**a**) *MSE* [25] and (**b**) accuracy of the *IM*-NNs versus the coefficient d assuming no process variation as well as no loading effect of memristive crossbar on the inverters VTC curves

of the NN in the ideal case. The MSE (accuracy) of the NNs under regression (classification) applications versus the value of the coefficient d is plotted in Fig. 12. The results reveal that when the coefficient d increases, the accuracy (MSE) of the NN increases (decreases). The rate of the changes, however, decreases when d is greater than 15. It should be noted that the variations of the inverter outputs increase when the coefficient d increases. In the RIM method, hence, a proper value for the coefficient d is determined by considering a tradeoff between the accuracy of the NN in the nominal case and the NN resiliency to the severe effects of variations. As a result, the coefficient d of the negative inverter was chosen as 15.

To justify the effect of the coefficient d on the accuracy of the trained NN, consider the VTC curves depicted in Fig. 13 assuming $d = 35$ similar to the PHAX method and $d = 15$ similar to the RIM method. In this figure, for the sake of illustration, it is assumed that $a = c = 0$ and $b = -V_{DD} = -0.25$. As shown in this figure, to have the ability to generate different output values by the inverters (i.e., achieving at least 90% of the inverter output swing), the swing of the *net* voltage should be at least 0.08 V (0.20 V) when d is equal to 35 Eq. (15). In other words, when $d = 15$, a larger range of *net* values should be generated to benefit from different output values produced by the inverters. Since the outputs of the lth layer are utilized as the inputs of the $(l + 1)^{st}$ layer, if the inverters of the lth layer cannot generate large output values, the situation becomes harder for the neurons of the $(l + 1)^{st}$ layer to generate large output values. In addition, in the output layer (L^{th} layer) of the classification NNs, it is expected that the selected (unselected) classes have output values $\cong V_{DD}$ ($\cong -V_{DD}$). Therefore, when $d = 15$, larger $|net|$ values are needed for the output layer requiring large values for the outputs of the previous layers and their corresponding weights. This is the reason for the accuracy degradation in *IM*-NNs when the VTC slope of the inverters decreases.

The flowchart of the RIM method is depicted in Fig. 14. As shown in this figure, the sizes of all the inverters, similar to the PHAX method, were chosen to be 5. Then, a grounded resistor was connected to the output node of the inverter, and the VTC curve of the inverter was extracted, while the resistance of the grounded resistor (R) was sweeping. Next, the value of coefficient d was plotted versus the R value. In the next step, the resistance which led to $d \approx 15$ was found and considered as

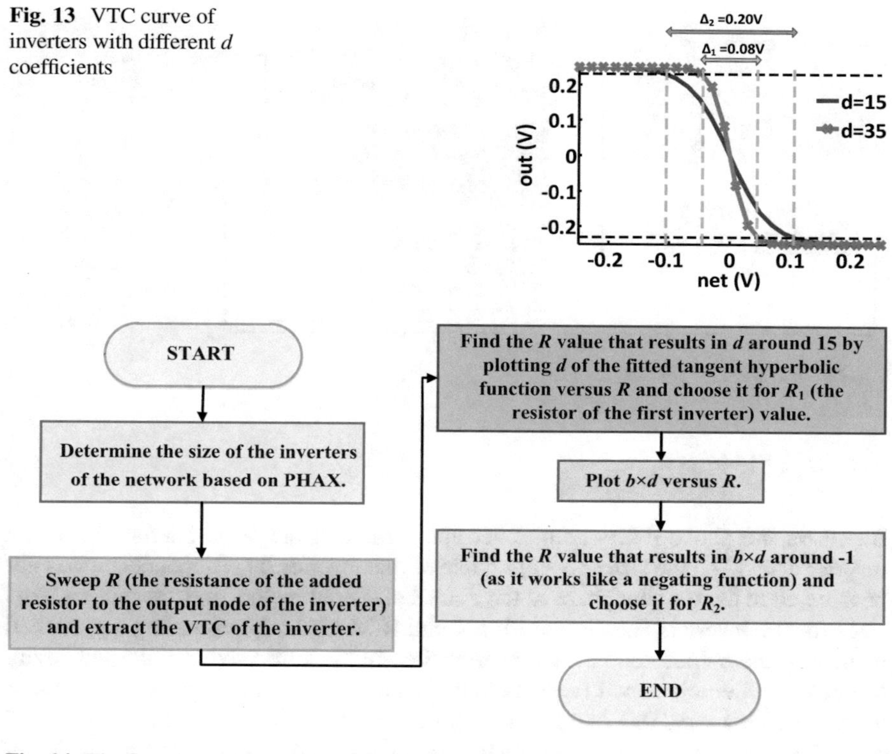

Fig. 13 VTC curve of inverters with different d coefficients

Fig. 14 The flowchart of the RIM method [25]

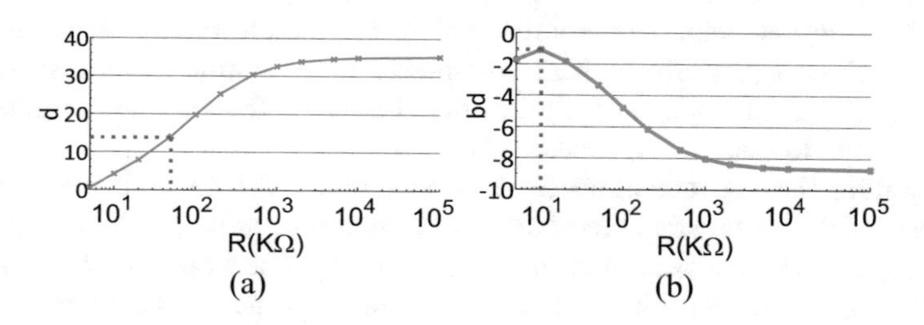

Fig. 15 (**a**) The coefficient d and (**b**) $b \times d$ values versus R [25]

the resistance of the grounded resistor connected to the output node of the negative inverter (R_1). Afterward, the value of $b \times d$ was plotted versus the value of R, and the resistance leading to $b \times d \approx -1$ was found and considered as the resistance of the grounded resistor connected to the output node of the positive inverter (R_2). As an example, the value of d and $b \times d$ versus R is plotted in Fig. 15 where the proper values for R_1 and R_2 are shown in these figures by dotted lines.

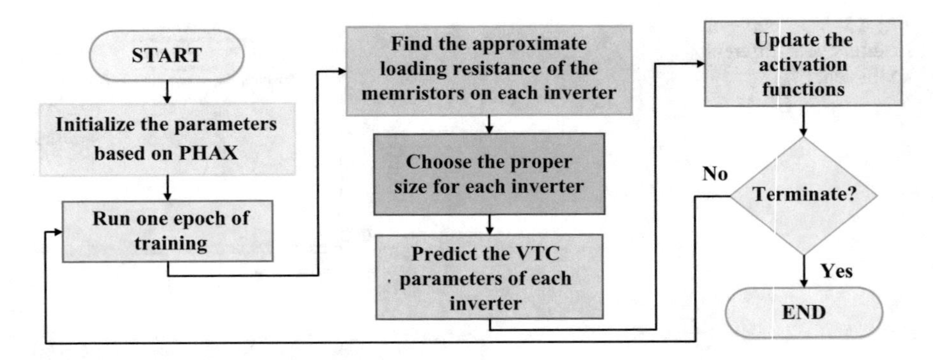

Fig. 16 The flowchart of LATIM [23]

5.3 *LATIM*[3]

Based on the above discussion, it became clear that the VTC coefficients of an inverter depend on the size and output load of that inverter. This dependency was not considered in the training phase of the PHAX and RIM methods. By assuming large sizes for the inverters (i.e., in the PHAX and RIM methods) or adding a grounded resistor to the output node of the inverter (i.e., in the RIM method), the sensitivity of the VTC coefficients with respect to the loading effect of the memristor crossbar decreased. However, this dependency still exists and can degrade the accuracy of the NN if it is not considered in the training phase of the network. In addition, by increasing the size of the inverters, the power consumption and area occupation of the neurons increase which are not desirable [18]. Therefore, sizes of the inverters should be chosen properly to have low-power consumption. This also requires modeling the output voltage of the inverters based on the sizes and output loads of the inverters to train the NN accurately.

To achieve the abovementioned goals, in the LATIM method, the equivalent resistance of the memristive crossbar was calculated approximately, and the sizes of the inverters were chosen to have a high accuracy. The flowchart of the LATIM method is shown in Fig. 16. In this method, first, the activation functions of all the neurons are considered the same, meaning that they were extracted without considering output loads for the inverters and assuming that the size of the inverters is equal to that of the PHAX method (i.e., 5).

Based on the obtained VTC coefficients, the activation functions are modeled, and one epoch of training is performed similar to the PHAX method. Next, the theta (θ) parameters are updated, and the conductance of the memristors is obtained based on Eq. (39). Afterward, the equivalent resistance of the memristive crossbar is calculated approximately, and the size of each inverter is chosen based on its

[3] Loading-aware offline training method for inverter-based memristive neural networks

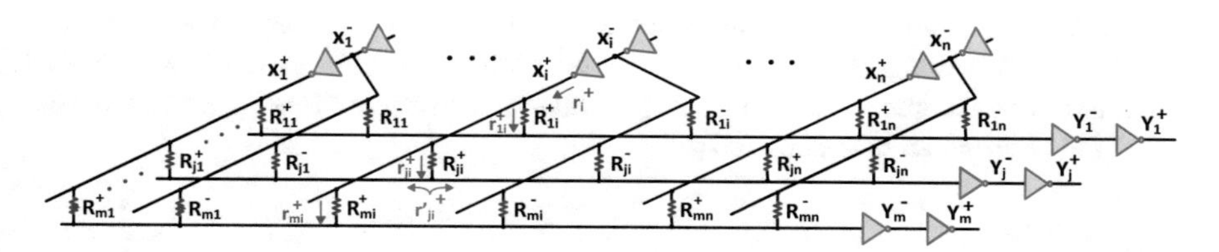

Fig. 17 An inverter-based memristive crossbar with n inputs and m outputs [23]

output load. In the next step, the VTC coefficients of each inverter are predicted using a pre-trained NN, and the activation functions of the neurons are updated. The abovementioned steps are repeated until the termination condition is satisfied. As an example, in [23], the number of epochs was considered as the termination condition for the training process.

To find the equivalent resistance of the memristive crossbar based on LATIM method, consider a crossbar with n inputs and m outputs (see Fig. 17). In this figure, r_i^+ represents the equivalent output load of the ith positive inverter which can be obtained from [23]

$$\frac{1}{r_i^+} = \sum_{j=1}^{m} \frac{1}{r_{ji}^+} = \sum_{j=1}^{m} \frac{1}{R_{ji}^+ + r'^+_{ji}}. \tag{46}$$

In addition, r'^+_{ji} can be calculated approximately as

$$\frac{1}{r'^+_{ji}} \approx \frac{1}{R_{ji}^-} + \sum_{p \neq i} \left(\frac{1}{R_{jp}^+} + \frac{1}{R_{jp}^-} \right). \tag{47}$$

Since r'^+_{ji} is the equivalent resistance of $(2n - 1)$ parallel branches, in LATIM, r'^+_{ji} is assumed to be considerably smaller than R_{ji}^+ and is neglected in the calculations of Eq. (46). Therefore, r_{ji}^+ is approximated as

$$r_{ji}^+ \approx R_{ji}^+, \tag{48}$$

which leads to

$$\frac{1}{r_i^+} \approx \sum_{j=1}^{m} \frac{1}{R_{ji}^+}. \tag{49}$$

The equivalent output load of the negative inverters (i.e., r_i^-) can be approximately calculated in a similar way.

It is worth noting that the weights and biases of an *IM* neuron are dependent, and the dependence limits the ability to select the weights and biases freely to achieve the maximum accuracy. Based on Eqs. (7) and (8), the sum of the weights and biases of an *IM* neuron can be calculated as

$$\mathrm{sum}_j = \sum_{i=1}^{n} \left(w_{j,i}{}^{+} + w_{j,i}{}^{-}\right) + \left(b_j^{+} + b_j^{-}\right) = \frac{\sum_{i=1}^{n} \left(\sigma_{j,i}{}^{+} + \sigma_{j,i}{}^{-}\right) + \left(\sigma b_j{}^{+} + \sigma b_j{}^{-}\right)}{\gamma_j} = 1. \tag{50}$$

This shows the dependence of the weights and biases of an *IM* neuron where increasing a weight (i.e., increasing $w_{j,\,i}{}^{+}$ by increasing $\sigma_{j,\,i}{}^{+}$) results in decreasing the other weights and biases. On the other hand, since the ratio of $w_{j,\,i}{}^{+}/w_{j,\,k}{}^{+}$ is equal to $\sigma_{j,i}^{+}/\sigma_{j,k}^{+}$, the maximum ratio of each two weights of the neuron equals to $\sigma_{max}/\sigma_{min}$ which is finite. This again limits our ability to select the weights arbitrarily. Let us assume that a neuron has n inputs and $\sigma_{max}/\sigma_{min} = k$. The maximum achievable weight occurs for the case where the conductance of the considered memristor is equal to σ_{max} and those of the others are σ_{min}. In this case, $\gamma = ((2n + 1)\sigma_{min} + \sigma_{max})$ and the maximum achievable weight is equal to $\sigma_{max}/((2n + 1)\sigma_{min} + \sigma_{max}) = k/((2n + 1) + k)$, while the other weights and biases are equal to $1/((2n + 1) + k)$. As an example, consider $n = 100$ and $k = 50$. In this case, the maximum achievable weight is ~0.2, while the other weights and biases are almost 0.004. This implies that having several large weights in an *IM* neuron makes the other weights very small. In addition, note that the inputs of the neurons are applied in the differential form (the positive and negative inputs have opposite signs). Generating large $|net|$ voltages by a neuron requires several large inputs having the same signs with corresponding large weights. Furthermore, the inputs with the opposite signs as well as their corresponding weights should be small. These requirements make the generation by a neuron rather difficult. The situation becomes worse when the neuron input swings decrease. This is due to the fact that the inputs of a neuron are multiplied by the weights to generate the net values. If the input swing decreases, it directly affects (reduces) the *net* swing. As the outputs of the inverters are modeled with $a + b \times \tanh(d \times (x - c))$ function, the output swing of the inverter is equal to $2b$. Also the outputs of the neurons of the lth layer are applied as the inputs of the next layer ($(l + 1)^{st}$ layer). Therefore, if the coefficient b of the neurons of the lth layer decreases, the input swing of the neurons of the $(l + 1)^{st}$ layer decreases, too. This situation may result in lower $|net|$ values which lead to smaller input swings for the neurons of the next layer. Consequently, inverters with small b coefficients or small output swings are not desirable. Based on Fig. 9, the coefficient b is a function of the output load and the size of the inverter. In addition, the value of b decreases when the size or the output load of the inverter decreases. By choosing proper sizes for the inverters based on their output load, however, one may achieve acceptable values for the coefficient b.

When the output load (R_M) of an inverter with the size of s is larger than a minimum resistance (denoted by RL_s), the values of coefficients a and b can be

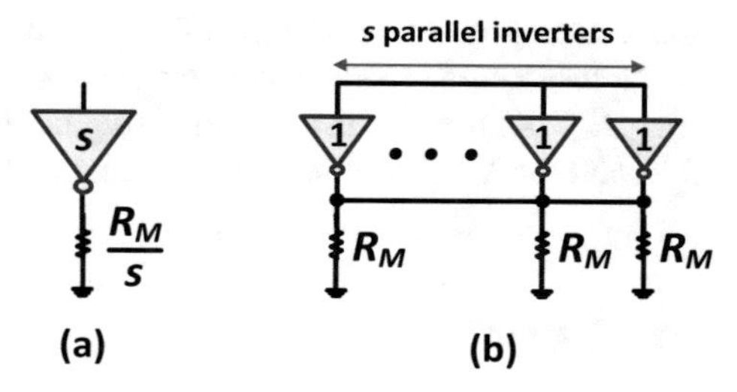

Fig. 18 (**a**) An inverter with the size of s and an output resistive load of R_M/s. (**b**) s inverters with the size of 1 and the output resistive load of R_M connected in parallel

considered constant (see Fig. 9). In addition, in this case, the value of b does not decrease which results in a high output swing for the inverters. To assume constant values for these coefficients, therefore, the output load of the inverter should be larger than RL_s. This limitation was used to choose proper sizes for the inverters in the LATIM method. The behavior of an inverter with the size of s and an output resistive load of R_M/s (similar to Fig. 18a) can be approximated by s inverters with the size of 1 (minimum-sized inverters) and the output resistive load of R_M connected in parallel (similar to Fig. 18b). Hence, RL_s was approximated as RL_1/s where RL_1 was defined as the resistance at which the approximation errors of coefficients a and b were small values (i.e., 1% and 2% of $2V_{DD}$ where V_{DD} $(-V_{DD})$ was the positive (negative) supply voltage of the inverter).

The output resistive load of the inverters was approximated based on Eq. (49). To have high-output swings for the inverters, this resistance should not be lower than RL_s. This constraint can be expressed as

$$r_i^- \geq RL_{s_i^-} = \frac{RL_1}{s_i^-}, \tag{51}$$

for the ith negative inverter with the size of s_i^- and approximate output resistance equal to r_i^-. This implied that to have the lowest power consumption while achieving acceptable output swings for the inverters, the sizes of the inverters should be chosen as

$$s_i^- = \left\lceil \frac{RL_1}{r_i^-} \right\rceil. \tag{52}$$

The next step after choosing the size of inverters is predicting the VTC coefficients of the inverters. Since the sizes of the inverters are chosen using Eq. (52), the output loads of the inverters are higher than RL_s. Consequently, constant values were considered for modeling coefficients a and b. As can be inducted from Fig. 9, however, the relation between the other coefficients, the output load, and the size of the inverter should be modeled with more complexity. For this purpose, two NNs, denoted by NN_c and NN_d, were trained to predict the values of coefficients c and d based on the output load and the size of the inverters. As a summary, after each training iteration of the LATIM method, the conductance of the memristors was updated, and the equivalent output resistance of the inverters was approximated using Eq. (49). Next, the proper sizes for the inverters were chosen based on Eq. (52), and the VTC coefficients were updated using NN_c and NN_d.

5.4 ERIM[4]

In this part, first, we review previous VTC modeling techniques and their weaknesses and then explain the ERIM approach for VTC modeling. The RIM method added a grounded resistor to the output node of the inverter to reduce the sensitivity of the VTC coefficients to the loading effect of the memristive crossbar. In other words, it lowered the VTC modeling error by reducing the memristor loading effect. The modeling accuracy in the LATIM method was improved by considering an approximation of the equivalent resistance of the memristive crossbar. In this method, the VTC coefficients, which are used to choose proper sizes for the inverters, are predicted more accurately. The loading effect of the memristive crossbar may not be fully modeled by considering only the equivalent resistance, and thus, other effects should be included. More specifically, since the inverter outputs are connected through the memristive crossbar, the loading effect becomes also dependent on the output voltages of other inverters of the same layer. Therefore, a more accurate loading effect modeling approach requires both equivalent resistance and equivalent voltage. In the ERIM method of [18], the Thevenin equivalent circuit of the network is utilized to more accurately model the loading effect on the output voltage of the inverters. In addition, as ERIM is an enhanced version of the RIM method, grounded resistors are also utilized in ERIM.

To illustrate the loading effect of the network on the output voltage of the inverters, assume that it is modeled by a Thevenin equivalent resistance and voltage denoted by R_M and V_{CM}, respectively (see Fig. 19a). The output voltage of the inverter for different R_M values assuming $V_{CM} = -0.05V$ is depicted in Fig. 19b. As was expected, the VTC slope of the inverter depends on the value of R_M (also see Eq. (47)). In addition, the output voltage of the inverter for different V_{CM} values assuming $R_M = 100K\Omega$ is depicted in Fig. 19c which shows that the VTC is vertically shifted due to the changes of V_{CM}. One may approximate the output voltage of an inverter (V_{out}) as [18]

$$V_{\text{out}} \approx f_n\left(V_{in}\right)|_{V_{CM}=0} + V_{CM}\frac{R_t}{R_t + R_M}, \tag{53}$$

where $R_t = r_p \| r_n$ and r_n (r_p) represents the resistance of the NMOS (PMOS) transistor of the inverter. The term $f_n\left(V_{in}\right)|_{V_{CM}=0}$ generates the VTC function

[4] Enhanced RIM

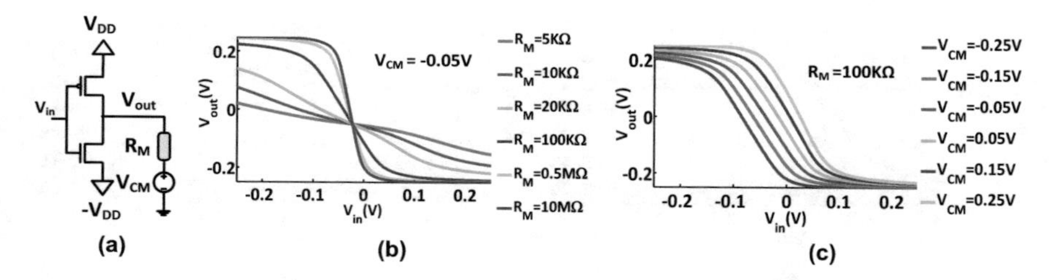

Fig. 19 (**a**) The circuit of an inverter with a resistive load and its VTC for different (**b**) RM values with VCM = −0.05 V and (**c**) VCM values with RM = 100 KΩ [18]

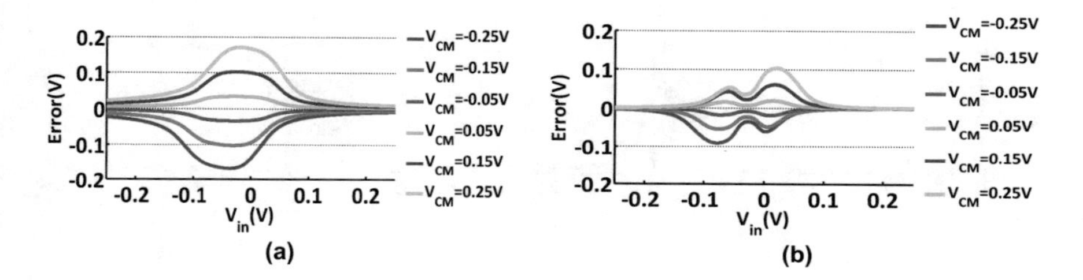

Fig. 20 The VTC approximation error of the (**a**) straight forward and (**b**) the formulation of Eq. (53) for an inverter with size = 2 and R_M = 100 KΩ [18]

considering only the approximate VTC slope which depends on the inverter output load (R_M) assuming $V_{CM} = 0$. The term $V_{CM}\frac{R_t}{R_t+R_M}$ generates the approximate vertical shift of the VTC due to V_{CM}. To demonstrate the superiority of this modeling approach compared to the straight forward method in which V_{out} is approximated with $f_n\left(V_{in}\right)|_{V_{CM}=0}$, the modeling error of these methods (the difference between the actual output voltage of the inverter and the one obtained using the mathematical model) is depicted in Fig. 20.

To approximate the vertical shift (see Eq. (53)), it is required to find the output resistance of the inverter (R_t). This resistance versus the input and output voltages of the inverter (with the inverter size as the running parameter) is plotted in Fig. 21. The output resistance versus the output voltage may be modeled with a simpler function (i.e., a parabolic one). The ERIM method used $A_s + B_s V_{\text{out}} + C_s {V_{\text{out}}}^2$ to model the characteristics of Fig. 21b which expresses R_t as a function of the output voltage and the size of the inverter [18]. The values of coefficients A_s, B_s, and C_s versus the size of the inverter (s) are depicted in Fig. 22. These coefficients were modeled by the function of $\alpha_\rho/(\beta_\rho + s)$ where ρ represents the considered coefficient (i.e., α_A represents the coefficient α of the fitted curve to A_s).

To calculate the amount of the vertical shift of the VTC curve due to the loading effect of the memristive crossbar (i.e., term $V_{CM}\frac{R_t}{R_t+R_M}$ in Eq. (53)), it is also required to find the V_{CM} and R_M. We can approximate R_M using Eq. (49) (similar to the LATIM method). The calculation of V_{CM} requires finding the Thevenin equivalent circuit branch connected to the output node of each inverter. Let us

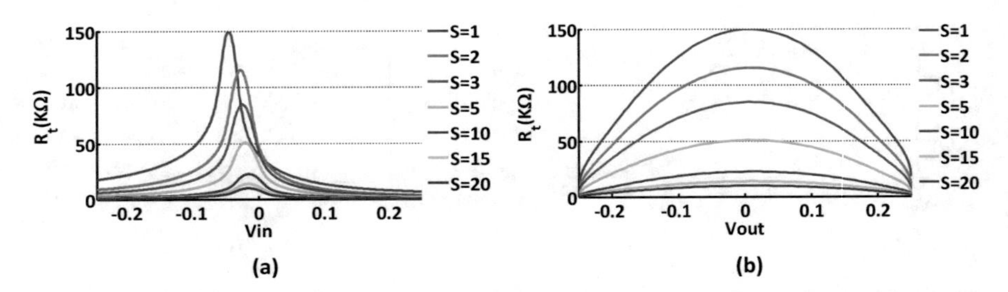

Fig. 21 Output resistance of inverters with different sizes based on the inverter (**a**) input voltage and (**b**) output voltage [18]

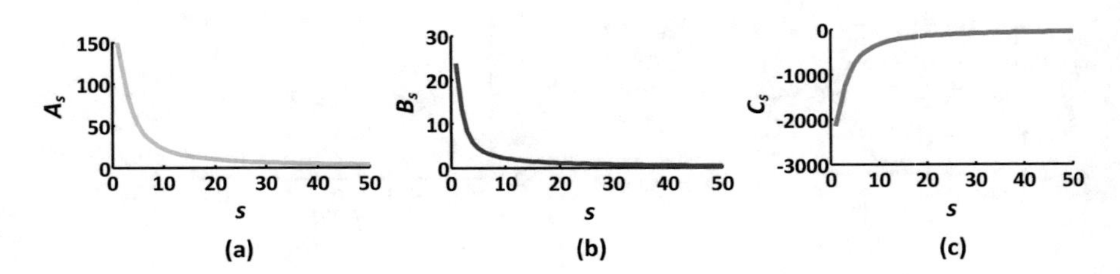

Fig. 22 Diagram of (**a**) A_s, (**b**) B_s, and (**c**) C_s versus the size of inverter (s) [18]

consider an inverter-based memristive crossbar with t inputs and u outputs as shown in Fig. 23a. The Thevenin equivalent circuit of each memristive branch connected to the output node of the inverters is depicted in Fig. 23b where Fig. 19a represents its Thevenin equivalent circuit. We can calculate R_M and V_{CM} as

$$R_M = \frac{1}{\sum_{j=1}^{u} \frac{1}{r_j}}, \tag{54}$$

and

$$V_{CM} = R_M \sum_{j=1}^{u} \frac{V_j}{r_j}. \tag{55}$$

Assume that r_{ji} and V_{ji} represent the values of r_j and V_j of the ith input inverter ($i \leq t$ and $j \leq u$). By comparing Figs. 17 and 23, one may find that r_{ji} can be approximated using Eq. (48) as

$$r_{ji} \approx R_{ji}. \tag{56}$$

Next, it is required to find V_{ji} which can be obtained by finding the open circuit voltage of each branch as shown in Fig. 23c. Therefore, only the ith input (X_i) is disconnected, and the other inputs are applied to the crossbar. In this case, the open

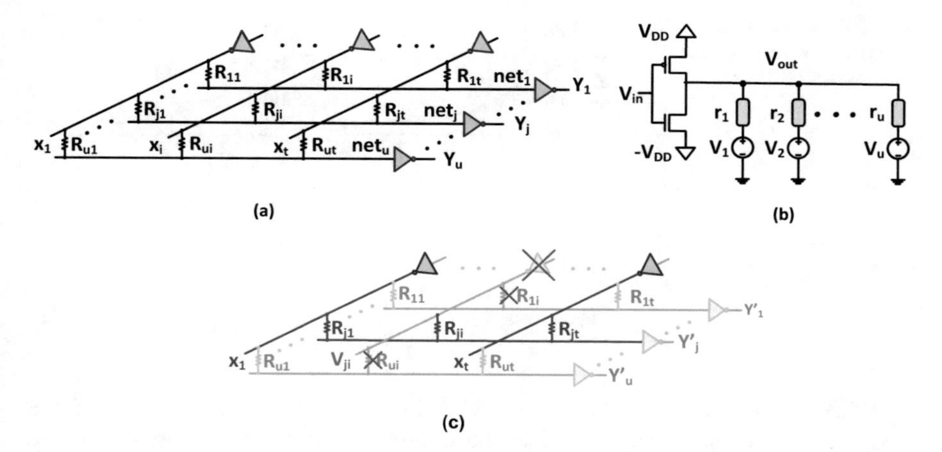

Fig. 23 (**a**) An inverter-based memristive crossbar with t inputs and u outputs. (**b**) Thevenin equivalent circuit of each memristive branch connected to the output node of each inverter. (**c**) Simplified circuit of a memristive crossbar to find the approximate value of VCM [18]

circuit voltage can be calculated as [18]

$$V_{ji} = \sum_{k \neq i} X_k \times \frac{\sigma_{jk}}{\gamma_{ji}}, \tag{57}$$

where $\gamma_{ji} = \sum_{k \neq i} \sigma_{jk}$. The net voltage of the jth output inverter (net_j), assuming that all inputs are applied to the crossbar, can be calculated as

$$net_j = \sum_{k=1}^{t} X_k \times \frac{\sigma_{jk}}{\gamma_j}, \tag{58}$$

where $\gamma_j = \sum_{k=1}^{t} \sigma_{jk}$. As the values of V_{ji} and net_j are almost similar, in the ERIM method, V_{ji} is approximately calculated as

$$V_{ji} \approx net_j. \tag{59}$$

The Thevenin equivalent resistance of the ith inverter ($R_{M,\,i}$) is obtained from

$$R_{M,i} \approx \frac{1}{\sum_{j=1}^{u} \frac{1}{R_{ji}}}, \tag{60}$$

and the Thevenin equivalent voltage of the ith inverter ($V_{CM,\,i}$) is calculated as

Fig. 24 The values of (**a**) R_n and (**b**) R_p versus the size of the inverter [18]

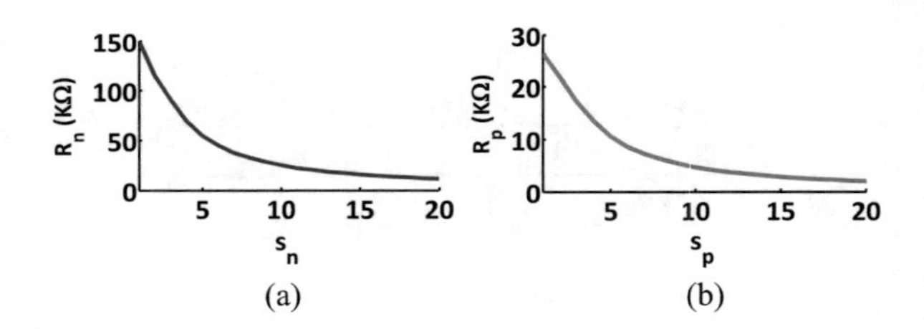

$$V_{CM,i} \approx R_{M,i} \sum_{j=1}^{u} \frac{net_j}{R_{ji}}. \tag{61}$$

ERIM is an enhanced version of RIM in which the loading effect of the memristive crossbar is considered in the training phase to increase the accuracy of the trained NN. Therefore, similar to the RIM method, the VTC slope of the negative and positive inverters should be controlled by adding the grounded resistors to the output node of the inverters to have low sensitivity to the variations of the circuit elements. Let us remind that the VTC slope of the inverter depends on the size of the inverter and the equivalent resistance connected to the output node of the inverter. Here, the sizes of the negative and positive inverters are denoted by s_n and s_p. In addition, R_n and R_p represent the equivalent grounded resistances connected to the output nodes of the negative and positive inverters leading to $d_n \approx 15$ and $b_p d_p \approx -1$, respectively (see Fig. 15). The values of R_n and R_p versus the size of the inverter (s) are plotted in Fig. 24. These resistances may be modeled using reciprocal functions (i.e., $R_n = A_n/(B_n + s_n)$) where the fitting coefficients (i.e., A_n and B_n) may be extracted using HSPICE and MATLAB simulations for the considered technology [18].

To find proper inverter sizes and the resistance of the grounded resistors (i.e., R_{Gn} and R_{Gp}) connected to the output node of the inverters, the equivalent resistive load of the memristive crossbar on the negative and positive inverters, which are denoted by R_{Mn} and R_{Mp}, respectively, may be calculated using Eq. (60). In this case, the equivalent resistance connected to the output node of the negative inverter can be calculated as

$$R_{eqn} = R_{Mn} \Big\| R_{Gn}. \tag{62}$$

To have $d_n \approx 15$ and $b_p d_p \approx -1$, the resistances R_{eq_n} and R_{eq_p} should be equal to R_n and R_p, respectively. In other words, the following equations should be satisfied [18]

$$R_n = \frac{1}{\frac{1}{R_{Mn}} + \frac{1}{R_{Gn}}}, \tag{63}$$

Therefore, $R_{Mn} > R_n$ which leads to

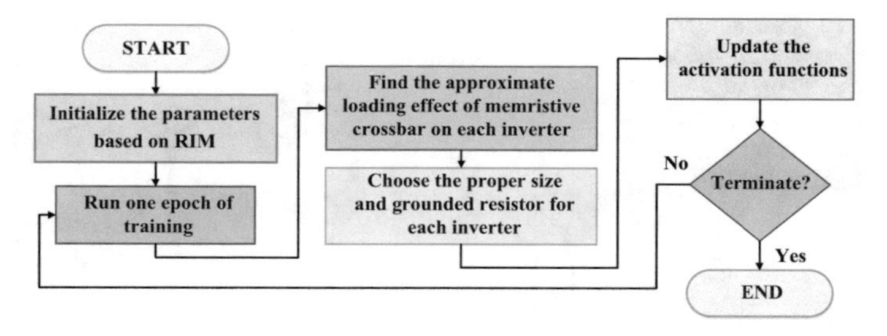

Fig. 25 The flowchart of the ERIM training method [18]

$$R_n = \frac{A_n}{B_n + s_n} < R_{Mn}. \tag{64}$$

Consequently, to have the lowest possible power consumption, the minimum inverter size that satisfy Eq. (64) can be obtained from

$$s_n = \left\lceil \frac{A_n}{R_{Mn}} - B_n \right\rceil. \tag{65}$$

Finally, R_{Gn} can be found using Eqs. (63) and (64). Similar equations were utilized to find the size and grounded resistor value of each positive inverter in the ERIM method.

The flowchart of the ERIM training method is depicted in Fig. 25. As shown in this chart, first, the size of the inverters and the resistance of their grounded resistors are selected similar to the RIM method. In other words, the sizes of all inverters are chosen as 5, and the grounded resistors are chosen such that $d_n \approx 15$ and $b_p d_p \approx -1$. Next, one epoch of training is performed similar to the RIM method, and the conductance of the memristors, the weights, and biases of the neurons are determined. Afterward, the Thevenin equivalent resistance and voltage of the memristive crossbar network on each inverter are found using Eqs. (60) and (61). Then, the size of each inverter and its grounded resistor are chosen based on Eqs. (65), (64), and (63). In the next steps, the activation functions are updated using Eq. (53), and the termination condition is checked to decide whether to continue or stop the training.

5.4.1 Implementation of Adjustable-Size Inverters

The size of the inverters of a chip cannot be changed after fabrication. Employing small-sized inverters would lead to smaller VTC slopes for a given output resistance. Also, as we observed in Fig. 12, the accuracy of the trained NN depends on the VTC slope of the inverters (larger slope would provide higher accuracy). On the other hand, utilizing large-sized inverters may increase the power consumption. To have

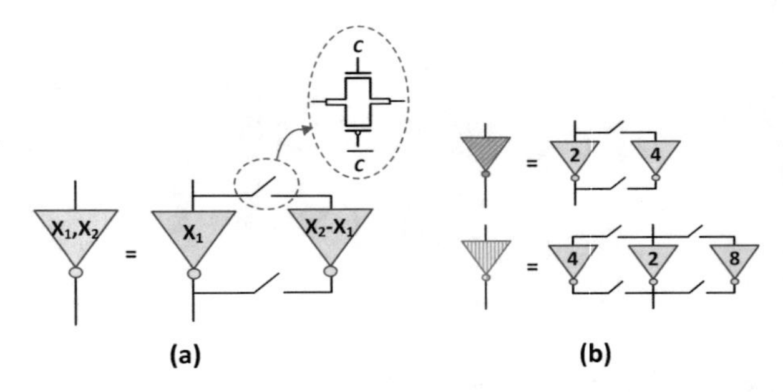

Fig. 26 (**a**) Implementation of an inverter with adjustable size. (**b**) An example of adjustable inverters composed of inverters with different sizes (modified from [18])

proper accuracy and power consumption, one may fabricate inverters with different sizes in the structure of an *IM*-based chip. To make it possible, in [18], it was proposed to utilize a bank of inverters with different sizes (i.e., with the sizes of 1, 2, 4, and 8) which can be connected to each other in parallel using transmission gates (TGs). This structure may satisfy the accuracy constraint while keeping the power consumption at the minimum possible level.

The structure of an adjustable-size inverter is depicted in Fig. 26a where two inverters with the size of X_1 and $X_2 - X_1$ are parallelized to construct an adjustable inverter. It can be utilized in two modes. In the first mode, the control signal (i.e., C shown in Fig. 26a) is logical "0" (or is equal to $-V_{DD}$), and the TG behaves as an off switch. The equivalent size of the adjustable inverter in this mode is equal to X_1. In addition, as the input node of the other inverter (with the size of $X_2 - X_1$) is float, the dynamic power consumption of this inverter is negligible. In the case where a small-sized inverter is required (with the size of X_1), the power consumption reduces as well. In the other mode where C is logical "1" (equal to V_{DD}), the TG behaves as an on switch, and the inverters are parallelized providing an inverter with equivalent size of X_2. The structures of two adjustable-size inverters composed of two and three inverters with the sizes of 2, 4, and 8 are depicted in Fig. 26b. The upper inverter can be utilized in two equivalent sizes of 2 and 6, while the bottom inverter can be utilized in four equivalent sizes of 2, 6, 10, and 14.

One may use the inverters shown in Fig. 26b in the structure of an *IM* crossbar. The structure of an *IM* crossbar composed of similar adjustable inverters is shown in Fig. 27a. For the NNs trained by the ERIM method, the VTC slope of the negative inverter is higher than the positive one which leads to the frequent use of larger sizes for the negative inverters (see Fig. 27a). Utilizing transmission gates increases the parasitic capacitance of the input and output nodes of the inverters which leads to higher delays. To handle this problem, one may employ a combination of fixed and adjustable size inverters in the structure of the crossbar, similar to the one shown in Fig. 27b. In this structure, the neurons which require small-sized inverters are mapped on the fixed-size inverters, while the neurons with larger-sized inverters are mapped on adjustable-size inverters.

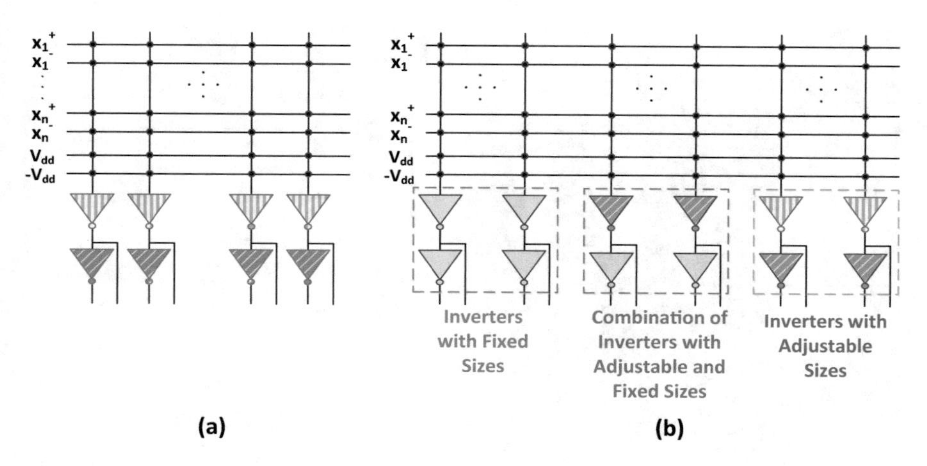

Fig. 27 The structure of a memristive crossbar composed of (**a**) similar adjustable size inverters, (**b**) a combination of fixed- and adjustable-size inverters (see Fig. 26b for the structures of the inverters) [18]

5.5 OCTAN[5]

Due to the complex mathematical model of *IM* neurons, the online training of these NNs using back propagation method, which is not a hardware-friendly task, becomes complicated. An on-chip training method, called OCTAN, updates the conductance of each memristor in an arbitrary direction and measures the error of the NN outputs [21]. If the error increases, the conductance of the memristor is updated in the reverse direction. Other approaches such as stochastic least mean squares (SLMS) [44, 45] and random weight change (RWC) [46] may also be utilized for training the NNs. In the SLMS method, only the weights of the output layer are updated, while the weights of the other layers are fixed random values [21]. In the RWC method, the weights of different layers are updated with constant values in a random direction [21]. Among the back propagation, SLMS, RWC, and OCATAN approaches, the back propagation (SLMS) has the highest (lowest) hardware complexity [21]. Due to the calculation of the derivative of the cost function (i.e., mean squared error (MSE)) with respect to the weights in the back propagation algorithm, this approach has the highest convergence speed among the methods. In contrast, in the RWC algorithm which relies on random changing of the weights, there is no guarantee that the random change will decrease the cost function. Hence, it may take several iterations to find the minimum point using this algorithm. OCTAN offers the simplicity of the RWC algorithm, while the updates in the conductance of the memristors are performed in a direction that leads to decrease

[5] On-chip training algorithm for the memristive neuromorphic circuits

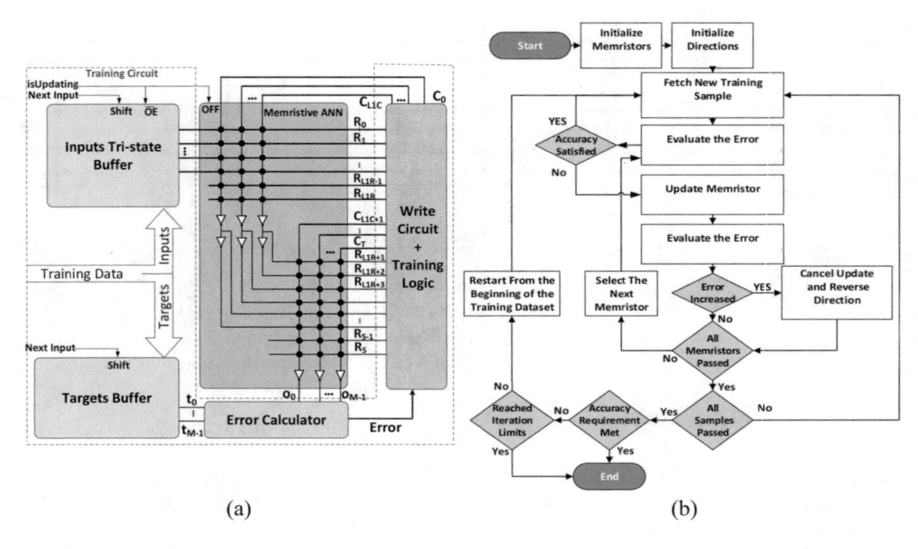

Fig. 28 (**a**) The overall view of OCTAN [21]. (**b**) The flowchart of OCTAN (modified from [21])

in the cost function in each iteration. Therefore, in comparison to RWC, it may take fewer iterations to find the minimum MSE.

The overall view of OCTAN, which includes an *IM*-NN and its online training circuit, is depicted in Fig. 28a. The training circuit includes two buffers, an error calculator unit, write circuit unit, and the training logic. The input and target buffers transmit the inputs and their corresponding target values one after another to the *IM*-NN. Next, the error calculator unit finds the absolute difference between the outputs of the NN and the target values and sends an error signal to the write circuit and training logic unit. In this unit, the conductance of the memristors is changed in an arbitrary direction (e.g., the conductance is added by $+\delta$), and the amount of the error for the considered sample is calculated again. If the current error is larger than the previously calculated error, the conductance of the memristor is changed in the opposite direction twice (added by -2δ). After updating the conductance of all the memristors, the next training sample should be applied to the *IM*-NN, and the steps are repeated until the termination condition is satisfied. The flowchart of the method is depicted in Fig. 28b.

6 Variation Mitigation Methods

In this section, state-of-the-art methods presented for reducing the severe effects of circuit element non-idealities on the accuracy of *IM*-NNs are explained. Note that RIM and ERIM methods, which reduce the NN accuracy degradation due to the non-idealities (by decreasing the VTC slope), were described in Sect. 5, and hence, they are not discussed here.

6.1 Variation-Aware Training (VAT)

As mentioned in Sect. 3, the output voltage of an *IM* neuron has the highest sensitivity to the horizontal shift of the inverter and buffer VTCs (coefficients $c^{-/+}$). In addition, the net voltage of the neuron may change due to the conductance variations of the memristors (see Eq. (5)). Using Eq. (9), one may write $\partial\left(y_j^{l\,+/-}\right)/\partial\left(net_j^l\right) = -\partial\left(y_j^{l\,+/-}\right)/\partial\left(c_j^{l\,+/-}\right)$. This implies that by reducing the sensitivity of the neuron output to the variations of coefficients $c^{+/-}$, the sensitivity to the conductance variations of the memristors is also reduced. Consequently, one may add a sensitivity term (ς) to the cost function of the training phase in which the sensitivity of the NN outputs to the variations of coefficients $c^{+/-}$ of different layers is considered when updating the NN weights [22]. Here, the cost function ($\mathcal{L}$) may be defined as

$$\mathcal{L} = \alpha J + \beta.\varsigma, \tag{66}$$

where the conventional cost function is denoted by J defined as the mean squared error (MSE) of the actual and expected NN output values for the regression applications and modified cross-entropy function for the classification applications [24]. Obviously, α and β determine the significance of J and ς terms during the training phase.

To obtain the mathematical interpretation of the sensitivity term, consider an *IM*-NN whose inputs and outputs are denoted by X and Y vectors with the sizes of $N^1 \times 1$ and $N_O \times 1$ where N^1 and N_O represent the number of the inputs and outputs of the NN. Furthermore, W^l_{total} ($=\left[W^{l+}, W^{l-}\right]$ where W^{l+}/W^{l-} refers to the positive/negative weight matrices) and WB^l_{total} represent the weight and bias matrices with the sizes of $N^{l+1} \times 2N^l$ and $N^{l+1} \times 1$, respectively. The element in the jth row and the ith column of W^{l+} (W^{l-}) is equal to $w^{l\ +}_{j,i}$ ($w^{l\ -}_{j,i}$), and the element in the jth row of WB^l_{total} is equal to $V_{DD}\left(wb_j^{l\,+} - wb_j^{l\,-}\right)$. Assuming that Y^l_{total} ($=\begin{bmatrix} Y^{l+} \\ Y^{l-} \end{bmatrix}$) with the size of $2N^{l+1} \times 1$ represents the output of the lth layer, NET^l_{total} with the size of $N^{l+1} \times 1$ may be obtained from

$$NET^l_{\text{total}} = W^l_{\text{total}} \times Y^{l-1}_{total} + WB^l_{\text{total}}, \tag{67}$$

where the element in the jth row of NET^l_{total} is equal to $net_j{}^l$. Furthermore, assuming that f^{l-} and f^{l+} model the VTC curve of the inverters and buffers, $Y^{l+/-}$ may be calculated as

$$Y^{l+/-} = f^{l+/-}\left(NET^{l}_{\text{total}}\right). \tag{68}$$

Assume that $A^{l+/-}$, $B^{l+/-}$, $C^{l+/-}$, $D^{l+/-}$, and $V^{l+/-}$ are vectors with the size of $N^{l+1} \times 1$ and the elements in their j^{th} rows are equal to $a_j^{l+/-}$, $b_j^{l+/-}$, $c_j^{l+/-}$, $d_j^{l+/-}$, and $v_j^{l+/-}$ $(j \le N^{l+1})$, respectively. Therefore, $Y^{l+/-}$ can be calculated as

$$Y^{l+/-} = A^{l+/-} + B^{l+/-} \circ \tanh\left(D^{l+/-} \circ \left(NET^{l}_{\text{total}} - C^{l+/-}\right)\right), \tag{69}$$

where $\circ$ represents the element-wise matrix multiplication. Let us define a sensitivity vector, called S^l. Here, the element in its jth row $(j \le N_O)$ may be calculated from [22]

$$s_j^l = \sum_{t=1}^{N^{l+1}} \left(\frac{\partial\left(y_j^L\right)}{\partial\left(c_t^{l+}\right)} + \frac{\partial\left(y_j^L\right)}{\partial\left(c_t^{l-}\right)} \right). \tag{70}$$

In other words, s_j^l represents the sensitivity of the jth output of the NN to the variations of coefficients $c^{+/-}$ of the lth layer. By substituting l with L in Eq. (70), one may find s_j^L as

$$s_j^L = \sum_{t=1}^{N_O} \frac{\partial\left(y_j^L\right)}{\partial\left(c_t^{L-}\right)}. \tag{71}$$

In the output layer of *IM*-NNs, only one inverter is utilized for the implementation of each neuron, and hence, no positive inverter is employed in the output layer and c_t^{L+} is not defined. In addition, $\partial\left(y_j^L\right)/\partial\left(c_t^{L-}\right) = -b_j^{L-} \times d_j^{L-} \times v_j^{L-}$ when $t = j$ and $\partial\left(y_j^L\right)/\partial\left(c_t^{L-}\right) = 0$ when $t \ne j$. Therefore,

$$s_j^L = -b_j^{L-} \times d_j^{L-} \times v_j^{L-}. \tag{72}$$

Consequently, S^L can be obtained from

$$S^L = -B^L \circ D^L \circ V^L. \tag{73}$$

For the other layers $(l < L)$, S^l can be found based on Eq. (70) as

$$S^l = \begin{bmatrix} \frac{\partial\left(y_1^L\right)}{\partial\left(c_1^{l\,+/-}\right)} + \cdots + \frac{\partial\left(y_1^L\right)}{\partial\left(c_{N^{l+1}}^{l\quad +/-}\right)} \\ \vdots \\ \frac{\partial\left(y_{N_O}^L\right)}{\partial\left(c_1^{l\,+/-}\right)} + \cdots + \frac{\partial\left(y_{N_O}^L\right)}{\partial\left(c_{N^{l+1}}^{l\quad +/-}\right)} \end{bmatrix}. \tag{74}$$

where $\partial\left(y_j^L\right)/\partial\left(c_t^{l+/-}\right) = \partial\left(y_j^L\right)/\partial\left(c_t^{l+}\right) + \partial\left(y_j^L\right)/\partial\left(c_t^{l-}\right)$. Taking into account that $h_j^L = \partial\left(y_j^L\right)/\partial\left(net_j^L\right) = b_j^{L^-} \times d_j^{L^-} \times v_j^{L^-}$, Eq. (74) can be rewritten as

$$S^l = H_{\text{total}}^L \circ \left(W_{\text{total}}^L \times \begin{bmatrix} \frac{\partial\left(y_1^{L-1^+}\right)}{\partial\left(c_1^{l\,+/-}\right)} + \cdots + \frac{\partial\left(y_1^{L-1^+}\right)}{\partial\left(c_{N^{l+1}}^{l\quad +/-}\right)} \\ \vdots \\ \frac{\partial\left(y_{NL}^{L-1^+}\right)}{\partial\left(c_1^{l\,+/-}\right)} + \cdots + \frac{\partial\left(y_{NL}^{L-1^+}\right)}{\partial\left(c_{N^{l+1}}^{l\quad +/-}\right)} \\ \frac{\partial\left(y_1^{L-1^-}\right)}{\partial\left(c_1^{l\,+/-}\right)} + \cdots + \frac{\partial\left(y_1^{L-1^-}\right)}{\partial\left(c_{N^{l+1}}^{l\quad +/-}\right)} \\ \vdots \\ \frac{\partial\left(y_{NL}^{L-1^-}\right)}{\partial\left(c_1^{l\,+/-}\right)} + \cdots + \frac{\partial\left(y_{NL}^{L-1^-}\right)}{\partial\left(c_{N^{l+1}}^{l\quad +/-}\right)} \end{bmatrix} \right), \tag{75}$$

where the element in the jth row of H_{total}^L is h_j^L. Assuming that

$$g_i^{l',l+} = \sum_{t=1}^{N^{l+1}} \left(\frac{\partial\left(y_i^{l'+}\right)}{\partial\left(c_t^{l+}\right)} + \frac{\partial\left(y_i^{l'+}\right)}{\partial\left(c_t^{l-}\right)} \right), \tag{76}$$

and

$$g_i^{l',l-} = \sum_{t=1}^{N^{l+1}} \left(\frac{\partial\left(y_i^{l'-}\right)}{\partial\left(c_t^{l+}\right)} + \frac{\partial\left(y_i^{l'-}\right)}{\partial\left(c_t^{l-}\right)} \right), \tag{77}$$

one may rewrite Eq. (75) as

$$S^l = H^L_{\text{total}} \circ \left(W^L_{\text{total}} \times \begin{bmatrix} g_1^{L-1,l+} \\ \vdots \\ g_{N^L}^{L-1,l+} \\ g_1^{L-1,l-} \\ \vdots \\ g_{N^L}^{L-1,l-} \end{bmatrix} \right) = H^L_{\text{total}} \circ \left(W^L_{\text{total}} \times G^{L-1,l}_{\text{total}} \right), \tag{78}$$

where $G^{l',l}_{\text{total}} = \begin{bmatrix} G^{l',l+} \\ G^{l',l-} \end{bmatrix}$ and the element in the ith row of $G^{l',l+/-}$ $(i \leq N^{l'+1})$ is equal to $g_i^{l',l+/-}$. To find S^l based on Eq. (75), it is required to find $\partial \left(y_i^{L-1^+} \right) / \partial \left(c_t^{l^+} \right)$ which can be calculated as

$$\frac{\partial \left(y_i^{L-1^+} \right)}{\partial \left(c_t^{l^+} \right)} = h_i^{L-1^+} \sum_{k=1}^{N^{L-1}} \left(w_{i,k}^{L-1^+} \times \frac{\partial \left(y_k^{L-2^+} \right)}{\partial \left(c_t^{l^+} \right)} + w_{i,k}^{L-1^-} \times \frac{\partial \left(y_k^{L-2^-} \right)}{\partial \left(c_t^{l^+} \right)} \right), \tag{79}$$

where $h_i^{L-1^+} = \partial \left(y_i^{L-1^+} \right) / \partial \left(net_i^{L-1} \right)$. Similarly, $\partial \left(y_i^{L-1^{+/-}} \right) / \partial \left(c_t^{l^{+/-}} \right)$ may be obtained. Therefore,

$$g_i^{L-1,l+} = h_i^{L-1^+} \sum_{k=1}^{N^{L-1}} \left(w_{i,k}^{L-1^+} \times g_k^{L-2,l^+} + w_{i,k}^{L-1^-} \times g_k^{L-2,l^-} \right), \tag{80}$$

where

$$g_k^{L-2,l^+} = \sum_{t=1}^{N^{l+1}} \left(\frac{\partial \left(y_k^{L-2^+} \right)}{\partial \left(c_t^{l^+} \right)} + \frac{\partial \left(y_k^{L-2^+} \right)}{\partial \left(c_t^{l^-} \right)} \right), \tag{81}$$

and

$$g_k^{L-2,l^-} = \sum_{t=1}^{N^{l+1}} \left(\frac{\partial \left(y_k^{L-2^-} \right)}{\partial \left(c_t^{l^+} \right)} + \frac{\partial \left(y_k^{L-2^-} \right)}{\partial \left(c_t^{l^-} \right)} \right). \tag{82}$$

Consequently, $G^{L-1,l}_{\text{total}}$ can be calculated similar to Eq. (78) as

$$G^{L-1,l}_{\text{total}} = H^{L-1}_{\text{total}} \circ \left(W^{L-1}_{\text{total}} \times G^{L-2,l}_{\text{total}} \right), \tag{83}$$

where $H^{l'}_{\text{total}} = \begin{bmatrix} H^{l'+} \\ H^{l'-} \end{bmatrix}$ and the element in the jth row of $H^{l'^{+/-}}$ $(j \leq N^{l'+1})$ is equal to $h_j^{l'+/-} = \partial \left(y_j^{l'+/-} \right) / \partial \left(net_j^{l'} \right)$. Thus, $G^{l',l}_{\text{total}}$ can be recursively calculated as

$$G^{l',l}_{\text{total}} = H^{l'}_{\text{total}} \circ \left(W^{l'}_{\text{total}} \times G^{l'-1,l}_{\text{total}} \right). \tag{84}$$

The amounts of $\partial \left(y_i^{l+} \right) / \partial \left(c_t^{l+} \right)$, $\partial \left(y_i^{l-} \right) / \partial \left(c_t^{l+} \right)$, $\partial \left(y_i^{l+} \right) / \partial \left(c_t^{l-} \right)$, and $\partial \left(y_i^{l-} \right) / \partial \left(c_t^{l-} \right)$ become zero for $i \neq t$. Thus, $g_i^{l,l+} = \partial \left(y_i^{l+} \right) / \partial \left(c_i^{l+} \right) = -\partial \left(y_i^{l+} \right) / \partial \left(net_i^l \right) = -h_i^{l+}$ and $g_i^{l,l-} = \partial \left(y_i^{l-} \right) / \partial \left(c_i^{l-} \right) = -\partial \left(y_i^{l-} \right) / \partial \left(net_i^l \right) = -h_i^{l-}$. In other words, $G^{l,l}_{\text{total}} = -H^{l}_{\text{total}}$.

The procedure of calculating S^l based on Eqs. (78) and (84) is depicted in Fig. 29. These calculations are also valid when a batch of training samples (with n_s samples) is used in the training phase. In this case, X, Y, $Y^{l'}_{\text{total}}$, $H^{l'}_{\text{total}}$, $G^{l',l}_{\text{total}}$, and S^l are matrices with the size of $N^1 \times n_s$, $N_O \times n_s$, $2N^{l'+1} \times n_s$, $2N^{l'+1} \times n_s$, $2N^{l'+1} \times n_s$, and $N_O \times n_s$, respectively.

Finally, the sensitivity term (ς) used in Eq. (66) can be obtained from

$$\varsigma = \sum_{l=1}^{L} \sum_{j=1}^{N_O} \sum_{p=1}^{n_s} \left| s^l_{j,p} \right| \tag{85}$$

where $s^l_{j,p}$ represents the sensitivity of the jth output of the NN to the variations of coefficients of the lth layer for the pth training sample.

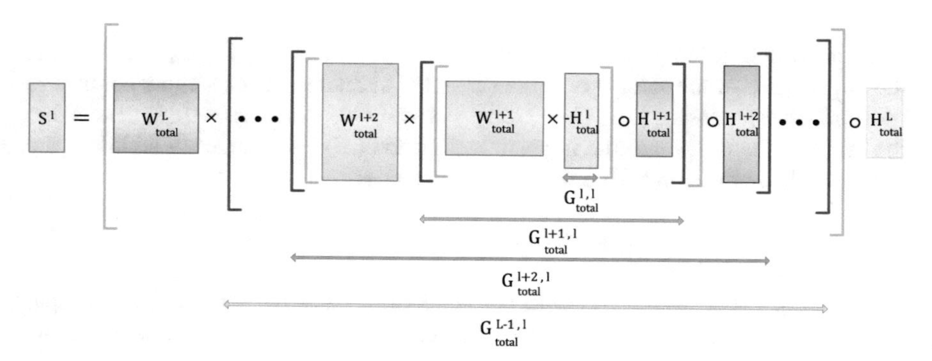

Fig. 29 The matrix calculations for the generation of S^l [22]

6.2 INTERSTICE[6]

As mentioned in Sect. 3, the sensitivity of an *IM* neuron output to the conductance variations of the memristors as well as the characteristic variations of the inverters may be mitigated by decreasing the parameter v or forcing the inverters to operate in the low-gain regions of their VTCs. In classification applications, the output voltage of the NN is expected to be nearly equal to V_{DD} ($-V_{DD}$) for the selected (unselected) classes. In other words, it is expected that the inverters of the output layer operate in the low-gain regions of their VTCs, which results in smaller values for v parameters (see Fig. 6) and lower sensitivity to the non-idealities of circuit elements. In regression applications, however, the output voltage of the NN varies in the range of $[-V_{DD}, +V_{DD}]$ which may lead to larger values for the v parameters. These facts lead us to employing an *IM*-based classification NN to predict the output range of a regression application and then approximate the final output based on the predicted range [24]. The method is called INTERSTICE.

In INTERSTICE (K), the total output range of the regression problem is divided into $2K - 1$ subranges, and a classification NN with K output classes is trained to predict the subrange to which each applied sample belongs. To train the classification NN, it is required to label the training samples based on their target values (O_T). The labels determine whether the sample belongs to the ith class (i.e., $C_i = 1$) or not (i.e., $C_i = 0$). To clarify this further, assume that $[a, b]$ represents the total output range of the regression problem and the classification NN has K output classes. In this case, the labels for each sample may be calculated as

$$C_i = \begin{cases} 1 & \text{if} \quad \frac{b-a}{4K} \max(4i - 5, 0) \leq O_T - a < \frac{b-a}{4K} \min(4i + 1, 4K) \quad \text{for} \quad 1 \leq i \leq K \\ 0 & \text{Otherwise} \end{cases} \tag{86}$$

If the sample belongs to only one class (i.e., $C_i = 1$ and $C_{j \neq i} = 0$), it means that the target value of the sample is in the $(2i - 1)^{st}$ subrange. If the sample belongs to two consecutive classes (i.e., $C_i = C_{i+1} = 1$ and $C_{j \neq (i \, or \, i+1)} = 0$), it means that the target value of the sample is positioned in the $(2i)^{th}$ subrange. In addition, the output of the regression problem is approximated as

$$I_{APX,j} = a + (j \times (b - a) / 2K), \tag{87}$$

where j $(1 \leq j \leq 2K - 1)$ represents the index of the selected subrange. The output generator (OG) unit is responsible to generate the approximate output based on the selected classes.

Now, assume that the total output range of the regression problem is $[a, b]$ and a classification NN with four output classes is utilized to predict the subrange to

---

[6] Inverter-based memristive neural network discretization

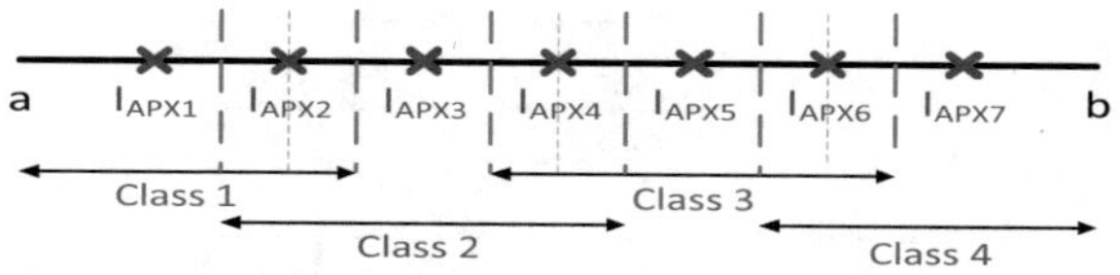

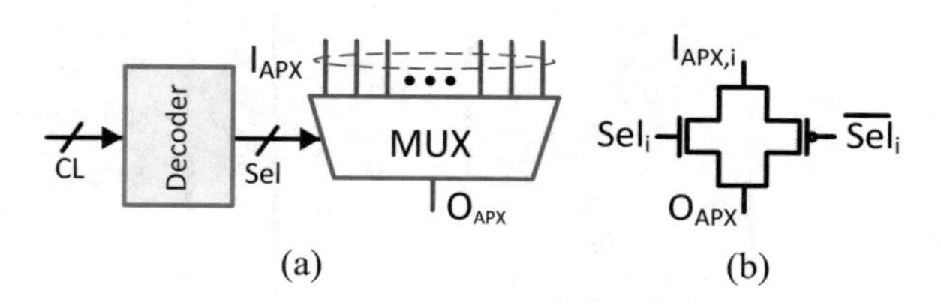

Fig. 30 The ranges of the classes and the values of the approximated outputs for INTERSTICE (4) method [24]

Fig. 31 The internal structure of the (**a**) OG module and (**b**) the TG-based multiplexer [24]

which the regression output belongs. In this case, the total output range (i.e., $[a, b]$) is divided into seven subranges, and the possible approximate output values can be determined based on Eq. (87) for $1 \leq j \leq 7$ (see Fig. 30). As an example, assume that only the second class of the classification NN is selected (i.e., $C_2 = 1$). This shows that the sample belongs to the third subsection, and the OG unit should pass I_{APX3} as the approximate output value. Consider another case in which the *third* and the *fourht* classes are selected which shows that the sample belongs to the *sixth* subrange. In this case, I_{APX6} should be considered as the approximate output value (see Fig. 30).

Furthermore, the internal structure of the OG unit which consists of a decoder and a transmission gate (TG)-based multiplexer is depicted in Fig. 31. The inputs of this unit (CL) are the buffered outputs of the classification NN. As shown in Fig. 31a, the decoder generates the *Sel* signal based on the selected classes by the classifier. The *Sel* signal is then applied to the multiplexer to select and pass the appropriate $I_{APX,j}$ value to the output (O_{APX}).

Even though in INTERSTICE method it is expected that only one class or two consecutive classes are selected, it is probable that in some cases the NN mistakenly selects none of the classes or more than two classes. In these cases, however, the OG unit should be able to select one of the $I_{APX,j}$ values as the final output value. The OG unit may use three rules to generate the final output [24]. The rules are presented in the following (see Fig. 32).

1. When no class or all of the classes are selected (all of CL_k are 0 or 1), the midpoint of the total range ($(a + b)/2$) is considered as O_{APX}. This rule is indicated by the symbol ♠ in Fig. 32.
2. When two nonconsecutive classes are selected by the classifier, the average of their midpoints is considered as O_{APX}. This rule is denoted by the symbol ♣ in Fig. 32.
3. When m classes are selected ($2 < m < K$), the decoder finds the largest chain of consecutive selected classes. The average of the midpoints of these classes is considered as O_{APX}. When none of the selected classes is consecutive, the

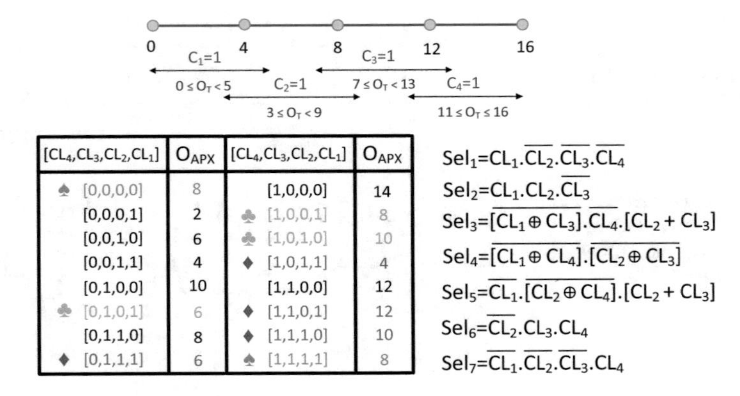

[CL4,CL3,CL2,CL1]	OAPX	[CL4,CL3,CL2,CL1]	OAPX
♠ [0,0,0,0]	8	[1,0,0,0]	14
[0,0,0,1]	2	♣ [1,0,0,1]	8
[0,0,1,0]	6	♣ [1,0,1,0]	10
[0,0,1,1]	4	♦ [1,0,1,1]	4
[0,1,0,0]	10	[1,1,0,0]	12
♣ [0,1,0,1]	6	♦ [1,1,0,1]	12
[0,1,1,0]	8	♦ [1,1,1,0]	10
♦ [0,1,1,1]	6	♠ [1,1,1,1]	8

Fig. 32 The *Sel* signals values and the O_{APX} for the case where the number of classes is four and the range of the target value is [0, 16] [24]

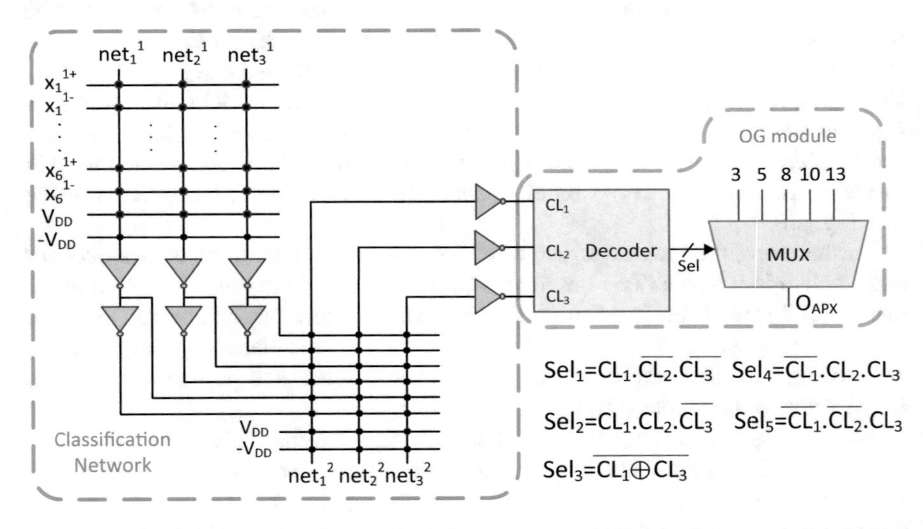

Fig. 33 The circuit diagram of INTERSICE (3) for a network with six inputs and three hidden neurons [24]

average of the midpoints of all the selected classes is considered as O_{APX}. This rule is indicated by the symbol ♦ in Fig. 32.

In Fig. 33, the hardware implementation of INTERSTICE method for a two-layer *IM*-NN implemented using INTERSTICE (3) method is shown. In this figure, it is assumed that the total output range of the regression problem is [0,16] and the final output (O_{APX}) is a 4-bit digital signal. The probable approximate output values ($I_{APX,j}$) become 3, 5, 8, 10, and 13 (see Eq. (87)), and the OG module selects one of these values based on the *Sel* signals which are determined by the decoder unit based on the selected classes by the classification NN (*CL* signals).

7 Comparison of Different Training Methods

In this section, the efficacies of different training methods used for *IM*-NNs are evaluated using MATLAB, Python, and HSPICE simulations under different regression and classification applications. Furthermore, the performances of different variation mitigation methods are studied. The training procedure is performed using MATLAB and Python (Keras and TensorFlow libraries) simulations. The electrical parameters of the networks (e.g., delay and power) and the accuracies of the trained NNs are measured using HSPICE simulations using a 90 nm technology and the memristor model of [47]. The employed memristor has a maximum and minimum conductance values of 8 μ℧ and 0.12 μ℧, respectively [21]. In addition, the positive and negative voltage sources of the network are considered to be 0.25 V and −0.25 V.

7.1 *Training Methods*

The characteristics of the *IM*-NNs in the ideal condition, extracted by an in-house simulator developed by Python programing, are reported in Table 1 where SePHAX, SeRIM, and SeINTERSTICE represent the cases where the sensitivity term is added to the cost function of the training phase of the PHAX, RIM, and INTERSTICE methods, respectively. In addition, the network configuration represents the number of inputs, hidden neurons, and outputs of the NN. The results show that by combining VAT with other training methods, the accuracy (*Acc*) for classification applications, and mean absolute error (*MAE*) for the regression application degrade in the ideal condition. The value of $v^{l^{+/-}}$ $(l \leq 3)$, however, decreases in most of the cases potentially leading to higher resiliency of the NN accuracy to the non-idealities of the circuit elements. In addition, by employing the RIM method, the maximum achievable accuracies for the MNIST and Fashion MNIST datasets decrease compared to those of the PHAX caused by the lower VTC slope of the inverters in the RIM method.

The MSE, delay, power, and energy consumption of the NNs (obtained by HSPICE simulations) trained by PHAX, RIM, and INTERSTICE methods under regression applications are given in Table 2 (extracted from [24] and [25]). In addition, PHAX (ADC) shows the case where the inverter-based memristive ADCs with 4-bit resolution is utilized as the output interfaces of the NNs trained by the PHAX method. In addition, in the training phase of the PHAX (ADC) and INTERSTICE cases, the whole conductance range of the memristors is not utilized to reduce the loading effect of the memristive crossbar on the output voltage of the NN. More specifically, $\sigma_{\max}$ is assumed to be 1 μ℧ instead of 8 μ℧ in Eq. (39) (used in the training phase of these cases).

The results of Table 2 reveal that the NNs trained by RIM method can generate the outputs with lower delays compared to those of the INTERSTICE and PHAX

Table 1 The characteristics of the *IM*-NNs trained based on PHAX, RIM, INTERSTICE, and their combination with VAT method under different benchmarks in the ideal condition [22]

Benchmark	Training method	Network configuration	$\left(\text{Acc}/\text{MAE}\right)_{train}$	$\left(\text{Acc}/\text{MAE}\right)_{test}$	v^{1-}	v^{1+}	v^{2-}	v^{2+}	v^{3-}
California housing	PHAX	8-8-4-1	0.043	0.053	0.63	0.17	0.70	0.09	0.86
	SePHAX		0.060	0.066	0.15	0.02	0.18	0.01	0.89
	RIM		0.047	0.056	0.78	0.71	0.93	0.89	0.83
	SeRIM		0.051	0.052	0.78	0.70	0.84	0.74	0.85
	INTERSTICE	8-12-8-4	0.047	0.047	0.62	0.11	0.54	0.11	0.43
	SeINTERSTICE		0.053	0.054	0.37	0.10	0.09	0.01	0.49
MNIST	PHAX	784-30-40-10	97.9	95.7	0.70	0.13	0.55	0.05	0.03
	SePHAX		93.4	93.2	0.66	0.10	0.35	0.04	0.04
	RIM		91.9	92.2	0.65	0.54	0.44	0.31	0.13
	SeRIM		90.5	90.9	0.47	0.38	0.34	0.22	0.10
Fashion MNIST	PHAX	784-30-40-10	92.3	87.0	0.71	0.14	0.54	0.06	0.06
	SePHAX		86.6	84.6	0.60	0.10	0.37	0.05	0.08
	RIM		85.8	84.2	0.65	0.53	0.46	0.34	0.13
	SeRIM		83.2	82.0	0.46	0.38	0.35	0.24	0.14

Table 2 Comparing the delay, power, energy, and MSE of the NNs trained by PHAX, INTERSTICE, and RIM approaches in the considered regression applications

Benchmark	Training method	Network configuration	Delay (nS)	Power (μW)	Energy (fJ)	MSE ($\times 10^{-3}$)
Blackscholes	PHAX	6 – 6 – 1	3.9	11.0	43.0	4
	PHAX (ADC)	6 – 6 – 1	15.1	14.8	223	6.2
	INTERSTICE (2)	6 – 2 – 2	5.0	4.4	22	6.6
	INTERSTICE (3)	6 – 4 – 3	5.0	9.1	46	4.9
	INTERSTICE (4)	6 – 6 – 4	5.3	14.3	76	3.3
	RIM	6 – 5 – 1	1.6	22.8	36.5	3.3
FFT	PHAX	1 – 8 – 2	4.3	10.6	45.5	9.7
	PHAX (ADC)	1 – 8 – 2	16.2	18.5	300	6.9
	INTERSTICE (2)	1 – 2 – 4	6.0	7.6	46	9.7
	INTERSTICE (3)	1 – 5 – 6	6.4	14.9	95	5.5
	INTERSTICE (4)	1 – 6 – 8	7.1	18.7	133	4.6
	RIM	1 – 5 – 2	2.8	27.4	76.7	3.7
K-means	PHAX	6 – 8 – 1	2.8	15.6	43.7	9
	PHAX (ADC)	6 – 8 – 1	16.2	21.8	353	1.6
	INTERSTICE (2)	6 – 3 – 2	4.4	6.1	27	2.5
	INTERSTICE (3)	6 – 3 – 3	5.3	7.5	40	2.6
	INTERSTICE (4)	6 – 7 – 4	4.7	14.1	66	1.6
	RIM	6 – 6 – 1	1.5	28.1	42.1	0.4
Sobel	PHAX	9 – 8 – 1	3.8	10.3	39.2	28.5
	PHAX (ADC)	9 – 8 – 1	13	15.9	207	18
	INTERSTICE (2)	9 – 6 – 2	3.9	11.9	46	10.4
	INTERSTICE (3)	9 – 8 – 3	3.7	15.7	58	10.5
	INTERSTICE (4)	9 – 9 – 4	4.0	19.1	76	12.8
	RIM	9 – 5 – 1	1.4	21.4	29.9	17.2

methods. Due to the added grounded resistors, however, the power consumptions of the NNs trained by RIM method become higher than those of the ones trained by PHAX and INTERSTICE methods.

As an example, the power consumption (delay) of the NNs trained by the RIM method is, on average, 113% (51%) higher (lower) than that of the ones trained by the PHAX method. Furthermore, it should be stated that the NNs trained by INTERSTICE can generate digital outputs without requiring ADCs as the output interfaces significantly reducing their delays and power consumptions compared to those of the PHAX (ADC) case. As an example, the delay and power consumption of the NNs trained by INTERSTICE are, on average, 67% and 31% smaller than those of the PHAX (ADC) case. In addition, except for Sobel, in all of the considered benchmarks, the MSEs of the NNs trained by the RIM method are smaller than those of the other methods.

To compare the efficacy of different training methods in classification applications, *IM*-NNs were trained by the PHAX, RIM, ERIM, and LATIM methods using Python-based simulator and the electrical parameters as well as the accuracy of the trained NNs were obtained using HSPICE simulations ([18] and [23]). The results are presented in Table 3 in which *OE* is the mean absolute difference between the voltage of the output layer outputs extracted by Python and HSPICE simulations. The results reveal that all of the training methods have acceptable accuracy when the size of the NN is small (i.e., for IRIS and BCW datasets). For larger-sized NNs (i.e., for MNIST and Fashion MNIST datasets), however, the efficacy of the PHAX method diminishes significantly, while RIM, ERIM, and LATIM still have acceptable accuracies. In the PHAX method, the VTCs of all the inverters and buffers of the NN are considered to be the same without taking into account the loading effect of the memristor crossbars. In the RIM method, the effect of memristor crossbars on the output voltage of the inverters is reduced by adding the grounded resistor, while in the ERIM and LATIM methods, this loading effect is mathematically modeled to find the output voltage of the inverters with a higher accuracy. In addition, due to the lower VTC slope of the inverters and buffers in the RIM and ERIM methods, the accuracies of the NNs trained by these methods are smaller than those of PHAX and LATIM in the ideal condition.

Furthermore, *OE* may be used as a criterion to compare the efficacy of different training methods in modeling the behavior of the NN. In other words, smaller *OE* values show that the mathematical model of the NN used in the training phase can estimate the output voltage of the NN with a higher accuracy. The results of Table 3 show that the modeling errors of the ERIM and LATIM methods are significantly smaller than those of PHAX and RIM as the output voltages of the inverters are modeled with a higher accuracy in the former methods. In addition, the power consumptions of the ERIM and LATIM methods are considerably smaller than that of the RIM method due to the proper sizing of the inverters.

As mentioned in Subsection 5.2, in *IM*-NNs with smaller d coefficients (i.e., in the RIM and ERIM methods with $d = 15$ which leads to a larger transition region in the VTC (see Fig. 13)), the *net* values should be varied in larger ranges to benefit from the whole output swing of the inverters. This is translated to larger

Table 3 Comparison of the electrical parameters as well as the accuracy of the NNs trained by PHAX, RIM, ERIM, and LATIM methods

Dataset	Training method	Network configuration	Train samples		Test samples				
			Acc (%)	OE (mV)	Acc (%)	OE (mV)	Delay (ns)	Power (μW)	Energy (fJ)
IRIS	PHAX	4 – 5 – 3	97.5	7	96.7	7	4.0	5.8	23.1
	RIM		95.8	8	96.7	8	1.5	30.5	45.8
	ERIM		95.8	10	96.7	9	1.1	9.6	10.5
	LATIM		98.3	8	100	7	3.0	3.7	11.2
BCW	PHAX	10 – 5 – 2	98.5	4	96.3	4	3.1	6.2	19.3
	RIM		97.4	3	96.3	3	1.2	29.3	35.1
	ERIM		97.6	7	95.6	7	1.0	9.0	9.0
	LATIM		98.2	7	96.3	8	1.8	2.8	5.1
MNIST	PHAX	784 – 30–40 – 10	28.0	106	31.7	107	2.1	164.7	345.9
	RIM		89.8	45	88.9	46	1.6	315.7	505.1
	ERIM		91.1	12	89.2	14	1.3	110.2	143.3
	LATIM		93.6	7	92.3	8	2.3	77.8	396.5
Fashion MNIST	PHAX	784 – 30–40 – 10	26.6	263	24.5	258	2.1	174.6	366.7
	RIM		83.0	39	82.1	39	1.4	304.7	426.6
	ERIM		83.6	12	82.6	12	1.4	106.4	149.0
	LATIM		86.1	9	83.7	9	1.9	158.4	306.0

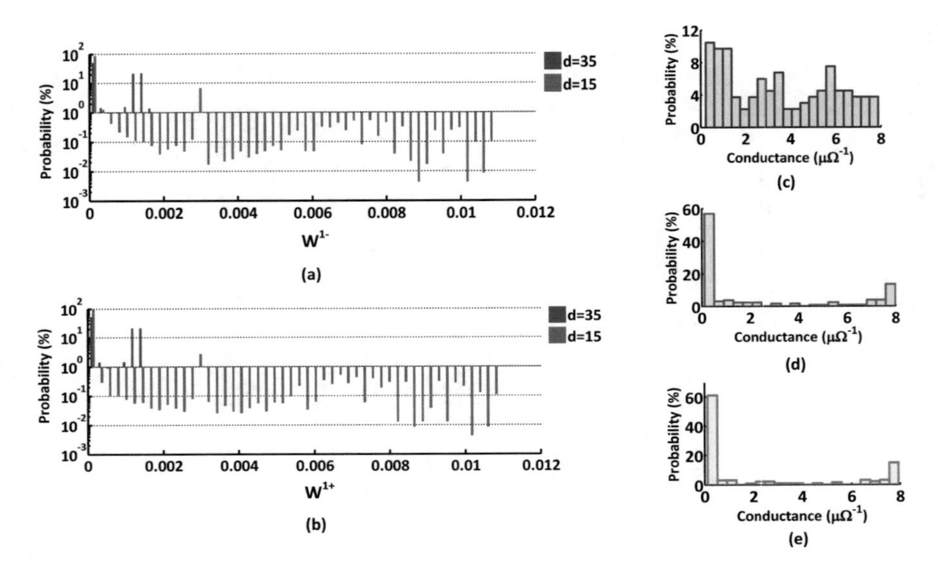

Fig. 34 The distributions of the (**a**) negative, (**b**) positive weights of the first layer of *IM*-NNs under the MNIST benchmark for different *d* values. The distributions of memristor conductance values of the NNs trained by the (**c**) PHAX, (**d**) RIM, and (**e**) ERIM methods for the BCW dataset [18]

values for some of the weights. Since the sum of weights should be equal to 1, the other weights become very small. Consequently, the conductance values of the memristors stick to σ_{max} or σ_{min} during the training phase to achieve large or very small weights. When d is large (i.e., in PHAX where $d = 35$ which leads to a smaller transition region in the VTC), the whole output swing of the inverters can be reached even when *net* voltage changes in a small range. Thus, it is not necessary to have very large weights letting the memristors take different conductance values between σ_{max} and σ_{min}. As an example, the distribution of the weights of the first layer of the *IM*-NNs under the MNIST benchmark is plotted in Fig. 34a and b, showing that the weights take larger values when d is smaller. In addition, the distribution of the conductance values of the memristors of the NNs trained by PHAX, RIM, and ERIM (for BCW dataset) are plotted in Fig. 34c, d, and e. The figure shows that the memristor conductance values of the NNs trained by RIM and ERIM are frequently equal to σ_{max} or σ_{min}.

It is worth noting that the delay of a path is inversely proportional to the conductance of the memristors that convey the signal. As shown in Fig. 34, in the RIM and ERIM methods, most of the conductance values are almost equal to $0.12\ \mu\Omega^{-1}$ ($=\sigma_{min}$) or $8\ \mu\Omega^{-1}$ ($=\sigma_{max}$). It should be noted that the memristors with the conductance of σ_{min} have considerably a lower impact on the *net* voltage of the inverters compared to the ones with the conductance of σ_{max}. Therefore, the memristors with the conductance of $8\ \mu\Omega^{-1}$ are more critical for determining the

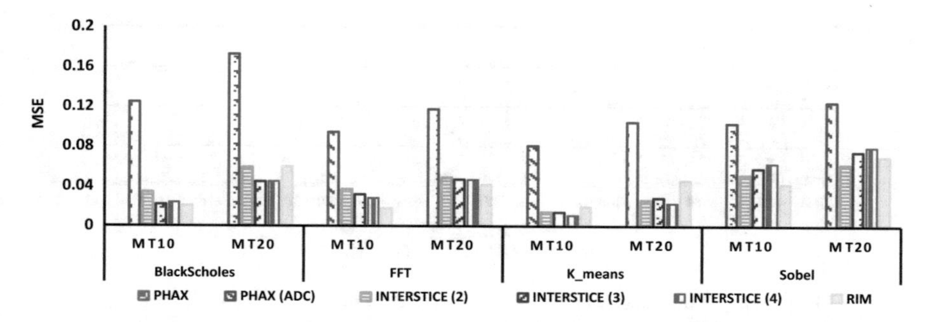

Fig. 35 Worst-case MSE of the *IM*-NNs trained by different methods considering transistors and memristors non-idealities

delay of the NNs trained by the RIM and ERIM methods. As Fig. 34 shows, the conductance values of the memristors of the NNs trained by PHAX are frequently in the range of [1 $\mu\Omega^{-1}$, 8 $\mu\Omega^{-1}$] conveying the fact that the *net* voltage of the inverters of these NNs depends on almost all of the inputs of the neurons which may have conductance values lower than 8 $\mu\Omega^{-1}$. Therefore, as the delay of the NN is inversely proportional to the conductance of the memristors that affect the *net* voltages of the inverters, it is expected that the delay of the NNs trained by PHAX becomes larger than that of the RIM and ERIM methods which is also in accordance with the results of Tables 2 and 3.

7.2 *IM-NN Accuracy in the Presence of Variations*

To evaluate the effect of non-idealities on the performance of the trained NNs, random variations with Gaussian distributions are applied to the conductance of the memristors as well as the width, length, and threshold voltage of the transistors. The results of 500 simulations are used to measure the mean (μ_{MSE}) and standard deviation (σ_{MSE}) of the MSE of the networks under the considered non-idealities scenarios. The worst case MSE of each network measured as $\mu_{\text{MSE}} + 3\sigma_{\text{MSE}}$ is shown in Fig. 35. In this figure, MT10 and MT20 represent the cases where μ/σ of the applied Gaussian random variations is equal to 10% and 20%, respectively. The results reveal that the NNs trained by the INTERSTICE and RIM methods have considerably higher resiliency to the variations. As an example, the worst case MSE of the NNs trained by RIM (INTERSTICE (4)) method is, on average, 77% (69%) and 60% (62%) smaller than those of PHAX and PHAX (ADC) under MT10 and MT20 conditions, respectively.

The accuracies of the NNs (for classification applications) under the presence of memristors and transistors non-idealities as well as the ideal case are plotted in Fig. 36. The ideal or nominal case is denoted by Nom in this figure, while Mα (Tα) represents the case where μ/σ of the applied Gaussian random variations to the

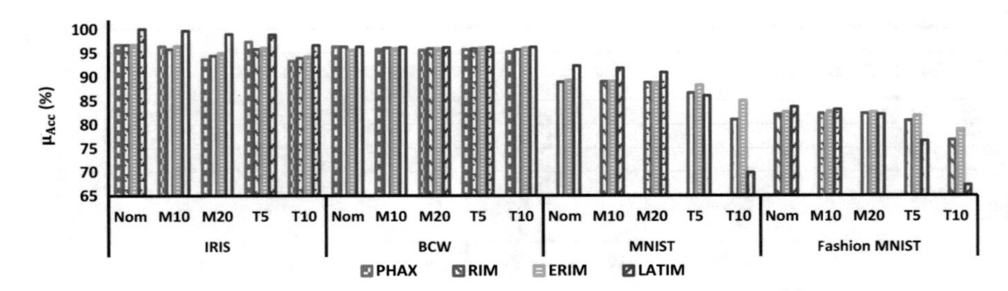

Fig. 36 The means of the accuracies of the NNs trained by the PHAX, RIM, ERIM, and LATIM methods for the cases of the nominal and variation conditions for different benchmarks

conductance of the memristors (characteristics of the transistors) is equal to $\alpha\%$. In addition, as the accuracies of the NNs trained by the PHAX method are small, their results are not included in this figure. As shown in Fig. 36, when the size of the NN is small (i.e., for IRIS and BCW datasets), the NNs trained by the LATIM method have the highest accuracy in all of the non-ideal conditions compared to the other training methods due to their higher accuracies in the ideal condition. For larger-sized NNs (i.e., for MNIST and Fashion MNIST datasets), the NNs trained by the LATIM method have the highest accuracy when considering memristors non-idealities except for the M20 case under the Fashion MNIST dataset in which ERIM has a better performance. In addition, larger-sized NNs trained by the ERIM (LATIM) method have the highest (lowest) accuracy when considering transistors non-idealities. It should be noted that based on the mathematical analysis of Sect. 3 as well as the simulation results shown in Fig. 36, memristor conductance variations have small impacts on the output voltage as well as the accuracy of the NN. Hence, the LATIM method which has a higher accuracy in the ideal condition yields a better accuracy in the presence of memristor conductance variations (i.e., M10 and M20 conditions).

To mitigate the severe effects of the transistors non-idealities on the accuracy of the NN, it is required to employ variation mitigation approaches such as RIM and ERIM. As the results of Fig. 36 reveal, ERIM (LATIM) leads to the highest (lowest) accuracy in the presence of transistors non-idealities for larger NNs. In the LATIM method, the equivalent resistance of the memristors is calculated approximately and used to find the output voltage of the inverters. This led to a higher accuracy of the NN in the ideal condition compared to those of RIM and ERIM. In the latter methods, the VTC slopes of the inverters were reduced by adding the grounded resistors to the output node of the inverters, providing higher resiliency of the NN outputs to the non-idealities of the circuit elements.

To evaluate the efficacy of the VAT method in mitigating the variations, the *MAE*s and accuracies of the NNs are obtained using Python simulations considering the variations for the memristor conductance values as well as transistors parameters. The degradations of these parameters compared to the ideal condition are plotted

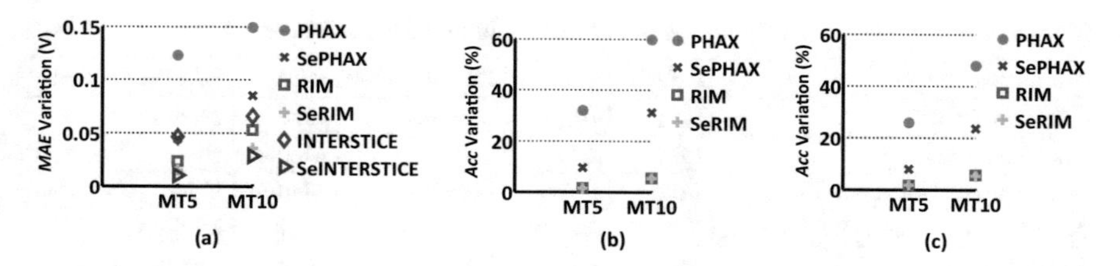

Fig. 37 (**a**) The *MAE* variation of the *IM*-NNs under California Housing, (**b**) *Acc* variation of the *IM*-NNs under MNIST benchmark, and (**c**) *Acc* variation of the *IM*-NNs under Fashion MNIST benchmark considering inverters and memristor non-idealities [22]

in Fig. 37. As shown in this figure, by combining VAT with the other methods, the performance of the NN in the presence of variations can be improved.

7.3 Comparing Different Training Methods

The presented simulation results and mathematical analysis provide us with some insights about the characteristics of different training methods for *IM*-NNs. Results are summarized in Table 4. As seen in this table, the PHAX method has the lowest complexity, modeling accuracy, and NN accuracy in both of the ideal and non-ideal conditions due to considering similar coefficients for modeling the VTC of all inverters of the NN. The VAT method has a high training complexity and a moderate NN accuracy in the non-ideal condition due to the calculation of the sensitivity term. As the VTC modeling used in this method is similar to PHAX, the accuracy of the trained NN in the ideal condition is low. In the INTERSTICE method, the need to the ADC as the output interfaces of the NNs is eliminated which leads to small delays and power consumptions for the networks trained by this approach. In the presence of non-idealities, this approach has a higher accuracy compared to that of the PHAX method due to the utilization of a classification NN for predicting the output range of the regression application. The VTC modeling of this method is similar to PHAX giving rise to a low modeling accuracy. Since the difference between the output voltage of the selected and unselected classes is large, the low modeling accuracy of this method does not degrade the accuracy of the trained NN significantly (moderate NN accuracy in the ideal condition). In addition, this approach has a higher training complexity compared to that of PHAX due to the labeling process of the input data and employing larger-sized NNs.

The NNs trained by the RIM method have the highest power consumptions due to utilization of the low resistance grounded resistors in the structure of the network. In addition, the technique has a moderate modeling accuracy as well as a moderate NN accuracy in the ideal condition due to the suppression of the sensitivity of the VTC coefficients to the loading effect of the memristor crossbars. Moreover, in the case of non-idealities, the NNs trained by RIM have high accuracies due to the low VTC

Table 4 Comparing the characteristics of different training methods for *IM*-NNs

Training method	PHAX	VAT	INTERSTICE	RIM	LATIM	ERIM
Complexity	Low	High	Medium	Low	Medium	High
Modeling accuracy	Low	Low	Low	Medium	High	High
NN accuracy (ideal condition)	Low	Low	Medium	Medium	High	Medium
NN accuracy (nonideal condition)	Low	Medium	Medium	High	Medium	High
NN delay	Medium	NA	Low	Low	Medium	Low
NN power consumption	Medium	NA	Low	High	Low	Medium

NA: Not reported in [22]

slope of the inverters mitigating the effects of non-idealities on the output voltage of the inverters. Furthermore, the delay of the NNs trained by the RIM method is small due to the sparse distribution of the memristors conductance values (Fig. 34).

In the LATIM method, the equivalent resistance of the memristor crossbars is taken into account to predict the VTC coefficients of the inverter. This gives rise to a moderate training complexity and a high modeling accuracy. Since the VTC slope of the inverters does not reduce considerably, the NNs trained by this approach have the highest accuracies in the ideal case and moderate accuracies in the case of non-idealities. In addition, due to choosing proper sizes for the inverters, the NNs have moderate delays and low power consumptions.

The training complexity and modeling accuracy of the ERIM method are the highest among the other methods due to utilization of the Thevenin equivalent circuit of the memristor crossbar for modeling the output voltage of the inverters. In addition, the accuracies of the NNs trained by this method are moderate in the ideal condition and high in the case of non-idealities due to the low VTC slopes of the inverters. Furthermore, the delays of the NNs become small due to the sparse distribution of the memristor conductance values. Moreover, the power consumptions of the NNs are moderate owing to the reduction of the power overhead of the grounded resistors using proper-sized inverters.

7.4 Opportunities for Future Research

Based on the results of Table 1, the variation mitigation approaches presented in the literature for *IM*-NNs (i.e., VAT and RIM) reduce the maximum achievable NN accuracy in the ideal condition due to the added sensitivity term or the lower VTC slope of the inverters. In addition, the INTERSTICE method is only applicable for regression applications. This makes the design of a variation mitigation method that does not degrade the accuracy of the NN in the ideal condition an important open problem. As a promising solution, one may propose a peripheral circuit for calibrating the VTC of the inverters after the chip fabrication. The important challenges in this case are the energy and speed overheads of the peripheral circuit and its calibration accuracy which should be considered during the circuit design phase. Moreover, the utilization of IM neurons in the structure of binary NNs (BNNs) would be another interesting research topic which has not been considered in the previous works. In fact, BNNs are expected to have lower vulnerability to the variations. Furthermore, all of the previous approaches regarding *IM*-NNs are implemented in the CMOS technology. It is possible that the *IM*-NNs offer better performance in the presence of non-idealities in other device technologies such as gate-all-around field-effect transistors (GAAFETs). In addition, the performances of the *IM* neurons in the presence of non-idealities in the other NN structures such as convolutional NNs (CNNs) and recurrent NNs (RNN) have not been investigated in any previous works. This can be an excellent research direction for future. Moreover, developing CAD tools for mixed-signal structures would be another interesting research topic in which the routing and placement of the transistors and memristors in a 3D structure and their challenges are taken into account. Finally, the idea of the VAT method and the mathematical analysis presented in Sect. 3 can be extended to other memristive structures to improve their resiliency against non-idealities of the circuit elements.

References

1. Huang, Z., Du, X., Chen, L., Li, Y., Liu, M., Chou, Y., Jin, L.: Convolutional neural network based on complex networks for brain tumor image classification with a modified activation function. IEEE Access. **8**, 89281–89290 (2020)
2. Luo, Y., Yu, S.: Accelerating deep neural network in-situ training with non-volatile and volatile memory based hybrid precision synapses. IEEE Trans. Comput. **69**(8), 1113–1127 (2020)
3. Si, X., Chen, J.J., Tu, Y.N., Huang, W.H., Wang, J.H., Chiu, Y.C., Wei, W.C., Wu, S.Y., Sun, X., Liu, R., Yu, S., Liu, R.S., Hsieh, C.C., Tang, K.T., Li, Q., Chang, M.F.: 24.5 A twin-8T SRAM computation-in-memory macro for multiple-bit CNN-based machine learning. In: *Proceedings of IEEE International Solid- State Circuits Conference (ISSCC)*, San Francisco, CA, USA, pp. 396–398 (2019)
4. Chen, Y., Krishna, T., Emer, J.S., Sze, V.: Eyeriss: an energy-efficient reconfigurable accelerator for deep convolutional neural networks. IEEE J. Solid-State Circuits. **52**(1), 127–138 (2017)
5. Vahdat, S., Kamal, M., Afzali-Kusha, A., Pedram, M.: TOSAM: an energy-efficient truncation- and rounding-based scalable approximate multiplier. IEEE Trans. Very Large Scale Integr. VLSI Syst. **27**(5), 1161–1173 (2019)
6. Vahdat, S., Kamal, M., Afzali-Kusha, A., Pedram, M.: LETAM: a low energy truncation-based approximate multiplier. Comput. Electr. Eng. **63**, 1–17 (2017)
7. Yao, P., Wu, H., Gao, B., Tang, J., Zhang, Q., Zhang, W., Yang, J.J., Qian, H.: Fully hardware-implemented memristor convolutional neural network. Nature. **577**, 641–646 (2020)
8. LeCun, Y.: The MNIST database of handwritten digits. http://yann.lecun.com/exdb/mnist/
9. Lee, E.H., Wong, S.S.: Analysis and design of a passive switched-capacitor matrix multiplier for approximate computing. IEEE J. Solid-State Circuits. **52**(1), 261–271 (2017)
10. Tripathi, A., Arabizadeh, M., Khandelwal, S., Thakur, C.S.: Analog neuromorphic system based on multi input floating gate MOS neuron model. In: Proceedings of *IEEE International Symposium on Circuits and Systems (ISCAS)*, Sapporo, Japan, pp. 1–5 (2019)

11. Ansari, M., Fayyazi, A., Banagozar, A., Maleki, M.A., Kamal, M., Afzali-Kusha, A., Pedram, M.: PHAX: physical characteristics aware Ex-Situ training framework for inverter-based memristive neuromorphic circuits. IEEE Trans. Comput. Aided Des. Integr. Circuits Syst. **37**(8), 1602–1613 (2018)
12. Yeo, I., Chu, M., Gi, S., Hwang, H., Lee, B.: Stuck-at-fault tolerant schemes for memristor crossbar array-based neural networks. IEEE Trans. Electron Devices. **66**(7), 2937–2945 (2019)
13. Chen, J., Pan, W.Q., Li, Y., Kuang, R., He, Y.H., Lin, C.Y., Duan, N., Feng, G.R., Zheng, H.X., Chang, T.C., Sze, S.M., Miao, X.S.: High-precision symmetric weight update of memristor by gate voltage ramping method for convolutional neural network accelerator. IEEE Electron Device Lett. **41**(3), 353–356 (2020)
14. Krestinskaya, O., Salama, K.N., James, A.P.: Learning in memristive neural network architectures using analog backpropagation circuits. IEEE Trans. Circuits Syst. I Regul. Pap. **66**(2), 719–732 (2019)
15. Krestinskaya, O., James, A.P.: Binary weighted memristive analog deep neural network for near-sensor edge processing. In: *Proceedings of 18th International Conference on Nanotechnology (IEEE-NANO)*, Cork, Ireland, pp. 1–4 (2018)
16. Khodabandehloo, G., Mirhassani, M., Ahmadi, M.: Analog implementation of a novel resistive-type sigmoidal neuron. IEEE Trans. Very Large Scale Integr. VLSI Syst. **20**(4), 750–754 (2012)
17. Hasan, R., Taha, T.M., Yakopcic, C.: A fast training method for memristor crossbar based multi-layer neural networks. Analog Integr. Circ. Sig. Process. **93**(3), 443–454 (2017)
18. Vahdat, S., Kamal, M., Afzali-Kusha, A., Pedram, M.: Loading-aware reliability improvement of ultra-low power memristive neural networks. IEEE Trans. Circuits Syst. I Regul. Pap. **68**(8), 3411–3421 (2021)
19. Fayyazi, A., Ansari, M., Kamal, M., Afzali-Kusha, A., Pedram, M.: An ultra low-power memristive neuromorphic circuit for internet of things smart sensors. IEEE Internet Things J. **5**(2), 1011–1022 (2018)
20. BanaGozar, A., Maleki, M.A., Kamal, M., Afzali-Kusha, A., Pedram, M.: Robust neuromorphic computing in the presence of process variation. In: *proceedings of Design, Automation & Test in Europe Conference & Exhibition (DATE)*, Lausanne, pp. 440–445 (2017)
21. Ansari, M., Fayyazi, A., Kamal, M., Afzali-Kusha, A., Pedram, M.: OCTAN: an on-chip training algorithm for memristive neuromorphic circuits. IEEE Trans. Circuits Syst. I Regul. Pap. **66**(12), 4687–4698 (2019)
22. Vahdat, S., Kamal, M., Afzali-Kusha, A., Pedram, M.: Reliability enhancement of inverter-based Memristor crossbar neural networks using mathematical analysis of circuit non-idealities. IEEE Trans. Circuits Syst. I Regul. Pap. **68**(10), 4310–4323 (2021)
23. Vahdat, S., Kamal, M., Afzali-Kusha, A., Pedram, M.: LATIM: loading-aware offline training method for inverter-based memristive neural networks. IEEE Trans. Circuits Syst. II Express Briefs. **68**(10), 3346–3350 (2021)
24. Vahdat, S., Kamal, M., Afzali-Kusha, A., Pedram, M.: INTERSTICE: inverter-based memristive neural networks discretization for function approximation applications. IEEE Trans. Very Large Scale Integr. VLSI Syst. **28**(7), 1578–1588 (2020)
25. Vahdat, S., Kamal, M., Afzali-Kusha, A., Pedram, M.: Offline training improvement of inverter-based memristive neural networks using inverter voltage characteristic smoothing. IEEE Trans. Circuits Syst. II Express Briefs. **67**(12), 3442–3446 (2020)
26. Chen, X., Jiang, J., Zhu, J., Tsui, C.: A high-throughput and energy-efficient RRAM-based convolutional neural network using data encoding and dynamic quantization. In: in Proceedings of *23rd Asia and South Pacific Design Automation Conference (ASP-DAC)*, Jeju, pp. 123–128 (2018)
27. Shakiba, F.M., Zhou, M.: Novel analog implementation of a hyperbolic tangent neuron in artificial neural networks. IEEE Trans. Ind. Electron. **68**(11), 10856–10867 (2020). https://doi.org/10.1109/TIE.2020.3034856
28. Liu, X., et al.: RENO: a high-efficient reconfigurable neuromorphic computing accelerator design. In: *Proceedings of the 52nd ACM/EDAC/IEEE Design Automation Conference (DAC)*, pp. 1–6 (2015)

29. Chua, L.: Memristor-the missing circuit element. IEEE Trans. Circuits Theory. **18**(5), 507–519 (1971)
30. Strukov, D.B., Snider, G.S., Stewart, D.R., Williams, R.S.: The missing memristor found. Nature. **453**(7191), 80–83 (2008)
31. Li, B., Yan, B., Liu, C., Li, H.H.: Build reliable and efficient neuromorphic design with memristor technology. In: Proceedings of the 24th Asia and South Pacific Design Automation Conference, pp. 224–229 (2019)
32. Pouyan, P., Amat, E., Hamdioui, S., Rubio, A.: RRAM variability and its mitigation schemes. In: Proceedings of *26th International Workshop on Power and Timing Modeling, Optimization and Simulation (PATMOS)*, pp. 141–146 (2016)
33. Hasan, R., Taha, T.M., Yakopcic, C.: On-chip training of memristor crossbar based multi-layer neural networks. Microelectron. J. **66**, 31–40 (2017)
34. Pham, K.V., Nguyen, T.V., Tram, S.B., Nam, H.K., Lee, M.J., Choi, B.J., Truong, S.N., Min, K.S.: Memristor binarized neural networks. J. Semicond. Technol. Sci. **18**(5), 568–577 (2018)
35. Pham, K.V., Tran, S.B., Nguyen, T.V., Min, K.S.: Asymmetrical training scheme of binary-memristor-crossbar-based neural networks for energy-efficient edge-computing nanoscale systems. Micromachines. **10**(2), 141–154 (2019)
36. Rajendran, J., Karri, R., Rose, G.S.: Improving tolerance to variations in memristor-based applications using parallel memristors. IEEE Trans. Comput. **64**(3), 733–746 (2015)
37. Liu, C., Hu, M., Strachan, J.P., Li, H.: Rescuing memristor-based neuromorphic design with high defects. In: *Proceedings of 54th ACM/EDAC/IEEE Design Automation Conference (DAC)*, Austin, TX, pp. 1–6 (2017)
38. Li, C., Hu, M., Li, Y., Jiang, H., Ge, N., Montgomery, E., Zhang, J., Song, W., Davila, N., Graves, C.E., Li, Z., Strachan, J.P., Lin, P., Wang, Z., Barnell, M., Wu, Q., Williams, R.S., Yang, J.J., Xia, Q.: Analogue signal and image processing with large memristor crossbars. Nat. Electron. **1**, 52–59 (2018)
39. Jin, S., Pei, S., Wang, Y.: A variation tolerant scheme for memristor crossbar based neural network designs via two-phase weight mapping and memristor programming. Futur. Gener. Comput. Syst. **106**, 270–276 (2020)
40. Pham, K.V., Nguyen, T.V., Min, K.S.: Partial-gated memristor crossbar for fast and power-efficient defect-tolerant training. Micromachines. **10**(4), 245 (2019)
41. Liu, B., Li, H., Chen, Y., Li, X., Huang, T., Wu, Q., Bernell, M.: Reduction and IR-drop compensations techniques for reliable neuromorphic computing systems. In: *Proceedings of IEEE/ACM International Conference on Computer-Aided Design (ICCAD)*, San Jose, CA, pp. 63–70 (2014)
42. Li, B., Wang, Y., Chen, Y., Li, H.H., Yang, H.: ICE: inline calibration for memristor crossbar-based computing engine. In: *Proceedings of Design, Automation and Test in Europe Conference & Exhibition (DATE)*, Dresden, pp. 1–4 (2014)
43. Lou, Q., Gao, T., Faley, P., Niemier, M., Hu, X.S., Joshi, S.: Embedding error correction into crossbars for reliable matrix vector multiplication using emerging devices. In: *Proceedings of the ACM/IEEE International Symposium on Low Power Electronics and Design*, pp. 139–144 (2020)
44. Merkel, C., Kudithipudi, D.: A stochastic learning algorithm for neuromemristive systems. In: Proc. 27th IEEE Int. Syst.-Chip Conf. (SOCC), Las Vegas, NV, USA, pp. 359–364 (2014)
45. Gokmen, T., Onen, M., Haensch, W.: Training deep convolutional neural networks with resistive cross-point devices. Front. Neurosci. **11**, 538 (2017)
46. Hirotsu, K., Brooke, M.A.: An analog neural network chip with random weight change learning algorithm. In: *Proc. Int. Conf. Neural Netw. (IJCNN),* Nagoya, Japan, vol. 3, pp. 3031–3034 (1993)
47. Yakopcic, C., Taha, T.M., Subramanyam, G., Pino, R.E.: Generalized memristive device SPICE model and its application in circuit design. IEEE Trans. Comput. Des. Integr. Circuits Syst. **32**(8), 1201–1214 (2013)

AI-Based Hardware Security Methods for Internet-of-Things Applications

Jaya Dofe and Wafi Danesh

1 Introduction

The Internet is going through a new stage in which billions of smart objects, "things" that sense and interact with the physical world, are connected in homes, industry, hospitals, cities, and farms, to name a few. These connected objects—the Internet of Things (IoT)—is changing the world around us, being stitched into the very fabric of our everyday lives. IoT brings extraordinary possibilities for improvements in various domains like smart cities and grids, healthcare, wearable devices, robotic systems, and numerous other systems. IoT is gradually becoming an integral part of the betterment of our personal and professional lives. While the benefits of IoT are undeniable, it is a double-edged sword. An IoT ecosystem is constantly subjected to changes and threats at various levels. IoT devices allocate the majority of energy and computation resources for normal functionality, and incorporating security features becomes extremely challenging [1]. Coupled with the short time-to-market and fierce competition among device manufacturers, security has become an afterthought [2] and has not been prioritized as a crucial metric.

Security and trust are paramount considerations while designing IoT systems. Unlike security threats in the traditional Internet, which are relegated to the digital sphere, attacks on IoT systems directly impact the physical world. IoT systems and applications can be made more secure by utilizing cryptography to communicate between the physical and cyber worlds. Even though conventional cryptographic countermeasures are computationally intensive, several IoT devices

J. Dofe (✉)
California State University Fullerton, California, CA, USA
e-mail: jdofe@fullerton.edu

W. Danesh
University of Missouri, Kansas City, MO, USA
e-mail: wdhv3@mail.umkc.edu

A. Iranmanesh (ed.), *Frontiers of Quality Electronic Design (QED)*,
https://doi.org/10.1007/978-3-031-16344-9_10

have incorporated lightweight embedded cryptographic cores for authentication and information processing. IoT systems use sensors and other smart devices. As all the applications require fundamental security, cryptography is deployed as part of an attempt to secure them [3]. However, if the encryption algorithms are not implemented correctly, they can compromise the applications. It is a huge challenge that researchers have been addressing over the decades. A prominent side-channel attack (SCA) that breaks an encryption system's security by exploiting the information leaked from the physical devices is a rising threat in IoT applications [4, 5]. Current IoT studies show that adversaries can easily acquire side-channel information, which is hard to prevent because leakages from devices are inevitable [6]. Side channels in IoT systems may arise from timing information, sensor data, or traffic rates between devices prevalent in our everyday lives. Due to IoT devices' inherent characteristics and limitations, designing hardware security solutions is complex and nontrivial. Henceforth, the future of IoT applications will rely on the ability to safeguard hard-to-secure, resource-sparse devices effectively.

The current state of security of Internet-of-Things (IoT) devices therefore challenges conventional security protocols. Many IoT device manufacturers prioritize keeping their devices low cost and small in size, with low battery usage and computation power, making traditional security methods unsuitable. In this situation, end users have to trade-off between the security and performance of the device. This trade-off is causing IoT devices to become vulnerable to side-channel attacks. Much published research discusses IoT security and challenges[7–9]. Most existing literature focuses on secure IoT infrastructure creation and implementation, authentication, trust management, and attack in different IoT layers. The main focus of this chapter is to review the hardware security techniques implemented in the face of threats such as side-channel attacks and hardware Trojan (HT) insertions. We dedicate a part of this chapter to studying unified countermeasures for side-channel attacks for IoT applications. We also give an extensive overview of the machine learning-based approaches used to develop countermeasures against hardware attacks in IoT applications. ML algorithms generate a *model* from *historical* data, in order to learn the underlying pattern and generalize new, unseen data. ML-based approaches can perform prediction or classification by learning the underlying pattern. In the context of HT detection, ML approaches learn from various circuit characteristics and parametric data to determine if an IC design is HT infected or not [10–13]. One of the significant advantages of ML-based HT detection is the automation of the process of HT detection.

As a final note, we also propose a new security paradigm for IoT security, namely, 3D integration for IoT devices. 3D integration technology provides various advantages such as heterogeneous integration, split manufacturing, and combining disparate technologies for the IoT platform such as MEMS sensors, RF transmitters, energy harvesters, etc. It makes 3D integration the choice for developing secure IoT platforms and devices. The stacked-layer architecture of 3D ICs makes them suitable for deploying a number of security features against side-channel attacks or HT insertion.

The rest of this chapter is organized as follows. Section 2 presents an overview of hardware attacks. Section 3 discusses generic countermeasures for side-channel attacks and HT insertions. It also covers the countermeasures for these attacks in IoT domain. The unified approaches unique to physical attacks and HT attacks are discussed in Sect. 4. Section 5 introduces 3D technology and corresponding usage of ML for 3D integration. We provide a detailed overview of ML-based approaches for HT detection in general and IoT systems in particular in Sect. 6. Section 7 outlines the details of 3D integration and the benefit of using it to secure IoT devices. Finally, Sect. 8 concludes the paper.

2 Hardware Attacks

The emerging hardware threats arise because of the globalized IC supply chain. There are multiple stages within the supply chain that can be manipulated by a potential adversary to perform the attacks. These diverse hardware attacks can be broadly classified in the following categories.

2.1 IP Piracy

IP piracy is the illicit or unlicensed use of an intellectual property (IP). The semiconductor industry is increasingly relying on a hardware IP-based design approach, in which reusable, pre-verified hardware modules are combined to produce a complex system that performs as expected. An attacker can steal valuable hardware IPs in the form of register-transfer-level (RTL) representations ("soft IP"), gate-level designs directly implementable in hardware ("firm IP"), or the GDSII design database ("hard IP") and market them as legitimate IPs. Hardware IP reuse in the design of systems-on-chip (SoCs) is a common practice in the silicon industry because it drastically saves design time and cost. IP piracy can occur at any point in the supply chain for integrated circuits. Designers, third-party IP (3PIP) vendors, and SoC integrators at the design, synthesis, and verification stages could potentially pirate the IP. Untrusted foundries may overbuild IP cores in the fabrication stage and resell them under a different brand name to make a profit. Hardware IPs bought from untrustworthy third-party manufacturers may contain a variety of security and integrity flaws. An attacker inside an IP design house can purposefully implant a malicious circuit or design alteration to undermine system security.

2.2 Reverse Engineering

The process of identifying an IC's structure, design, and functionality is known as reverse engineering. Different types of reverse engineering include product teardowns, system-level analysis, process analysis, and circuit extraction. One can use reverse engineering to (1) determine the device technology, (2) extract the gate-level netlist, and (3) infer chip functionality. Several techniques and tools have been developed to facilitate reverse engineering. Traditionally, it has been a legal method for teaching, assessing, and evaluating mask work processes under the US Semiconductor Chip Protection Act. Reverse engineering, on the other hand, is a two-edged sword. Reverse engineering techniques could be used to pirate ICs. Reverse engineering attacks can be carried out at many levels of abstraction in the supply chain, depending on the attacker's goals [14].

2.3 Counterfeiting

A counterfeit semiconductor component is an illegal forgery or imitation of the original component. Counterfeiting is often performed by one of the many entities in the semiconductor supply chain, including new product vendors or secondary (recycled) IC vendors. In recent years, because of technological advances in 3D packaging, fake ICs are hard to distinguish from real ones. Counterfeit ICs are a serious threat to the IC supply chain. Computers, telecommunications, automotive electronics, and military systems are all affected by counterfeiting attacks. As counterfeiters get more sophisticated, counterfeit chips are becoming more difficult to detect.

2.4 Hardware Trojans

Maliciously altering a circuit's hardware is one of the most insidious ways of attacking it. A hardware Trojan (HT) is built by embedding hidden functionality into a hardware design discretely. This insertion can happen at any point in the IC supply chain, and it can have catastrophic consequences for the final design. Such Trojans can perform a wide range of functions, including denial-of-service attacks, which provide designers with a programmable kill switch, and concealed data leaks, exposing sensitive information. HTs pose a direct danger to the IoT, which is already susceptible. HTs, unlike software Trojans, cannot be removed simply by updating the firmware, making them extremely dangerous and difficult to remove. As a result, HT detection is a critical step in ensuring the chips used in IoT are safe. The simple structure of HT is shown in Fig. 1.

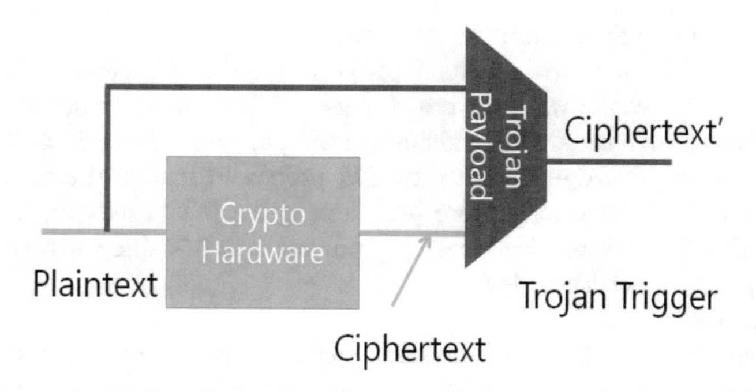

Fig. 1 Simple example of hardware Trojan

2.5 *Side-Channel Attacks*

A physical attack is one of the most prominent and influential methods in the hands of adversaries in the hardware security sector. Physical attacks are when the attacker has physical access to the device being targeted. These threats can assist an attacker in infiltrating the IoT. Invasive vs. noninvasive attacks and active vs. passive attacks are the two main types of physical attacks. Invasive attacks necessitate tampering with the target device, whereas noninvasive attacks do not. If the adversary actively influences the device's behavior, it's either an active attack or a passive observation of information leakage. The scope of side-channel attacks has shifted drastically with the introduction of mobile devices. Initially, attackers required physical access to the device. However, similar attacks are more common in IoT systems and can be performed easily.

Side-channel analysis (SCA) attacks [4, 5, 15] aim to retrieve the secret key in the cryptosystems by analyzing physical parameters like power, delay, or electromagnetic emission of the IC that runs security-critical applications.

- **Power Analysis Attacks**:
  Kocher et al. proposed power analysis attacks that take advantage of hardware implementation of cryptographic algorithm [16]. Power-based SCA attacks have been thoroughly explored. They take advantage of the correlation between the cryptosystem's power usage and the assumed crypto key to recover the secret key used. Simple, differential, and correlation power analysis are three prevalent power analysis attacks.
- **Timing Attacks**:
  Kocher [17] developed this attack in 1996 as well. To divulge secret information, it exploits the data-dependent execution time. Because cryptosystems use conditional branches in their algorithms and performance optimization, the execution of cryptographic system takes varying amounts of time to process different inputs.

- **Electromagnetic Side-Channel Attacks**:
  Electromagnetic side-channel attack [18] is also a valuable source of information that is available whenever a system is in use. This attack does not require device manipulation to measure side-channel leakage, and hence, it is noninvasive. Because of the ready availability of EM probes to launch the attack, electromagnetic SCA is becoming more prevalent in the IoT paradigm. In contrast to power SCA, this assault is increasingly prevalent in IoT since adversaries do not require physical access to devices.
- **Fault Attacks**:
  A fault attack is an attack on a physical, electronic device (e.g., smartcard, HSM, USB token) that involves straining the device using an external means (e.g., voltage, light) to induce faults that cause the system to fail. An adversary can use fault attacks to cause the device to defeat security safeguards or to retrieve secret information by exploiting faulty outputs. A fault attack can defeat the advanced encryption standard (AES) implementation with only a pair of fault-free and faulty ciphertexts, according to the research [19]. Manipulation of the external clock or power inputs, as well as the use of electromagnetic disturbances, are two of the most popular approaches to carrying out a fault attack. This type of attack is easy to perform as it requires only a motivated attacker with mid-level expertise and low-cost equipment. Thus, these fault injection techniques should be considered a severe threat to IoT systems.

3 Countermeasures Against Side-Channel Attacks and Hardware Trojans Insertion

3.1 Generic Countermeasures for SCA

For power-based side-channel attacks, the main objective of countermeasure is to make the power consumption of a device as independent as possible to the intermediate values of a cryptographic algorithm. The general countermeasures for AES include either hiding or masking the data. The goal of hiding [20, 21] is to cover up a correlation between the power traces and the intermediate values. Hiding deceives the power traces by randomizing power consumption in a device or flattening the power consumption to make all operations look similar. For the masking technique, the goal is to conceal data by adding/multiplying random numbers to the intermediate values in the encryption process to ward off potential attackers [21]. The challenge becomes implementing the countermeasures without reducing the speed, increasing the power consumption, or increasing the area of the cryptographic algorithm beyond reasonable limits.

Some of the countermeasures proposed against electromagnetic SCA include signal strength reduction techniques like shielding or signal information reduction using noise insertion [22]. Recently, Das et al. used white-box modeling [18] to develop a low-overhead generic circuit-level countermeasure against electromagnetic side-channel attacks. An electromagnetic equalizer is proposed in [23], where on-chip power grid impedance is adjusted to flatten the current waveform.

A common approach to protecting the cryptographic core from timing attacks is to ensure that its behavior is never data-dependent. The sequence of cache accesses or branches does not depend on either the key or the plaintext. Paper [24] proposed to perform rescheduling of instructions so that each encryption round will consume constant time independent of the cache hits and misses. Another way is to induce noise in all events to prevent exploitation of timing information [25]. One beneficial way to make time attacks challenging is to desynchronize the execution of sensitive parts by using random waits, dummy instructions, jitter on clocks, etc., as much as possible. The most cost-effective approach against FA attacks is modifying the cryptographic device's design to detect injected faults. Traditional fault detection methods for cryptosystems exploit information redundancy, spatial redundancy, or time redundancy to detect faults [26]. Survey paper [27] presented countermeasures against fault injection attacks, including algorithmic changes, sensors and shields, and fault detection or correction techniques.

3.2 *Generic Countermeasures Against Hardware Trojan Insertions*

Existing HT detection approaches are classified as presilicon and postsilicon detection. Presilicon detection is used to validate 3rd Party IP (3PIP) cores. It includes functional validation, structural analysis, and formal verification. Postsilicon detection is further categorized into nondestructive and destructive detection methods. Nondestructive methods include functional testing accelerated Trojan activation followed by testing analysis on side-channel signals (e.g., power, delay, temperature, or electromagnetic profiles) [28]. In destructive techniques, reverse-engineering is used, which involves depackaging an IC followed by scanning electron microscope (SEM) image reconstruction and analysis. HT detection in a very large-scale integrated system is comparable to finding a needle in a haystack. Hence, such methods are very high cost and time-consuming [29]. None of the existing HT detection techniques guarantee the full coverage of HTs. Therefore, the research [30] suggest embedding HT prevention methods during the design phase provides a more effective potential approach as a countermeasure against HTs.

3.3 *Countermeasures Against Physical Attacks in IoT*

Side-channel information may arise from timing information, sensor data, or data traffic prevalent in everyday lives. Current IoT studies show that adversaries can easily acquire side-channel information, which is hard to detect because leakages are inevitable; hence, tackling these attacks is of utmost importance [4–6]. The IoT devices are intended to be small and convenient. The traditional countermeasures against power attacks reduce the signal-to-noise ratio, which may be expensive to implement for lightweight IoT applications. The attenuated signature AES is proposed in [5] to resist power analysis attacks with reduced overhead. This approach implements AES in a signature attenuating hardware, making the variations in AES current highly suppressed. A false key-based AES engine that utilized wave dynamic differential logic (WDDL) is presented in [31] as a countermeasure against CPA attacks. The false round keys generated by the constant intermediate value are added to the original round keys to disguise the correlation between the dynamic power consumption profile and the actual key. As the area and power overhead of the proposed technique is negligible compared to the unprotected AES, this method fits IoT devices. Kai Yang et al. presented a flexible FPGA virtualization approach [32] to prevent the FPGA-based system from timing attacks. This method's masking and architectural diversity make it challenging to obtain the required information to carry out the successful timing attacks.

3.4 Countermeasures Against Hardware Trojans in IoT

In article [33], Guo et al. propose an HT detection technique that makes use of chip temporal thermal information and self-organizing map (SOM) neural networks to distinguish and isolate Trojan-infected chips with the Trojan-free chips. This method detects HT with a high probability even if HT is inserted into the block, which is best for hiding Trojans. A nonconventional approach is proposed in [34] in which switching activities are amplified to increase Trojan visibility and detect possible Trojans. This method guarantees the background noise does not mask Trojan activities. An HT detection method for IoT networks is presented in which the features of a clock tree are extracted and used as a signature to identify and detect Trojans. We discuss ML-based approaches thoroughly in Sect. 6.

4 Unified Countermeasures for IoT

As mentioned earlier, IoT devices have a constrained power budget, and hence, it is imperative to design unified countermeasures that can address multiple attacks simultaneously. The paper [35] propose strategies that could be used for the design-specific targets, specifically for lightweight IoT applications. The first method is to use a maximum distance separable linear layer to incorporate diffusion and fault space transformation that helps to protect against classical cryptanalysis and differential fault attacks. The second strategy exploits modified transparency order metrics to select from different S-box implementations that guide the adequate refresh rate for the mask to defeat the differential power attacks with the same

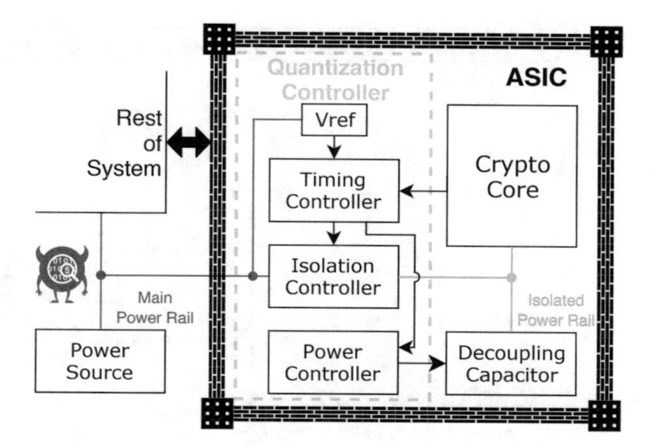

Fig. 2 Secure processor using quantization controller [36]

resistance. Cipher-dependent nibble-wise shuffling was proposed in their third method to enhance the side-channel resistance.

An embedded trusted platform module is proposed in [36] to address a variety of side-channel attacks, including power, timing, fault, and power-glitching attacks. This work makes use of a quantized controller as shown in Fig. 2 that sits between a security-critical core and the rest of the system. A controller uses integrated decoupling capacitors to create uniform power and timing footprints. The inherent implementation of the controller allows control where the computer processor receives its power. During security-critical processes, it can switch the processor's power source from the main power rail to the controller's internal storage capacitors, invisible to attackers. This allows the power traces to become unreadable with the proper implementation. A core design is to leverage on-demand isolation to allow side-channel protection from a software-level decision, making the method effective in real time to accommodate IoT design.

Recently, authors Das et al. used white-box modeling [18] to develop a low-overhead generic circuit-level countermeasure called STELLAR—Signature aTtenuation Embedded CRYPTO with Low-Level metAl Routing against electromagnetic and power side-channel attacks shown in Fig. 3. This approach utilizes the local lower-metal layers to route the crypto core with a signature suppression circuit, reducing the leakage reaching the top metal layer.

In research [37], authors proposed a concurrent software approach to resist the side-channel and fault attacks. This countermeasure is generic and applicable to any byte-size cipher. It utilizes larger data path of 32-bit or 64-bit microcontroller units to carry out parallel byte-sliced encryption. As depicted in Fig. 4, the same data byte D1 is cloned four times and encrypted using a fake key (K_F) twice and true key (K_T) twice. This arrangement will generate the correlated algorithmic noise to protect against SCA as both computations operate parallel on the same data but using two different keys. The same approach helps detect the fault injection attack because of duplicated results from both the fake and correct key computation to detect any anomalies.

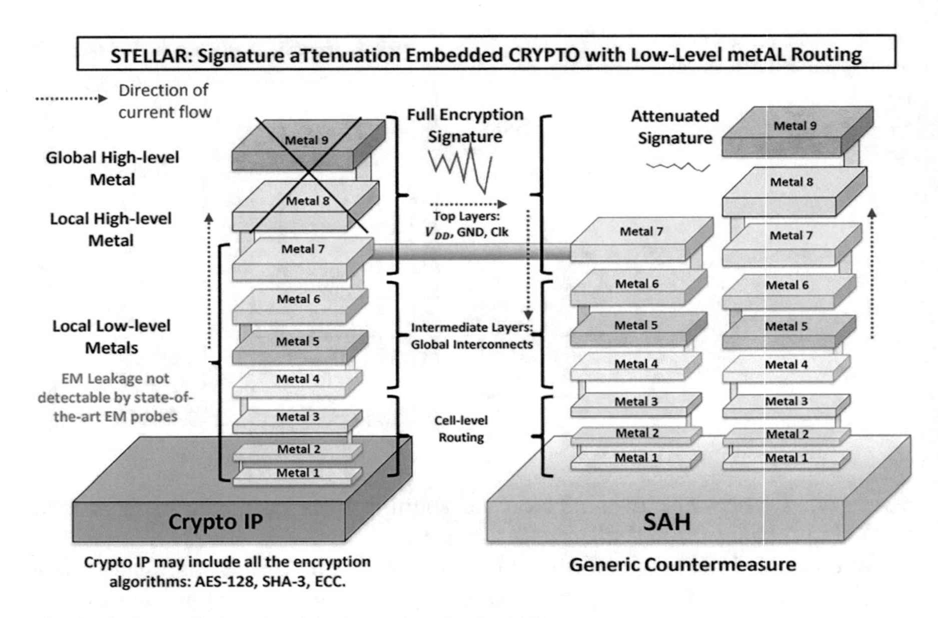

Fig. 3 Stellar technique for side-channel protection [18]

Fig. 4 Combined cand FA Countermeasure [37]

In study [28, 38], authors proposed to integrate a dynamic masking technique with an error control code-based error deflection mechanism to thwart power analysis and fault attacks simultaneously. This method generates the masking vector from the intermediate state register in runtime, which changes over time. The main working principle of this method is in Fig. 5. The technique is implemented for AES cipher, but it could be easily extended for other ciphers. In the method, the intermediate state value is encoded before processing and also uses dynamic masking that changes the masking value at runtime. This arrangement fails the power model modification according to a guessed masking vector.

An on-chip waveform measurement (OCM) technique is exploited in [39] that protect against physical side-channel attacks. The on-chip latch comparator resonator senses the proximate antennas using magnetic coupling. The OCM captures the voltage substrate waveforms when a laser hits the substrate detecting the fault attacks. When OCM detects the antenna or laser presence, the cryptographic chip forces are immediately halted or transitioned to a dummy state.

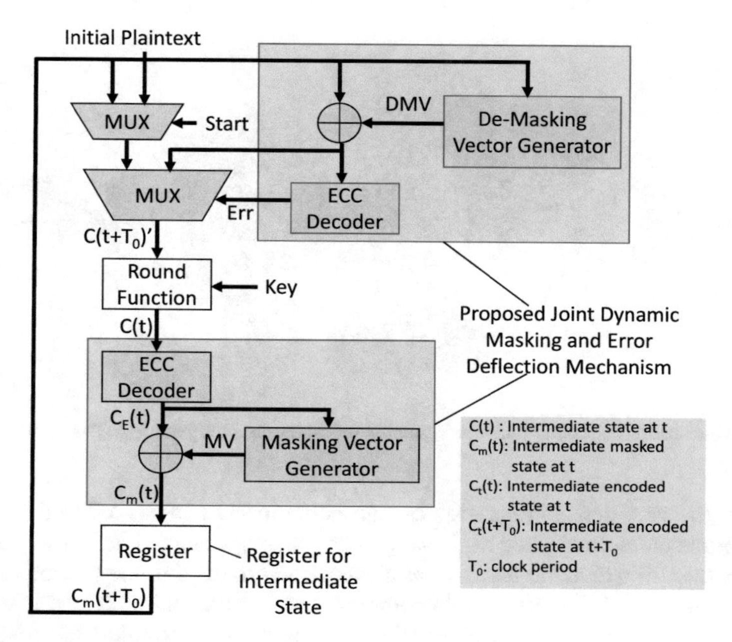

Fig. 5 Dynamic masking plus error deflection method

Sensors, actuators, and data collection devices are connected through the Internet in IoT applications. At least one data computation chip is needed to analyze the massive amount of data and make a corresponding decision. In paper [40], a dynamic permutation method is proposed to protect the processing unit from hardware threats, more precisely, HT insertion and side-channel analysis attacks simultaneously. The overview of the method is shown in Fig. 6. The incoming data from the sensor will be changed because of dynamic permutation. Hence, the attacker cannot successfully execute a predefined trigger condition. The cryptographic module or random number generator can control the dynamicity. A new permutation pattern will be requested if the cryptographic module detects an invalid message. With this framework, the processing node is obfuscated with a dynamic feature. It prevents a Trojan attack and also changes the power profile over time. As a result, the proposed dynamic permutation makes it more difficult to retrieve the crypto key based on the power analysis.

5 3D ICs and Machine Learning

For 2D ICs, the extreme level of scaling of transistor size, as approximately predicted by Moore's law, has allowed for very dense IC designs to be fabricated on the planar structure of a single layer. However, with the limits of Moore's law being

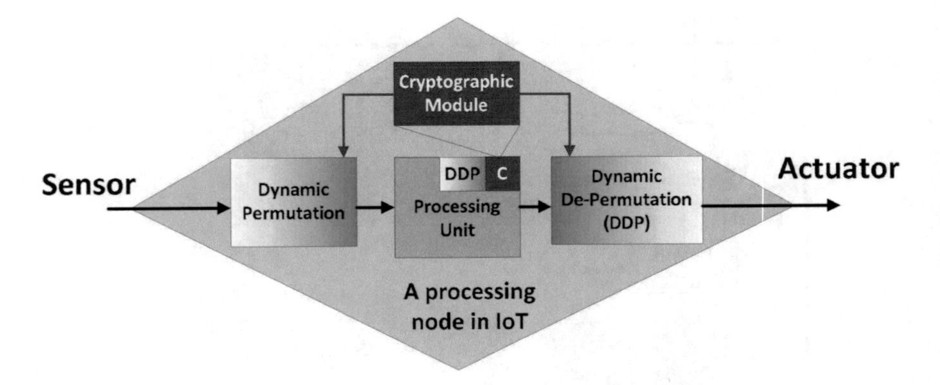

Fig. 6 Obfuscated processing unit with dynamic permutation for a generic IoT network

reached, the very dense integration of transistors in 2D ICs has led to degradation in performance metrics, such as delay, power consumption, and heat dissipation. In essence, there is a demand for an alternative paradigm to continue fabricating more complex integrated circuits. In this context, 3D ICs performing device integration in a 3D architecture offer an alternative paradigm for continuing the fabrication of more complex IC designs and increasing device integration even further. 3D integration is performed in 3D ICs by vertically stacking several layers, often with heterogeneous functionality, and connecting them, in general, using Through Silicon Vias (TSV), which are electrical interconnects etched into the silicon. As a result, significant improvements are achieved, such as reductions in interconnect and wire length, in chip area, and superior electrical performance.

As compared to the relative improvements, 3D IC fabrication gives rise to new critical design challenges [41], such as the design of TSVs, mitigation of chip hotspots, development of appropriate CAD tools for 3D IC, and formulation of optimization algorithms for 3D ICs, among others. Machine learning (ML) and, more recently, deep learning (DL) algorithms have emerged as key tools to solve many design challenges. The use of ML in electronic design automation (EDA) stretches back almost three decades and, in the context of 3D ICs, has been used for inter-die variation modeling, design space exploration (DSE), placement, and routing optimization among other purposes. In [42], Sandeep et al. used a nonlinear regression ML technique, called multivariate adaptive regression splines (MARS), and developed an efficient model for two-tier 3D IC designs. In this model, input parameters extracted from the place-and-route (P & R) database are fed to the MARS model. The developed model is integrated into a full-chip variation-aware 3D IC physical design flow and further verified with detailed Monte Carlo simulations on three different benchmark circuits. A 16% improvement was obtained for critical path delay under variations.

In the context of design space exploration, several notable studies have explored the use of ML algorithms and techniques to solve the design optimization issues arising in 3D ICs [43–48]. Das et al. [43] used an online ML algorithm called STAGE to perform design space exploration for optimizing the orientation of planar and vertical communication links in small-world (SW) network-based 3D Network-on-Chip (NoC) architectures and improve energy efficiency. Upon experimental verification, the optimized 3D SW NoC designs outperform their 3D MESH counterparts, with an average of 35% energy-delay-product (EDP) improvement over 3D MESH for the nine PARSEC SPLASH2 benchmarks. In [44], Sung et al. implement a Bayesian optimization (BO) algorithm using Gaussian Process (GP) to perform multivariate optimization in analyzing both the electrical and thermal performance of 3D integrated systems. On average, the results show an improvement of 4.4%, 31.1%, and 6.9%, respectively, in temperature gradient, CPU time, and skew using machine learning. A design for a 3D NoC for heterogeneous manycore platforms is proposed by Biresh et al. [47], where an ML-based multi-objective optimization (MOO) technique called MOO-STAGE is proposed. The MOO-STAGE algorithm uses a supervised learning approach that uses past search experience to optimize the DSE problem for the 3D NoC. The results show a 9.6% better Energy-Delay Product (EDP) on average, at nearly iso-temperature conditions compared to a thermally optimized design for 3D heterogeneous NoCs, by joint consideration of multiple requirements (latency, throughput, temperature, and energy). Hakki et al. [45] proposed a new BO global optimization algorithm, called Two-Stage Bayesian Optimization (TSBO), to minimize the clock skew for 3D IC designs. Compared with a widely used nonlinear solver (fmincon in MATLAB) and high-performance algorithm, IMGPO, TSBO resulted in faster convergence, being 3.96 times faster than 'fmincon' and 3.76 times faster than IMGPO. In [48], the author proposes MOO-STAGE and outlines the design considerations of two 3D IC architectures, TSV-based 3D heterogeneous manycore systems and monolithic 3D (M3D)-based NoC, that need to be made and where MOO-STAGE can play a pivotal role in DSE. Das et al. [46] propose using the STAGE to perform the DSE of an M3D-enabled energy-efficient NoC architecture.

For placement and routing (P&R) optimization in 3D IC designs, some of the notable methods [49, 50] attempt to extend 2D P&R for 3D IC designs by performing bin-based tier partitioning. This leads to a less than optimal solution as these methods do not account for the topology of the 3D architecture and underlying technology used. In [51], Lu et al. uses TP-GNN, an unsupervised graph-learning-based tier partitioning framework, to properly account for technology and design-related parameters in 3D IC P&R. A hierarchy-aware graph transformation algorithm is first devised to convert the original netlist (hypergraph) into an edge-contracted clique-based graph. Afterward, graph neural networks (GNNs) are leveraged to perform graph representation learning. The goal is to construct a node representation that captures each node's design characteristics related to tier partitioning. At the final stage, the weighted k-means algorithm is used to perform area-balanced partitioning based on the learned representation for each cell. Experimental results with OpenPiton, a RISC-V-based multi-core system, show 27.4%, 7.7%, and 20.3% improvements in performance, wirelength, and energy-per-cycle, respectively.

ML approaches have been used in several other notable applications for 3D IC designs. In [52], Hyeok et al. use a two-layer feedforward neural network to perform the optimization of the 14-nm Fully-Depleted (FD) Silicon On Insulator (SOI) FET structures to obtain the best DC performance (on/off current ratios) with the average errors of prediction and optimization found to be within 5% tolerance. Subhajit et al. [53] used ML methods to perform thermal validation for 3D ICs by implementing the obtained temperature prediction model to generate thermally aware test schedules to reduce the test times for standard ITC'02 benchmarks. In [54], Lang et al. propose a fast stress analysis method for runtime usage by training an artificial neural network (ANN) offline, in a supervised manner, using thermal data around each TSV as input and stress information around each TSV as output. Yong et al. [55] presented a dynamic thermal management method for 3D ICs with a time-dependent power map using the tier-specific microfluidic heat sink (MFHS) and ML-based control method. For the ML-based control method presented, a Bayesian optimization (BO) approach was initially applied, followed by ANN when the calculation complexity of the BO increased with more data. In [56], Pentapati et al. use a decision tree learning model, XGBoost, to provide a more accurate prediction of 3D parasitics for obtaining a well-optimized monolithic 3D (M3D) IC design. As commercial EDA tools such as Compact-2D are built for 2D ICs, for M3D IC designs, a pseudo-3D design is first implemented and then split into two dies, which are routed independently to create the M3D design. In the pseudo-3D design stage, accurate estimation of 3D wire parasitics is crucial for optimizing the final M3D IC design. Experimental results show a 2.9× and 1.7× smaller root mean square error in the resistance and capacitance predictions for the proposed XGBoost method.

6 Securing IoT Infrastructure Using Artificial Intelligence (AI) and Machine Learning (ML)

The expansion of IoT systems to various application domains has consequently brought about critical security challenges. Due to the heterogeneous, low-power, resource-constrained nature of IoT systems, conventional security measures such as authentication, access control, network, and access security become infeasible in deployment. Furthermore, usage of low-cost, low-power, low-latency components such as FPGAs can enable backdoors into IoT networks through bitstream reverse engineering and remote dynamic partial reconfiguration [57, 58]. An attacker is therefore presented with new, multiple attack vectors to insert HT into the deployed IoT system. Taking all of this into account, the security paradigm in IoT systems rapidly changes as new vulnerabilities arise from IoT deployment in novel application environments, which are termed zero-day attacks. In this context, AI/ML approaches have been extensively used as a countermeasure for zero-day attacks, particularly for IoT network intrusion detection [59–61]. This section provides an overview of the usage of AI/ML in HT detection in general and focusing on using AI/ML techniques for HT detection in IoT networks. We also briefly discuss the emerging area of 3D integrated IoT devices, the potential for HT insertion in said context, and potential countermeasures against them.

6.1 Hardware Trojan Detection Using AI and ML

Over the last decade, machine learning (ML) methods have been used extensively among HT detection approaches. Due to the increasing complexity of IC designs and the expanding range of applications, ML approaches provide a learning-based methodology for HT detection that can adapt to the dynamic nature of the threat scenario. ML methods can learn patterns from high-dimensional feature spaces and perform pertinent feature extraction to differentiate between HT-free and HT-infected designs. In this context, the ML algorithms and approaches used for HT detection can be categorized as follows:

- **Supervised Learning**
  - Supervised ML approaches require *historical* data to train algorithms to perform either prediction (provide a continuous-valued output) or classification (provide a target label). A domain expert labels the training data, and in practice, the lack of domain expertise can be a critical bottleneck in adopting supervised learning approaches. Among the supervised ML methods for HT detection are artificial neural networks [62], support vector machines [10], Bayesian classifiers [63], extreme learning machines [64], decision trees [65], and k-nearest neighbors [66]. Supervised ML models, in general, are robust against noise and can handle cases where only a few features are available. However, these methods require golden designs for training and reference and are prone to over- and under-fitting.
- **Unsupervised Learning**
  - Unsupervised ML techniques attempt to discover the relationships between the data points without labels for the target classes. In HT detection, the unsupervised ML techniques used are, in general, clustering methods, where unlabeled data points are grouped according to some user-defined criteria [11, 29, 67, 68]. These techniques require no golden designs for training and are scalable with the available training dataset size. As unsupervised ML techniques are generally used for clustering, they have poor classification performance and are sensitive to noise and the number of clusters chosen.
- **Dimensionality Reduction**
  - Dimensionality reduction and feature selection techniques are used to select the feature space attributes that have the most impact on the output. Often,

dimensionality reduction methods are used to reduce the dimension of the feature space in preparation for supervised ML algorithms. Among the dimensionality reduction and feature selection methods used for HT detection are genetic algorithms (GA) [11], principal component analysis (PCA) [12], and two-dimensional PCA [67]. In general, these techniques increase the accuracy and other performance metrics for HT detection. A consequence of this improved performance is that features of HT may be discarded as redundant. In addition, multiple iterations of this method may be needed for selecting the appropriate number of features for a specific case.

- **Design Optimization and Model Enhancement**
  - The performance of ML methods can be enhanced by the use of optimization algorithms such as adaptive iterative optimization algorithm (AIOA) [13], multi-objective evolutionary algorithm (MOEA) [69], and particle swarm optimization (PSO) algorithm [70], among others. HT detection is generally used to improve the design-for-security (DFS) strategies to prevent HT insertion in a given design. These techniques, however, are limited to simple combinational circuits and require multiple iterations to obtain an optimum (or near optimum) solution.

With regard to ML countermeasures for HT defense, they can be categorized into *four* types: HT detection, design-for-security (DFS), bus security, and secure architecture. We provide a brief overview of each of these countermeasures.

- **HT Detection**
  - Methods for HT detection are used primarily for design verification to verify the presence of malicious modifications in designed or fabricated ICs. In three key areas, ML-based approaches for HT detection have had extensive usage: (1) reverse engineering, (2) circuit feature analysis, and (3) side-channel analysis. In reverse engineering, microscopic imagery of every layer of the IC is obtained to reconstruct the original design. ML-based approaches used in reverse engineering [29, 71, 72] have a major advantage in reducing the total number of steps in the reverse engineering process. Circuit feature analysis extracts functional or structural features of the IC from gate-level netlists, which are quantified and analyzed for HT detection [10, 73–75]. These methods reduce the size of feature space and automate the process of HT-Net classification. The perturbations in circuit parameters such as power, path delay, temperature, and electromagnetic (EM) radiation profiles are analyzed to distinguish an HT-infected IC from the golden IC for side-channel analysis. ML techniques improve the side-channel analysis methods by improving the parameters' signal-to-noise ratio (SNR) [12, 64, 67, 76–78].

- **Design-for-Security (DFS)**
  - DFS techniques enable either improving the trustworthiness of IC designs or a DFS that exploits security design strategies to enhance the trustworthiness of IC designs or assist in HT detection, prevention, and monitoring via on-chip modules. ML countermeasures for DFS include (1) Trojan detection assistance, (2) implantation prevention, and (3) trusted library. In Trojan detection assistance, on-chip modules are embedded in the design to either identify anomalies due to HT or increase the sensitivity of HT detection at both design and test time. ML techniques are integrated with these on-chip modules to improve accuracy of HT detection [11, 62, 63, 79, 80]. Implantation prevention techniques implement either design obfuscation or layout-filler approaches to protect ICs/IPs from HT insertion and activation, reverse engineering, or theft. ML methods have been incorporated in the EDA design and testing domains [69]. In trusted library generation, a golden reference library, meaning trusted datasets that include golden side-channel fingerprints and circuit features, or enhanced training models for HT detection, is generated. ML techniques have had significant usage in the construction of said golden libraries [13, 67, 81].
- **Bus Security**
  - Bus security measures are enacted to ensure the security of on-chip communication between multiple cores in a multiprocessor SoC environment. An attacker can intercept private communication, manipulate communication data, interrupt normal operations, and perform DoS attacks by altering router or linker behavior. ML approaches have been used in bus security to detect anomalous on-chip traffic behavior due to HT [82–84]. As these ML countermeasures require a training dataset, unknown attacks may bypass detection.
- **Security Architecture**
  - Secure architecture countermeasures have been extensively researched to protect a design from architectural-level HT, and ML techniques have been incorporated into many of these approaches. ML techniques, such as one-class artificial neural network (ANN), have been used for trust evaluation [85] and for defending against HTs that compromise confidentiality [86]. Both of these approaches require on-chip modules to implement the required ML technique. Trust evaluation approaches require a golden model for reference, whereas confidentiality protection does not have such requirements.

As can be observed, ML approaches can automate the process of HT detection and, in general, can improve the detection accuracy. In addition to the application of ML techniques for development of HT countermeasures in IC designs, several deep learning (DL) approaches have been implemented in HT countermeasure development in recent years. DL approaches are a subcategory of ML techniques, which are implemented using neural networks with multiple layers (more specifically, more than one hidden layer), allowing them to be able to extract hidden patterns and features from both structured and unstructured data (such as images, text, and speech). For HT detection, DL approaches have been used for advanced image processing on IC images and information extraction from netlists among other applications. Kulkarni and Xu [87] perform transfer learning using a ResNet architecture with 34 layers to perform optical inspection on IC images in order to classify them as HT infected, defective, or functioning correctly. A novel convolutional neural network (CNN) architecture, deep Siamese CNN (DSCNN), uses a few-shot learning approach to implement an HT detection algorithm from IC layout images obtained from partial reverse engineering (RE) process [88]. In [89], the authors propose using a stochastic reinforcement learning framework to reduce test generation complexity and automate effective test vector generation. The proposed approach considers both rare and testable signals using a combination of Sandia Controllability/Observability Analysis Program (SCOAP) measurement and dynamic simulation, to improve trigger coverage of suspicious regions. Yu et. al [90] apply natural language processing (NLP) for the first time to extract features from gate-level netlists, which are used for training long short-term memory (LSTM) and CNN in order to perform HT detection. To perform HT detection, the authors in [91] represent the hardware design as a graph and use, for the first time, a graph neural network (GNN) to learn circuit behavior and generate data flow graphs (DFG) for the register-transfer level (RTL) codes.

6.2 Hardware Trojan Detection in IoT Systems Using AI and ML

The expansion of IoT systems to various application domains has provided attackers with opportunities to generate new attack vectors to bypass conventional security features such as authentication, access control, and network and access security. Compounding this threat is the fact that the IC design cycle is geographically distributed to reduce costs, which enables the involvement of untrusted third-party actors. As a result, hardware security threats such as HTs have emerged as a critical security threat. In this scenario, ML-based countermeasures for HT detection specifically tailored to IoT systems have gained much research attention. Hossein et. al [92] used ML classifiers for malware detection in resource-constrained embedded systems. The classifiers were trained for real-time malware detection using data generated from a number of real malware threats. A random forest classifier trained using noninvasive measurements of current, voltage, and power attributes of the IoT device was proposed to detect both covert channel attacks and power depletion attacks [93]. Baibhab et. al [94] proposed using a deep neural network (DNN)-based authentication method for IoT devices that uses the inherent variations in the RF signals of an RF physically unclonable function (RF-PUF) for training. The area of ML-based HT detection in IoT systems is an emerging research area and continues to receive significant research focus from both the academic and industrial communities.

7 Leveraging 3D Integration for Hardware Security in IoT Devices

An emerging area for IoT applications has been the integration of 3D IC technology for IoT device construction. The fact that security is not the main functionality of an IoT device means that an even lesser portion of its computing power is available for security. Security measures implemented in traditional computers, such as cryptography, present a challenge from this context when applied in IoT devices. Further, due to the heterogeneity of devices, the power budget may not be enough to implement sophisticated security features in IoT paradigm. Many studies have shown that side-channels in IoT devices are easy to obtain and hard to defend against; hence, addressing side-channel leakage is crucial. Although various threats challenge IoT security, the root of trust starts from the hardware [95]. High-level approaches cannot stop these attacks without trusted and authenticated IoT devices. As many IoT devices are small in size, low in computation capabilities, and powered by low-capacity batteries, we need to rethink the trusted environment for IoT.

Three-dimensional (3D) integration [96] is an emerging technology to ensure the growth in transistor density and performance expected for future ICs. 3D integration and similar forms of die-level integration provide novel design methodologies to increase transistor density, reduce interconnect distances, and integrate additional system components. 3D integration covers various technologies, from interposer-based 2.5D methodology to monolithic sequential integration, but 3D stacked die-level integration, based on microbumps and Through-Silicon Vias (TSV), is widely seen as one of the most promising technologies for meeting future needs. In this methodology, separate dies are fabricated using standard lithography, TSVs are added (during or after lithography), individual dies or wafers are thinned, and 3D stacks are formed through alignment and bonding. The trend in 3D packaging technologies is shown in Fig. 7. 3D integration has attracted significant attention to developing diverse computing platforms such as high-performance processors, low-power systems-on-chip (SoCs), and portable devices during the past two decades. Recently, wireless approach has been introduced to replace TSVs with fabricated inductive coupling links (ICLs) in each of the layers of a 3D IC, providing

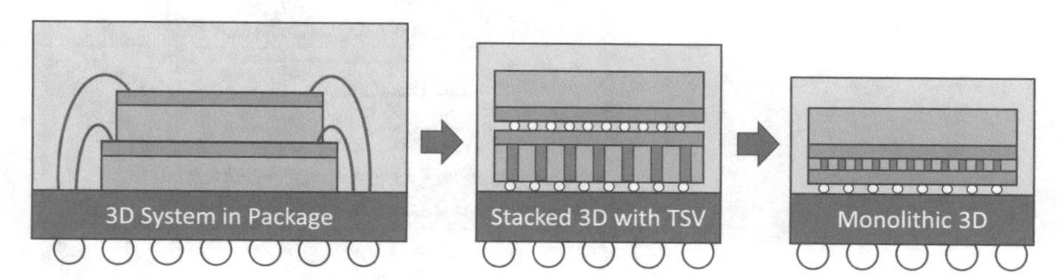

Fig. 7 3D packaging technology is expected to transition from current wire bond SiP to high-bandwidth TSV-based die stacking, and eventually to full monolithic integration

wireless 3D integration [97]. ICLs enable contactless 3D integration, wherein AC data or power transmission between the layers is performed by electromagnetic (EM) coupling between planar inductors fabricated in the back-end-of-line (BEOL) interconnect layers of each die.

Low-power consumption, small form factor, and multifunctionality are required for embedded devices in loT, and heterogeneous 3D integration can provide these attributes. Hybrid integration technology of complementary metal-oxide–semiconductors, microelectromechanical systems (MEMS), and photonic circuits for optoelectronic heterogeneous integrated systems on an LSI wafer is developed in [98]. 3D integration is not yet used in IoT devices. 3D integrated circuits (3D ICs) include several heterogeneous layers in a stacked architecture in the chip layout and provide a promising paradigm for secure, heterogeneous 3D integration suitable for IoT devices. 3D technology provides various advantages such as heterogeneous integration [99], split manufacturing [100, 101], disparate technologies for IoT like MEMS sensors [102], and others.

3D integration provides the following benefits for their application in the IoT paradigm. The overview of the 3D structure for IoT devices is shown in Fig. 8 [103].

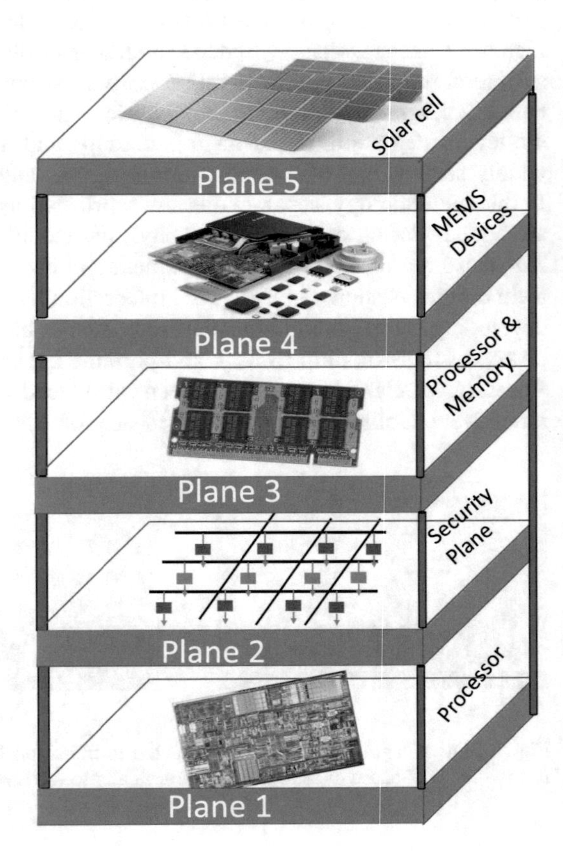

Fig. 8 3D structure for IoT devices

1. **Separate security plane using 3D stack**: Sherwood et al. [104] introduce a novel architecture using a separate control plane, stacked using 3D integration that provides security mechanisms to protect the design from explicit and implicit channels of information leakage. 3D will provide much higher integration, bringing multiple CPUs, memory blocks, and cryptographic engines together. Hence, the side-channel information will become noisy, making the attacks very challenging. If the control (security) plane is placed in the middle stack of 3D IC for fault prevention, it will be unlikely to inject reliable faults to carry out successful fault attacks.
2. **Shielding side-channels with 3D stacking**: In this approach, authors utilize intrinsic characteristics of 3D chip and dynamic shielding to hide the security-related activities on the chip [105]. They propose to use a microcontroller unit to produce complementary activity patterns, dynamically thwarting side-channel information leakage.
3. **Intrinsic power distribution network (PDN) noise to defeat SCA in 3D ICs**: In this work, the authors demonstrate that 3D PDN introduces noise to the power profile of a crypto unit that depends on the load switching activities, PDN topology, and crypto module deployment in the 3D chip. Using real 3D PDNs and through-silicon-vias (TSVs) models, we performed quantitative experimentation to exploit intrinsic noise to defeat the side-channel attacks [106, 107]. In the method shown in Fig. 9, we divide the crypto unit into four subunits (we can use multiple). Each subunit is driven by a local supply voltage V DDi ($i = 1, 2, 3, 4$). We utilize a crossbar to connect the local VDD pins with the PDN nodes close to four power TSVs. Due to the nonuniform switching activities in every 3D plane, each TSV passes a unique voltage from other 3D planes to the plane carrying the crypto unit. The effect of parasitic resistance and capacitance (RC) of the metal wire between the power grid and the local VDD pin further increases the variance of the four VDDs for the crypto unit.
4. **Energy harvesting using solar cell**: Many IoT devices will be battery operated or self-powered. 3D integration provides an opportunity to use alternate forms of

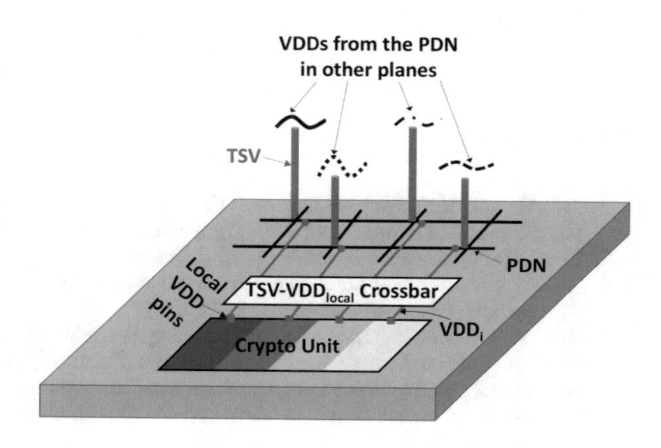

Fig. 9 Proposed countermeasure against CPA attacks in 3D ICs

energy like solar, electromagnetic, thermal, and others, because of its heterogeneous nature.

8 Discussion

The Internet-of-Things (IoT) has radically changed how we interact with the world around us. IoT devices have had widespread applications in many fields of production and social living, such as healthcare, energy and industrial automation, and military application, to name a few. While the benefits of IoT systems are undeniable, they have disadvantages as well. Conventional security measures are infeasible in IoT deployments due to their low-power, heterogeneous, and resource-constrained nature. In addition, due to the geographical redistribution of the IC design cycle and the involvement of untrusted entities in the design process, hardware attacks in IoT systems are becoming a serious threat. The heterogeneous nature of IoT systems introduces multiple attack vectors that an attacker can exploit for hardware attacks, especially side-channel attacks and hardware Trojan insertions. A great deal of research focuses on software, network, and cloud security concerning IoT; however, hardware security in these devices has been overlooked. In this chapter, we have discussed the existing countermeasures for individual side-channel attacks (SCA) and HT insertions and unified hardening approaches that benefit IoT devices because of area footprint and power constraints. Machine learning-based hardware security countermeasures that have been extensively used in SCA attack countermeasures and HT detection have been thoroughly discussed. Furthermore, we proposed to use 3D integration technology for developing a secure IoT platform. 3D integration technology provides various advantages such as heterogeneous integration, high-performance, low-power, multifunctionality, and small form factor integration, making it the best choice for developing secure IoT platforms. We hope this chapter stimulates the research community in academia and industry to further investigate the new hardware security approaches and improve existing ones.

References

1. Ray, S., Jin, Y., Raychowdhury, A.: The changing computing paradigm with Internet of Things: a tutorial introduction. IEEE Design Test **33**(2), 76–96 (2016)
2. Bastos, D., Shackleton, M., El-Moussa, F.: Internet of Things: a survey of technologies and security risks in smart home and city environments. In: Living in the Internet of Things: Cybersecurity of the IoT—2018, pp. 1–7 (2018)
3. Yan, Y., Oswald, E., Tryfonas, T.: Exploring potential 6LoWPAN traffic side channels. Cryptology ePrint Archive, Report 2017/316 (2017). https://ia.cr/2017/316
4. Workshop Report by Guru Prasadh Venkataramani and Patrick Schaumont: NSF Workshop on side and covert channels in computing systems (2019). https://www2.seas.gwu.edu/~guruv/workshop-report.pdf. Accessed 5 Jan 2021

5. Das, D., Maity, S., Nasir, S.B., Ghosh, S., Raychowdhury, A., Sen, S.: High efficiency power side-channel attack immunity using noise injection in attenuated signature domain. In: 2017 IEEE International Symposium on Hardware Oriented Security and Trust (HOST), pp. 62–67 (2017)
6. Stout, W.M.S., Urias, V.E.: Challenges to securing the Internet of Things. In: 2016 IEEE International Carnahan Conference on Security Technology (ICCST), pp. 1–8 (2016)
7. Sicari, S., Rizzardi, A., Grieco, L., Coen-Porisini, A.: Security, privacy and trust in Internet of Things: the road ahead. Comput. Netw. **76**, 146–164 (2015)
8. Al-Omary, A., Othman, A., AlSabbagh, H.M., Al-Rizzo, H.: Survey of Hardware-Based Security support for IoT/CPS Systems (2018)
9. Roman, R., Zhou, J., Lopez, J.: On the features and challenges of security and privacy in distributed Internet of things. Comput. Netw. **57**, 2266–2279 (2013)
10. Hasegawa, K., Oya, M., Yanagisawa, M., Togawa, N.: Hardware trojans classification for gate-level netlists based on machine learning. In: 2016 IEEE 22nd International Symposium on On-Line Testing and Robust System Design (IOLTS), pp. 203–206 (2016)
11. Karimian, N., Tehranipoor, F., Rahman, M.T., Kelly, S., Forte, D.: Genetic algorithm for hardware trojan detection with ring oscillator network (ron). In: 2015 IEEE International Symposium on Technologies for Homeland Security (HST), pp. 1–6 (2015)
12. Liu, Y., Jin, Y., Nosratinia, A., Makris, Y.: Silicon demonstration of hardware trojan design and detection in wireless cryptographic ICS. IEEE Trans. Very Large Scale Integr. Syst. **25**(4), 1506–1519 (2017)
13. Xue, M., Wang, J., Hu, A.: An enhanced classification-based golden chips-free hardware trojan detection technique. In: 2016 IEEE Asian Hardware-Oriented Security and Trust (AsianHOST), pp. 1–6 (2016)
14. Rostami, M., Koushanfar, F., Rajendran, J., Karri, R.: Hardware security: threat models and metrics. In: 2013 IEEE/ACM International Conference on Computer-Aided Design (ICCAD), pp. 819–823 (2013)
15. Das, D., Sen, S.: Electromagnetic and power side-channel analysis: advanced attacks and low-overhead generic countermeasures through white-box approach. Cryptography **4**(4), 30 (2020)
16. Kocher, P., Jaffe, J., Jun, B.: Differential power analysis. In: Advances in Cryptology—CRYPTO' 99, (Berlin, Heidelberg), pp. 388–397. Springer, Berlin Heidelberg (1999)
17. Kocher, P.C.: Timing Attacks on Implementations of Diffie-Hellman, RSA, DSS, and Other Systems. In: Advances in Cryptology—CRYPTO '96, pp. 104–113. Springer, Berlin, Heidelberg (1996)
18. Das, D., Nath, M., Chatterjee, B., Ghosh, S., Sen, S.: STELLAR: a generic EM side-channel attack protection through ground-up root-cause analysis. In: 2019 IEEE International Symposium on Hardware Oriented Security and Trust (HOST), pp. 11–20 (2019)
19. Tunstall, M., Mukhopadhyay, D., Subidh Ali, S.: Differential fault analysis of the advanced encryption standard using a single fault, pp. 224–233 (2011)
20. Fritzke, A.: Obfuscating Against Side-Channel Power Analysis Using Hiding Techniques for AES (2012)
21. Mangard, S., Oswald, E., Popp, T.: Power analysis attacks—revealing the secrets of smart cards (2007)
22. Agrawal, D., Archambeault, B., Rao, J.R., Rohatgi, P.: The EM side-channel(s). In: Cryptographic Hardware and Embedded Systems—CHES 2002, 4th International Workshop, Redwood Shores, CA, USA, August 13–15, 2002, Revised Papers, vol. 2523 of Lecture Notes in Computer Science, pp. 29–45. Springer, Berlin (2002)
23. Wang, C., Cai, Y., Wang, H., Zhou, Q.: Electromagnetic equalizer: an active countermeasure against EM side-channel attack. In: 2018 IEEE/ACM International Conference on Computer-Aided Design (ICCAD), pp. 1–8 (2018)
24. Jayasinghe, D., Ragel, R., Elkaduwe, D.: Constant time encryption as a countermeasure against remote cache timing attacks. In: 2012 IEEE 6th International Conference on Information and Automation for Sustainability, pp. 129–134 (2012)

25. Ge, Q., Yarom, Y., Cock, D., Heiser, G.: A survey of microarchitectural timing attacks and countermeasures on contemporary hardware. J. Cryptogr. Eng. **8**, 1–27 (2018)
26. Mozaffari-Kermani, M., Reyhani-Masoleh, A.: Concurrent structure-independent fault detection schemes for the advanced encryption standard. IEEE Trans. Comput. **59**(5), 608–622 (2010)
27. Barenghi, A., Breveglieri, L., Koren, I., Naccache, D.: Fault injection attacks on cryptographic devices: theory, practice, and countermeasures. Proc. IEEE **100**(11), 3056–3076 (2012)
28. Yu, Q., Zhang, Z., Dofe, J.: Proactive Defense Against Security Threats on IoT Hardware, ch. 18, pp. 407–433. Wiley, London (2020)
29. Bao, C., Forte, D., Srivastava, A.: On reverse engineering-based hardware trojan detection. IEEE Trans. Comput.-Aided Design Integr. Circuits Syst. **35**(1), 49–57 (2016)
30. Yu, Q., Dofe, J., Zhang, Z.: Exploiting hardware obfuscation methods to prevent and detect hardware trojans. In: 2017 IEEE 60th International Midwest Symposium on Circuits and Systems (MWSCAS), pp. 819–822 (2017)
31. Yu, W., Köse, S.: A lightweight masked AES implementation for securing IoT against CPA attacks. IEEE Trans. Circuits Syst. I: Regul. Pap. **64**(11), 2934–2944 (2017)
32. Yang, K., Park, J., Tehranipoor, M., Bhunia, S.: Robust timing attack countermeasure on virtual hardware. In: 2018 IEEE Computer Society Annual Symposium on VLSI (ISVLSI), pp. 148–153 (2018)
33. Guo, S., Wang, J., Chen, Z., Li, Y., Lu, Z.: Securing IoT space via hardware trojan detection. IEEE Internet Things J. **7**(11), 11115–11122 (2020)
34. Jedari, E., Rashidzadeh, R.: A hardware trojan detection method for IoT sensors using side channel activity magnifier. IEEE Internet Things J. **9**(6), 4507–4517 (2021)
35. Patranabis, S., Roy, D.B., Chakraborty, A., Nagar, N., Singh, A., Mukhopadhyay, D., Ghosh, S.: Lightweight design-for-security strategies for combined countermeasures against side channel and fault analysis in IoT applications. J. Hardw. Syst. Secur. **3**(2), 103–131 (2019)
36. Moukarzel, M., Eisenbarth, T., Sunar, B.: Leech: a side-channel evaluation platform for IoT. In: 2017 IEEE 60th International Midwest Symposium on Circuits and Systems (MWSCAS), pp. 25–28 (2017)
37. Aerabi, E., Papadimitriou, A., Hely, D.: On a side channel and fault attack concurrent countermeasure methodology for MCU-based byte-sliced cipher implementations. In: 2019 IEEE 25th International Symposium on On-Line Testing and Robust System Design (IOLTS), pp. 103–108 (2019)
38. Dofe, J., Pahlevanzadeh, H., Yu, Q.: A comprehensive FPGA-based assessment on fault-resistant AES against correlation power analysis attack. J. Electron. Test. **32**(5), 611–624 (2016)
39. Nagata, M.: On-chip protection of cryptographic ICs against physical side channel attacks: invited paper. In: 2019 IEEE 13th International Conference on ASIC (ASICON), pp. 1–4 (2019)
40. Dofe, J., Frey, J., Yu, Q.: Hardware security assurance in emerging IoT applications. In: 2016 IEEE International Symposium on Circuits and Systems (ISCAS), pp. 2050–2053 (2016)
41. Shanthi, J., Rajaram, S., et al.: Machine learning optimization techniques for 3d IC physical design. In: Handbook of Research on Emerging Trends and Applications of Machine Learning, pp. 47–61. IGI Global (2020)
42. Samal, S.K., Chen, G., Lim, S.K.: Machine learning based variation modeling and optimization for 3d ICs (2016)
43. Das, S., Doppa, J.R., Kim, D.H., Pande, P.P., Chakrabarty, K.: Optimizing 3d NoC design for energy efficiency: a machine learning approach. In: 2015 IEEE/ACM International Conference on Computer-Aided Design (ICCAD), pp. 705–712 (2015)
44. Park, S.J., Bae, B., Kim, J., Swaminathan, M.: Application of machine learning for optimization of 3-d integrated circuits and systems. IEEE Trans. Very Large Scale Integr. Syst. **25**(6), 1856–1865 (2017)
45. Torun, H.M., Swaminathan, M.: Black-box optimization of 3d integrated systems using machine learning. In: *2017 IEEE 26th Conference on Electrical Performance of Electronic Packaging and Systems (EPEPS)*, pp. 1–3 (2017)

46. Das, S., Doppa, J.R., Pande, P.P., Chakrabarty, K.: Monolithic 3d-enabled high performance and energy efficient network-on-chip. In: *2017 IEEE International Conference on Computer Design (ICCD)*, pp. 233–240 (2017)
47. Joardar, B.K., Kim, R.G., Doppa, J.R., Pande, P.P., Marculescu, D., Marculescu, R.: Learning-based application-agnostic 3d NoC design for heterogeneous manycore systems. IEEE Trans. Comput. **68**, 852–866 (2019)
48. Lee, D., Das, S., Kim, D.H., Doppa, J.R., Pande, P.P.: Design space exploration of 3d network-on-chip: a sensitivity-based optimization approach. J. Emerg. Technol. Comput. Syst. **14**(3), 1–26 (2018)
49. Ku, B.W., Chang, K., Lim, S.K.: Compact-2d: a physical design methodology to build commercial-quality face-to-face-bonded 3d ICs. In: Proceedings of the 2018 International Symposium on Physical Design, ISPD '18, (New York, NY, USA), pp. 90–97. Association for Computing Machinery (2018)
50. Panth, S., Samadi, K., Du, Y., Lim, S.K.: Shrunk-2-d: a physical design methodology to build commercial-quality monolithic 3-d ICs. IEEE Trans. Comput.-Aided Design Integr. Circuits Syst. **36**(10), 1716–1724 (2017)
51. Lu, Y.-C., Pentapati, S.S.K., Zhu, L., Samadi, K., Lim, S.K.: Tp-GNN: a graph neural network framework for tier partitioning in monolithic 3d ICs. In: Proceedings of the 57th ACM/EDAC/IEEE Design Automation Conference, DAC '20. IEEE Press (2020)
52. Yun, H., Yoon, J.-S., Jeong, J., Lee, S., Choi, H.-C., Baek, R.-H.: Neural network based design optimization of 14-nm node fully-depleted SOI FET for SoC and 3DIC applications. IEEE J. Electron Devices Soc. **8**, 1272–1280 (2020)
53. Chatterjee, S., Roy, S.K., Giri, C., Rahaman, H.: Machine learning based temperature estimation for test scheduling of 3d ICs. In: 2020 IEEE International Test Conference India, pp. 1–8, (2020)
54. Zhang, L., Wang, H., Tan, S.X.-D.: Fast stress analysis for runtime reliability enhancement of 3d IC using artificial neural network. In: 2016 17th International Symposium on Quality Electronic Design (ISQED), pp. 173–178 (2016)
55. Li, Y.-S., Yu, H., Jin, H., Sarvey, T.E., Oh, H., Bakir, M.S., Swaminathan, M. and Li, E.-P.: Dynamic thermal management for 3-d ICs with time-dependent power map using microchannel cooling and machine learning. IEEE Trans. Comp. Packag. Manuf. Technol. **9**(7), 1244–1252, 2019.
56. Pentapati, S.S.K., Ku, B.W., Lim, S.K.: ML-based wire RC prediction in monolithic 3d ICs with an application to full-chip optimization. In: Proceedings of the 2021 International Symposium on Physical Design, ISPD '21, (New York, NY, USA), pp. 75–82. Association for Computing Machinery (2021)
57. Danesh, W., Banago, J., Rahman, M.: Turning the table: using bitstream reverse engineering to detect FPGA trojans. J. Hardw. Syst. Secur. **5**(3), 237–246 (2021)
58. Johnson, A.P., Patranabis, S., Chakraborty, R.S., Mukhopadhyay, D.: Remote dynamic partial reconfiguration: a threat to internet-of-things and embedded security applications. Microprocess. Microsyst. **52**, 131–144 (2017)
59. Thomas, L., Bhat, S.: Machine learning and deep learning techniques for IoT-based intrusion detection systems: a literature review. Int. J. Manag. Technol. Soc. Sci. **6**(2), 296–314 (2021)
60. Asharf, J., Moustafa, N., Khurshid, H., Debie, E., Haider, W., Wahab, A.: A review of intrusion detection systems using machine and deep learning in internet of things: challenges, solutions and future directions. Electronics **9**, 1177 (2020)
61. Tsimenidis, S., Lagkas, T., Rantos, K.: Deep learning in IoT intrusion detection. J. Netw. Syst. Manag. **30**, 1–40 (2022)
62. Liu, Y., Volanis, G., Huang, K., Makris, Y.: Concurrent hardware trojan detection in wireless cryptographic ICs. In: 2015 IEEE International Test Conference (ITC), pp. 1–8 (2015)
63. Chen, X., Wang, L., Wang, Y., Liu, Y., Yang, H.: A general framework for hardware trojan detection in digital circuits by statistical learning algorithms. IEEE Trans. Comput.-Aided Design Integr.Circuits Syst. **36**(10), 1633–1646 (2017)

64. Wang, S., Dong, X., Sun, K., Cui, Q., Li, D., He, C.: Hardware trojan detection based on ELM neural network. In: 2016 First IEEE International Conference on Computer Communication and the Internet (ICCCI), pp. 400–403 (2016)
65. Lodhi, F.K., Hasan, S.R., Hasan, O., Awwadl, F.: Power profiling of microcontroller's instruction set for runtime hardware trojans detection without golden circuit models. In: Design, Automation Test in Europe Conference Exhibition (DATE), 2017, pp. 294–297 (2017)
66. Lodhi, F.K., Abbasi, I., Khalid, F., Hasan, O., Awwad, F., Hasan, S.R.: A self-learning framework to detect the intruded integrated circuits. In: 2016 IEEE International Symposium on Circuits and Systems (ISCAS), pp. 1702–1705 (2016)
67. Nowroz, A.N., Hu, K., Koushanfar, F., Reda, S.: Novel techniques for high-sensitivity hardware trojan detection using thermal and power maps. IEEE Trans. Comput.-Aided Design Integr. Circuits Syst. **33**(12), 1792–1805 (2014)
68. Cakır, B., Malik, S.: Hardware trojan detection for gate-level ICs using signal correlation based clustering. In: 2015 Design, Automation Test in Europe Conference Exhibition (DATE), pp. 471–476 (2015)
69. Marcelli, A., Restifo, M., Sanchez, E., Squillero, G.: An evolutionary approach to hardware encryption and trojan-horse mitigation. In: Design, Automation Test in Europe Conference Exhibition (DATE), 2017, pp. 1593–1598 (2017)
70. Wang, C., Zhao, S., Wang, X., Luo, M., Yang, M.: A neural network trojan detection method based on particle swarm optimization. 2018 14th IEEE International Conference on Solid-State and Integrated Circuit Technology (ICSICT), pp. 1–3 (2018)
71. Bao, C., Forte, D., Srivastava, A.: On application of one-class SVM to reverse engineering-based hardware trojan detection. In: Fifteenth International Symposium on Quality Electronic Design, pp. 47–54 (2014)
72. Li, W., Wasson, Z., Seshia, S.A.: Reverse engineering circuits using behavioral pattern mining. In: 2012 IEEE International Symposium on Hardware-Oriented Security and Trust, pp. 83–88 (2012)
73. Zhou, E.-R., Li, S.-Q., Chen, J.-H., Ni, L., Zhao, Z.-X., Li, J.: A novel detection method for hardware trojan in third party ip cores. In: 2016 International Conference on Information System and Artificial Intelligence (ISAI), pp. 528–532 (2016)
74. Hasegawa, K., Yanagisawa, M., Togawa, N.: Trojan-feature extraction at gate-level netlists and its application to hardware-trojan detection using random forest classifier. In: 2017 IEEE International Symposium on Circuits and Systems (ISCAS), pp. 1–4 (2017)
75. Hoque, T., Cruz, J., Chakraborty, P., Bhunia, S.: Hardware IP trust validation: learn (the untrustworthy), and verify. In: 2018 IEEE International Test Conference (ITC), pp. 1–10 (2018)
76. Li, J., Ni, L., Chen, J., Zhou, E.: A novel hardware trojan detection based on bp neural network. In: 2016 2nd IEEE International Conference on Computer and Communications (ICCC), pp. 2790–2794 (2016)
77. Jap, D., He, W., Bhasin, S.: Supervised and unsupervised machine learning for side-channel based trojan detection. In: 2016 IEEE 27th International Conference on Application-specific Systems, Architectures and Processors (ASAP), pp. 17–24 (2016)
78. Iwase, T., Nozaki, Y., Yoshikawa, M., Kumaki, T.: Detection technique for hardware trojans using machine learning in frequency domain. In: 2015 IEEE 4th Global Conference on Consumer Electronics (GCCE), pp. 185–186 (2015)
79. Dong, C., He, G., Liu, X., Yang, Y., Guo, W.: A multi-layer hardware trojan protection framework for IoT chips. IEEE Access **7**, 23628–23639 (2019)
80. Shanyour, B., Tragoudas, S.: Detection of low power trojans in standard cell designs using built-in current sensors. In: 2018 IEEE International Test Conference (ITC), pp. 1–10 (2018)
81. Liu, Y., Huang, K., Makris, Y.: Hardware trojan detection through golden chip-free statistical side-channel fingerprinting. In: Proceedings of the 51st Annual Design Automation Conference, DAC '14, (New York, NY, USA), pp. 1–6. Association for Computing Machinery (2014)

82. Kulkarni, A., Pino, Y., Mohsenin, T.: SVM-based real-time hardware trojan detection for many-core platform. In: 2016 17th International Symposium on Quality Electronic Design (ISQED), pp. 362–367 (2016)
83. Madden, K., Harkin, J., McDaid, L., Nugent, C.: Adding security to networks-on-chip using neural networks. In: 2018 IEEE Symposium Series on Computational Intelligence (SSCI), pp. 1299–1306 (2018)
84. Kulkarni, A., Pino, Y., French, M., Mohsenin, T.: Real-time anomaly detection framework for many-core router through machine-learning techniques. J. Emerg. Technol. Comput. Syst. **13**, 1–22 (2016)
85. Jin, Y., Maliuk, D., Makris, Y.: Post-deployment trust evaluation in wireless cryptographic ICs. In: 2012 Design, Automation Test in Europe Conference Exhibition (DATE), pp. 965–970 (2012)
86. Guha, K., Saha, D., Chakrabarti, A.: RTNA: securing SOC architectures from confidentiality attacks at runtime using ART1 neural networks. In: 2015 19th International Symposium on VLSI Design and Test, pp. 1–6 (2015)
87. Kulkarni, A., Xu, C.: A deep learning approach in optical inspection to detect hidden hardware trojans and secure cybersecurity in electronics manufacturing supply chains. Front. Mech. Eng. **7**, 709924 (2021)
88. Sharma, R., Sharma, G.K., Pattanaik, M.: A few shot learning based approach for hardware trojan detection using deep Siamese CNN. In: 2021 34th International Conference on VLSI Design and 2021 20th International Conference on Embedded Systems (VLSID), pp. 163–168 (2021)
89. Pan, Z., Mishra, P.: Automated test generation for hardware trojan detection using reinforcement learning. In: 2021 26th Asia and South Pacific Design Automation Conference (ASP-DAC), pp. 408–413 (2021)
90. Yu, S., Gu, C., Liu, W., O'Neill, M.: Deep learning-based hardware trojan detection with block-based netlist information extraction. IEEE Trans. Emerg. Topics Compu. (2021).
91. Yasaei, R., Yu, S.-Y., Al Faruque, M.A.: GNN4TJ: graph neural networks for hardware trojan detection at register transfer level. In: 2021 Design, Automation Test in Europe Conference Exhibition (DATE), pp. 1504–1509 (2021)
92. Sayadi, H., Makrani, H.M., Randive, O., PD, S.M., Rafatirad, S., Homayoun, H.: Customized machine learning-based hardware-assisted malware detection in embedded devices. In: 2018 17th IEEE International Conference on Trust, Security and Privacy in Computing and Communications/12th IEEE International Conference on Big Data Science and Engineering (TrustCom/BigDataSE), pp. 1685–1688 (2018)
93. Mohammed, H., Odetola, T.A., Hasan, S.R., Stissi, S., Garlin, I., Awwad, F.: (hiadiot): Hardware intrinsic attack detection in internet of things; leveraging power profiling. In: 2019 IEEE 62nd International Midwest Symposium on Circuits and Systems (MWSCAS), pp. 852–855 (2019)
94. Chatterjee, B., Das, D., Maity, S., Sen, S.: RF-PUF: Enhancing IoT security through authentication of wireless nodes using in situ machine learning. IEEE Internet Things J. **6**(1), 388–398 (2019)
95. Rostami, M., Koushanfar, F., Karri, R.: A primer on hardware security: models, methods, and metrics. Proc. IEEE **102**(8), 1283–1295 (2014)
96. Lu, J.-Q.: 3-D hyperintegration and packaging technologies for micro-nano systems. Proc. IEEE **97**(1), 18–30 (2009
97. Dofe, J., Danesh, W.: LC-physical unclonable function in wireless 3d IC for securing internet of things devices. In: 2021 IEEE 34th International System-on-Chip Conference (SOCC), pp. 67–70 (2021)
98. Lee, K.-W., Noriki, A., Kiyoyama, K., Fukushima, T., Tanaka, T., Koyanagi, M.: Three-dimensional hybrid integration technology of CMOS, MEMS, and photonics circuits for optoelectronic heterogeneous integrated systems. IEEE Trans. Electron Devices **58**(3), 748–757 (2011)

99. Dofe, J., Gu, P., Stow, D., Yu, Q., Kursun, E., Xie, Y.: Security threats and countermeasures in three-dimensional integrated circuits, pp. 321–326 (2017)
100. Xie, Y., Bao, C., Liu, Y., Srivastava, A.: 2.5D/3D integration technologies for circuit obfuscation. In: 2016 17th International Workshop on Microprocessor and SOC Test and Verification (MTV), pp. 39–44 (2016)
101. Dofe, J., Yu, Q., Wang, H., Salman, E.: Hardware security threats and potential countermeasures in emerging 3d ICs. In: Proceedings of the 26th Edition on Great Lakes Symposium on VLSI, GLSVLSI '16, (New York, NY, USA), pp. 69–74. Association for Computing Machinery (2016)
102. Wang, Z.: 3-D integration and through-silicon vias in MEMS and microsensors. J. Microelectromech. Syst **24**, 1211–1244 (2015)
103. Dofe, J., Nguyen, A., Nguyen, A.: Unified countermeasures against physical attacks in internet of things—a survey. In: 2021 IEEE International Symposium on Smart Electronic Systems (iSES), pp. 194–199 (2021)
104. J. Valamehr, T. Huffmire, C. Irvine, R. Kastner, C. Koc, T. Levin, T. Sherwood: A qualitative security analysis of a new class of 3-D integrated crypto co-processors, vol. 6805, pp. 364–382 (2012)
105. Gu, P., Li, S., Stow, D., Barnes, R., Liu, L., Xie, Y., Kursun, E.: Leveraging 3D technologies for hardware security: opportunities and challenges. In: 2016 International Great Lakes Symposium on VLSI (GLSVLSI), pp. 347–352 (2016)
106. Dofe, J., Yu, Q.: Exploiting PDN noise to thwart correlation power analysis attacks in 3D ICs. In: 2018 ACM/IEEE International Workshop on System Level Interconnect Prediction (SLIP), pp. 1–6 (2018)
107. Zhang, Z., Dofe, J., Yu, Q.: Improving power analysis attack resistance using intrinsic noise in 3D ICs. Integration **73**, 30–42 (2020)

Enabling Edge Computing Using Emerging Memory Technologies: From Device to Architecture

Arman Roohi, Shaahin Angizi, and Deliang Fan

1 Introduction and Motivations

1.1 Von-Neumann vs. Non-Von-Neumann Architectures

In the past decades, the amount of data required to be processed by computing systems has been dramatically increasing to exascale (10^{18} bytes/s or flops) [1, 2]. However, the incapacity of modern computing platforms to deliver both energy-efficient and high-performance computing solutions leads to a gap between meets and needs [3]. Unfortunately, with current Boolean logic and Complementary Metal Oxide Semiconductor (CMOS)-based computing platforms, such a gap will keep widening mainly due to limitations in both *devices* and *architectures*. First, at the device level, CMOS Boolean systems' computing efficiency and performance are beginning to stall due to approaching the end of Moore's law. Because of reaching its power wall, i.e., huge leakage power consumption limits the performance growth when technology scales down [1, 4]. For example, the highest power efficiency of modern CPU and GPU systems is only ~10GFLOPS/W [5], which is challenging to improve substantially in the predictable scaled technology node. Second, at

A. Roohi (✉)
School of Computing, University of Nebraska-Lincoln, Lincoln, NE, USA
e-mail: aroohi@unl.edu

S. Angizi
Department of Electrical and Computer Engineering, New Jersey Institute of Technology, Newark, NJ, USA
e-mail: shaahin.angizi@njit.edu

D. Fan
School of Electrical; Computer and Energy Engineering, Arizona State University, Tempe, AZ, USA
e-mail: dfan@asu.edu

A. Iranmanesh (ed.), *Frontiers of Quality Electronic Design (QED)*,
https://doi.org/10.1007/978-3-031-16344-9_11

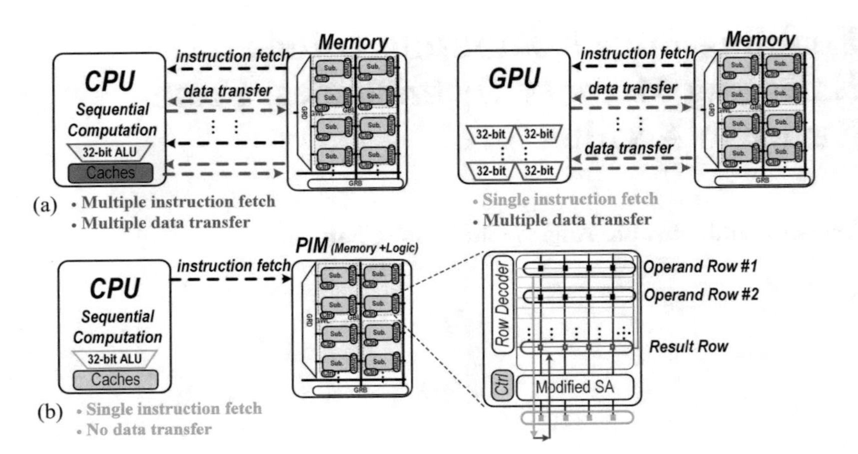

Fig. 1 (**a**) General von-Neumann computing architecture in CPU and GPU vs. (**b**) processing-in-memory architecture [6]

the architecture level, as depicted in Fig. 1a, today's computers are based on Von-Neumann architecture with separate computing and memory units connecting via buses, which leads to the memory wall issue. This bottleneck imposes long memory access latency, limited memory bandwidth, energy-hungry data transfer, and immense leakage power for holding data in volatile memory [6, 7]. This comes from the fact that there is a massive number of instruction fetch and data transfer between computing and memory units. Therefore, there is a great need to leverage innovations from both device and architecture to build intelligent, reconfigurable, energy-efficient, and high-performance computing platforms integrating memory and logic to break the existing memory and power walls.

The processing-in-memory (PIM) architecture, a potentially viable way to solve the memory wall challenge, has been well explored [8, 9]. The key concept behind PIM, as depicted in Fig. 1b, is to embed logic units within memory to process data by leveraging the inherent parallel computing mechanism and exploiting large internal memory bandwidth. It could lead to remarkable savings in off-chip data communication energy and latency. Ideally, the PIM architectures must be capable of performing bulk bit-wise operations that are needed in many big data applications [10–12]. Generally, at the sub-array level, a PIM holds the operand rows, e.g., #1 and #2 shown in Fig. 1b in two target rows of the memory. The PIM's row decoder simultaneously activates the target rows by receiving a particular instruction from the CPU side. It performs the bit-wise logic function between all the bit-cells in two rows, storing two operands. This could be achieved by modifying memory components at Sense Amplifiers (SA) level [10, 13, 14], memory bit-cell level [15], or even adding combinational circuits after SA [13, 14, 16, 17]. The proposals for exploiting SRAM-based [18] PIM architectures can be found in recent literature. However, PIM in the context of main memory (DRAM-based [9]) has drawn much more attention in recent years, mainly due to larger memory capacities and off-chip data transfer reduction as opposed to SRAM-based PIM. However, existing DRAM-based PIM architectures have significant shortcomings, e.g., high refresh/leakage power, multi-cycle logic operations, overwritten operand data, operand locality, etc. The PIM architecture has become even more intriguing when integrated with emerging nonvolatile memory (NVM) technology, such as Phase Change Memory (PCM) [19] and resistive RAM (ReRAM) [8]. ReRAM and PCM offer more packing density ($\sim 2-4\times$) than DRAM and appear to be competitive alternatives to DRAM. However, they suffer from slower and more power-hungry writing operations than DRAM [19]. Magnetic RAM (MRAM) technology in emerging NVM technologies is another promising high-performance candidate for last-level cache and main memory due to its ultralow switching energy, nonvolatility, superior endurance, excellent retention time, high integration density, and compatibility with CMOS technology. Meanwhile, MRAM technology is in the process of commercialization [20]. Hence, PIM in the context of different NVMs, without sacrificing memory capacity, can open a new way to realize efficient PIM paradigms [10, 17].

1.2 Normally Off Computing Systems

CMOS scaling challenges have inspired considerable advancements in reduced-power datapath designs. Practical techniques to reduce dynamic energy consumption, such as low-voltage operation, clock gating, and efficient RTL design, have been widely successful [21, 22]. Nonetheless, an increasing number of modern intelligent systems from many-core dies to Internet-of-Things (IoT) components, making the standby power dissipation of such systems a critical issue, especially under the deep-scaling impacts of CMOS process technology. For this reason, various state-of-the-art *Normally-off Computing (NoC)* techniques have been developed, which provides promising features such as zero standby power consumption during idle time, instant wake-up time, and resilience to power failure [23–25]. Hence, nonvolatile elements, including nonvolatile memories (NVMs) and nonvolatile flip-flops (NV-FFs), have received increasing attention because of their utility in designing an NoC architecture [26]. Various hardware-assisted approaches for normally off computing have recently been promulgated [27]. For instance, in [28], all of the conventional FFs are replaced by NV-FFs, while in [27], many small NV memory arrays are utilized to backup and restore data. Although NV elements offer the desirable feature of nonvolatility, their advantages are achieved at the cost of increased write-power consumption. Hence, a comprehensive datapath synthesis strategy is essential. In previous approaches, the roles and costs of the additional middleware and checkpointing operations[1] needed have been prominent [29, 30]. In

[1] Almost all the previous checkpointing techniques suffer from data movement overhead, new programming paradigms, and internal and external consistency.

addition to the overheads resulting from the checkpointing operations themselves, existing approaches may suffer from leakage occurring between the checkpointing operations made to nonvolatile backup storage [31].

2 Emerging Magnetic RAM (MRAM) Technology

Recent experiments and fabrication of nanomagnets demonstrate the ability to switch the magnetization using ultrasmall current-induced Spin-Transfer Torque (STT) or spin-orbit torque (SOT) with high-speed (sub-nanosecond), long-endurance (10 years), and less than fJ/bit memory write energy (close to SRAM) [32]. Various nanoscale spintronic devices have been explored to realize nonvolatile storage devices for MRAM applications. They include but not limited to Magnetic Tunnel Junction (MTJ) [33], Domain Wall Motion (DWM) device [34], SOT-MTJ memory device [35, 36], and Skyrmions. Several companies, including IBM [37] and Everspin [20], are developing MRAM chips for next-generation universal NVM systems. In early 2016, Everspin announced 256 Mb STT-MRAM chips based on MTJ with interface speed similar to DRAM and was planning 1Gb chips in the near future [20]. Toshiba and SK Hynix codeveloped a 4-Gbit STT-MRAM chip prototype and demonstrated it at IEDM 2016 [38]. In [39], a field-free switching SOT-MRAM on a 300 mm wafer was demonstrated with reliable sub-ns switching and CMOS-compatible processes. In [40], a SOT-MRAM achieving 60-MHz write and 90-MHz was fabricated under a 55-nm CMOS process, and then the first successful example of large-capacity SOT-MRAM fabrication (4 kB) on a single wafer is shown in [40]. With the significant advancement of fabrication technology and commercialization progress, MRAM is becoming a next-generation universal NVM technology, with potential applications in both last-level cache and main memory. The comparison also reveals that the latency of the STT-MRAM is sufficient to meet the requirements of the last-level caches in the high-performance computing domain, which operate around 100 MHz clock frequency [41]. It will significantly change the state-of-the-art memory hierarchy due to its nonvolatility, zero leakage power in un-accessed bit-cell, high integration density (2$\times$ more than SRAM), excellent endurance ($\sim 10^{15}$ cycles [42]), and compatibility with the CMOS fabrication process (back end of the line) [33].

2.1 STT-MRAM

A typical Magnetic Tunnel Junction (MTJ) structure [43, 44], as shown in Fig. 2a, consists of two ferromagnetic layers with a tunnel barrier sandwiched between them. Due to the Tunnel MagnetoResistance (TMR) effect [32, 44], the resistance of MTJ is high when the magnetizations of two ferromagnetic layers are in an antiparallel state or vice versa. The *TMR ratio* is defined as $(R_{AP}\text{-}R_P)/R_P$, which may vary

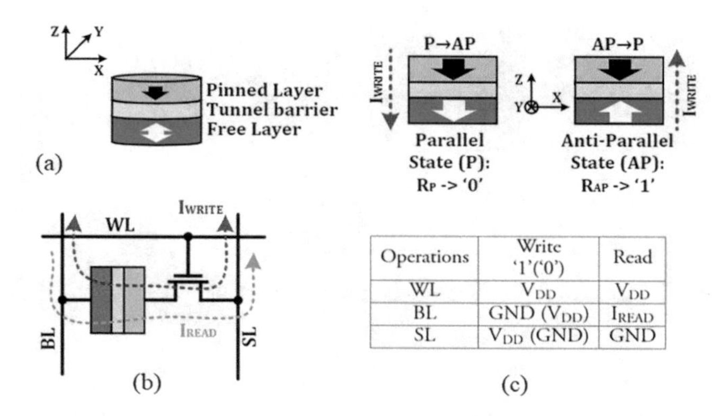

Operations	Write '1'('0')	Read
WL	V_{DD}	V_{DD}
BL	GND (V_{DD})	I_{READ}
SL	V_{DD} (GND)	GND

Fig. 2 (**a**) Device structure of conventional Magnetic Tunnel Junction (MTJ) in parallel and antiparallel states, with Spin-Transfer Torque (STT) switching scheme. (**b**) 1T1R STT-MRAM. (**c**) Biasing condition for memory operations

from 10% to 400% depending on materials and temperature. R_{AP} and R_P denote antiparallel and parallel MTJ's resistance, respectively. Thus, data are stored and determined by the magnetization direction in the free layer, which can be flipped through current-induced STT. Note that the MTJ with Perpendicular Magnetic Anisotropy (PMA) is used in this chapter. The 1T1R memory bit-cell is widely used in the typical MRAM design, as depicted in Fig. 2b, which is controlled by Bit Line (BL), Word Line (WL), and Source Line (SL). The biasing conditions of memory read/write are presented in Fig. 2c. For both memory read and write operations, the WL is enabled, which turns on the access transistor. The corresponding WL is activated using a Memory Row Decoder (MRD) to write data in a memory cell. Then appropriate voltage difference (Fig. 2c) is applied to the corresponding BL and SL using the Write Driver (WD) connected to them (the write current path is shown in Fig. 2b), leading to MTJ resistance in High-R_{AP} (/Low-R_P). For memory read, a sensing current (I_{READ}) is applied on the BL and consequently generates a sensing voltage, which can be detected by a Sense Amplifier (SA).

For the STT-MRAM modeling in this chapter, the Non-Equilibrium Green's Function (NEGF) and Landau-Lifshitz-Gilbert (LLG) equation are used before the circuit-level simulation. The magnetization dynamics of MTJ's Free Layer-FL (m) can be modeled as [45–47]:

$$\frac{dm}{dt} = -|\gamma| m \times H_{\text{eff}} + \alpha\left(m \times \frac{dm}{dt}\right) + |\gamma|\beta(m \times m_{\text{p}} \times m) - |\gamma|\beta\epsilon'(m \times m_{\text{p}}) \tag{1}$$

$$\beta = |\frac{\hbar}{2\mu_0 e}| \frac{I_c P}{A_{\text{MTJ}} t_{\text{FL}} M_s} \tag{2}$$

where $\hbar$ is the reduced plank constant, γ is the gyromagnetic ratio, I_c is the charge current flowing through MTJ, t_{FL} is the thickness of the free layer, ϵ' is the second Spin transfer torque coefficient, and H_{eff} is the effective magnetic field. P is the

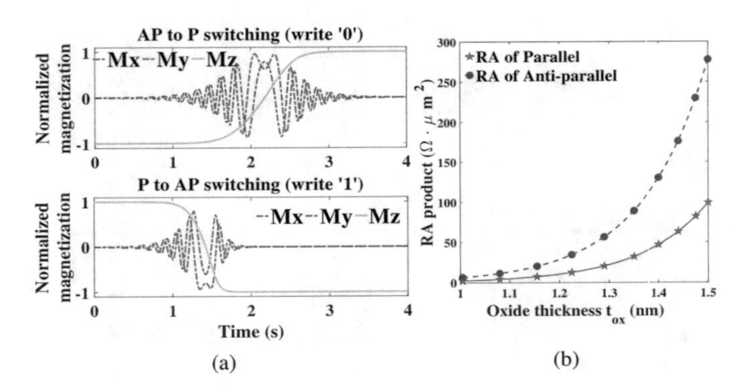

Fig. 3 (**a**) The normalized magnetization switching in *x*-, *y*- and *z*-axis. (**b**) The Resistance-Area product w.r.t the thickness of MTJ tunnel oxide (t_{ox})

effective polarization factor, A_{MTJ} is the cross-sectional area of MTJ, and m_P is the unit polarization direction. Figure 3a shows the normalized magnetization dynamics of the free layer in x-, *y*-, and *z*-axis when performing the STT-MRAM write scheme as described earlier.

Based on the simulation parameters listed in Table 1, the magnetization dynamic from the LLG equation can provide the relative angle θ between the magnetization of Pinned Layer-PL ($\hat{z}$) and Free Layer-FL (m). Therefore, the real-time conductance of MTJ (G_{MTJ}) is given by:

$$G_{MTJ} = \frac{G_P + G_{AP}}{2} + \frac{G_P - G_{AP}}{2} \cos\theta \tag{3}$$

where G_P and G_{AP} are the conductance of MTJ in parallel ($\theta = 0$) and antiparallel ($\theta = 180$) configurations. Both G_P and G_{AP} are obtained from the atomistic level simulation framework based on Non-Equilibrium Green's Function (NEGF) [48], while the Resistance-Area product with respect to the thickness of MTJ tunnel oxide is shown in Fig. 3b.

2.2 *SOT-MRAM*

As shown in Fig. 2b, in the typical STT-MRAM design, only one access transistor is used for both memory write and read. Due to the intrinsic device physics and structure, this suffers several limitations, including long write latency (>10-ns), high write current (>2 MA/cm^2), and thus considerable writing power and area (due to large transistor sizing). Moreover, shared read and write paths causing read-write conflict; asymmetric writing of data '0' and '1' due to different spin polarization factor of fixed and free ferromagnetic layers; reliability concern due to tunnel oxide breakdown in large write voltage [3, 49].

Table 1 Simulations parameters for MTJ

Parameter	Value
Free layer dimension $(W \times L \times t)_{FL}$	$65 \times 65 \times 2\,\text{nm}^3$
Polarization factor, P	0.4
Gilbert Damping Factor, α	0.007
Saturation Magnetization, M_s	850 kA/m
Oxide thickness, t_{ox}	1.5 nm
Resistance-Area product, RA_p / TMR	$10.58\,\Omega \cdot \mu\text{m}^2$/171.2%
Supply voltage	1 V
CMOS technology	45 nm
STT-MRAM cell area	48F^2
Access transistor width	9F
Cell aspect Ratio	1.34

In order to address the above limitations of STT-MRAM, the recent application of SOT has been explored to switch the adjacent MTJ free layer magnetization (i.e., programming MTJ resistance) much more energy efficient in I/FM/HM structure (I: insulator, FM: ferromagnet, and HM: heavy metal) [50]. Figure 4a presents the device structure of SOT-MTJ,[2] which is an MTJ mounted on a heavy metal substrate. When electrons flow through the nonmagnetic heavy metal substrate (in the ±y direction) with strong spin-orbit coupling, the electrons with the reverse direction of rotation accumulate on the opposite surfaces of HM. Thus, a pure spin current (I_s) in the ±z direction is generated, which exerts an SOT on the adjacent FM and switches the magnetization. The relationship between the generated spin current (I_s) and the applied charge current (I_c) can be expressed as:

$$I_s = P_{she}(\sigma \times I_c) \tag{4}$$

$$P_{she} = \frac{I_s}{I_c} = \frac{A_{FM}}{A_{HM}}\theta_{sh}\left(1 - sech\left(\frac{t_{HM}}{\lambda_{sf}}\right)\right) \tag{5}$$

where P_{she} is spin Hall injection efficiency. σ is the electron spin polarization, transverse to both the spin current and charge current directions. A_{FM} is the area of the adjacent FM area, and A_{HM} is the cross-sectional area of HM in the direction of current flow. θ_{sh} is the spin Hall angle, which is defined as the ratio of generated spin current density to the applied charge current density. t_{HM} is the thickness of HM substrate, and λ_{sf} is the spin-flip length. Recently, large spin Hall angle was experimentally demonstrated in different heavy metal materials, such as Pt [51], β-Ta [52], β-W [53], and CuBi alloys. High magnetization switching speed (<1 ns) of SOT-MTJ is achieved mainly due to larger spin injection efficiency than the conventional MTJ with an STT-switching scheme. Therefore, choosing a SOT-

[2] Note: SOT-MTJ and SHE-MTJ are used interchangeably in this chapter book.

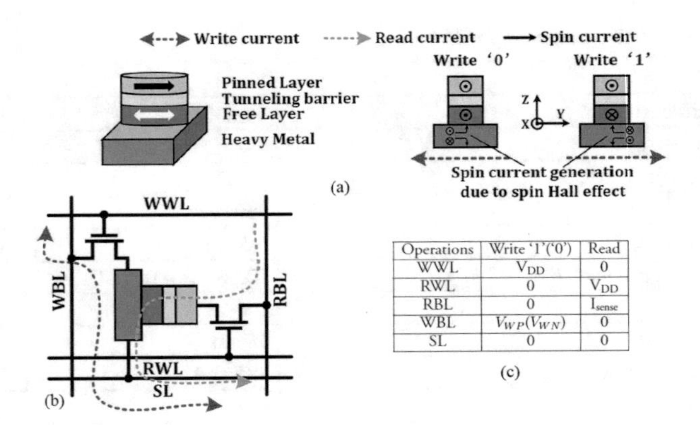

Operations	Write '1'('0')	Read
WWL	V_{DD}	0
RWL	0	V_{DD}
RBL	0	I_{sense}
WBL	$V_{WP}(V_{WN})$	0
SL	0	0

Fig. 4 (**a**) The stacking device structure of MTJ and heavy metal substrate, which uses spin-orbit torque-induced magnetization switching scheme. (**b**) Bit-cell schematic of SOT-MRAM with two access transistors (1R/1W). (**c**) Biasing condition for memory operations

induced switching scheme is much more efficient as the next-generation MRAM design. Figure 4b shows the corresponding 2T1R SOT-MRAM bit-cell design with separated write and read access transistors, correspondingly controlled by Write Bit Line (WBL), Write Word Line (WWL), Read Bit Line (RBL), Read Word Line (RWL), and the shared Source Line (SL). The memory read and write biasing conditions are presented in Fig. 4c. WWL is pulled high for memory write, which turns on the write access transistor. Then, to write '1' (or '0'), a positive voltage V_{WP} (or negative voltage V_{WN}) is applied to WBL with SL connected to ground. For memory read, RWL is set to V_{DD}, and the read access transistor is switched on. A sensing current (I_{sense}) flowing through SOT-MTJ consequently generates a sensing voltage (V_{sense}) on RBL, which can be detected by the SA.

The magnetization dynamics of SOT-MTJ's FL (m) can be also modeled by the modified LLG equation, which can be mathematically described as:

$$(1+\alpha^2)\frac{dm}{dt} = -|\gamma|\mu_0 m \times H - \alpha|\gamma| m \times m \times H - m \times m \times \frac{I_s}{qN_s} + \alpha m \times \frac{I_s}{qN_s} \tag{6}$$

where α is Gilbert damping factor, γ is the gyromagnetic ratio, and μ_0 is the vacuum permeability. H is the effective field, which includes dipolar coupling, demagnetization, thermal noise, and anisotropy fields. $N_s = M_s V/\mu_B$ is the number of spins, μ_B is Bohr magneton, and M_s and V are the saturation magnetization and volume of the ferromagnet, respectively. The simulation parameters are listed in Table 2. To realize the desired 1 ns switching speed, about 130 μA writing current is required, which leads to 1 V and −0.35 V for V_{WP} and V_{WN}, respectively. The magnetization dynamic from the LLG equation can provide the relative angle θ between the magnetization of PL and FL. Therefore, the real-time conductance of MTJ (G_{MTJ}) is given by the Eq. (3), where again both G_P and G_{AP} are obtained from the atomistic level simulation framework based on Non-Equilibrium Green's Function (NEGF) [48]. The Resistance-Area product with respect to the thickness of MTJ tunnel oxide is listed in Table 2.

Table 2 Simulation parameters for SOT-MTJ

Parameter	Value
Free layer dimension,$(W \times L \times t)_{FM}$	$60 \times 40 \times 2\,\text{nm}^3$
SHM dimension, $(W \times L \times t)_{HM}$	$60 \times 80 \times 2\ \text{nm}^3$
Demagnetization Factor, D_x, D_y, D_z	0.066, 0.911, 0.022
Spin flip length, λ_{sf}	1.4 nm
Spin Hall angle, θ_{sh}	0.3
Gilbert Damping Factor, α	0.007
Saturation Magnetization, M_s	850 kA/m
Oxide thickness, t_{ox}	1.2 nm
Resistance-Area product, RA_p / TMR	$10.58\ \Omega \cdot \mu\text{m}^2$ / 171.2%
Supply voltage	1 V
CMOS technology	45 nm
SOT-MRAM cell area	69F^2
Access transistor width	4.5F
Cell aspect Ratio	1.91

3 Enabling Data-Intensive Computing Paradigm

3.1 General Processing-in-Memory Structure

The general memory organization to realize PIM in NVMs is shown in Fig. 5 [54, 55]. The main memory chip is basically divided into multiple banks. Each bank consists of multiple memory matrices (mats). Banks within the same chip typically share I/O and buffer, and banks in different chips work in a lock-step manner. The mats are connected to a Global Row Decoder (GRD) and a shared Global Row Buffer (GRB). Each mat consists of multiple computational memory sub-arrays (i.e., PIM-enhanced sub-array) connected to a GRD and GRB.

According to the application type and physical address of operands within memory, the PIM's Controller (Ctrl) can configure the computational sub-arrays to perform data-parallel inter-sub-array computations. Every two computational sub-arrays share a Local Row Buffer (LRB) as well as a Digital Processing Unit (DPU) to further process the data (if necessary) in specific applications, as will be discussed later. Figure 6 gives an overview of the PIM's acceleration steps. Assume input tensors A and B (that can belong to various applications) are initially stored in Data Banks of the memory. In the first step, either raw data or preprocessed data (by DPU) are mapped into the computational sub-arrays in specific mats. In the second step, parallel computational sub-arrays are designed to handle the computational load employing PIM techniques, perform bulk bit-wise operations between tensors, and generate the output. The results at this step can be considered the ultimate output in data encryption or graph processing applications. Additionally, the generated data can be further processed by DPU to generate the output for neural network-based applications.

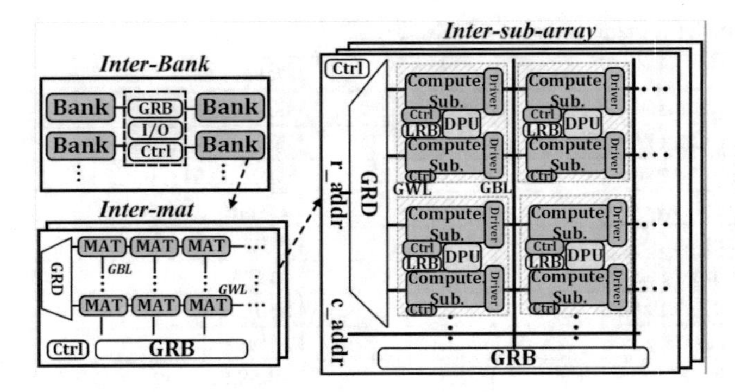

Fig. 5 The overall PIM architecture

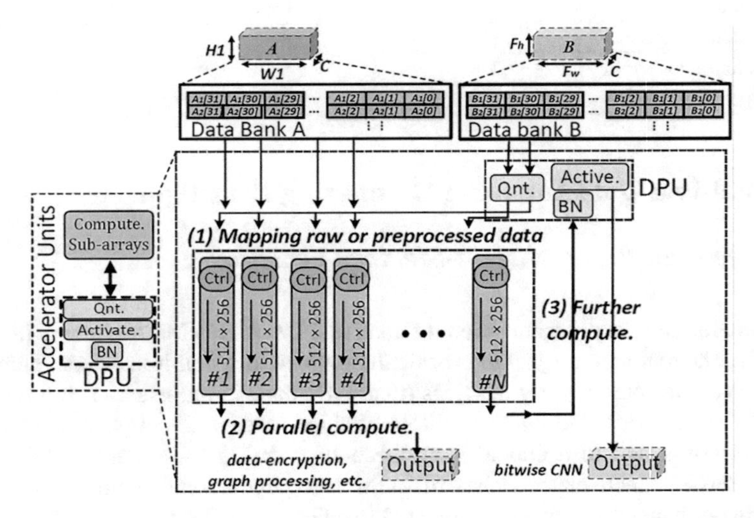

Fig. 6 The PIM's acceleration steps. The size of the computational sub-arrays could be tailored

3.2 Circuit-Level Exploration: Evolution of the MRAM-Based PIM Platforms

3.2.1 Basic PIM Supporting (N)AND, (N)OR

In emerging resistive NVMs, like MRAM and ReRAM, the data are stored in resistive states of memory cells as discussed in Sects. 2.1 and 2.2. In the traditional NVM's read operation, one selected memory cell will be activated and compared with a reference resistance through memory SA to read out data value. Therefore, firstly, the corresponding WL(/RWL) is activated using the Memory Row Decoder (MRD) and the corresponding BL(/RBL) is connected to the SA using the Memory Column Decoder (MCD) (the read current path is shown in

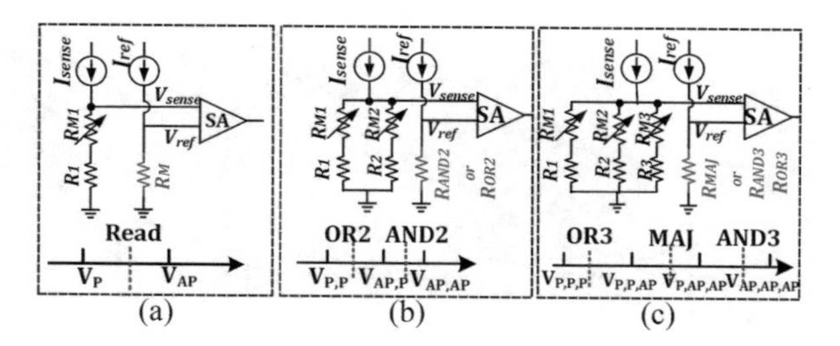

Fig. 7 The idea of voltage comparison between V_{sense} and V_{ref} for (**a**) memory read, (**b**) two-input in-memory logic, and (**c**) three-input in-memory logic. Note that R_{Mi} and R_i denote the equivalent resistance of the nonvolatile component and selecting transistor, wire, etc., respectively

Fig. 2b). The idea of voltage comparison for memory read is depicted in Fig. 7a, and a single cell is addressed to generate a sense voltage (V_{sense}), which will be compared with memory mode reference voltage activated by an enable signal EN_M ($V_{sense,P} < V_{ref,M} < V_{sense,AP}$). Now, if the path resistance is higher (/lower) than R_M (memory reference resistance), i.e., R_{AP} ($/R_P$), then the SA produces high (/low) voltage indicating logic '1' (/'0'). Note that one SA per BL(/RBL) is considered in the whole chapter to maximize the output bandwidth.

With a careful study of this operation, new peripheral circuits are designed such that multiple resistive memory cells (i.e., data operands) could be activated and sensed simultaneously, leading to different parallel resistive levels at the SA side. In this way, by carefully selecting different reference resistance levels, various Boolean logic outputs could be intrinsically 'read out' based on input operand data in the memory array.

The first idea was relatively simple [13, 56], where every two bits stored in the identical column could be selected and sensed simultaneously, as depicted in Fig. 8a. The MRD was modified to support the multiline enable of this function by combining two single-line enable decoders with their outputs connected to OR gates. To activate computing, the current path, shown in Fig. 8a for the first column, RWL1 and RWL2 are activated by the MRD while SL1 and SL2 are grounded, and all the other WLs and SLs are kept deactivated. The MCD/CD activates the RBL1 to be connected to the SA. Now the sense (read) current is applied to RBL1. With that, the equivalent resistance voltage of such parallel-connected SOT-MRAMs (m1 and m2) and their cascaded access transistors can be compared with a specific reference voltage generated by SA. Through selecting different reference resistances by new enable signals (EN_M, EN_{AND}, EN_{OR}) as shown in SA box in Fig. 8a, the SA can perform basic memory and in-memory Boolean functions (i.e., (N)AND2 and (N)OR2). For (N)AND2 operation, R_{ref} is set at the midpoint of $R_{AP}//R_P$ ('1','0') and $R_{AP}//R_{AP}$ ('1','1') as shown in Fig. 7b. Thus only when both of the selected MRAM bit-cells are in an antiparallel state (i.e., binary input: '1', '1'), the output is high, whereas the output is low. Similarly, for (N)OR2 operation, R_{ref} is set at the midpoint of $R_P//R_P$ and $R_P//R_{AP}$. Only when the two selected MRAM bit-cells

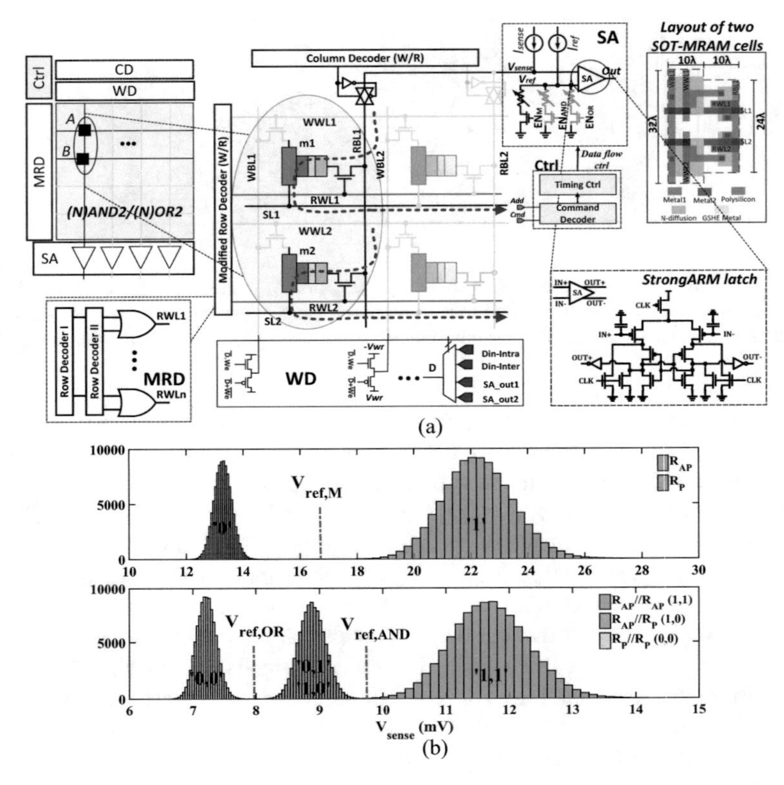

Fig. 8 (**a**) Presented PIM sub-array architecture based on SOT-MRAM supporting (N)AND, (N)OR functions with peripherals [13, 56]. The layout of two adjacent SOT-MRAM cells is also indicated. (**b**) Monte-Carlo simulation result of the sense voltage (V_{sense}) distribution

are in the parallel state (i.e., binary input: '0', '0'), the output is low, whereas the output is high.

To validate the sense circuit's variation tolerance, we have performed a Monte-Carlo simulation with 100,000 trials. A $\sigma = 5\%$ variation is added on the Resistance-Area product (RA_P) and a $\sigma = 10\%$ process variation is added on the TMR. The simulation result of sense voltage (V_{sense}) distributions in Fig. 8b shows the sense margin of in-memory computing. It will be reduced by increasing the logic fan-in (i.e., number of parallel memory cells). It is worth pointing out that this design does not necessarily rely on NVM technology or cell structure. As long as the technology is based on resistive cells, i.e., PCM and ReRAM, the presented SA can readily perform in-memory computation. Based on our experiments, leveraging PCM and ReRAM cells (with higher ON/OFF ratio) leads to a significantly larger read margin than SOT-MRAM, which further translates to much higher reliability even by activating more number of rows (e.g., up to 64-row operation for PCM [10]). Therefore, it is possible to use other emerging NVMs to achieve a better read margin. Notwithstanding, PCM and ReRAM consume more power than SOT-MRAM if converted to the PIM platform. This is mainly because of their relatively higher writing power, which inevitably causes overall power increase when dealing with complex real-world applications requiring massive intermediate operand data write-back into memory.

While the presented PIM design in Fig. 8a could implement any in-memory Boolean logic functions based on universal NAND2/NOR2 functions, it requires multiple cycles. The operation's result has to be written back into the memory after each memory cycle. Such write-back operation reduces the platform's performance and energy efficiency in computationally intensive big data applications and eventually may fade the PIM advantages. This motivated us to move forward and design reconfigurable complete PIM platforms supporting more Boolean functions.

3.2.2 Reconfigurable Complete PIM Supporting X(N)OR

In [14, 50], an enhanced and reconfigurable PIM platform on top of the previous design is presented. In the new design, every RBL is routed to a Modified Sense Amplifier (MSA), as shown in Fig. 9. The new MSA consists of two sub-SAs and three reference resistors compared to the first design with one SA in Sect. 3.2.1. Every two bits stored in the identical column can be selected with the MRD and sensed simultaneously, as shown in Fig. 9a. Again, the equivalent resistance of such parallel SOT-MRAMs and their cascaded access transistors is compared with MSA's programmable reference. In the new design, through selecting two reference resistances (i.e., EN_{AND}, EN_{OR}), two sub-SAs can operate simultaneously to realize two basic in-memory Boolean functions, i.e., (N)AND2 and (N)OR2 at the same time, as shown in Fig. 9b. This provides more flexibility to the PIM to implement more complex logic functions through combining the outputs. The X(N)OR2 logic can be realized with two sub-SA's outputs (AND2 and NOR2 logic) with an extra CMOS NOR2 gate after the outputs in the MSA. As depicted in Fig. 9b, the operation of such sense circuit is determined by the control signals (EN_{AND}, EN_M, EN_{OR}), while the desired result is acquired by the select signal (SEL) of the output multiplexer [14]. It is noteworthy that only one SA is used during (N)AND2/(N)OR2/memory read operation to reduce the power consumption of sensing. Parallel computing/read is implemented by using one SA per bit-line.

Figure 10 depicts the transient simulation result of the sense circuit under a 2 ns period clock signal (CLK), which takes the data stored in MRAM1 (m1) and MRAM2 (m2) as inputs. When CLK is high, the sense amplifier is in the pre-charge phase, and the output is reset to '0'. When CLK is low, the sense amplifier is in the sampling phase and generates logic computation results depending on the reference voltage configuration. V_{cmp} plots the comparison between sense voltage (V_{sense}) and two reference voltages, i.e., V_{ref1} and V_{ref2}. Again, V_{ref1} is set to $(V_{AP,AP}+V_{AP,P})/2$, and V_{ref2} is set to $(V_{P,P}+V_{AP,P})/2$, for performing AND2 and OR2, respectively.

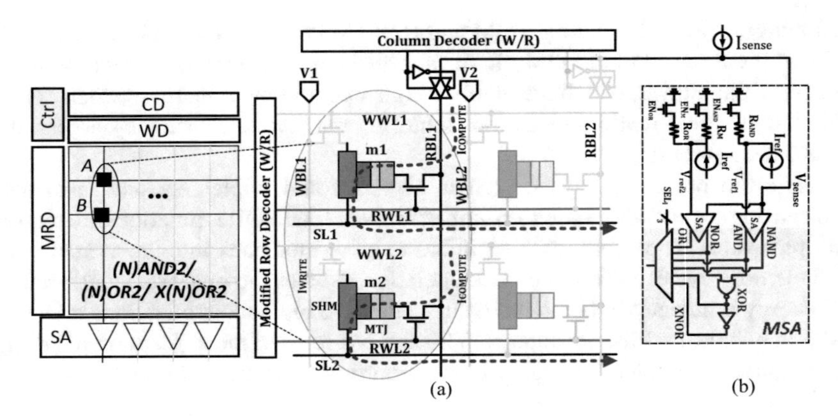

Fig. 9 (**a**) Presented in-memory processing sub-array architecture based on SOT-MRAM supporting (N)AND, (N)OR, X(N)OR functions [14, 50], (**b**) Modified Sense Amplifer with two sub-SAs and three reference resistors

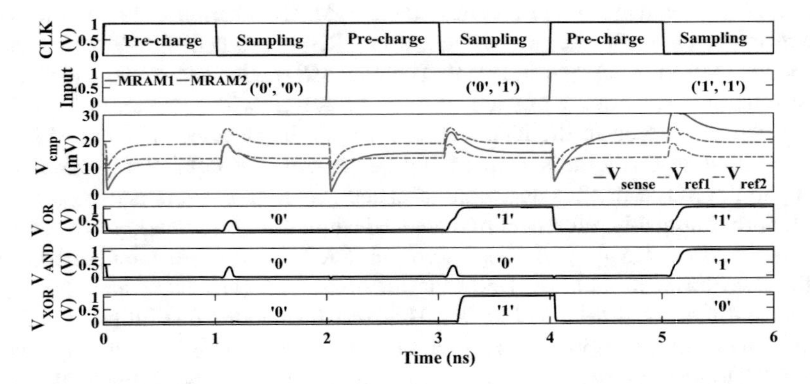

Fig. 10 Transient simulation results of in-memory computing operations (i.e., AND, OR and XOR) [50]

3.2.3 Reconfigurable PIM Supporting Two-Cycle In-Memory Addition

Aiming to provide more flexibility and reconfigurability for the PIM platforms, a new PIM sub-array architecture based on STT-MRAM, named MRIMA, was presented in [54]. This in-memory circuit design, as depicted in Fig. 11a, mainly consists of Write Driver (WD), MRD, MCD, and SA (Fig. 11b) and can be adjusted by Ctrl unit (Fig. 11b) to work in a dual mode that performs memory write/read and bit-line computing. The presented reconfigurable SA, as depicted in Fig. 11b, consists of two sub-SAs and totally six reference-resistance branches that can be selected by enable bits (EN_M, EN_{OR3}, EN_{OR2}, EN_{MAJ}, EN_{AND3}, EN_{AND2}) by the sub-array's Ctrl to realize the memory and computation schemes as tabulated in Table 3. Such reconfigurable SA could implement memory read and one-threshold-

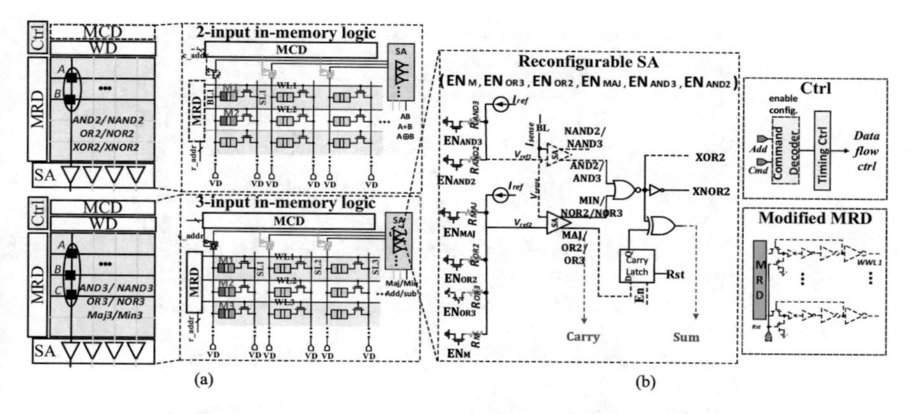

Fig. 11 The MRIMA's sub-array architecture [54]: (**a**) Block level scheme and STT-MRAM realization of 2-input and 3-input in-memory logic methods, (**b**) Peripherals of computational sub-arrays to support computation

Table 3 Configuration of MRIMA's enable bits for different functions

Ops.	read/NOT	(N)OR2/NOR2	(N)AND2	X(N)OR2	MAJ/MIN	(N)OR3	(N)AND3
EN_M	1	0	0	0	0	0	0
EN_{OR2}	0	1	0	1	0	0	0
EN_{AND2}	0	0	1	1	0	0	0
EN_{OR3}	0	0	0	0	0	1	0
EN_{AND3}	0	0	0	0	0	0	1
EN_{MAJ}	0	0	0	0	1	0	0

based logic functions on top of the discussed bit-line computing scheme by activating one enable at a time. For instance, by setting EN_{AND2} to '1', (N)AND2 logic can be readily implemented between operands located in the same bit-line. Meanwhile, by activating two enables at a time, e.g., EN_{OR2}, EN_{AND2}, two logic functions can be simultaneously implemented and further used to generate two-threshold-based logic functions like X(N)OR2, as in Sect. 3.2.2. Here, we elaborate on the main functions supported by MRIMA.

Fast Row Copy (FRC) MRIMA's FRC mechanism needs consecutive memory read and write operations. In the first half-cycle, the source row is activated by sub-array's MRD and readout to LRB (shown in Fig. 5); in the second half-cycle, the data stored in the buffer is written back to the destination row.

Two-Input In-Memory Logic (IML2x) The computational sub-array of MRIMA is designed to perform bulk bit-wise in-memory logic operations between two or three operands located in the same bit-line. The IML2x is essentially the same as the two-input PIM operation in the previous designs, where every two bits stored in an identical column can be selected employing the MRD and sensed simultaneously, as depicted in Fig. 11a. The equivalent resistance of such parallel-connected STT-

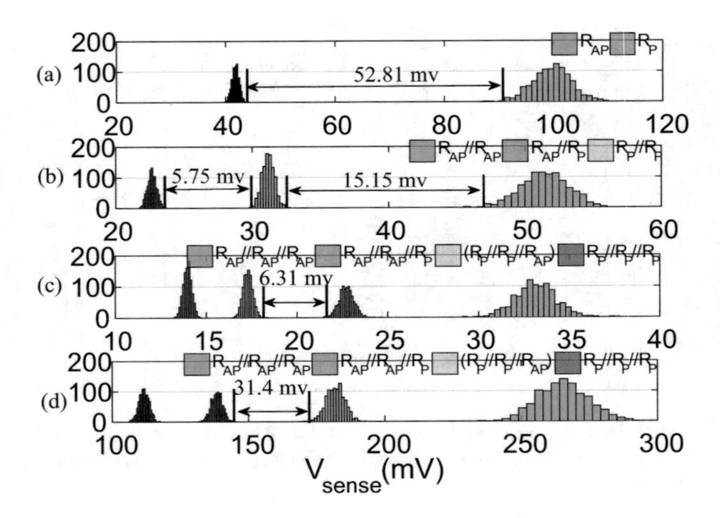

Fig. 12 Monte-Carlo simulation of V_{sense} (with RAP/TMR=2%/5% - t_{ox}=1.5 nm) for (**a**) memory read, (**b**) IML2x, (**c**) IML3x when $I_{sense} = 6.6\,\mu A$, and (**d**) IML3x when $I_{sense} = 18\,\mu$A [54]

MRAMs and their cascaded access transistors is compared with a programmable reference by SA. Through selecting different reference resistances (R_{AND2}, R_{OR2}), the SA can perform basic two-input in-memory Boolean functions (i.e., (N)AND2 and (N)OR2) in a single memory cycle.

Three-Input In-Memory Logic (IML3x) In the IML3x, every three cells located in an identical column can be selected by MRD and sensed simultaneously to realize three-input logic functions (i.e., (N)AND3, (N)OR3, MAJ/MIN). For instance, consider the data organization shown in Fig. 11a, where A, B, and C operands correspond to M1, M2, and M3 memory cells, respectively, and the computational sub-array can perform majority function ($AB + AC + BC$) by setting EN_{MAJ} to '1'. As shown in Fig. 7c, to perform MAJ operation, R_{MAJ} is set at the midpoint of $R_P//R_P//R_{AP}$ ('0','0','1') and $R_P//R_{AP}//R_{AP}$ ('0','1', '1'). A comprehensive study on the MRIMA's sensing circuit's variation tolerance is done by running the Monte-Carlo simulation with 10,000 trials. A $\sigma = 2\%$ variation is added to the RAP, and a $\sigma = 5\%$ process variation (typical MTJ conductance variation [3]) is added on the TMR. The simulation result of V_{sense} distributions in Fig. 12 shows the sense margin for memory read, IML2x, and IML3x. It can be seen that sense margin gradually reduces when increasing the number of fan-ins. To avoid logic failure and guarantee the output's reliability, we limited the number of sensed cells to three. Such sense margin could be even improved by either increasing the sense current or oxide thickness (t_{ox}), but obviously by sacrificing the operation's energy efficiency [54].

Two-Cycle In-Memory Addition In addition to the abovementioned single-cycle logic operations, MRIMA's sub-array can perform addition/subtraction (add/sub) operation quite efficiently. In the full-adder Boolean logic, the carryout can be directly produced by MAJ function (Carry in Fig. 11b) just by setting EN_{MAJ} to '1'. Accordingly, a carry latch is inserted to store intermediate carry outputs used in the summation of the next bits. Meanwhile, Sum output can be obtained by inserting a two-input XOR gate in the reconfigurable SA, taking the latch output and in-memory XOR2 output as the inputs. Now, assume A, B, and C operands (in Fig. 11a), IML2x and IML3x are able to generate Sum (/Difference) based on XOR3 and Carry (/Borrow) bits and perform parallel multi-bit addition operation.

System Integration While MRIMA is meant to be an independent, high-performance, and energy-efficient accelerator, it needs to be exposed to programmers and system-level libraries to utilize it. From a programmer's perspective, MRIMA is more of a third-party accelerator connected directly to the memory bus or through PCI-Express lanes rather than a memory unit. Accordingly, the programs are translated at install time to the MRIMA hardware instruction set tabulated in Table 4. The micro and control transfer instructions are not shown in the table. The MRIMA commands/instructions can be directly copied/written to predefined memory-mapped address ranges. For example, defined in the memory type range registers (MTRRs) or programmed through writing to memory-mapped I/O regions are allocated through a simple device driver to do initialization/cleanup for required software memory structures. Note that the first approach can potentially bring more performance gains than the latter; accessing MRIMA as an I/O device can incur significant overheads due to interrupts and page faults (in the shared memory model). In contrast, a memory-mapped MRIMA scheme can cause significant contentions in the memory bus if the processor executes memory-intensive applications simultaneously. Choosing the scheme of integrating MRIMA is left to system architects based on their workloads and use-cases. In both schemes for integrating MRIMA, the commands/instructions that MRIMA architecture accepts are similar and based on the ISA.

Table 4 The basic instructions of MRIMA

Opcode		Operation	Function
FRC		$B \leftarrow A$	Copy row A to Row B
IML2x	IML21	$A.B$	AND2/NAND2
	IML22	$A + B$	OR2/NOR2
	IML23	$A \oplus B$	XOR2/XNOR2
IML3x	IML31	$A.B.C$	AND3/NAND3
	IML32	$A + B + C$	OR3/NOR3
	IML33	$AB + AC + BC$	MAJ/MIN

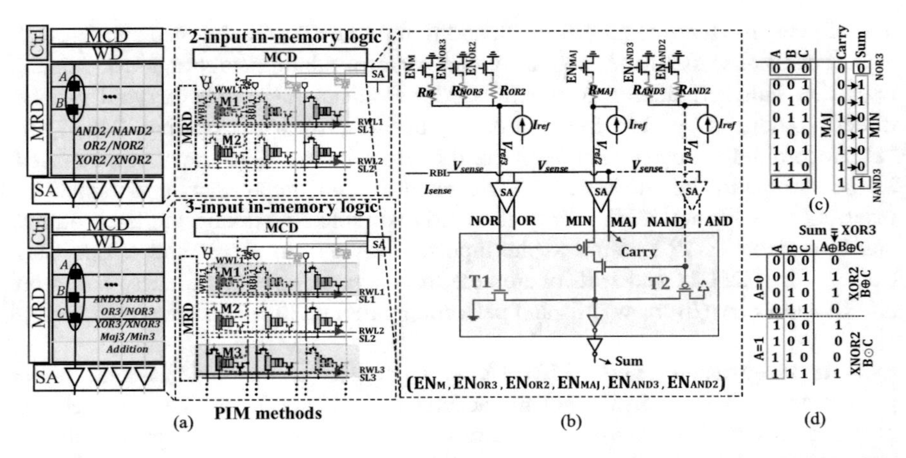

Fig. 13 (**a**) Block level scheme of computational sub-array and SOT-MRAM realization of 2-input and 3-input in-memory logic methods in GraphS [55], (**b**) Reconfigurable SA, (**c**) Truth table of addition operation implementation, (**d**) Truth table for realizing X(N)OR2

3.2.4 Reconfigurable PIM Supporting One-Cycle In-Memory Addition

The GraphS's reconfigurable SA[3] [55], as depicted in Fig. 13b, consists of three sub-SAs and totally six reference-resistance branches that can be selected by enable bits (EN_M, EN_{OR3}, EN_{OR2}, EN_{MAJ}, EN_{AND3}, EN_{AND2}) by the sub-array's Ctrl to realize the memory and computation schemes as tabulated in Table 5. Such reconfigurable SA could again implement memory read and one-threshold-based logic functions only by activating one enable at a time. Meanwhile, by activating two or three enables at a time, two or three logic functions can be simultaneously implemented and further used to generate complex logic functions like X(N)OR3, as explained accordingly. GraphS supports both IML2x and IML3x operations. In IML3x, every three cells located in an identical column can be selected by MRD and sensed simultaneously to realize three-input majority/minority functions (MAJ/MIN) in a single sensing cycle. Consider the data organization shown in Fig. 13a where A, B, and C operands correspond to M1, M2, and M3 memory cells, respectively, and the computational sub-array can perform $AB + AC + BC$ Boolean function by setting EN_{MAJ} to '1'.

Besides, with careful observation on the full-adder (FA) truth table, we realized that in six out of eight possible input combinations, Sum output could be directly obtained by inverted Carry signal as shown in Fig. 13c. Keep this fact in mind that FA's Carry can be produced by MAJ function; the presented reconfigurable SA can implement such Sum output readily by MIN (majority-not) function. As depicted in Fig. 13b–c, the Sum signal is directly connected to the MIN output. However,

[3] A variation of this design is named PIM-Aligner [57].

Table 5 Configuration of enable bits for different functions

Ops.	Read	(N)OR2	(N)AND2	MAJ/MIN	(N)OR3	(N)AND3	Add/XNOR3/X(N)OR2
EN_M	1	0	0	0	0	0	0
EN_{OR2}	0	1	0	0	0	0	0
EN_{AND2}	0	0	1	0	0	0	0
EN_{OR3}	0	0	0	0	1	0	1
EN_{AND3}	0	0	0	0	0	1	1
EN_{MAJ}	0	0	0	1	0	0	1

for two extreme cases, i.e., (0,0,0) and (1,1,1), the MIN signal is disconnected and Sum can be respectively implemented by NOR3 (T1:ON, T2:OFF → Sum='0') and NAND3 functions (T1:OFF, T2:ON → Sum='1'). This is realized by adding two pass transistors in the MIN function path. Note that, considering the fact that Sum output is the XOR3 function, the presented reconfigurable SA can also implement two-input and three-input XOR functions, without imposing additional XOR gates like previous works [9, 18, 54, 58] as shown in Fig. 13d. Now, assume A, B, and C as input operands (in Fig. 13a), and IML3x can generate Sum(/Difference) and Carry(/Borrow) bits in a single cycle.

3.3 *Convolutional Neural Networks (CNN) Acceleration: Analog or Digital PIM Approach?*

3.3.1 CNN Terminology

CNN is a machine learning classifier that takes an image as input and then computes the probability that image features belong to a sort of output class. Typically, a CNN consists of several convolutional layers and pooling layers followed by Fully-Connected layers (FC) as shown in Fig. 14. Note that it has been proven that convolutions could equivalently implement FC layers [59, 60]. Figure 14 also shows a visualization of the convolutional layer of CNN where each layer receives a set of features organized in multichannel as input (Input fmaps). It applies kernels (filters) by performing high-dimensional convolutions, i.e., Multiplication-and-Accumulation (MAC), and then produces the features (Output fmaps) for the next layer. The dimensions of both fmaps (input/output) and kernels are 4D (multiple 3D structures), and a batch of input fmaps is typically processed by multiple 3D kernels. After convolution, a nonlinear activation function, such as ReLU, will be applied to the results. By considering the shape parameters listed in Table 6, the computation of one convolutional layer can be defined as follows:

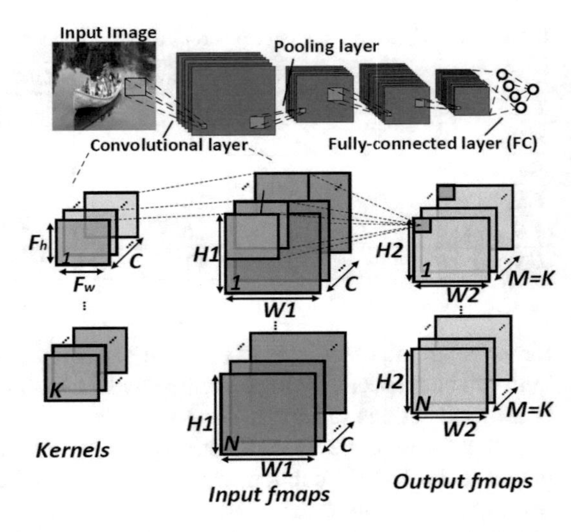

Fig. 14 Visualization of inference (a.k.a. forward propagation) in CNN

Table 6 Shape parameters of a convolutional layer

Shape parameter	Description
Input fmaps dimension	$W1 \times H1 \times C$
3-D fmaps batch size (input/output)	N
No. of 3D kernels	K
Spatial extent of kernels	$F_w \times F_h \times C$
Stride	S
No. of zero padding	P
Output fmaps dimension	$W2 \times H2 \times M$

$$O[n][k][x][y] = ReLU\left(B[k] + \sum_{i=0}^{F_h-1} \sum_{j=0}^{F_w-1} \sum_{z=0}^{C-1} I[n][z][U_x+i][U_y+j]W[k][z][i][j]\right),$$

$$0 \leq n < N, 0 \leq k < K, 0 \leq x < W2, 0 \leq y < H2 \tag{7}$$

where O, B, I, and W are the matrices representing output fmaps, Bias, input fmaps, and kernels, respectively. W2/H2 dimensions can be achieved as $W2 = (W1 - F_w + 2P)/S + 1$ and $H2 = (H1 - F_h + 2P)/S + 1$.

For CNN acceleration in memory, analog resistive crossbar memory, as one of the most popular memory array structures, has drawn significant interest due to its high memory accessing bandwidth and in situ computing capability. More importantly, its current-mode weighted summation operation intrinsically matches the dominant MAC in the artificial neural network, making it one of the most promising candidates as the primary computing unit for neural network accelerator design. For example, ISAAC [61] architecture improves throughput and energy by 14.8× and 5.5×, respectively, relative to a well-known ASIC architecture. PipeLayer [62] achieves the speed-up and energy saving of 42.45× and 7.17×, respectively, compared with a GPU platform on average. However, many nonideal

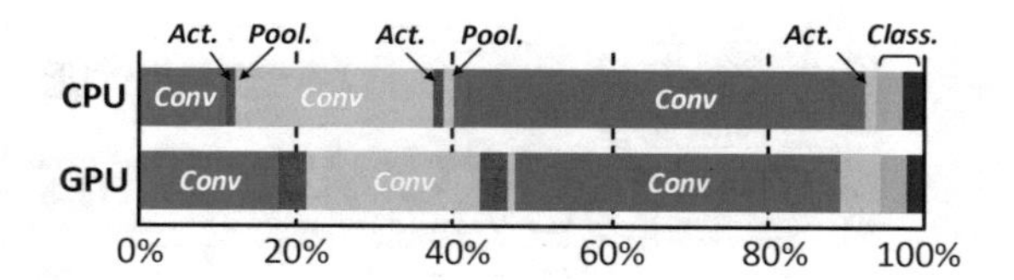

Fig. 15 Execution time of a sample CNN for scene labeling on CPU and GPU [65]

effects, such as IR-drop (i.e., wire resistance), Stuck-At-Fault (SAF), thermal noise, and random telegraph noise, are limiting the progress of hardware implementation of large-scale CNNs on ReRAM crossbar-based accelerators [63]. Many recent works have investigated such issues with either hardware or software solutions [64] (Fig. 15).

As an alternative solution to realize massive MAC and memory operations in CNN deployments, researchers have come up with weights and/or activations to be quantized/binarized in the forward propagation [59]. These modifications convert the conventional MAC operation to much simpler bulk bit-wise operations (based addition/subtraction [66, 67] or comparison [14]) that can be accelerated in the content of digital memories. For example, Neural Cache [18], as an SRAM-based platform, improves inference latency by 18.3× over the state-of-the-art multicore CPU (Xeon E5) and 7.7× over server-class GPU. DRISA [4], as a DRAM-based platform, employs 3T1C- and 1T1C-based computing mechanisms and achieves 7.7× speed-up and 15× better energy-efficiency over GPUs for CNN accelerations. CMP-PIM [14] as an MRAM-based platform achieves ∼10× better energy-efficiency compared to CNN-ReRAM accelerators. While the respective benefits of the aforementioned acceleration-in-memory approaches (i.e., analog and digital) are well known, it still lacks cross-technology comparison and analysis.

3.3.2 Evaluation Framework

Various data-intensive applications with distinct workload sizes and memory access patterns are expected to benefit from processing-in-memory in both cache and main memory levels; selecting the right design for a particular application is complex. Besides, by choosing a PIM design, it is imperative to establish uniform evaluation conditions to make an impartial choice between available design options. To perform the cross-technology comparison among aforementioned PIM techniques, we developed a comprehensive bottom-up cross-layer framework [63, 68] shown in Fig. 16.

1. For *Device level* modeling, the device parameters are first extracted from different assessments and models. The Non-Equilibrium Green's Function (NEGF) and Landau-Lifshitz-Gilbert (LLG) equations are used to model STT-MRAM and SOT-MRAM bitcell (indicated under MRAM in Fig. 16) [46]. Large numbers of physical parameters are integrated into the compact model to achieve a good agreement with experimental measurements. The default ReRAM and SRAM

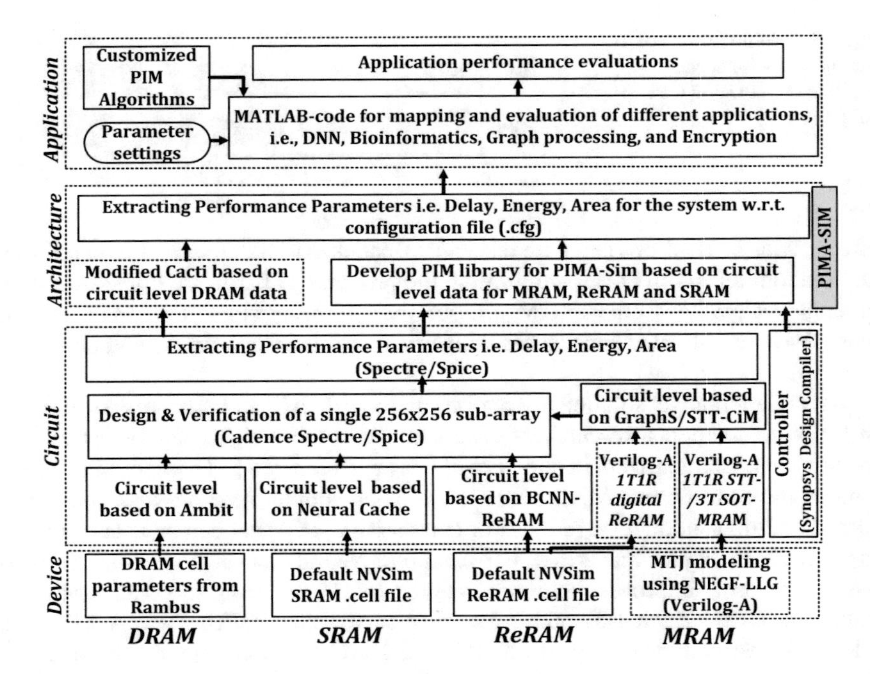

Fig. 16 The bottom-up evaluation framework developed for PIM platform evaluation

cell configurations of NVSIM [69] are considered for evaluation. Moreover, DRAM cell parameters are taken from Rambus [70] and scaled. Rambus has been developed to evaluate a wide variety of DRAM architectures such as a typical 55 nm DDR3 with respect to power consumption [70].

2. For *Circuit level* simulation, the memory sub-array with peripheral circuity (SA, MRD, MCD, etc.) could be implemented based on a particular PIM style for each technology on top of the device level data. For CNN acceleration, the GraphS [55] PIM style is used for SOT-MRAM and digital ReRAM implementations; STT-CiM [71] as the STT-MRAM design, BCNN-ReRAM [72] design for analog ReRAM crossbar, Neural Cache [18] design for SRAM, and Ambit [9] design for DRAM are accordingly used. The memory sub-arrays are simulated in Cadence Spectre with 45 nm NCSU Product Development Kit (PDK) library [73] to verify the PIM's circuit functionality and achieve the circuit performance parameters. The memory controller circuits for all platforms are synthesized by Design Compiler [74] with the same 45 nm industry library.
3. For *Architecture level*, a PIM support evaluation tool is developed for the NVSIM [69] named PIMA-SIM as shown in Fig. 17. NVSIM [69] was built as a circuit-level model for NVM performance and supporting various NVM technologies including STT-RAM, PCRAM, ReRAM, and conventional NAND Flash. The model has been successfully validated against industrial NVM prototypes [69]. PIMA-SIM also models the timing, energy, and area of various PIM technolo-

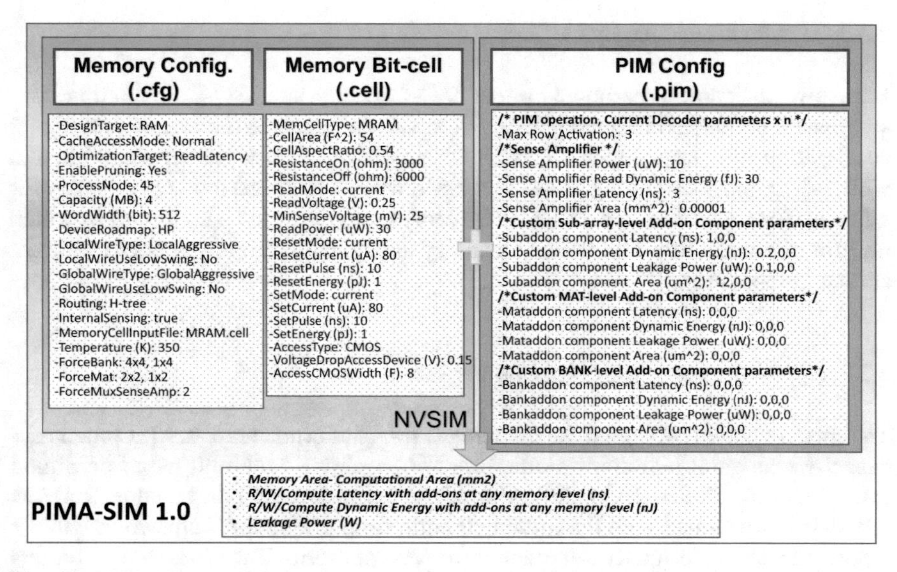

Memory Config. (.cfg)
Memory Bit-cell (.cell)
PIM Config (.pim)
-DesignTarget: RAM
-CacheAccessMode: Normal
-OptimizationTarget: ReadLatency
-EnablePruning: Yes
-ProcessNode: 45
-Capacity (MB): 4
-WordWidth (bit): 512
-DeviceRoadmap: HP
-LocalWireType: LocalAggressive
-LocalWireUseLowSwing: No
-GlobalWireType: GlobalAggressive
-GlobalWireUseLowSwing: No
-Routing: H-tree
-InternalSensing: true
-MemoryCellInputFile: MRAM.cell
-Temperature (K): 350
-ForceBank: 4x4, 1x4
-ForceMat: 2x2, 1x2
-ForceMuxSenseAmp: 2
-MemCellType: MRAM
-CellArea (F^2): 54
-CellAspectRatio: 0.54
-ResistanceOn (ohm): 3000
-ResistanceOff (ohm): 6000
-ReadMode: current
-ReadVoltage (V): 0.25
-MinSenseVoltage (mV): 25
-ReadPower (uW): 30
-ResetMode: current
-ResetCurrent (uA): 80
-ResetPulse (ns): 10
-ResetEnergy (pJ): 1
-SetMode: current
-SetCurrent (uA): 80
-SetPulse (ns): 10
-SetEnergy (pJ): 1
-AccessType: CMOS
-VoltageDropAccessDevice (V): 0.15
-AccessCMOSWidth (F): 8
/* PIM operation, Current Decoder parameters x n */
-Max Row Activation: 3
/*Sense Amplifier */
-Sense Amplifier Power (uW): 10
-Sense Amplifier Read Dynamic Energy (fJ): 30
-Sense Amplifier Latency (ns): 3
-Sense Amplifier Area (mm^2): 0.00001
/*Custom Sub-array-level Add-on Component parameters*/
-Subaddon component Latency (ns): 1,0,0
-Subaddon component Dynamic Energy (nJ): 0.2,0,0
-Subaddon component Leakage Power (uW): 0.1,0,0
-Subaddon component Area (um^2): 12,0,0
/*Custom MAT-level Add-on Component parameters*/
-Mataddon component Latency (ns): 0,0,0
-Mataddon component Dynamic Energy (nJ): 0,0,0
-Mataddon component Leakage Power (uW): 0,0,0
-Mataddon component Area (um^2): 0,0,0
/*Custom BANK-level Add-on Component parameters*/
-Bankaddon component Latency (ns): 0,0,0
-Bankaddon component Dynamic Energy (nJ): 0,0,0
-Bankaddon component Leakage Power (uW): 0,0,0
-Bankaddon component Area (um^2): 0,0,0
NVSIM
PIMA-SIM 1.0
Memory Area- Computational Area (mm2)
R/W/Compute Latency with add-ons at any memory level (ns)
R/W/Compute Dynamic Energy with add-ons at any memory level (nJ)
Leakage Power (W)

Fig. 17 PIMA-SIM as a PIM support evaluation tool developed to model the timing, energy, and area of various PIM technologies

gies. This tool offers the same flexibility in memory configuration in terms of bank/mat/subarray organization and peripheral circuitry design as NVSIM while supporting PIM-level configurations. PIMA-SIM can be configured using three configuration files. At the cell level, it uses NVSIM's .cell file to save the device-circuit level info. The architecture level uses NVSIM's .cfg file to configure the memory organization and optimization target. In addition, as depicted in Fig. 17, at the PIM level, PIMA-SIM's .pim file is designed to save the PIM-level parameters. The PIM libraries are accordingly developed for each platforms on top of NVSIM [69] and Cacti [75] based on device/circuit level data. Accordingly, the performance data (i.e., latency, energy, and area) could be extracted for different PIM platforms w.r.t. a single input memory configuration file (.cfg).

4. For *Application level* simulations, a behavioral-level simulator is developed in Matlab. It takes architecture-level results and the presented customized in-memory algorithm for various big data applications to calculate the latency, energy, and area that different PIM platforms spend on them. It has a mapping optimization framework to maximize the performance w.r.t. the available resources.

3.3.3 Performance Analysis

Here, two different experiments under ISO-Capacity and ISO-Computation constraints are conducted to quantitatively compare and analyze the analog and digital acceleration-in-memory approaches for CNNs. ISO-Capacity denotes a condition in which all memory technologies are developed with an identical memory capacity for a fair performance benchmarking. ISO-Computation denotes a condition in which all PIM platforms based on various memory technologies are leveraged to realize a similar computation.

3.3.3.1 ISO-Memory-Capacity Comparison

The performance of digital and analog PIM platforms with an ISO-memory-capacity constraint is initially studied. A 32 Mb, single bank unit based on digital (SOT-MRAM, STT-MRAM, ReRAM, SRAM, and DRAM) and analog ReRAM crossbar is developed with the presented bottom-up evaluation framework. Table 7 reports eleven performance parameters for each platform. The observations on this experiment are listed below.

Area The area metric was divided into two parts: memory die area (*M*) and computational area (*C*), which includes controller, modified decoder, SA, 8-bit ADC for the relevant analog ReRAM crossbar, etc. In terms of memory die area, the digital PIM platforms impose a relatively larger area than analog ReRAM crossbar except for STT-MRAM design [71]. However, if we take the computational area into account, the ReRAM crossbar consumes 2.5 mm^2, which is much larger than that of digital counterparts, such as digital ReRAM (0.4 mm^2). Accordingly, a memory to computational area ratio as M/C can be defined. The M/C ratio equals 23.53 for SOT-MRAM-based PIM, while the analog ReRAM crossbar shows a ratio of 1.33. The low M/C ratio of the ReRAM crossbar is the consequence of sizeable peripheral circuit overhead, such as buffers and DAC/ADC, which contributes more than 85% of the computational area [72]. Furthermore, according to the results reported in Fig. 7, the STT-MRAM and SRAM platforms occupy the smallest and the largest overall area, respectively, compared to other PIM counterparts.

Latency As listed in Table 7, the analog ReRAM crossbar achieves the shortest read latency (1.48 ns) as compared with digital platforms. Still, it has the longest write latency (20.9 ns). The SOT-MRAM platform achieves the shortest write latency compared to other technologies and has a higher endurance (10^{10}–10^{15}) compared to ReRAM-based platforms.

Energy Based on Table 7, SOT-MRAM and STT-MRAM platforms consume the smallest write dynamic energy among all the NVM platforms due to their intrinsically low-power device operation. At the same time, SRAM achieves the smallest read and write energy compared to all the platforms. The analog ReRAM crossbar

Table 7 Per operation estimation results for different PIM designs. In the Area part, M denotes memory die area, and C denotes computation area overhead. (iso-capacity: 32Mbit-single Bank, Data Width: 512-bit)

	Digital					Analog
Metrics	SOT-MRAM†	STT-MRAM‡	ReRAM†	SRAM*	DRAM§	ReRAM**
Nonvolatility	Yes	Yes	Yes	No	No	Yes
Area (mm^2)	M: 7.06 C: 0.3	M: 2.14 C: 0.3	M: 3.92 C: 0.4	M: 10.38 C: 0.5	M: 4.53 C: 0.04	M: 3.34 C: 2.5
Read latency (ns)	2.85	1.90	1.65	2.9	3.4	1.48
Write latency (ns)	2.59	5.29	19.8	2.7	3.4	20.9
Read dynamic energy (nJ)	0.57	0.37	0.76	0.34	0.66	0.38
Write dynamic energy (nJ)	0.66	0.67	2.9	0.38	0.66	2.7
(N)AND/(N)OR computation energy (nJ)	~0.64	~0.46	~1.13	~0.59	~0.75	1.96 per MAC
Full-adder computation energy (nJ)	~1.92	~1.59	~3.4	~1.18	~11.25	
Leakage power (mW)	550	410.2	362.4	5243	335.5	587.6
Endurance	~10^{10} – 10^{15}	~10^{10} – 10^{15}	~10^5 – 10^{10}	Unlimited	10^{15}	~10^5 – 10^{10}
Data overwritten issue	No	No	No	No	Yes	No

† Implemented based on [55].
‡ Implemented based on [71].
* Implemented based on [18].
§ Implemented based on [9].
** Implemented based on [72]

achieves a close-to-SRAM read dynamic energy, consuming a considerable write dynamic energy. Computational energy is measured based on the PIM's capability to perform (N)AND/(N)OR and full-adder functions for digital platforms. As seen from Table 7, the STT-MRAM [71] and SRAM [18] PIM respectively consume the smallest computational energy compared to different technologies to perform various operations, where SOT-MRAM stands as the third most energy-efficient platform. Although the DRAM PIM design based on Ambit [9] consumes 0.75 nJ to perform (N)AND/(N)OR-based TRA mechanism, it requires over 14 memory cycles to perform the addition operation to avoid overwriting data. It leads to much higher energy consumption compared to other platforms. The computational energy per MAC was reported for the analog crossbar, comparable with the addition operation in the digital SOT-MRAM platform. The digital ReRAM and DRAM can be observed as relatively more power-efficient platforms regarding leakage power consumption. Moreover, the SRAM platform consumes $\sim$14.5$\times$ and $\sim$9$\times$ more power than digital and analog ReRAM, respectively.

3.3.3.2 ISO-Computation Comparison

The performance of the digital and analog PIM platforms was further explored for CNN acceleration. Hereby, we took the classical LeNet-5 as a simple example to perform the handwritten digit classification task with the MNIST dataset. For correctly mapping the target CNN into the PIM, offline training of the LeNet-5 network was conducted with weight and activation quantization, following the methods presented in [59]. A description model of each platform based on the data reported in Table 7 was then employed in the application-level CNN simulator. For fair hardware comparison, the bit-width configuration of [1:8] for [Weight: Activation] was selected, although ReRAM crossbar-based accelerator supports higher weight bit-width ($>$1 bit) with better CNN performance (i.e., classification accuracy in the experiment). No quantization was applied in the first and last layer of CNN, and the PIM-based accelerator also handled the full-precision computations. For the sake of simplicity, the estimated performance results (area, energy, latency) of convolutional layers are only reported.

Area Contrary to the approach used to report the area in Table 7, we leverage the method presented in [14, 72] to report the results. Specifically, we consider the area overhead due to computation by calculating the number of crossbars or sub-arrays. Table 8 reports the area for digital and analog PIM platforms by dividing it into the memory and logic parts. We observe that the digital ReRAM and STT-MRAM platforms require the smallest area than other platforms, respectively, mainly due to their single transistor cell structure. It is noteworthy that the DRAM platform has one of the least die areas due to its single-transistor cell and owns the least computational area under ISO-capacity constraint because of almost unchanged peripheral circuitry (1% as listed in Table 7). However, it requires access to multiple sub-arrays to avoid overwriting data problems and fitting the network simultaneously, resulting in a larger area requirement compared to NVMs. As for the analog crossbar platform, the logic part contributes $\sim$4$\times$ more than the memory area. Overall, it imposes a larger area than that of other digital NVM platforms due to matrix splitting and extra-large add-on area overhead [8].

Latency Table 8 summarizes each platform's latency required to process the convolutional layers of the CNN. According to the table, the SRAM platform is the fastest, with 0.7 ms latency. This mainly comes from its short read and write latency and fast two-cycle addition scheme [18]. Besides, we observe that the SOT-MRAM platform achieves 0.9 ms latency and stands as the second-fastest platform. The

Table 8 Estimated row performance of various PIMs without parallelism techniques

	Digital					Analog
Parameters	SOT-MRAM	STT-MRAM	ReRAM	SRAM	DRAM	ReRAM
Area (mm^2) (memory + logic)	0.018 ∼(0.0172 + 0.0008)	∼0.012 ∼(0.011 + 0.0008)	0.0097 ∼(0.009 + 0.0007)	0.64 ∼(0.608 + 0.032)	0.16 ∼(0.158 + 0.002)	0.06 ∼(0.011 + 0.049)
Energy (μJ) (write-back+read-based Ops)	0.85 ∼(0.31+0.54)	0.78 ∼(0.25+0.53)	1.9 ∼(0.75+1.15)	1.6 ∼(0.42+1.18)	2.1 ∼(0.8+1.3)	13.5 ∼(0+13.5)
Latency (ms)	0.9	1.8	1.3	0.7	13.5	5.8

DRAM platform shows a long latency mainly due to the excessive copy operations needed to avoid overwriting data. The analog crossbar needs 5.8 ms to process the convolutional layers.

Energy Table 8 also reports the energy consumption of different platforms. It can be seen that SOT-MRAM- and STT-MRAM-based platforms save 15.8× and 17.3× energy compared to the analog crossbar. In addition, the volatile digital memories consume much smaller energy than that of the analog platform. Therefore, from an energy-saving standpoint, digital PIM platforms could be a better choice than analog crossbar. Note that, for PIM platforms, all operands are assumed to be stored in memory. Unlike traditional computation, an extra intermediate data write-back is needed, which affects the overall energy and latency. Based on this, we split the reported energy into write-back and read-based logic operations energy. The write-back energy involves the energy required to write the weights or inputs into PIM plus the energy needed to write the computation results back to the memory for computation in the next layer. The read-based operation energy involves the read and bit-line computing energy. The analog crossbar [72] can accomplish the MAC operation without writing back the intermediate data; that is why we omit the write-back energy for this platform.

4 Enabling Reliable and Resilient Computing Paradigm

In this section, a power failure resilient/analysis design approach, as a cross-layer method from device level to architectural level, is developed, as shown in Fig. 18. It uses a nonvolatility feature in Spintronics when selectively inserted into the implementations concerning power failure/analysis resiliency and performance overhead. In step ❶, a physics-based and compact model of novel spin-based devices is

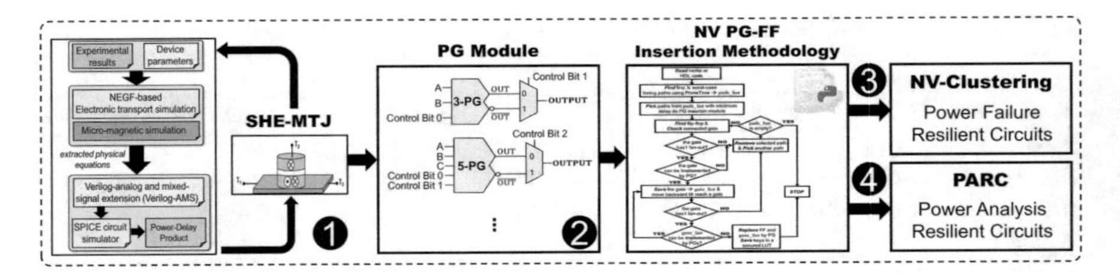

Fig. 18 Cross-layer design exploration on spin-based designs to attain power failure and power analysis-resilient architectures

constructed. The trade-off between write energy and retention time for various switching energy barriers is investigated. To do so, Matlab, Verilog-A, and SPICE models are developed, enabling straightforward integration with VLSI circuits in SPICE-friendly platforms. In step, ❷, a dual-mode spin-based polymorphic gate (PG) with built-in logic and memory features is efficiently designed and analyzed using the developed Majority Gate (MG). The PGs' libraries contain a functionally complete set of Boolean logic gates. A standardized methodology referred to as NV-clustering is developed for targeted insertion of PG modules as new compact means to increase the functionality of pipeline registers ❸. In step ❹, new algorithms are developed and added to NV-clustering to address the secure-computation demands, specifically in the presence of power analysis attacks.

4.1 MG-Based Synthesis and Optimization Research Tool

As it can be perceived, the unifying computational mechanism underlying all of these TMR-based devices is an accumulation-mode operation that enables the realization of majority logic functions as basic computational building blocks [76, 77]. Therefore, first, we developed an MG-based synthesis and optimization research tool. In our approach, a Genetic Algorithm (GA) scheme has been used to design logic circuit networks that are implemented by majority gates [78, 79]. GAs algorithms are one of the most popular optimization tools due to their ability to optimize any objective function regardless of the gradient or higher derivatives of the objective function. We develop a SHE-based Synthesis and Optimization Routine and Tool (SORT). It produces an optimized MG-based implementation regarding the Boolean expression. A combination of three-input and five-input MGs are considered the primary building blocks to optimize the designs. This combination of majority gates includes either design based on only one type of MGs (three-input or five-input) or designs including both types of the MG. The logical functions of three-input MG and five-input MG are expressed by Eq. (8). It represents the output for an MG with n inputs, where n is always an odd number. The MG outputs "1" if and only if more than (n-1)/2 of the inputs are "1" and vice versa.

$$\mathbf{M(A, B, C)} = \mathrm{AB + AC + BC}$$

$$\mathbf{M(A, B, C, D, E)} = \mathrm{ABC+ABD+ABE+ACD+ACE+ADE+BCD+BCE+BDE+CDE} \tag{8}$$

A tree structure is used to represent the chromosomes. In this structure, the root and the inner nodes of the tree are either a Majority or an Inverter specified with the Maj. and Inv., respectively. The algorithm starts with a population including 500 chromosomes. So to both three- and five-input MGs, of the effects be taken into account, a linear abstraction of fan-in as its cost function is considered, which has been defined as below:

$$f(C_i) = \frac{N(m, C_i)}{|m|} + \frac{1}{N(r, C_i)} + \frac{1}{Nodes(C_i)} \tag{9}$$

where $N(,)$ is a function that calculates the number of minterms in the first parameter implemented by the second one, m contains the minterms to be implemented, $|m|$ is the size of m and has been added for scaling issues, and r is the rest of minterms that should not be implemented. The algorithm stops when no improvement in fitness function happens during more than 20 generations or the total number of generations exceeds 1500. To provide experimental evidence to study how the combination of three-input and five-input majority gates improves the performance of traditional design methods, the presented optimization procedure is implemented in Python, and results are illustrated in Table 9.

As shown in Fig. 19, SORT is comprised of two modules: (1) *Developed Genetic Algorithm (GA) optimization unit* realizes a tree-structured Boolean expression according to the optimization criterion. The tree structure is constructed of a combination of inverters and MGs with varying numbers of inputs, and (2) *netlist generator* develops a nodal circuit topology according to the generated tree structure of the target Boolean expression. The SPICE circuit simulation tool leverages the produced optimized netlist and a SHE device model to validate the functionality and estimate power and delay metrics of the realized SHE-based Boolean logic circuit. The netlist generator research emphasizes algorithmic subtree methods to collapse an optimized MG graph based on the MG device libraries developed. The netlist generator outputs a SPICE syntax compatible file that can be utilized by the circuit simulation toolchain in conjunction with the SHE model library to synthesize the desired Boolean circuit. Hence, this developed GA-driven research synthesis tool is used to extract an optimized netlist for standard majority logic-based gate libraries. Our optimization methodology for spin-based NoC circuits is described, as shown in Fig. 20a. Spin-based components are utilized for storing and computing, whereas CMOS-based elements are used for implementing logic in storage elements and conducting the read operation. Required sensing scheme is provided by PCSA, which generates both output (OUT) and invert of the output ($\overline{\mathrm{OUT}}$). Hence, the intrinsic structure of the presented spin-based NoC cell includes one MG, which provides a functionally complete unit. Thus, in our proposed

Table 9 Optimization of three standard functions

Functions	Previous works using 3MG	Presented approach using combination of three and five MGs
1. BCD+ABC+ABD+ACD	$\mathbf{M}$(B,C,$\mathbf{M}$(D,A,0))	$\mathbf{M}$(A,B,C,D,0)
2. $\overline{A}.\overline{B}.\overline{C}$+ABC	$\mathbf{M}(\mathbf{M}(\overline{C}, A, 1), \mathbf{M}(C, B, 0), \overline{\mathbf{M}(A, B, 1)})$	$\mathbf{M}(\mathbf{M}(A, B, C, 0, 0), 1, \overline{\mathbf{M}(A, B, C, 1, 1)})$
3. A.B.C.D	$\mathbf{M}$($\mathbf{M}$(A,B,0),0,$\mathbf{M}$(0,C,D))	$\mathbf{M}$(0,$\mathbf{M}$(A,B,0),0,C,D)

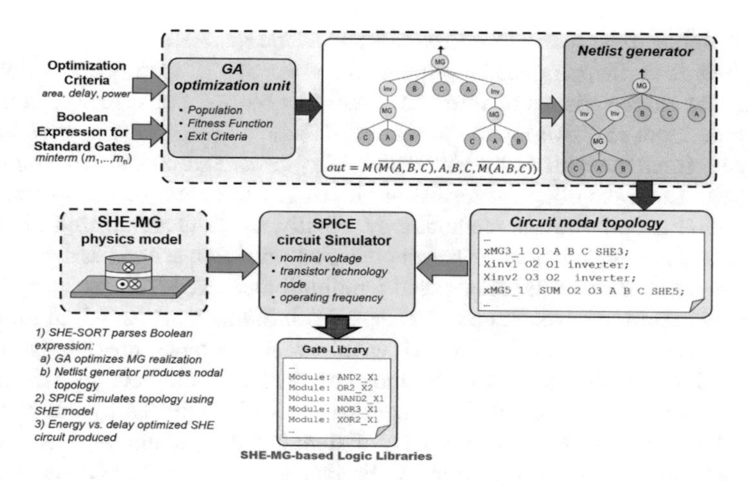

Fig. 19 Presented MG synthesis approach to realize SHE-based Boolean logic, including SHE-MG based gate libraries

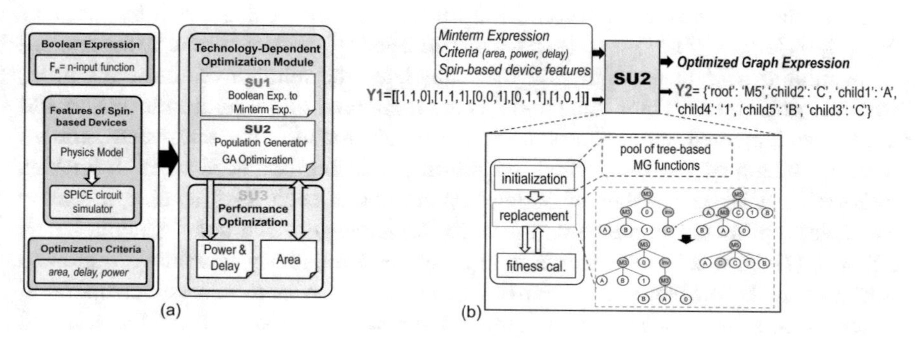

Fig. 20 (**a**) Schematic of the presented evolutionary approach to realize MG-based NoC circuit and (**b**) operations of F1 and F2 blocks for A·B+C in technology-dependent optimization process

optimization methodology, the implementation cost of the inverter gate is equal to zero. Our presented evolutionary approach includes two levels of optimization to reduce the convergence time: *technology-dependent optimization* and *performance optimization*.

4.1.1 Technology-Dependent Optimization

In the first level of the optimization, shown in Fig. 20a, GAs are utilized to optimize the implementation of a Boolean logic expression in terms of area, delay, or power. It leverages the spin-based device characteristics as inputs to achieve a semi-optimized implementation. First, a transforming unit, which is *Synthesis Unit 1* (SU1), decomposes a Boolean expression into its minterms. The generated minterm expression is applied to a mapping and optimization unit, SU2, along with optimization criteria and characteristics of spin-based building blocks. For instance, in a design with three-input and five-input spin-based MGs as building blocks, first MGs are separately implemented, and their related delay, area, and power consumption are measured. Then, the obtained results are leveraged to define their implementation cost within the optimization methodology. Finally, the GAs are utilized to optimize a Boolean logic implementation based on the optimization criteria and the obtained implementation cost of the spin-based building blocks in SU3. The mapping and optimization unit involves three main steps: (1) *Initialization*: An initial set of tree-based structures are created, in which each parent can have three or five random children. Each of the trees is a chromosome, and the complete set is called the initial population. The GA convergence time could be adjusted by the population size and range of chromosome variety. Extending the population size increases the variety of chromosomes, which is limited to some upper bounds. However, this extension leads to an increase in the total processing time of GA. (2) *Fitness Evaluation*: To evolve the population toward better solutions, the fitness of each chromosome is evaluated. Therefore, a fitness function is defined to assign a fitness value to the chromosomes. Herein, the fitness function is expressed by $f(t_i) = N(m, t_i)/(lengthoft_i) + 1/N(r, t_i) + 1/(numberofgate)$, where m is the applied input minterms; $N()$ is a function that calculates the number of minterms in m, which is implemented by t_i tree; and r is the remainder of the minterms that should not be implemented. As it can be seen, the fitness function has an inverse relation with the length of the tree, which results in producing balanced trees. It enables performing a larger number of parallel operations at each level leading to power and delay optimized implementations. (3) *Replacement*: The code generates new offspring(s) from selected parents with a defined probability to achieve improved solutions to the problem. The subtree has been chosen as the crossover operator, which selects two nodes and exchanges their subtrees rooted from the selected nodes. The mutation operation is applied to avoid the algorithm being trapped in a local optimum. Tournament selection has been utilized to select the parents for crossover and mutation operators. The algorithm stops when no improvement in fitness function happens after more than 100 generations. The output of this mapping and optimization unit is an optimized graph expression, as shown in Fig. 20b. Figure 21a illustrates the evolutionary approach leveraged in the presented technology-dependent optimization methodology.

4.1.2 Power and Delay Optimization

Due to the nature of spin-based devices, increasing the input current decreases the operation's delay at the expense of increased power consumption. As it was mentioned earlier, AND/OR gates can be readily implemented by majority gates. The disjunction operator (OR) has larger power consumption than the conjunction operator (AND) due to a higher number of ON transistors that leads to the higher

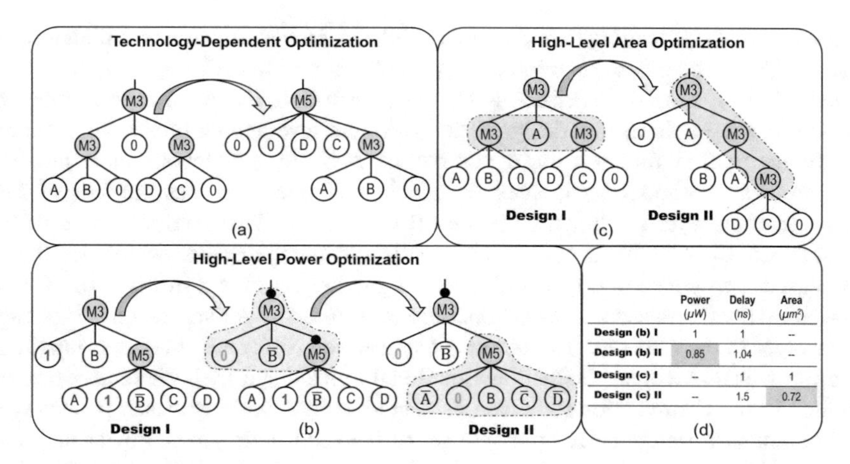

	Power (μW)	Delay (ns)	Area (μm^2)
Design (b) I	1	1	--
Design (b) II	0.85	1.04	--
Design (c) I	--	1	1
Design (c) II	--	1.5	0.72

Fig. 21 (**a**) Technology-dependent optimization for F= A.B.C.D, (**b**) power optimization for F= A+B+C+D, (**c**) area optimization for F= A·(B+CD), and (**d**) comparison results for designs in (**b**) and (**c**)

input current. Since the implementation cost of an inverter is equal to zero in our optimization methodology, disjunction operators and conjunction operators can be replaced according to the well-known De Morgan's law without any redundancy cost. Hence, a third functional unit (SU3) (Fig. 20a) is added to the optimization tool, which replaces the OR (AND) operations by AND (OR)-inverter operations within the logic implementation to reduce power (delay). The algorithm first takes the optimized tree obtained by SU2. Then, it executes a pre-order traversal scheme to visit a node, check its value, and updates it recursively. All trees or subtrees with a root labeled M3 or M5 are examined to find any leaf with value "1". Then, it replaces "1" with "0" and inverts all remaining leaves with the same parent. Finally, it uses the OUT signal instead of OUT to invert the whole tree or subtree. An example of a power-optimized implementation of (A+B+C+D) expression and its corresponding normalized simulation results are shown in Fig. 21b,d, respectively.

4.1.3 Area Optimization

In the generated implementations, each MG node requires one PCSA. Therefore, the number of required PCSAs for each layer depends on the number of MG nodes existing in that layer. On the other hand, PCSAs can be shared between different layers. Thus, the required PCSAs for implementing an NoC circuit equals the maximum number of MG nodes utilized in any spintronic layer. However, according to the fitness function described previously, trees with a balanced structure have a larger fitness value. Although the balanced tree structure, e.g., shown in Design I of Fig. 21c, provides an optimized implementation in terms of delay or power consumption, it requires a larger number of PCSAs due to having more MG nodes in the second layer leading to higher area overhead. Hence, for area optimization, we have modified the fitness function to $f(t_i) = N(m, t_i)/(lengthoft_i) + 1/N(r, t_i) + 1/(numberofgate) + 1/(nMG + 1)$, where nMG is the maximum number of PCSAs in the implemented design. The procedure leveraged a breadth-first search technique to find the maximum number of MGs in one level. Therefore, the optimization methodology creates an unbalanced tree with less number of MG nodes in each layer as shown in Design II of Fig. 21c. Thus, only a single PCSA is required to implement the A·(B+CD) Boolean expression, which results in decreased area consumption while increasing delay, as shown in Fig. 21d. This is caused by the increased sequential operations required to deliver the output of each logic layer to the next one. To implement spin-based NoC cells, the three-input and five-input SHE-MGs are defined as functional blocks, and their characteristics are applied to the optimization tool. The presented evolutionary approach is leveraged to implement a functionally complete set of Boolean logic gates. Power and area optimization resulted in an identical implementation for each Boolean function, while the delay optimization generated a different implementation.

4.2 *Power Failure Resilient: NV-Clustering Design Methodology*

On top of the optimized MG-based cells, we developed a standardized methodology to synthesize optimized NV architectures, which is referred to as NV-Clustering [80, 81]. NV-Clustering selectively collects together compatible Boolean logic functions and state holding functions, as depicted in Fig. 22. It utilizes (1) Logic-Embedded FFs (LE-FFs) as NV storage elements that also serve as computational elements, (2) a methodology for utilizing the developed cells to achieve robust intermittent

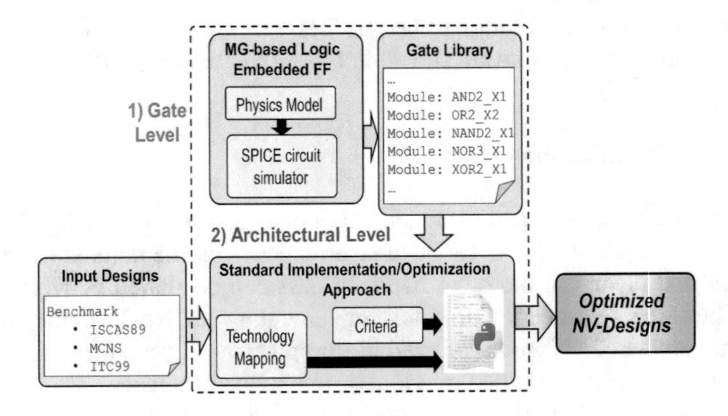

Fig. 22 Optimized NV implementations using NV-Clustering methodology diagram

operation, and (3) a constraint-based optimization step considering the area, power, and delay to realize a preferred NV-enhanced datapath design.

4.2.1 Logic-Embedded FF (LE-FF) Design

The presented LE-FF is composed of a spin MG-based master latch and a CMOS-based slave latch, as shown in Fig. 24a. An LE-FF has three different modes: store mode, in which the write operation to NVM is performed; *standby mode*, in which the power is disabled; and *sense mode*, in which the stored data in NVM is read. After power-up, the data is restored into the slave latch. Therefore, due to its nonvolatility, the entire design can be power gated without incurring vulnerability to the datapath. It can compute operation and store value during the first cycle, whereas the output, Q, is propagated during the second cycle. The circuit implementation of a three-input SHE-based LE-FF is depicted in Fig. 23a. LE-FF functionality in the presence of power failure and power-up situations is depicted in Fig. 23b, which verifies the desired forward progress of the design's operation while supporting the intermittent operation.

Our proposed LE-FF has two significant features in comparison to the previously presented NV-FF designs: (I) in addition to storing a value with nearly zero standby power, similar to the other NV-FFs, the LE-FF design is capable of computing rudimentary Boolean expressions intrinsically, resulting in area, complexity, and power reduction. Figure 24b shows a two-input OR, which is connected to an NV-FF and its equivalent implementation using LE-FF. Figure 24c summarizes all possible Boolean expression, which can be implemented using three- and five-input LE-FFs. Their implementation capacities might be enhanced by leveraging larger MGs. Moreover, (II) by using LE-FFs, the implemented designs have lower sensitive time to power failures. It is determined by the duration of signal propagation between two NV elements, including (1) input registers and an NV-FF, (2) two NV-FFs, or

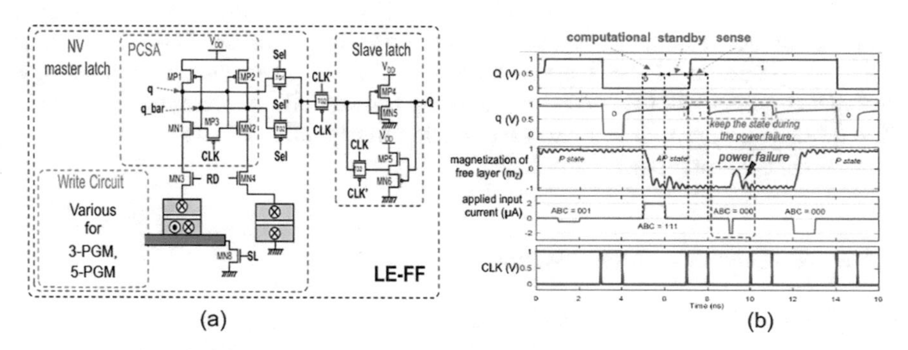

Fig. 23 Circuit-level design of proposed three-input SHE-based LE-FF and (**b**) transient response for three different input ABC= "001", "111", and "000" in presence of power failure. Three different modes are shown: (1) store mode, (2) standby mode, and (3) sense mode

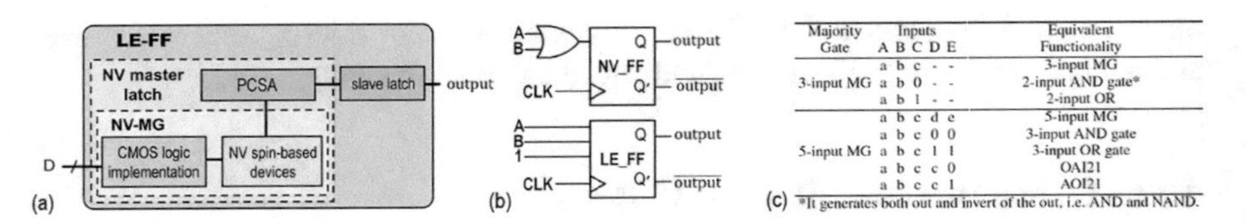

Majority Gate	A	B	C	D	E	Equivalent Functionality
3-input MG	a	b	c	-	-	3-input MG
	a	b	0	-	-	2-input AND gate*
	a	b	1	-	-	2-input OR
5-input MG	a	b	c	d	e	5-input MG
	a	b	c	0	0	3-input AND gate
	a	b	c	1	1	3-input OR gate
	a	b	c	c	0	OAI21
	a	b	c	c	1	AOI21

*It generates both out and invert of the out, i.e. AND and NAND.

Fig. 24 (**a**) Schematic of proposed MG-based LE-FF, and (**b**) different implementations using NV-FF (top), and proposed LE-FF (bottom). (**c**) Boolean Expressions using three and five-input MGs

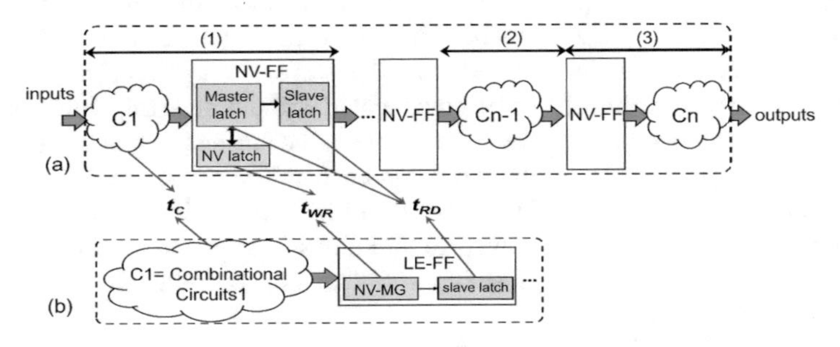

Fig. 25 All three sensitive time durations for (**a**) NV-FF based implementation and for (**b**) proposed implementation approach, in which C1(b) < C1(a)

(3) an NV-FF and output registers, in which if a power failure occurred, data will be lost and rebooting required. Figure 25 depicts all three possible durations. The vulnerability interval is expressed by $t_s = t_{WR} + t_{RD} + t_D$, where, t_{WR} is the write operation time for the NV element; t_{RD} is the switching time of CMOS-based latches, e.g., a master latch; and t_C is the required time for combinational circuits before storing into NV-FFs. In a datapath, the summation of all obtained sensitive time is considered a design vulnerability time (DVT), implying that a design with a smaller DVT provides higher tolerance to power failure. Hence, replacing cones of gates and NV-FFs with LE-FFs will reduce DVT, increasing failure robustness, which improves redundant restart efficiency. To design optimized NV architectures using the proposed LE-FF, there is a need to develop a systematic methodology, which incorporates all LE-FF features to design power failure-tolerant architectures. The developed approach leverages the maximum capability of LE-FFs in terms of replacement and implementation steps.

4.2.2 NV-Clustering Methodology

The proposed NV-Clustering methodology takes a hardware description language (HDL) representation of a datapath and MG-based gate modules as its inputs and produces an optimized NV-enhanced datapath. NV-Clustering was constructed in Python, according to the control flow illustrated in Algorithm 1. Its three primary procedures are (1) `find_gate(X)` that finds a gate generating the output **X**, (2)

Algorithm 1 NV-Clustering Methodology

```
 1: procedure MAIN()
 2:     Input: Hardware Description Language (HDL) code
 3:     Output: optimized HDL code
 4:     find all FFs and update FF_list
 5:     for FF in FF_list do
 6:         if size (create_cone (FF)) > 1 then
 7:             replace cone_gates by MG_FF
 8:         else
 9:             replace cone_gates by NV_FF
10:         end if
11:         update HDL code
12:     end for
13: end procedure
14: procedure CREATE_CONE()
15:     Input: a combinational gate, i.e. G
16:     Output: list of gates connected to a FF, i.e. cone_gates
17:     input_list = find_input (G)                    ▷ return list of G's input
18:     for item in input_list do
19:         if criterion #3 or criterion #4 is violated then
20:             input_list.remove (item)
21:         end if
22:     end for
23:     for item in input_list do
24:         tmp_gate = find_gate (item)                ▷ return gate with item as its input
25:         cone_gates.append (tmp_gate)
26:         if check (cone_gates) then
27:             create_cone (tmp_gate)
28:         else
29:             cone_gates.remove (tmp_gate)
30:         end if
31:     end for
32:     return cone_gates
33: end procedure
34: procedure CHECK()
35:     Input: cone of gates
36:     Output: Boolean expression
37:     if criterion #1 or criterion #2 is violated then
38:         return FALSE
39:     end if
40:     return TRUE
41: end procedure
```

`find_input(Y)` that finds all the primary inputs of gate `Y`, and (3) `check(Z)` that validates correctness according to the following circuit-level criteria regarding gate list `Z` :

Criterion #1: All gates in list `Z` are implemented by exactly one LE-FF. Rationale: Whereas each LE-FF requires one clock cycle for computation and to ensure the functional correctness of the design, the list `Z` including a cone of combinational logic gates and a master latch has a tight bound to occur within one clock cycle. Hence, the use of more than one MG for complex functions could increase the propagation delay enough to violate timing constraints. Hence, all elements in the list should be implemented using one LE-FF.

Criterion #2: Fan-out of every gate in list **Z** cannot exceed one. Rationale: Whereas LE-FFs realize sequential designs, outputs are obtained after a delay of two clock cycles, one for computation/storing (master layer) and one for reading (slave layer). If a computational circuit connects to more than one gate, then two gates that are driven require output1 and output2 as their inputs without any delay. Therefore, implementing a clockless design is permitted if the combinational function has a fan-out of one driven into a single sequential block. In addition to the abovementioned conditions, two more crucial considerations will be checked:

Criterion #3: An item in the *input_list* should not be a primary input. Rationale: If the input port is one of the primary design inputs, it cannot be an output of a gate; hence, it is removed from the input list.

Criterion #4: The item should not be an FF's output. Rationale: If the FF's output is in the *input_list*, the possible cone gates contain two FFs that requires two clock cycles instead of one, which causes a timing violation. Hence, the input is removed from the list. Therefore, due to the timing criterion, each cone gates should include only one FF and one (several) combinational gate(s).

If all criteria are satisfied, a cone of gates including all gates connected to an FF is replaced by precisely one LE-FF. Otherwise, the FF is replaced by a logic-free NV-FF. Then, the HDL code is updated based on the changes. These steps are performed for all FFs in the candidate design. Finally, the optimized HDL code is produced. To exemplify the functionality of the proposed methodology, the s27 circuit from the ISCAS-89 benchmark is analyzed, as shown in Fig. 26a. The following steps are performed:

(1) All FFs are listed, FF_list = {FF#1, FF#2, FF#3}

For FF#1:

STEP 1. **create_cone** (FF#1) is invoked. Next, **find_input** (FF#1) only returns **X1** as primary inputs and neglects the clock input. Thus, it satisfies both C3 and C4 conditions. This implies **input_list** = {**X1**}.

STEP 2. List of inputs has only one item. The **find_gate** (**X1**) function returns **INV1**, in which **X1** is its output. The **cone_gates** list is updated with **INV1**, thus **cone_gates** = {**INV1**}.

STEP 3. The function **check** (cone_gates) returns TRUE because **INV1** satisfies criteria C1 and C2. Therefore, **INV1** is retained in the cone.

STEP 4. Function **create_cone** (INV1) is invoked which performs all steps 1, 2, and 3. The **find_input**(INV1) returns **X2**, after checking criteria. Next, **input_list** is updated to {**X2**}. Thus, **find_gate**(X2) returns **OAI21** gate, which is appended in the **cone_gates** list as {**INV1, OAI21**}. Meanwhile, **check**(cone_gates) is still TRUE, whereas all gates can be implemented by one MG, simultaneously and each gate has fan_out of one. Thus, **create_cone**(OAI21) is invoked.

STEP 5. Invoking **create_cone**(OAI21) implies that **input_list** equals to {**X3, X4, X5**}. Meanwhile, **X3** violates the C4 condition corresponding to the output of the FF, so it is removed from the input list. Moreover, **X4** violates the C3 condition, the primary input of the circuit, thus **input_list=X5**. Accordingly, **find_gate**(X5) returns **INV2**, thus the revised set of cone_gates= {INV1, OAI21, INV2}. Since cone_gates satisfies C1 and C2 criteria, **check**(cone_gates) returns TRUE. Hence, **create_cone**(INV2) is invoked.

STEP 6. Invocation of **create_cone**(INV2) generates input_list=**X6**. However, X6 violates criterion C3. Then **cone_gates** returns to the main procedure. If its cardinality exceeds one, then the replaceable combinational gates are specified in **cone_gates** while the FF is replaced by an LE-FF. Otherwise, the FF is replaced by a conventional NV-FF. In this case, the HDL code becomes updated accordingly.

For FF#2:

STEP 1. Initially, **create_cone**(FF#2) is invoked, thus **find_input**(FF#2) returns **Y1**, which satisfies criteria C3 and C4. Accordingly, input_list = {**Y1**}.

STEP 2. The **find_gate(Y1)** function returns NOR1. The cone_gates set is updated such that cone_gates=NOR1.

STEP 3. The function **check**(cone_gates) returns TRUE whereas NOR1 satisfies criteria C1 and C2. Therefore, NOR1 is retained in the cone.

STEP 4. Function **create_cone**(NOR1) is invoked such that input_list={Y2,Y3}. However, Y2 and Y2 violate criteria C2 whereas both gates fan-out of 2. Thus, cone_gates is returned to the main procedure and because it is non-null, whereby the FF and NOR gates become replaced by LE-FF. The HDL code is updated accordingly.

For FF#3:

STEP 1. Procedure **create_cone**(FF#3) is invoked resulting in **find_input** (FF#3) returning Z1. It violates criterion C2. Thus, **cone_gates** is returned to the main procedure, and because of an empty list, the FF is replaced by a conventional NV-FF. Whereas FF_list is empty, it outputs the optimized HDL code. The optimized schematic for s27 is shown in Fig. 26b, which is discussed below.

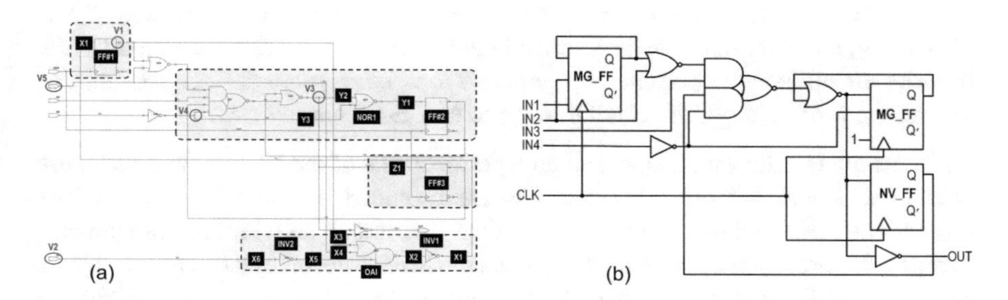

Fig. 26 (**a**) s27 schematic with highlighted FFs and (**b**) optimized LE-FF based design

Table 10 NV-Clustering gate equivalent reduction

ISCAS 89	Circuit function	Latch	Gate equivalent		Improvement %
			Baseline	NV-Clustering	
s27	Logic	3	10	8	20
S298	PLD	8	119	49	59
S349	4-bit multiplier	15	161	102	36
S400	TLC	21	164	144	12
S420	Fractional multiplier	16	218	152	30
S526	TLC	21	193	83	57
S820	PLD	5	289	259	10
S838	Fractional multiplier	32	446	329	26
S1196	Logic	18	529	459	13
S1423	Logic	74	657	396	40
S15850	Logic	534	9772	8942	8
S38584	Logic	1426	19,253	12,504	35

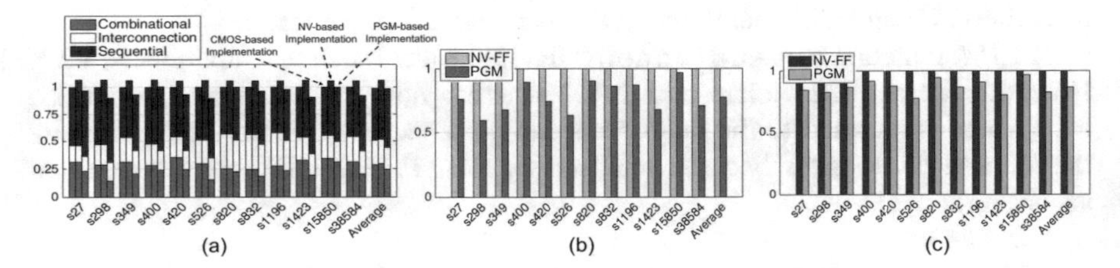

Fig. 27 Normalized (**a**) area, (**b**) power, and (**c**) delay, compared to CMOS and NV-FF based implementations

4.2.3 Simulation Results

In this section, performance characteristics, including power, delay, and area of NV-Clustering, are elaborated on large-scale benchmark circuits. The generated LE-FF libraries are utilized in a commercial synthesis tool, i.e., Synopsys Design Compiler, to map the produced optimized HDL code to an LE-FF based design.

Area Analysis The gate counts and area performance of the ISCAS-89 benchmark circuits with and without NV-Clustering are provided in Table 10 and Fig. 27a, respectively. All building blocks, including functional and buffer components, except FFs, are counted as gate equivalent. Whereas no gates are clustered with an NV-FF realization, its number of gate equivalents is identical to a CMOS-only realization. Meanwhile, NV-Clustering leverages LE-FFs, which can implement one (or a set of) Boolean function (s). For instance, benchmark circuit s1423 has 657 gates, reduced to approximately 60% of the original number of gates, 396, in the LE-FF implementation. Figure 27a depicts the total area of ISCAS-89 circuits, including the interconnection, combinational, and sequential components regarding these different implementations. For combinational circuits, NV-FF and

CMOS implementations occupy a similar area as mentioned above. However, from a sequential point of view, implementing NV-FFs and LE-FFs requires additional peripheral circuits such as write and read circuits, which can incur area overhead. Hence, the NV-FF implementation occupied the largest area among the implementations. On the other hand, a reduction in the equivalent gate count decreases the area of both combinational and interconnection components. Owing to the back-end process vertical integration of spintronic devices, the area of NV elements can be significantly reduced; hence, LE-FF implementations indicate the least area consumption. As shown in Fig. 27a, the area overhead of ISCAS-89 benchmark circuits using LE-FF shows an average 15% area reduction over NV-FF realization.

Power Analysis Figure 27b depicts the power consumption regarding the combinational blocks. Generally, our implementations illustrate an excellent amount of power reduction for all benchmarks. However, due to the constraints in implementing large logic functions using three- and five-input MGs, the differences between the two implementations are insignificant in some benchmark circuits. Although this issue can be readily addressed by developing larger MGs with a higher number of inputs, increasing the number of inputs also increases the complexity of the MG. Figure 27b depicts an average of 22% power reduction using the NV-Clustering average for ISCAS-89 benchmark circuits.

Delay Analysis The optimized RTL Verilog HDL codes for the benchmarks are synthesized using a Synopsys Design Compiler, and then worst-case timing paths are obtained through applying STA on compiled netlists using Synopsys PrimeTime. The obtained results regarding benchmarks are shown in Fig. 27c. The delay is directly proportional to the number of combinational components. If the number of FFs is minimal and the number of replaced combinational blocks is maximum, then the delay is reduced to the greatest possible extent. As shown in Fig. 27c, the delay reduction for selected ISCAS-89 benchmark using NV-Clustering exhibits an average of 14% over NV-FF. It is worth noting that the obtained results herein are at the gate level, and physical design parameters are not considered.

Resumption Overhead In energy harvesting systems, the power supply has a limited capacity. On the one hand, in a CMOS-based design, if the system is powered down, then volatile memories lose data, and up to a few milliseconds [82] is necessary to restore information after a new power-up. Furthermore, this charge/discharge cycle, which is an intrinsic characteristic of energy harvesting devices, may occur hundreds of times per second. The system might consume its entire power supply capacity to restore to the initial states. On the other hand, although NV-FF and LE-FF implementations provide power failure-tolerant designs, the required power consumption of write operations for nonvolatile elements remains an issue. Hence, due to the capacity limitation of power supplies and the aforementioned issues, various conditions should be considered to choose between CMOS-based or NV-based implementations. Two main conditions are (1) a total number of completed operations and (2) a power failure rate. According to the equality $ConstantPowerSupply = \sum_i^m n_i \times P_i$, where m is the total number of operations, n is the number of operation i, and P is the required power consumption of operation i, in a low/free power failure situation, volatile CMOS-based implementations perform more operations than nonvolatile-based designs. However, in an environment with a high occurrence rate of power failure, CMOS's number of completed tasks is excessively reduced, which degrades the overall system performance. Since the power supply capacity and power consumption of each operation are constant, the usage of an NV approach is affordable if the power failure rate is relatively high, which can disable CMOS-based designs' functionalities.

Thus, there are two potential scenarios to be considered. *Scenario #1* corresponds to the case when **intermittency is absent**, in which power failure did not occur during the processing interval under observation. *Scenario #2* represents the case in which **intermittency is present**. Considering *Scenario #1*, the application of MTJs in memory device applications [43, 44], the retention time, $\tau = \tau_0 exp(\Delta/kT)$, is arranged to be 10–15 years by choosing a thermal barrier, Δ, between 40 and 60 kT. On the other hand, the critical spin current is linearly proportional to the thermal barrier, Δ. Thus, for applications that do not require retention times of years, we investigate via simulation the reduction of the thermal barrier of nanomagnets employing uniaxial anisotropy and other possibilities such as lowering their volume or their saturation magnetization. This ultimately reduces the charge currents that are required for write operations, which can result in significant energy improvement due to the quadratic relationship between the Ohmic (I^2R) losses and the input write currents. Therefore, LE-FFs using SHE-MTJ devices with 30 kT energy barriers are investigated to achieve retention times ranging from minutes to hours while providing at least 50% energy reduction. Figure 28 shows the power-delay-product (PDP) values for the two scenarios. In the intermittency-absent condition, the obtained PDP results for CMOS-based designs are relatively lower than the other implementations because of CMOS's high-speed/low-power switching, whereas in the intermittency-present scenario for various ISCAS-89, ITC-99, and MCNC benchmark circuits, the results exhibit an average of 14%, 12%, and 4% PDP improvements, respectively, for LE-FF ($\Delta = 40$ kT)-based designs compared to NV-FF-based implementations. Further PDP improvements can be achieved by using low-energy barrier SHE-MTJ devices ($\Delta = 30$ kT) within LE-FFs at the cost of shorter retention times. However, in the energy-harvesting-powered IoT devices, retention time in the range of days and hours could be sufficient to achieve proper functionality. Thus, leveraging SHE-MTJ devices with 30 kT energy barrier in intermittency occurred situations provides up to 12%, 48%, and 39% average PDP improvements compared to CMOS-based designs, NV-FF-based designs, and LE-FF-based implementations with $\Delta = 40$ kT, respectively, without incurring any area overhead. It is worth noting that the results provided herein are obtained at the gate level, and physical design parameters are not considered within the document space available.

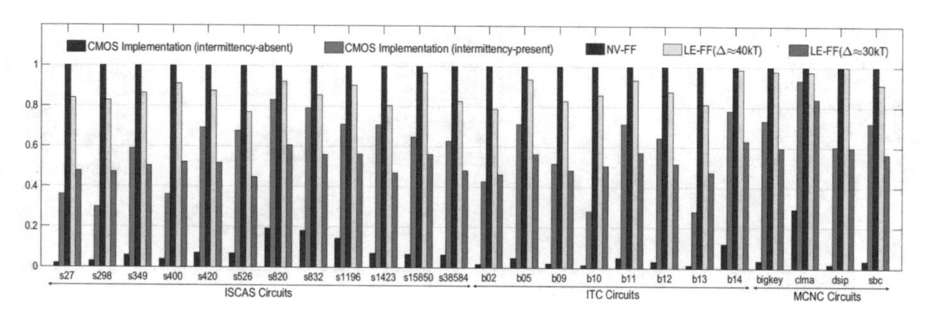

Fig. 28 Normalized PDP compared to NV-FF-based implementations for *intermittency-absent* and *intermittency-present* scenarios

4.3 Power Analysis Resilient: PARC Design Methodology

Our approach [83–85] is inspired by power masking approaches as a possible power analysis countermeasure. In this method, two completely separate units are utilized, where the inputs are stored in registers and operate similarly with different power profiles. The selector building block includes a true random number generator (TRNG) that connects to a multiplexer, leveraged to enable one of the two functional blocks. Because of the random behavior of the selector, the power consumption of this design will change randomly. These designs suffer from area overhead, $\sim 2\times$ larger than the original design, and a narrow range in power profiles for masking power. Moreover, implementing CMOS-based power maskable units with reconfigurability features imposes area and power overhead. Therefore, herein, a standardized methodology to synthesize optimized PAA-resilient architectures, referred to as Power Analysis-Resilient Circuit (**PARC**), is developed. It leverages PGMs as programmable building blocks, which offer advantages in evolvable, intelligent, and security-critical applications. PARC determines where PGMs should be inserted to provide robust coverage at minimal overhead. It is worth noting that the randomizing process is performed using ultradense and energy-aware spin-based TRNG. PARC methodology incorporates various metrics of PAA, including the number of required samples and a correlation between the secret key and the sample to design a PAA-tolerant circuit for small footprint IoT devices.

4.3.1 PARC Design Methodology

Figure 29 depicts the design flow of the PARC design methodology used to synthesize power analysis-resilient circuits. The proposed approach is described in Algorithm 2, which is developed in Python including three main procedures: (1) **`Search()`**, (2) **`Insertion()`**, and **`Analyze()`**.

First, in step ❶, hardware description language (HDL) code of a design such as VHDL or Verilog is taken by **`Search`** procedure, which utilizes Synopsys

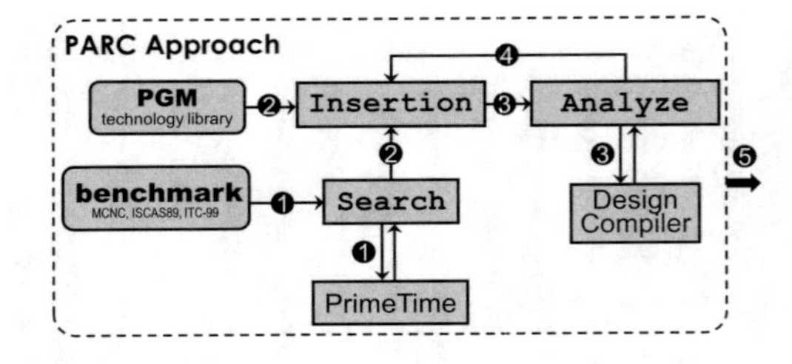

Fig. 29 Systematic PG insertion methodology to elevate the immunity of designs in the presence of PAA with minimal performance overhead

Algorithm 2 PARC methodology

```
1: Input: Hardware Description Language (HDL) code, PGM
2: Output: Power analysis-resilient architectures HDL code
3: procedure SEARCH ()
4:     delay_list [] ← PrimeTime()                              ▷ find delay
5:     update P_path_list ← prioritize delay_list []            ▷ sort paths based on their delays
6: end procedure
7: procedure INSERTION ()
8:     for i ← 1 to length (P_path_list) do
9:         update (gates_list)                                  ▷ all gates in P_path_list [i]
10:        for j ← 1 to length (gates_list) do
11:            if check (design criteria) then                  ▷ can be replaced
12:                replace (input_list[j])
13:            else
14:                Break
15:            end if
16:            m_HDL ← update (HDL code)
17:            Analyze(m_HDL)
18:        end for
19:    end for
20: end procedure
21: procedure ANALYZE ()
22:    PDP and Area ← DesignCompiler (m_HDL)
23:    Y [pdp_N, area_N] ← Normalize (PDP and Area)
24:    Compute (ND)                                             ▷ calculates ND w.r.t. inserted PGMs
25:    Compute (EoD)                                            ▷ calculates EoD based on Eq. 1
26:    return EoD
27: end procedure
```

PrimeTime to obtain worst-case timing paths and marks all gates within these paths as low replacement priority. Subsequently, in addition to PGMs, the prioritized list of paths is applied to the **Insertion** procedure. In step ❷, a cone of gates including combinational logic gates is selected and replaced by the PGMs using the NV-Clustering approach presented in our previous work [80]. Then in step ❸, **Analyze** procedure takes the modified HDL code generated by step ❷ and applies it within a commercial synthesis tool, i.e., Synopsys Design Compiler and SPICE circuit simulators. With respect to the original implementation, O, the *Effectiveness of Design* (EoD) values are computed in step ❹:

$$Y_{PDP} = \frac{PDP_M}{PDP_O}, \quad Y_{Area} = \frac{Area_M}{Area_O},$$

$$EoD = \frac{ND}{Y_{PDP} \times Y_{AREA}} \tag{10}$$

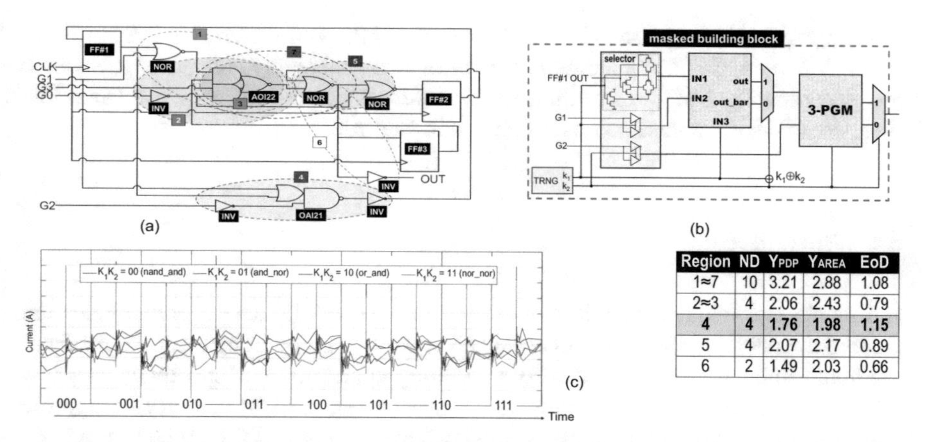

Region	ND	Y_{PDP}	Y_{AREA}	EoD
1≈7	10	3.21	2.88	1.08
2≈3	4	2.06	2.43	0.79
4	**4**	**1.76**	**1.98**	**1.15**
5	4	2.07	2.17	0.89
6	2	1.49	2.03	0.66

Fig. 30 (**a**) s27 schematic from ISCAS-89 benchmark circuit and possible replacement regions, (**b**) PGM-based design and EoD results for the selected regions, and (**c**) power traces results for K_1K_2 combinations

where PDP_{M} and $Area_{\mathrm{M}}$ are power-delay-product and area of the modified circuit, respectively. *ND* is the maximum number of possible configurations of the new design that produce the correct output. The procedure including all steps is performed *N* times, calculated by Eq. (11) to maximize the EoD value. Higher EoD values have higher PAA resiliency. It means higher EoD escalates the required number of power profiles/traces, which is expressed by $2^{nk} \times 2^{nin}$, where *nk* is the number of key bits, and $nk = \lceil \sqrt{ND} \rceil$ and *nin* is the number of input bits for the inserted building blocks.

$$N = \sum_{m=1}^{n} (n - m + 1) = \frac{n(n+1)}{2} \tag{11}$$

where, n = number of gates in one path and m = number of the selected gate(s) to be replaced by PGM(s). Finally, in step ❺, the optimized power analysis-resilient HDL code is produced.

To exemplify PARC operation and functionality, circuit s27 from the ISCAS-89 benchmark is analyzed as shown in Fig. 30a. Since it has nine different paths with a maximum of four logic gates in one path, the procedure will run ten times to determine the optimized EoD. The selection process is performed using a CMOS-based MUX, which chooses the proper PGM's outputs as the final output. The EoD results for different selected cone regions are listed in Fig. 30b. For instance, region 1 (or 7) consists of a two-input NOR and a four-input AND_OR_Inverter22 (AOI22). Although it can generate ten different structures (maximum number of equivalent logic realization) with an identical function, it requires four 3-input PGMs, which impose ~3.2× and ~2.9×, power and area overhead, respectively, compared to the original implementation, whereas region 4 shows the highest EoD value, which can be implemented by four equivalent structures, including

NAND_AND, AND_NOR, OR_AND, NOR_NOR, with different power profiles. Since $ND = 4$, this implementation requires two enable keys (en_{K1}, en_{K2}), which are generated by TRNG, resulting in 32 different power profiles for all possible three-input design combinations, as shown in Fig. 30c.

5 Conclusion

In this chapter book, magnetic random-access memory (MRAM) components are studied to enable in-memory processing and gate-level pipelining essential for data-/compute-intensive tasks and energy harvesting applications, respectively. Regarding data-/compute-intensive applications, new customized in-memory computing algorithms and mapping methods were developed to convert the crucial iteratively used functions to bit-wise PIM-supported functions. Moreover, a generic and comprehensive evaluation framework was also presented to analyze the performance quantitatively. About NoC systems, the design methodology was first extended to realize the targeted insertion PGMs within the VLSI implementations to make them resilient against power failure. Then PARC as an extension of NV-Clustering was developed as a power-masked synthesis method in the presence of power analysis side-channel attack. Due to the exciting accelerator achievements and normally off computing units' implementations, the presented frameworks and methodologies are promising schemes for resource-constrained edge devices.

References

1. Wang, Y., Yu, H., Ni, L., et al.: An energy-efficient nonvolatile in-memory computing architecture for extreme learning machine by domain-wall nanowire devices. IEEE Trans. Nanotechnol. **14**(6), 998–1012 (2015)
2. Fact sheet: Big data across the federal government (2012) [Online]. Available:
3. Fong, X., Kim, Y., Yogendra, K., et al.: Spin-transfer torque devices for logic and memory: Prospects and perspectives. IEEE Trans. Comput. Aided Des. Integr. Circuits Syst. **35**(1), 1–22 (2016)
4. Li, S., Niu, D., Malladi, K.T., et al.: Drisa: A dram-based reconfigurable in-situ accelerator. In: Proceedings of the 50th Annual IEEE/ACM International Symposium on Microarchitecture, pp. 288–301. ACM, New York (2017)
5. Li, B., Gu, P., Shan, Y., et al.: Rram-based analog approximate computing. IEEE Trans. Comput. Aided Des. Integr. Circuits Syst. **34**(12), 1905–1917 (2015)
6. Angizi, S.: Processing-in-memory for data-intensive applications, from device to algorithm. Ph.D. dissertation. Arizona State University, New York (2021)
7. Cheng, M., Xia, L., Zhu, Z., et al.: Time: A training-in-memory architecture for memristor-based deep neural networks. In: 2017 54th ACM/EDAC/IEEE Design Automation Conference (DAC), pp. 1–6. IEEE, New York (2017)
8. Chi, P., Li, S., Xu, C., et al.: Prime: a novel processing-in-memory architecture for neural network computation in reram-based main memory. In: ACM SIGARCH Computer Architecture News, vol. 44(3), pp. 27–39. IEEE Press, New York (2016)

9. Seshadri, V., Lee, D., Mullins, T., et al.: Ambit: In-memory accelerator for bulk bitwise operations using commodity dram technology. In: 2017 50th Annual IEEE/ACM International Symposium on Microarchitecture (MICRO), pp. 273–287. IEEE, New York (2017)
10. Li, S., Xu, C., Zou, Q., et al.: Pinatubo: A processing-in-memory architecture for bulk bitwise operations in emerging non-volatile memories. In: 2016 53nd ACM/EDAC/IEEE Design Automation Conference (DAC), pp. 1–6. IEEE, New York (2016)
11. He, Z., Angizi, S., Parveen, F., Fan, D.: Leveraging dual-mode magnetic crossbar for ultra-low energy in-memory data encryption. In: Proceedings of the on Great Lakes Symposium on VLSI 2017, pp. 83–88 (2017)
12. Angizi, S., Roohi, A., Taheri, M., Fan, D.: Processing-in-memory acceleration of mac-based applications using residue number system: A comparative study. In: Proceedings of the 2021 on Great Lakes Symposium on VLSI, pp. 265–270 (2021)
13. Angizi, S., He, Z., Parveen, F., Fan, D.: Imce: Energy-efficient bit-wise in-memory convolution engine for deep neural network. In: 2018 23rd Asia and South Pacific Design Automation Conference (ASP-DAC), pp. 111–116. IEEE, New York (2018)
14. Angizi, S., He, Z., Rakin, A.S., Fan, D.: Cmp-pim: an energy-efficient comparator-based processing-in-memory neural network accelerator. In: Proceedings of the 55th Annual Design Automation Conference, p. 105. ACM, New York (2018)
15. Yin, S., Jiang, Z., Seo, J.-S., Seok, M.: Xnor-sram: In-memory computing sram macro for binary/ternary deep neural networks. IEEE J. Solid-State Circuits **55**(6), 1733–1743 (2020)
16. Roohi, A., Angizi, S., Fan, D., DeMara, R.F.: Processing-in-memory acceleration of convolutional neural networks for energy-effciency, and power-intermittency resilience. In: 20th International Symposium on Quality Electronic Design (ISQED), pp. 8–13. IEEE, New York (2019)
17. Roohi, A., Sheikhfaal, S., Angizi, S., et al.: Apgan: Approximate gan for robust low energy learning from imprecise components. IEEE Trans. Comput. **69**(3), 349–360 (2019)
18. Eckert, C., Wang, X., Wang, J., et al.: Neural cache: Bit-serial in-cache acceleration of deep neural networks, pp. 383–396 (2018)
19. Lee, B.C., Ipek, E., Mutlu, O., Burger, D.: Architecting phase change memory as a scalable dram alternative. In: ACM SIGARCH Computer Architecture News, vol. 37(3), pp. 2–13. ACM, New York (2009)
20. Everspin announces sampling of the world's first 1-gigabit mram product (2016). [Online]. https://www.everspin.com
21. Baumann, A., Jung, M., Huber, K., et al.: A mcu platform with embedded fram achieving 350na current consumption in real-time clock mode with full state retention and 6.5 μs system wakeup time. In: 2013 Symposium on VLSI Circuits (VLSIC), pp. C202–C203. IEEE, New York (2013)
22. Chien, T.-K., Chiou, L.-Y., Lee, C.-C., et al.: An energy-efficient nonvolatile microprocessor considering software-hardware interaction for energy harvesting applications. In: 2016 International Symposium on VLSI Design, Automation and Test (VLSI-DAT), pp. 1–4. IEEE, New York (2016)
23. Senni, S., Torres, L., Sassatelli, G., Gamatie, A.: Non-volatile processor based on mram for ultra-low-power iot devices. ACM J. Emerg. Technol. Comput. Syst. (JETC) **13**(2), 17 (2017)
24. Senni, S., Torres, L., Benoit, P., et al.: Normally-off computing and checkpoint/rollback for fast, low-power, and reliable devices. IEEE Magn. Lett. **8**, 1–5 (2017)
25. Prenat, G., Jabeur, K., Vanhauwaert, P., et al.: Ultra-fast and high-reliability sot-mram: From cache replacement to normally-off computing. IEEE Trans. Multi-Scale Computing Systems **2**(1), 49–60 (2016)
26. Bishnoi, R., Oboril, F., Tahoori, M.B.: Non-volatile non-shadow flip-flop using spin orbit torque for efficient normally-off computing. In: 2016 21st Asia and South Pacific Design Automation Conference (ASP-DAC), pp. 769–774. IEEE, New York (2016)
27. Khanna, S., Bartling, S.C., Clinton, M., et al.: An fram-based nonvolatile logic mcu soc exhibiting 100% digital state retention at vdd = 0 v achieving zero leakage with < 400-ns wakeup time for ulp applications. IEEE J. Solid State Circuits **49**(1), 95–106 (2014)

28. Sakimura, N., Tsuji, Y., Nebashi, R., et al.: 10.5 a 90 nm 20 mhz fully nonvolatile microcontroller for standby-power-critical applications. In: 2014 IEEE International Solid-State Circuits Conference Digest of Technical Papers (ISSCC), pp. 184–185. IEEE, New York (2014)
29. Ransford, B., Sorber, J., Fu, K.: Mementos: System support for long-running computation on rfid-scale devices. In: ACM SIGARCH Computer Architecture News, vol. 39(1), pp. 159–170. ACM, New York (2011)
30. Lucia, B., Ransford, B.: A simpler, safer programming and execution model for intermittent systems. ACM SIGPLAN Not. **50**(6), 575–585 (2015)
31. Shi, K., Howard, D.: Challenges in sleep transistor design and implementation in low-power designs. In: Proceedings of the 43rd annual Design Automation Conference, pp. 113–116. ACM, New York (2006)
32. Zhao, H., Glass, B., Amiri, P.K., et al.: Sub-200 ps spin transfer torque switching in in-plane magnetic tunnel junctions with interface perpendicular anisotropy. J. Phys. D. Appl. Phys. **45**(2), 025001 (2011)
33. Rowlands, G., Rahman, T., Katine, J., et al.: Deep subnanosecond spin torque switching in magnetic tunnel junctions with combined in-plane and perpendicular polarizers. Appl. Phys. Lett. **98**(10), 102509 (2011)
34. Roohi, A., Zand, R., DeMara, R.F.: A tunable majority gate-based full adder using current-induced domain wall nanomagnets. IEEE Trans. Magn. **52**(8), 1–7 (2016)
35. Rakin, A.S., Angizi, S., He, Z., Fan, D.: Pim-tgan: A processing-in-memory accelerator for ternary generative adversarial networks. In: 2018 IEEE 36th International Conference on Computer Design (ICCD), pp. 266–273. IEEE, New York (2018)
36. Roohi, A., Zand, R., Fan, D., DeMara, R.F.: Voltage-based concatenatable full adder using spin hall effect switching. IEEE Trans. Comput. Aided Des. Integr. Circuits Syst. **36**(12), 2134–2138 (2017)
37. Gallagher, W.J., Parkin, S.S.: Development of the magnetic tunnel junction mram at ibm: From first junctions to a 16-mb mram demonstrator chip. IBM J. Res. Dev. **50**(1), 5–23 (2006)
38. Chung, S.-W., Kishi, T., Park, J., et al.: 4gbit density stt-mram using perpendicular mtj realized with compact cell structure. In: 2016 IEEE International Electron Devices Meeting (IEDM), pp. 27–1. IEEE, New York (2016)
39. Garello, K., Yasin, F., Hody, H., et al.: Manufacturable 300 mm platform solution for field-free switching sot-mram. In: 2019 Symposium on VLSI Circuits, pp. T194–T195. IEEE, New York (2019)
40. Natsui, M., Tamakoshi, A., Honjo, H., et al.: Dual-port field-free sot-mram achieving 90-mhz read and 60-mhz write operations under 55-nm cmos technology and 1.2-v supply voltage. In: 2020 IEEE Symposium on VLSI Circuits, pp. 1–2. IEEE, New York (2020)
41. Sakhare, S., Perumkunnil, M., Bao, T.H., et al.: Enablement of stt-mram as last level cache for the high performance computing domain at the 5 nm node. In: 2018 IEEE International Electron Devices Meeting (IEDM), pp. 18–3. IEEE, New York (2018)
42. Kan, J., Park, C., Ching, C., et al.: Systematic validation of 2x nm diameter perpendicular mtj arrays and mgo barrier for sub-10 nm embedded stt-mram with practically unlimited endurance. In: 2016 IEEE International Electron Devices Meeting (IEDM), pp. 27–4. IEEE, New York (2016)
43. Slaughter, J., Rizzo, N., Janesky, J., et al.: High density ST-MRAM technology. In: 2012 IEEE International Electron Devices Meeting (IEDM), pp. 29–3. IEEE, New York (2012)
44. Slaughter, J., Nagel, K., Whig, R., et al.: Technology for reliable spin-torque mram products. In: 2016 IEEE International Electron Devices Meeting (IEDM), pp. 21–5. IEEE, New York (2016)
45. Donahue, M.J.: Oommf user's guide, version 1.0. *-6376* (1999)
46. Fong, X., Gupta, S.K., Mojumder, N.N., et al.: Knack: A hybrid spin-charge mixed-mode simulator for evaluating different genres of spin-transfer torque mram bit-cells. In: 2011 International Conference on Simulation of Semiconductor Processes and Devices, pp. 51–54 (2011)

47. Zand, R., Roohi, A., Fan, D., DeMara, R.F.: Energy-efficient nonvolatile reconfigurable logic using spin hall effect-based lookup tables. IEEE Trans. Nanotechnol. **16**(1), 32–43 (2016)
48. Panagopoulos, G., Augustine, C., Roy, K.: A framework for simulating hybrid mtj/cmos circuits: Atoms to system approach. In: 2012 Design, Automation and Test in Europe Conference and Exhibition (DATE), pp. 1443–1446. IEEE, New York (2012)
49. Huai, Y.: Spin-transfer torque mram (stt-mram): Challenges and prospects. AAPPS Bull. **18**(6), 33–40 (2008)
50. He, Z., Zhang, Y., Angizi, S., Gong, B., Fan, D.: Exploring a SOT-MRAM based in-memory computing for data processing. IEEE Trans. Multi-Scale Comput. Syst. **4**(4), 676–685 (2018)
51. Liu, L., Moriyama, T., Ralph, D., Buhrman, R.: Spin-torque ferromagnetic resonance induced by the spin hall effect. Phys. Rev. Lett. **106**(3), 036601 (2011)
52. Liu, L., Pai, C.-F., Li, Y., et al.: Spin-torque switching with the giant spin hall effect of tantalum. Science **336**(6081), 555–558 (2012)
53. Pai, C.-F., Liu, L., Li, Y., et al.: Spin transfer torque devices utilizing the giant spin hall effect of tungsten. Appl. Phys. Lett. **101**(12), 122404 (2012)
54. Angizi, S., He, Z., Awad, A., Fan, D.: Mrima: An mram-based in-memory accelerator. IEEE Trans. Comput. Aided Des. Integr. Circuits Syst. **39**(5), 1123–1136 (2019)
55. Angizi, S., Sun, J., Zhang, W., Fan, D.: Graphs: A graph processing accelerator leveraging sot-mram. In: 2019 Design, Automation and Test in Europe Conference and Exhibition (DATE), pp. 378–383. IEEE, New York (2019)
56. Angizi, S., Fan, D.: Imc: energy-efficient in-memory convolver for accelerating binarized deep neural network. In: Proceedings of the Neuromorphic Computing Symposium, pp. 1–8 (2017)
57. Angizi, S., Sun, J., Zhang, W., Fan, D.: Pim-aligner: a processing-in-mram platform for biological sequence alignment. In: 2020 Design, Automation and Test in Europe Conference and Exhibition (DATE), pp. 1265–1270. IEEE, New York (2020)
58. Angizi, S., Sun, J., Zhang, W., Fan, D.: Aligns: A processing-in-memory accelerator for dna short read alignment leveraging sot-mram. In: 2019 56th ACM/IEEE Design Automation Conference (DAC), pp. 1–6. IEEE, New York (2019)
59. Zhou, S., Wu, Y., Ni, Z., et al.: Dorefa-net: Training low bitwidth convolutional neural networks with low bitwidth gradients. arXiv preprint arXiv:1606.06160 (2016)
60. Rastegari, M., Ordonez, V., Redmon, J., Farhadi, A.: Xnor-net: Imagenet classification using binary convolutional neural networks. In: European Conference on Computer Vision, pp. 525–542. Springer, Berlin (2016)
61. Shafiee, A., Nag, A., Muralimanohar, N., et al.: ISAAC: A convolutional neural network accelerator with in-situ analog arithmetic in crossbars. ACM SIGARCH Computer Architecture News **44**(3), 14–26 (2016)
62. Song, L., Qian, X., Li, H., Chen, Y.: Pipelayer: A pipelined reram-based accelerator for deep learning. In: 2017 IEEE International Symposium on High Performance Computer Architecture (HPCA), pp. 541–552. IEEE, New York (2017)
63. Angizi, S., He, Z., Reis, D., et al.: Accelerating deep neural networks in processing-in-memory platforms: Analog or digital approach? In: 2019 IEEE Computer Society Annual Symposium on VLSI (ISVLSI), pp. 197–202. IEEE, New York (2019)
64. Jain, S., Sengupta, A., Roy, K., Raghunathan, A.: RX-CAFFE: Framework for evaluating and training deep neural networks on resistive crossbars. arXiv preprint arXiv:1809.00072 (2018)
65. Cavigelli, L., Magno, M., Benini, L.: Accelerating real-time embedded scene labeling with convolutional networks. In: Proceedings of the 52nd Annual Design Automation Conference, pp. 1–6 (2015)
66. Angizi, S., He, Z., Fan, D.: Dima: a depthwise cnn in-memory accelerator. In: 2018 IEEE/ACM International Conference on Computer-Aided Design (ICCAD), pp. 1–8. IEEE, New York (2018)

67. Roohi, A., Taheri, M., Angizi, S., Fan, D.: Rnsim: Efficient deep neural network accelerator using residue number systems. In: IEEE/ACM International Conference on Computer-Aided Design (ICCAD), pp. 1–9. IEEE, New York (2021)
68. Reis, D., Gao, D., Angizi, S., et al.: Modeling and benchmarking computing-in-memory for design space exploration. In: Proceedings of the 2020 on Great Lakes Symposium on VLSI (2020), pp. 39–44
69. Dong, X., Xu, C., Xie, Y., Jouppi, N.P.: NVSIM: A circuit-level performance, energy, and area model for emerging nonvolatile memory. IEEE Trans. Comput. Aided Des. Integr. Circuits Syst. **31**(7), 994–1007 (2012)
70. DRAM Power Model. https://www.rambus.com/energy/
71. Jain, S., Ranjan, A., Roy, K., Raghunathan, A.: Computing in memory with spin-transfer torque magnetic RAM. IEEE Trans. Very Large Scale Integr. VLSI Syst. **26**(3), 470–483 (2018)
72. Tang, T., Xia, L., Li, B., et al.: Binary convolutional neural network on rram. In: 2017 22nd Asia and South Pacific Design Automation Conference (ASP-DAC), pp. 782–787. IEEE, New Your (2017)
73. (2011) Ncsu eda freepdk45. [Online]. http://www.eda.ncsu.edu/wiki/FreePDK45:Contents
74. Synopsys, Inc., Synopsys design compiler, product version 14.9.2014 (2014)
75. Chen, K., Li, S., Muralimanohar, N., et al.: CACTI-3DD: Architecture-level modeling for 3d die-stacked dram main memory. In: Design, Automation and Test in Europe Conference and Exhibition (DATE), 2012, pp. 33–38. IEEE, New York (2012)
76. Behin-Aein, B., Datta, D., Salahuddin, S., Datta, S.: Proposal for an all-spin logic device with built-in memory. Nat. Nanotechnol. **5**(4), 266 (2010)
77. Nikonov, D.E., Bourianoff, G.I., Ghani, T.: Proposal of a spin torque majority gate logic. IEEE Electron Device Lett. **32**(8), 1128–1130 (2011)
78. Roohi, A., Menbari, B., Shahbazi, E., Kamrani, M.: A genetic algorithm based logic optimization for majority gate-based qca circuits in nanoelectronics. Quantum Matter **2**(3), 219–224 (2013)
79. Roohi, A., Zand, R., DeMara, R.F.: Synthesis of normally-off boolean circuits: An evolutionary optimization approach utilizing spintronic devices. In: 2018 19th International Symposium on Quality Electronic Design (ISQED), pp. 49–54. IEEE, New York (2018)
80. Roohi, A., DeMara, R.F.: Nv-clustering: Normally-off computing using non-volatile datapaths. IEEE Trans. Comput. **67**(7), 949–959 (2018)
81. Roohi, A., DeMara, R.F.: IRC cross-layer design exploration of intermittent robust computation units for IoTs. In: 2019 IEEE Computer Society Annual Symposium on VLSI (ISVLSI), pp. 354–359. IEEE, New York (2019)
82. Kimura, H., Fuchikami, T., Maramoto, K., et al.: A 2.4 pj ferroelectric-based non-volatile flip-flop with 10-year data retention capability. In: , 2014 IEEE Asian Solid-State Circuits Conference (A-SSCC), pp. 21–24. IEEE, New York (2014)
83. Roohi, A., DeMara, R.F.: PARC: A novel design methodology for power analysis resilient circuits using spintronics. IEEE Trans. Nanotechnol. **18**, 885–889 (2019)
84. Roohi, A., DeMara, R.F., Wang, L., Köse, S.: Secure intermittent-robust computation for energy harvesting device security and outage resilience. In: 2017 IEEE International Conference on Advanced and Trusted Computing (ATC), pp. 1–5. IEEE, New York (2017)
85. Roohi, A., Zand, R., DeMara, R.F.: Logic-encrypted synthesis for energy-harvesting-powered spintronic-embedded datapath design. In: Proceedings of the 2018 on Great Lakes Symposium on VLSI, pp. 9–14 (2018)

IoT Commercial and Industrial Applications and AI-Powered IoT

Khaled Ahmed Nagaty

1 Introduction

Internet of Things is a disruptive technology that is commonly available and easily accessible. It connects heterogeneous devices with each other through sending and receiving information in different formats to reach a common goal [1]. The main goal of IoT devices is to sense data and interact with the environment [2]. Companies use IoT to gather information about customers to understand customers' needs and preferences, and in the same time IoT personalizes customer's products and services and customizes them to the user's needs and preferences. Therefore, many companies and industries in various fields in our daily lives adopt IoT because it helps them automate processes, reduce labor costs, and "increase productivity, save time, optimize cost, optimize human resource, predict maintenance, and provide a lot of comfort to human life." The IoT also reduces waste of resources by monitoring the utilization of these resources, hence improving the quality of products and service delivery. The IoT is composed of physical objects called things. Sensors, software, and communication technologies connect devices and exchange information over the Internet. The devices of IoT systems may range from ordinary devices such as home appliances to complex sensor networks in various industries such as weather forecast, or military. The sensors and devices in IoT systems collect data from the environment and send the data to the cloud through the Internet. When the data gets to the cloud, data processing could be done and finally the information is sent to the end user, which could be another IoT device. IoT systems are characterized by heterogeneity, dynamism, autonomy, extensiveness, privacy, and security [2]. For heterogeneity, an IoT system may be composed of different

K. A. Nagaty (✉)
The British University in Egypt, El-Sherouk City, Cairo, Egypt
e-mail: khaled.nagaty@bue.edu.eg

A. Iranmanesh (ed.), *Frontiers of Quality Electronic Design (QED)*,
https://doi.org/10.1007/978-3-031-16344-9_12

hardware devices, network infrastructure, and processing applications, and they need to connect and exchange information. For dynamism, IoT systems need to keep their correct behavior independent of changes occurring in the environment. For autonomy, IoT devices must be capable of making decisions without human intervention. For extensiveness, as the number of devices estimated to connect to the Internet in 2021 is 35 billion devices, an infrastructure and platform are required to manage such large number of connections. As IoT systems are connecting to the Internet to exchange information, this may let them vulnerable to cyberattacks especially if these systems are collecting sensitive data such as military sensor networks or healthcare systems. Therefore, IoT systems must guarantee privacy and security to protect their collected data. The global market for IoT can be segmented based on technologies, components, applications, end users, and geography. In this chapter, we will consider the IoT market segment based on applications. There are many applications of IoT technologies in our real lives; this is because IoT can be customized to almost any field of applications that can produce data about monitoring the environment, its operations, and activities. IoT applications can be commercial or industrial. The most important commercial applications of IoT are wearables, healthcare, traffic monitoring, hospitality, retail, maintenance management, and digital marketing. The most important industrial applications are manufacturing, agriculture, water supply, smart cities, financial services, energy, supply chain, transportation, telematics, and building automation. Figure 1 shows the IoT global market from 2018 to 2023. According to global data, IoT technology global market reached 130 billion dollars in 2018. It is estimated to reach 318 billion dollars in 2023, at a 20% compound annual growth rate [3].

This chapter is organized as follows: Sect. 2 is dedicated for IoT commercial applications, Sect. 3 is dedicated for industrial IoT applications, Sect. 4 is dedicated for IoT data analytics, Sect. 5 is dedicated for IoT security risks and threats, Sect. 6 is dedicated AI-powered IoT, and finally Sect. 7 is dedicated for conclusion.

2 IoT Commercial Applications

Commercial IoT applications improve experiences of customers, patients, and guests in different places such as hospitals, markets, hotels, and restaurants through more efficient monitoring of operations in smart buildings and smart offices. It improves company's insight into retail business and allow them to make real-time decisions to target potential customers with appropriate messages. The most common commercial IoT applications are the following.

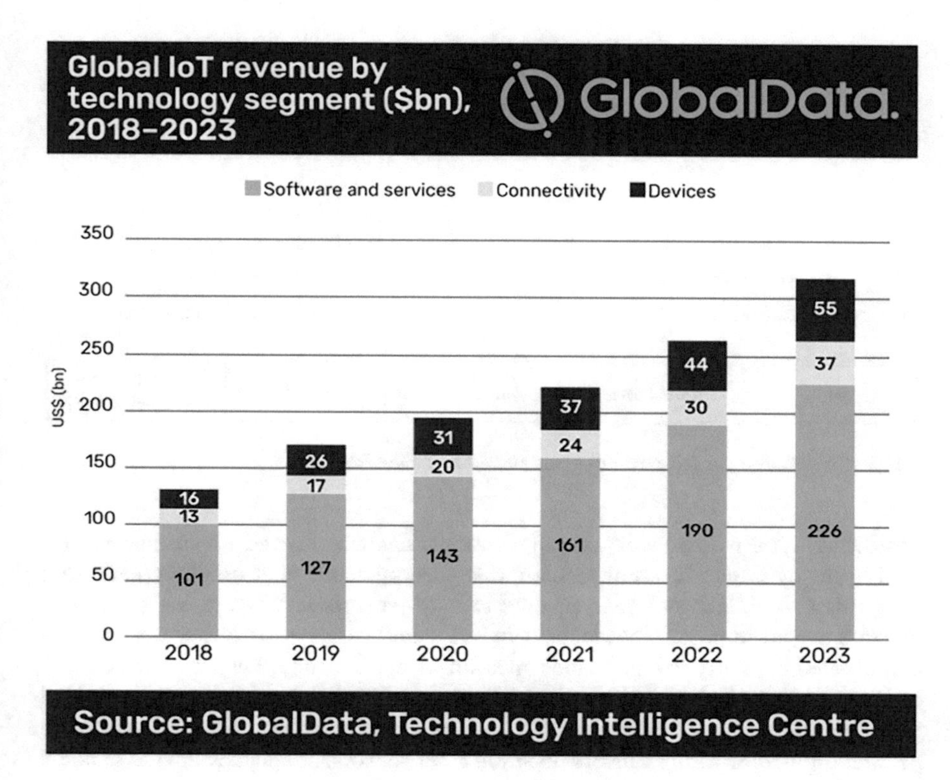

Fig. 1 Forecast End-user Spending on IoT Solutions Worldwide from 2017 to 2025 (in billions of dollars)

2.1 *Healthcare*

IoT healthcare systems provide flexibility and save much time than traditional healthcare systems [4]. Families, patients, hospitals, physicians, and insurance companies can benefit from the applications of IoT in healthcare. Lives of healthcare providers become safer and easier, costs are reduced, and healthcare services for patients are ultimately improved. IoT enables medical providers to detect patients' commitment to medical plans and provide them immediate help in case of emergency. For instance, monitoring noncritical patients, providing assisted ambient living (AAL) for elderly persons, and rehabilitation after physical injury save hospital resources for patients that are more critical. Patients who live far away from hospitals could not reach the hospital on time, so providing them medical advice using the IoT healthcare system could be a lifesaver. However, Internet disconnection is one of the major risks of IoT healthcare systems [5]. On the other side, patients can learn about their health records and can communicate with their medical providers. This section discusses the major components of IoT

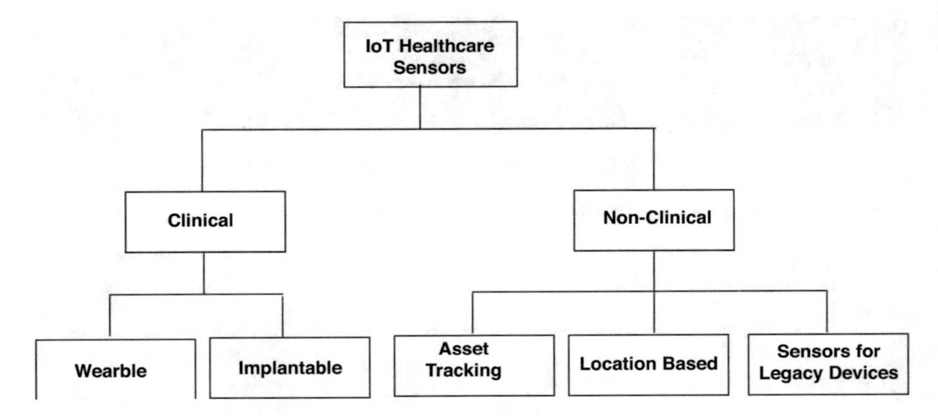

Fig. 2 Classification of IoT healthcare sensors. (As adapted from [6])

healthcare systems, services and applications, challenges, and opportunities. IoT healthcare systems are composed of many components that communicate with each other to collect and analyze data to help patients and healthcare providers. Sensors are an important component of IoT healthcare systems. Sensors must be recalibrated regularly to ensure their measurement efficiency. They are divided into two categories: clinical and nonclinical sensors. The clinical sensors are divided into wearable sensors and implantable sensors. Nonclinical sensors are divided into asset/equipment tracking sensors, location asset sensors, and legacy device sensors. Figure 2 shows the classification of healthcare sensors.

2.1.1 Clinical Sensors

2.1.1.1 Wearable Sensors

They are useful within healthcare for monitoring the bodies of patients and collecting physiological and movement data. Wearable sensors are deployed on parts of medical interest on the patient's body.

Pulse Sensors

It reads heart pulse of a patient. The pulse sensor can be placed on the patient's wrist, chest, earlobe, or fingertips. Figure 3 shows a pulse sensor.

Respiratory Rate Sensors

Respiratory sensors measure respiratory rate by detecting variations in chest movement per minute. The sensing system can be placed around the chest with a

Fig. 3 Pulse sensor

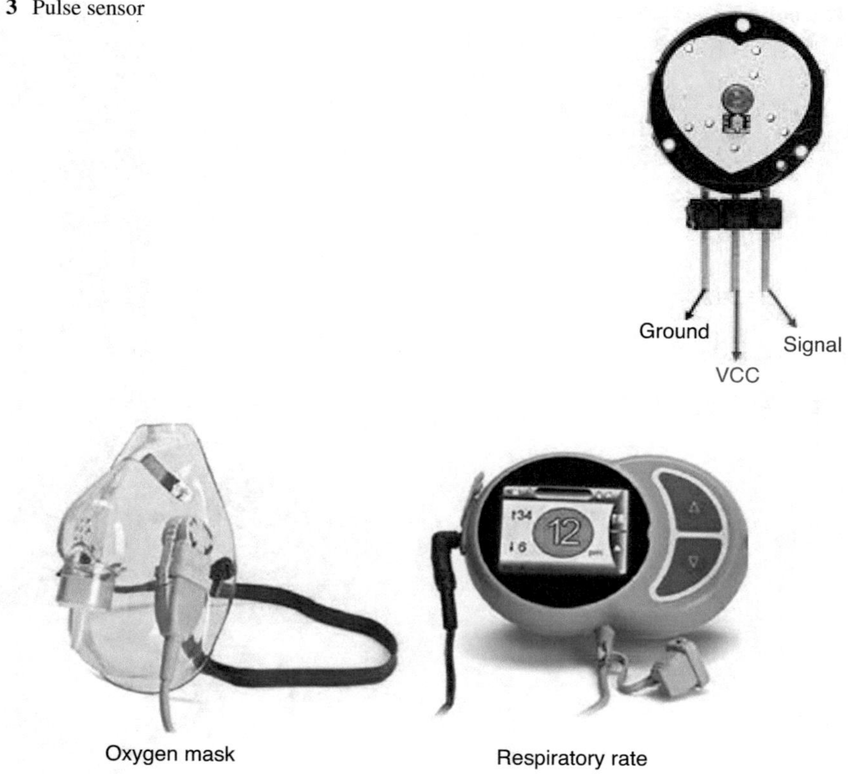

Fig. 4 Respiratory rate sensor [7]

strap. Figure 4 shows a respiratory rate sensor composed of an oxygen mask and a respiratory rate device.

Body Temperature Sensors

An accurate body temperature depends on how far the sensor position from the human body. Infrared (IR) body temperature sensors are noncontact sensors and can be placed close to the patient's forehead, earlobe, or skin. Figure 5 shows a body temperature sensor.

Blood Pressure Sensors

IoT blood pressure monitoring system (IBPMS), for example, Raspberry Pi, can remotely monitor patient's blood pressure [8]. The IBPMS reads the data and sends it to both Telegram and Gmail applications [8]. Figure 6 shows blood pressure sensor.

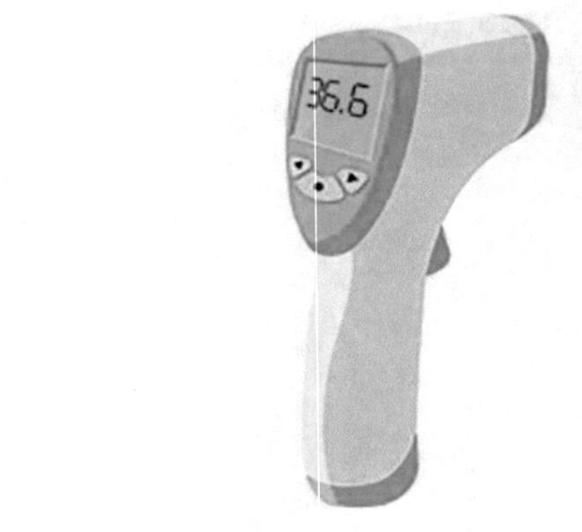

Fig. 5 Body temperature sensor

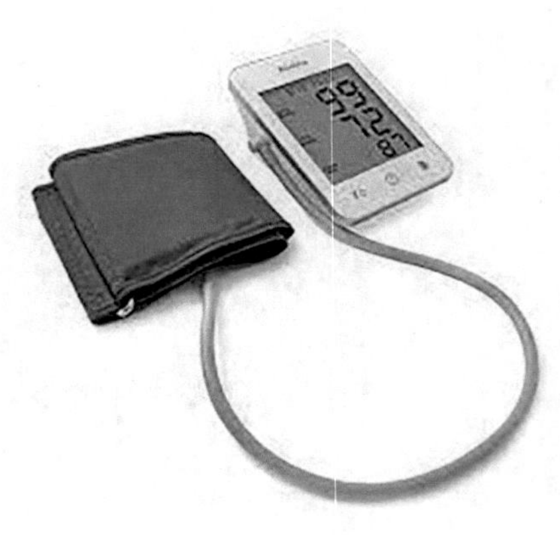

Fig. 6 Blood pressure sensor

Fig. 7 Pulse oximetry sensor

Pulse Oximetry Sensors

Oxygen saturation in the patient's blood can be measured using pulse oximeter sensors. This noninvasive device is composed of a red light source, infrared source, photo detectors, and a probe to transmit light through a translucent, pulsating arterial bed, typically an earlobe or fingertip. Figure 7 shows a pulse oximetry sensor.

2.1.1.2 Implantable Sensors

These sensors are inserted into the patient's body for diagnosis, treatment, and long-term monitoring. Such implantable sensors allow health monitoring systems to detect changes in the health conditions of the patients whether they are conscious or not to provide them with immediate treatment. For example, to design better prosthetics, implantable strain sensors can be incorporated into orthopedic prosthetics to specify the forces acting on those joints. Patients with high risk of excessive clotting or impeded blood flow implantable cardiovascular flow and pressure sensors can provide early warning of excessive clotting or impeded flow, and implantable neurostimulators can treat muscular and neurological damage [9].

2.1.2 Nonclinical Sensors

IoT devices can help track physician's location and people inside hospitals, find the nearest ambulance in case of emergency, track assets to achieve operational efficiency and compliance with hygiene standards, and provide real-time information for logistics [6].

2.2 Tourism and Hospitality

The tourism industry is highly affected by digital transformation and diffusion of disruptive technologies such as IoT. IoT technologies help increase customers' satisfaction in tourism and hospitality while reducing operational costs at the same time. IoT devices provide guests with automated guest check-in, smart hotel rooms with IoT-enabled door locks, motorized curtains, smart TVs which have positive impact on customers' satisfaction. IoT security and safety enable safer hotel stays and increased peace of mind for guests. Housekeeping staff can gain from IoT hotel systems to know when a guest room is occupied and when it is ready to be cleaned. Maintenance department can receive quick reports on malfunctioning gadgets such as burned-out lightbulbs or plumbing leaks to fix them as quickly as possible to minimize costs to fix problems and increase guests' satisfaction. The staff can detect vacant rooms and minimize electricity by dimming lights, turning off lights and A/C units, and tracking equipment with asset tracking systems.

2.3 Retail Industry

IoT technologies can play a fundamental role in retail industry. RFID is one of the most common forms of IoT used in retailing [12]. Data obtained from reading the tags attached to items and products by radio-frequency identification (RFID) is analyzed via IoT data analytics tools which allow retailers to obtain more valuable information on sales update and customers' purchasing patterns. Retailers can enhance customer experience by creating ideal shopping atmosphere using smart self-checkout cart where customers pick an item off a shelve, scan barcode of the item, drop it in the cart, and, when shopping is finished, pay directly on the cart. IoT reduce costs by tracking lost carts and baskets, tracking shipments in real time to prevent spoilage, keep products protected in transit, prevent theft or loss, adjust the air conditioning based on how many people are coming and going, and dim light switches when a store is less occupied. IoT retail uses sensors to monitor customer satisfaction, food safety, and sales opportunities in real time. IoT technologies allow retailers to have deep insight in the supply chain and track assets. Based on customers' behavior and demographics collected by IoT systems, customized products can be delivered to customers and placed in the right places. IoT inventory systems eliminate downtime at warehouse and uphold timely deliveries, and using smart shelves, retailers can monitor stock levels and guarantee products' availability on shelves.

2.4 Digital Marketing

IoT technology provides marketers with more insights on customer's usage of products. IoT technology finds patterns in product usage which allow digital marketers to predict exact demand and understand the daily lifestyle of customers and may read the customer's mind set. This makes 100% of advertisements are aligned with the customers' needs, interests, behaviors, buying patterns, individual preferences, and past purchases. Digital marketers can better understand their audience, determine the elements affect purchasing patterns, analyze customers' behavior in real time, determine markets where a particular product will sell best, and target potential customers with promotion messages and personalized advertisements. For example, if the milk in a customer's smart fridge in his/her smart home is about to end, then the connected smart fridge will record for buying a new bottle of milk and sends a purchasing order to the store. As IoT technology becomes more advanced, it learns the customer's behavior in consuming milk and calculate the number of days the bottle of milk will be consumed. Such smart fridge can warn the customer when the milk is approaching to an end through a text message to the customer's smartphone before sending a purchasing order to the store. As a result, advertisers can predict when customers will replace products, which type of products they want to accurately target potential customers, determine the type of marketing campaign, determine potential messages that will actively engage customers, and improve the marketing campaign. IoT in social media allows digital marketers to automate market data gathering, creating posts with contents they are sure that the audience will want to see and sharing them with potential customers.

3 IoT Commercial Applications

3.1 Agriculture

IoT in agriculture uses remote sensors embedded into plants and fields to collect data about the physical properties of soil and environment to help make decisions for increasing crop production which is essential for sustainable food security. Data analysis helps farmers monitor their crops, organize the irrigation process, promptly diagnose plant diseases to minimize the use of pesticides, help decision-makers to have better insights, and develop management plans to save time and money. Using IoT in agriculture is one step closer to smart agriculture where farms will be self-dependent, thus making the right decisions to increase productivity [10]. IoT sensors used in smart agriculture are classified into two categories: field and climate sensors. The IoT agriculture system is composed of three layers: the first layer is data collection layer that collects field data and climate data and send over the Internet to the data analysis layer. The data analysis layer contains tools to analyze the collected data and historic stored data to predict the parameters required for making decisions in the upcoming days. For example, in rescheduling field irrigation, data about soil moisture is collected from sensors inserted into the soil closer to the root zone of the plants. Climatic data such as temperature and wind speed collected from wind speed sensors, sunshine data, and field data is sent to a dedicated server in the cloud to be analyzed and decision is made based on a threshold value [11]. The decision on whether to start or stop the water pump or increase or reduce the amount of water served to the field is sent to the water pump regulator. The framer or human expert can override the decisions of the IoT agriculture system. Figure 8 shows IoT agriculture sensors which can be categorized into field sensors and climate sensors.

3.1.1 Field Sensors

3.1.1.1 Soil Temperature

It is the measure of warmth in the soil, i.e., how hot or cold the soil is, the air may be warm, but the soil may be cool. So, soil temperature is necessary because it is a very important physical property that affects the germination of seeds and plant growth. It controls the speed of chemical reactions and biological activities in the soil; most soil organisms work better at warm soil temperature. Factors that influence soil temperature are climate, season, water levels, soil color, plant cover, compost and manure, and soil moisture. An instant-read thermometer used for cooking is used to measure soil temperature by putting the thermometer's probe as deep as possible into the soil to get a precise reading of the soil temperature. Figure 9 shows a soil temperature sensor.

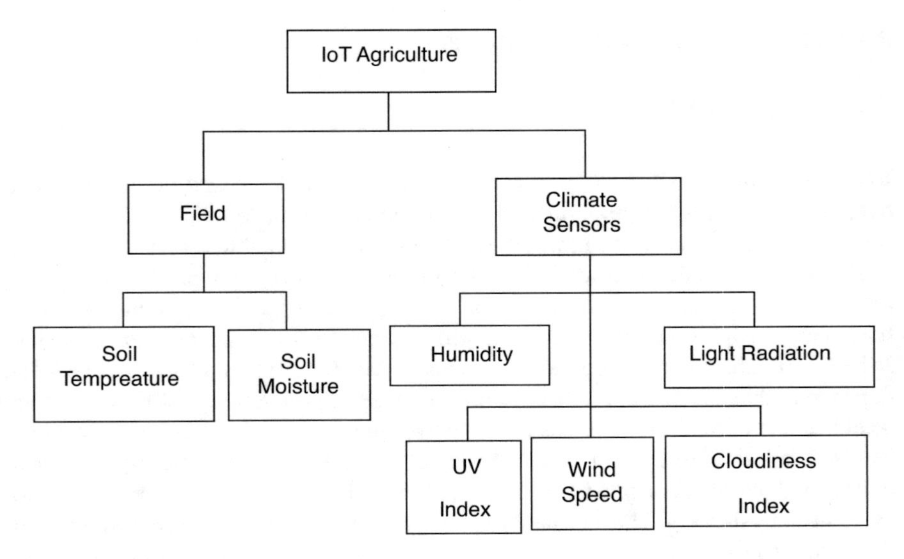

Fig. 8 Classification of IoT agriculture sensors

Fig. 9 Soil temperature sensor

3.1.1.2 Soil Moisture

Soil moisture is the water between the spaces of soil particles; it dissolves minerals and nutrients the plants need and absorb to grow. Soil moisture controls water exchange and heat energy between land surface and atmosphere through evaporation and plant transpiration. Precipitation, temperature, and soil characteristics affect soil moisture. Soil moisture sensors measure the volumetric water content at the root zone in the soil to manage the irrigation systems to use less water. Soil moisture is measured using tensiometers that measure soil moisture tension. Capacitive soil moisture sensor determines the amount moisture in the soil by measuring changes in capacitance to determine the water content of soil. Soil moisture sensors are inserted deep in the soil at the plants' root zone. Figure 10 shows a soil moisture sensor.

Fig. 10 Soil moisture sensor

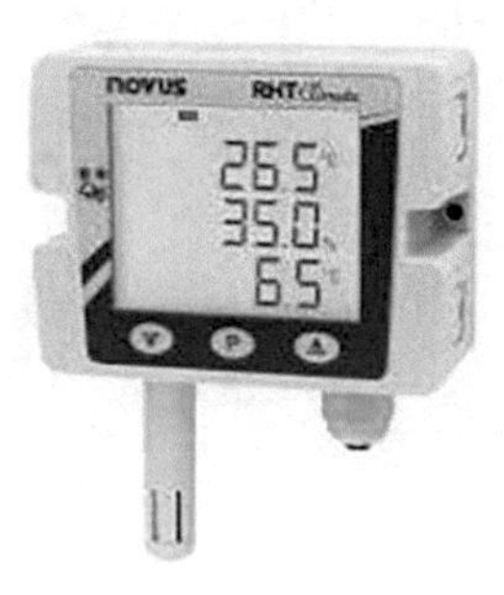

Fig. 11 Humidity sensor

3.1.2 Climate Sensors

3.1.2.1 Humidity

Humidity sensors or hygrometers measure moisture in the air. They combine relative humidity (RH) measurement and air temperature (T) to provide accurate measurement of dew point and absolute humidity (AH). Relative humidity is an important factor for comfort, it measures the ratio of moisture in the air to the highest amount of moisture at a particular air temperature. Humid air is subject to less daily temperature variation than dry air because humid air takes longer to heat up and cool off. Under high humidity maximum shade temperature rarely exceeds 38 °C, while under dry condition, a maximum of 54 °C is possible. Minute changes in the atmosphere are monitored using humidity sensors to calculate humidity in the air. Humidity sensors are placed in home heating, ventilating, and air conditioning systems. In addition, they are also used in offices, cars, industrial spaces, museums, greenhouses, and meteorology stations to forecast weather. Figure 11 shows a humidity sensor.

3.1.2.2 Ultraviolet (UV) Index

UV index sensor provides an accurate measurement of the ultraviolet radiation index (UVI) from sunlight. UV radiation boosts the intensity of photosynthesis processes and facilitates the production of chlorophyll and nutrient which strengthens the

Fig. 12 Ultraviolet sensor

Fig. 13 Photo-resistor sensor

plants. UV affects the life cycle of plants as it can speed up the germination process for starting seeds when grown indoors. UV sensors help identify risks associated with different levels of UV exposure. UV index may be different from one place to another; it is affected by a number of factors including time of day, cloud cover, altitude, and more. Monitoring UV radiation in agricultural field allows farmers to take better precautions to improve the growth of agricultural crops and increase productivity. Figure 12 shows an ultraviolet sensor.

3.1.2.3 Light Radiation

A light sensor measures the radiant energy or illuminance that exists in a very narrow range of frequencies basically called "light." It outputs an electric signal which indicates the intensity of daylight or artificial light. Illuminance decreases as the distance from light source increases, so light sensors can be used to gauge relative distance from the source. Photo-resistors, photodiodes, and phototransistor are types of light sensors. Figure 13 shows a photo-resistor sensor.

3.1.2.4 Cloudiness Index

Cloudiness or cloud cover index refers to the part of the sky covered by clouds when observed from a specific location. Sunshine duration is inversely proportion to cloud cover, i.e., the least cloudy locales are the sunniest ones and vice versa. Variations in daily temperature are affected by cloud cover buffering which lowers

Fig. 14 Ceilometer

the daytime high but raises the nighttime low. Growth and yields are adversely affected by high daytime temperatures which cause pollen sterility and blossom drop, while hot nights can reduce crop yields. Cloudiness is estimated in terms of how many eighths of the sky are covered in cloud. This measure rates from 0 Oktas which is complete clear sky to 8 Oktas which is completely overcast. Figure 14 shows a ceilometer.

3.2 Oil and Gas Mining

Smart personal protective equipment (PPE) embeds tracking devices, sensors, and monitors in clothes worn by workers who may face specific hazards because of their working environment. PPEs provide better safety and long-term cost savings in mining fields through early prevention of health issues, hazardous situations, or the exposure to danger zones. Data collected by these sensors and monitors is analyzed to provide insights on any harms a worker may be susceptible to. PPE sensors can be infused or interwoven in the fabric so that they cannot be removed by laundering or they have built-in electronic devices. In oil and field mining, these PPEs monitor worker's temperature to ensure they are not overheating or developing hypothermia and monitor air toxicity in mine fields to provide real-time safety for the workers. Personal protective equipment includes items such as safety glasses, gloves, earplugs, shoes, muffs, respirators, hard hats or coveralls, vests, and full body suits.

3.2.1 Smart Pipelines

Engineers and operators of oil and gas pipelines must ensure optimal performance of pipelines for continuous flow operations 24 $\times$ 7. We need to ensure cost-effective maintenance of old pipelines network that covers thousands of kilometers across international borders to mitigate risk of flow disruption. Old pipelines operate inefficiently, are vulnerable to damages due to environmental reasons, and require more labor to inspect each kilometer of the pipeline on regular basis that make the cost to maintain high. IoT sensors, meters, and diagnostic devices such as lasers and ultrasonic and acoustic sensors can contribute to networks of smart pipelines. They collect data on pressure, flow, and compressor conditions and report movement, corrosion, leakage, or impact to the pipelines. Data collected from pipelines are analyzed using advanced analytical tools to enhance the existing engineering capabilities of the operating teams and allow them to formulate maintenance and repair options on-demand, with the help of potential risk probability and impact calculators. Accurate and timely information allows the operating teams to be more proactive and ensure better pipeline management.

3.2.2 Lone Worker Monitoring

IoT lone worker tracking system offers an easy way to monitor workers in hazardous environment to ensure safety of the workers, tracking them via GPS to show where they are and to which direction they are heading. Moreover, workers can report emergency situations and talk to emergency professionals.

3.2.3 Safety and Security

Safety is the set of protective measures that must be taken to detect risks, assess safety, and prevent accidents at the workplace. Safety IoT devices collect data that is required to assess health hazards in the work field and identify, evaluate, and prioritize risks. Safety IoT devices include systems like door locks, surveillance cameras, smart safes, access control systems, fire alarm systems, and similar devices mostly used to secure a location or prevent a hazard.

3.3 Wearables

IoT wearables, also known as “wearables,” is a category of network-connected devices that can be implanted in the user’s body, embedded in clothing, tattooed on the skin, or worn as accessories. Wearables are disruptive technologies that can collect data, track activities, and customize experiences to users’ requirements and desires. Wearable technologies are growing fast and expect to influence our social, economic, and legal norms. Wearable devices track employees’ performance and monitor their health and location inside the premises. Besides, the wearables can also report collisions and falls, thereby improving the safety of the operations. Fitness trackers and body-mounted sensors such as accelerometers, gyroscopes,

Fig. 15 Fitness tracker

Fig. 16 Smartwatch

magnetic sensors, and their combinations, smart jewelries and smartwatches, are the prevailing trends in today's wearable market.

3.3.1 Fitness Trackers

They are devices that are worn typically as wristbands to measure vital parameters such as heart beat rate and monitor the number of steps taken each day, how long you spent sleeping in each sleep stage, how much calories you have burned, and how long you spent working out. Gathered data can be sent to your smartphone to help you monitor changes over time. Some trackers have GPS to track user's locations during running, biking, or walking especially for old people who may have Alzheimer's disease. Figure 15 shows a typical fitness tracker.

3.3.2 Smartwatches

They are watches with computer capabilities such as calculations, translations, digital media playing, and game playing. They have mobile operating systems with apps similar to mobile apps that include digital maps, schedulers, calculators, personal organizers, Wi-Fi, and Bluetooth connectivity. Figure 16 shows a typical smart watch.

Fig. 17 Tri-axial accelerometer

3.3.3 Accelerometers

They are devices that can measure the acceleration of an object. Figure 17 shows a tri-axial accelerometer, which provides simultaneous measurements in three orthogonal axes x, y, and z to analyze all vibrations being experienced by a structure. Tri-axial accelerometers have three crystals that are placed so that each one reacts to vibration in a different axis. The output has three signals; each one represents the vibration for one of the three axes. In IoT, tri-axial accelerometers detect shake, orientation, tap, double tap, tilt, fall, positioning, motion, shock, or vibration.

3.4 Smart Cities

The main components of smart cities are data collection, data exchange, data storage, and data analysis [13]. A huge amount of information is collected daily through millions of connected IoT devices such as sensors and meters that collect and analyze data to better monitor and use the infrastructure and improve maintenance. IoT technology can develop efficient methods to minimize operational costs of public utilities and services, manage traffic, reduce pollution, maintain people safety and city cleanness, increase productivity, and allow quick response to people's needs [14]. Collected data is transmitted to the cloud through Wi-Fi networks, 4G or 5G technologies for storage. Data is stored in the cloud in formats that make it usable for data analysis. In the data analysis stage patterns are extracted and inferences are obtained from the stored data to guide decision-makers. Simple analysis for basic decision-making could be enough, while more detailed and deep analysis for heterogeneous data is required for more complex decisions. Smart city contains smart homes, smart schools, smart hospitals, smart power plants, smart wastewater management, smart agriculture, smart transportation, smart health, smart environment, smart governance, and more. IoT technology uses smart device and sensor networks to collect and analyze data to eliminate risks, avoid damages, and reduce costs. Traffic authorities can use this information to build safe road strategies and predict the outcomes and efficacy of specific measures and precautions that promote optimal safety. IoT technologies collect data and provide

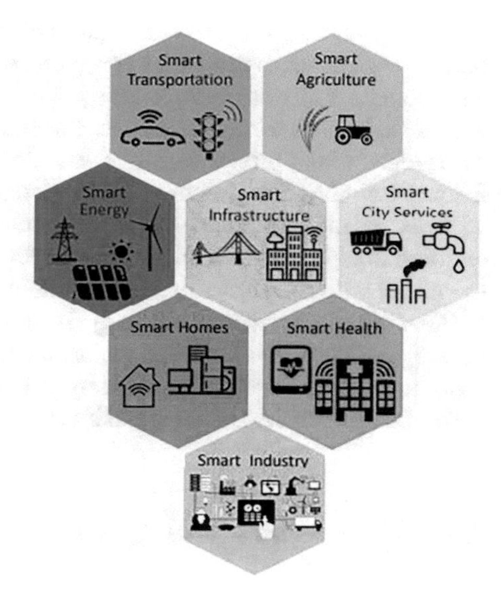

Fig. 18 Smart city components

deep insight on what drivers do to hold them accountable and encourage them to adopt safer habits to reduce road accidents and protect them from collisions and casualties that can happen. Figure 18 shows smart city components [13].

3.4.1 Environmental Monitoring

IoT sensors are deployed at many points in the cities to collect accurate data on the environment to guide us on how to interact with the environment and put plans to improve the quality of life in cities. Major applications of IoT in environmental monitoring include weather monitoring, endangered species protection, water quality, air quality, waste monitoring, and more.

3.5 Smart Buildings

They are digitization of buildings to provide people living in buildings a safe, efficient, comfortable, and convenient environment. Billions of devices are now installed and connected all over the world, thus enabling smart buildings to communicate with their owners, tenants, occupants, and maintenance teams. Smart buildings use IoT sensors to monitor, maintain, and control everything in the building such as lighting, humidity, occupancy, smart elevators, ventilation, shading, security, Co_2 monitoring to identify poorly ventilated areas in the building, and more. Smart buildings use integrated systems to share, exchange information to facilitate collaboration in order to manage resources in cooperative way to improve building efficiency, optimize resource use, enhance security, reduce operating costs, and monitor and troubleshoot easily. Monitoring smart building 24 × 7 registers events such as abnormal activities, fire breakouts, or security breaches and helps management take proper care of the building for now and in the future. To build a smart building, the property owner should consider having a powerful wireless networking infrastructure. Distributed antenna system is the key component of building IoT systems as it allows emergency responders such as police or firefighters to interact with each other in case of building fires, earthquakes, or natural disasters. Cellular phone networks enhance the mobility of cellular devices to increase coverage of the whole building. A data analytics software can help the management team understand the data collected by IoT sensors and let them be more flexible to make the right decisions in certain conditions and cost constraints.

3.6 Maintenance Management

IoT-based predictive maintenance keeps track of the operating conditions of equipment and machines which make it easier and more efficient to monitor, maintain, and optimize asset utilization for better availability and performance especially in remote locations. Attaching IoT sensors to assets and service items excludes human errors and unnecessary visits to remote locations, ensure accuracy and availability of usable data, and gain better visibility into assets through real-time monitoring and receiving automatic alerts, notifications, and reports on time between failures, when operating conditions are out of specification, mean time to repair, and key performance indicators. Instead of waiting for a failure, technicians and mechanics can see equipment failures in real time during the breakdown and determine the object's exact location that needs maintenance; this helps technicians to predict machine failure and identify which parts to be replaced. Technicians can receive problem description, list of spare parts needed to fix the asset, options for repairs, and recommended actions to take which make them effective decision-makers. Maintenance will only be performed if it is required, thus reducing the costs of labor and spare parts. This will empower managers to keep productivity at maximum and the cost of repairs and downtime at minimum. Predictive maintenance reports contain failure data, system operating conditions at the time of failure in addition to previous repair data from the enterprise asset management (EAM). These allow manufacturers improve the quality of their products, optimize spare parts stocks, reduce downtime, control maintenance budgets, and increase customers' satisfaction.

3.7 Water Supply

IoT water supply systems monitor water quality in real time, conserve water supplies, and enable cities to function efficiently. An IoT smart water sensor tracks water quantity in the storage reservoir to turn on the water pump to refill the reservoir and switch off the water pump when it reaches the maximum level. IoT technology monitors water flow across the building to optimize water distribution among tenants and monitors water pressure to detect water leakage or wear of water pipes or equipment to reduce water wastage and maintain acceptable water pressure [15]. IoT smart water sensors track water quality by measuring the physical and chemical properties of the water such as temperature, pH, and turbidity [16]. Managers at different points of water supply chain use data collected by IoT water sensors to receive key insights into the changing conditions of water resources and equipment and become able to take on-demand data-driven corrective measures.

3.8 Manufacturing

IoT technology automatically connects machines, tools, and sensors on floor of the factory to provide production engineers and managers with the information they produce to monitor equipment and track parts in real time during the assembly and supply chain processes. With a granular visibility into the production process, managers can make more informed and smarter decisions to ensure reliability, compliance, and safety to optimize productivity. IoT technology provides data on how products are used that could be fed back in real time to manufacturers who can iteratively correct and rapidly design improvement, which improves prediction of demand, enables faster time to market, and enhances customer satisfaction. Implementing IoT in manufacture allows for more efficient energy saving as sensors can help managers determine places of waste, boost equipment efficiency, predict failures, and detect issues of compliance with quality assurance. Equipment failure is the main reason for poor production and poor-quality products, which result in more sales return, poor customer satisfaction, and reduced customers' trust in the products and finally damage the brand reputation. Moreover, repairing defective products consumes more resources and increases the production costs. High-quality products reduce costs and wastes, enhance customer experience, and increase product sales. Maintenance can be done based on machine needs at an exact time not on historical data or guessing because manufacturers may not know about machine faults which may cause production problems.

3.9 Transportation

IoT-based transportation has improved the conventional operations of transportation through embedding sensors, actuators, and other IoT devices. IoT devices collect data about the environment and transmit it for predictive analysis to make decisions within real time. A telematics device is an instrument that can be installed in a vehicle to record accurate up-to-date real-time information about location, idling time, tire pressure, fuel consumption that has better impact on the environment, vehicle activity, and driver behavior including driving style, alerts for harsh acceleration, how fast you brake, and the distance you drive. With telematics data, the operation managers can ensure that a vehicle is on its route. The managers can take adjustment actions if a vehicle has drifted from the optimal route based on a specific threshold. IoT-based systems help managers to better plan for journeys, monitor traffic congestion and vehicle's load, react quickly to traffic accidents, and improve safety by tracking vehicle location in case the vehicle has been stolen. Traffic management is a main application of IoT-based transportation. Traffic management includes smart parking, traffic lights, and smart accident assistance. An IoT-based smart parking system sends data through web/mobile application about free and occupied parking places. An IoT-based real-time traffic monitoring system dynamically handles traffic signals based on traffic density. Smart accident assistance automatically detects an *accident* and notify the nearest emergency unit. IoT technology facilitates tolling and ticketing processes where modern vehicles are equipped with IoT devices and can be sensed a kilometer away from the tolling station, which is correctly identified, and the barrier lifted for the vehicle to pass through. Smart vehicles are connected to the Internet and communicate with each other to prevent collisions and allow smooth traffic. In public transport management, IoT smart transportation can be widely used in automated ticketing and fare collection. It helps public transport operators and transit agencies to monitor vehicles routes, waiting times, and schedules, estimate the overall fleet performance, and provide tools to analyze and interpret the real-time data collected over short time periods. The IoT technology allows public transport systems to better serve its customers and create better customer satisfaction that leads to increased ridership. IoT devices installed in the buses of a fleet provide passengers in digital bus stops with accurate real-time arrival information to decrease passengers' average waiting time. IoT technologies allow for better communication with passengers through text messages on their mobile phones, which increases customers' satisfaction. Transit agencies can monitor passengers' behavior and travel patterns and send them personalized information on their mobile phones with updates on routes, closure of stations re-routing of buses, or delays. Figure 19 shows a vehicle OBD II Dongle telematics device with a vehicle GPS tracking.

Fig. 19 Telematics device

3.10 *Warehouses*

Smart warehousing is essential for the profitability of any business as it allows the company to optimize its operations and stay competitive in hard competition markets and volatile global economy. The management should adopt IoT technologies to determine the best layout and configuration of the warehouse to ensure optimum utilization of storage space to maintain a seamless workflow at its fullest efficiency and improve movement and fulfillment of goods through the warehouse. IoT smart warehouse wastes no resources and provides visibility into the flow of outgoing and incoming supply chain. Connected sensors track materials from ordering until the shipment reaches the end customer or third-party logistics (3PL) warehouses. IoT-based warehouse tracks the equipment and products more quickly and accurately, thus making the movement of products faster, and tracks the quantity and quality of goods by monitoring the temperature and location of the cargo within the warehouse in real time, which reduces food spoilage and results in increased profits and reduces management costs. IoT can help the warehouse management to calculate time, infrastructure, and budget needed to scale up the warehouse storage space using data collected by IoT devices. IoT-based warehouse collects inventory data to forecast the workload based on seasonal changes of demand and broadcast inventory information to warehouse managers to inform them of low stock, displaced products, unsuitable temperature, theft, and more. IoT robots in smart warehouses move independently and utilize sensors and cameras to help humans pick and pack products faster. IoT devices continuously run without feeling tired, which eliminates fatal human errors and reduces operation costs.

4 IoT Security and Privacy Issues

IoT devices connecting over the Internet are growing exponentially which causes a wide variety of potential concerns that relate to security and privacy. Adding more IoT devices to the Internet increases the vulnerability of connected IoT systems, which makes security needs in the heart of any decision to adopt IoT technology. IoT devices include surveillance cameras, drones, home appliances, smart home devices, monitors, sensor networks that can transmit data, smart toys, routers, and

Internet gateways. There are potential drawbacks to use IoT technology because most of these devices are developed with little attention to data protection and access control has opened gates for the hackers and attackers to invade IoT devices. Cyber-criminals are targeting IoT devices; file-less malware is the most obvious attack on IoT devices. It does not rely on physical files that can be transferred and stored on a victim machine that allows it to evade antivirus software. This means that it leaves little evidence behind it and can only be detected by sophisticated security applications. File-less attacks are spread via botnet that detects vulnerable applications. Rule-based detection can be able to detect malicious execution of commands. Machine learning techniques are now widely spread to detect file-less malware by studying the behavior of malwares [17]. In healthcare, unapproved access to IoT wearable devices can cause changes in the data collected by these devices, which may put the life of a patient at risk. In smart cities, safety penetrations IoT devices are risky and could pose risks on individuals' safety. In agriculture, safety attack on sensors can cause change in reading important information such as soil humidity, temperature, and more, which affect crop production. Hackers focused their efforts last year with the following tools.

4.1 IoT Malware

IoT malware families such as Aidra, Bashlite, and Mirai scan the ports of IoT to locate exposed ports and acquire default credentials on these devices to launch distributed denial-of-service (DDoS) attacks or gain access to IoT devices [18, 19]. IoT malware detection is either non-graph-based or graph-based detection methods. Non-graph-based methods classify a binary file as malicious or benign by extracting static features which are either high-level or low-level features. High-level features include operation code, which is a single instruction executed by the CPU and describes the behavior of an executable file. Strings are usually a sequence of characters stored in ASCII or Unicode format in an executable file. Printable strings contain valuable information such as IP address, URL, etc. to determine whether an executable file is malicious or not [20]. Low-level features include elf (executable and linkable format) file header which contains important information for malware detection and grayscale image where each executable file is converted to binary strings and combined into 8-bit vectors that represent hex value from 00 to FF [19]. Graph-based IoT malware detection methods include control flow graph (CFG) which is a directed graph that represents all possible execution paths of a program. Each block is represented by a vertex and each edge represents the control flow between basic blocks. In [21] the control flow graph (CFG) is used to demonstrate the differences and similarities between IoT malware and Android malware. Control flow graph (CFG) methods achieved 99.66% in detecting IoT malware using a dataset of 6000 malware and benign samples. Federated learning is used to detect malware affecting IoT devices. In [22] a deep learning method is proposed to detect the Internet of Battlefield Things (IoBT) malware and achieved 98.37% accuracy rate and 98.59% precision rate. In [23] a framework for IoT malware detection is proposed that employs federated learning to train and evaluate supervised and unsupervised models without sharing critical data. This framework consists of a client side that is deployed on the RAN SLICING Edge Nodes or in the CLOUD SLICING Fog Nodes and a server side that is deployed in fog/cloud. The results showed that a lot of research is required to reach satisfying results.

4.2 Encrypted Threats

It is an encryption malware that helps attackers escape the secure socket layer (SSL) protocol and invade IoT networks with the intention of stealing data. Ransomware is an example of encryption malware that uses encryption to encrypt victim's information at ransom. Attackers encrypt a user's or organization's critical information so that they cannot access databases, applications, or files until a demanded ransom is paid. If attackers received the demanded ransom, information will be decrypted and owners can get use of it. Ransomware spreads quickly over networks and copies itself on other servers to encrypt files and databases, thus paralyzing entire organization. Ransomware threat is growing very fast generating very big revenue for hackers and paralyzing entire organizations by encrypting their critical information, thus causing a major damage [24]. IoT ransomware is capable of paralyzing the entire network of physical devices by controlling IoT devices, hitting on all IoT security aspects including authentication, integrity, and availability which cause financial losses and could be life threatening. Attacks occur on real-time IoT devices such as healthcare devices, smart vehicles, autopilots, etc. Attacks on these IoT devices are launched from multiple devices because they do not own user interfaces [25]. Botnet, malvertisement, or social engineering are the major methods for ransomware penetration [26]:

(i) Botnet: In any IoT network, botnets are incubators of all IoT malware including IoT ransomware. IoT malware penetrates the IoT network with the help of botnets, which may cause distributed DoS attacks or flooding attacks [27].
(ii) Social engineering: An attacker deceives the IoT network by acting as authorized users, to gain access to the IoT networks and steal sensitive information.
(iii) Malvertisement: Content delivery network (CDN) that appears to be benign can broadcast malware to be installed on IoT devices.

4.3 Perception Layer

It is responsible for data acquisition, so it is also called a sensor layer. This layer is vulnerable against many security attacks.

4.3.1 Node Capture

It is a serious attack against user authentication schemes through which an intruder can gain an unauthorized access of the IoT network. The attacker can perform various operations on the network such as modifying the memory content, modifying computation, or forging messages sent by the gateway to legitimate users and gain additional knowledge by interacting with the captured slave node to reveal cryptographic keys and try to break security. A captured node can make arbitrary queries on behalf of the attacker such as denial-of-service (DoS) attack against availability [28].

4.3.2 Replay Attack

It is a form of network attack where an attacker eavesdrops on a secure network communication to intercept it. The attacker can delay valid data maliciously or captures it and resends it to mislead the receiver as the messages or data appear authentic. The receiver will do what the attacker wants. The IoT replay attack reproduces a signal to control an IoT device to make spoofing and launch DoS attack.

4.3.3 Malicious Node

An intruder can add a malicious or infected node to existing network. Most malicious nodes can launch different attacks based on tampering, retransmission, and discarding methods [29]. Figure 20 shows an invisible node attack launched by malicious node C located between two legitimate nodes A and B that are indirectly connected. The malicious node C repeats the signals and messages between A and B to make them think they are directly connected. This way, malicious node C impersonates node A to node B and vice versa. Figure 21 shows a stolen identity attack launched by malicious node C which can steal authentication credentials such as cryptographic keys from node A. If malicious node C outraces legitimate node in updating the stolen credentials, then the credentials of the legitimate node will not be valid anymore. If a malicious node controls the legitimate node, it can abuse the trust relationship built with other legitimate nodes [30].

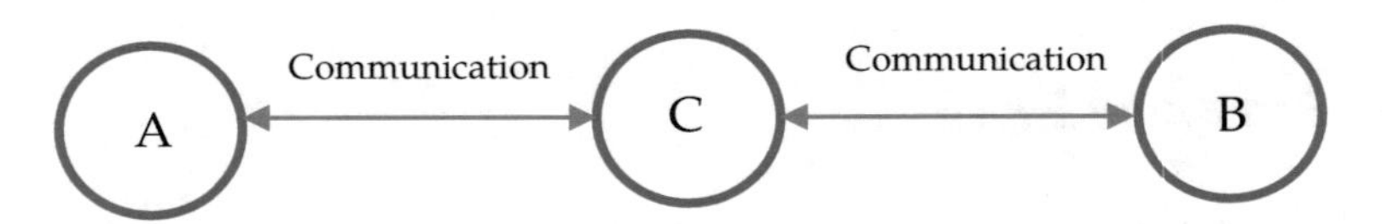

Fig. 20 Invisible node attack [30]

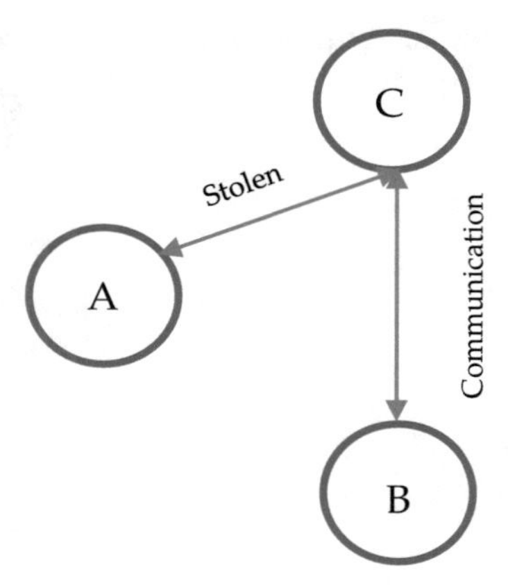

Fig. 21 Stolen identity attack [30]

4.4 Network Layer

This is a transmission layer that performs networking and routing by handling various networking devices. As it carries a large amount of information, it is a target for several types of attacks which may cause network congestion. The main types of attacks a network layer is susceptible to are authentication and integrity attacks [31]. The most common attacks on the network layer are:

4.4.1 DDoS (Distributed Denial-of-Service) Attack

It expends network resources that make services unavailable for actual users [32]. This attack attempts to prevent users from accessing the network services such as emails, portals, or other resources such as printers and storage devices. It floods the network with huge number of redundant traffic to make its resources unavailable, slowing down performance or even crashing the system [31].

4.4.2 Man-in-the-Middle Attack

In this attack, an intruder comes between the sender and receiver to intercept exchanged communication to steal personal information or impersonate both parties, which creates a real threat to confidentiality and integrity [33]. For example, an attacker can intentionally change the temperature recorded by an IoT sensor, malfunction a working device, and drive the whole process to failure [31].

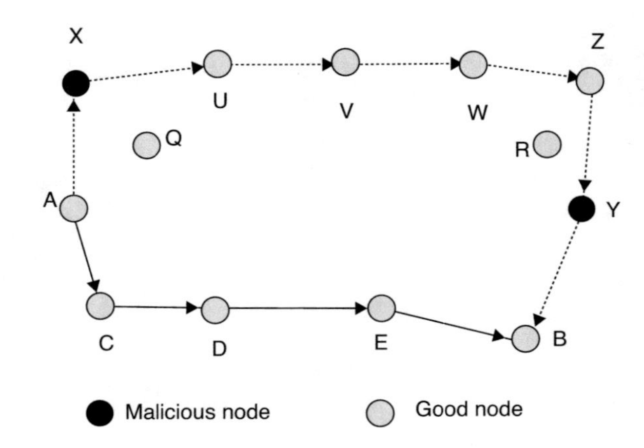

Fig. 22 Wormhole attack [34]

4.4.3 Spoofing Attack

Spoofing is easily launched when security is violated on a shared IoT network where IoT devices are sharing the same network with protected resources. When an IoT device on a shared network is hacked, the intruder can easily get access to the protected resources. Spoofing happens by impersonating an identity of a genuine IoT device using fake Internet protocol (IP), MAC addresses, or both to claim to be another genuine IoT device and gain illegal access to the IoT network. The intruder can then launch denial-of-service attacks or man-in-the-middle attacks to steal credential information and take control of genuine devices.

4.4.4 Wormhole Attack

It is one of the most challenging and severe internal attacks on IoT routing. It is very effective in attacking any protocol even with encrypted traffic [34]. Wormhole attack can insert information in wrong routes or topology information to make other nodes in the network believe they are closer to other nodes, which may be not true and can cause problems to the routing algorithm. Figure 22 shows a wormhole attack that forwards data through a tunnel from a compromised node to another malicious node at the other end of the network.

In Fig. 22 nodes *X* and *Y* are malicious nodes that formed a tunnel from *X* to *Y* to exchange packets. *X* falsely advertises that to reach *B*, a shortest path is through *X* but physically *B* is at far distance from node *X*. So, if node *A* wants to send a packet to node *B* on the other side of the network, the packet takes more time to reach the destination which is considered one of the wormhole attack symptoms [34].

4.4.5 Black Hole Attack or Drop Attack

It is a denial-of-service attack where an aggressive node displays itself as having the shortest route to the destination node using its routing protocol. The malicious node replies to the route requests before any actual node replies, thus creating a fake route. The malicious node acts as black hole and it intercepts the packets and discards them instead of relaying them, disrupting the communication between the nodes of the network without their knowledge [35]. The malicious node can launch this attack randomly or against a particular node at specific dates and times.

4.4.6 Sybil Attack

This attack is very destructive to sensor networks because a malicious node tries to gain illegal influence on the network by creating multiple fake identities that appear to be a real and unique identity to the outside. This malicious node can change the information reaching other nodes, generate false reports, and send spam messages. There are two types of Sybil attacks, namely, direct and indirect. In direct Sybil attack, honest nodes directly influenced a Sybil node, while in the indirect Sybil attacks, the nodes that directly communicate with Sybil nodes influence the honest nodes.

4.4.7 Sinkhole Attack

It is a routing attack in IoT networks. An attacker compromises a node in the network to launch attacks. This node tries to attract all the traffic from neighbor nodes by advertising fake routing update. Examples of sinkhole attacks are selective forwarding attack, acknowledging spoofing attack, dropping or altering routing information, and sending bogus information to base station [36]. In Fig. 23, node *M* launches sinkhole attack in tiny AODV. Node *A* sends RREQ to nodes *B*,*C*, and *M*. However, node *M* instead of broadcasting to node *E* just as nodes *B* and *C* do to node *D* replies back RREP to node *A*. Then node *A* will reject nodes *B* and *C* and forward packets to *M* because nodes *A* and *B* are very far to node *F* than node *M*.

4.4.8 Malicious Code Injection

It is the oldest known web application attack vector; SQL injection is one of these attacks. An attacker gains control of a working node in a network by injecting it with a malicious code. Firstly, the attacker probes the application that can accept untrusted data. By exploiting data input vulnerabilities such as data format, allowed number of characters, and amount of expected data, the attacker can launch denial-of-service attacks, resulting in loss of data integrity and loss of data, and compromise or even shut down the whole network.

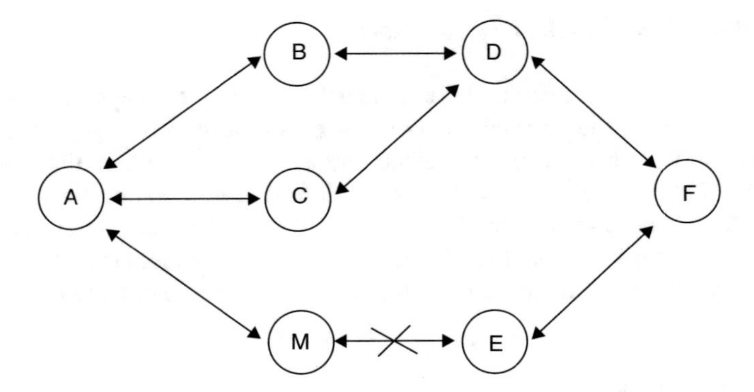

Fig. 23 Sinkhole in tiny AODV protocol (Teng and Zhang, [37]

4.5 Application Layer

4.5.1 Cross-Site Scripting

This is a dangerous injection attack where the attacker can change the content of the application [38]. In cross-site scripting (XSS) attacks, malicious scripts are injected into benign and trusted websites. The browser under attack could not know that the injected script must not be trusted and will execute it. Accordingly, the malicious script can gain access to cookies, session keys, or other sensitive information, redirecting the victim to web content controlled by the attacker, or performing other malicious operations on the victim's machine.

Cross-site scripting (XSS) attacks occur when [39]:

1. Input data enters into a web application through untrusted sources, most frequently a web request.
2. Data is within a dynamic content that is sent to a web user without being checked for malicious content.

There are three types of cross-site scripting (XSS) attacks: stored, blind, and reflected. In stored attacks the injected script is permanently stored on the target servers. The victim retrieves the malicious script when it requests the information stored on the server. Blind cross-scripting attacks occur when the attacker's malicious script is saved on the server. The attacker submits a feedback form that contains the malicious script. When the server admin opens the attacker's submitted form, the attacker's script is executed. Reflected attacks are sent to victims through different routes such emails or another website. When the victim clicks on a malicious link, browses a malicious site, or submits a specially crafted form, the injected code transfers to the vulnerable website, which reflects the attack back to the victim's browser. The victim's browser then executes the transferred malicious code.

4.5.2 Privacy and Confidentiality

It is an important issue that must be carefully handled. Unauthorized access of sensitive data is a serious attack. Endless IOT applications of smart health, smart houses, smart farming, self-drive vehicles, intelligent networks, and more require privacy [14]. Botnet attacks are another type of IOT attacks where botnets are used to launch attacks. These attacks may include malicious activities such as data theft, credential theft, unauthorized access, data theft, and distributed denial-of-service (DDoS) attacks. IoT and wearable devices are always sensing, collecting, and communicating data, which challenge the traditional social privacy and legal norms. Privacy and data breach are both significant concerns to most businesses, because it interrupts the work flow, activities, and network services [40]. Wearables generate a massive amount of data that is collected and analyzed and can be shared by several parties without its owner's knowledge. Individuals wearing wearable devices may not have control on these devices and could not approve or reject to whom this data is transferred, with whom it is shared, or how long it will be retained. This data can be sensitive such as medical, fitness, or personal health information that can be used for marketing purposes or used by insurance companies to increase their premiums or by employers for jobs-related issues [41].

5 IoT Data Analytics

Data analytics is an important component of IoT solution. It allows finding patterns, conducting forecasts, and integrating machine learning algorithms and predictive analysis and finds out insights from collected IoT data. IoT data analytics use data analysis tools to analyze huge data volumes generated by connected IoT devices to extract valuable information that can be used to improve processes, operations, and services. Without data analysis the purpose of IoT systems becomes operation automation not operation optimization. This makes data analysis processes be done by human experts which may not be available for most organizations and causes valuable data not to be used. IoT data analytics can detect data trends within collected IoT data. They can highlight expectations and deviations from normal trends or normal performance. The continuous analysis of gathered IoT data provides the management with continuous feedback on equipment performance to ensure that all equipment is running with high efficiency. Data analytic tools do not replace human experts but provide them with more insight into the ways to optimize the efficiency of equipment. Managers can put plans to cut operational costs and maintenance costs as they have the required information to predict costly breakdown of equipment and achieving strategic goals. Decision-makers can be confident that their choices are based on accurate and complete information obtained from real-time reports and alerts. Huge amount of IoT data from multiple disjoint resources that belong to different systems may have different structures that reduce data reconciliation and leads to inaccurate and incomplete data analysis, which is considered a fundamental issue in IoT data analysis. Data not accessible by IoT data analytics tools or inaccurate collected data affect the accuracy of the generated reports.

6 AI-Powered IoT

The main role of IoT devices or sensors is to collect data. The traditional IoT processing sends data to a cloud server for processing to extract valuable information that helps managers have more insight into this data to make the right decisions. AI plays an important role in fast and accurate data analysis to improve the outcomes for users and service providers, maintain confidentiality of data and privacy, and provide security of IoT devices from cyberattacks. AI can create a predictive model which captures the information in your organization which can be used to examine data in real time [42]. The following steps help build a predictive model:

- Define the business to be analyzed, scope of analysis, and the desired output.
- Predictive models intensively depend on data. Streams of data collected or created by IoT sensors or devices that are embedded into machines are communicated to servers to be stored and ready for analysis.
- IoT data may be incomplete or contain noise which requires data preprocessing. Data preprocessing include data cleaning and complete missing data. IoT devices can aggregate and analyze data before it is transmitted to the server for ultimate analysis and action.

However, sending data collected by IoT devices to cloud server for analysis has issues with privacy, security, latency, storage, and efficiency. Data is vulnerable against man-in-the-middle attack where data can be intercepted by malicious people. So, keeping data on the IoT devices for processing improves security and privacy. Most of the data collected by some IoT devices such as surveillance cameras may be useless as nothing could happen most of the days which wastes valuable storage. Therefore, embedding intelligent systems in IoT devices allows these devices to collect data when necessary which reduces the required storage and the amount of data to be transmitted to the cloud server. Data transmission between IoT devices and the cloud server depends on the Internet. Latency occurs because of slow Internet connection where a long time is required to transmit data to the cloud server for analysis and return the output to the receiver. Edge computing is a possible solution for latency where processing of data should take place on the edge device. Edge devices have low memory, limited power, and low computation power; therefore, applications that are lightweight and more computationally efficient are required. Empowering edge devices with lightweight applications allows all data to be processed locally on these devices. With the advent of tiny machine learning (TinyML), it is possible to enable powerful artificial intelligence (AI) algorithms such as image understanding, voice recognition, hand gesture recognition, pose estimation, speech analysis, sequence analysis, and more to run efficiently on

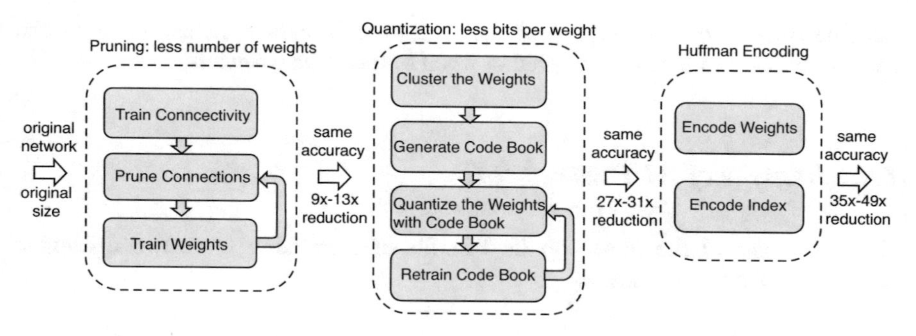

Fig. 24 The three-stage compression pipeline: pruning, quantization, and Huffman coding [43]

low-power mobile and IoT devices. TinyML is a type of machine learning that compresses deep learning networks to embed into tiny hardware such as IoT devices or mobile devices. TinyML algorithms are trained on desktop computers or in cloud servers as traditional machine learning algorithms; post-training is where TinyML starts. Deep compression is the main operation in the post-training phase. Figure 24 shows the three-stage pipeline for network deep compression: network pruning, quantization, and Huffman encoding [43]:

- *Network pruning:* reduces the number of weights by 10×
- Quantization: further improves the compression rate, between 27× and 31×
- *Huffman encoding*: reduces more the network size by storing the data in a maximum efficient way

The IoT devices that are powered with AI bring many benefits to businesses, for example, formal and informal activities to control situations and implement changes, designing and producing services specialized for customer's needs, and requirement and cognitive automation which automates low-level processes without human intervention. IoT-powered AI can discover patterns hidden in collected data which could not appear on devices or sensors, analyze it before transmitting it to other devices, predict risks and automate preventive actions, and determine redundant or time-consuming operations to eliminate or improve them to enhance the efficiency of the system. Machine learning combined with AI can be used to predict outcome and estimate parameters to improve accuracy. Fujitsu analyzes data collected by wearable devices to ensure that safety precautions are implemented by its workers [44]. Google analyzes the data collected by heat sensors in its data centers to reduce costs [45]. People can communicate with IoT devices using natural language processing which helps people to efficiently operate these devices. Robots in manufacturing are empowered with AI-powered sensors that make them more intelligent; self-driving vehicles are the best examples for hybrid IoT and AI as they can predict pedestrians' behavior and other neighboring vehicles. They can learn from each trip and become more intelligent to make appropriate actions. For example, they can predict weather conditions and road circumstances and learn to become smarter to make appropriate actions. In retail industry, smart cameras can observe customers' behavior to predict when to reach the cashiers.

6.1 Benefits of AI-Powered IoT

Here are some of the most popular benefits of combining these two disruptive technologies to the businesses.

6.1.1 Boosting Operational Efficiency

AI in IoT crunches the constant streams of data and detects nondeceptive patterns on simple gauges. In addition, machine learning coupled with AI can predict the operation conditions and detect the parameters to be modified to ensure ideal outcomes. Hence, intelligent IoT offers an insight into which processes are redundant and time-consuming and which tasks can be fine-tuned to enhance efficiency. Google, for example, brings the power of artificial intelligence into IoT to reduce its data center cooling costs [46].

6.1.2 Better Risk Management

Pairing AI with IoT helps businesses to understand as well as predict a broad range of risks and automate for the prompt response. Thereby, it allows them to better handle financial losses, employees' safety, and cyber threats. Fujitsu, for example, ensures worker safety by engaging AI for analyzing data sourced from connected wearable devices.

6.1.3 Triggering New and Enhanced Products and Services

NLP (natural language processing) is getting better at allowing people to communicate with devices. Undeniably, IoT and AI together can directly create new products or enhance existing products and services by enabling the business to rapidly process and analyze the data. Rolls Royce, for example, plans to leverage AI technologies in the implementation of IoT-enabled airplane engine maintenance amenities. Indeed, this approach will support to spot patterns and discover operational insights [47].

6.1.4 Increase IoT Scalability

IoT devices range from mobile devices and high-end computers to low-end sensors. However, the most common IoT ecosystem includes low-end sensors, which offer floods of data. AI-powered IoT ecosystem analyzes and summarizes the data from one device before transferring it to other devices. As such, it reduces large volumes of data to a handy level and allows connecting a large number of IoT devices. This is called scalability.

6.1.5 Eliminates Costly Unplanned Downtime

In some sectors like offshore oil and gas and industrial manufacturing, equipment breakdown can result in costly unplanned downtime. The predictive maintenance with AI-enabled IoT allows you to predict the equipment failure in advance and schedule orderly maintenance procedures. Hence, you can avoid the side effects of downtime. Deloitte, for example, finds the following results with AI and IoT [48]:

- 20–50% reductions in their time invested in maintenance planning
- 10–20% increase in equipment availability and uptime
- 5–10% reduction in maintenance costs

6.1.6 Smart Thermostat

A user can check and manage the temperature from anywhere using an integrated thermostat with a smartphone based on the work schedule and temperature preferences.

Overall, IoT coupled with AI technology can lead the way to the advanced level of solutions and experience. To obtain better value from your network and transform your business, you should integrate AI with incoming data from the IoT devices.

7 Conclusion

All IoT systems architecture use IoT sensors connected to things through a network to collect data. Smart IoT devices can preprocess the data before transferring it to the data center or the cloud for analysis and storage. Every IoT system is composed of the same four components: devices, connectivity, platform, and an application. This chapter discussed the most famous IoT commercial and industrial applications with emphasis on the IoT devices used in each application. As many of IoT devices and sensors are connected and communicate with their data center or cloud using the Internet, the lack of security increases the vulnerability of these devices against cyberattacks. Also, personal information can be leaked using the IoT sensors which violates users' privacy. Hence, IoT security and privacy issues are important issues which were discussed in this chapter. Finally, the conventional integration between AI and IoT devices was discussed where IoT devices or sensors collect data and send it to the cloud for intelligent applications to analyze this data and send the results to decision-makers. The next AI revolution is to embed intelligent applications in the IoT devices that are characterized with low resources so that data processing can be done locally. Therefore, tiny machine learning approach was discussed. Finally, the benefits of AI-powered IoT were presented.

References

1. Costa, B., Pires, P.F., Delicato, F.C., Li, W., Zomaya, A.Y.: Design and Analysis of IoT Applications: A Model-Driven Approach. In: 2016 IEEE 14th Intl Conf on Dependable, Autonomic and Secure Computing, 14th Intl Conf on Pervasive Intelligence and Computing, 2nd Intl Conf on Big Data Intelligence and Computing and Cyber Science and Technology Congress(DASC/PiCom/DataCom/CyberSciTech), Auckland, New Zealand, vol. 2016, pp. 392–399. https://doi.org/10.1109/DASC-PICom-DataCom-CyberSciTec.2016.81
2. Rusu, I.C.A.R.C.: Commercial and Industrial Internet of Things Applications with the Raspberry Pi. Apress (2020)
3. https://www.trialog.com/en/iot-systems-and-interoperability/ (Accessed on 03/14/2022).
4. Jayashankara, M., Udmale, S.S., Pandey, A.K., Singh, R.S.: IoT-based data analytics for the healthcare industry techniques and applications. Intell. Data-Centric Syst., 9–29 (2021)
5. Baker, S., Xiang, W., Atkinson, I.: Things for Smart Healthcare: Technologies, Challenges, and Opportunities. IEEE Access (2018)
6. Naresh, V.S., Pericherla, S.S., Murty, P.S.R., Reddi, S.: Internet of Things in healthcare: Architecture, applications, challenges, and solutions. Comput Syst Sci Eng. **6**, 411–421 (2020)
7. Anand, G., Heuss, L.: Feasibility of breath monitoring in patients undergoing elective colonoscopy under propofol sedation: A single-center pilot study. World J Gastrointest. Endosc. **6** (2016). https://doi.org/10.4253/wjge.v6.i3.82
8. Hashim, N., Norddin, N., Idris, F., Yusoff, S.N.I.M., Zahari, M.: IoT blood pressure monitoring system. Indonesian J. Elect. Eng. Comp. Sci. **19**(3), 1384–1390. ISSN: 2502–4752 (2020). https://doi.org/10.11591/ijeecs.v19.i3.pp1384-1390
9. Nelson, B.D., et al.: Wireless technologies for implantable devices. Sensors (Basel, Switzerland). **20**(16), 4604 (2020). https://doi.org/10.3390/s20164604
10. Ratnaparkhi, S., Khan, S., Arya, C., Khapre, S., Singh, P., Diwakar, M., Shankar, A.: Smart agriculture sensors in IOT: A review. Mater. Today: Proceed.. (In Press)
11. Sushanth, G., Sujatha, S.: IOT based smart agriculture system. IEEE (2018)
12. Caro, F., Sadr, R.: The Internet of Things (IoT) in retail: Bridging supply and demand. Business Horizons. **62**, 47–54 (2019)
13. Syed, A.S., Sierra-Sosa, D., Kumar, A., Elmaghraby, A.: IoT in smart cities: A survey of technologies, practices and challenges. Smart Cities. **4**, 429–475 (2021). https://doi.org/10.3390/smartcities4020024
14. Raghuvanshi, A., Singh, U.K.: Internet of Things for smart cities- security issues and challenges. Mater. Today: Proceed. https://doi.org/10.1016/j.matpr.2020.10.849
15. Natividad, J.G., Palaoag, T.D.: IoT based model for monitoring and controlling water distribution. Int. Conf. Inform. Technol. Digit. Appl., IOP Conf. Series: Mater. Sci. Eng. **482**, 012045. IOP Publishing (2019). https://doi.org/10.1088/1757-899X/482/1/012045
16. Daigavane, V.V., Gaikwad, M.A.: Water quality monitoring system based on IOT. Adv. Wireless Mobile Commun.., ISSN 0973–6972. **10**(5), 1107–1116 (2017)

17. https://csrc.nist.gov/publications/detail/sp/800-61/rev-2/ (Accessed on 07/01/20).
18. Ngo, Q.-D., Nguyen, H.-T., Le, V.-H., Nguyen, D.-H.: A survey of IoT malware and detection methods based on static features. **6**(4), 280–286 (2020)
19. Ngo, Q.-D., Nguyen, H.-T., Le, V.-H., et al.: A survey of IoT malware and detection methods based on static features. ICT Exp. **6**(4), 280–286 (2020)
20. Plu, T.N., Hoang, L.H., Touan, N.N., Tho, N.D., Binh, N.N.: CFDVex: A novel feature extraction method for detecting cross-architecture IoT malware. In: Proceedings of the Tenth International Symposium on Information and Communication Technology, pp. 248–254 (2019)
21. Alasmary, H., et al.: Graph-based comparison of IoT and android malware. In: Proceedings of International Conference on Computational Social Networks, pp. 259–272 (2018)
22. Azmoodeh, A., et al.: Robust malware detection for Internet of (Battlefield) things devices using deep eigenspace learning. IEEE Trans. Sustain. Comput., 88–95 (2018)
23. Rey, V., Sánchez, P.M.S., Celdrán, A.H., Bovet, G.: Federated learning for malware detection in IoT devices. Comp. Netw. **204**(26), 108693 (2022)
24. https://www.trellix.com/en-us/security-awareness/ransomware/what-is-ransomware.html (Accessed on 05/09/2022).
25. Wani, A., Sathiya, R.: Ransomware protection in IoT using software defined networking. Int. J. Elect. Comp. Eng. **10**(3), 3166–3174
26. Bertino, E.: Botnets and internet of things security. Computer. **50**(2), 76–79 (2017)
27. Azmoodeh, A., Dehghantanha, A., Conti, M., Choo, K.K.R.: Detecting crypto-ransomware in IoT networks based on energy consumption footprint. J. Ambient Intell. Human. Comput. **9**(4), 1141–1152 (2018)
28. Butun, I., Osterberg, P., Song, H.: Security of the internet of things: Vulnerabilities, attacks and countermeasures. https://arxiv.org/pdf/1910.13312.pdf
29. Li, B., Ye, R., Gao, G., Liang, R., Liu, W., Ken Cai, E.: A detection mechanism on malicious nodes in IoT. Comp. Commun. **151**(1), 51–59 (2020)
30. Jiang, J., Han, G., Zhu, C., Dong, Y., Zhang, N.: Secure localization in wireless sensor networks: A survey. J. Commun. **6**(6), 460–470 (2011)
31. Deep, S., Zheng, X., Jolfaei, A., Yu, D., Ostovari, P., Bashir, A.K.: A survey of security and privacy issues in the Internet of Things from the layered context. https://arxiv.org/pdf/1903.00846.pdf (Accessed on 05-19-2022).
32. Prabhakar, S.: Network security in digitalization: Attacks and defence. Int. J. Res. Comput. Appl. Robot. **5**, 46–52 (2017)
33. Conti, M., Dragoni, N., Lesyk, V.: A survey of man in the middle attacks. IEEE Commun. Surv. Tutor.
34. Bhosale, S.D., Sonavane, S.S.: Wormhole attack detection in internet of things. Int. J. Eng. Technol. **7**(2.33), 749–751 (2018)
35. Fazeldehkordi, E., Amiri, I.S., Akanbi, O.A.: A study of blackhole attack solutions. Syngress. (2016)
36. Kibirige, G.W., Sanga, C.: A survey on detection of sinkhole attack in wireless sensor network. https://arxiv.org/ftp/arxiv/papers/1505/1505.01941.pdf#:~:text=Sinkhole%20attack%20is%20a%20type,drops%20or%20altered%20 routing%20information.
37. Teng, L., Zhang, Y.: Secure routing algorithm against sinkhole attack for mobile wireless sensor network, in computer modeling and simulation, in proceedings of 2010. ICCMS'10. Second IEEE Int. Conf. Comp. Model. Simul. **4**, 79–82 (2010)
38. Gupta, S., Gupta, B.B.: Cross-site scripting (XSS) attacks and defense mechanisms: Classification and state-of-the-art. Int. J. Syst. Assur. Eng. Manage. **8**, 512–530 (2017). https://doi.org/10.1007/s13198-015-0376-0
39. https://owasp.org/www-community/attacks/xss/ (Accesed on 06/18/2022).
40. Tawalbeh, L.'a., Muheidat, F., Tawalbeh, M., Quwaider, M.: IoT privacy and security: Challenges and solutions. Appl. Sci. **10**, 4102 (2020). https://doi.org/10.3390/app10124102
41. Thierer, A.D.: The internet of things and wearable technology: Addressing privacy and security concerns without derailing innovation. Richmond J. Law Technol. **XXI**(2) (2015)

42. Nelson, J.W. (editor), Jayne, F., Mary, A. H. (Co-editors): Using Predictive Analytics to Improve Healthcare Outcomes, Wiley (2021)
43. Han, S., Mao, H., Dally, W.J., Deep compression: Compressing deep neural networks with pruning, trained quantization and Hufman coding, 4th international conference on learning representations, *ICLR 2016*, San Juan.
44. https://www.fujitsu.com/au/imagesgig5/IoT_solutions_UBIQUITOUSWARE_Digital_ Solutions.pdf (Accessed on 05/06/2022).
45. https://static.googleusercontent.com/media/www.google.com/en//corporate/datacenter/dc-best-practices-google.pdf (Accessed on 05/06/2022).
46. https://www.clariontech.com/blog/ai-and-iot-blended-what-it-is-and-why-it-matters (Accessed on 06/18/2022).
47. https://www.rolls-royce.com/country-sites/sea/discover/2021/tapping-ai-technologies-to-create-solutions-of-tomorrow.aspx (Accessed on 05/06/2022).
48. https://www.clariontech.com/blog/ai-and-iot-blended-what-it-is-and-why-it-matters (Accesed on 06/18/2022).

Hardware and System Security: Attacks and Countermeasures Against Hardware Trojans

Konstantinos Liakos, Georgios Georgakilas, and Fotis Plessas

1 Introduction

Every year, more and more innovative technology-based applications are being developed and implemented on a professional and personal level. A significant percentage of these applications are based on the Internet of Things (IoT) and artificial intelligence (AI). These technologies have a lot of advantages and make our life easier, giving us the ability to automate tasks and to have access to information of our data from any device, anytime and from anywhere. But advantages have disadvantages; a fault in the system can corrupt all the devices, and the devices can increase the potential for a hacking attack both of which enhance the need for more advanced hardware security (HS) systems.

IoT devices consist of complex sophisticated integrated circuits (ICs). These ICs due to their continuous need and in combination with economic reasons are outsourced from design companies for their fabrication to third-party foundry companies. Furthermore, ICs are based on their development of intellectual property (IP) cores from untrusted third-party vendors. These third-party companies are not always trustworthy. All these are responsible for the insertion into ICs of powerful hardware viruses, known as HTs. HT infection is a crucial present problem in the world of electronics, with the potential to spread in the next years, providing a serious threat both technologically and socially.

HTs are related to circuit modifications that occur during the design and/or manufacturing phases. Because of the sophistication of modern circuits, HTs can be added at any stage of IC development and stay idle until activated by a range of

K. Liakos · G. Georgakilas · F. Plessas (✉)
University of Thessaly, Volos, Greece
e-mail: kliakos@e-ce.uth.gr; ggeorgakilas@e-ce.uth.gr; fplessas@e-ce.uth.gr

A. Iranmanesh (ed.), *Frontiers of Quality Electronic Design (QED)*,
https://doi.org/10.1007/978-3-031-16344-9_13

activation mechanisms. HTs are related to unexpected IC faults, circuit damage, and sensitive information loss regardless of encryption level [1].

As a result, establishing well-designed and efficient HS countermeasures against HTs are critical for the development of more dependable and trustworthy integrated circuits and IoT devices.

The question that quickly comes to mind is, who gains from the insertion of HTs into ICs? A rival, for example, may implant an infected circuit into another company's electronic component to discredit it, reducing its market share, consumer confidence, and profitability. Another HT use case is the use of HT cyber warfare to damage military equipment and infrastructure between countries [2].

Ideally, any unwanted alteration applied to an IC should be detected by pre-silicon verification/simulation and post-silicon testing. But the verification or simulation phase of a circuit needs the golden model of it. This may not always be provided, particularly for IP-based designs in which IPs would originate from third-party manufacturers. Furthermore, a complex design is generally never suitable for exhaustive testing. Mainly, ICs design testing at the post-silicon phase can be provided via conventional side-channel analysis (SCA) [3–5] and logic testing (LT) [6, 7] approaches. However, these approaches have limitations and are efficient ware combined. Another method for the verification of the ICs design at the post-silicon phase consists the reverse engineering (RE) [8, 9].

The development of countermeasure methods against HTs consists of a complex and costly process, because of the stealthy nature and the wide variety of possible HT instances that a competitor may use. Ideally, the development of countermeasure methods against HTs must be a combination of the effectiveness against HTs and the cost designers can afford. Specifically, these methods can be used as countermeasures against HTs that may need extra circuitry and/or significant adjustment to post-silicon on the test setup. As a result, the cost of the methods against HTs is a combination of the needed area, test period, and setup charge overhead. The key issue for the approaches against HTs is the search for solutions to minimize this cost.

The purpose of this chapter book is to provide readers with a thorough understanding of what are HTs, their structure, models, and assault categories. The chapter also refers to the techniques used as countermeasures against HTs and demonstrates their evolution over time, through a historical review. Furthermore, there presents an overview of artificial intelligence (AI) and, in particular, AI terminology and definitions, tasks, and learning analyses, as well as the most essential learning models against HTs. Finally, through this book chapter, the readers will be able to learn how to identify an AI problem, create their dataset, proceed to feature selection, select a learning model, and build their own AI models against HTs.

2 IC Supply Chain

To have a thorough grasp of the issue of HTs, the difficulty of preventing its contagion, and the challenges of detecting them, while ensuring the smooth operation of ICs, we must first have a good understanding of the modern circuit production chain and especially the production chain of the application-specific integrated circuits (ASICs). ASIC production chain consists of two stages, pre- and post-silicon stages. The pre-silicon stage is the circuit design period and consists of four steps: register transfer level (RTL), gate-level netlist (GLN), placement and routing (P&R), and graphic design system II (GDSII). The post-silicon stage is the fabrication period of the circuit and consists of the SCA phase. Specifically, the RTL phase describes the specifications that the circuit will have through the usage of a hardware design language (HDL) like Verilog or VHDL. When IC design and integration are completed at RTL, the design must be synthesized to a GLN. GLN is characterized as the logic synthesis phase, and RTL is translated to GLN. The logic synthesis phase is done via professional electronic design automation (EDA) tools, e.g., Cadence Genus Synthesis Solution and Synopses Design Compiler NXT [10]. These tools provide the area, power, and timing analysis of the circuit. Also, the throughput and efficiency of the designed circuit could be derived as well. The last phase is the P&R and is known as the physical design phase where the layout level is created via the GLN and is produced in the final GDSII of the circuit.

HT attacks are divided into four general groups for the pre-silicon stage (Fig. 1), i.e., RTL, GLN, P&R, and GDSII, as well as, fabrication and testing/assembly for the post-silicon stage. In the pre-silicon attacks, the attacker aims to gain full access to the source code, design files, or compromise design tools and scripts, to develop a modified IC representation without altering the source code, while, in post-silicon attacks, the attacker aims to add or remove components from the designed circuit through reverse engineering, modification of the layout geometry, or measurement of the IC (Fig. 1).

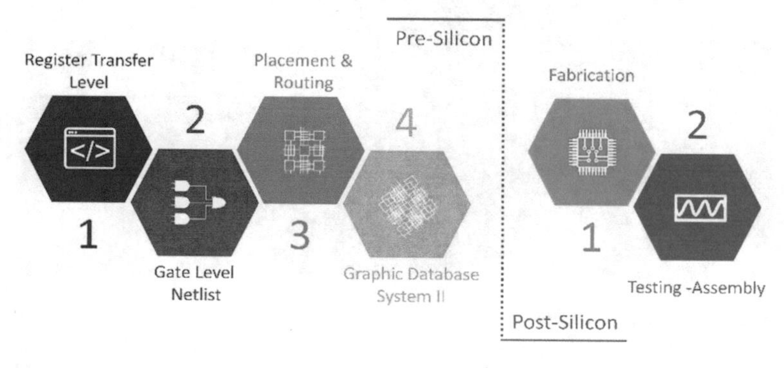

Fig. 1 IC supply chain and HTs insertion in pre- and post-silicon stages

3 HT Structure

The typical structure of an HT has two mechanisms, triggers and payloads (Fig. 2). Triggers are related to rare signals or events [11] and payloads with the activation of malicious functions. The goal of an HT is to remain stealthy – to be undetectable during design simulation or testing and to be activated under rear conditions. An HT is activated when the rare signal or event appears and through the payload mechanism.

4 HT Models

HTs are designed to be undetectable; their structure is consisted of a trigger and a payload mechanism and can be implemented in all pre- and post-silicon phases of the IC production chain. Another characteristic of HTs is their logic models. Logic models are associated with the trigger mechanism and especially how the rare signal or event will activate the trigger mechanism. HTs are designed to have two logic models, a combinational or a sequential [11]. In combinational logic models, the trigger mechanism is activated from a set of simultaneous rare signals or events (Fig. 3a) and in sequential logic models from a series of rare events or signals (Fig. 3b).

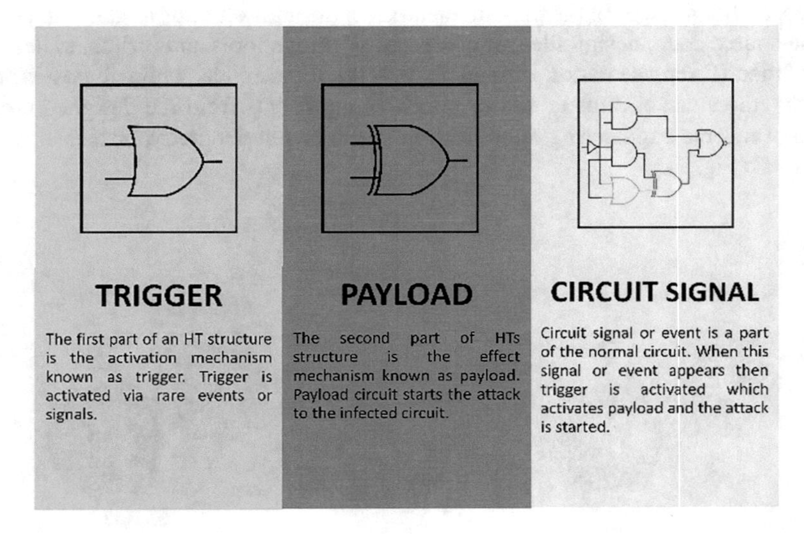

TRIGGER
The first part of an HT structure is the activation mechanism known as trigger. Trigger is activated via rare events or signals.
PAYLOAD
The second part of HTs structure is the effect mechanism known as payload. Payload circuit starts the attack to the infected circuit.
CIRCUIT SIGNAL
Circuit signal or event is a part of the normal circuit. When this signal or event appears then trigger is activated which activates payload and the attack is started.

Fig. 2 Hardware Trojan structure

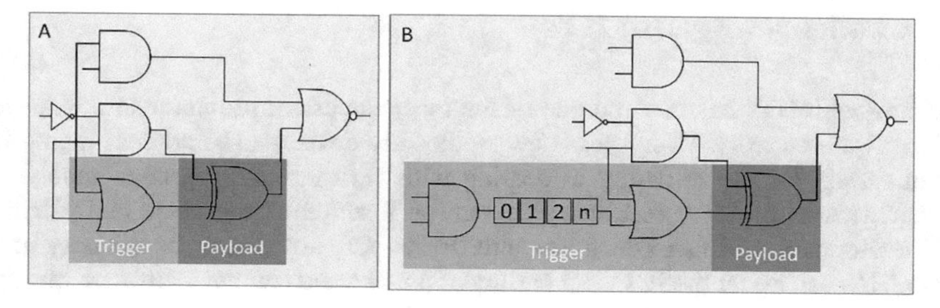

Fig. 3 Concept graph presenting (**a**) combinational and (**b**) sequential model logic

5 HT Attacks

The purpose of HTs is to affect the normal functioning of the infected circuit. Thus, the HT attacks can be divided into two types of attacks: those aimed at destroying the device, known as general-purpose processors attacks, and those aimed at leaking sensitive information, known as cryptographic engine attacks. Cryptographic engine attacks aim at the crypto engine of the infected circuit through various attack mechanisms and leak encrypted information. General-purpose processor attacks aim at the lower levels of the processor, kernel, memory, and secret keys and degrade the system or destroy it completely. These types of HTs can be triggered under rare signals or events and disable the safe start mechanism of the infected circuit [12, 13].

6 HT Taxonomy

There is no formal taxonomy for HTs. Each study has its taxonomy structure. Tehranipoor et al. [14] presented a taxonomy of HTs based on three main characteristics of HTs, action, activation, and, physical. Physical characteristics are considered to be the type, size, or structure of an HT. Activation characteristics are divided into external and internal activation mechanisms of an HT, and action characteristics are considered the types of HT attack on the infected circuit. Karri et al. [15] developed a taxonomy approach for HTs, based on five characteristics: abstraction level, insertion phase, activation mechanism, effect, and localization, while Bhunia et al. [11] developed a taxonomy approach based on trigger and payload mechanisms.

7 Challenges Against HTs

Dealing with HTs has become one of the most important problems in the science of hardware security. Every year new studies are developed to address them. The main reason for the difficulty in dealing with HTs is mainly a large number of different cases of HT infections. HTs can be inserted at any stage and phase of IC development and can attack any unit of the ICs, processors, or memory units. Also, HTs can affect the ICs via a variety of attacks and can have different physical layouts.

8 Structure and Purpose of the Chapter

Next it presented an overview of artificial intelligence (AI), the tasks of learning, and the most significant types of learning models, to be able for the readers to understand the philosophy of AI. In Sect. 3 the approaches as countermeasures against HTs, a historical throwback, the study's trend, and the function of each subcategory and category with tables are presented. Section 4 presents in depth the methodology of how an AI model is "built," and Sect. 5 presents the conclusions of the chapter.

The main purpose of this chapter is for readers to understand the function of HT viruses and to know the basic philosophy of machine learning, because the science of AI is becoming more and more widespread in all research fields, including this field. And according to the research trends, it is increasingly being applied as a solution for dealing with HT viruses. Another aim of the chapter is for readers to understand the main categories of countermeasures used to deal with viruses as well as to learn the fundamental steps of AI, to be able to create their models against HTs.

9 An Overview of Artificial Intelligence

Every year, terms such as AI, machine learning (ML), and deep learning (DL) are increasingly mentioned in our everyday life. This happens because a technology trend is the development and use of AI-based technologies on a professional or personal level. As a result, the meaning of these terms has been lost. It is important to understand that all these terms are part of the AI scientific field.

In this section, a detailed reference is made to the science of AI. Specifically, in this section of the book, the differences between the AI, ML, and DL terms are presented. It also details the learning tasks of AI, such as supervised and unsupervised learning. Furthermore, a plethora of learning models and algorithms are discussed exhaustively. The purpose of this section is for readers to be able to distinguish the differences between the AI, ML, and DL, as well as to comprehend how each learning model works and when their algorithms are applied.

9.1 Artificial Intelligence Term

The term AI-first was introduced in 1956 by John McCarthy through an academic conference. McCarthy defined AI as the science of making intelligent machines. AI can be defined as the scientific field that aims to teach machines to think without the need for human intervention. AI consists of a broad area of computer science and can be categorized into three main categories, AI-narrow, AI-general, and AI-super. AI-narrow is goal-oriented and has been programmed to complete a single task. AI-general allows machines to learn and apply their intelligence to solve any problem by mimicking human intellect and/or behaviors, and in AI-super machines are capable of outperforming even the best humans in terms of intelligence.

9.2 Machine Learning Term

ML term was introduced in 1959 by Samuel et al. [16], and it was defined as the scientific field that allows machines to learn without being strictly programmed. Specifically, ML consists of a subset of AI that uses statistical learning algorithms for the development of smart systems. Without being explicitly programmed, ML-based systems can learn and improve on their own. The ML algorithms can be categorized into three main categories, supervised, unsupervised, and semi-supervised learning.

9.3 Deep Learning Term

DL is a subset of ML techniques utilizing multiple layers of training with more reliable performance and fastest speed. The DL technique was inspired by the way a human brain analyzes information. DL-based systems consist of interrelated layers for the classification or prediction of information. Figure 4 presents in brief the differences between AI, ML, and DL.

10 Tasks of Learning

AI, ML, or DL algorithms can be categorized into three learning task categories, supervised, unsupervised, and semi-supervised learning. The main difference is that

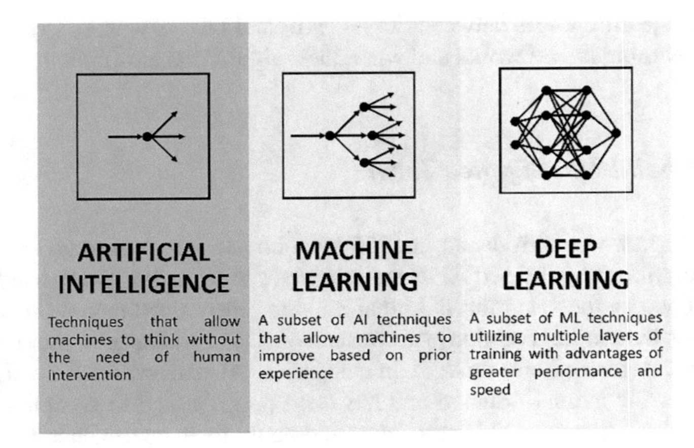

Fig. 4 Artificial intelligence vs machine learning vs deep learning

supervised learning uses labeled data to help in prediction, while unsupervised does not. Semi-supervised learning uses data mixed with labeled and unlabeled examples. However, there are some distinctions between the three techniques, as well as key areas where one surpasses the others. In this section the differences between the three learning tasks are presented.

10.1 Supervised Learning

Supervised learning uses datasets with labeled samples as inputs and outputs for the development of an ML- or DL-based model. Supervised learning can be used as a solution for two categories of problems, classification or regression. In the classification problems, a labeled dataset is split into two sets, the training and test set for the development of a model. The purpose is for the model to be able to classify with high performance the samples of the test set. For example, a classic supervised classification learning problem is the classification of original from spam emails. Furthermore, in regression problems, the model's purpose is to comprehend the connection between the dataset's dependent and independent variables using a labeled dataset. Regression models are useful for predicting numerical values based on various data samples, such as sales revenue estimates for a certain business. Figure 5 presents a typical figure of supervised learning.

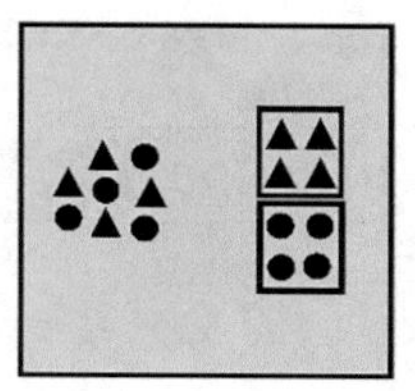

Fig. 5 Supervised learning

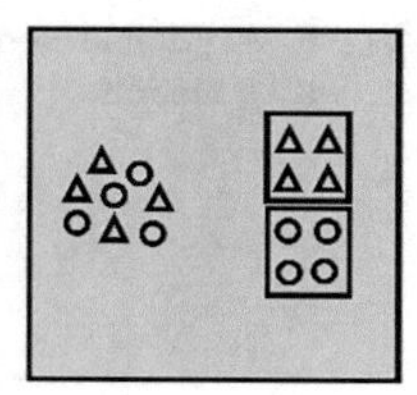

Fig. 6 Unsupervised learning

10.2 Unsupervised Learning

Unsupervised learning uses datasets with unlabeled samples as inputs and outputs for the development of an ML- or DL-based model. In unsupervised learning-based models from the dataset, it derives patterns between the features, and when the model analyzes new data, it can classify the new samples into a class, based on the already learned feature patterns. Unsupervised learning can be used as a solution for clustering or dimensionality reduction problems. In the clustering problems, the aim of the model is via an unlabeled dataset to group the dataset. In dimensionality reduction problems, the aim of the model is to convert the higher-dimension dataset into lesser dimensions without losing information, to reduce the poor performance which is produced from the datasets with a large number of features. Figure 6 presents a typical figure of unsupervised learning.

10.3 Semi-supervised Learning

Semi-supervised learning uses datasets with mixed samples like labeled and unlabeled samples as inputs and outputs for the development of an ML- or DL-based model. There is a desirable prediction problem, but the model must learn the structures to arrange the data and produce predictions. Classification and regression are two common semi-supervised problems. Unsupervised and semi-supervised learning may be more tempting options because relying on domain expertise to label data accurately for supervised learning can be time-consuming and costly. In Fig. 7 a typical figure of semi-unsupervised learning is presented.

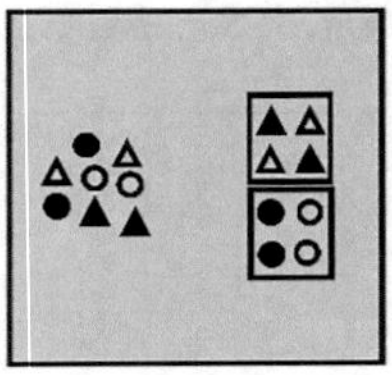

Fig. 7 Semi-supervised learning

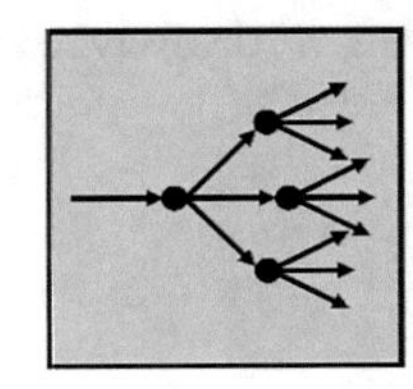

Fig. 8 Artificial neural networks models

11 Learning Models

11.1 Artificial Neural Network Models

The functionality of the human brain inspired the development of artificial neural networks (ANNs). ANNs emulate complicated tasks like recognition, classification, decision-making, and pattern generation [17]. The human brain is made up of billions of neurons that communicate and process information. Based on the same philosophy, an ANN is a simplified model of this structure. Specifically, ANNs consist of three categories of layers: input, hidden, and output layers. Input layers fed the dataset into the system. Hidden layers produce the learning of the model, and the decision/prediction is given from the output layer. ANNs are supervised models which are used to solve regression and classification problems. The most common ANNs-based algorithms are perceptron [18], multilayer perceptron [19], back-propagation [20], resilient back-propagation [21], and counter propagation algorithms [22]. Also, other common ANN algorithms are radial basis function networks [23], Kohonen networks [24], Hopfield networks [25], generalized regression networks [26], autoencoder [27], adaptive-neuro fuzzy inference systems [28], extreme learning machines [29], and self-adaptive evolutionary extreme learning machines [30]. In Fig. 8 a typical structure of an ANN model is presented.

11.2 Bayesian Models

Bayesian models (BM) are a type of probabilistic graphical model where the analysis is carried out using Bayesian inference. This model belongs to the domain of supervised learning and can be used to solve classification or regression problems.

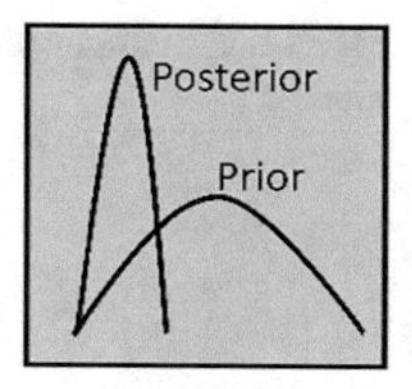

Fig. 9 Bayesian models

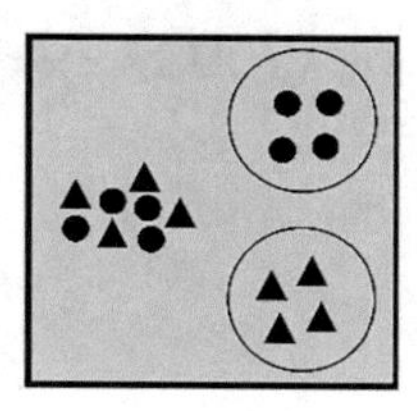

Fig. 10 Clustering models

Some of the most common BM-based algorithms are the Bayesian network [31], Bayesian belief network [32], naive Bayes [33], multinomial naive Bayes [34], and Gaussian naive Bayes [35]. In Fig. 9 a typical figure of a Bayesian model is presented.

11.3 Clustering Models

Clustering-based models [36] are typical applications of unsupervised learning models. These types of models are used to find natural groupings of data, known as clusters. Common clustering algorithms are the K-means [37], hierarchical clustering [38], and the expectation-maximization algorithm [39]. In Fig. 10 a typical structure of a cluster-based model is presented.

11.4 Computer Vision Models

Computer vision (CV) models aim to understand information from digital images or videos. Automatic extraction, analysis, and interpretation of meaningful information from a picture or sequence of images are all part of CV-based models. It entails the creation of a theoretical and computational foundation for autonomous visual comprehension. Some of the most common algorithms are HRNet- OCR [40], FixEfficientNet [41], and EfficientDet [42]. In Fig. 11 a typical structure of a CV model is presented.

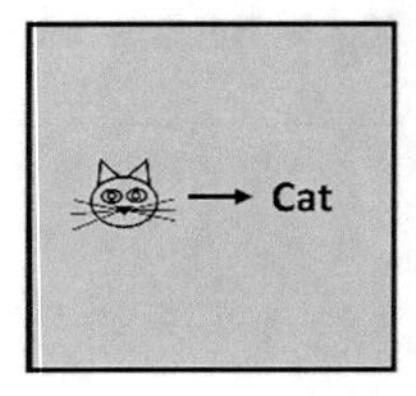

Fig. 11 Computer vision models

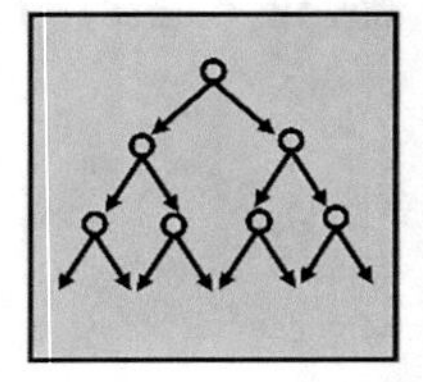

Fig. 12 Decision tree models

11.5 Decision Tree Models

Decision tree (DT) models are based on a tree-like architecture [43] and solve regression or classification problems. The dataset is grouped into smaller homogeneous subsets, known as subpopulations, while a relative tree graph is generated at the same time. Each branch of the tree structure represents the result of a distinct pairwise comparison on a given property, and each internal node represents a separate pairwise comparison on a given attribute. The leaf nodes, which follow the path from the root to the leaf, convey the process' ultimate prediction. Common DT-based algorithms are regression and classification trees [44], chi-square automatic interaction detector [45], and the iterative dichotomiser [46]. In Fig. 12 a typical structure of a DT model is presented.

11.6 Deep Learning Models

Deep learning (DL) or deep neural networks (DNNs) [47] consist of a modern version of ANNs. DL-based models are the new era of AI, while more and more models are developed based on them. As in ANNs, so in DL models consist of three categories of layers, input, multiple hidden, and output layers. The significant difference between ANNs is the usage of multiple processing layers which can learn complex data representations via multiple levels of abstraction. Furthermore, one more advantage of DL-based models is that the feature extraction can be performed by the model itself. These models can be used for supervised, unsupervised, and semi-supervised learning. The most common DL-based algorithms are convolutional neural networks [48], deep Boltzmann machines [49], deep belief networks [50], autoencoders [51], recurrent neural networks [52], and long- and short-term memory networks [53]. In Fig. 13 a typical structure of a DNN model is presented.

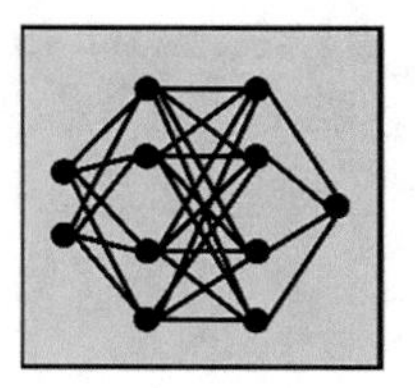

Fig. 13 Deep learning models

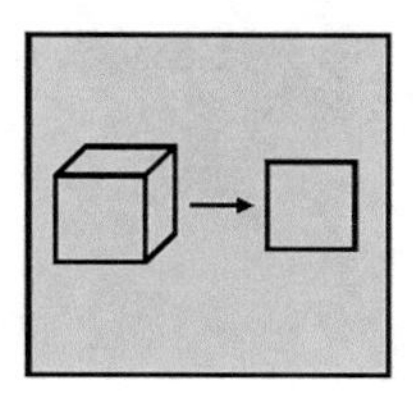

Fig. 14 Dimensionality reduction models

11.7 Dimensionality Reduction Models

Dimensionality reduction (DR)-based models convert the original higher-dimensional dataset into a lower-dimensional representation to preserve as much information from the original data as feasible and to reduce the poor performance which is produced from the datasets with a large number of features. DR-based models can be used for supervised and unsupervised learning types and usually are applied to solve regression problems. The most common DR-based algorithms are principal components [54], partial least squares [55], and linear discriminant [56]. In Fig. 14 a typical structure of a DR model is presented.

11.8 Ensemble Learning Models

By building a linear combination of simpler base learners, ensemble learning (EL) models are aimed to improve the prediction performance of a particular statistical learning or model-fitting approach. EL-based models or multiple-classifier systems enable the hybridization of hypotheses that were not produced by the same base learner, producing improved outcomes in the case of high variety among the single models. Typically, the DT architecture is used in EL-based models. Common EL-based algorithms are AdaBoost [57], boot-strap aggregating [58], boosting technique [59], gradient boosting machines [60], and random forest [61]. In Fig. 15 a typical structure of an EL model is presented.

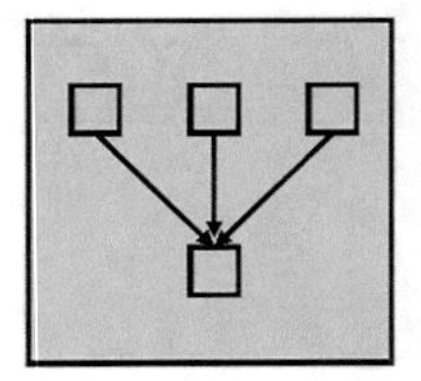

Fig. 15 Ensemble learning models

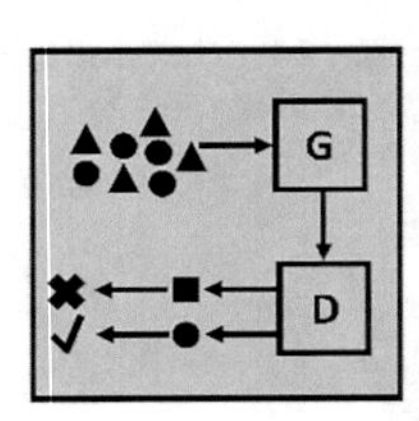

Fig. 16 Generative learning models

11.9 Generative Learning Models

Generative learning (GL) models aim to generate new synthetic samples. A typical GL model consists of two neural networks, the generative network and the discriminative network. The generative network learns how to produce new synthetic samples according to the initial dataset and the discriminative network distinguishes the generated from the initial original samples. GL-based models mostly are used to generate new samples in art, video games, and advertising. Common GL-based algorithms are generative adversarial networks (GANs) [62], conditional generative adversarial networks (CGANs) [63], Wasserstein generative adversarial networks (WGANs) [64], Wasserstein conditional generative adversarial networks (WCGANs) [65], StyleGAN [66], and CycleGAN [67]. In Fig. 16 a typical structure of a GL model is presented.

11.10 Instance-Based Models

Instance-based (IB) models are memory-based models that learn from the comparison of new cases to instances in the training dataset. These models generate hypotheses based on the information supplied from the data. Also, IB-based models generate regression or classification predictions only via specific instances, while these models do not adhere to a set of abstractions. The fundamental downside of IB-based models is that they get more complex as more data is collected. K-nearest neighbor [68], vector quantization [69], locally weighted [70], support vector machines [71], and self-organizing map [72] are the most common IB-based algorithms. In Fig. 17 a typical structure of an IB model is presented.

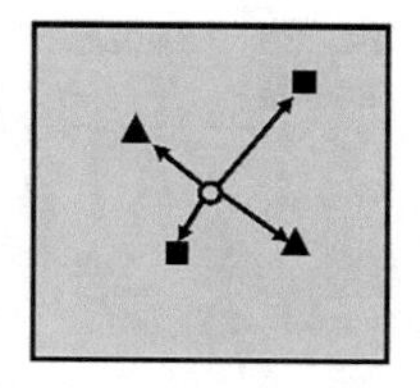

Fig. 17 Instance-based models

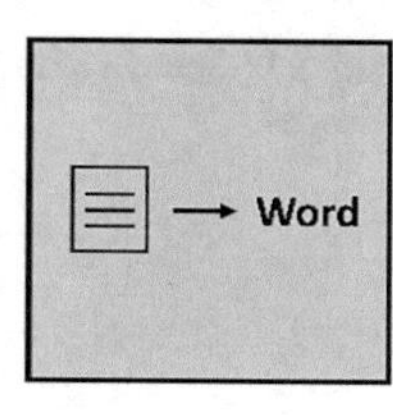

Fig. 18 Natural language processing models

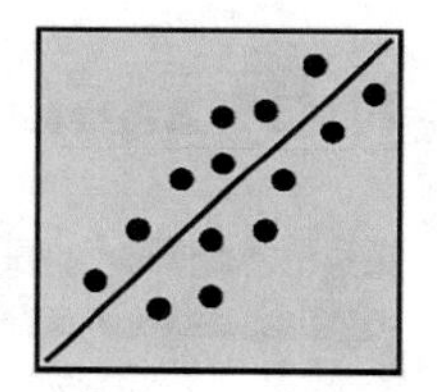

Fig. 19 Regression models

11.11 Natural Language Processing Models

Automatic summarizing of the major points in a text or document is achieved using natural language processing (NLP) models. Also, these algorithms are used to classify text into specified classes or categories or to organize information through this. The most common NLP-based algorithms are BERT [73] and XLNet [74]. In Fig. 18 a typical function of an NLP model is presented.

11.12 Regression Models

The role of a regression learning model is to produce an output result based on known input values. Linear regression [75], logistic regression [76], ordinary least squares regression [77], cubist [78], and locally estimated scatterplot smoothing [79] are the most common regression-based algorithms. In Fig. 19 a typical structure of a regression model is presented.

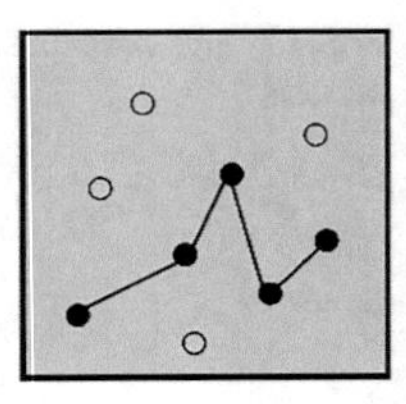

Fig. 20 Regularization models

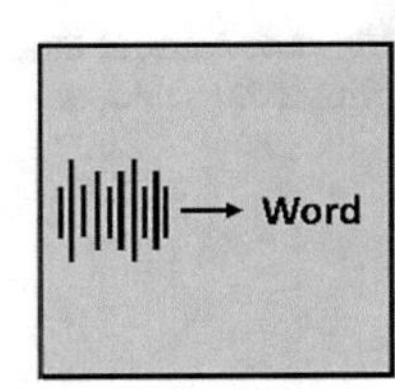

Fig. 21 Speech recognition models

11.13 Regularization Models

Regularization models consist of an extension of regression models. The purpose of regularization-based models is through a penalize technique to simplify complex models to simpler performance models. Common regularization algorithms are ridge regression [80], least absolute shrinkage and selection operator [81], and least-angle regression [82]. In Fig. 20 a typical structure of a regularization model is presented.

11.14 Speech Recognition Models

Speech recognition (SR) models or voice recognition models are used in speech recognition technology to convert voice to text. SR-based models work as follows: they break down a speech audio file into individual sounds, analyzing each sound to find the most suitable word that matches the language and converting these sounds into text. Most common SR-based algorithms are ContextNet [83], LiGRU[84], and ResNet [85]. In Fig. 21 a typical function of an SR model is presented.

12 AI History Timeline

As can be observed from Fig. 22, the first algorithms were created in 1950 to develop simple AI models to solve basic mathematical problems. Moreover, from 1950 to 1970, an increase in the development of new algorithms can be observed, while from 1980 to 2000, there is a sharp decline. The main reason was the need to solve increasingly complex mathematical problems, combined with the lack of computational resources. This led to a lack of interest in this field of research.

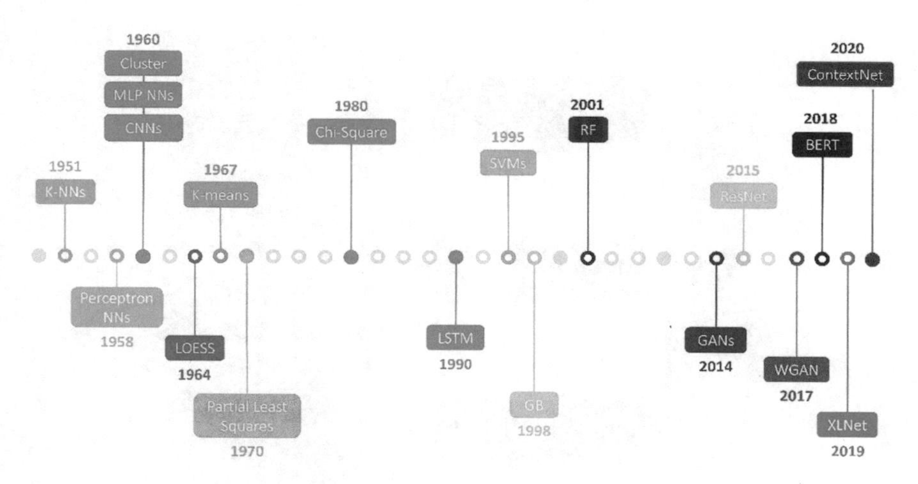

Fig. 22 ML and DL algorithms history timeline

While it is observed that since 2014, the period in which computing resources have increased, more sophisticated algorithms are being developed to solve more complex problems, such as computer vision, natural language processing, and speech recognition problems.

13 Countermeasures Against HTs

HTs can be inserted at any stage and phase of IC development, can attack any unit of the ICs, can affect the ICs via a variety of attacks, and can have different physical layouts. For these reasons in this book, the countermeasures against HTs are categorized into three major categories, SCA-based approaches, ML-based and simulation approaches, and auxiliary approaches (Fig. 23).

14 Historical Throwback

Historically, the first research attempt that mentioned and studied the existence of HTs in ICs was presented by Agrawal et al. [86] in 2007. The authors have developed the first detection approach based on SCA-based power analysis. In 2009, Chakraborty et al. [87] developed the first method for HT detection based on LT. In 2012, Salmani et al. [88] proposed the first PF approach. In 2014 introduced by Bao et al. [89], the first ML-based approach for the post-silicon stage. In 2015, Ngo et al. [90] proposed an RM approach. Lastly, in 2016, the detection of HTs at GLN was proposed by Hasegawa et al. [91], while in 2022 proposed “GAINESIS” [92],

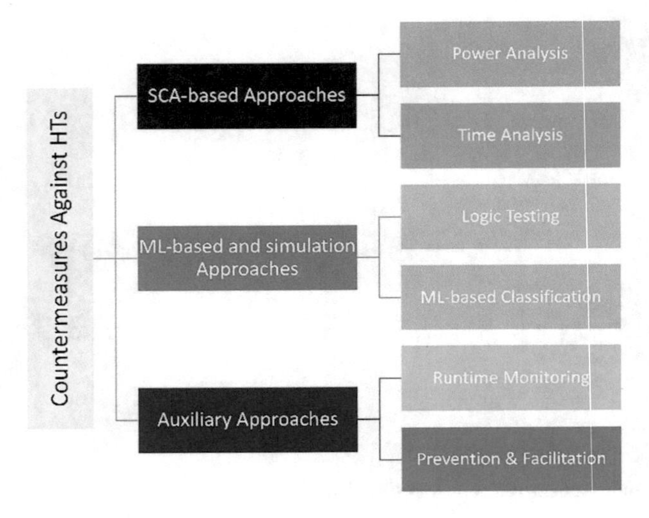

Fig. 23 Categorization of countermeasures approaches against HTs

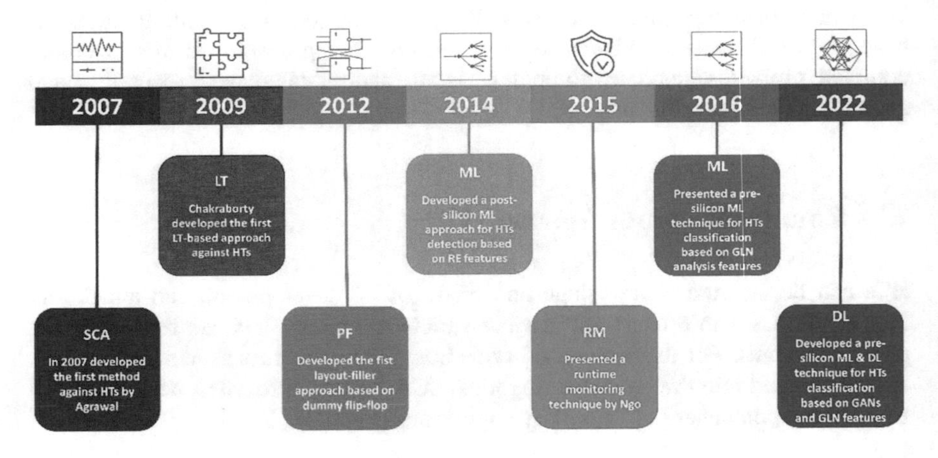

Fig. 24 Countermeasures against HTs history timeline

the first GAN-based approach for the synthesis of new generated samples for GLN. In Fig. 24 a history timeline for countermeasures against HTs is presented.

15 Studies Trend

In Fig. 25 the popularity of each subcategory over the years can be observed. Specifically, from 2007 to 2013, most of the studies focused on the development

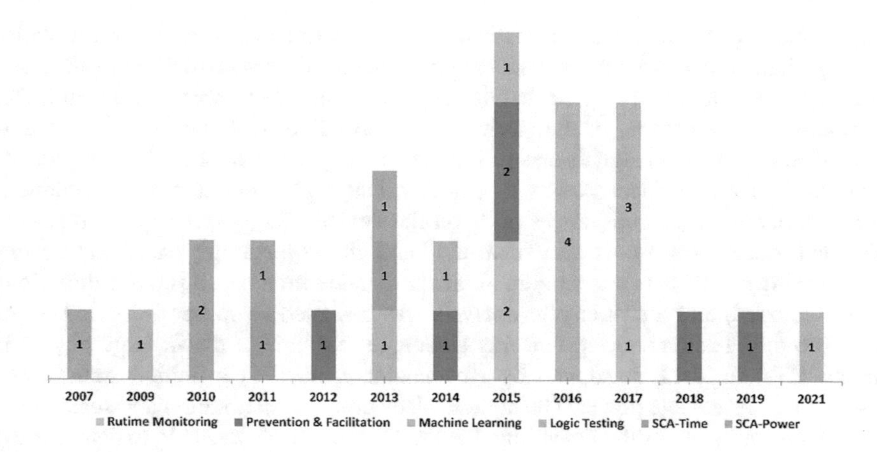

Fig. 25 Countermeasures trend

of methods for the detection of HTs based on SCA power and time analysis. In 2012, the first auxiliary-based study appears. And the golden era of auxiliary-based approaches was 2015 when the majority of these studies are developed. The first ML-based approach was introduced in 2014. But in 2016 and 2017, there is a sharp increase in the development of such methods. As regards the LT simulation approaches, the first study was presented in 2009, and other such approaches have been developed over time.

16 Side-Channel Analysis-Based Approaches

SCA-based approaches aim to secure ICs for the SCA phase of the post-silicon stage of ICs. These approaches use techniques based on side-channel analysis features to detect alterations of physical characteristics like area, time, and power caused by HTs. If the original SCA values of an IC differ, then the circuit is infected. This is because when an HT is partially or fully activated, the original infected circuit shows more interrupting activity compared to the original normal circuit because consumes extra power.

16.1 SCA Power Analysis-Based Approaches

The first study which mentioned the existence of HTs was presented in 2007 by Agrawal et al. [93], and it was an SCA-based approach. Specifically, the authors developed a method for the detection of large or small physical layout HTs based on SCA of transient current characteristics. In the study [94], the authors proposed an approach for HT detection via static current characteristics. They measured static current characteristics at numerous locations over the 2D surface of the chip simultaneously. The outcomes revealed that this approach can effectively detect small HTs. Furthermore, the authors of the study [95] proposed an SCA-based method via a power supply transient signals analysis. A power supply transient analysis approach was used to assess local power supply transient signal measurements obtained from numerous individual power ports on the device. The power supply transient signals for each power port were measured, and the power supply transient of each surrounding power port was examined. Then, to decrease noise, a signal calibration was employed, and a scatter plot analysis was designed to successfully detect an HT. The final results showed that this technique was able to detect large physical layout HTs. In 2011 developed by Koushanfar et al. [96] a unified framework based on SCA leakage power. The authors also combined calibration and sensitivity analysis techniques for the detection of HTs. This approach was able to detect with low process overhead, large in size HTs. In the study [3], the authors proposed a multiple-parameter SCA-based approach for the detection of HTs. They used and combined dynamic current and maximum frequency analysis features for HTs detection. The results showed that their approach was able to detect varying types and sizes of HTs. In Table 1 a summary of SCA-based power analysis approaches is presented.

Table 1 Summary of approaches in SC-based power analysis

Authors	Observed features	Functionality	Effectiveness	Benchmark
[93]	Transient supply current	Detection of HTs in ICs, based on side-channel information analysis via transient current	Large and small HTs	RSA circuit
[94]	Quiescent supply current	Detection of HTs based on the analysis of a chip's IDDQS	Small HTs	N/A
[95]	Transient supply current	Detection of HTs via sensitivity analysis of power signal	Large HTs	ISCAS 85 benchmark circuit: C499
[96]	Delay, quiescent supply current, transient supply current	Detection of HTs in ICs based on gate-level characterization and multi-parameter measurements	Large HTs	ISCAS 85 benchmark circuits: C8, C499, C432, C1355, C3450
[3]	Transient supply current, maximum operating frequency	Detection of HTs, based on dynamic current and maximum operating frequency	Varying types and sizes of HTs	Xilinx FPGA: Virtex-II XC2V500

Table 2 Summary of approaches in SCA-based time analysis

Authors	Observed features	Functionality	Effectiveness	Benchmark
[97]	Power, delay	Detection of HTs, based on the analysis of power and delay	Large and small HTs	Xilinx FPGA circuit: Virtex XUP-V2Pro
[98]	Transition, delay	Detection of HTs based on clock sweeping and delay-based detection	Small HTs	ISCAS 89: S38417

16.2 SCA Time Analysis-Based Approaches

In 2011, Lamech et al. [97] developed an SCA-based on-time analysis features approach. Specifically, the authors combined SCA delay and power features for the detection of HTs. The experimental results showed that their method was able to detect large- and small-sized HTs. In 2013, Xiao et al. [98] developed an approach based on clock sweeping and SCA delay characteristics. They used a combination of path delay fault patterns with a clock sweeping transition technique for the detection of HTs in a circuit. The results demonstrated that their approach was capable of detecting HTs of modest size. In Table 13.2 a summary of SCA-based time analysis approaches is presented.

17 ML and Simulation-Based Approaches

The purpose of ML-based methods is to classify infected with HTs from uninfected circuits. These types of approaches developed ML-based classifiers for the classification of HTs in different phases of ICs development. Simulated methodologies such as logic testing techniques generate tests to activate HTs and propagate the HT payload to principal outputs for comparison with the golden circuit. The difficulty with these methods is coming up with effective assays to activate HTs. In this section, are presented ML- and simulated-based approaches as countermeasures against HTs.

17.1 Logic Testing Simulation Approaches

Previously, LT simulation approaches aim to generate effective tests aiming to activate and discover the stealthy nature of HTs. Because of their stealthy nature, HTs are difficult to distinguish. Randomly generated tests are not effective, therefore LT-based simulation approaches aim to generate guided tests for HT activation and detection. In 2009, Chakraborty et al. [7] suggested a method based on LT simulation as a countermeasure against HTs. Specifically, they developed an LT approach named MERO. Multiple excitations of unusual logic situations at internal nodes were used to construct test patterns. The findings demonstrated that this method was capable of detecting small HTs. In 2011, Waksman et al. [99] developed an LT-based framework named FANCI. They used Boolean functional analysis features to generate test patterns for HT activation. The findings demonstrated that this method could detect infected circuits with a low percentage of false positive rate. In the study [100], the authors developed an LT-based simulation technique named VeriTrust for HT detection at the design phase based on HT trigger inputs. VeriTrust technique consisted of a traced and a checker. The tracer parsed verification tests to identify trigger signals containing inactive entries, while the checker examined these signals to determine which are associated with HTs. The findings demonstrated that this method could detect various types and sizes of HTs. In Table 13.3 a summary of LT simulation approaches is presented.

Table 3 Summary of LT simulation approaches

Authors	Observed features	Functionality	Effectiveness	Benchmark
[7]	Nodes	Detection of HTs based on test pattern generation and multiple excitations of rare logic conditions at internal nodes	Small HTs	ISCAS 85: C2670, C3540, C5315, C6288, C7552 ISCAS 89: S13207, S15850, S35932
[99]	Wires	Detection of HTs based on Boolean functional analysis	HTs and IPs	ISCAS 89: S15850, S35932, S38417
[100]	Netlists	Identification of HTs at the design stage, based on the detection of trigger inputs	Different types and sizes of HTs	ISCAS 89: S15850, S35932, S38417, S38584 Microcontrollers: MC8051, LEON3

17.2 ML-Based Approaches

ML-based approaches aim to detect an infection with HT circuit from an uninfected normal circuit. In these approaches models which can classify infected from normal circuits or used as reverse engineering or side-channel analysis methods for the detection of HTs in a circuit are developed. Specifically, for the pre-silicon stage proposed ML-based classifiers for the classification of infected and normal circuits at GLN phase, based on netlist, area and power analysis features. ML-based methods that work as reverse engineering techniques and ML-based methods trained via side-channel analysis features were developed for HT detection at the post-silicon stage.

For the pre-silicon phase, in 2016, Hasegawa et al. [91] proposed an SVM-based model for the classification of infected from normal circuits. Specifically, the authors developed an SVM-based model for the classification of HTs at the GLN phase of the pre-silicon stage. For the training of the model, a dataset consisted of GLN-based features like nets and gates of the circuits was used. The results showed that this approach was able to classify effectively the infected with HTs from normal nets. The same group [101] proposed another ML-based model. They developed an RF-based model which was trained via GLN-based area features, like a number of flip-flops and multiplexors before and after for each net. The results showed that the RF-based model was effective for the classification of the two classes. In 2018, Inoue et al. [102] proposed an SVM-based model in a combination of GLN-based area features for the classification of HTs at the GLN phase of the pre-silicon stage of IC development. The SVM-based model was trained via area features like the number of logic gates and flip-flops for each net of the infected and normal circuits. The final results proved the validity of the method. In the study [103], the authors developed six ML-based models for the classification of HTs at the GLN phase. Specifically, they developed and compared six ML-based models which were trained via a dataset consisting of GLN-based area, power, and time analysis features from infected and normal circuits. The features consisted of area features like the number of cells, nets, ports, and power features like the number of total switching and combinational power of each normal and infected circuits. The experimental results showed that their GB-based model was able to classify effectively the normal from HTs circuits.

ML-based approaches were developed for the detection of HTs at the post-silicon stage. For the post-silicon phase in 2014, Bao et al. [89] developed an ML-based model as a reverse engineering approach for the detection of HTs. Specifically, they trained an SVM classifier based on high-resolution images features from golden and infected with HTs circuit layouts. The simulation results showed that the SVM-based classifier was able to classify the two classes efficiently. The same group in the study [9] proposed a K-means-based clustering model. The K-means-based model has been developed again via high-resolution images features from golden circuits and of three types modifications based on the golden circuits which consisted of the infected circuits. Another post-silicon detection approach was developed in 2016 by Jap et al. [104]. Specifically, the authors developed an SVM-based model for the detection of HTs. The model was trained from a dataset consisting of SCA-based time features like leakage from normal and infected circuits. Another study with ML and SCA techniques was proposed by Xue et al. [105]. In this study, the authors developed an SVM-based model for the detection of HTs at the post-silicon stage. The model was trained via a dataset that consisted of SCA-based power features and specifically transient power supply features of normal and infected circuits. The experimental results showed that this method was able to detect with effectiveness the infective from normal circuits. Wang et al. [106] proposed another SCA-based method in combination with ML techniques for the detection of HTs in the post-silicon phase. Specifically, they developed an ELM-based model which was trained from a dataset consisting of dynamic power features from infected and normal circuits. In the study [107], the authors developed an SVM-based model for the detection of HTs via SCA power features. Specifically, they developed an SVM-based model which was trained via a data set consisting of SCA-based power consumption waveforms features from infected and normal circuits and given. The experimental results proved the validity of the method. Liu et al. [108] proposed another SCA-based in combination with an ML-based model approach for the detection of HTs in the post-silicon phase. They developed an SVM-based model which was trained via SCA wireless transmission power waveform features from HT-free and infected circuits. The results showed that their method was able to detect effectively wireless transmission power signals produced from HTs. In Table 13.4 a summary of ML-based approaches is presented.

18 Auxiliary Approaches

The purpose of the auxiliary approaches is to enhance the effectiveness of the detection techniques against HTs for the pre-silicon or post-silicon stage. Auxiliary approaches can be categorized into two categories, the runtime monitoring approaches and the prevention-facilitation approaches.

When HTs are triggered, runtime monitoring systems try to limit the catastrophic impacts of these infections. These methods are focused on finding ostensibly undetected attacks and their consequences from time-delayed HT activation. These approaches develop algorithms that can use finite state machines to investigate the behavior of signals of interest or construct and perform many functionally identical tests to identify HT attacks. Furthermore, due to their concurrent execution on the circuit, these approaches can identify comparable HTs or bypass HTs, simulating software HTs. In addition, utilizing verification tests, runtime monitoring systems can discover underutilized circuitry and mark it as a suspect. Suspicious circuitry is then replaced by a software logic exception, allowing the system's regular operation to bypass the HTs.

Prevention-facilitation approaches try to enhance the difficulty of HT insertion into ICs, primarily during the design process, or to make detection procedures easier. Prevention-facilitation approaches use hardware security techniques like

Table 4 Summary of ML-based approaches

Authors	Observed features	Benchmark	Algorithm	Results
[91]	Features extracted from known gate-level netlists, like LGFi, FFi, FFo, PI, and PO	Trust-HUB: RS232-T1000, RS232-T1600, S15850-T100, S35932-T100, S35932-T300, S38417-T100, S38417-T300, S38584-T100, S38584-T300	SVM	80%–100% TPR
[101]	Features extracted from gate-level netlists	Trust-HUB: RS232-T1000, RS232-T1200, RS232-T1300, RS232-T1400, RS232-T1500, S15850-T100, S35932-T100, S35932-T300, S38417-T100, S38417-T200, S38417-T300, S38584-T100	RF	74.6% F-measure
[102]	Features extracted from netlists, like LGFi, FFi, FFo, PI, and PO	Trust-HUB: RS232-T1000, RS232-T1100, RS232-T1200, RS232-T1300, RS232-T1400, RS232-T1500, RS232-T1600	SVM	Type A: 58.9% accuracy Type B: 69.5% accuracy Type C: 65.1% accuracy
[103]	Features from area, power, and time analysis through the DC compiler tool	Trust-HUB: Aall benchmarks	GB	100% F1-score
[89]	High-resolution images from IC golden layouts	ISCAS 89: S27, S298, S280, S15850, S38417 ITC 99: B18	SVM	90% accuracy
[9]	Trojan-free IC golden layout images and three types of modifications produced based on these images, Trojan addition, deletion, and parametric	ISCAS 89: S27, S298, S280, S15850, S38417 ITC 99: B18	K-means	Trojan-free: 99.23% accuracy Trojan addition: 100% accuracy Trojan deletion: 100% accuracy Trojan parametric: 98.86% accuracy

(continued)

Table 4 (continued)

Authors	Observed features	Benchmark	Algorithm	Results
[104]	Features extracted from side-channel analysis to leakage of the chip based on-time samples	Xilinx FPGA circuit: Spartan-6	SVM	N/A
[105]	Features extracted from the transient power supply currents (IDDT) of each simulated IC and a Trojan-free or Trojan-inserted indicator	ISCAS 89: S38417, S35932	SVM	Trojan-inserted ICs known: 100% accuracy Trojan-inserted ICs unknown: 98% accuracy
[106]	Features from side-channel analysis, dynamic power consumption	N/A	ELM	90% success rate
[107]	Features from converted power consumption waveform into the frequency domain	N/A	SVM	72.72% accuracy
[108]	Features consist of transmission power measurements for six ciphertext blocks transmitted by each of 40 Trojan-free circuits	Trojan-free: TSMC microcontroller: 0.35-μm technology Trojan-I and Trojan-II: Created two HTs, which leak the secret key of a wireless cryptographic IC consisting of an advanced encryption standard (AES) core and an ultra-wideband (UWB) transmitter (TX)	SVM	0/10 FP and 0/80 FN

obfuscation, layout-filler, and path-delay fingerprinting to enhance the detection of HTs. The obfuscation technique alters the circuit's transition mode, allowing it to function in two separate modes: regular and obfuscated. The regular model generates the circuit's expected output, but the obfuscated mode causes the circuit to fail in specific input patterns. The usage of this technique makes it more difficult to install a malicious circuit into a system. Layout-filler approaches are used to limit the insertion of extra components by filling the vacant areas in a circuit with filler cells. These solutions, however, are incapable of preventing the malicious conversion of a transistor set or the addition of a circuit that does not require additional layout space. Another method for detecting HTs is to use synthesis algorithms based on the path-delay fingerprint. These techniques improve the HT detection probability by minimizing the maximum delay and shortest path of the circuits.

18.1 Runtime Monitoring Approaches

In 2015, Ngo et al. [90] proposed a runtime monitoring method for the detection of HTs. Specifically, they developed an assertion approach for identifying and validating high-level important behavioral invariants through an integrated circuit and hardware property checker. The findings showed that this method could detect HTs in circuits with different system overhead and adjust the protection levels accordingly. In the study [109], the authors developed a general methodology based on runtime monitors for the identification and detection of HT attacks through burst mode communication. They developed a runtime monitor technique based on an examination of susceptible routes. The statistical and experimental analysis revealed that this strategy had a minimal area and power overhead when compared to previous monitor methods and could be employed without the need for additional IP module information. Furthermore, the authors in the study [110] developed three low-overhead runtime approaches based on power/thermal features of infected and normal circuits for the detection of HTs. The first approach was a sensor-based approach based on thermal features produced from the thermal sensors. In the second approach, a filter known as the Kalman filter for the tracking of circuits' thermal profile was used. The third approach combined the Kalman filter with leakage power features of the circuits to track the thermal profiles. The simulation results verified that all the approaches were able to detect HTs effectively. In Table 13.5 the summary of RM approaches is presented.

18.2 Prevention and Facilitation Approaches

An obfuscation-based technique was developed by Kamali et al. [111]. The authors developed an obfuscation-based method via embedded key features for the protection of ICs against HT attacks. The simulation results demonstrated that their technology could successfully safeguard ICs. The same group in the study [112] proposed again an obfuscation-based method for the defense of IP-piracy and reverse engineering approaches via the replacement of parts of logic design with programmable logic routing blocks. In 2012, Salmani et al. [88] developed an improving HT detection technique based on an analysis of the transition generation time and dummy flip-flop insertion. Specifically, the authors developed a method

Table 5 Summary of RM approaches

Authors	Observed features	Functionality	Benchmark
[90]	Critical behavioral invariants	Configurable security monitor	Microcontroller circuit: LEON3
[109]	Handshaking protocol features	Configurable security monitor	Trust-HUB: AES-T100, AES-T1000, AES-T1100, AES-T1200, AES-T1300, AES-T1400, AES-T1500, AES-T200, AES-T2000, AES-T2100, AES-T300, AES-T400, AES-T500, AES-T600, AES-T700, AES-T800, AES-T900
[110]	Thermal and power profiles	Variant-based parallel execution	Trust-HUB: AES-T1700, BasicRSA-T200, MC8051-T300, MC8051-T400, MC8051-T600, RS232-T400, RS232-T900, S38417-T300, PIC16F84-T100, PIC16F84-T200

based on dummy multiplexors to be able to remove rare trigger conditions, reduce transition generation time, and increase the activity of HTs for the detection of HTs. In the study [113], the authors proposed a layout-filler based on a dummy circuit insertion technique against HT attacks. This technique is identified and replaced the unused resources of a circuit with dummy logic cells. Experimental results showed that the proposed study was effective for field programmable gate arrays (FPGAs) with no cost on power or performance. In 2014, Nejat et al. [114] proposed an approach for enhancing HT detection based on a combination technique of an effective test-vector selection scheme with a path-delay fingerprinting. The basic concept behind this procedure was to test the circuit at a gamma of frequencies. Each path was examined at a clock cycle with a period equal to the path's delay. According to the results, this technique improves the detection of HTs with low area overhead. The same group in the study [115] developed a path-delay fingerprinting-based approach for HT detection. Specifically, they developed a logic-level synthesis

Table 6 Summary of PF approaches

Authors	Observed features	Functionality	Benchmark
[111]	Several embedded key numbers	Obfuscation	ISCAS 85: C2670, C3540, C5315, C6288, C7552
[112]	Fully programmable logic and routing blocks	Obfuscation	ISCAS 85: C432, C499, C880, C1355, C1908, C2670, C3540, C5315, C7552
[88]	Features based on average clock cycles per transition	Dummy circuit insertion	ISCAS 89: S38417
[113]	Low-level dummy logics (LLDLs)	Layout filler	Xilinx FPGA circuit: Virtex-II
[114]	Features based on path-delay fingerprinting	Improvement of HT detection based on path-delay fingerprinting and an effective test-vector selection scheme	ISCAS 89: S713, S1423, S5378, S13207, S35932
[115]	Features based on path-delay fingerprinting	Enhance HTs detection based on the improvement of the path-delay fingerprinting technique via a logic-level synthesis retiming algorithm	ISCAS 89: S208, S344, S1196, S1238, S1494, S9234, S13207, S38417

retiming algorithm that shortened for each node of a circuit the connection paths to minimize the communication delay. The results showed that the shorted paths improve the detection of HTs. In Table 13.6 the summary of PF approaches is presented.

19 Build Your Model Against HTs

This section outlines the processes required for the development of an ML- or DL-based model. In particular, all of the critical steps required to develop an ML- or DL-based model are presented and analyzed. The significance of the dataset is mentioned as well as the actions required to prepare it before usage. In addition, examples from datasets that were used for the development of ML-based models against HTs are provided. Also, the significance of the features contained in a dataset, as studies and the features based on for the development of models against

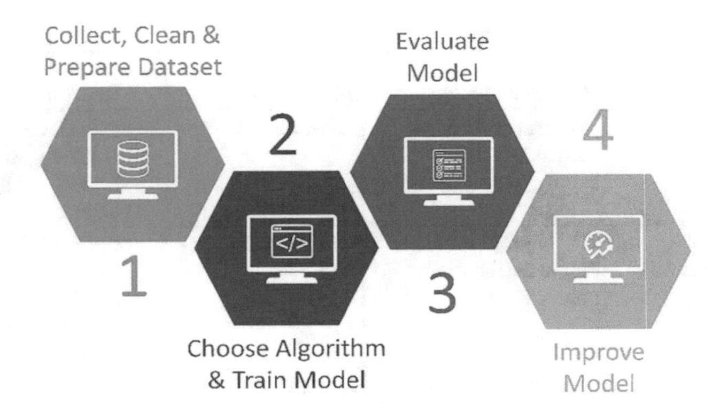

Fig. 26 Steps for the development of an ML- or DL-based model

HTs, are mentioned. Furthermore, the most essential ML-based algorithms which were utilized for the development of ML-based models against HTs are presented. GL-based algorithms for the generation of synthetic datasets are also presented in the case there is an imbalance problem with the dataset. Finally, all the necessary metrics for the evaluation of the ML-based algorithms are mentioned. The goal of this section is to inform readers about the ML-based models that have been developed as countermeasures against HTs and to enable them to develop their models.

It must be mentioned that developing a model based on the principles of ML or DL is a costly process in both time and computing power. Depending on the problem, the size of the dataset, the size, type, and quantity of features contained in the dataset as well as the algorithms and the set of parameters that will be used and combined for the development of the new ML- or DL-based model, time and computing power can vary significantly from model to model. Until today, ML-based models need significantly more time for training and testing than the DL-based models. The reason is that ML-based algorithms for the development of a model are built to use the central processing unit (CPU) and not the graphics processing unit (GPU). On the other, DL-based algorithms can use either CPU or GPU for the training and evaluation of the development model. In Fig. 26 the steps for the development of an ML- or DL-based model are presented.

19.1 Dataset

Every year more and more ML-/DL-based approaches are developed as countermeasures against HTs. These approaches are aimed at classifying or detecting circuits infected with HTs from normal uninfected circuits. Also, some approaches are used to enhance the classification or detection methods. The development of these types of approaches needs a quality dataset that will contain a sufficient number of quality samples and features to be able to train the ML-/DL-based model more efficiently.

Dataset can be divided into three categories: structured, unstructured, and semi-structured. Structured data is data that follows a pre-defined data model and is thus easy to analyze. Structured data is presented in a tabular format, including relationships between rows and columns. Excel files and SQL databases are common examples of structured data. Each of them has sortable organized rows and columns. Unstructured data is information that lacks a predefined data model or is not organized in a specific way. Common examples of unstructured data include text, images, video, or audio files. Semi-structured data is a type of structured data that does not follow the rules of structured data. However, tags or other markers are used to distinguish semantic pieces and enforce hierarchies of records and fields inside the data. Examples of semi-structured data include JSON and XML files.

The dataset plays a significant role in the creation of a robust ML- or DL-based model. Specifically, the dataset before being used for the development of a model must be cleared from unnecessary values and organized. For example, the dataset must be checked for consistency, cleared of zeros and/or unspecified values, and labeled where needed. An unreliable dataset like a dataset with imbalanced samples per class leads to the development of unreliable models. A type of unreliable model is a model that was taught to overclassify a class compared with another class. Due to the lack of samples for a class, the model has learned to underclassify this class compared with the other one.

It should be noted that each sample or feature provides a quantifiable piece of data that may be analyzed. The features which are included in a dataset can vary widely depending on the problem which is analyzed. The basic building elements of a dataset are features. The quality of the features in a dataset has a significant influence on the quality of the insights gained during model construction. For the development of a model, the developer needs to understand the goals of the project and select the appropriate feature values for the training of the model. Feature selection and feature engineering are two strategies for increasing the quality of a dataset's features. These techniques require extensive user experience for proper application. For the creation of the model, the features which will be used must be scaled. Scale techniques alter features by scaling them to a certain range. The most common scale methods are standard and min-max scale methods. The standard scaler assumes that data within each feature is normally distributed and scales it so that the distribution is centered around 0 with a standard deviation of 1. Centering and scaling are performed individually on each feature by computing the necessary statistics on the training set samples. The min-max scaler scales and translates each feature separately such that it is inside the training set's defined range, e.g., between zero and one. If there are negative values, this scaler reduces the data to a range of -1 to 1. Below are presented datasets that were built and used for the training of models as countermeasures against HTs.

It should be noted that, in the case of HT challenges, the datasets typically consist of two classes, the class with the infected and the class with the noninfected circuits, and most studies for the development of an ML-based model against HTs used circuits from a free access benchmark suit named as Trust-HUB [116, 117]. The study [89] was the first study that proposed an SVM-based model for the detection of HTs based on the reverse engineering (RE) method. For the development of the SVM-based model, a dataset consisting of two classes was used. The first class consisted of high-resolution images of golden circuits and the second class with images from three types of modifications based on the golden circuits and were used as infections. For the classification of free and infected circuits at the GLN phase, Hasegawa et al. [101] proposed an RF-based classification model. The dataset consisted of area features from free and infected circuits. Also, the same group [91] proposed an SVM-based classifier developed based on a dataset consisting of free and infected circuit area features like gates, nets, etc. In the study [103], authors developed and compared six ML-based classifiers for the classification of infected and normal circuits at GLN.

19.2 Training of Our ML- and DL-Based Models

The next step in the development of an ML-based model includes the selection of a suitable ML-based algorithm for the training of the model. For the development of an ML-based model, there is often more than one algorithm that can be used. The most important criterion for selecting the most suitable algorithm for its model is the type of problem to be addressed. According to this criterion, more than one algorithm which is indicated as a solution for the problem can be chosen. Another criterion consists of the structure of the dataset which will be used for the training and evaluation of the model. According to the features of the dataset, there may be a need to choose other types of algorithms. Also, it is significant to know the complexity and the speed of each algorithm, because each algorithm needs specific computing power, according to the parameters used for the development of a model. There is a case where the model cannot be built due to a lack of computer power. It should be noted that using more sophisticated algorithms does not always result in the best outcomes.

One of the most crucial steps in the development of machine learning models is the training step. Each training phase involves updating the weights and biases. Training a model simply entails learning/determining good values for all of the weights and biases based on the samples in our dataset. In supervised learning, a model can be developed based on labeled data samples, but in unsupervised ML, inferences from unlabeled data can be attempted. For training, a set of hyperparameters is utilized, and the weights must be updated to get better outcomes from cycle to cycle. As the number of training steps increases, more accurate outcomes may be expected. The typical procedure is to train the model with the default parameters of each algorithm and then depending on the problem and the data that are used, we make changes to the parameters in order to optimize the model. According to Section 3, the most common algorithms for the training of ML-based models as countermeasures against HTs are presented.

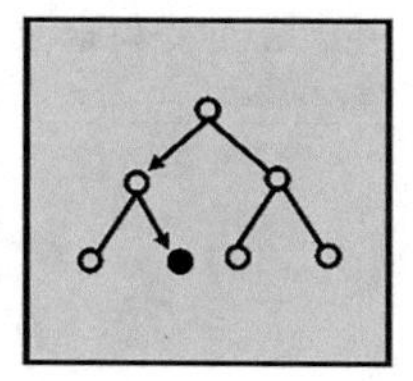

Fig. 27 GB-based algorithm

19.2.1 Gradient-Boosting Algorithm

Gradient boosting (GB) [60], models are components of ensemble learning algorithms that rely on a group choice using inefficient prediction models known as decision trees. During the boosting step, each new tree is based on a modified version of the original dataset. To begin, GB constructs a decision tree and assigns equal weight to each observation. Following the initial tree evaluation, the weights for easy-to-classify observations fall, while the weights for difficult-to-classify data increase. The next tree develops on the weighted data, seeking to improve on the predictions of the first tree. The new model is a cross between the first and second trees. The classification error is computed, and a third tree is constructed to predict the corrected residuals. This procedure is repeated for a certain number of iterations until convergence is achieved. The final ensemble model's forecast is the weighted sum of all previous model iterations' projections. The most common hyperparameters for the training of GB-based models are learning rate, number of estimators, max features, and max tree depth. A number of estimators consist of the total number of sequential trees to be modeled. Max tree depth parameter controls the depth of the individual trees. And max features parameter is the number of features that will be used for the best split of the model. In Fig. 27 a typical structure of a GB algorithm is presented.

19.2.2 K-Nearest Neighbor Algorithm

The K-nearest neighbor (KNN) [68] is a kind of IB learning that may be used to solve supervised regression and classification problems in a straightforward manner. The KNN algorithm is predicated on the premise that similar entities occur in close proximity. In other words, similar things are close to one another. KNN is based on the concept of similarity (also known as distance, proximity, or closeness) in calculating the space between graph nodes. There are several techniques for estimating distance, and depending on the situation, one approach may be preferred. The KNN algorithm is initially loaded with the training dataset, which is commonly referred to as x, and their goal values, which are referred to as y. Goal values y needs to be classified from the model. The distance between the sample whose goal value is intended to categorize is then computed for each data sample, and k is initialized to a desirable number of neighbors. The query example's index and distance are then added to an ordered list of indices and distances, and the list is sorted in ascending

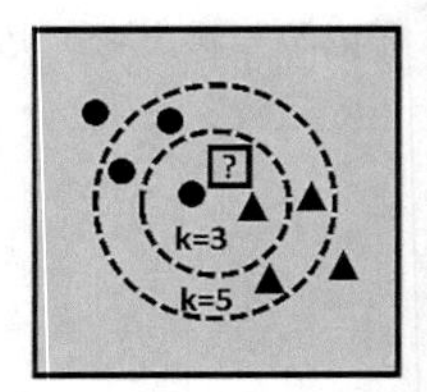

Fig. 28 KNN-based algorithm

order (from smaller to bigger), with the distance as the order criterion. Finally, the first k elements from the sorted list are chosen, and the labels of the chosen k entries are obtained. As a result, the form of the k labels may be returned. Some of the most often used hyperparameters for the training of a KNN-based model are leaf size, the number of neighbors, and weight metrics. Leaf size parameter is the maximum number of points a node can hold. The number of neighbors is used to return indices of and distances to the neighbors of each point. Using a training set, the weights parameter is utilized to approximate the ideal degree of effect of various attributes. Relevant characteristics are given a high weight value when successfully applied, while irrelevant features are given a weight value near zero. In Fig. 28 a typical structure of a KNN algorithm is presented.

19.2.3 Multilayer Perceptron Algorithm

ANNs are built based on the human brain. Based on the philosophy of ANNs, the algorithm multilayer perceptron (MLP) consists of a feedforward ANN that generates a set of outputs from a set of inputs. Specifically, an MLP [19] is a neural network that links numerous layers in a directed graph, which means that the signal path between nodes is only one way. Each node, aside from the input nodes, has a nonlinear activation function. MLP is frequently utilized for supervised learning tasks. Common hyperparameters for an MLP model are hidden layer sizes, solver, activation, max iterations alpha, and learning rate. Hidden layer size is used for the creation of the hidden layers. The hidden layers are produced according to the size value. Furthermore, the hidden layer merely generates layers of mathematical functions, each of which is designed to create an output particular to an intended outcome. The solver parameter represents a stochastic gradient descent-based optimizer for optimizing the computation graph's parameters. An activation hyperparameter consists of an activation function that describes how the weighted sum of the input is converted into an output from a network layer node or nodes. An iteration is the number of times a batch of data is processed by the algorithm. In the context of neural networks, this refers to the forward and backward passes. As a result, each time you run a batch of data through the ANN, you complete an iteration. The alpha parameter is a regularization term, also known as a penalty term, that limits the size of the weights to prevent overfitting. Increasing alpha may reduce high variance by promoting lower weights, which results in a decision boundary plot with fewer curvatures. The learning rate, in particular,

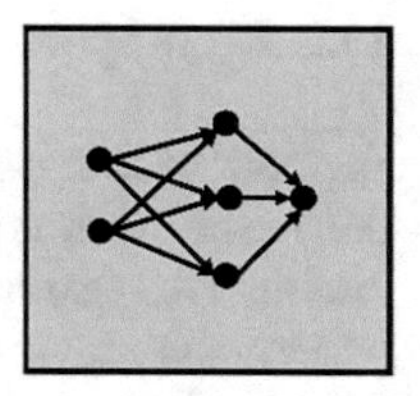

Fig. 29 MLP-based algorithm

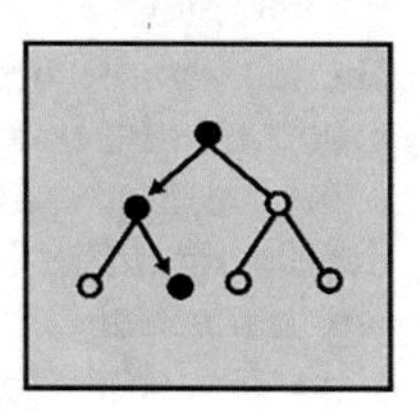

Fig. 30 RF-based algorithm

is an adjustable hyperparameter used in neural network training that has a tiny positive value, typically in the range of 0.0 to 1.0. The learning rate determines how quickly the model adapts to a new situation. It could be the model's most essential hyperparameter. In Fig. 29 a typical structure of an MLP algorithm is presented.

19.2.4 Random Forest Algorithm

A random forest (RF) is a collection of DTs. The main notion behind this strategy is that combining several learning models improves the overall output. To accomplish the precision and reliability of the prediction, RF constructs numerous decision trees and blends them. In this approach, it eliminates overfitting by generating random subsets of the features, constructing smaller trees from these subsets and combining them to improve overall performance. RF assigns a sample to the class with the most "votes" in each subtree. While developing the trees, RF makes the model more random. When splitting a node, it scans for the best element from a random group of characteristics rather than looking for the most significant feature. As a result, there is a wide range of possibilities, which leads to a more comprehensive model. Some of the most common hyperparameters for the training of an RF-based model are max features, the number of estimators, min sample leaf, and max depth. The max feature parameter specifies the maximum number of features that RF is permitted to attempt in a single tree. For example, if the total number of variables is 100, only 10 of them can be used in a single tree. The number of estimators refers to the number of trees that are built before computing the maximum voting or prediction averages. A larger number of trees improves performance but necessitates more computing power. The min sample leaf parameter specifies the minimal number of samples that must be present at a leaf node. The maximum depth parameter reflects the maximum depth of each tree in the forest. The more splits there are in the tree, the more information it collects about the data. In Fig. 30 a typical structure of an RF algorithm is presented.

19.2.5 Support Vector Machine Algorithm

Support vector machine (SVM) is a binary problem-solving method. SVMs use the kernel technique dot product to turn the input feature space into a higher-dimensional feature space. The sample distance of each dataset to a specific dividing hyperplane may be calculated. Margin is defined as the shortest distance between the samples and the hyperplane. A hyperplane, or dividing curve between various classes, can be used to divide the altered data. The best hyperplane optimizes the profit margin. Its purpose is to categorize a fresh sample by calculating its distance from the hyperplane. SVMs, which are based on global optimization, deal with overfitting difficulties that arise in high-dimensional spaces, making them interesting in a variety of applications [118, 119]. Most used SVM algorithms include the support vector regression [120], least-squares SVM [121], and successive projection algorithm-SVM [122]. In other terms, an SVM is a linear separator that focuses on building the biggest feasible margin in a hyperplane. Its purpose is to categorize a fresh sample by calculating its distance from the hyperplane. The hyperplane is a single line that divides two classes in a two-dimensional feature space. An SVM cannot linearly classify data in a multi- dimensional feature space if the data are nonlinearly separable. It employs the kernel method in this situation. The basic idea is that the new multidimensional feature space may contain a linear decision boundary that was not linear in the original feature space. Commonly used SVM hyperparameters are C, gamma, and kernel. The C parameter informs the SVM optimizer how much you don't want to misclassify each training example. When C is big, the optimization will choose a smaller-margin hyperplane if it performs a better job of properly categorizing all of the training points. The gamma parameter specifies how far the influence of a single training example extends, with low values suggesting "far" and high values indicating "close." The gamma parameters may be thought of as the inverse of the model's radius of effect for samples selected as support vectors. A kernel function is a way of taking data as input and transforming it into the needed form for processing. The term "kernel" is chosen because the collection of mathematical functions utilized in SVM provides a window through which data can be manipulated. In Fig. 31 a typical structure of an SVM algorithm is presented.

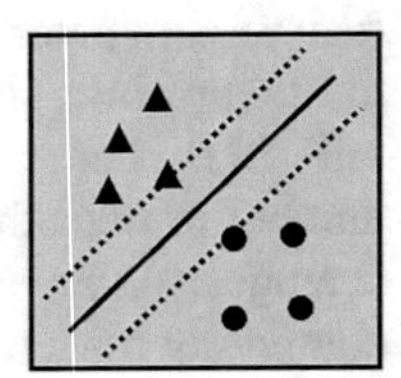

Fig. 31 SVM-based algorithm

19.2.6 GAN Algorithm

GL algorithms aim to generate new synthetic samples, and they can be applied as a solution for the imbalanced datasets. This section is mentioned GL-based algorithms which can be used for the synthesis of new samples for database cases such as normal and infected circuits. For the development of GL-based models, as many models as the number of classes that are contained in the dataset must be developed. Then, depending on the algorithm which will be used, there may be a need to apply some clustering algorithms. With the use of the clustering algorithms, the user will be able to cluster each given class into sub-classes to be able to use the class label as an extra feature.

GANs are based on CNNs and consist of a complex DL algorithm. GANs were designed and introduced by Goodfellow et al. [62] in 2014. GAN algorithms consisted of two models which are trained simultaneously by an adversarial process. The two models are a generator ("the artist") and the discriminator ("the art critic"). The purpose of the generator is to learn and to create samples that look real like the training samples. On the other hand, the discriminator aims to learn and to distinguish the real from the fake samples. During the training, the generator becomes better and can synthesize samples that look real, while the discriminator becomes better at distinguishing them from the real samples. The process finishes when the discriminator can no longer distinguish real samples from fakes.

19.2.7 CGAN Algorithm

Another GL-based algorithm for the synthesis of new samples is the CGANs [63]. CGANs are close to the philosophy of the GANs. The only difference is that CGAN uses an extra feature for the training of the model which is the class labels.

19.2.8 WGAN Algorithm

WGANs were developed by Arjovsky et al. [64] in 2017. WGANs are based on the main idea of GANs with the difference that for the synthesis of generated samples, WGANs use as an extra feature the Wasserstein distance. An improved version of WGANs is the algorithm WCGANs. Specifically, WCGANs were developed in 2018 by Qin et al. [65]. WGANs have the same function as WGANs with the difference that for the synthesis of new generated samples, they use as an extra feature the label of the classes, from the training set.

Common hyperparameters for the training of a GL-based model are learning rate, data dimensional, activation, kernels, optimizers, layer dense, and according to the algorithm the class labels of the dataset. Learning rate controls how efficiently the algorithm descends the gradient descent by evaluating each tree's contribution to the ultimate outcome. Data dimensional is the number of features that will be considered to determine the optimal split. The activation hyperparameter consists of an activation function that defines how the weighted sum of the input is turned into an output from a node or nodes in a network layer. Kernels are used in convolutional layers to extract features. They are essentially filters that are applied to the input data. They are implemented as matrices, with the kernel "moving" above the input data and the dot product between the kernel and the sub-region of the input matrix below it being calculated at each step, with the result being a matrix of the dot products. Optimizers are algorithms or methods that are used to change the neural network's attributes such as weights and learning rate to reduce losses. By minimizing the function, optimizers are used to solve optimization problems. Layer dense in any neural network is deeply connected to the layer before it, which indicates that the neurons in the layer are connected to every neuron in the layer before it. In artificial neural network networks, this is the most widely utilized layer.

19.3 Evaluation

Once it was completed – the steps of data collection and preparation and trained the model – it is time for the evaluation of the model. For the evaluation of the model, a test set which mainly consisted of 20% of the total dataset is used, and the samples of this set are unknown to the model. For example, in the case of HT classification, the test set consisted of unknown infected and free circuits features which the model will process for the first time and needs to classify.

19.3.1 Metrics for Classification ML-Based Algorithms

According to the ML-based methods presented, the problem of HTs consisted of a classification problem. In this section the metrics which are used for the evaluation of a classification model are presented.

To evaluate the performance of ML classification, algorithms used specific metrics: accuracy, precision, recall or sensitivity, specificity, 1-specificity, and F1 score. These metrics for their evaluation need four values: true positive (TP), false positive (FP), true negative (TN), and false negative (FN) values. TP and TN values consist of the correct predictions for the positive and the negative class, while FP and FN values consist of the error predictions for the positive and negative classes, respectively. According to each study, TP values can be used to indicate the infected circuits or the normal circuits, respectively.

Accuracy is defined as the number of correct predictions divided by the total number of predictions (1). Accuracy consists of the simplest of metrics and is not the best metric for the evaluation of a model. For example, when the dataset consisted of imbalanced classes, the model will be highly accurate which is wrong, because the model will predict all samples as the most frequent class. Therefore, the users need to look at class-specific performance metrics.

$$\text{Accuracy} = (\text{TP} + \text{TN}) / (\text{TP} + \text{TN} + \text{FP} + \text{FN}) \tag{1}$$

Precision consists of one of these metrics and is defined as the total number of TP values divided by the total number of all positive values (2).

$$\text{Precision} = \text{TP}/ (\text{TP} + \text{FP}) \tag{2}$$

On the other hand, recall or sensitivity is defined as the total number of TP values divided by the total number of TP and FN values (3). Recall can be characterized as the true positive rate (TPR).

$$\text{Recall} = \text{TP}/ (\text{TP} + \text{FN}) \tag{3}$$

Specificity is defined as the total number of TN values divided by the total number of TN and FP values (4) and can be characterized as the true negative rate (TNR).

$$\text{Specificity} = \text{TN}/ (\text{TN} + \text{FP}) \tag{4}$$

1-Sensitivity is defined as the total number of FP values divided by the total number of TN and FP values (5).

$$\text{1-Specificity} = \text{FP}/ (\text{TN} + \text{FP}) \tag{5}$$

Lastly, F1 is the harmonic mean of Precision and Recall and is defined by the multiplication of Precision by Recall and then by number two divided by the product of Precision and Recall (6).

$$\text{F1 Score} = 2 \left(\text{Precision}^{*}\text{Sensitivity}\right) / (\text{Precision} + \text{Sensitivity}) \tag{6}$$

19.3.2 Metrics for the Evaluation of GL-Based Algorithms

As proposed in the previous chapter, the differences between the GL algorithms for the synthesis of new samples are in use or not of extra features. For this reason, the evaluation metrics are different between the mentioned algorithms. First, the metrics for GANs and CGANs algorithms and next for the Wasserstein-based algorithms are introduced.

To evaluate the performance of GANs and CGANs algorithms, loss functions are used. Specifically, GAN- and CGAN-based models use two types of loss functions for their evaluation, a generator and a discriminator loss function. The generator loss function evaluates how effectively the generator tricked the discriminator. For example, the discriminator will classify the fake samples as real when the generator performs well. On the other hand, the discriminator loss function evaluates how well the discriminator can distinguish real from fake samples. As a result, for the calculation of loss functions in GAN and CGAN algorithms, the minmax loss (7) is used. E_x is the expected value for real sample instances, and E_z is the expected value over all random inputs to the generator. $D_{(x)}$ is the discriminator's estimate of the probability that real data instance x is real. Specifically, $D_{(x)}$ is the output for a real instance at the discriminator. $G_{(z)}$ is the output when given noise z, at the generator (z is random noise based on a bell curve from a Gaussian distribution and produces sample values selected by the generator). $D(G_{(z)})$ is the output for a fake instance at the discriminator.

$$\text{Minmax Loss} = E_x\left[\log\left(D(x)\right)\right] + \mathrm{E}_z\left[\log\left(1 - D\left(G(z)\right)\right)\right] \tag{7}$$

On the other hand, WGAN and WCGAN algorithms do not classify instances but produce an output number with values between 0 and 1. For the evaluation of the discriminator, the Wasserstein discriminator loss metric (8) is used and for the generator, the Wasserstein Generator Loss metric (9). On WGANs and WCGANs, the aim of the discriminator is to increase the output between the real and fake instances. $D(x)$ is the discriminator's output for real instances, $G(z)$ is the generator's output for the given noise z, and $D(G(z))$ is the discriminator's output for fake instances.

$$\text{Wasserstein Discriminator Loss} = D(x) - D\left(G(z)\right) \tag{8}$$

$$\text{Wasserstein Generator Loss} = D\left(G(z)\right) \tag{9}$$

19.4 Hyperparameter Tuning

Once completed, the evaluation step, according to the performance of the model based on the aforementioned metrics, may need to optimize the model based on a set of hyperparameters. To train a model based on ML or DL, a set of default hyperparameters is used. If the model evaluation is not effective, combinations with the hyperparameter sets must be made to enable optimization of the model. A classic technique is to give each hyperparameter a list of values and then combine all the hyperparameter values together to find the most effective evaluation result.

20 Languages, Frameworks, and Tools

The development of an ML-based or DL-based model has required the use of tools. There is a huge gamma of tool combinations. For the development of a model, it is important to know three basic things: the programming language, the framework, and the environment that will be used and combined for the implementation of the model. In the world, literature has reported a variety of combinations of programming languages, frameworks, and development tools.

The most common programming languages for the development of ML- or DL-based models are Python [123] and R [124]. Python is a general-purpose programming language, which means it can be used to develop a wide range of applications and is not specialized for any particular problem. R, on the other hand, is a programming language as well as a free software environment for statistical computation and graphics. It is commonly used by statisticians and data miners to create statistical applications and do data analysis.

There is a great variety of frameworks that are differentiated according to the user. The most common in use are, Tensorflow [125], Keras [126], and PyTorch [127] which mostly are used for the development of DL-based models. Specifically, TensorFlow is a free and open-source machine learning and deep learning software library. It may be used for a variety of tasks; however, it is most commonly employed for DNN training and inference. Keras, on the other hand, is a Tensor-Flow-based deep learning API built-in Python. It was created to allow for quick experimentation. It is critical to be able to get from concept to outcome as quickly as feasible when conducting research. For ML-based models, the most usable framework is the Scikit-Learn. Scikit-Learn is a robust library for ML in Python. It offers a set of efficient tools for ML and statistical modeling, including classification, regression, clustering, and dimensionality reduction, via a Python interface. OpenCV [128] is another important framework for the construction of CV-based models. It is an open-source library that is particularly helpful for computer vision applications such as video analysis, CCTV footage analysis, and picture analysis.

The most usable development tools for the creation of a model are Jupyter Notebook [129], PyCharm, and Microsoft Visual Studio Code. The Jupyter Notebook is an open-source web tool that allows data scientists to create and share documents that contain live code, equations, computational output, visualizations, and other multimedia elements, as well as explanatory text. PyCharm is a Python-integrated development environment (IDE) that provides a wide range of necessary tools for Python developers. These tools are tightly integrated to offer a pleasant environment for productive Python, web, and data science development. Microsoft Visual Studio Code is a simplified code editor that supports development processes such as debugging, task execution, and version control. It tries to give only the tools required by a developer for a speedy code-build-debug cycle, leaving more complicated processes to full-featured IDEs. A not-so-common tool is Spyder, which is an open-source cross-platform IDE for scientific programming in the Python language. Another useful tool is Anaconda [130]. Anaconda is a Python and

R programming language distribution for scientific computing, such as data science, ML applications, large-scale data processing, and predictive analytics, to simplify package management and deployment using virtual environments. Furthermore, Anaconda enables the creation of virtual environments which installed all the necessary programming languages, frameworks, and development tools.

21 Conclusions

Despite substantial research efforts over the years to build scalable and automated security validation methodologies, there are still many hurdles to designing secure and trustworthy ICs. There is currently no one-size-fits-all answer to HT attacks. ICs are often built to perform multiple functions, which introduces significant differences between them because even minor changes might affect their entire operation. These variables significantly increase the difficulty of designing HT prevention/detection systems that can be deployed uniformly on ICs intended for a variety of purposes. Furthermore, the sophistication of ICs is continually increasing, making the bulk of present HT countermeasures outdated.

Classical methodologies were established more than a decade ago, setting the framework for meticulously planning and strategizing HT detection with SCA-based and simulation-based techniques like LT. Despite their accuracy, these analytical frameworks lacked scalability and generalizability across a wide range of circuit types and sizes. The need for fresh ideas to overcome the barriers immediately became apparent, paving the route for the ultimate adoption of ML. However, as strong as ML might be, it should not be considered a panacea because it comes with its own set of idiosyncrasies and issues.

The most crucial stage in developing a fundamentally robust ML model is the creation of a training dataset that embeds the underlying variability of the queried domain. Experts in the area would quickly consider getting data from small-sized circuits and using it to train an ML system for detecting HTs in large-sized circuits as wishful thinking. Furthermore, the process of feature extraction and subsequent selection is always constrained by the art of robust dataset development, particularly in the case of ML algorithms that lack inherent methods for reducing the side effects of redundant or non-informative features. This area of machine learning is heavily reliant on domain specialists, who must commit substantial time and effort to collect and assess the features that will be used to train the ML algorithms.

Unfortunately, the bulk of previous ML-based research attempting to address the HT detection problem failed to overcome the aforementioned limitations. The authors presented ML algorithms trained on a tiny number of samples generated by circuits covering a narrow range of types and sizes, unavoidably resulting in overfitting and models that fail to compete with sophisticated opponents. Furthermore, the generalizability of these models to new circuits would be called into doubt because it was never tested, even via subsequent studies that sought to push the boundaries even further. Furthermore, except for one research, none of the prior studies used industry-level software in the process of reviewing circuit designs and extracting features/values that were as close to their printed equivalent as feasible. Instead, they opted for open-source solutions, which include a trade-off between financial/time cost and precision in measured feature values. Except for a few studies that performed feature selection before training, ML was largely regarded as a black box, with little effort expended in explaining what drives the algorithm's conclusions, a process that identifies feature combinations that are significant for the classification objective. The lack of publicly accessible Trojan-free and trojan-infected IC designs is a serious restriction in the HT detection field. Even though Trust-Hub contains over 1000 Trojan-infected designs, they are all based on a tiny number of Trojan-free circuits. Because adversaries already have access to the HT detection training data, they know what degree of resistance to expect; nevertheless, the scientific community is unable to keep up with the level of complexity observed in HT assaults due to a lack of plentiful and easily accessible data.

The focus on GLN is a recurring theme in existing ML-based studies, most likely due to the ease of access and feature extraction for this particular IC design phase. Because Trojans may be injected at any point of the process, this severely limits research into how HTs can be discovered throughout the IC production process. Furthermore, the great majority of Trust-Hub circuit designs are "friendly" to FPGAs, leaving the ASIC HT detection environment mostly untouched.

We believe that technological developments in the twenty-first century are forcing the HT detection sector to rapidly evolve. However, the path is hampered by ambiguity about how to properly use the benefits of ML, a paucity of publicly available data, intellectual property rights, and the fact that industrial-scale IC design is not, by definition, a conventional research area. Existing HT detection studies have contributed significantly to publicizing the socio-economical direct and indirect repercussions of HT insertion in devices critical to our society's homeostasis. Our understanding of how HTs operate has developed substantially over the years, laying a solid basis from which we can evaluate the most recent technical developments and seek to address the ever-growing cluster of new challenges that lie ahead.

One of these challenges is bridging the gap between research data that is scattered throughout multiple ID design phases. A new study must encompass the full IC production chain, which is made up of phases with various characteristics, features that may be retrieved, and challenges to avoid. The accumulation of study data that comply with this approach framework will help us understand how HT works and how it is dependent on the design phase chosen as the insertion point. Knowledge transfer within disciplines of HT detection research, such as the combination of techniques concentrating on various manufacturing phases, modes of operation, or even a mix of software-oriented and hardware-only solutions, can push the frontiers of HT prevention even farther.

The future of HT countermeasure approaches comprises meticulous and planned procedures for the production of diversified and scalable solutions that integrate cutting-edge AI algorithms and features extracted software specifically intended for industrial-precision level. On the other hand, any research that uses ML should commit a significant portion of its resources to probe the mechanisms underlying the ML algorithm's judgments. Such holistic approaches should prioritize explainable AI since it is capable of condensing the information obtained by these complicated algorithms and highlighting problems in specific design phases, features, methodologies, software, and production pipelines.

Robust HT countermeasures will limit socio-economic losses from malicious hardware to a minimum, but disclosing high-order linkages between subcomponents of the IC manufacturing chain may have a disruptive influence on how the IC Industry is distributed internationally.

References

1. Bhunia, S., et al.: Protection against hardware trojan attacks: Towards a comprehensive solution. IEEE Des. Test. **30**(3), 6–17 (2013). https://doi.org/10.1109/MDT.2012.2196252
2. Mitra, S., Wong, H.S.P., Wong, S.: The Trojan-proof chip. IEEE Spectrum. (2015). https://doi.org/10.1109/MSPEC.2015.7024511
3. Narasimhan, S., et al.: Hardware trojan detection by multiple-parameter side-channel analysis. IEEE Trans. Comput (2013). https://doi.org/10.1109/TC.2012.200
4. Amelian, A., Borujeni, S.E.: A Side-Channel Analysis for Hardware Trojan Detection Based on Path Delay Measurement. J. Circuits. Syst. Comput. (2018). https://doi.org/10.1142/S0218126618501384
5. He, J., Zhao, Y., Guo, X., Jin, Y.: Hardware Trojan detection through Chip-free electromagnetic side-channel statistical analysis. IEEE Trans. Very Large Scale Integr. Syst. (2017). https://doi.org/10.1109/TVLSI.2017.2727985
6. Nourian, M.A., Fazeli, M., Hely, D.: Hardware Trojan detection using an advised genetic algorithm based logic testing. J. Electron. Test. Theory Appl. (2018). https://doi.org/10.1007/s10836-018-5739-4
7. Chakraborty, R.S., Wolff, F., Paul, S., Papachristou, C., Bhunia, S.: MERO: A statistical approach for hardware Trojan detection (2009). https://doi.org/10.1007/978-3-642-04138-9_28
8. Sklavos, N., Chaves, R., Di Natale, G., Regazzoni, F.: Hardware security and trust: Design and deployment of integrated circuits in a threatened environment. 2017.
9. Bao, C., Xie, Y., Liu, Y., Srivastava, A.: Reverse engineering-based hardware trojan detection. In: The Hardware Trojan War: Attacks, Myths, and Defenses (2017)
10. Synthesis, C., Script, E., Design, C.: Synopsys design compiler tutorial. Technology (2002)
11. Bhunia, S., Hsiao, M.S., Banga, M., Narasimhan, S.: Hardware trojan attacks: Threat analysis and countermeasures. Proceedings of the IEEE. (2014). https://doi.org/10.1109/JPROC.2014.2334493
12. Hicks, M., Finnicum, M., King, S.T., Martin, M.M.K., Smith, J.M.: Overcoming an untrusted computing base: Detecting and removing malicious hardware automatically. (2010). https://doi.org/10.1109/SP.2010.18
13. King, S.T., Tucek, J., Cozzie, A., Grier, C., Jiang, W., Zhou, Y.: Designing and implementing malicious hardware (2008)
14. Tehranipoor, M., Koushanfar, F.: A survey of hardware trojan taxonomy and detection. IEEE Design Test Comp. (2010). https://doi.org/10.1109/MDT.2010.7
15. Karri, R., Rajendran, J., Rosenfeld, K., Tehranipoor, M.: Trustworthy hardware: Identifying and classifying hardware trojans. Computer (Long. Beach. Calif). (2010). https://doi.org/10.1109/MC.2010.299

16. Samuel, A.L.: Some studies in machine learning using the game of checkers. IBM J. Res. Dev (2000). https://doi.org/10.1147/rd.441.0206
17. McCulloch, W.S., Pitts, W.: A logical calculus of the ideas immanent in nervous activity. Bull. Math. Biophys. (1943). https://doi.org/10.1007/BF02478259
18. Rosenblatt, F.: The perceptron: A probabilistic model for information storage and organization in the brain. Psychol. Rev. (1958). https://doi.org/10.1037/h0042519
19. Pal, S.K., Mitra, S.: Multilayer perceptron, fuzzy sets, and classification. IEEE Trans. Neural Networks. (1992). https://doi.org/10.1109/72.159058
20. Kelley, H.J.: Gradient theory of optimal flight paths. ARS J. (1960). https://doi.org/10.2514/8.5282
21. Riedmiller, M., Braun, H.: Direct adaptive method for faster backpropagation learning: The RPROP algorithm (1993). https://doi.org/10.1109/icnn.1993.298623
22. Hecht-Nielsen, R.: Applications of counterpropagation networks. Neural Networks (1988). https://doi.org/10.1016/0893-6080(88)90015-9
23. Broomhead, D., Lowe, D.S.: Multivariable functional interpolation and adaptive networks. Complex Sys. **2**, 321–355 (1988)
24. Melssen, W., Wehrens, R., Buydens, L.: Supervised Kohonen networks for classification problems. Chemom. Intell. Lab. Syst. **83**(2), 99–113 (2006). https://doi.org/10.1016/j.chemolab.2006.02.003
25. Hopfield, J.J.: Neural networks and physical systems with emergent collective computational abilities. Proc. Natl. Acad. Sci. U. S. A. (1982). https://doi.org/10.1073/pnas.79.8.2554
26. Specht, D.F.: A general regression neural network. IEEE Trans. Neural Networks. (1991). https://doi.org/10.1109/72.97934
27. Liou, C.Y., Cheng, W.C., Liou, J.W., Liou, D.R.: Autoencoder for words. Neurocomputing (2014). https://doi.org/10.1016/j.neucom.2013.09.055
28. Jang, J.S.R.: ANFIS: Adaptive-network-based fuzzy inference system. IEEE Trans. Syst. Man Cybern. (1993). https://doi.org/10.1109/21.256541
29. Bin Huang, G., Zhu, Q.Y., Siew, C.K.: Extreme learning machine: Theory and applications. Neurocomputing. (2006). https://doi.org/10.1016/j.neucom.2005.12.126
30. Cao, J., Lin, Z., Bin Huang, G.: Self-adaptive evolutionary extreme learning machine. Neural Process. Lett (2012). https://doi.org/10.1007/s11063-012-9236-y
31. Hasman, A.: Probabilistic reasoning in intelligent systems: Networks of plausible inference. Int. J. Biomed. Comput. (1991). https://doi.org/10.1016/0020-7101(91)90056-k
32. Neapolitan, R.E.: Models for reasoning under uncertainty. Appl. Artif. Intell. (1987). https://doi.org/10.1080/08839518708927979
33. Ligeza, A.: Artificial intelligence: A modern approach. Neurocomputing. **9**(2), 215–218 (1995). https://doi.org/10.1016/0925-2312(95)90020-9
34. Ali, K., Jamali, A., Abbas, M., Ali Memon, K., Aleem Jamali, A.: Multinomial naive Bayes classification model for sentiment analysis. IJCSNS Int. J. Comput. Sci. Netw. Secur. (2019)
35. Ontivero-Ortega, M., Lage-Castellanos, A., Valente, G., Goebel, R., Valdes-Sosa, M.: Fast Gaussian Naïve Bayes for searchlight classification analysis. Neuroimage (2017). https://doi.org/10.1016/j.neuroimage.2017.09.001
36. Tryon, R.C.: Communality of a variable: Formulation by cluster analysis. Psychometrika (1957). https://doi.org/10.1007/BF02289125
37. Lloyd, S.P.: Least squares quantization in PCM. IEEE Trans. Inform. Theory. (1982). https://doi.org/10.1109/TIT.1982.1056489
38. Johnson, S.C.: Hierarchical clustering schemes. Psychometrika (1967). https://doi.org/10.1007/BF02289588
39. Dempster, A.P., Laird, N.M., Rubin, D.B.: Maximum likelihood from incomplete data via the EM algorithm. J. Royal Stat. Soc., Series B. **39**(1), 1–22 (1977). https://doi.org/10.1111/j.2517-6161.1977.tb01600.x
40. Yuan, Y., Chen, X., Chen, X., Wang, J.: Segmentation transformer: Object-contextual representations for semantic segmentation. arXiv Prepr. (2021)

41. Touvron, H., Vedaldi, A., Douze, M., Jégou, H.: Fixing the train-test resolution discrepancy (2019)
42. Tan, M., Pang, R., Le, Q.V.: EfficientDet: Scalable and efficient object detection (2020). https://doi.org/10.1109/CVPR42600.2020.01079
43. Belson, W.A.: Matching and prediction on the principle of biological classification. Appl. Stat. (1959). https://doi.org/10.2307/2985543
44. Breiman, L., Friedman, J. H., Olshen, R. A., Stone, C. J.: Classification and regression trees (2017)
45. Kass, G.V.: An exploratory technique for investigating large quantities of categorical data. Appl. Stat. (1980). https://doi.org/10.2307/2986296
46. Hormann, A.M.: Programs for machine learning part I. Inf. Control (1962). https://doi.org/10.1016/S0019-9958(62)90649-6
47. Lecun, Y., Bengio, Y., Hinton, G.: Deep learning. Nature. (2015). https://doi.org/10.1038/nature14539
48. Milosevic, N.: Introduction to convolutional neural networks (2020)
49. Salakhutdinov, R., Hinton, G.: Deep Boltzmann machines (2009)
50. Hua, Y., Guo, J., Zhao, H.: Deep Belief Networks and deep learning (2015). https://doi.org/10.1109/ICAIOT.2015.7111524
51. Vincent, P., Larochelle, H., Lajoie, I., Bengio, Y., Manzagol, P.A.: Stacked denoising autoencoders: Learning useful representations in a deep network with a local denoising criterion. J. Mach. Learn. Res. (2010)
52. Medsker, L.R., Jain, L.C.: Recurrent neural networks design and applications. J. Chem. Inf. Model. (2013)
53. Hochreiter, S., Schmidhuber, J.: Long short term memory. Neural computation. Neural Comput. (1997)
54. Pearson, K.: LIII. On lines and planes of closest fit to systems of points in space. London, Edinburgh, Dublin Philos. Mag. J. Sci. (1901). https://doi.org/10.1080/14786440109462720
55. Leguina, A.: A primer on partial least squares structural equation modeling (PLS-SEM). Int. J. Res. Method Educ. (2015). https://doi.org/10.1080/1743727x.2015.1005806
56. Sarkar, P.: What is LDA: Linear discriminant analysis for machine learning. Knowledge Hut. (2019)
57. Schapire, R.E.: Explaining adaboost. In: Empirical Inference: Festschrift in Honor of Vladimir N. Vapnik (2013)
58. Breiman, L.: Bagging predictors. Mach. Learn. (1996). https://doi.org/10.1007/bf00058655
59. R. E. Schapire, "A brief introduction to boosting," 1999.
60. Friedman, J.H.: Greedy function approximation: A gradient boosting machine. Ann. Stat. (2001). https://doi.org/10.1214/aos/1013203451
61. Breiman, L.: Random forests. Mach. Learn (2001). https://doi.org/10.1023/A:1010933404324
62. Goodfellow, I., et al.: Generative adversarial networks. Commun. ACM. (2020). https://doi.org/10.1145/3422622
63. Mirza, M., Osindero, S.: Conditional Generative Adversarial Nets Mehdi. arXiv1411.1784v1 [cs.LG] 6 Nov 2014 Cond. (2018)
64. Arjovsky, M., Chintala, S., Bottou, L.: Wasserstein GAN Martin. arXiv:1701.07875. (2017)
65. Qin, S., Jiang, T.: Improved Wasserstein conditional generative adversarial network speech enhancement. Eurasip J. Wirel. Commun. Netw (2018). https://doi.org/10.1186/s13638-018-1196-0
66. Karras, T., Laine, S., Aila, T.: A style-based generator architecture for generative adversarial networks (2019). https://doi.org/10.1109/CVPR.2019.00453
67. Zhu, J.Y., Park, T., Isola, P., Efros, A.A.: Unpaired image-to-image translation using cycle-consistent adversarial networks (2017). https://doi.org/10.1109/ICCV.2017.244
68. Fix, E., Hodges, J.L.: Discriminatory analysis. Nonparametric discrimination: Consistency properties. Int. Stat. Rev./Rev. Int. Stat. **57**(3), 238 (1989). https://doi.org/10.2307/1403797

69. Kohonen, T.: Statistical pattern recognition Revisited. In: Advanced Neural Computers (1990)
70. Atkeson, C.G., Moore, A.W., Schaal, S.: Locally weighted learning. Artif. Intell. Rev. (1997). https://doi.org/10.1007/978-94-017-2053-3_2
71. Cortes, C., Vapnik, V.: Support-vector networks. Mach. Learn (1995). https://doi.org/10.1023/A:1022627411411
72. Kohonen, T.: The self-organizing map. Neurocomputing. **21**(1–3), 1–6 (1998). https://doi.org/10.1016/S0925-2312(98)00030-7
73. Devlin, J., Chang, M.-W., Lee, K., Google, K.T., Language, A.I.: BERT: Pre-training of deep bidirectional transformers for Language understanding. Naacl-Hlt. **2019** (2018)
74. Yang, Z., Dai, Z., Yang, Y., Carbonell, J., Salakhutdinov, R., Le, Q. V.: XLNet: Generalized autoregressive pretraining for language understanding (2019)
75. Park, K., Rothfeder, R., Petheram, S., Buaku, F., Ewing, R., Greene, W.H.: Linear regression. In: Basic Quantitative Research Methods for Urban Planners (2020)
76. Cox, D.R.: The regression analysis of binary sequences. Journal of the Royal Statistical Society, Series B. **21**(1), 238–238 (1959). https://doi.org/10.1111/j.2517-6161.1959.tb00334.x
77. Hutcheson, G., Hutcheson, G.: Ordinary least-squares regression. In: The SAGE Dictionary of Quantitative Management Research (2014)
78. Quinlan, J.R.: Learning with Continuous Classes (1992)
79. Cleveland, W.S.: Robust locally weighted regression and smoothing scatterplots. J. Am. Stat. Assoc. (1979). https://doi.org/10.1080/01621459.1979.10481038
80. Hoerl, A.E., Kennard, R.W.: Ridge regression: Biased estimation for nonorthogonal problems. Technometrics (1970). https://doi.org/10.1080/00401706.1970.10488634
81. Tibshirani, R.: Regression shrinkage and selection via the lasso. J. R. Stat. Soc. Ser. B (1996). https://doi.org/10.1111/j.2517-6161.1996.tb02080.x
82. Efron, B., et al.: Least angle regression. Ann. Stat (2004). https://doi.org/10.1214/009053604000000067
83. Han, W., et al.: ContextNet: Improving convolutional neural networks for automatic speech recognition with global context. (2020). https://doi.org/10.21437/Interspeech.2020-2059
84. Ravanelli, M., Brakel, P., Omologo, M., Bengio, Y.: Light gated recurrent units for speech recognition. IEEE Trans. Emerg. Top. Comput. Intell. (2018). https://doi.org/10.1109/TETCI.2017.2762739
85. He, K., Zhang, X., Ren, S., Sun, J.: Deep residual learning for image recognition (2016). https://doi.org/10.1109/CVPR.2016.90
86. Agrawal, D., Baktir, S., Karakoyunlu, D., Rohatgi, P., Sunar, B.: Trojan detection using IC fingerprinting. Proc. – IEEE Symp. Secur. Priv., 296–310 (2007). https://doi.org/10.1109/SP.2007.36
87. Chakraborty, R.S., Wolff, F., Paul, S., Papachristou, C., Bhunia, S.: MERO: A statistical approach for hardware Trojan detection. Lect. Notes Comput. Sci. (including Subser. Lect. Notes Artif. Intell. Lect. Notes Bioinformatics). **5747 LNCS**, 396–410 (2009). https://doi.org/10.1007/978-3-642-04138-9_28
88. Salmani, H., Tehranipoor, M., Plusquellic, J.: A novel technique for improving hardware Trojan detection and reducing Trojan activation time. IEEE Trans. Very Large Scale Integr. Syst. **20**(1), 112–125 (Jan. 2012). https://doi.org/10.1109/TVLSI.2010.2093547
89. Bao, C., Forte, D., Srivastava, A.: On application of one-class SVM to reverse engineering-based hardware Trojan detection (2014). https://doi.org/10.1109/ISQED.2014.6783305
90. Ngo, X.T., Danger, J.L., Guilley, S., Najm, Z., Emery, O.: Hardware property checker for run-time Hardware Trojan detection. 2015 Eur. Conf. Circuit Theory Des. ECCTD. **2015**, 1–4 (2015). https://doi.org/10.1109/ECCTD.2015.7300085
91. Hasegawa, K., Oya, M., Yanagisawa, M., Togawa, N.: Hardware Trojans classification for gate-level netlists based on machine learning (2016). https://doi.org/10.1109/IOLTS.2016.7604700

92. Liakos, K.G., Georgakilas, G.K., Plessas, F.C., Kitsos, P.: GAINESIS: Generative artificial intelligence NEtlists SynthesIS. Electron. **11**(2) (2022). https://doi.org/10.3390/electronics11020245
93. Agrawal, D., Baktir, S., Karakoyunlu, D., Rohatgi, P., Sunar, B.: Trojan detection using IC fingerprinting (2007). https://doi.org/10.1109/SP.2007.36
94. Aarestad, J., Acharyya, D., Rad, R., Plusquellic, J.: Detecting trojans through leakage current analysis using multiple supply pad IDDQs. IEEE Trans. Inf. Forensics Secur. (2010). https://doi.org/10.1109/TIFS.2010.2061228
95. Rad, R., Plusquellic, J., Tehranipoor, M.: A sensitivity analysis of power signal methods for detecting hardware trojans under real process and environmental conditions. IEEE Trans. Very Large Scale Integr. Syst. (2010). https://doi.org/10.1109/TVLSI.2009.2029117
96. Koushanfar, F., Mirhoseini, A.: A unified framework for multimodal submodular integrated circuits trojan detection. IEEE Trans. Inf. Forensics Secur. (2011). https://doi.org/10.1109/TIFS.2010.2096811
97. Lamech, C., Rad, R.M., Tehranipoor, M., Plusquellic, J.: An experimental analysis of power and delay signal-to-noise requirements for detecting trojans and methods for achieving the required detection sensitivities. IEEE Trans. Inf. Forensics Secur. (2011). https://doi.org/10.1109/TIFS.2011.2136339
98. Xiao, K., Zhang, X., Tehranipoor, M.: A clock sweeping technique for detecting hardware trojans impacting circuits delay. IEEE Des. Test. **30**(2), 26–34 (2013). https://doi.org/10.1109/MDAT.2013.2249555
99. Waksman, A., Suozzo, M., Sethumadhavan, S.: FANCI: Identification of stealthy malicious logic using boolean functional analysis. (2013). https://doi.org/10.1145/2508859.2516654
100. Zhang, J., Yuan, F., Wei, L., Liu, Y., Xu, Q.: VeriTrust: Verification for hardware trust. IEEE Trans. Comput. Des. Integr. Circuits Syst. (2015). https://doi.org/10.1109/TCAD.2015.2422836
101. Hasegawa, K., Yanagisawa, M., Togawa, N.: Trojan-feature extraction at gate-level netlists and its application to hardware-Trojan detection using random forest classifier (2017). https://doi.org/10.1109/ISCAS.2017.8050827
102. Inoue, T., Hasegawa, K., Yanagisawa, M., Togawa, N.: Designing hardware trojans and their detection based on a SVM-based approach. (2017). https://doi.org/10.1109/ASICON.2017.8252600
103. Liakos, K.G., Georgakilas, G.K., Plessas, F.C.: Hardware Trojan classification at gate-level netlists based on area and power machine learning analysis (2021). https://doi.org/10.1109/ISVLSI51109.2021.00081
104. Jap, D., He, W., Bhasin, S.: Supervised and unsupervised machine learning for side-channel based Trojan detection. In: Proceedings of the International Conference on Application-Specific Systems, Architectures and Processors, vol. 2016-Novem, pp. 17–24 (2016). https://doi.org/10.1109/ASAP.2016.7760768
105. Xue, M., Wang, J., Hux, A.: An enhanced classification-based golden chips-free hardware Trojan detection technique. (2017). https://doi.org/10.1109/AsianHOST.2016.7835553
106. Wang, S., Dong, X., Sun, K., Cui, Q., Li, D., He, C.: Hardware Trojan detection based on ELM neural network. 2016 1st IEEE Int. Conf. Comput. Commun. Internet, ICCCI 2016. **7**, 400–403 (2016). https://doi.org/10.1109/CCI.2016.7778952
107. Iwase, T., Nozaki, Y., Yoshikawa, M., Kumaki, T.: Detection technique for hardware Trojans using machine learning in frequency domain. 2015 IEEE 4th Glob. Conf. Consum. Electron. GCCE 2015, 185–186 (2016). https://doi.org/10.1109/GCCE.2015.7398569
108. Liu, Y., Jin, Y., Nosratinia, A., Makris, Y.: Silicon demonstration of hardware Trojan design and detection in wireless cryptographic ICs. IEEE Trans. Very Large Scale Integr. Syst. (2017). https://doi.org/10.1109/TVLSI.2016.2633348
109. Khalid, F., Hasan, S.R., Hasan, O., Awwad, F.: Runtime hardware Trojan monitors through modeling burst mode communication using formal verification. Integration. **61**(October 2017), 62–76 (2018). https://doi.org/10.1016/j.vlsi.2017.11.003

110. Bao, C., Forte, D., Srivastava, A.: Temperature tracking: Toward robust run-time detection of hardware Trojans. IEEE Trans. Comp. Des. Integr. Circuits Syst. **34**(10), 1577–1585 (2015). https://doi.org/10.1109/TCAD.2015.2424929
111. Mardani Kamali, H., Zamiri Azar, K., Gaj, K., Homayoun, H., Sasan, A.: LUT-lock: A novel LUT-based logic obfuscation for FPGA-Bitstream and ASIC-hardware protection (2018). https://doi.org/10.1109/ISVLSI.2018.00080
112. Kamali, H.M., Azar, K.Z., Homayoun, H., Sasan, A.: Full-lock: Hard distributions of SAT instances for obfuscating circuits using fully configurable logic and routing blocks (2019). https://doi.org/10.1145/3316781.3317831
113. Khaleghi, B., Ahari, A., Asadi, H., Bayat-Sarmadi, S.: FPGA-based protection scheme against hardware trojan horse insertion using dummy logic. IEEE Embedded Systems Letters. **7**(2), 46–50 (2015). https://doi.org/10.1109/LES.2015.2406791
114. Nejat, A., Shekarian, S.M.H., Saheb Zamani, M.: A study on the efficiency of hardware Trojan detection based on path-delay fingerprinting. Microprocess. Microsyst. **38**(3), 246–252 (2014). https://doi.org/10.1016/j.micpro.2014.01.003.
115. Shekarian, S.M.H., Saheb Zamani, M.: Improving hardware Trojan detection by retiming. Microprocess. Microsyst. (2015). https://doi.org/10.1016/j.micpro.2015.02.002
116. Salmani, H., Tehranipoor, M., Karri, R.: On design vulnerability analysis and trust benchmarks development (2013). https://doi.org/10.1109/ICCD.2013.6657085
117. Shakya, B., He, T., Salmani, H., Forte, D., Bhunia, S., Tehranipoor, M.: Benchmarking of hardware trojans and maliciously affected circuits. J. Hardw. Syst. Secur. (2017). https://doi.org/10.1007/s41635-017-0001-6
118. Suykens, J.A.K., Vandewalle, J.: Least squares support vector machine classifiers. Neural Processing Letters. **9**(3), 293–300 (1999). https://doi.org/10.1023/A:1018628609742
119. Chang, C.C., Lin, C.J.: LIBSVM: A library for support vector machines. ACM Trans. Intell. Syst. Technol. (2011). https://doi.org/10.1145/1961189.1961199
120. Smola, A.J., Schölkopf, B.: A tutorial on support vector regression. Stat. Comp. (2004). https://doi.org/10.1023/B:STCO.0000035301.49549.88
121. Suykens, J.A.K., Van Gestel, T., De Brabanter, J., De Moor, B., Vandewalle, J.: Basic methods of least squares support vector machines. In: Least Squares Support Vector Mach., pp. 71–116 (2002)
122. Galvão, R.K.H., et al.: A variable elimination method to improve the parsimony of MLR models using the successive projections algorithm. Chemom. Intell. Lab. Syst. (2008). https://doi.org/10.1016/j.chemolab.2007.12.004
123. Van Rossum, G., Drake, F.L.: Python reference manual (2006)
124. R Core Team: R: A language and environment for statistical computing. In: R Foundation for Statistical Computing, Vienna, Austria. 2020 (2020)
125. Abadi, M., et al.: TensorFlow: A system for large-scale machine learning (2016). https://doi.org/10.5555/3026877.3026899
126. Chollet, F.: Keras. J. Chem. Inf. Model (2013)
127. Paszke A., et al.: PyTorch: An imperative style, high-performance deep learning library (2019)
128. Bradski, G.: The OpenCV library. Dr. Dobb's J. Softw, Tools (2000)
129. Kluyver, T., et al.: Jupyter notebooks—A publishing format for reproducible computational workflows. (2016). https://doi.org/10.3233/978-1-61499-649-1-87
130. Anaconda: Anaconda Software Distribution. Computer software. Vers. 2-2.4.0. Anaconda, Nov. 2016. Web. Anaconda Soft. Distrib. Comp. Soft. (2016)

FPGA Security: Security Threats from Untrusted FPGA CAD Toolchain

Sandeep Sunkavilli, Zhiming Zhang, and Qiaoyan Yu

1 Introduction

Field programmable gate arrays (FPGAs) have gained popularity over the years and slowly made their way into advanced applications like machine learning, artificial intelligence, cloud services, military, and aerospace. Due to their flexibility in programming, FPGAs have become prevalent in system prototyping, hardware implementation for low-volume products, replacing obsolete components in legacy systems, and implementing hardware security modules [1, 2].

Due to the increase in market share and features like field programmability, FPGAs have become a target for attackers. FPGAs are subjected to traditional security threats like Trojan insertion, side-channel analysis, reverse engineering, and information leakage through a covert channel. The majority of research efforts on FPGA security includes inserting hardware Trojans [3, 4], reverse engineering intellectual property (IP) by decomposing or decrypting bitstream files [5], side-channel analysis attacks [6, 7], and using counterfeit devices [8] to degrade system performance. In existing literature the underlying FPGA CAD tool is considered trusted, and the investigation is performed typically on the stand-alone system as shown in Fig. 1. In a stand-alone system, FPGA users have physical access to the FPGA board and design suite. The regulations implemented in the design flow are simple.

Section 2 presents two types of FPGA CAD tools. Section 3 introduces traditional and emerging security threats in FPGA CAD tools and illustrates practical attacks. Section 4 proposes a new security threat landscape. Section 5 concludes this chapter and highlights future research directions.

S. Sunkavilli · Z. Zhang · Q. Yu (✉)
University of New Hampshire, Durham, NH, USA
e-mail: Sandeep.Sunkavilli@unh.edu; qiaoyan.yu@unh.edu

A. Iranmanesh (ed.), *Frontiers of Quality Electronic Design (QED)*,
https://doi.org/10.1007/978-3-031-16344-9_14

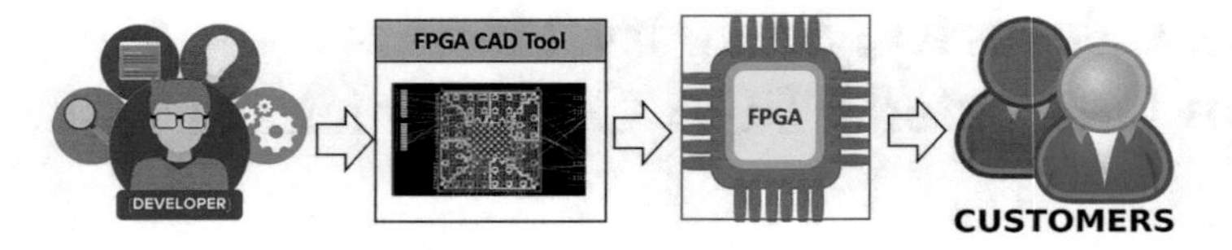

Fig. 1 Traditional FPGA utilization model [9]

2 Commercial and Open-Source FPGA CAD Tools

FPGA is a reprogrammable device that can be programmed with a different design from time to time. FPGA CAD tools play the most important role in FPGA functioning. The primary job of an FPGA CAD tool is to convert a design file written in hardware description language into a bitstream file, which configures the FPGA. FPGA CAD tool runs three major steps with the HDL file provided by the designer: synthesis, implementation, placement, and routing. With advancements in FPGA CAD tools, they provide rich simulation features such as post-synthesis, and implementation analysis, and also implement countermeasures like IP encryption, bitstream encryption, and design isolation. FPGA CAD tools are also providing IP modules to help designers with complex designs. Two major types of FPGA CAD tools, open-source FPGA CAD tools (e.g., VTR and Symbiflow) and commercial FPGA CAD tools (e.g., Xilinx ISE, Xilinx Vivado, Altera Quartus), are used in FPGA deployment. The FPGA design flow in commercial FPGA design suites can be found in FPGA handbooks and literature [10].

Different from commercial FPGA software, open-source FPGA CAD tools are developed to investigate the impact of various FPGA architectures and CAD algorithms on FPGA configuration. Due to their reliability and no cost, researchers and low-income businesses prefer open-source FPGA CAD tools. VTR and Symbiflow are the two popular and widely used open-source FPGA CAD tools. VTR [11] is an open-source tool that generates .net, .place, and .route files as the end product. VTR requires two input files: one is a hardware description file (Verilog), and the other is an architecture description file (EARCH.xml). As shown in Fig. 2, VTR tool chain is comprised of ODIN-II, ABC, and Versatile Place and Route (VPR) tools. ODIN-II is a synthesis and elaboration tool that takes the Verilog HDL file as an input file and synthesis netlist file. ABC takes the netlist file synthesized by ODIN as input and perform logic optimization, technology mapping, and produces two .blif files. The file generated by the ABC tool is the final and complete netlist file for the input Verilog file. VPR is the final tool in the VTR tool chain. VPR performs packing, placement, routing, and timing analysis. The information of packing, placement, and routing is stored in .net, .place, and .route respectively.

Symbiflow is an end-to-end open-source FPGA CAD tool that can generate programmable bitstream file for a given Verilog and architecture file. The tools used in the Symbiflow toolchain are Yosys, ABC, VPR, nextpnr, and open FPGA assembler as shown in Fig. 2. Yosys is similar to the ODIN-II used in VTR.

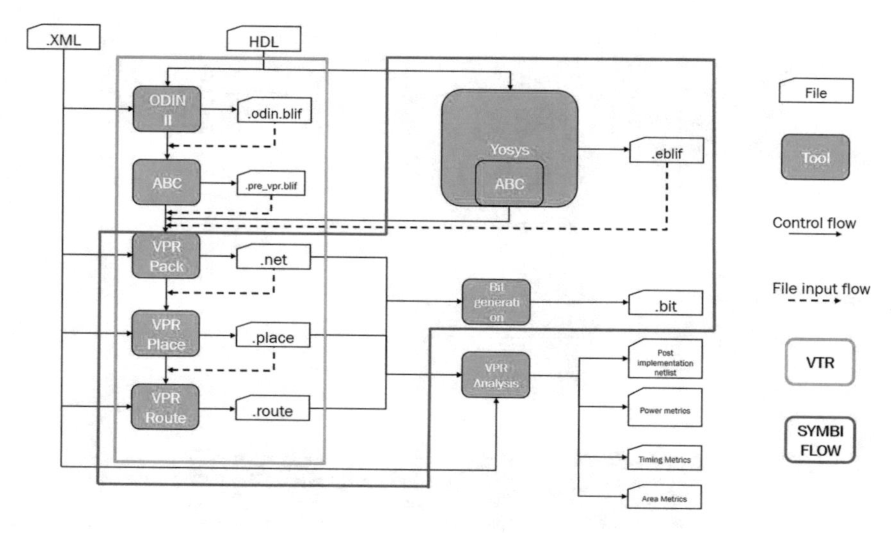

Fig. 2 Open-source FPGA CAD flow [12]

Yosys performs synthesis and converts a Verilog file into a netlist (.eblif) file. To perform placement and routing, Symbiflow is equipped with two different tools. For Xilinx-7 series FPGAs, VPR is used and for Lattice iCE40 FPGAs, nextpnr is used. Currently, Symbiflow supports only Xilinx-7 series, Lattice iCE40, and Lattice ECP5 FPGAs. PnR tools generate .fasam file, and VPR also generates .net, .place, and .route files in addition to fasam file. The fasam file is used to generate the bitstream file. Symbiflow also allows performing analysis on a critical path.

3 Security Threats from FPGA CAD Tools

In existing FPGA security literature, FPGA CAD tools are typically assumed trusted. Unfortunately, more and more attack examples [12] show that security threats from FPGA CAD tools are practical and imperative to address. As shown in Fig. 3a, our attack model assumes that the FPGA deployment engineers, in-house designs, the bitstream downloading channel, and procedure are trusted. The untrusted phase interested in this work is the FPGA configuration, especially the design mapping, place, and route stages. The attacks originated from malicious software mounted on top of the original FPGA design suite for SRAM FPGAs, as shown in Fig. 3b. The FPGA design suite may not be malicious initially, but advanced attackers could exploit the vulnerability of the FPGA design suite to implant malicious software into the original suite through software upgrading. We believe that the FPGA design suite will be propagated through a computer network or retailers, so the integrity of the software may be sabotaged by advanced attackers.

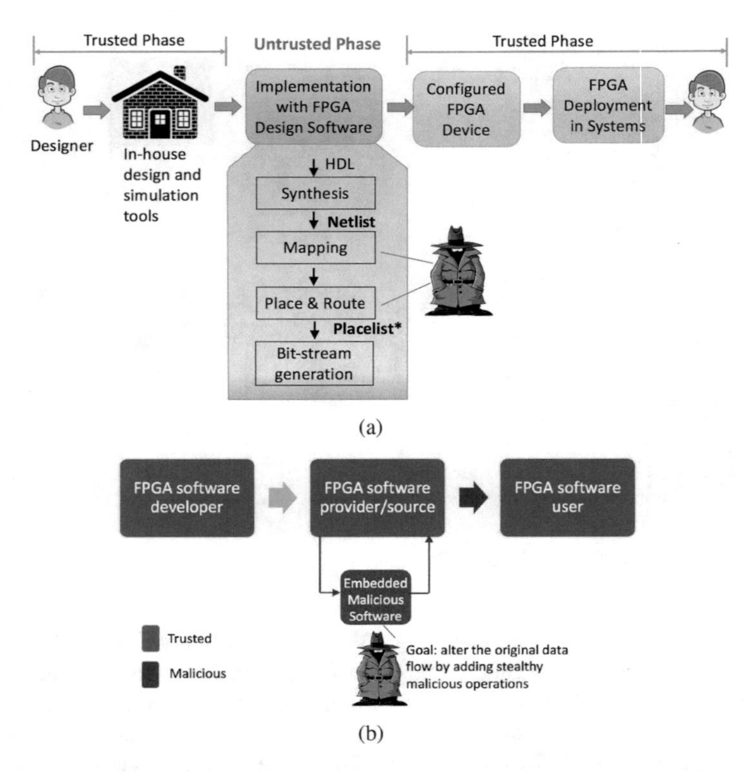

Fig. 3 Contaminated FPGA design suite leading to a stealthy modification on the placelist for an FPGA device. (**a**) Software compromising stage and (**b**) malicious software add-on in the supply chain of FPGA tools

3.1 Security Threats in Commercial FPGA CAD Tool

3.1.1 Attacks on Xilinx ISE

Figure 4 depicts the design flow for a Xilinx FPGA design suite. There are three potential attack surfaces for maliciously implanted FPGA tools to land on. We use Xilinx ISE 14.1 as an example in the following discussion. In the step of mapping, an attacker could introduce additional I/O pins, exchange the existing I/O pin connection, and modify the slew rate and the voltage level of I/O pins. As the tampered mapping output *_map.ncd** is not readable (unless the FPGA design suite provides a program like *ncd2xdl* to read back the native circuit description file), it is not easy to notice the modification performed by the malicious FPGA software. More tampering on the FPGA configuration can be done in the step of place and route (*PAR*) than in the mapping stage because all the LUTs, flip-flops, SRAM blocks, and interconnects are specifically designated on the FPGA die. The attack on the stage of bitstream generation is mainly for the purpose of IP piracy,

Fig. 4 Attack surfaces on the Xilinx FPGA design flow. The rectangles represent the output file from each step. The file with the symbol of * is an output file modified by the malicious FPGA software

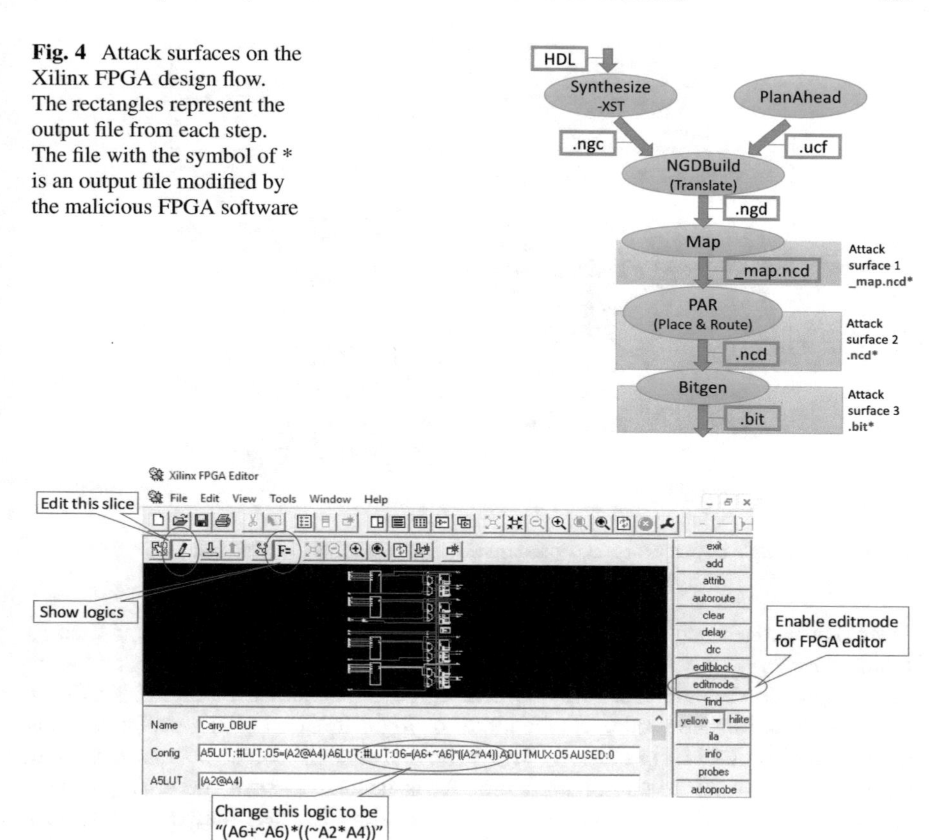

Fig. 5 An example of practical attack performed through the FPGA editor tool available in the Xilinx ISE 14.1 design suite [10]

which is out of the scope of this chapter. Our work focuses on the first two attack surfaces shown in Fig. 4.

We successfully modified the configuration of the target slice through the *FPGA editor* tool from Xilinx. Figure 5 shows the graphic interface. In the edit mode of the FPGA editor, we changed the logic configuration after the PAR stage and then re-did bitstream generation. The attack can also be performed via XDL file editing followed by the command *xdl2ndc*. All attack actions here can be implemented in a malicious FPGA software implanted in the original FPGA design suite.

3.1.2 Attacks on Altera Quartus

The Altera FPGA design suite, Quartus, leaves similar back doors for attackers to insert hardware Trojans. The security vulnerability of Quartus is in the process of

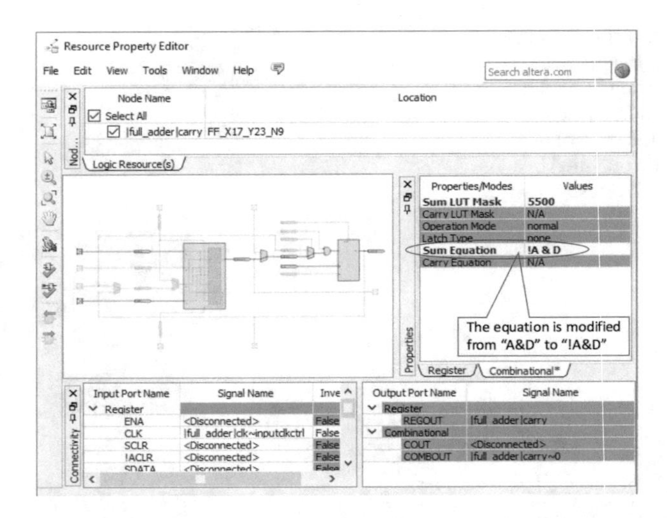

Fig. 6 An example of practical attack performed through Quartus Chip Planner [10]

placement and routing *Fitter*, like *PAR* in the Xilinix ISE. Attackers can, in theory, manipulate the entire FPGA configuration if they control *Fitter* or access and alter the design file that the tool *Fitter* is dealing with. As shown in Fig. 6, attackers can change buffer slew rate, I/O standard, or logic function of the design via the Quartus built-in tool *Chip Planner*. The malicious changes can be done after design compilation, and *no* re-compilation process is needed to save the changes. The attacks performed through *Chip Planner* are stealthy because they do not disturb the functional module in a format of the hardware description language and the constraint settings.

3.1.3 Attack Surfaces Induced by Integrating Countermeasures to Commercial CAD Tools

Encryption and isolation are two general categories of FPGA security countermeasures. Here we are going to analyze the potential risks of those techniques and demonstrate some case studies using Xilinx Vivado. Like Xilinx ISE, Xilinx Vivado is also an FPGA suite developed by Xilinx. Xilinx Vivado is the latest FPGA design suite developed by Xilinx which provides all the features provided by ISE and a few additional countermeasures at different levels of design flow.

Security Vulnerability in IP Encryption The goal of IP encryption is to protect the design IP modules from counterfeiting and reverse engineering. The access to the encrypted IP is based upon the access rights defined by the IP owner. Two important files in this encryption are an encrypted IP file and a key file. Any mismatch in the key of the CAD tool will block the user from using the IP. If

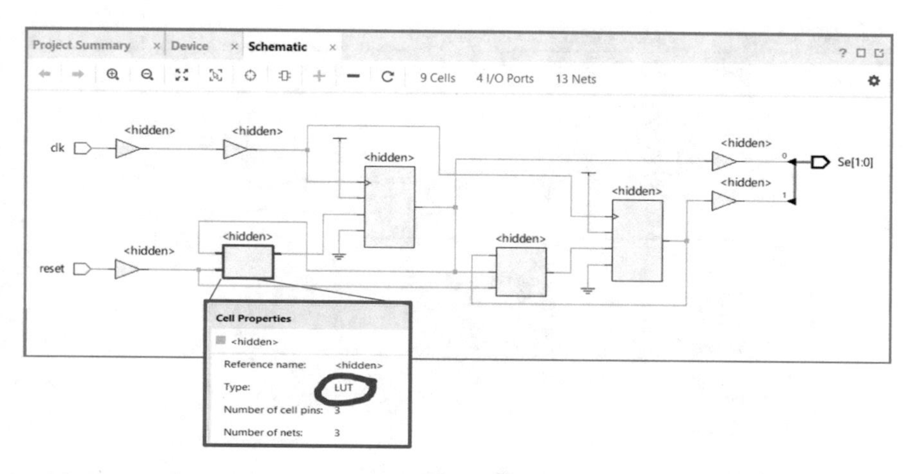

Fig. 7 Diagram of an instantiated encrypted counter [12]

the trustworthiness of the FPGA CAD tool is not guaranteed, there is a potential risk of key leaking. IP encryption still suffers from the risk of piracy even with a secured key. Even though IP is encrypted, the design net and pin names are still visible to the IP user as shown in Fig. 7. IP users could leverage this fact to trace the signals and precisely target the critical signals. The schematic view of an encrypted IP will also reveal the FPGA components used in the physical implementation of the module. As seen in Fig. 7, the hidden component used in the encrypted module is a LUT. It is also possible to find the hierarchy of the encrypted module instanced in a design. Figure 8 shows a partially encrypted linear-feedback shift register (LFSR), where the encrypted submodules d1, d2, d3, and d4 are described in the format of encrypted netlist *dff.vp*. This example shows that we can trace the relationship between the encrypted IPs and other submodules in the system design.

Security Vulnerability in Design Isolation Fence is a dedicated empty FPGA area that prohibits any logic from being implemented. The purpose of the fence is to isolate modules, thus thwarting the crosstalk attack. Figure 9a depicts an example of fence-based isolation. Xilinx introduces a concept called Isolation Design Flow (IDF) to physically and logically isolate modules with the fence. The fence-based isolation requires a slight modification to the conventional FPGA design flow. As shown in Fig. 9b, the two grey steps are added to the RTL elaboration and design synthesis phases. The property *HD.ISOLATE* directs the tool to create an isolated region. Six design rule check (DRC) rules [13] employed in Vivado can examine provenance and violations on I/O banks, package pins, floor planning, placement, and routing. However, the errors reported by DRC do not stop the designer from deploying the design into the FPGA device; instead DRC only warns the designer to be aware of the violations.

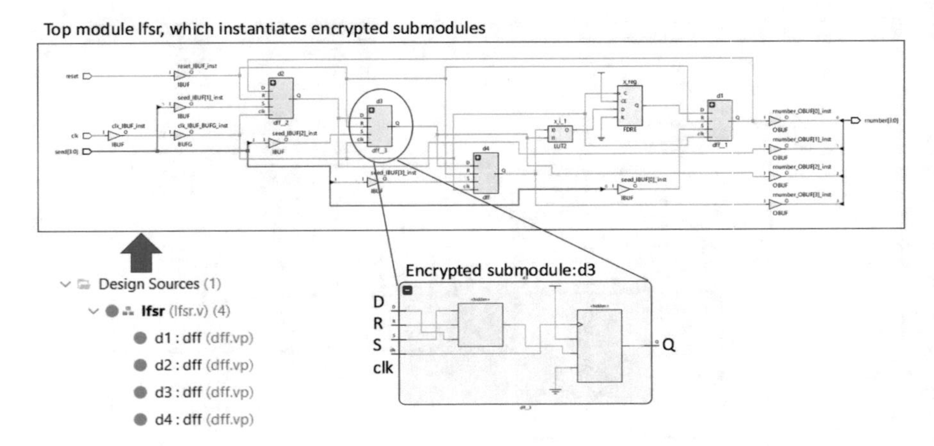

Fig. 8 Hierarchical view of a partially encrypted LFSR [12]

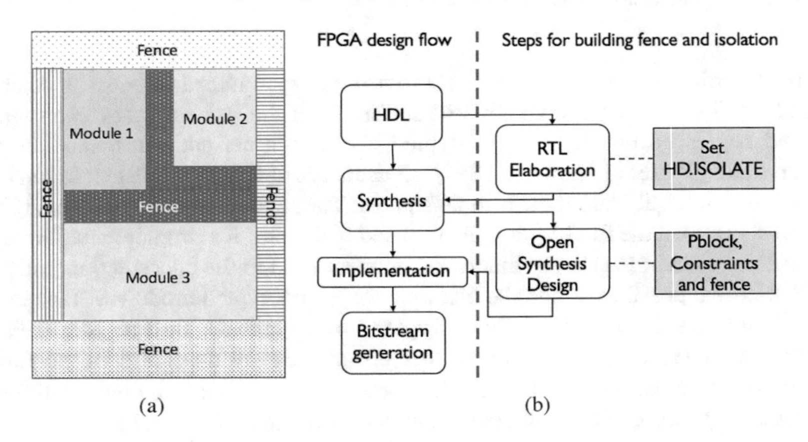

Fig. 9 Isolation design flow (IDF) in FPGA design. (**a**) Fence implementation and (**b**) modified floor planning in the design flow [12]

The isolation fence only prevents hardware Trojan insertion in the fence area, but it cannot thwart malicious modification of the design due to a few limitations. Limitation 1: Exception on global signals, trusted routing is the key for design isolation, and global signals are exempted. In trusted routing, each input is driven by one source, and each output only drives one load. This cannot be the case for global signals since they have to drive more than one input. This exemption could be exploited by attackers to disguise the interconnect for a Trojan as global signals. Limitation 2: Exception on non-isolated modules, a non-isolated module is not passed through the DRC check and can still communicate freely with isolated modules. According to the IDF rules, a module without the HD.ISOLATE property will not be protected by a fence, and thus it can be placed in any of the isolated regions (p-blocks). In the design shown in Fig. 10a, only M1, M2, and M3 with

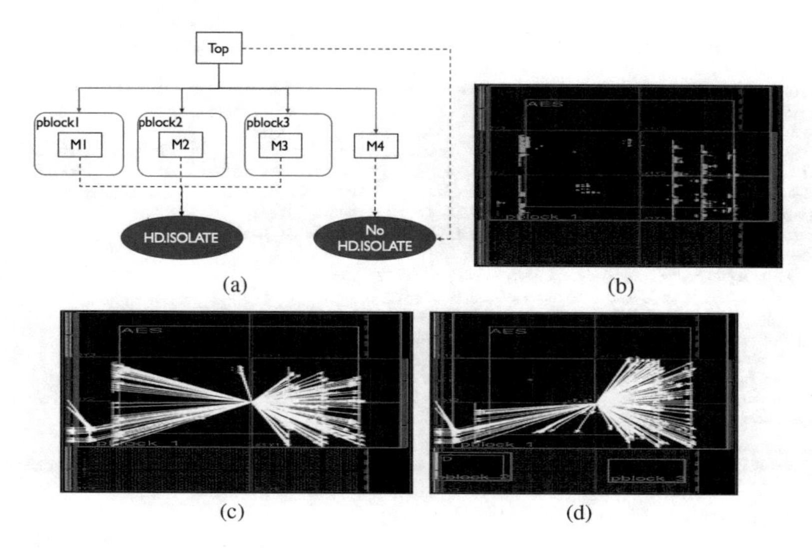

Fig. 10 An example for exception on non-isolated modules. (**a**) Top module overview of an IDF design, post-implementation device view for AES (**b**) without Trojans, with (**c**) a hardware Trojan in a single isolation p-block, and with (**d**) a hardware Trojan in one of the three p-blocks [12]

the isolation property will be configured in the three isolated FPGA regions, p-block1, p-block2, and p-block3, respectively. The remaining logic in the top module, including M4, can be placed in any of the three p-blocks above. Limitation 3: no protection mechanism at routing.

The instructions for fence implementation, isolation region creation, HD.ISOLATE property setting, and isolation region assignment for specific modules are stored in the constraints file. Fig. 11 illustrates a snapshot for a constraints file, which is used in the process of synthesis and implementation. To place a Trojan in a design, modifying the constraints file is a vital step. Any personal/tool who has access to the constraints file can make changes to the constraint setting, as there is no integrity check available for the constraints file. Limitation 4: unutilized resources. The FPGA resources allocated for an isolation region are not only used for the module to be isolated but also accessible for the remaining logic without isolation property in the top module. As the prediction of the size of the isolation region for the module under protection is not always precise, there are unutilized FPGA resources available for attackers to implement hardware Trojans.

We performed a case study in Vivado using design isolation limitations to implement a covert channel, which aims for leaking key information from an AES encryption module. The baseline is a top module only carrying an AES unit. The isolation property is set to the AES unit to form an isolated region (p-block1). The output after implementation is shown in Fig. 10b. Next, we introduced a hardware Trojan to the top module for the purpose of leaking the secret key from the AES unit.

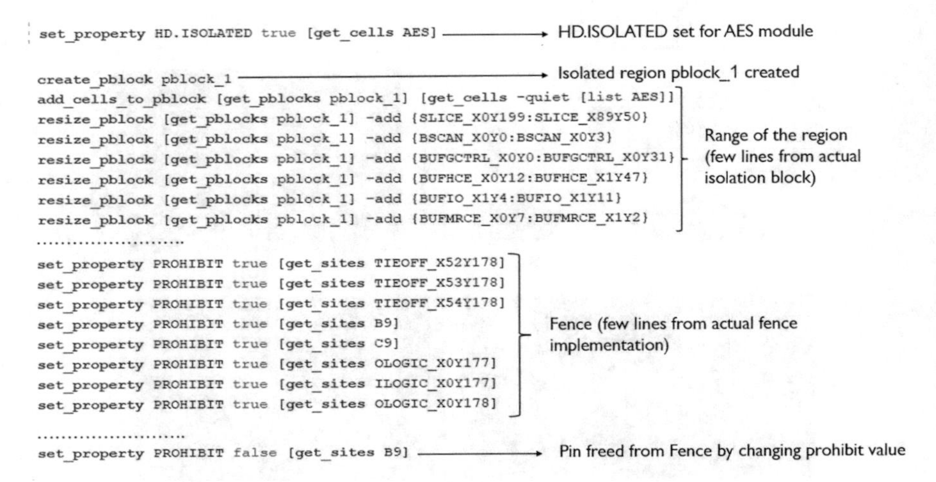

set_property HD.ISOLATED true [get_cells AES]
HD.ISOLATED set for AES module
create_pblock pblock_1
Isolated region pblock_1 created
add_cells_to_pblock [get_pblocks pblock_1] [get_cells -quiet [list AES]]
resize_pblock [get_pblocks pblock_1] -add {SLICE_X0Y199:SLICE_X89Y50}
resize_pblock [get_pblocks pblock_1] -add {BSCAN_X0Y0:BSCAN_X0Y3}
resize_pblock [get_pblocks pblock_1] -add {BUFGCTRL_X0Y0:BUFGCTRL_X0Y31}
resize_pblock [get_pblocks pblock_1] -add {BUFHCE_X0Y12:BUFHCE_X1Y47}
resize_pblock [get_pblocks pblock_1] -add {BUFIO_X1Y4:BUFIO_X1Y11}
resize_pblock [get_pblocks pblock_1] -add {BUFMRCE_X0Y7:BUFMRCE_X1Y2}
Range of the region (few lines from actual isolation block)
....
set_property PROHIBIT true [get_sites TIEOFF_X52Y178]
set_property PROHIBIT true [get_sites TIEOFF_X53Y178]
set_property PROHIBIT true [get_sites TIEOFF_X54Y178]
set_property PROHIBIT true [get_sites B9]
set_property PROHIBIT true [get_sites C9]
set_property PROHIBIT true [get_sites OLOGIC_X0Y177]
set_property PROHIBIT true [get_sites ILOGIC_X0Y177]
set_property PROHIBIT true [get_sites OLOGIC_X0Y178]
Fence (few lines from actual fence implementation)
....
set_property PROHIBIT false [get_sites B9]
Pin freed from Fence by changing prohibit value

Fig. 11 A snapshot of a constraints file in IDF [12]

As the Trojan is the circuit under protection, we did not set the isolation property for the Trojan. Figure 10c indicates that the Trojan is successfully placed in the isolated region for AES. The center of the white interconnection is the Trojan. We further increased the number of p-blocks in the top module. As shown in Fig. 10d, we can insert the Trojan to the p-block where the AES unit is located. For both cases shown in Fig. 10c and d, no IDF DRC error is reported to disclose the occurrence of the Trojan even though the entire IDF for the fence implementation is followed. To assure the success of the Trojan insertion in the case study, we modified the pin prohibit values from *true* to *false* by altering constraints shown in the last line in Fig. 11.

3.2 Security Threats in Open-Source FPGA CAD Tool

VTR and Symbiflow are analyzed to examine unique security vulnerabilities in open-source FPGA CAD tools. We demonstrate the potential attack surfaces that can be used to exploit for practical attacks. The two open-source CAD design suites are all comprised of various tools to support multistep FPGA design flow, in which different intermediate files are generated. Our analysis indicates that the security threats in open-source FPGA CAD tools are originated from those unprotected intermediate files.

Maliciously modifying the intermediate files will let the attacker to insert hardware Trojans, modify design logic, and compromise the integrity of the CAD tool. To leverage the intermediate files to perform attacks, adversaries need to understand the structure and the information stored in the intermediate files. The intermediate files generated in VTR tool chain are .blif, .net, and .place files. The intermediate files generated in Symbiflow toolchain are .eblif which is similar to .blif file in VTR, .net, .place, .route, and .fasam.

A Berkeley Logic Interchange Format file (.blif) describes the logic level hierarchical circuit in a textual format. This file holds the netlist information in textual format. Figure 12a shows an example of a .blif file for a design under attack. The .blif file consists of four important parts: *.model* (the module name, e.g., Test_M1), *.inputs* (all the input pins for the module), *.outputs* (all the output pins), and *.names* (the complete list of signals involved in a particular output logic e.g., N4).

Net file holds the information provided by FPGA architecture file and .blif file. The file consists of block names, subblocks, instances, modes, and clocks. The information inside the blocks is populated based on the architecture file.

A .place file holds the information of position of the blocks in FPGA fabric. These blocks are defined in the .net file. The position of these blocks are defined by X and Y coordinates. The X and Y coordinates can be changed and re-route the input and output pins to the places where attacker can probe and develop covert channels later.

3.2.1 Potential Attacks on VTR

After understanding the file format of each tool in the VTR toolchain, three potential attack surfaces are shown in Fig. 13: *.blif*, *.net*, and *.place* files after *ABC*, *Net*, and *Place*, respectively. Three practical attacks can be performed to alter input ports, output ports, and logic truth tables of the FPGA design. With the help of these attack surfaces, three potential attacks can be performed with a malicious open-source FPGA CAD tool. The three potential attacks are attacking inputs, attacking outputs, and attacking the logic truth table.

A .blif file can be modified to create new input and output pins and also add new logic to the original function of the circuit. A new input pin creation attack is successful only when the attacker changes .inputs and .names as shown in Fig. 12a. A fourth column is added to the lines 10 to 12. The VTR tool checks for the file integrity when the .blif file is sent to VPR for packing, placement, and routing. The attack successfully creates a new input pin as shown in Fig. 12c. We can see that the attack introduced a new input pin (we have four blue blocks now, instead of three) when compared with the baseline circuit before the attack shown in Fig. 12b. The number of output pins (in red) remains the same, but their position is shifted. A similar attack procedure can be followed on .blif to add new output pins for the design. The same attack can be performed by tampering with the .net file by adding a block similar to the one in Fig. 12d. The attacks via the .net file are more stealthy than the attack via the .blif file.

Another modification on .blif is to sabotage the original logic truth table. This is practical for an attacker who has a good understanding of the design under attack. The truth table is defined in the .blif file under the line starting with the keyword *names*. Theoretically, one can remove/add one row or revise the logic in the original

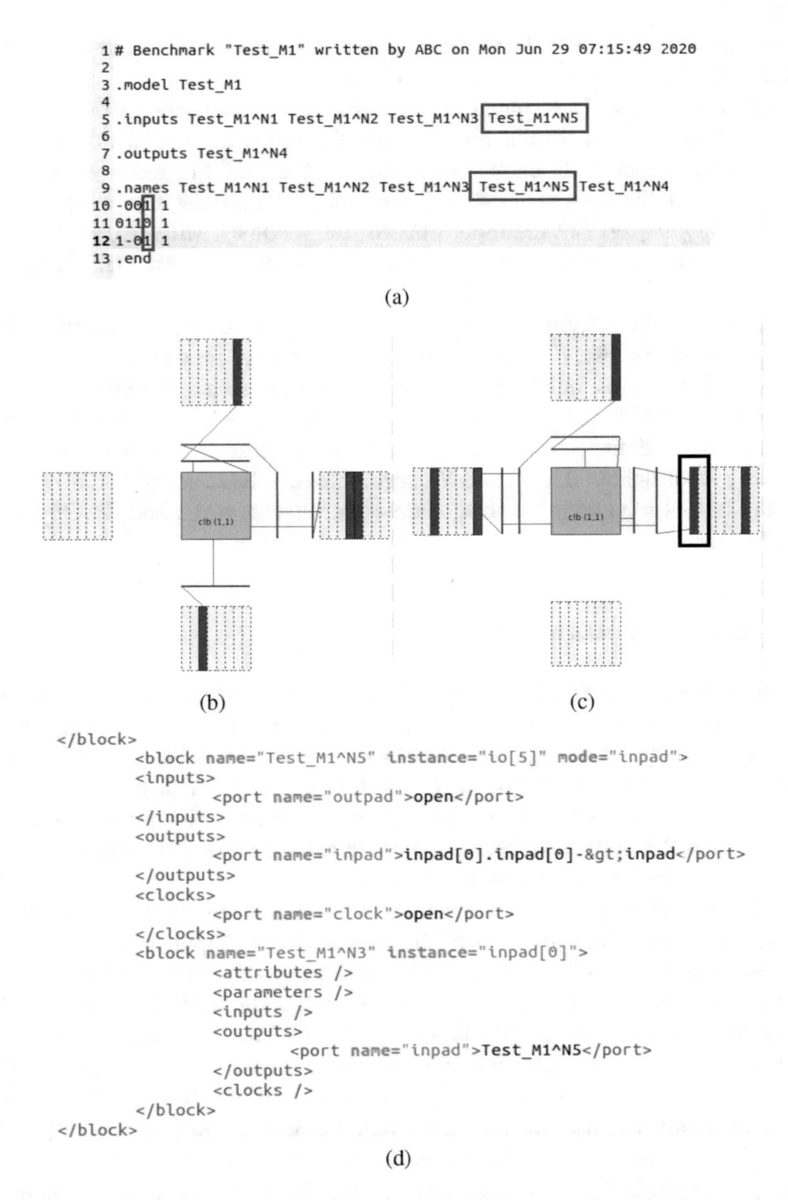

Fig. 12 Implementation of the input attack on the .blif and .net files. (**a**) Tampered .blif file (modified portion highlighted in red boxes), (**b**) graphical view of the mapped circuit before attack, (**c**) graphical view of the mapped circuit after attack, and (**d**) equivalent modification on the .net file [9]

table in the attack. Due to the built-in integrity check in VTR, the output of VPR will only be accepted if the attack on the logic truth table removes some rows, instead of

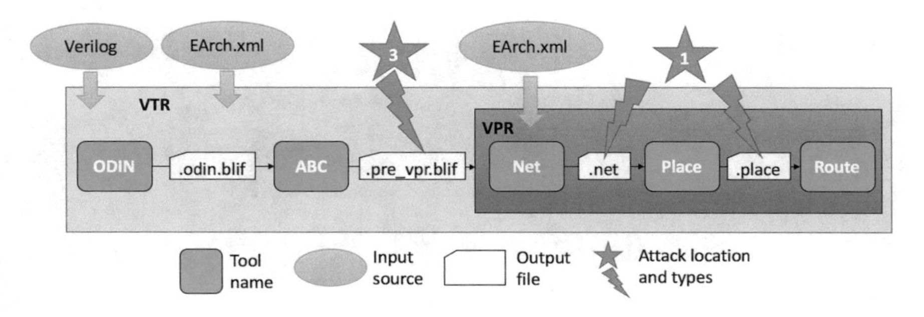

Fig. 13 Overview of attack surfaces on VTR [9]

```
 1 # Benchmark "Test_M1" written by ABC on Mon Jun 29 07:15:49 2020
 2
 3 .model Test_M1
 4
 5 .inputs Test_M1^N1 Test_M1^N2 Test_M1^N3
 6
 7 .outputs Test_M1^N4 Test_M1^N5
 8
 9 .names Test_M1^N1 Test_M1^N2 Test_M1^N3 Test_M1^N4
10 -00 1
11 011 1
12
13 .end
```

(a)

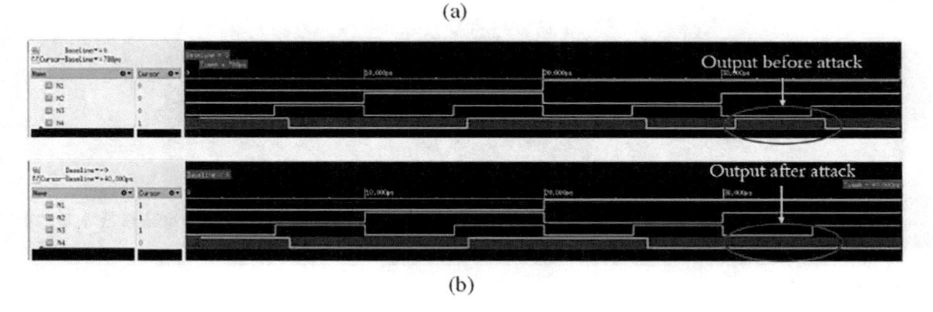

(b)

Fig. 14 Implementation of the logic attack on .blif file. (**a**) Modified .blif file due to the attack on logic description and (**b**) waveform showing the change on logic table leading to tampered output [9]

adding new rows. The same baseline example used in Fig. 12a is used to implement the attack on logic description. As shown in Fig. 14a, the logic expression on line 12 is removed. Consequently, the output of N4 is altered by the attack. The red circles in Fig. 14b highlight the change in the outcome of the proposed attack. Another interesting observation we notice is VTR does not have the capability to check if the logic description is modified. Functional verification (conducted in other tools) is necessary to detect the attack on the logic truth table.

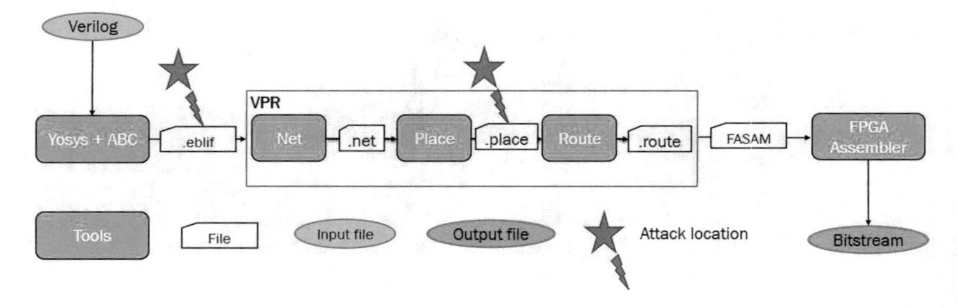

Fig. 15 Overview of attack surfaces on Symbiflow [9]

```
#block name       x        y        subblk  bloc
#----------       --       --       ------  ----
$abc$153$auto$iopadmap.cc:361:execute$91
$auto$iopadmap.cc:321:execute$90             2
N2                2        114      0       #2
N3                2        120      0       #3
N1                2        104      0       #4
```

(a)

```
#block name       x        y        subblk  blo
#----------       --       --       ------  ---
$abc$153$auto$iopadmap.cc:361:execute$91
$auto$iopadmap.cc:321:execute$90             2
N2                2        114      0       #2
N3                2        120      0       #3
N1                2        124      0       #4
```

(b)

Fig. 16 Implementation of an input attack on .place file. (**a**) Original .place file before attack and (**b**) the .place file after attack [9]

3.2.2 Potential Attacks on Symbiflow

Symbiflow has more capabilities than VTR and is capable of generating executable bitstreams. After thoroughly examining the intermediate files generated in Symbiflow, as shown in Fig. 15, two attack surfaces were identified: *.eblif* (generated by Yosys) and *.place* files (produced by the Place tool in VPR). Extended Berkely

Logic Interchange Format (.eblif) file is similar to a .blif file except for a few structural changes. Any attack performed on the .place file in the VPR tool can also be performed in Symbiflow since it is embedded in the Symbiflow toolchain.

The attacks implemented on the .blif file in VTR can be implemented on the .eblif file as well since they are the same files with small structural changes. Tampering with the .place file is a basic attack that can be realized in Symbiflow. The place file holds the information on the position of blocks in FPGA fabric. As shown in Fig. 16, the Y coordinates of a block N1 are changed from 104 to 124. Unlike VTR, Symbiflow does not have a graphical view for analysis, but it can report holding and

Table 1 Timing slack affected by the attacks on Symbiflow [9]

Case ID	Y coordinates	Holding time	Setup time
Original	104	2.214 ns	2.418 ns
Attack 1	124	2.418 ns	−3.011 ns
Attack 2	134	2.214 ns	−3.424 ns
Attack 3	144	2.214 ns	−3.631 ns
Attack 4	154	2.214 ns	−3.913 ns

Table 2 Timing results for different circuits after attacking .place file [9]

		Holding time		Setup time	
2*Circuit	2*.place file	Critical	Noncritical	Critical	Noncritical
3*S298	Normal operation	3.046 ns	1.731 ns	−4.201 ns	−0.464 ns
	Under attack	3.046 ns	1.415 ns	−4.185 ns	−0.637 ns
	Under attack	3.046 ns	1.371 ns	−4.185 ns	−0.464 ns
3*S15850	Normal operation	−0.585 ns	0.75 ns	−8.966 ns	−3.046 ns
	Under attack	−0.708 ns	0.345 ns	−8.966 ns	−2.938 ns
	Under attack	−0.908 ns	0.343 ns	−8.966 ns	−2.923 ns

setup time for the generated bitstream. As shown in Table 1, the modified N1 block in the .place file have a different holding and setup time. The change in holding and setup time could be positive or negative. A negative change is difficult to detect since the attack does not influence the worst-case delay. The attack on the place file is further extended to two benchmark circuits s298 and s15480. The results of the attack are shown in Table 2. The attacks have negligible impact on the critical-path delay but could result in large changes on the noncritical paths.

3.2.3 Practical Attacks Using Open-Source FPGA CAD Tools

Two practical attacks are demonstrated one on LFSR (Linear Feedback Shift Register) and the other on AES (Advanced Encryption Standard) crypto module using malicious open-source FPGA CAD tools. For realizing the practical attack, it is assumed that the CAD tool is already mounted with a malicious library.

LFSR is used to generate random numbers for cryptographic modules. In LFSR depending on the feedback paths, random number sequences are shifted from serially connected registers. Basic attacks demonstrated in VTR are used to perform the practical attack on 8-bit LFSR. Post the practical attack, the generated random numbers are confined to a limited range. This shows that with a practical attack, we can affect the randomness of LFSR.

Figure 17 shows the schematic of the LFSR design. This 8-bit LFSR can produce 255 different random numbers. Through the malicious tool VTR, we can successfully implement the pin creation attack and logic attack. The pin creation attack adds a new output pin in the .blif file for the LFSR to leak the random numbers being generated. Post pin creation attacks, the schematic of the LFSR circuit is

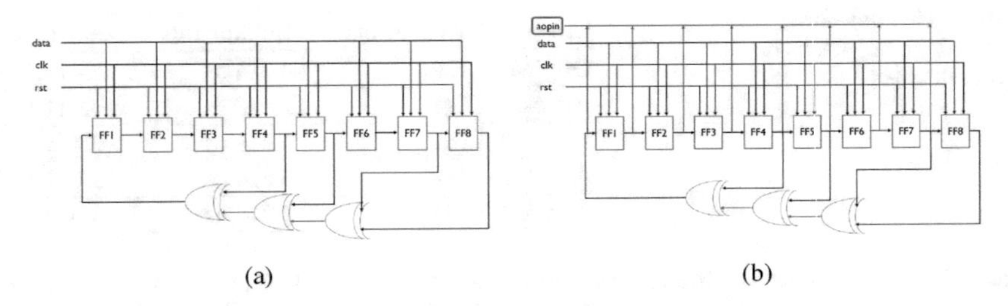

Fig. 17 Impact of FPGA CAD attacks on an 8-bit LFSR circuit schematic. (**a**) LFSR before attack and (**b**) LFSR after attack [9]

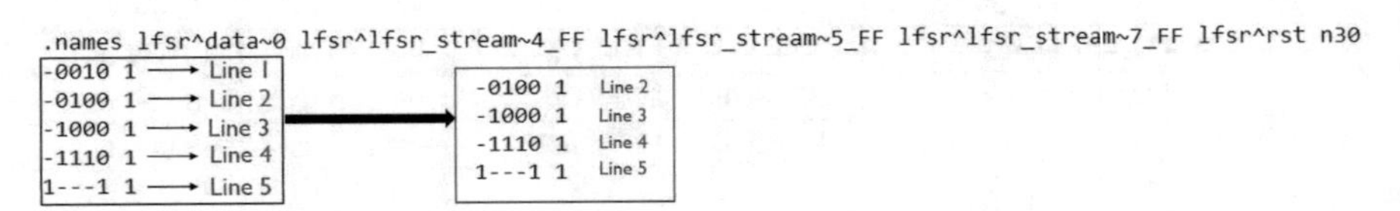

Fig. 18 LFSR feedback logic described in the .blif file. Note that the attack from the FPGA CAD tool removes Line 1, a part of the feedback loop logic [9]

changed to as shown in Fig. 17b. The new output pin, *aopin*, is added to manipulate the feedback logic and leak the random numbers being generated. The logic attack attacks the normal function of LFSR. Figure 18 shows how the logic specifying feedback logic is modified by the malicious FPGA CAD tool.

The consequence of the practical attack on LFSR is illustrated in Fig. 19. As shown in Fig. 19a, the dynamic range of the random numbers generated by the tampered LFSR is significantly smaller compared to the original LFSR. Moreover, the diversity of the random numbers due to the FPGA CAD attack is decreased dramatically. As shown in Fig. 19b, the LFSR suffering from the FPGA CAD attack only generates a limited number of distinct random numbers; in contrast, the LFSR without the attack can produce evenly distributed random numbers in the range of 0 and 255.

A malicious FPGA CAD tool (Symbiflow) can also implement a covert channel in an AES encryption module as shown in Fig. 20a. The malicious internal library consists of a Trojan that can modulate the secret key used in the encryption module. In this attack, the CAD tool scans the .eblif file for the AES module and then adds Trojan logic to the .eblif file. Two more files are modified along with the .eblif file for successful generation of Trojan-infested bitstream. The targeted FPGA board used here is a Digilent Nexys Artix-7 FPGA board. The constraints file (.pcf) and the make file are modified to assign all the ports to the FPGA board and to pass the required instructions to generate the bitstream file. All the old .net, .place, .route, .fasam, and .bit files are removed from the directory. Figure 20b shows the normal operation of the AES encryption module. Figure 20c and d shows the information leaked through the covert channel implemented through Symbiflow.

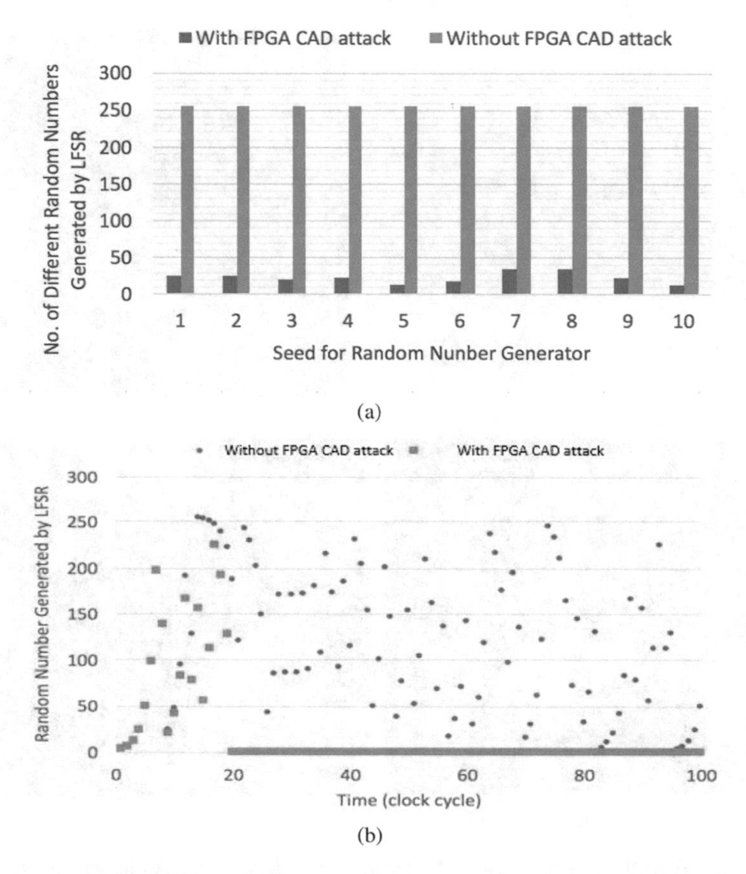

Fig. 19 Impact of pin addition and logic modification attacks from the malicious VTR on the random numbers generated by the 8-bit LFSR [9]

3.2.4 Generalized Attack Flow in Open-Source FPGA CAD Tools

Both the open-source FPGA CAD tools, VTR and Symbiflow use different interfaces in the process of design compilation and bitstream generation. In the security threat analysis, we were able to abstract common steps that a typical attack will use and summarize them in Fig. 21. In step 1, an attack performed via FPGA CAD tools will selectively incorporate the user constraints, either ignoring the user's constraints or stealthily adding new constraints, so that the FPGA configuration could be modified covertly. In step 2, attackers need to fully understand the format of the intermediate output files (e.g., .blif) and foresee what changes on the intermediate files can fulfill the intended attack purpose. The core attacks on FPGAs will take place in step 3. One could modify I/O and also alter the Boolean expressions indicated by the hardware description language. Step 4 is the most challenging one as the success of bitstream generation requires the modified intermediate files to be acceptable in other phases of the CAD flow. Attackers need to collaboratively adjust the malicious modifications so that the intermediate files

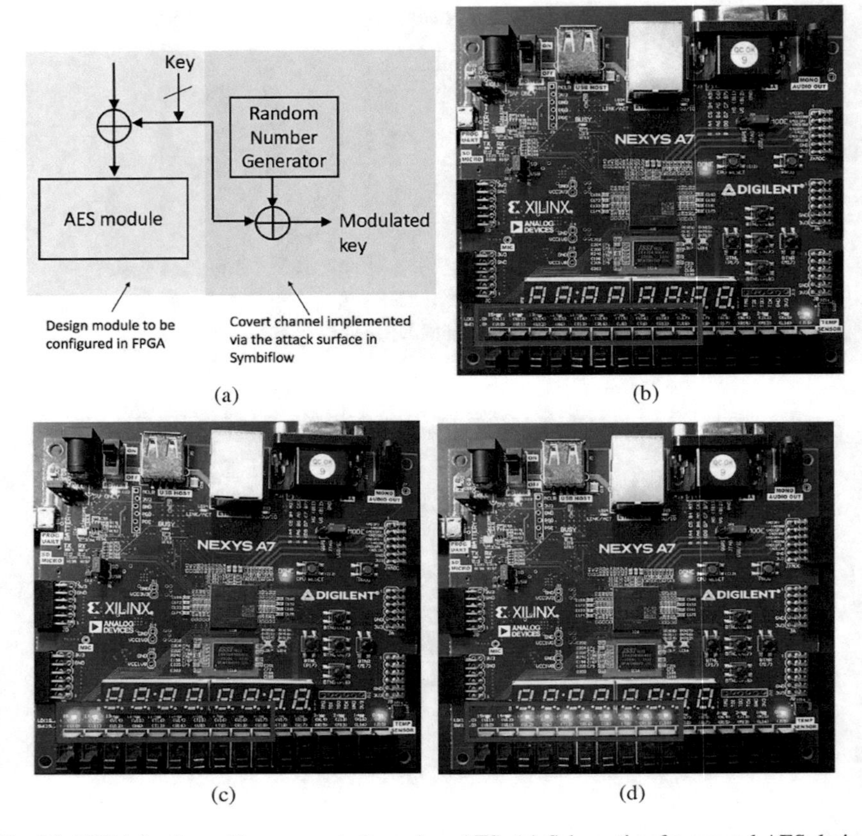

Fig. 20 FPGA implementing a covert channel on AES. (**a**) Schematic of tampered AES design, (**b**) FPGA running the normal AES operation, and (**c**, **d**) FPGA running the AES with a covert channel leaking the crypto key information at different moments [9]

are synchronized in the entire FPGA configuration flow. In an open-source FPGA CAD tool, the fact that all the intermediate stages whose input or output files are open for editing will introduce potential attack surfaces.

4 New Security Threat Landscape

4.1 New FPGA Utilization Model

The increase in demand for FPGAs and their ability to perform computation-intensive tasks has forced FPGAs and their vendors to change the utilization model as shown in Fig. 22. To increase the flexibility and to facilitate different applications

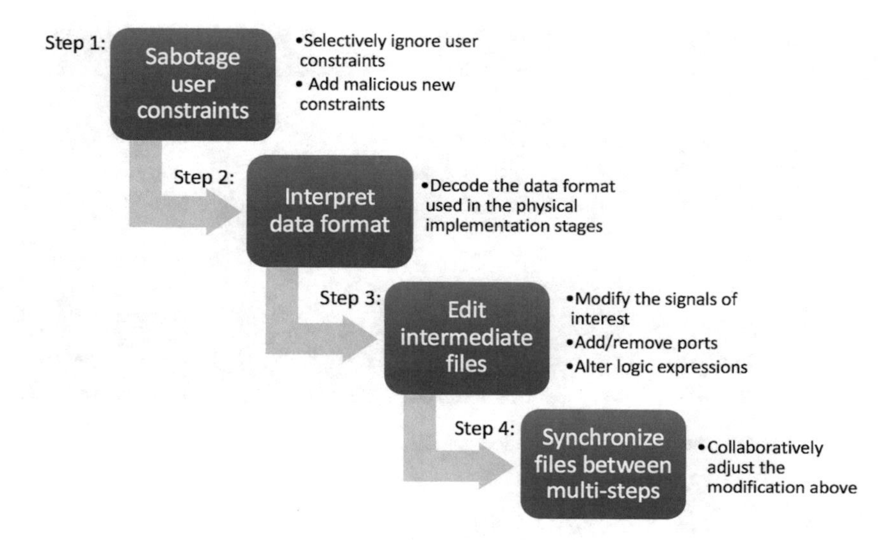

Fig. 21 Proposed general attacks on open-source FPGA CAD tools [9]

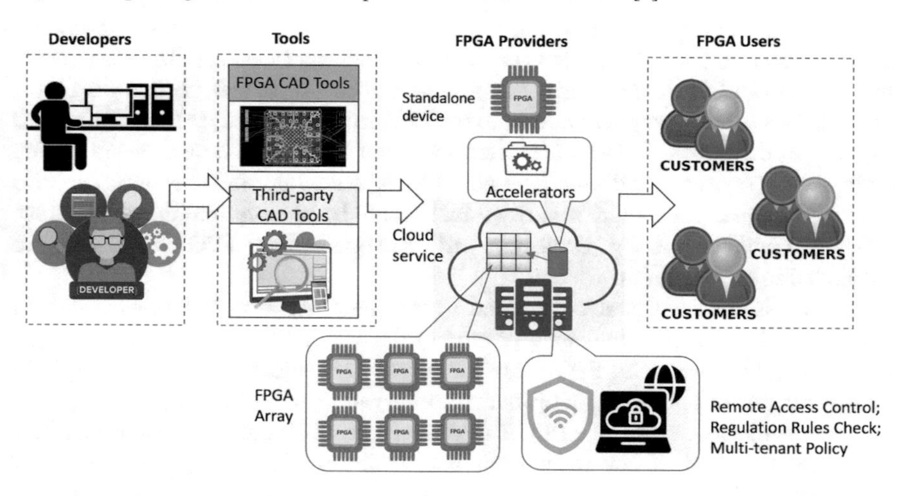

Fig. 22 Emerging FPGA utilization model [12]

in FPGAs, FPGA CAD tools are modified to support third-party IPs and plug-ins, which could make CAD tools vulnerable to security threats. Third-party developers and FGPA CAD tools have made their way into the FPGA design flow to ease the development. The ability to perform complex mathematical calculations has made FPGAs work as accelerators in the cloud and data centers. A compromised FPGA design could affect an entire data center or user applications.

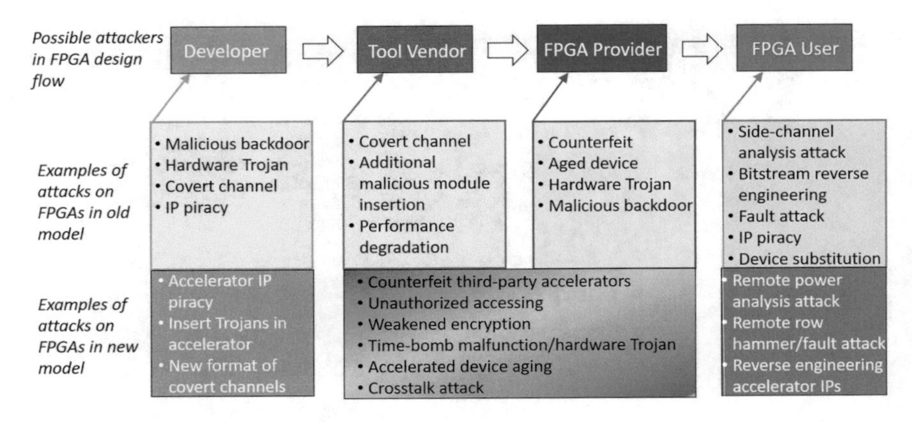

Fig. 23 Proposed new security threat landscape [12]

4.2 New FPGA Security Challenges

An accurate security threat landscape is needed to find attack surfaces and develop countermeasures for CAD tools against those attack surfaces. As discussed earlier there is an old FPGA use model and a New FPGA use model. In the old FPGA utilization scenarios, *physical access* to the FPGA devices or CAD tools is required to execute the attack. Attacks are often localized either in the device and design suite vendor or in the end user. The new FPGA use model involves more entities than the old one. Thus, more security threats could be brought into the design flow. This trend will become severer as more and more open-source FPGA CAD tools are employed in the design flow.

The most common attacks reported in the literature are (1) hardware Trojan insertion to build a covert channel or modify the original function at the developer end [3, 14], (2) counterfeit FPGA devices [8], and (3) side-channel analysis (SCA) attacks via power consumption or thermal observation at the FPGA end user [6, 7]. Few works disclose that some FPGA CAD tools can be tampered with to facilitate the implementation of covert channels and Trojan insertion [10, 15]. No matter which kind of attacks mentioned above is performed, collaborative attacks are less likely to happen because of the limited knowledge sharing among multiple sectors in the FPGA design flow. In cloud-based FPGA computing, the physical access to the CAD tools and FPGA devices are often prohibited; now, the new attack format could be a *synergy effort* deployed in the tools and devices. As a result, the attack model is shifted. Figure 23 summarizes the threat models induced by old and new FPGA utilization models. As shown in Fig. 23, the attacks in the new FPGA utilization model from the tool vendors and FPGA providers may be combined. Open-source FPGA CAD tools are not customized for a particular FPGA chip; the architecture information and essential characteristics of the FPGA of interest should be available for the CAD tool.

The accessibility of intermediate files will further facilitate the occurrence of new attacks. For example, more counterfeit third-party accelerators [16] could be employed in the design flow. The diverse sources of IPs will challenge the authorized access control. The security strength provided by IP encryption will be weakened as well. Information leaking via crosstalk and thermal covert channels [17, 18] could be more prevalent than those in the old FPGA use model. In addition, when we enter the era of FPGA-as-a-service, local SCA attacks are upgraded to remote versions, such as remote power analysis [19], remote row hammer attack [20], and remote fault attack [21].

The new FPGA utilization model urges us to expand the threat model for the new FPGA development flow. Our summary shown in Fig. 23 may not include all new attack examples reported in the existing literature. We hope that our work can inspire more engineers and researchers in the FPGA security community to enhance their knowledge and foresee the potential risks that we are facing in the new FPGA development scenarios. On the other hand, there are new opportunities for countermeasure design in open-source FPGA CAD tools. Various security features could be added to the conventional design flow by modifying and extending the vendor tools to secure FPGA systems against the new threats.

4.3 Comprehensive Summary of Attack Surfaces

Figure 24 summarizes the potential attack surfaces and the aims of possible attacks. The open-source tools often have a transparent flow for how the user constraints are employed in synthesis and floor planning. Some constraints defined by FPGA users could be tampered with or simply muted such that the pre-defined protection mechanisms are nullified. Second, more intermediate files in open-source CAD tools (e.g., *.blif*, *.net* and *.place* used in VTR and Symbiflow) are readable and editable before bitstream generation. The interfaces for third-party IP integration are more diverse than those in the commercial tools. FPGA system developers could leverage the interfaces to weaken the encryption on the licensed IPs or build covert channels for information leaking. Last but not least, the loosely protected toolchain will suffer from Trojan insertion and bitstream tampering [15, 22].

5 Conclusion and Future Research Directions

The new features in commercial FPGA CAD tools, open-source FPGA CAD tools, and cloud-based FPGA services gradually change the traditional FPGA design and use model. More and more entities are involved in the development flow of FPGA systems. The newly emerged FPGA use model poses new and unique challenges to FPGA security. This book chapter complements the existing surveys on the attacks from FPGA developers and users by investigating the potential security

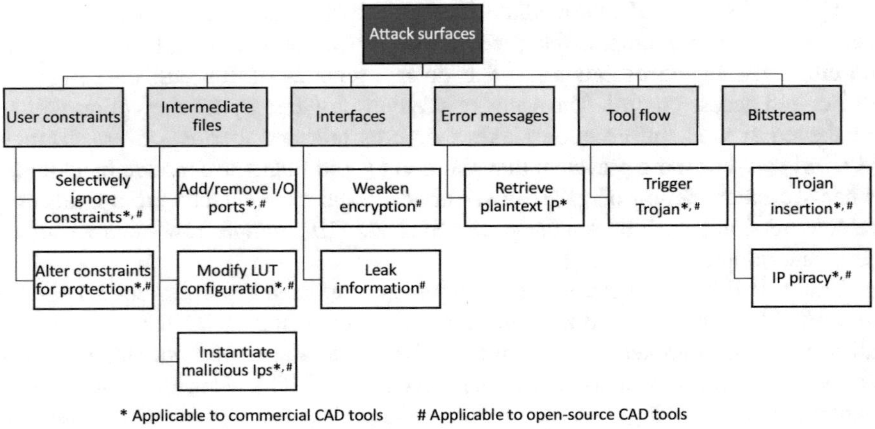

Fig. 24 Attack surfaces on commercial and open-source FPGA CAD tools [12]

threats from the commercial and open-source FPGA CAD tools. A comprehensive landscape for the new security threats is proposed in this work. Furthermore, this work analyzes the new security vulnerabilities induced by the integration of defense methods into the typical FPGA design flow. Several case studies are provided accordingly to inspire researchers to reconsider the new security issues that may occur in the deployment of countermeasures into FPGA CAD tools, especially when we implement more FPGA applications in the new FPGA utilization model.

Acknowledgments This work is partially supported by the National Science Foundation awards CNS-1652474 and CNS-2022279.

References

1. Hallmans, D., Sandström, K., Nolte, T., Larsson, S.: A method and industrial case: replacement of an FPGA component in a legacy control system. In: *2015 IEEE 13th International Conference on Industrial Informatics (INDIN)*, pp. 208–214 (2015)
2. Zhang, Z., Njilla, L., Kamhoua, C., Kwiat, K., Yu, Q.: Securing FPGA-based obsolete component replacement for legacy systems. In: *Proc. ISQED18*, pp. 401–406 (2018)
3. Gundabolu, S., Wang, X.: On-chip data security against untrustworthy software and hardware IPs in embedded systems. In: *Proc. ISVLSI'18*, pp. 644–649 (2018)
4. Zhang, J., Qu, G.: Recent attacks and defenses on FPGA-based systems. ACM Trans. Reconfigurable Technol. Syst. **12**(3), 1–24 (2019)
5. Zhang, T., Wang, J., Guo, S., Chen, Z.: A comprehensive FPGA reverse engineering tool-chain: from bitstream to RTL code. IEEE Access. **7**, 38379–38389 (2019)
6. Thoonen, M.: Hardening FPGA-based AES implementations against side channel attacks based on power analysis. B.S. thesis, University of Twente (2019)
7. Wei, L., Luo, B., Li, Y., Liu, Y., Xu, Q.: I know what you see: power side-channel attack on convolutional neural network accelerators. In: *Proc. ACSAC'18*, pp. 393–406 (2018)

8. Dogan, H., Forte, D., Tehranipoor, M.M.: Aging analysis for recycled FPGA detection. In: *Proc. DFT'14*, pp. 171–176 (2014)
9. Sunkavilli, S., Zhang, Z., Yu, Q.: Analysis of attack surfaces and practical attack examples in open source fpga cad tools. In: *2021 22nd International Symposium on Quality Electronic Design (ISQED)*, pp. 504–509 (2021a)
10. Zhang, Z., Njilla, L., Kamhoua, C.A., Yu, Q.: Thwarting security threats from malicious FPGA tools with novel FPGA-oriented moving target defense. TVLSI. **27**(3), 665–678 (2019)
11. Murray, K.E., Petelin, O., Zhong, S., Wang, J.M., Eldafrawy, M., Legault, J.-P., Sha, E., Graham, A.G., Wu, J., Walker, M.J.P., Zeng, H., Patros, P., Luu, J., Kent, K.B., Betz, V.: VTR 8: high-performance CAD and customizable FPGA architecture modelling. ACM Trans. Reconfigurable Technol. Syst. **13**(2), 1–55 (2020)
12. Sunkavilli, S., Zhang, Z., Yu, Q.: New security threats on fpgas: From fpga design tools perspective. In: *2021 IEEE Computer Society Annual Symposium on VLSI (ISVLSI)*, pp. 278–283 (2021b)
13. Pitaka, S.: Isolation design flow for Xilinx 7 series FPGAs or Zynq- 7000 SoCs (Vivado Tools). In: *Xilinx XAPP1222* (2020)
14. Mal-Sarkar, S., Krishna, A., Ghosh, A., Bhunia, S.: Hardware Trojan attacks in FPGA devices: threat analysis and effective countermeasures. In: *Proc. GLSVLSI'14*, pp. 287–292 (2014)
15. Chakraborty, R.S., Saha, I., Palchaudhuri, A., Naik, G.K.: Hardware Trojan insertion by direct modification of FPGA configuration bitstream. IEEE Des. Test. **30**(2), 45–54 (2013)
16. Turan, F., Verbauwhede, I.: Trust in FPGA-accelerated cloud computing. CSUR. **53**(6), 1–28 (2020)
17. Provelengios, G., Ramesh, C., Patil, S.B., Eguro, K., Tessier, R., Holcomb, D.: Characterization of long wire data leakage in deep submicron FPGAs. In: *Proc. FPGA'19*, pp. 292–297 (2019)
18. Tian, S., Szefer, J.: Temporal thermal covert channels in cloud FPGAs. In: *Proc. FPGA'19*, pp. 298–303 (2019)
19. Schellenberg, F., Gnad, D.R., Moradi, A., Tahoori, M.B.: An inside job: remote power analysis attacks on FPGAs. In: *Proc. DATE'18*, pp. 1111–1116 (2018)
20. Krautter, J., Gnad, D.R., Tahoori, M.B.: FPGAhammer: remote voltage fault attacks on shared FPGAs, suitable for DFA on AES. In: *TCHES*, pp. 44–68 (2018)
21. Alam, M.M., Tajik, S., Ganji, F., Tehranipoor, M., Forte, D.: RAM-jam: remote temperature and voltage fault attack on FPGAs using memory collisions. In: *Proc. FDTC'19*, pp. 48–55 (2019)
22. Moradi, A., Schneider, T.: Improved side-channel analysis attacks on xilinx bitstream encryption of 5, 6, and 7 series. In: *Proc. COSADE'16*, pp. 71–87 (2016)

DoS Attack Models and Mitigation Frameworks for NoC-Based SoCs

Mitali Sinha, Sidhartha Sankar Rout, and Sujay Deb

1 Introduction

Increasing demand for high performance and energy efficiency have led to integration of more and more processing cores within a single System-on-Chip (SoC). Furthermore, a surge of interest is also observed in the recent computing systems for employing heterogeneous SoCs, which comprise of different processing elements with varied computing abilities [1, 2]. These systems are suitable for real-time applications in the domains like national security, traffic monitoring, autonomous vehicles, etc. As the number of Intellectual Property (IP) blocks increases, Network-on-Chip (NoC) has gained traction as an effective interconnection platform due to its ability to provide high-bandwidth and low-latency communication [3]. NoCs are typically comprised of routers that are connected to their neighbor nodes through the wired links. Each IP block is attached to a router node through the network interface (NI) and communicate with other on-chip modules through the wired communication network. Besides the wired NoCs, multiple emerging interconnection architectures like wireless NoC (WNoC) [4], Photonic NoCs [5], etc., are also proposed to meet the needs of diverse traffic patterns and improve system performance. Although the shared and distributed infrastructure of the NoC provides effective data communication, it can also introduce various security threats within the system. These security threats get even more aggravated when the third-

M. Sinha (✉)
Department of Computer Science Engineering, Indraprastha Institute of Information Technology Delhi, Delhi, India
e-mail: mitalis@iiitd.ac.in

S. S. Rout · S. Deb
Department of Electronics and Communication Engineering, Indraprastha Institute of Information Technology Delhi, Delhi, India
e-mail: sidharthas@iiitd.ac.in; sdeb@iiitd.ac.in

A. Iranmanesh (ed.), *Frontiers of Quality Electronic Design (QED)*,
https://doi.org/10.1007/978-3-031-16344-9_15

party IP blocks are integrated together with other on-chip modules in the SoC. This is because the lack of access to the design details of third-party IPs makes them difficult to validate. Even with the design details, it is infeasible to perform exhaustive explorations of millions of logic elements to detect a possible security threat due to design flaws or any malicious modifications like Hardware Trojans (HT). Such security threats can lead to information leakage, data alteration, system performance degradation, and memory corruption.

A Denial-of-Service (DoS) attack on an interconnection network is an attack that curtails a network's ability to provide appropriate services to the legitimate on-chip modules. A malicious IP (MIP) can trigger a DoS attack by either disrupting the normal working of the NoC or holding the shared NoC resources. Here, we consider flooding-based DoS attacks triggered by an MIP on the NoC. An MIP triggers a flooding attack by injecting continuous random packets into the network resulting in higher communication latency. As a result, it manipulates the perceived availability of on-chip resources by other legitimate users and creates a DoS attack. For instance, an MIP can generate a large number of useless read/write requests to the shared memory banks located at distant nodes. This will thwart the network traffic and degrade the performance of other benign cores. Along with the targeted victim node, other intermediate nodes are also congested as they fall on the DoS attack path. Consequently, when a source node tries to access the destination node residing on the attack path, it will face higher latency due to congestion on its communication path. As a result, the affected nodes starve for and thus degrades the overall application performance. Therefore, it is of utmost importance to accurately detect the DoS attacks at an early stage and prevent the system from severe performance degradation. Once an attack is detected, the source of the attack needs to be localized and appropriate actions need to be taken to restore the system to a healthy state. Since the primary goal of multicore system design is to improve the performance of the applications running on the SoC, it is imperative to address the DoS attacks in NoCs that lead to the overall system performance degradation. This particular chapter primarily discusses the state-of-the-art flooding-based DoS attack models in the NoCs and the corresponding frameworks for detecting such attacks along with the countermeasures.

A widely used approach for detecting such flooding-based DoS attacks is to monitor the traffic pattern and distinguish the network behavior in case of an attack and non-attack scenario. A threshold is set for the chosen network parameters based on empirical observations, and attack scenario is flagged out in case of any threshold violations [6]. However, during real-time operations, different applications coexist at different times over the entire execution period. To take advantage of such varied workload condition, a number of runtime optimization methods have been proposed [7–9], which provide utmost performance for the applications running on the SoC. As a result of these optimizations, a wide variation is observed in system states and network-level activities of an SoC during its runtime. Such dynamic environment of SoCs imposes a challenge to set appropriate thresholds for the chosen network parameters for attack detection. Therefore, there is a need to employ efficient techniques that will accurately distinguish the NoC traffic behavior between a DoS attack and a non-attack scenario. Furthermore, the inherent nature of NoCs, allowing multiple parallel communication and resource sharing, makes it extremely difficult to locate the source of the attack. In the cases where multiple MIPs are present in the network, it becomes ever more difficult to distinguish attack and non-attack scenarios, which highlights the importance for a robust MIP localization framework.

Along with the traditional wired NoCs, multiple emerging interconnection platforms like WNoC and Photonic NoCs are being explored to further improve system performance by providing long-range on-chip communication. However, an MIP can exploit these emerging interconnection networks to introduce vulnerabilities within the system and lead to DoS attacks. Since the long-range communication channels are shared among multiple contenders to transfer their data, an MIP can hold the channel unauthorizedly to introduce DoS attacks or can eavesdrop or spoof to leak crucial information. These result in significant performance degradation or compromising the system security. The integration of different third-party untrusted IPs and difficulty in detecting small HT circuits on the multicore SoCs increase the probability of unauthorized channel access. Therefore, it is imperative to design countermeasures for such kind of DoS attacks that can disrupt the regular communication flow and drastically degrade the system performance.

The remainder of the chapter is organized as follows. Section 2 discusses different threat models in NoC. It provides a brief description on the existing threat models in both wired and wireless NoCs. An overview of DoS attack detection and localization frameworks is presented in Sect. 3. An example DoS attack detection and localization framework for wired NoC-based SoC is discussed in Sects. 3.1 and 3.2. An example DoS attack detection framework along with countermeasures for wireless NoC-based SoCs is presented in Sect. 4. Finally, Sect. 5 concludes the work along with insights into the future research opportunities in this domain.

2 Threat Model

This section presents the various state-of-the-art threat models proposed for DoS attacks in NoC-based system architectures. Firstly, the attack models in wired NoCs are presented, followed by the attack models in wireless NoC-based systems.

2.1 Wired NoC Threat Model

The DoS attack models in wired NoCs can be broadly divided into network resource blocking attack and bandwidth consumption attack. In case of the network resource blocking attacks, the network services are denied by blocking or disrupting the functionality of critical resources leading to their unavailability for the legitimate users, whereas the aim of the bandwidth consumption attacks is to exhaust the network bandwidth by excessive link usage leading to bandwidth reduction for the legitimate users.

The DoS attack proposed in [10] demonstrates a network resource blocking attack, where the critical router components of the NoC are targeted. The router micro-architecture is modified in [10] to incorporate the control and data path of an HT within the router's pipeline stages and manipulate its functionality. The "arbitration" and "allocation" stages of router micro-architectures are modified to disrupt the crossbar traversal slots for the packets generated from (or targeted to) a victim node. In the arbiter stage, the HT de-prioritizes the packets generated from (or targeted to) a victim node and delay the packets at a given router. Similarly, the requests for crossbar allocation is suppressed at the allocation stage, denying services for the specific targeted packets, leading to a DoS attack. While the network resource blocking attacks can create adverse situation, the bandwidth consumption attacks are even more disruptive. The authors in [6, 11, 12] presents such attacks where an HT-infected MIP node generates a large amount of useless and frequent packets targeted to a Victim IP (VIP) node, creating high network congestion. The perceived availability of the on-chip network bandwidth by other legitimate users are manipulated, leading to a flooding-based DoS attack. For instance, a malicious node can generate a large number of useless read/write requests to a shared memory block residing in a distant part of the NoC. This will thwart the network traffic and result in higher network congestion on the attack path and significantly increase the network latency. The other on-chip IPs generating requests for the same/nearby memory blocks face high congestion in the network and are denied appropriate services. Hence, the domination of the malicious traffic over the legitimate traffic disrupts regular network packet communication along with the targeted VIP node, which results in overall system performance degradation. Such bandwidth consumption-based DoS attacks become more hazardous in the presence of multiple malicious nodes within the NoC. The authors in [13–15] present such attacks, where multiple MIP nodes target single or multiple VIP nodes and create a Distributed DoS attack (DDoS attack). Under such scenarios, the network is highly dominated by the useless packets generated by multiple malicious nodes, which make the legitimate traffic to starve for network bandwidth. Figure 1 shows the various scenarios, where the malicious nodes create a bandwidth consumption-based DoS attack. As shown in the figure, along with the targeted victim node, other router nodes residing on the attack path also gets affected and result in extreme performance degradation of the overall system. We consider a comprehensive threat model that encompasses different scenarios of a flooding-based DoS attack as follows:

Number of MIPs/TVIPs A single MIP can initiate an attack on a single TVIP (Fig. 1a) or multiple TVIPs (Fig. 1b). Multiple MIPs can also initiate attacks on a single TVIP or multiple TVIPs (Fig. 1c, d, respectively).

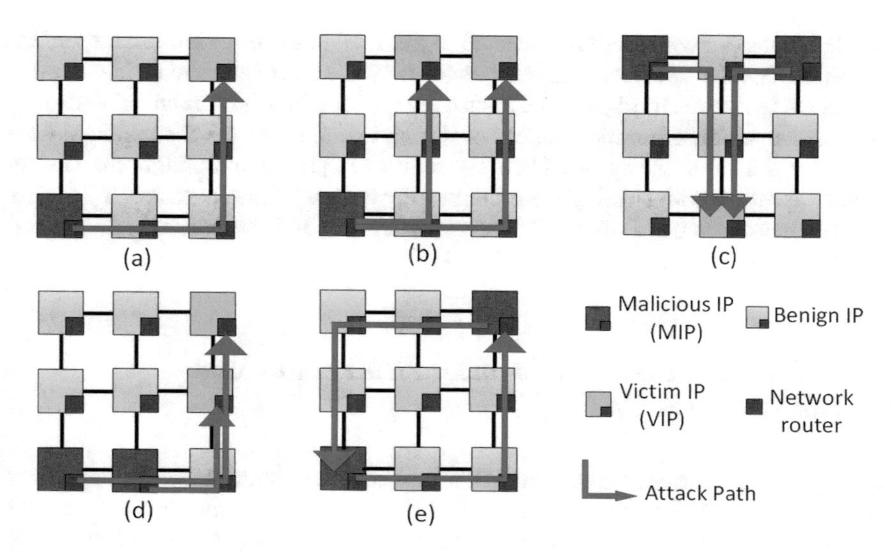

Fig. 1 Example scenarios of multiple MIP/VIP placement

Placement of MIPs/TVIPs Different placements of MIPs/TVIPs can create different scenarios, where the attack paths can partially or completely overlap with each other or form a loop. Figure 1 shows a few illustrative scenario.

Coordinated and Uncoordinated Attack Multiple MIPs can cooperate and mount a coordinated attack. For instance, in Fig. 1e, a coordinated attack is created by exchanging frequent packets between two MIPs. In an uncoordinated attack, the MIPs target TVIPs independently without any cooperation.

2.2 *Wireless NoC Threat Model*

In case of WNoC, both the network resource blocking and bandwidth consumption-based DoS attacks are possible. Multiple existing research works highlight bandwidth consumption attack [16, 17] and provide solution for it. Authors in [16] focus on the DoS attack that is created by the excessive transfer of useless data over wireless channel. The work in [17] highlights the occurrence of burst error on the received data bits due to the presence of interference on the wireless channel caused by the persistent jamming-based DoS attack. The network resource blocking-based DoS attack in case of WNoC is shown by unauthorized holding of the wireless channel [18, 19]. The Wireless Interfaces (WIs) on the network share the wireless channel for data transmission and reception. While multiple WIs access the same wireless channel to transfer their data, a malicious WI can hold the channel unauthorizedly. This results in blocking of wireless channel for the healthy wireless hubs (network nodes connected to the WIs) and creates DoS in the system operation. A WI is considered to be malicious if it has an HT present inside it or it is connected to a rogue third-party IP. Such a malignant wireless hub can either disturb the channel access arbitration pattern or it might manipulate the channel hold time to introduce DoS to the system [18]. The authors in [19] also highlight the DoS in WNoC by bandwidth stealing from the healthy nodes. With regard to WNoC, this chapter focuses on the designing of threat model for wireless channel blocking and its countermeasures as discussed in Sect. 4.

3 DoS Attack Detection and Localization in wired NoC-Based SoCs

The profound security threats from MIPs have triggered multiple works to address flooding-based DoS attacks in SoCs. In [15], design guidelines were given for mesh-based multicore systems to mitigate the effects of DoS attacks, whereas [20] presents security verification methods for NoC micro-architectures. The authors in [10] relies on using extra packets to perform security checks, which requires multiple windows of checking and packet injection, further delaying network communication. A central monitoring unit is proposed in [12] to detect unnatural traffic conditions based on average bandwidth violations, which is computed empirically at the design time. In [6], a DoS attack detection method is proposed by statically profiling a defined network traffic to distinguish between the attack and non-attack scenarios. Once an attack is detected, it is imperative to localize the MIPs and take necessary actions to prevent any further system degradation. But the inherent nature of NoCs, allowing multiple parallel communication and resource sharing, makes it extremely challenging to locate the source of the attack. The presence of multiple MIPs makes it even more onerous and signifies the need for a robust MIP localization framework. An effort is made in [21] and [14] to localize MIPs in NoC-based systems. However, their approaches need multiple checking and communication between the network nodes, which will further increase the complexity of MIP localization. Moreover, the methods proposed in [21] and [14] rely on a single network parameter, whose threshold is set through static profiling, to identify an MIP within the system. However, in real scenarios, a varied range of network traffic is observed due to the dynamic nature of applications running on complex SoCs. Hence, tuning a network parameter through static profiling might set it to an erroneous threshold and result in an inefficient method generating numerous false predictions. While including multiple parameters may decrease the number of false predictions, it will substantially increase the challenge of setting those parameters to appropriate thresholds. To address these issues, the authors in [11, 13] proposed machine learning (ML) approach for detecting and localizing flooding-based DoS attacks in NoC-based SoCs. By tuning multiple network parameters to appropriate thresholds, the ML algorithms help in accurate

MIP detection and localization. The following sections discuss such ML-based detection and localization frameworks in detail.

3.1 Attack Detection Framework

This section describes an efficient flooding-based DoS attack framework using ML algorithms [11] for accelerator-rich heterogeneous SoCs, which comprises of a few general-purpose CPU cores, a number of specialized hardware accelerators, and memory subsystems, all connected together by an NoC as shown in Fig. 2. The CPU cores allocate the compute-intensive part of the applications onto the accelerators and prepare the accelerator-specific data in memory. The accelerators are initiated by the respective CPU cores through system calls (e.g., ioctl calls). Whenever a task is completed, the accelerator informs the corresponding CPU by raising an interrupt signal. Figure 3 shows the flow of the attack detection framework comprising of a two-step method to detect a flooding-based DoS attack in such an accelerator-rich heterogeneous SoC. In the first step, a first level sanity check is performed by the CPU cores to flag a possible flooding attack based on precomputed accelerator execution time (AET) and interrupt generation. To accurately differentiate between an attack and a non-attack scenario, an ML-based method is employed as a second step of detection. Once a flag is raised in the first step, the ML classifier is triggered to perform binary classification of flooding-based DoS attack by inspecting the current system behavior. The rationale behind employing a two-step framework is

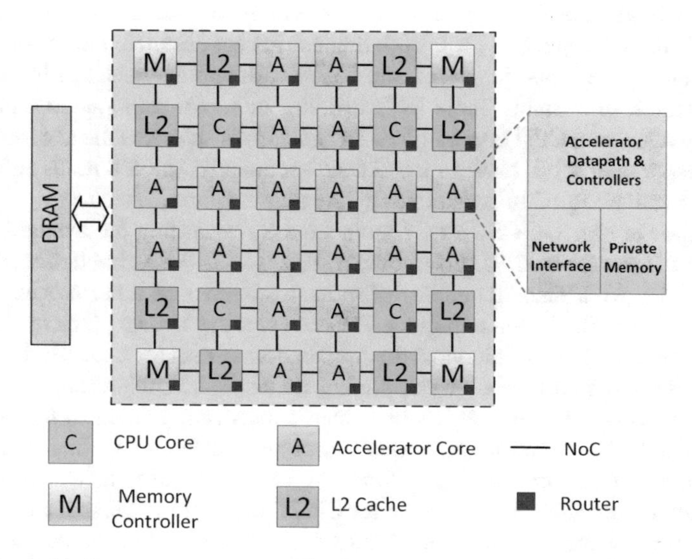

Fig. 2 A typical heterogeneous system architecture

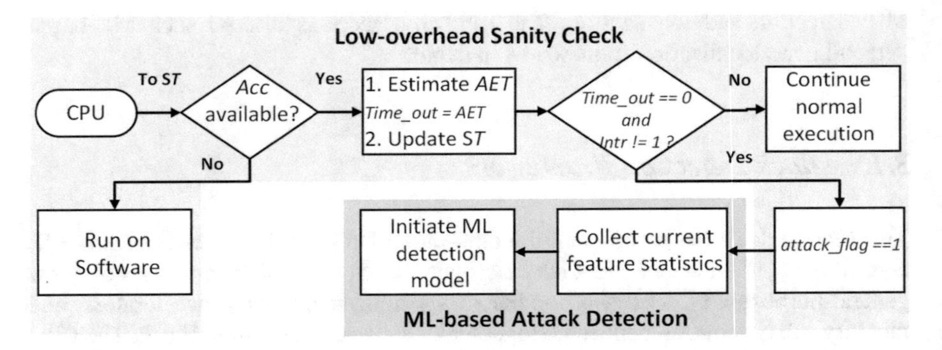

Fig. 3 Flow of the attack detection framework. Here, *Acc* Accelerator, *ST* Status table, *Intr* Interrupt, *AET* Accelerator execution time

to reduce runtime computation overhead by avoiding frequent ML invocation. In the following sections, the attack detection framework is described in detail.

3.1.1 First-Level Sanity Check

To facilitate the first-level sanity check, the CPU cores compute AET of the target accelerator before off-loading an application kernel. Here, AET comprises of (i) an accelerator's computation time and (ii) average network latency. An accelerator's computation time is data-dependent and is estimated using linear regression as a function of input data size, whereas the average network latency is estimated using an analytical model given in [22]. Both the metrics are calculated at the design time with representative input data sizes and benchmark applications. All the estimated data are stored in a Status Table (ST) residing in an on-chip shared memory. At runtime, whenever a CPU core initiates an accelerator, it stores the accelerator's ID and corresponding AET back in the ST. Consequently, the ST holds information about the availability of an accelerator at any point of time.

As shown in Fig. 3, a CPU first enquires the ST regarding the availability of an accelerator. If available, CPU sets a counter, *Time out*, to the estimated AET for each accelerator invoked. Then, it updates ST, prepares input data, and starts execution of the accelerator while decrementing the *Time out* counter at each clock cycle. In case of unavailability of an accelerator, the application is executed locally in the CPU core. As discussed earlier, after completing its task, an accelerator must generate an interrupt to update the CPU, which then subsequently updates the ST. If the CPU does not receive an interrupt signal from the accelerator within the estimated execution time (i.e., $intr! = 1$ and $Time\ out == 0$), it raises a flag to indicate a possible attack scenario. The rationale for CPU flag generation is that an accelerator must be starving for data accesses due to network congestion, which delayed its interrupt generation. Once a flag is raised, the second-level check is executed that employs an ML model for accurate attack detection.

3.1.2 Machine Learning for Attack Detection

The fundamental notion behind ML-based attack detection is to accurately distinguish the system behavior in an attack and non-attack scenario. To achieve this, the system state and network statistics are used as features for training an ML classifier. Although deploying an ML-based solution may lead to accurate attack detection, an inefficient design may introduce computational and performance overheads. One design approach is to have a centralized ML model in which feature data need to be collected from all the network nodes. The other approach is to have a decentralized method where the ML model resides in every network node and takes the decision locally based on the feature data of that node. Such a decentralized approach results in huge computational and/or hardware overheads due to multiple copies of ML classifiers running simultaneously. On the other hand, a centralized approach might incur performance overhead due to feature data collection from each network node.

The authors in [11] adopt a centralized ML-based approach to provide an accurate and low-cost solution while using a hierarchical data collection method to address the performance overheads. The entire network is logically divided into multiple clusters, where a CPU acts as a cluster head and facilitates feature data collection. The ML classifier runs on a CPU and is triggered only when there is a flag raised by it in the first-level sanity check, indicating possible security violations. After raising a flag, the CPU sends a broadcast message instructing the cluster heads to start feature data aggregation. Each cluster head collects feature data from every node belonging to that particular cluster. Once the data is collected, average feature values are calculated at each cluster head and transferred to the CPU core running the ML classifier. Finally, after receiving data from all the network clusters, the average feature values are calculated for the entire network and fed to the ML model running in the CPU core. To facilitate steady forwarding of feature data through the likely compromised network, each node groups them into one feature packet and prioritizes those packets to avoid any unwanted delay. A priority bit is employed to distinguish a feature packet from the regular network packets.

The framework used two classes of features to train the ML classifiers for attack detection: *Network State Features* and *System State Feature*. The *Network State Features* capture the variation in on-chip communication behavior over the execution period. The *System State Feature* captures the system's current state to understand the expected network load at any given time.

(1) Network State Features

Inter-Packet Interval (IPI) In a non-attack scenario, there is an appreciable time between any two consecutive packet reception at a network node. On the contrary, the Inter-packet Interval (*IPI*) time becomes very small in case of a flooding-based DoS attack. As a result, *IPI* is a potential feature candidate to distinguish between attack and non-attack traffic.

Router-Buffer Waiting Time (RWT) It refers to the time interval for which a packet waits in the buffer of a router due to the unavailability of its output ports or congestion at the input port of the downstream router. This waiting time significantly increases in an attack scenario and serves as a good metric for attack detection. Similar to *IPI* calculation along with a timestamp counter, *RWT* is computed for incoming packets at each network router.

Packet Delivery Ratio (PDR) *PDR* refers to the ratio of packets received by a destination to those generated by a source node. In comparison to a non-attack scenario, *PDR* value will drastically decrease during a flooding attack as more number of packets will get dropped before reaching the destination nodes. To compute *PDR*, each source node router holds the information of the number of packets generated along with their targeted destination nodes. With that knowledge, and based on the number of retransmission requests, *PDR* is determined for each node.

Global Average Delay (GAD) It represents the average latency incurred by the entire on-chip communication network. In an attack scenario, the network exhibits high congestion as compared to a non-attack scenario. This results in extended communication latency during a flooding attack and hence drastically increase the Global Average Delay (GAD). This signifies the use of GAD as an effective feature to distinguish between attack and non-attack traffic. GAD is readily available from the simulation tool, Noxim[23] employed in the experimental evaluation.

(2) System State Feature

Number of Active Accelerators (#AAcc) As discussed earlier, in an accelerator-rich SoC, the CPU cores request from a pool of accelerators to off-load the entire or part of its application. While multiple accelerators are assigned various tasks, the rest of the inactive accelerators are turned off to reduce power consumption. The number of ingress or egress packets generated in the network vastly depends upon the number of accelerators active at a given point of time. This influences all the *Network State Features* due to the increased packet density on the network. Considering the number of active accelerators as a feature will drive the ML classifier to set appropriate thresholds leading to better prediction of attack scenarios. As a result, the classifier will significantly reduce the number of false positives. This quantifies the importance of including *#AAcc* as a feature to accurately detect an ongoing flooding attack. The information regarding the number of currently active accelerators is collected from the ST and fed to the ML classifier for detection.

3.1.3 Results and Analysis

To evaluate the attack detection framework, gem5-Aladdin [1] and Noxim [23] simulators are used. The detection framework is implemented in Noxim with the configurations presented in Table 1. The application traces are obtained from gem5-

Table 1 System configuration and simulation setup

Component	Configuration
Processing cores	4×86 CPU cores; customized fixed-function accelerator cores, 500 MHz frequency
Memory	32 KB private and 2MB shared on-chip memory; 4 GB off-chip DRAM
NoC	8X8 2D mesh, XY routing, wormhole switching, 1-cycle link latency, 3-cycle router latency
ML models	KNN, RF, SVM, DT, and ANN

Table 2 Detection performance of classifiers

ML models	Accuracy	F1 score	Precision	Recall
KNN	0.855	0.855	0.856	0.856
RF	0.974	0.970	0.970	0.971
SVM	0.638	0.621	0.631	0.630
DT	0.929	0.928	0.928	0.928
ANN	0.891	0.890	0.890	0.890

Aladdin and fed to Noxim. The MachSuite [24] and MiBench[25] benchmark suites are used for the evaluation.

Detection Accuracy This section presents the offline detection accuracy achieved by the ML classifiers on the dataset. The detection results are presented by averaging the outcomes of fivefold cross validation. Table 2 presents the offline accuracy results of the ML classifiers used in the attack detection framework. It also shows the robustness of the classifiers with regard to precision, recall, and F1-score metrics. The F1-score metric gives the harmonic mean between the precision and recall values, which provides a statistical measure of performance. It is observed that the attack detection accuracy ranges from 0.63 to 0.97 for all the classifiers. A higher value of accuracy and robustness metric represents higher performance. A relatively low accuracy of 0.63 given by the SVM classifier suggests that the data-space considered for training the classifier is not linearly separable. However, the KNN model performed fairly well with an accuracy of 0.85. This indicates that the classifier was able to cluster the data into attack and non-attack classes with regard to the considered feature space. In the case of neural network (ANN), a detection accuracy of 0.89 was reported. DT and RF classifiers performed extremely well with an attack detection accuracy of 0.92 and 0.97, respectively. This is due to the nonlinear nature of DT and RF classifiers that allow them to achieve higher accuracy by constructing multiple linear boundaries for detection. The RF classifier performs even better by considering a pool of decision tree outcomes and is used for further experimental analysis.

Feature Importance Figure 4 shows the importance of all the features considered in the evaluation of the framework in terms of prediction accuracy. Here, each feature is taken in isolation as well as all possible combination of features, and

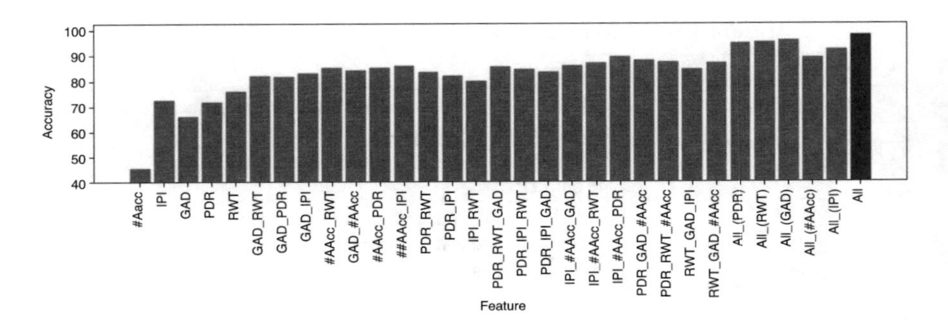

Fig. 4 Feature importance (accuracy of RF for binary classification with each feature in isolation and in combination). We also plot the accuracy with all features at the rightmost bar of the figure (labeled as 'All'). Here, "$All\_(Feature\_name)$" denotes all features except *"Feature_name"*

Table 3 On-the-fly detection performance of classifiers

ML models	Accuracy	F1 score	Precision	Recall
KNN	0.832	0.832	0.831	0.831
RF	0.965	0.965	0.965	0.964
SVM	0.598	0.591	0.590	0.590
DT	0.909	0.908	0.909	0.9098
ANN	0.861	0.862	0.862	0.862

the RF classifier is run that reported the best binary-classification accuracy in the experimental evaluation. A prediction accuracy ranging from 45.25% to 76.58% is observed while considering each feature in isolation. Although having relatively less accuracy in isolation, a combination of multiple feature metrics to detect an attack scenario significantly increased the overall prediction accuracy.

On-the-Fly Detection Accuracy The evaluation results provided so far have been obtained offline by applying the detection algorithms on collected data samples. Now, the ML classifiers are integrated within the simulation environment. The flooding attack is simulated by increasing packet injection from a particular network node. Whenever the flag is raised in the first-level sanity check, the ML model is triggered that resides within the corresponding CPU node. In this case, the classifiers are trained offline on the collected data samples while the inference is carried out at the runtime. As a result, the ML classifiers encounter a variation in runtime network behavior, which helps in evaluating their effectiveness to make accurate attack detection for dynamic runtime network traffic. The on-the-fly detection accuracy of different ML classifiers is presented in Table 3. All the classifiers show similar results as compared to the offline detection accuracy presented in Table 2. Among all the classifiers considered for evaluation, RF classifier provides the best performance in terms of detection accuracy. From these observations, it can be concluded that such an ML-based attack detection framework is suitable for a heterogeneous SoC environment where the network traffic exhibits dynamic load behavior.

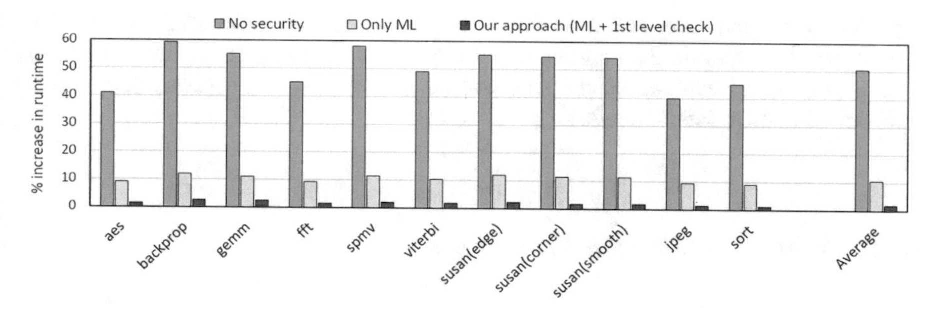

Fig. 5 Accelerator runtime in three scenarios: (i) absence of any security mechanism, (ii) ML-only-based detection, and (iii) two-step ML-based detection: first-level sanity check and second-level ML detection

Performance Evaluation In this section, the impact of the attack detection framework on the system performance is evaluated. While the framework will be able to curtail the performance degradation of the accelerators, it might also incur traffic overhead under non-attack scenarios. The performance degradation (i.e., increase in runtime) of all the accelerator cores is observed in three scenarios: (i) absence of any security mechanism, (ii) ML-only-based detection, and (iii) two-step ML-based detection: first-level sanity check and second-level ML detection. In the first scenario, the system has no built-in security measures, whereas in other scenarios, there is a security mechanism to detect a flooding attack. In the second scenario, the ML classifier is assumed to be active after every fixed time interval throughout the application execution time. As a result, feature data is continuously collected at each interval from all the network nodes and fed to the ML classifier. Figure 5 shows the increase in runtime of the accelerators in all the three scenarios as compared to a baseline non-attack scenario. As shown in the figure, the absence of any security mechanism significantly affects application performance leading to an increase in accelerator runtime of up to 59.21% as compared to the baseline. The ML-only-based method is able to accurately detect a flooding attack and prevent the accelerators from experiencing significant performance degradation. However, it incurs an average runtime overhead of 10.58% due to the additional on-chip communication involved in regularly collecting feature data from all the network nodes. On the contrary, the two-step ML-based detection framework incurs a minimal runtime overhead of 1.84% on average, as compared to the baseline. This is because the first-level sanity check eliminates the need for invoking the ML classifier on regular intervals. As a result, the communication overhead incurred in collecting the feature data for every ML invocation is drastically reduced. Finally, the hierarchical feature data collection further minimizes the overall runtime overhead by allowing cluster-based data aggregation as compared to collecting data from every node as done in ML-only scenario. The first-level sanity check prevents frequent ML invocations and reduces the system performance overhead. However, the first-level sanity check may introduce a few false positives, which will invoke the second-level ML detection in the non-attack scenarios. This will result in unnecessary feature data collection and incur traffic overheads in such cases. In the experimental evaluation, a traffic overhead of 0.31% is observed due to this unwanted feature data collection.

3.2 Attack Localization Framework

This section presents Sniffer, an ML-based framework for localizing MIPs in case of the flooding-based DoS attacks in NoC-based SoCs. An overview of the ML-based MIP localization framework is presented in Fig. 6. Once a VIP detects an attack, it initiates the localization process by inspecting the congestion status of its router's incoming ports. The notion behind this analysis is that the router port falling on the attack path will exhibit high congestion in an attack scenario as compared to a non-attack scenario. Sniffer employs a machine learning approach to accurately distinguish the congestion status of a given router port in attack and non-attack scenarios. If the ML model flags the status as an attack, the VIP creates a probing packet and forwards it to the neighbor node residing in the direction of the suspected port. Consequently, each router will inspect its incoming ports and forward the probing packet accordingly. The local IP port at each node is also inspected, and the Node ID is stored as an MIP if the model flags its status as an attack. In addition, Sniffer also checks if a node is traversed before and stops the process to indicate the presence of a loop. Hence, by traversing back in the opposite direction of the attack path, Sniffer successfully localizes all the MIPs.

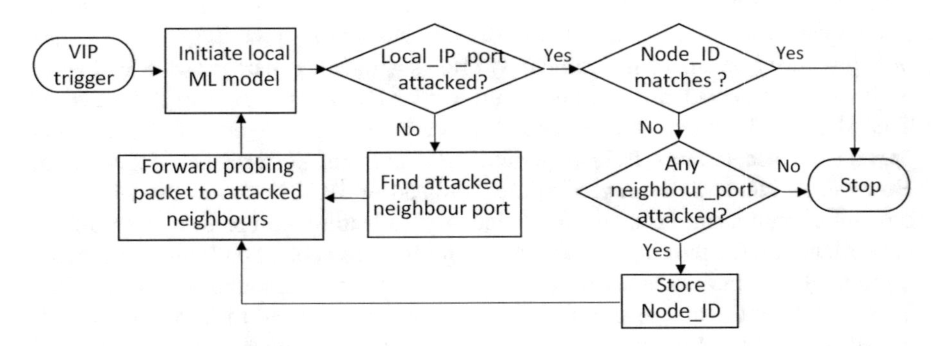

Fig. 6 Overview of the MIP localization framework

3.2.1 Machine Learning for Localization

A fundamental step in Sniffer is to accurately determine the congestion status of a given router port. If a router port's congestion status is falsely classified as an attack scenario, it may increase localization time and generate unnecessary probing packets into the network. Hence, it is crucial to accurately distinguish the network traffic behavior for attack and non-attack scenarios. A common approach used in the literature, [21] and [14], is to employ a single network parameter, whose threshold is set by static profiling, to flag anomalies in traffic behavior. However, the heterogeneous SoCs experience varied network-level activities throughout its execution due to various runtime optimizations [7–9] and coexistence of different applications at different times. As a result, setting the threshold of a network parameter through static profiling for determining a port's congestion status will lead to inefficient classification. Although using multiple parameters has been shown to increase the accuracy in various scenarios [26], it will significantly increase the complexity of the entire process. Firstly, there can be a number of network parameters that have the potential to help in distinguishing the congestion status of a router's port. One needs to meticulously explore the system or network characteristics and boil down to a set of parameters that can most accurately check the congestion status. Secondly, while tuning a single parameter through static profiling can be error-prone, tuning multiple parameters to achieve a single goal would dramatically increase the complexity and make static profiling infeasible. An ML model is able to tune multiple parameters to appropriate thresholds for congestion detection. Therefore, it eliminates the need for onerous and error-prone tuning of parameters by static profiling and provides a suitable solution for varied network traffic of heterogeneous SoCs.

Features Multiple network features are meticulously selected to train the ML model to provide accurate congestion status. These features capture the variation in network-level activities throughout the applications' execution period.

Buffer Waiting Time (BWT) BWT is defined as the time interval for which a flit waits in the buffer of a network router. A flit will wait in the router buffer if its output port is unavailable or the input port of the downstream router is congested. Since the routers face high congestion in an attack scenario, a flit will be waiting for a significant time interval as compared to a non-attack scenario. As a result, BWT is a potential feature candidate to accurately detect the congestion status of a router port.

Inter-Flit-Interval (IFI) In a non-attack scenario, a router port observes an appreciable time period between reception of any two consecutive network flits. However, in a flooding-based attack scenario, a router port residing on the attack path observes a large number of incoming flits within a given time interval, and IFI becomes significantly small. While BWT gives a good indication for an attack scenario, IFI further helps to capture the cases where a large number of packets are destined for the same output port, resulting in higher BWT in non-attack scenarios. Hence, IFI serves as a good metric for detecting the congestion status of a given router port.

Virtual Channel Occupancy (VCO) VCO refers to the amount of virtual channel (VC) space occupied by the incoming flits at each network router. In case of an attack scenario, the VCs of a router on the attack path will be heavily filled due to a large number of incoming flits. This signifies the use of VCO as an effective metric to distinguish between attack and non-attack scenarios.

Data Collection To train the ML model, data samples are collected for both attack and non-attack scenarios. Noxim [23], a cycle-accurate network simulator, is used to facilitate data collection for the experimental evaluation. The simulator is operated in two modes, namely, attack and non-attack modes. In non-attack mode, there are no MIPs, and the system experiences regular network-level activities, while in the attack mode, the MIPs target one or more VIPs and inject a large number of frequent packets into the network. The higher packet generation rate of the MIPs affects the network bandwidth and increases congestion, resulting in a flooding-based DoS attack. Since, in a heterogeneous system, the traffic density varies depending on the placement and type of IPs, the ML model is trained separately for each network node to precisely capture its communication behavior. The models are trained offline on the collected data samples to reduce computational overheads. In the experiments, the evaluation is carried out on a 64-node system, while the ML models can be further trained for localizing MIPs in larger system sizes. Multiple runs of different benchmarks are considered, each for 20,000 cycles during data collection (details of system configurations and benchmarks are given in Sect. 3.3). A total of 80,640 data samples are collected, 40,320 for attack and 40,320 for non-attack scenarios.

3.2.2 Algorithm for MIP Localization

Algorithm 1 describes the localization process of Sniffer. During the localization process, one of the three scenarios is encountered at each node as follows. In the case of a non-malicious node, only an incoming neighbor port (here, incoming neighbor port refers to the current node's incoming ports connected to the neighbor router nodes) will be congested, which signifies that the node is on an attack path, whereas an MIP will experience only a congested incoming local IP port and other incoming neighbor ports will undergo regular network activities. However, in the case of multiple MIPs, there may be scenarios where one of the MIPs resides on the attack path of the other (Fig. 1d) or a loop is formed due to a coordinated attack (Fig. 1e). In such scenarios, the MIPs will experience congestion in both incoming local IP port and incoming neighbor port. Upon VIP triggering, the localization process starts, and final decisions are taken at each node based on the three scenarios. The steps involved in the MIP localization algorithm are described as follows.

Algorithm 1: MIP localization algorithm

```
1  MIP[]=Null; port_attacked[]=Null
2  Upon VIP trigger:
3  do in pipeline
4      foreach in_port from N,E,S,W direction do
5          /*Perceptron check congestion in current node*/
6          if in_port == attack then
7              Mark in_port as attack
8  do in pipeline
9      foreach neighbour in N,E,S,W direction do
10         /*Perceptron check congestion for its outgoing port towards current node*/
11         if out_port == attack then
12             Mark out_port as attack
13         if prob_model(in_port, out_port)==attack then
14             port_attacked += direction
15 if local_IP_port ≠ attack then
16     Forward probing packet
17 else
18     if port_attacked == Null then
19         MIP += Node_ID
20         Stop & return MIP
21     else
22         if Node_ID not traversed then
23             MIP += Node_ID
24             Forward probing packet
25         else
26             Stop & return MIP
27 Reset port_attacked
28 Goto next node with probing packet & Repeat from step 3
```

N=North, E=East, S=South, W=West

- At a current node, its local perceptron checks the congestion status of all the incoming router port (steps 4–7). If the incoming neighbor port is congested ($in\_port == attack$), it is marked as under attack (steps 6–7). At the same time, a signal is sent to the neighbor nodes to check the congestion status of their outgoing port toward the current node.
- All the neighbor nodes check the congestion status on the corresponding output port by using their locally residing perceptron hardware. If any neighbor node finds its outgoing port congested ($out\_port == attack$), it marks it as under attack and communicates this information back to the current node (steps 9–12). The perceptrons' decisions of both current and neighbor nodes for each direction are fed to the probabilistic model, which finally declares a router port as under attack or non-attack scenario (steps 13–14).

- If the incoming local IP port of current node undergoes regular network activities ($local\_IP\_port \neq attack$), it indicates that the node is on the attack path and not the attacker. In such scenarios, a probing packet is forwarded to the neighbor node in the direction of the port under attack (steps 15–16).
- If the incoming local IP port is under attack and incoming neighbor ports undergo regular activities, then the localization process stops and current $Node\_ID$ is flagged as the MIP (steps 18–20).
- If both the incoming local IP port and neighbor ports are under attack, the framework first checks if the current node is already traversed. If the current node is not traversed before, the Node ID is stored in a probing packet and forwarded to the neighbor node in the direction of the port under attack (steps 21–24). If the current node is traversed before, it implies that Sniffer has encountered a complete loop due to a coordinated attack by multiple MIPs. As a result, the localization process stops and all the $Node_ID$s stored in the probing packet are flagged as MIPs (step 26).

In the case of a single MIP or multiple MIPs with no closed loop, Sniffer stops as soon as the probing packets flag the attackers. However, in the case where a closed loop is encountered or the attack paths overlap, Sniffer extracts all the Node IDs inserted in the probing packet and flags them as MIPs, which were creating a coordinated attack. In scenarios where one MIP targets multiple VIPs, the MIP will be localized by the probing packet that takes minimum time to reach the MIP node. When the MIP gets localized, the probing packets generated from other VIPs do not find any router port under attack and get dropped. Once localized, the operating system shuts down all the MIPs to prevent any further attacks and the system is restored to allow the legitimate IPs to undergo regular network activities.

3.2.3 Walk-Through Example

This section presents the MIP localization process of Sniffer with single/multiple MIPs/VIPs that include different scenarios as shown in Fig. 1. The localization process is first demonstrated using a walk-through example for a single MIP and VIP scenario. A transaction-level representation for the walk-through example with a 16-node system is presented in Fig. 7. Later, the working of the framework works in the presence of multiple MIPs/VIPs is described. Sniffer is able to work in conjunction with any existing DoS attack detection methods where a system is flagged when experiencing a flooding attack. To illustrate the localization process, an example flooding-based DoS attack detection method presented in [6] is considered. After a flooding attack is mounted on the system, one or more VIPs raise a flag indicating that the system is under attack. Whenever an attack flag is raised, Sniffer starts the MIP localization process from the VIPs that generate the attack flag.

Single MIP and VIP Figure 7.a shows the attack path through which an MIP (Node 12) creates a flooding attack targeting a VIP node (Node 6). Node 6 flags the system as under flooding attack and the localization process is initiated.

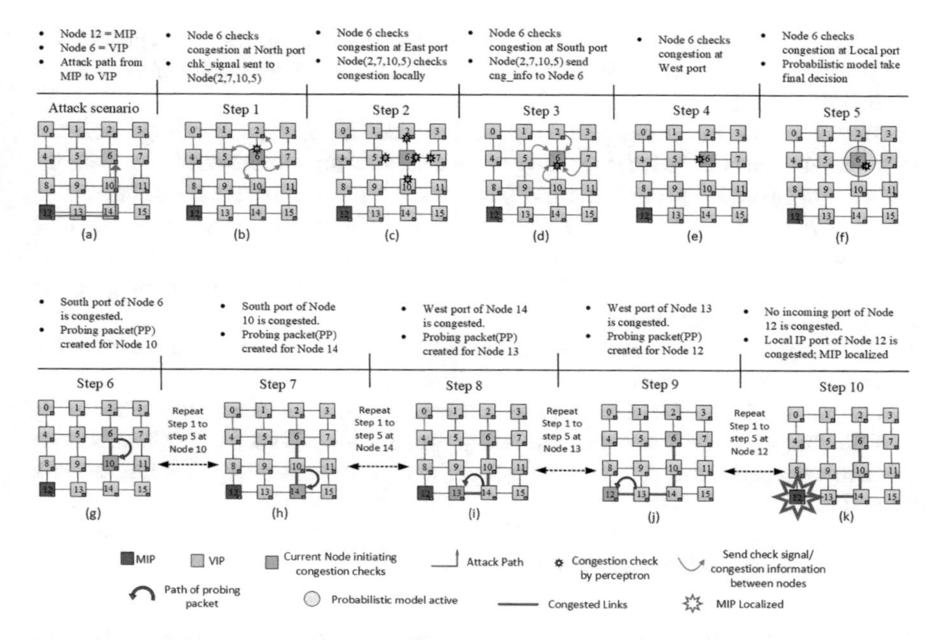

Fig. 7 A walk-through example for MIP localization by Sniffer with an example attack scenario (Node 12 = MIP, Node 6 = VIP)

- In Step 1 (Fig. 7b), Node 6 checks if its incoming port in the north direction is congested or not using its locally residing perceptron model. At the same time, it also sends a signal (chk_signal) to its immediate neighbor nodes (2, 7, 10, and 3) to start the collective decision-making on the congestion status at each port.
- In Step 2 (Fig. 7c), after completing the congestion check at north direction, the perceptron model at Node 6 checks its incoming port at the east direction for congestion. The local perceptrons at Node 2, Node 7, Node 10, and Node 3 also check the congestion status of their corresponding output ports in the direction of the received chk_signal.
- In Step 3 (Fig. 7d), Node 6 subsequently makes a congestion check at its incoming south port while Node 2, Node 7, Node 10, and Node 3 send their computed congestion information (cng_info) back to Node 6.
- In Step 4 (Fig. 7e), Node 6 checks its incoming port in the west direction for determining its congestion status. It is to be noted that Sniffer employs a single perceptron hardware at each router node for low overhead implementation. So the perceptron at the current node starts the checking of the congestion status of the incoming ports one after another. There is no specific rule for the order in which the ports need to be checked. For this example case, the checks on incoming neighbor ports are done in N-E-S-W order.
- In Step 5 (Fig. 7f), Node 6 makes a congestion check at its incoming Local IP port to determine if Node 6 is an attacker. After receiving all the congestion

check results from its local perceptron and neighbor nodes' perceptrons, the probabilistic models at Node 6 take the final decision for congestion at all its incoming ports.

- If any of the incoming port is marked as under attack, a probing packet is generated toward the node residing at the corresponding direction. In the example scenario, Node 6 generates a probing packet for Node 10 as shown in Fig. 7g. Once Node 10 receives the probing packet, it performs all the steps done at Node 6 (Steps 1–5) to check the congestion status of its incoming ports. Eventually, the incoming south port of Node 10 is found to be congested, and the probing packet is forwarded toward Node 14.
- Similarly, steps 1–5 are repeated at Node 14 and subsequently at Node 13 to find the direction of the probing packet, as shown in Fig. 7h and i.
- As Node 12 receives the probing packet, it starts its congestion checks by following steps 1–5, as shown in Fig. 7j. Since Node 12 is the only attacker node in this example case, all its incoming neighbor ports will undergo regular network activities, and its incoming local IP port will be congested. Hence, no probing packet is generated, Sniffer flags Node 12 as an MIP node, and the localization process stops (as indicated in Fig. 7k).

Multiple MIPs/VIPs Figure 1b–e show scenarios where multiple MIPs/VIPs coexist within the system. Under such scenarios, multiple MIPs can target the same or different VIPs to create a flooding-based DoS attack. They can also mount an effective coordinated attack where multiple MIPs form a closed loop in the network.

In the case where multiple MIPs follow different attack paths to target different VIPs, Sniffer parallelly starts the localization process from the corresponding VIP nodes. A probing packet starts from each VIP node and follows all the steps described in the walk-through example (Fig. 7). Each probing packet reaches the corresponding MIP node by backtracking and traversing through all the nodes in the attack path. In the case where the attack paths overlap (e.g., Fig. 1c and d), the probing packets are merged and forwarded toward the corresponding node residing at the direction of the congested port. In the case of a closed loop formed by multiple MIPs (Fig. 1e), whenever any one of nodes flags the presence of an attack, Sniffer starts the localization process and follows the steps shown in the walk-through example. However, the probing packets might keep on traversing through the nodes in a cycle as all the nodes will always find one of their incoming ports under attack. To overcome this situation, Sniffer stores the Node IDs of the attacker nodes in the probing packets. Whenever it encounters a node that is traversed before, Sniffer identifies a loop and stops the localization process. It then marks all the Node IDs stored in the probing packets as the MIPs. Therefore, Sniffer is able to successfully localize multiple MIPs under different MIP/VIP placements and attack scenarios.

If the DoS attack is not handled timely, the network resources will quickly get exhausted as the malicious IPs will keep on injecting frequent useless packets into the network. If the network gets too congested due to the malicious injection of packets, it creates a problem for the movement of probing packets, which is required for communicating between the network nodes and tracing back the attack path. Typically, such scenarios are handled by the attack detection frameworks that are responsible for timely detection of an attack scenario within the network. For instance, our attack detection framework (presented in Sect. 3.1) proposed a two-level attack detection. The first-level sanity check helps in timely attack detection as well as lowering the overhead of frequent second-level ML invocation. The use of hierarchical trace data collection for second-level ML detection was also introduced to reduce further network overheads and help in seamless trace data collection. By using such mechanisms for timely attack detection, the framework reduces the possibility of further network degradation, which is also demonstrated by the experimental results in Sect. 3.1.3. Once an ongoing attack is detected timely, the Sniffer localization framework can start the localization process immediately. Since the detection framework detects the attack before the network gets completely clogged, the probing packets are able to move into the network and successfully localize the malicious IPs. Therefore, in this work, we assume that the probing packets will find free network resources to traverse through the localization path. However, in a rare worst-case scenario, where the detection framework is not able to detect a timely attack, the network resources can get completely clogged due to severe and prolonged attack. To handle such a situation, we propose that the network router will drop packets residing in a virtual channel of the corresponding port, which is requested by the probing packet. This will allow a steady forwarding of probing packets and provide efficient MIP localization.

3.3 Results and Analysis

This section discusses the experimental setup and evaluation of the MIP localization framework. To evaluate the performance of Sniffer, a cycle-accurate network simulator, Noxim [23], is employed. The system is tested using real-world heterogeneous benchmarks from *Rodinia* [27] benchmark suite. All the application-level traces are obtained from gem5-gpu [28] and fed to Noxim, which emulates the network communication behavior. Table 4 summarizes the simulation setup and system configurations.

Performance of Perceptron Model The performance of perceptrons in checking the port congestion status is evaluated in terms of true positives (TP) and true negatives (TN), i.e., correct classification of congestion status as attack and non-attack, respectively, and false positives (FP) and false negatives (FN), i.e., incorrect classification of congestion status as attack and non-attack, respectively. A higher value of TP and TN indicates that the perceptrons can more accurately classify the congestion status, whereas a lower FP and FN represents the robustness of the classification results. Several MIP and VIP pairs with different placements for each benchmark application are considered. Table 5 presents the perceptron accuracy results (average accuracy among all the network nodes) for different benchmark

Table 4 System configuration and simulation setup

Component	Configuration
Topology	8X8 2D mesh, XY routing, 1-cycle link latency
Router	5 I/O ports, 1 VCs per port, 4 flit VC buffer, 32 bit flits, wormhole switching, 3-stage router, 2 flit packet
Processing cores	8×86 OOO CPU cores; 32 GPU cores
Memory	32 KB private L1 cache, 256 KB L2 cache, 8 L2 caches, 2 MB L3 cache; 4 GB DRAM
Workloads	Rodinia benchmark suite: Backprop, Gaussian, Hotspot, Kmeans, Lud, Mummergpu

Table 5 Performance of perceptron model

Benchmarks	TP (%)	FP (%)	TN (%)	FN (%)
Backprop	98.33	1.71	98.28	1.66
Gaussian	98.32	1.73	98.26	1.68
Hotspot	98.35	1.69	98.30	1.65
Kmeans	98.36	1.69	98.31	1.63
Lud	98.34	1.72	98.27	1.65
Mummergpu	98.35	1.66	98.33	1.64
Average	98.35	1.70	98.30	1.65

Table 6 Performance of MIP localization

Benchmarks	Localization accuracy
Backprop	96.739%
Gaussian	96.691%
Hotspot	96.773%
Kmeans	96.789%
Lud	96.722 %
Mummergpu	96.808%
Average	96.754%

applications. As shown in the table, the perceptron model is able to provide high accuracy with an average of 98.35% and 98.30% for TP and TN values, respectively. The FP and FN values are also very less with an average of 1.70% and 1.65%, respectively, across all benchmarks, signifying the robustness of the model. Hence, the perceptron models trained with carefully selected features provide high accuracy for localizing the MIPs.

Performance of MIP Localization In this section, the performance of Sniffer in correctly localizing MIPs within the system is evaluated. Real-world heterogeneous benchmarks from Rodinia benchmark suite [27] with multiple test cases are run. All the test cases comprise of different MIP and VIP pairs along with various placements of MIP and VIP nodes across the network. Multiple design considerations are included in the framework such that it provides a stable performance across different benchmarks. The node-level training, where each perceptron at a router node is trained on data collected from different heterogeneous benchmarks, ensures that

Table 7 Localization time in the presence of multiple MIPs

Test cases	Router pairs (MIP, VIP)	Localization time for each (MIP, VIP) pair (cycles)	Final localization time (cycles)
2 MIPs & 2 VIPs	(0, 14), (23, 42)	70, 80	80
2 MIPs & 1 VIP	(24, 7), (56, 7)	100, 140	140
3 MIPs & 1 VIP	(29, 22), (40, 22), (62, 22)	20, 50, 70	70
1 MIP & 2 VIPs	(23, 6), (23, 48)	30, dropped	30
2 MIPs (coordinated attack)[a]	(39, 1) coordinates with (1, 39)	(50, 100)	100

[a] Please refer to Sect. 3.2.3

each perceptron model gets a holistic view of various possible scenarios. Along with a carefully designed training method, the probabilistic model for collective decision-making strategy enhances the congestion checks at each router node, which further improves the performance of Sniffer in MIP localization. Table 6 shows the MIP localization accuracy of Sniffer for different benchmark applications. As given in the table, Sniffer provides high localization accuracy of 96.754% on average. It is also evident from the table that Sniffer is able to provide a consistent performance across the benchmarks, which makes the framework robust across different applications.

MIP Localization Time The MIP localization time of Sniffer depends on the time taken by the probing packet to traverse through the network nodes, starting from the node at which the localization process is initiated (VIP) to the node at which it is stopped (MIP), and the time taken for congestion checks at each node on the localization path. Figure 8 shows the localization time of Sniffer across various test cases, where each test case consists of a single MIP and VIP pair. As evident from the figure, Sniffer is able to localize the MIPs in a timely manner across all the test cases. The worst-case localization time for a 64-node system is found to be as low as 150 cycles in the experimental evaluation. The efficiency of Sniffer in localizing the MIPs in a timely manner is achieved primarily by the collective decision of perceptrons and prioritized probing packet transmission. In large system sizes, the search space for attack localization is vast, as there are a number of possible network nodes that can be a potential attacker. The collective decision of perceptrons significantly decreases the search space by allowing Sniffer to make congestion checks with high accuracy at each step of localization. Hence, Sniffer needs to traverse only through the nodes residing on the attack path and localize the MIPs in a timely manner. Furthermore, Sniffer forwards the probing packets with higher priority through the attack path, which leads to a faster MIP localization. From Fig. 8, it is observed that the time required to localize an MIP varies across different (MIP, VIP) pairs. The variations in the localization time are primarily due to the varied distance between each MIP and VIP pair. The localization time increases with the increase in hop distance between the MIP and VIP nodes, as shown in Fig. 8.

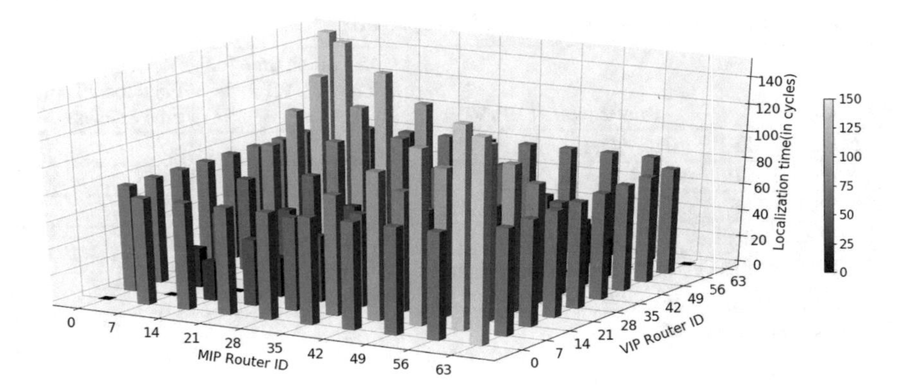

Fig. 8 MIP localization time

The localization time of Sniffer in the presence of multiple MIPs and/or VIPs is also analyzed. In the case where multiple MIPs target different VIPs present in the system, the time taken to localize all the MIPs is the maximum localization time across all the MIP and VIP pairs. In the case of a coordinated attack mounted by multiple MIPs, which forms a closed loop within the network, the localization time will depend on the time taken to complete one iteration of the loop and the congestion checks at the corresponding network nodes. If the system contains only one MIP that targets different VIPs, the total localization time will correspond to the minimum time taken by any one of the probing packets generated from all the VIPs to reach the MIP. This is because once the MIP is localized and the system is restored, the probing packets generated from other VIPs do not find any congestion in the router ports and get dropped. Table 7 shows the localization time for a (MIP, VIP) pair for each test case and the final localization time in the presence of multiple MIPs/VIPs. Here, a few illustrative test cases are shown, where each test case comprises of different MIP and VIP pairs.

As evident from the table, in the case of multiple MIPs targeting different VIPs, the final localization time corresponds to the maximum of the individual localization time of the (MIP, VIP) pairs in each test case, whereas in a single MIP and multiple VIP case, the probing packet with minimum hop distance localizes the MIP, and the probing packets generated from other VIPs get dropped. Lastly, under a coordinated attack[1], the final localization time of all the MIPs is found to be the time required to check one iteration of the closed loop under attack. For instance, a test case for two MIPs, where Node 39 and Node 1 create a coordinated attack, is shown in Table 7. In this case, Node 1 generates a probing packet, which traverses through the localization path to reach Node 39 and then back to Node 1. When the probing packet reaches Node 1, one iteration of the loop is completed and both the MIPs get localized. Hence, the final localization time is the time taken to complete one iteration of the loop, which is found to be 100 cycles in this case. In this scenario,

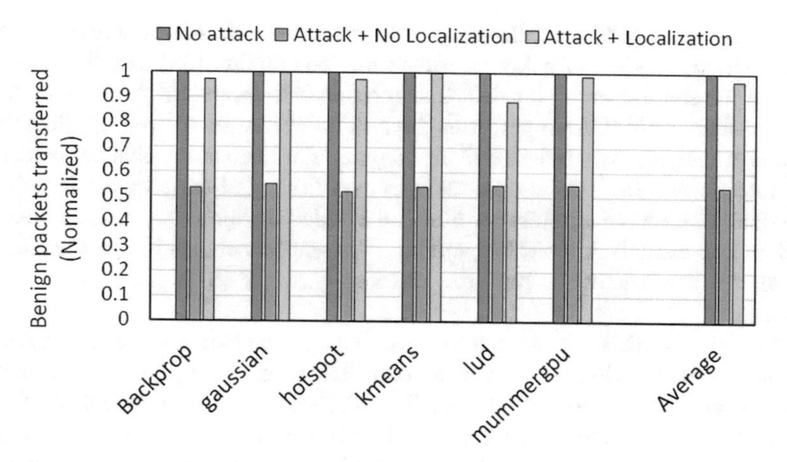

Fig. 9 Benign packets transferred in different scenarios

Node 39 will also generate a probing packet, apart from Node 1. However, the probing packet that will complete the loop first will localize both the MIPs.

Impact on Application Performance The network traffic is analyzed to understand the impact of Sniffer's localization process on applications running on the SoC. The number of benign packets successfully transmitted to their destination nodes are observed in three scenarios: (i) baseline system with no attack, (ii) system under attack with no MIP localization method, and (iii) system under attack with Sniffer deployed. To make a fair comparison, all the three scenarios are evaluated for the same time interval across all benchmarks. In an attack scenario with no MIP localization, as the network gets flooded with malicious packets, the applications experience performance degradation due to significant packet loss. This is also evident from Fig. 9, where the number of successfully transferred benign packets reduce to almost half as compared to the non-attack baseline scenario. However, in the third case, due to the presence of Sniffer, the system under attack gets restored timely and prevents further packet loss. Hence, Sniffer reduces the effect of flooding attack and curtails system performance degradation.

4 DoS Attack Detection and Localization in WNoC-Based SoCs

With the evolution of CMOS compatible millimeter-wave wireless interconnects [4], WNoC is emerged as a promising interconnect solution for the multicore systems. The use of WNoC is proved to be beneficial not only for the system performance gain [29] but also for faster debug [30], efficient cache coherence [31], high performance CNN acceleration [32], etc. Major WNoC architectures adopt single wireless channel for low overhead implementation [33]. All the hubs need to share the same channel for data transmission and reception. A Channel Access control Mechanism (CAM) performs the task of arbitration as well as allocation of channel among all the wireless hubs. At a particular instant, the channel is assigned to a specific hub for a fixed amount of access time, and thereby, the interference and contention on the wireless medium are avoided. In such a scenario, a malicious wireless hub can attack the CAM and hold the wireless channel in an unauthorized manner to create network resource blocking-based DoS attack as discussed in Sect. 2.2.

Traditional CAMs follow a token passing mechanism for the distribution of channel access to wireless hubs in a controlled manner [29, 34]. In this method, a token is circulated among all the existing wireless hubs and an associated WI gets the access of the channel till the time token is with it. Most of the token passing methods follow a fixed *Round Robin* arbitration and a timer-based approach [29], where each hub gets the channel access for a predetermined fixed period of time. For such conventional CAMs with static access pattern and time, a deviation in arbitration or access time is clearly visible to the system, and detecting as well as localizing a malicious hub is easy [19]. But to meet the stringent power and performance needs of modern-day systems, a dynamic CAM is desired [35, 36]. However, a malicious hub can exploit the varying channel access pattern and time in dynamic CAM to introduce vulnerability to the system. Moreover, such a dynamic CAM can be controlled in a centralized [35] or decentralized manner [36]. A centralized controller incurs significant overhead in transferring the control signals till all the WIs. Furthermore, the whole system would malfunction, if the attacker can intrude the centralized controller.

Decentralized method deploys simple controllers inside each of the wireless hubs. Such a controller decides the channel access time of the corresponding hub based on local load level [36]. Every wireless packet communication on such network informs the source address and the channel access time of the corresponding source WI to the remaining hubs using the broadcast capability of the WNoC. The packet header flit as shown in Fig. 10 is modified by the controller to embed the WI addresses and access time information. The *Source WI* and *Access Time* in the header flit of a transmitted packet indicate who and when will get the next access of the channel. Whenever a packet header is received by a hub, the receiver checks for the *Dest WI*, *Source WI*, and *Access Time*. If the *Dest WI* matches to its own address, then it starts receiving the payload of the packet. If the *Source*

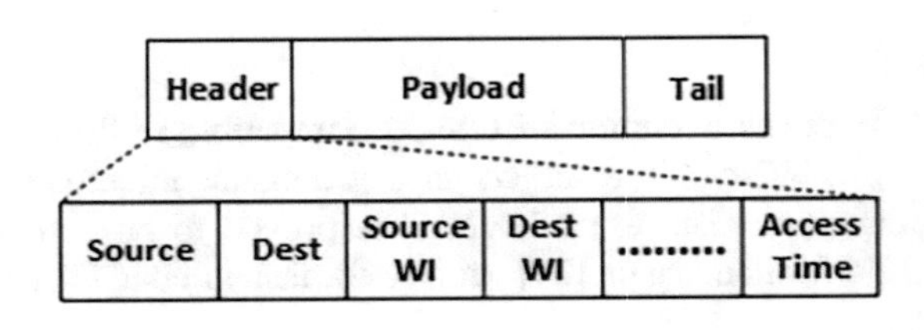

Fig. 10 Packet format transmitted over wireless channel

WI appears to be the address of the previous hub, then the hub interprets that it is the next to get the access of the channel and it holds the token if the *Access Time* is found to be zero.

This section particularly highlights the work done in [18] that adopts the decentralized approach for wireless channel allocation, establishes a threat model for such architecture, and finally reutilizes the same decentralized control infrastructure for establishing the defense mechanism.

4.1 Possible Threat Model

In a decentralized CAM, if any local controller is corrupted by some means, then the *Source WI* and *Access Time* information in packet header flit can be manipulated to hold the channel in an unauthorized manner. Presence of a simple HT circuit can easily be triggered to change the header flit values. Such flit manipulation can lead to DoS or spoofing attack. These attacks would result in poor utilization of wireless channel and would drive the victim hubs to starvation.

4.1.1 Attacks on WNoC CAM

The DoS on CAM refers to the unavailability of the wireless channel for a particular wireless hub whenever the channel is used by another hub in an illegitimate way. This can be done by a malicious hub by periodically changing its *Access Time* in the header flit of the transmitting packets with an intention not to release the wireless channel. The *Access Time* of the malignant hub would never be seen as zero, and the next wireless hub waiting for the channel would be misguided that the previous node has the rightful access of the channel. Another way of creating DoS is by changing the source address of the packet being transmitted on the wireless channel during its last access cycle. It can be done by a rough wireless hub by manipulating the *Source WI* in the packet header flit as the source address of the previous hub. This would mislead all the remaining hubs interpreting that the malicious one is the next to get the access of the channel. This way, the malicious hub would continue to hold the channel. In this case, since the malicious hub imitates as another hub to create the DoS, it is called as a spoofing attack.

4.1.2 Trojan Attack Activation

Both DoS and spoofing attack scenarios can be triggered by a simple HT circuit implanted inside the wireless hub as shown in Fig. 11. The circuit used is a sequential *Time-bomb Trojan* [37]. *Trojan Trigger* takes the inputs from the *Access Time Register* and generates the trigger signal when all bits of the register becomes zero. *Trojan Payload* modifies the *Load Enable* (LE_n) signal whenever there is a

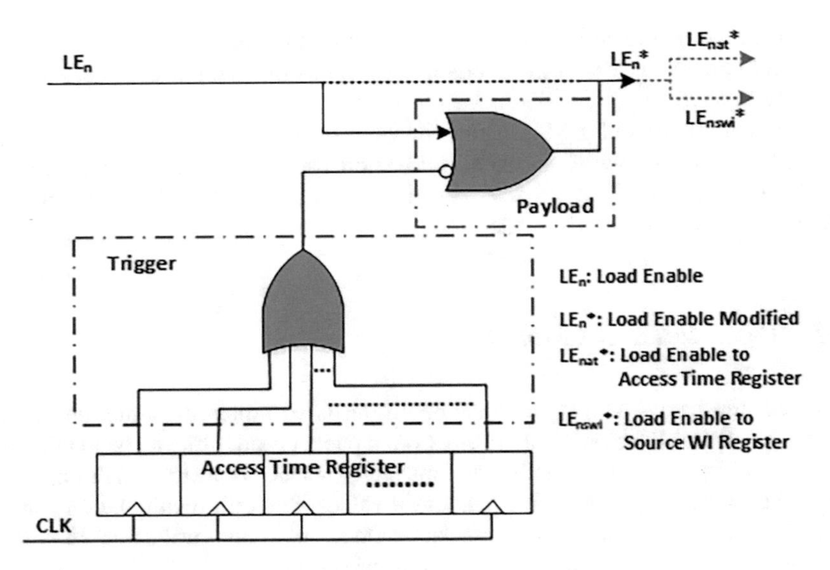

Fig. 11 Time-bomb Trojan in the malicious hub

trigger. During the *First Token Round* (As discussed in Sect. 4.2.1), LE_n signal in each of the hubs gets activated to load the *Access Time Register* and *Source WI Register*. Both these registers provide values to be updated in the corresponding fields of packet header flit before transmission. The value in *Access Time Register* is decremented based on the load release. The Trojan circuit exploits the decrement nature of the *Access Time Register* to generate a *Time-bomb Trigger* during the last channel access cycle of the corresponding hub in the *Second Token Round*. The *Payload* circuit modifies the LE_n to generate $LE_n{}^*$ that enables unauthorized loading to either *Access Time Register* ($LE_{nat}{}^*$) leading to DoS attack or *Source WI Register* ($LE_{nswi}{}^*$) leading to spoofing attack.

4.2 Security Countermeasures

The security countermeasures are established based on assigned access time and actual access time values. The decentralized CAM adopts a ranking-based channel allocation. Figure 12a demonstrates the architectural changes to the wireless hub by introducing a distributed ranking-based channel access controller (DRCAC) in WI and the security blocks such as *Majority Voter* and *Flag Generator* in network interface (NI). The NI is assumed to be completely secured and free from attacks. The NI also provides separate interfaces to the wired router (eNI) and to the WI (wNI).

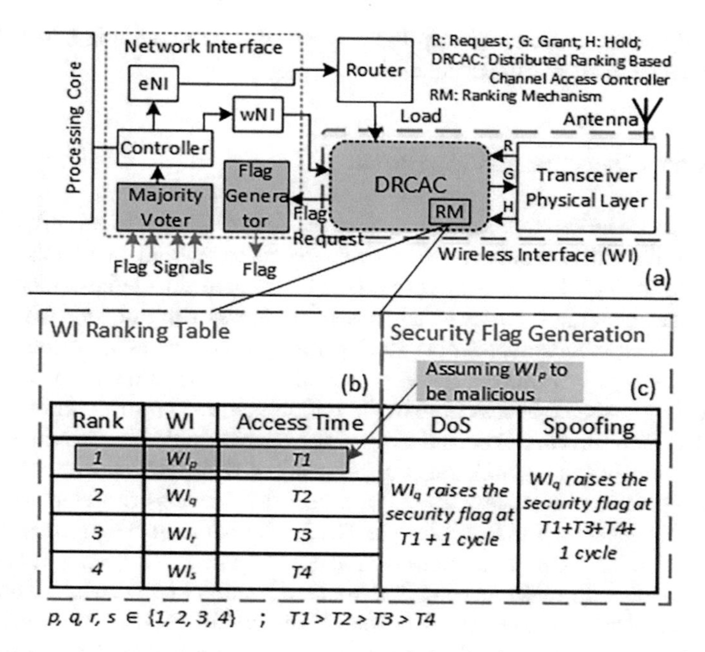

Fig. 12 (**a**) Ranking mechanism and security architecture. (**b**) Ranking table. (**c**) Security flag generation

4.2.1 Ranking Based CAM

For a wireless data transmission, the transceiver layer sends a request (R) to the controller (DRCAC) for gaining the access of the channel. The controller grants (G) the request whenever it gets the token, and the transceiver holds (H) the channel till the time WI has the token with it. The token is circulated on the network based on the ranking of the existing hubs. DRCAC implements the ranking mechanism and ranking table as shown in Fig. 12a and b. During the *First Token Round*, each WI broadcasts a packet with its access time value in a fixed *Round Robin* manner. Access time is calculated by the local DRCAC based on the current load density at the hub. At the end of the *First Token Round*, all the WIs have access time information of all other WIs, which provides consistency of information throughout the whole system and leads to coherent WI ranking computation. A *WI Ranking Table* based on access time values is prepared by the local controller inside each of the hubs as shown in Fig. 12b. The ranking is performed based on the descending order of the access time values. In the *Second Token Round*, the token transfer follows the ranking pattern in the ranking table. During the channel access by a particular WI, the *Source WI* information is extracted from the header flit of transmitted packet and is matched in the local ranking table to figure out who is the next WI that will get the token. Moreover, the access time information in the transmitted packet header indicates the amount of time remaining before the next WI will be able to access the channel.

4.2.2 Attack Detection and Correction

Any malicious wireless hub can manipulate the *Access Time* or *Source WI* information while transmitting a packet to hold the wireless channel in an unauthorized way as discussed in Sect. 4.1. To deal with this, a distributed self-defense mechanism is proposed that generates security flag based on access time of any WI as shown in Fig. 12c. The mechanism works during runtime and takes a corrective action whenever there is a violation between actual access duration by a WI and the corresponding assigned access time in the ranking table. For an example case, a 64 node, 4 WI, 2D mesh NoC is considered, whose ranking table is shown in Fig. 12b. It is assumed that WI_p is a malicious hub. Whenever WI_p holds the channel beyond T_1 by manipulating its *Access Time*, DoS security threat is detected by WI_q (at $T1 + 1$ *cycle*) as shown in Fig. 12c. The DRCAC module in WI_q requests its *Flag Generator* to raise a *DoS attack flag* and sends the same to all other WIs over the wired NoC paths. To make an unbiased attack detection and correction, a majority voting mechanism is implemented in the NI module of each of the wireless hub. The *DoS attack flag* is received by all other WI hubs present in the network and is evaluated locally based on the *Access Time* information present in the local ranking table. In case of a real violation, all other WIs would raise *Support flags*. All the flags would travel on the wired NoC path till the NI of the malicious hub. The associated *Majority Voter* module will collect all the flags and would disable the wNI if found guilty. This leaves the node with only the wired router for data transfer and suspends its wireless capability. In case of spoofing attack, the correction mechanism will be same with a small difference in the detection mechanism. In this scenario, the *spoofing attack flag* will be generated by WI_q at $T1 + T3 + T4 + 1$ *cycle*. This is because at $T1$ time, WI_p would manipulate the packet source address (*Source WI*) as WI_s to confuse the other wireless hubs. While receiving the packet and matching the source address, WI_q would misinterpret that WI_p is the next candidate to have the token. So it would not generate the flag at $T1+1$ *cycle*. To detect such scenarios, the maximum possible idle time of a WI during a token round is considered. In this case, when WI_q would not get the flag within $T1 + T3 + T4$ (maximum idle time of WI_q), it will raise a *spoofing attack flag* in the next cycle.

4.3 Experimental Setup and Results

The DoS and spoofing attacks discussed in this chapter degrade the system performance drastically due to the underutilization of wireless medium. With the proposed solution, the network blacklists the malicious WI and reconfigures the wireless network with the available healthy wireless hubs. This helps the system regain its performance close to the completely healthy one. A 2D mesh WNoC of size 8x8 nodes with 4 WIs is modeled on the network simulator Noxim [38]. An HT is implemented by modifying the WI module. The simulation results are generated by executing three synthetic applications and three workloads from SPLASH-2 benchmark suite [39]. Network traces for Noxim simulator is generated using a full system simulator Graphite tool [40]. Network architecture and simulation setup are shown in Table 8.

The experimental results are illustrated in Figs. 13 and 14. Simulations are executed for three system architectures such as (i) baseline system with four healthy hubs, (ii) malicious system with one attacked hub, and (iii) proposed reconfigured system with three healthy hubs. Figure 13 represents the normalized average wireless utilization, and Fig. 14 represents the normalized network throughput for all the three system architectures. In case of an attacked system, the wireless channel is captured by the malignant hub all the time leading to drastic reduction in wireless utilization. By detecting and eliminating the malicious hub, the system regains a considerably high wireless utilization, which is around 45% higher than the attacked system (shown in Fig. 13). Due to the poor utilization of wireless channel, the whole network throughput also gets affected during the attack. The proposed method reconfigures the network with the available healthy hubs, and the wireless channel is shared between them. This enables the system to achieve better network throughput,

Table 8 Network architecture and simulation setup

Component	Configuration
Topology	8×8 2D Mesh WNoC, 4 wireless hubs, XY routing
Router	5 I/O ports, 1 I/O wireless port (for wireless hub), 8 flit buffers, 8 flit packets, 32 bit flits
Wireless link	60 GHz carrier, 16 Gbps bandwidth
Workload	SPLASH-2—*Barnes, FFT, Radix* Synthetic—*Random, Transpose, Butterfly*

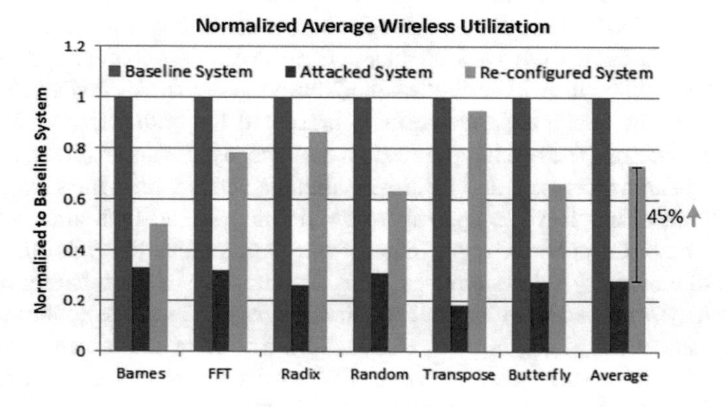

Fig. 13 Normalized average wireless utilization

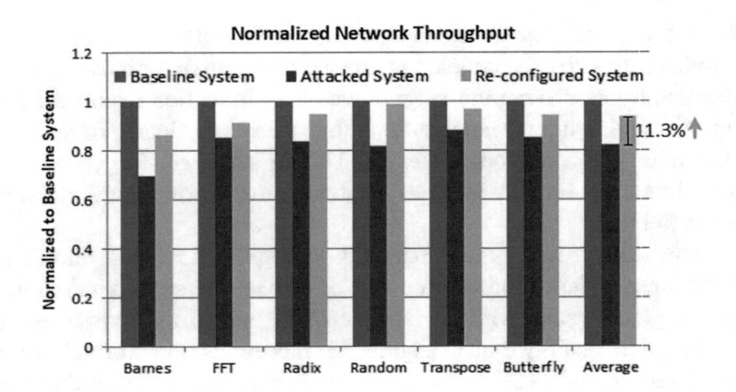

Fig. 14 Normalized network throughput

which is around 11.3% larger than the attacked system. The presented technique successfully detects the discussed DoS and spoofing attacks, and the suggested modification shows significant improvement over the attacked system with a very nominal hardware overhead. The associated overhead is due to the introduction of local ranking module and the security modules. The additional circuitry designed and synthesized in Synopsis design compiler with 65-nm technology-node occupies only 0.0036 mm^2 of area with 0.13 mW of power consumption, whereas the baseline WI designed in 65-nm CMOS process acquires 0.3 mm^2 of area with 36.7 mW of power consumption [4]. Thus, the area and power overhead per WI is found to be 1.2% and 0.36%, respectively.

5 Conclusion and Future Work

This chapter provides an overview of the DoS attack detection and localization frameworks for NoC-based SoCs. Along with the traditional systems with wired NoCs, it also discusses the DoS attacks and mitigation techniques for emerging interconnect architectures like wireless NoC-based systems. A detailed discussion on state-of-the-art example frameworks is presented for both wired and wireless NoC-based systems. These example frameworks mostly use sophisticated machine learning algorithms to detect and localize malicious nodes within the system.

As a future research work, apart from the flooding-based DoS attacks through the NoCs, the SoCs need to be prevented from DoS attacks through other shared resources like memory subsystems, etc. Furthermore, due to the inherent nature of NoCs, such systems can be vulnerable to other security attacks like spoofing, eavesdropping, packet tampering attacks, etc. Therefore, it is of the utmost importance to develop frameworks to make a secure SoC environment that can address such attacks and minimize application performance degradation.

References

1. Shao, Y.S., Xi, S.L., Srinivasan, V., Wei, G.Y., Brooks, D.: Co-designing accelerators and SOC interfaces using gem5-aladdin. In: 2016 49th Annual IEEE/ACM International Symposium on Microarchitecture (MICRO), pp. 1–12. IEEE, Piscataway (2016)
2. Sinha, M., Harsha, G.S., Bhattacharyya, P., Deb, S.: Design space optimization of shared memory architecture in accelerator-rich systems. ACM Trans. Design Autom. Electron. Syst. **26**(4), 1–31 (2021)
3. Benini, L., De Micheli, G.: Networks on chips: a new SOC paradigm. Computer **35**(1), 70–78 (2002)
4. Deb, S., Chang, K., Yu, X., Sah, S.P., Cosic, M., Ganguly, A., Pande, P.P., Belzer, B., Heo, D.: Design of an energy-efficient CMOS-compatible NoC architecture with millimeter-wave wireless interconnects. IEEE Trans. Comput. **62**(12), 2382–2396 (2012)
5. Shacham, A., Bergman, K., Carloni, L.P.: Photonic networks-on-chip for future generations of chip multiprocessors. IEEE Trans. Comput. **57**(9), 1246–1260 (2008)
6. Charles, S., Lyu, Y., Mishra, P.: Real-time detection and localization of dos attacks in NoC based SoCs. In: 2019 Design, Automation & Test in Europe Conference & Exhibition (DATE), pp. 1160–1165. IEEE, Piscataway (2019)
7. Reddy, B.K., Singh, A.K., Biswas, D., Merrett, G.V., Al-Hashimi, B.M.: Inter-cluster thread-to-core mapping and DVFS on heterogeneous multi-cores. IEEE Trans. Multi-Scale Comput. Syst. **4**(3), 369–382 (2017)
8. Iskandar, V., Salama, C., Taher, M.: Dynamic thread mapping for maximizing performance in power-efficient multi-core systems. In: 2018 13th International Conference on Computer Engineering and Systems (ICCES), pp. 230–235. IEEE (2018)
9. Attia, K.M., El-Hosseini, M.A., Ali, H.A.: Dynamic power management techniques in multi-core architectures: a survey study. Ain Shams Eng. J. **8**(3), 445–456 (2017)
10. JS, R., Ancajas, D.M., Chakraborty, K., Roy, S.: Runtime detection of a bandwidth denial attack from a rogue network-on-chip. In: Proceedings of the 9th International Symposium on Networks-on-Chip, pp. 1–8 (2015)
11. Sinha, M., Bhattacharyya, P., Rout, S.S., Prakriya, N.B., Deb, S.: Securing an accelerator-rich system from flooding-based denial-of-service attacks. IEEE Trans. Emerg. Topics Comput. **10**(2), 855–869 (2021)
12. Fiorin, L., Palermo, G., Silvano, C.: A security monitoring service for NoCs. In: Proceedings of the 6th IEEE/ACM/IFIP international conference on Hardware/Software Codesign and System Synthesis, pp. 197–202 (2008)
13. Sinha, M., Gupta, S., Rout, S.S., Deb, S.: Sniffer: a machine learning approach for dos attack localization in NoC-based SoCs. IEEE J. Emerg. Sel. Topics Circuits Syst. **11**(2), 278–291 (2021)
14. Charles, S., Lyu, Y., Mishra, P.: Real-time detection and localization of distributed dos attacks in NoC-based SoCs. IEEE Trans. Comput.-Aided Design Integr. Circuits Syst. **39**(12), 4510–4523 (2020)
15. Fang, D., Li, H., Han, J., Zeng, X.: Robustness analysis of mesh-based network-on-chip architecture under flooding-based denial of service attacks. In: 2013 IEEE Eighth International Conference on Networking, Architecture and Storage, pp. 178–186. IEEE, Piscataway (2013)
16. Ganguly, A., Ahmed, M.Y., Vidapalapati, A.: A denial-of-service resilient wireless NoC architecture. In: Proceedings of the Great Lakes Symposium on VLSI, pp. 259–262 (2012)
17. Vashist, A., Keats, A., Dinakarrao, S.M.P., Ganguly, A.: Securing a wireless network-on-chip against jamming-based denial-of-service and eavesdropping attacks. IEEE Trans. Very Large Scale Integr. Syst. **27**(12), 2781–2791 (2019)
18. Rout, S.S., Singh, A., Patil, S.B., Sinha, M., Deb, S.: Security threats in channel access mechanism of wireless noc and efficient countermeasures. In: 2020 IEEE International Symposium on Circuits and Systems (ISCAS), pp. 1–5. IEEE (2020)

19. Lebiednik, B., Abadal, S., Kwon, H, Krishna, T.: Architecting a secure wireless network-on-chip. In: 2018 Twelfth IEEE/ACM International Symposium on Networks-on-Chip (NOCS), pp. 1–8. IEEE, Piscataway (2018)
20. Boraten, T., DiTomaso, D., Kodi, A.K.: Secure model checkers for network-on-chip (NoC) architectures. In: 2016 International Great Lakes Symposium on VLSI (GLSVLSI), pp. 45–50. IEEE, Piscataway (2016)
21. Chaves, C.G., Azad, S.P., Hollstein, T., Sepúlveda, J.: Dos attack detection and path collision localization in NoC-based MpsoC architectures. J. Low Power Electron. Appl. **9**(1), 7 (2019)
22. Qian, Z., Juan, D.C., Bogdan, P., Tsui, C.Y., Marculescu, D., Marculescu, R.: A comprehensive and accurate latency model for network-on-chip performance analysis. In: 2014 19th Asia and South Pacific Design Automation Conference (ASP-DAC), pp. 323–328. IEEE, Piscataway (2014)
23. Catania, V., Mineo, A., Monteleone, S., Palesi, M., Patti, D.: Noxim: an open, extensible and cycle-accurate network on chip simulator. In: 2015 IEEE 26th International Conference on Application-specific Systems, Architectures and Processors (ASAP), pp. 162–163. IEEE, Piscataway (2015)
24. Reagen, B., Adolf, R., Shao, Y.S., Wei, G.Y., Brooks, D.: Machsuite: benchmarks for accelerator design and customized architectures. In: 2014 IEEE International Symposium on Workload Characterization (IISWC), pp. 110–119. IEEE, Piscataway (2014)
25. Guthaus, M.R., Ringenberg, J.S., Ernst, D., Austin, T.M., Mudge, T., Brown, R.B.: Mibench: a free, commercially representative embedded benchmark suite. In: WWC-4. 2001 IEEE International Workshop on Workload Characterization, 2001, pp. 3–14. IEEE, Piscataway (2001)
26. Xu, W., Trappe, W., Zhang, Y., Wood, T.: The feasibility of launching and detecting jamming attacks in wireless networks. In: Proceedings of the 6th ACM International Symposium on Mobile Ad Hoc Networking and Computing, pp. 46–57 (2005)
27. Che, S., Boyer, M., Meng, J., Tarjan, D., Sheaffer, J.W., Lee, S.H., Skadron, K.: Rodinia: a benchmark suite for heterogeneous computing. In: 2009 IEEE international symposium on workload characterization (IISWC), pp. 44–54. IEEE, Piscataway (2009)
28. Power, J., Hestness, J., Orr, M.S., Hill, M.D., Wood, D.A.: GEM5-GPU: a heterogeneous CPU-GPU simulator. IEEE Comput. Archit. Lett. **14**(1), 34–36 (2014)
29. Ganguly, A., Chang, K., Deb, S., Pande, P.P., Belzer, B., Teuscher, C.: Scalable hybrid wireless network-on-chip architectures for multicore systems. IEEE Trans. Comput. **60**(10), 1485–1502 (2010)
30. Rout, S.S., Deb, S., Basu, K.: Wind: an efficient post-silicon debug strategy for network-on-chip. IEEE Trans. Comput.-Aided Design Integr. Circuits Syst. **40**(11), 2372–2385 (2021)
31. Franques, A., Kokolis, A., Abadal, S., Fernando, V., Misailovic, S., Torrellas, J.: Widir: a wireless-enabled directory cache coherence protocol. In: 2021 IEEE International Symposium on High-Performance Computer Architecture (HPCA), pp. 304–317. IEEE, Piscataway (2021)
32. Sinha, M., Gade, S.H., Singh, W., Deb, S.: Data-flow aware CNN accelerator with hybrid wireless interconnection. In: 2018 IEEE 29th International Conference on Application-specific Systems, Architectures and Processors (ASAP), pp. 1–4. IEEE, Piscataway (2018)
33. Gade, S.H., Rout, S.S., Sinha, M., Mondal, H.K., Singh, W., Deb, S.: A utilization aware robust channel access mechanism for wireless NoCs. In: 2018 IEEE International Symposium on Circuits and Systems (ISCAS), pp. 1–5. IEEE, Piscataway (2018)
34. Kumar, A., Peh, L.S., Jha, N.K.: Token flow control. In: 2008 41st IEEE/ACM International Symposium on Microarchitecture, pp. 342–353. IEEE, Piscataway (2008)
35. Palesi, M., Collotta, M., Mineo, A., Catania, V.: An efficient radio access control mechanism for wireless network-on-chip architectures. J Low Power Electron. Appl. **5**(2), 38–56 (2015)
36. Rout, S.S., Chaudhari, V.I., Patil, S.B., Deb, S.: Rcas: critical load based ranking for efficient channel allocation in wireless NoC. In: 2019 32nd IEEE International System-on-Chip Conference (SOCC), pp. 21–26. IEEE, Piscataway (2019)
37. Chakraborty, R.S., Narasimhan, S., Bhunia, S.: Hardware trojan: threats and emerging solutions. In: 2009 IEEE International High Level Design Validation and Test Workshop, pp. 166–171. IEEE, Piscataway (2009)

38. Catania, V., Mineo, A., Monteleone, S., Palesi, M., Patti, D.: Cycle-accurate network on chip simulation with noxim. ACM Trans. Model. Comput. Simul. **27**(1), 4 (2016)
39. Woo, S.C., Ohara, M., Torrie, E., Singh, J.P., Gupta, A.: The splash-2 programs: characterization and methodological considerations. ACM SIGARCH Comput. Archit. News **23**(2), 24–36 (1995)
40. Miller, J.E., Kasture, H., Kurian, G., Gruenwald, C., Beckmann, N., Celio, C., Eastep, J., Agarwal, A.: Graphite: A distributed parallel simulator for multicores. In: HPCA-16 2010 the Sixteenth International Symposium on High-Performance Computer Architecture, pp. 1–12. IEEE, Piscataway (2010)

Defense against Security Threats with Regard to SoC Life Cycle

Usha Mehta and Jayesh Popat

1 Motivation

We are accompanied by billions of computational devices (such as secured encryption devices, IoT devices, etc.) in our daily lives. They are used mainly for collecting, monitoring, and understanding some of our private data, including sleep, place, and network of relationships. The new study estimates that approximately 50 billion "smart" and "connected" hardware devices will be in use by 2021, according to Cisco. These devices produce, process, and share a substantial amount of confidential details and information (which are referred to as "security assets" or "assets").

System-on-chip (SoC) architecture is responsible for the main computational tasks of these hardware devices. The SoC integrates the different components of a hardware device, such as processing units, memory, secondary storage, and I/O ports in a single integrated circuit. The SoC normally includes multiple secure components and confidential information (like encryption keys, software keys, digital rights management credentials, and config bits) that are required to be shielded from attackers [1].

Typically, hardware systems were deemed stable, trustworthy, and confidential, whereas software components (such as operating systems, programs, and/or user applications) were designed and developed on top of them. But hardware systems should no longer be regarded as trustworthy anymore nowadays since state-of-the-art research practices [2, 3] show that hardware systems are vulnerable against security threats. Hence, the security of the hardware system toward such threats is utmost needed.

U. Mehta (✉) · J. Popat
Institute of Technology, Nirma University, Ahmedabad, India
e-mail: usha.mehta@nirmauni.ac.in

A. Iranmanesh (ed.), *Frontiers of Quality Electronic Design (QED)*,
https://doi.org/10.1007/978-3-031-16344-9_16

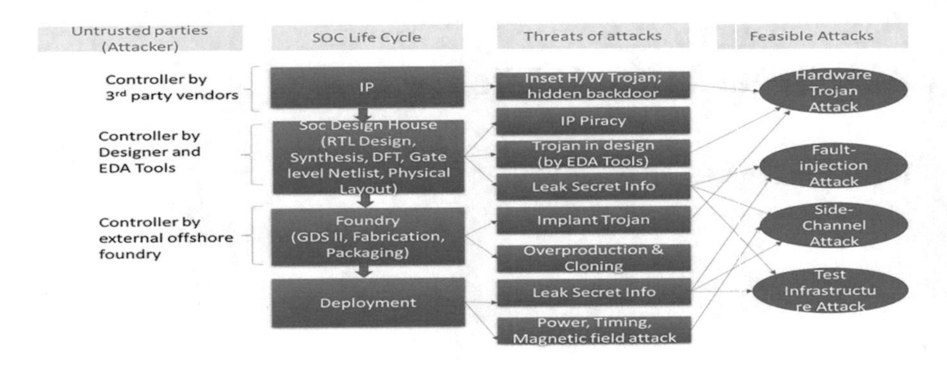

Fig. 1 Hardware security threats from multiple stages of SoC life cycle

2 Security Threats with Regard to SoC Life Cycle and Supply Chain

Several important problems are involved in securing the modern computing system. The first problem is the highly complex architecture of the SoC. System-on-chip (SoC) are most often built using the integration of predesigned hardware or software blocks (also known as intellectual properties (IPs)) that are linked together by communication protocols. The IPs are well-engineered blocks with high complexity and power, performance, and area efficient. The on-chip communication fabric is adding one more layer to complexity. Finally, more security-critical devices are incorporated in the SoC architecture, and the access of such devices is controlled by the complicated security protocols.

The protocols are developed by system architects and various SoC integration teams, and their specifications are refined and modified during the system design phase. It is therefore more difficult to build the on-chip security measure that thwart the unwanted entry or evolving security measure specifications specifically on demand of customer requirements.

Another source of the challenge is the life cycle and supply chain involved in the development of a modern computing device. Figure 1 illustrates the SoC life cycle, different hardware attacks, and the threats caused by attacks. However, the fast globalization of the SoC design, manufacturing, and supply process makes it more vulnerable to different security threats. The third-party predesigned IPs are widely used in this process, and the demand for the IPs has reached over 2.1 billion [4]. As the technology advanced by following Moore's law, the IPs and SoC integration process are becoming more and more complex. So it makes the SoC design process highly dependent on the IPs from third-party vendors for their integrities by considering them as a black box. It concerns about overall system security.

An attacker involved in the IP design may modify the design with malicious intentions or incorporate a hidden backdoor in the design. Further, the robust IP validation may not be ensured by small IP suppliers due to high time-to-market pressure. Moreover, the DfT (design-for-test) and DfD (design-for-debug) structures are incorporated during this SoC design process to ensure quality testing after fabrication. The attacker at the design house may exploit such test and debug interfaces or side channels like power/timing information to leak secret data from the security-critical IPs [5].

The SoC designs have become so much complex that design houses need to work with many parties to fulfill all the requirements of the market. Examples of different partners are third-party IP vendors; EDA tool developers, standard cell library, and designers. The EDA tools are used to optimize the design for power, area, and performance. These optimizations may lead to new threats [6]. A rogue designer can compromise the integrity of SoC design by inserting malicious hardware Trojan. The SoC design is outsourced to an untrusted partner for DfT insertion; these untrusted partners may reverse engineer the netlist and poses the threat of IP piracy.

Many SoC design houses have nowadays become fabless and outsourcing the designs to overseas untrusted foundries for fabrication. This is a serious concern in terms of design security since the foundry has full command over the entire design. It may lead to piracy, overproduction and cloning of the design or IP blocks. There is also a chance of implantation of a malicious hardware Trojan in the design which is used for tampering the design to work as a backdoor or compromise structural/functional parameters of a system.

After deployment, the hardware system is also attacked by exploiting existing test/debug infrastructure or noninvasive side channels to leak the secret data or magnetic field fault attacks for tampering the memory values to cause malfunction or DoS (denial-of-service).

Considering all of the above attacks on hardware, the roots of security and trust seem to be violated in modern computing portable devices. Hardware security refers to the security problems related to the underlying hardware. These hardware security and trust issues have drawn the attention of the researcher community [3, 7, 8]. The recent proposed solutions mainly focus on the information leakage from the cryptographic engine (security-critical IPs), Trojan attacks, and counterfeit IPs.

In this book chapter, we show various threat and attack models for SoC designs as well as their Ips, and we discuss several countermeasures approaches to address various security vulnerabilities in them.

The outline of this chapter is as follows. Section 2 presents the source of hardware security attacks at different stages of an SoC life cycle. Then, we review four threat models associated with the hardware security: side-channel attack (SCA), fault injection attacks, test infrastructure-based attacks, and hardware Trojan (HT) insertion in Sect. 3. Finally, Section 4 will cover the defense against those security threats, respectively.

3 Sources of Attacks in SoCs

Security issues may be applied during the SoC design and production phase. Prior to manufacturing, there are three main sources of threats: (1) design flaws, dishonest employee, and 3rd party IP cores from an untrusted facility; (2) unreliable EDA tools used for design and synthesis purpose; and (3) untrusted partners of the outsourced design for DfT and DfD insertion. During and post-manufacturing, the threats may be introduced by (1) untrusted fabrication facility and (2) invasive and side-channel attacks after shipping of the SoC. We have listed the sources of attacks in SoCs at different stages as follows.

3.1 Design Stage

The SoC design stage begins with outlining the specification in the natural as well as high-level languages. Then, they are converted into RTL with hardware description languages. Over the years, the complete SoC is designed in-house. However, the incorporation of IP cores from different vendors and outsourcing certain modules of the SoC to untrusted parties has become a trend in the SoC design due to the high time-to-market pressure and the sheer complexity of SoC.

The main source of attack may come from third-party IP cores. They may have already introduced vulnerabilities to compromise the SoC security. These vulnerabilities cause the malfunctioning of SoC design and ultimately used for leakage of confidential details to the attacker. These vulnerabilities may be instituted by unreliable designers (insiders). Since they have absolute visibility and control of the design description files, insider attacks are especially harmful. Furthermore, the SoC IP cores may also be heisted during the design phase. Stolen intellectual property would result in a forfeit of copyright for the IP vendor along with illegitimate production of the design instances. In addition, it will assist to discover known design flaws together with new means to assault the SoC.

3.2 Synthesis RTL to Layout

After completion of SoC architectural design and incorporation of IPs at RTL, the design must be synthesized. Various EDA tools [9, 10] can be used to conduct the synthesis procedure. While carrying out the lower abstraction level transformation, the tools are only bounded toward execution time, area and power consumption, and ignoring the security. During design optimization, EDA tools can introduce unintended security flaws. For instance, when performing the optimization of the controller architecture, the tool can add new don't care states to the existing Finite State Machine (FSM). The hypothesis is that the newly introduced states are not approachable from any other states, and the design functionality is not influenced by them. Recent findings, however, indicate that these vulnerabilities can be triggered by injections of faults [6, 11]. With the condition that the newly introduced states are linked with secured states of the design, then an attacker would be able to insert the faults to reach to don't care states along with secured states unauthentically.

Synthesized netlist must be translated into transistor-level netlist with the help of a standard-cell library. More often, these two (gate and transistor-level) netlists are outsourced to unreliable parties for activities like DFT incorporation, clock-tree synthesis, or placement and routing. Those parties can introduce extra gates/transistors to the netlist or alter the routing of the layout, resulting in the implantation of the malicious feature. Furthermore, it is possible to produce IP infringement, fake products, and copyright issues with the help of reverse-engineering the netlist.

3.3 *Fabrication and Manufacturing*

Once the layout is completed, it is submitted to an overseas fabrication facility to manufacture the SoC. Design houses ship their design to unreliable foundries because of the more expensive in-house manufacturing. An intruder in the fabrication facility may introduce harmful features within the SoC. IP infringement, cloning, recycling, or copyright issues may take place at the untrusted fabrication facility. The untrusted fabrication facility may not follow the terms and conditions specified in the agreement and can overproduce the chips to generate more revenue by supplying them illegally.

3.4 *In-Field Attacks*

Once the chip is shipped and functionally working in-field, there are still chances for different kinds of attacks. Trojan can be triggered and execute its planned attack or failure if it was introduced during the design or fabrication phase. There are various ways to trigger malfunction by introducing the fault in the chip (such as clock glitches, underpowering, heating or laser beam, etc.). The intruder may exploit noninvasive side channels to observe the device characteristics (such as execution time of operation, power consumption, EM radiation, etc.) and retrieve confidential details. Furthermore, the adversary with advanced equipment can remove the package and take high-quality images of the chip to perform an invasive attack to obtain the design details which leads to cloning and infringement of IP.

Moreover, the secret keys (stored inside read-only memory) are retrieved by probing the internal signals. Finally, the system with recycled chips is not trustworthy as they may have faults (might have failed in some tests).

Hence, it is significant to establish the reliability of the SoC which can be achieved through security verification and correction at each stage of the SoC life cycle.

4 Threat Model

Various threat models that can affect the security of SoC are discussed in the present section.

Figure 2 shows that system vulnerabilities comprise the highest percentage of hardware as well as software vulnerabilities that endanger the SoC security [11]. Discovering these vulnerabilities is exceptionally difficult because they are deceptive and hidden in nature. There are mainly four classes of attacks that cause such threat models and vulnerabilities in the SoC. They are as follows:

- Hardware Trojan attacks
- Side-channel attacks
- Fault injection attacks
- Test-infrastructure-based attacks

It is of utmost importance to defend against these attacks to ensure the security of the SoC. In the following subsection, we presented details of the abovementioned attacks.

4.1 Hardware Trojan Attacks

The hardware Trojans are referred to as malicious modifications made during the design or fabrication process by the insertion of malicious circuitry. Hence, the ICs

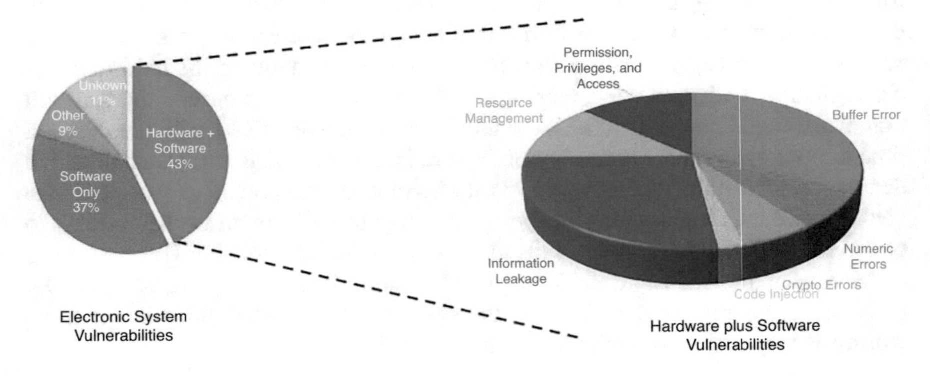

Fig. 2 Various categories of hardware and software vulnerability in an electronic system [11]

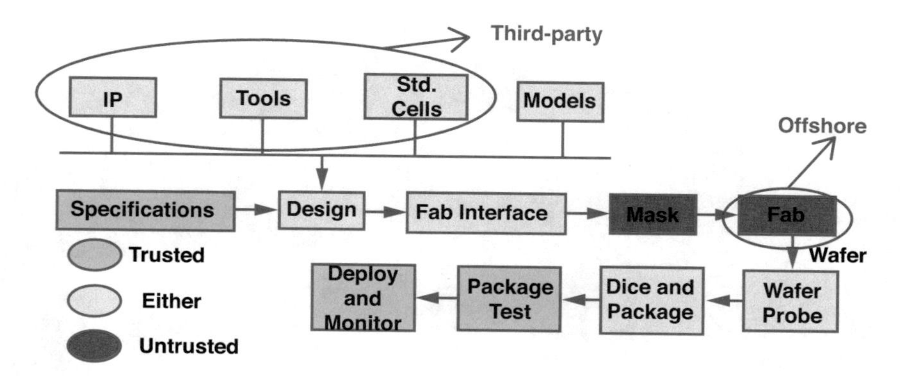

Fig. 3 DARPA's model of hardware security Threats [12]

may include hidden functions that work in rare conditions that are called "hardware Trojan." They are located in low controllability and observability positions [12–15]. So they are inactive most of the time and are triggered under very rare conditions at the internal nodes. Such alteration to the functionality of the chip may lead to disabling the entire chip, financial theft, or security vulnerabilities [12–14]. The US Military has published evidence of such malicious Trojans in ICs recently [15–17].

In the era of complex SoC designs and short time-to-market, third-party IP cores play a vital role in the VLSI industry. The SoC designs have become so much complex that fabless design houses need to work with many parties to fulfill all the requirements of the market. Examples of different partners are third-party IP vendors, CAD tool developers, standard cell library, hardware models, and overseas manufacturing facilities [12]. Therefore, hardware Trojans may be inserted during any step of the SoC design and development cycle process as illustrated in Fig. 3.

Hence, there is the utmost need for detecting, preventing, or tolerating such undesirable malicious changes in the chip. The discovery of hardware Trojan is feasible during pre-silicon verification, physical verification, or post-silicon testing [18].

However, pre-silicon verification is definitely not a decent decision for hardware Trojan discovery since there is a requirement for a golden model which is commonly not available in IP core-based SoCs. The thorough verification design isn't achievable too [18]. Further, the customary test and validation techniques don't work consistently for the detection of hardware Trojan as they are meant to check the functional hardware only, not the additional functionality meant for harmful activities.

After fabrication, the hardware Trojan can be detected using a destructive or nondestructive approach. The destructive technique involves reverse engineering of IC, while the nondestructive is carried out by verifying IC's function with a golden model. Destructive approaches are less recommended since the possibility of inserting a Trojan may be in a small population of ICs and they are expensive too [12].

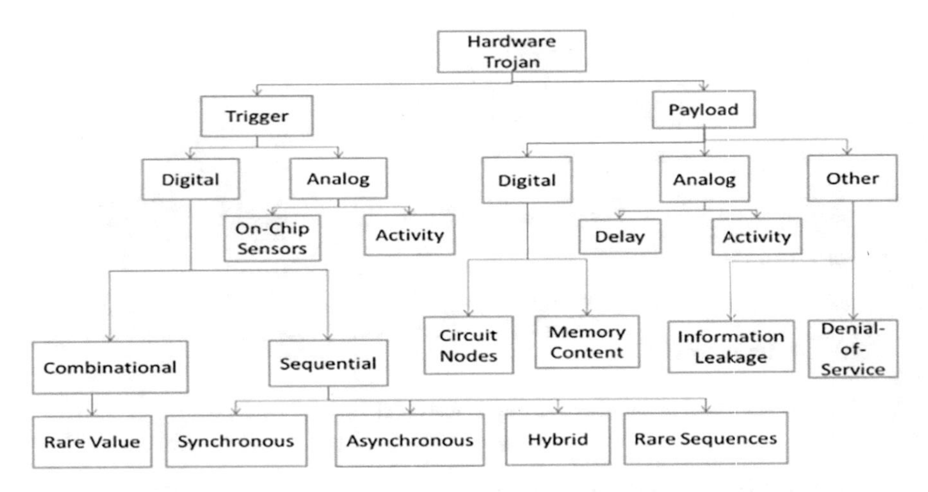

Fig. 4 Trojan taxonomy based on activation mechanism (trigger) and malicious effect (payload)

The attacker may embed the Trojan in the circuit so that it tends to be hard to detect such Trojan through the post-silicon test. To achieve such a situation, the triggering or activation of Trojan is only possible under very rare conditions. Hence, the erroneous output is only produced under the sequence of multiple rare conditions. The malfunctioning output node is referred to as a payload node. Hardware Trojans are categorized by trigger and payload approaches [19, 20] as shown in Fig. 4.

The trigger method is classified into two types: digital and analog. Digital triggering can be done in two ways: combinational trigger or sequential trigger. During combinational trigger, the rare triggering condition on the trigger node will produce an erroneous output value on the payload node.

While in the case of the sequential trigger, the erroneous output value is produced by sequentially applying rare node conditions multiple times to trigger the node synchronously or asynchronously. While in the case of digital payload Trojans, the circuit node may be affected and produce wrong value, or tampering in the memory data may happened. Analog payload Trojans are in the form of pull-up resistor or pull-down capacitor. They don't target changing the circuit functionality. However, they may change analog parameters of circuits like power, noise, and delay. Some of them may focus on increasing switching activity, and hence, the aging of the chip speeds up without modifying the functionality of the chip. The other types of payload Trojans include information leakage and denial of service [21–24]. Information Leakage involves the leaking of confidential and sensitive information through radio signals. The denial-of-service attack will not allow users to access their system functionality. In [2], the author compares the hardware Trojans based on their area overhead and activation probability. Trojans can be inserted into hardware design using various ways as listed below:

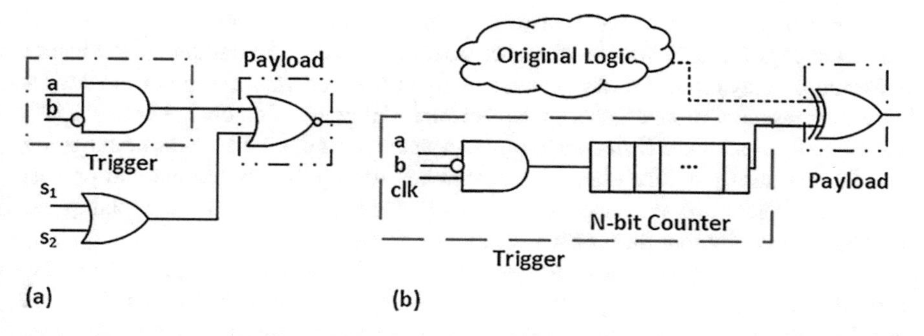

Fig. 5 (**a**) A combinational Trojan that can be triggered using rare condition $a = 1$ and $b = 0$. (**b**) A sequential Trojan that is triggered when rare condition $a = 1$ and $b = 0$ happens 2^N times, where N is the length of the counter [25]

- *Rare nodes*: An intelligent attacker can make the malicious circuit (Trojan) to be activated under very rare input condition at the internal nodes [25]. So, the trojan can remain hidden as much as possible unless specified input condition is met. As illustrated in Fig. 5a, the combinational Trojan can be activated by applying very rare input condition ($a = 1$ and $b = 0$). Contrarily, Fig. 1.5b depicts sequential Trojan which is activated only if the N-bit counter goes beyond the last counting value. So, it required to apply rare input condition ($a = 1$ and $b = 0$) 2^N times to activate the Trojan.
- *Rare branches*: The attacker may add Trojans in the RTL design in such a way that it is concealed in the rare branch and concurrent assign statements. Or else they may get detected using random constraint verification method.
- *Gate misplacement*: An attacker may replace the gate in the netlist which results in variation in the design and alter the specified function of the design. Further, this attack can cause the bit-change in the response as compared to golden response, and it results in the security vulnerability since there may be unwanted switching to secured states, false outputs, and DOS (denial-of-service). These attacks have very less impact on the physical features such as area and power. So, they are not easily identified through review of the design and random constraint verification.

4.2 Side-Channel Attacks

These attacks are called noninvasive attacks. They are commonly performed on data picked up from the secondary interfaces/side-channels of the crypto-device like electromagnetic radiation, execution time, and power consumption.

These attacks are performed to gain more information about design and be able to attack. For example, an attacker can guess some internal values or secret keys by measuring the execution time of various computations (note that “0” or “1” bits in a register can initiate different operations) [26]. Extracting side-channel information may require some knowledge about the internal structure of the design. However, some of these attacks such as differential power side-channel attacks [27] are black-box attacks. Unfortunately, side-channel analysis has a common issue, i.e., the sensitivity of side-channel signatures is susceptible to thermal and process variations. Therefore, the success of these attacks is determined by the quality and precision of equipment that is used for measurement.

Power-side channel attacks use the amount of power consumption and transient/dynamic current leakage to attack the design. A device like an oscilloscope can be used to collect power traces, and those traces are statistically analyzed using correlation analysis to derive secret information of the design. The different types of power analysis-based attacks include simple power analysis (SPA) [27], differential power analysis (DPA) [27], and correlation power analysis (CPA) [28, 29]. Therefore, it is very important to develop automated security validation methods that can identify power side-channel leakage. We need to detect the parts of a design that is responsible for power side-channel leakage in an automated fashion.

Two different types of EM analysis-based attacks are reported: simple electromagnetic analysis (SEMA) and differential electromagnetic analysis (DEMA). There are certain differences between power analysis and EM analysis-based attacks. Power analysis just uses the power dissipation of circuits, while EM analysis essentially centers around antenna placement on the device. Usually, the attacker, who performs the EM attacks, remains far from the device. For instance, amplitude demodulators are available at a very long distance from the circuit and still capable of performing the attack [30, 31].

4.3 Fault Injection Attacks

Many secured SoCs that execute specific cryptographic operations are typically believed to work securely while they are being used, and we never care of asking how the reliability of those operations relies on the trustworthiness of those SoCs that execute them. Despite this belief, hardware faults that occur in the time of the function of an SoC with crypto hardware have been seen to significantly affect the reliability. These defective behaviors or outputs can even be essential side channels and can significantly improve the sensitivity of a cipher to cryptanalysis. Fault attacks against encrypted hardware systems such as smart cards pose realistic and successful threats. Hence, we concentrate primarily here on the fault attacks on hardware systems.

The fault injection techniques that have been developed in order to alter maliciously the correct functioning of a computing device currently include variations in the power supply voltage level, injection of irregularities in the clock signal, radiation or electromagnetic (EM) disturbances, overheating the device, or exposing it to intense light.

Change in power supply, inducing accurate timing power spikes in the power supply or rise in temperature [32–34], will cause the transient fault to be injected. This fault will produce a single-bit error at the beginning and a multi-bit error as the power supply goes down further. This technique is applied to ARM9 processor IC [35, 36] and small ASIC of crypto devices [37, 38].

Wrong data bytes are stored in the memory by the processor by forcing it to process the next instruction in the earlier clock cycle [39, 40]. This is achieved by tampering the circuit clock signal with the help of external clock generator.

The crypto-chip may be subjected to high-energy optical sources such as a UV laser beam or a camera photo flash [41]. This way the chip can conduct or alter the logic state stored in the memory. Bit flipping is also possible by inducing EM field which causes eddy current to flow in the crypto-chip [42, 43].

For fault injection attacks, the adversary should have physical access to the device. Setup time violations can be performed by different fault injection methods. The main objective of the fault injection attack is to malfunction or tamper data stored in the memory.

4.4 Test-Infrastructure-Based Attacks

Integrated circuits are tested after fabrication for manufacturing defects to ensure high product quality. However, in current generation nanometer IC design and fabrication technology, ICs are incorporated with DFT (design-for-test) infrastructure so that it is quite easy to test the IC with high fault coverage and have diagnostic facility after fabrication. This is true for crypto-chips as well. The scan-chain insertion, built-in self-test, test compressors, etc. are the regular DFT features of any complex integrated circuit. The attack makes the benefit of such available test hardware to uncover the secret of crypto-devices.

Attack Principle

To bypass the chip's security, the scan-chain architecture may be exploited. Many of the scan flops contain chip's secret data while performing encryption algorithms. The attacker attacks the scan flops which store output from the encryption algorithm's interim operations. A scan-based attack uses the chip's PI/PO (primary input or output) and the SI/SO (scan in or scan out) easily accessible pins to retrieve the private key. This attack is carried out by applying differential cryptanalysis on AES cipher [44, 45].

There are mainly two types of test infrastructure-based attacks: (1) differential scan attack and (2) test-mode-only attack.

Attack Assumptions

Scan attack is carried out upon consideration of the following assumptions:

- The attacker has control over SE (scan-enable) pin or can access test infrastructure (scan-chains) using JTAG.

- The attacker is aware of the AES crypto algorithm.
- The execution time of the desired operation of the crypto algorithm is in the knowledge of the attacker. Thus, he can execute only one round of AES operation in normal mode before switching it to test mode.
- Although being aware of DFT structure (scan chains, space compaction/time compaction) implemented in the AES circuits, the attacker cannot have knowledge about detailed hardware implementation such as the number of scan-chains, test compression ratio, etc.
- The scan chain does not include the registers holding the secret key of the AES crypto circuit.

4.4.1 Differential Scan Attack (DSA)

This is a traditional scan-based attack. The attacker may exploit the test infrastructure of the AES cipher. As mentioned in Chap. 1, the scan-based AES cipher includes the round register in the scan path. Hence, the attacker can observe the content of the round register by changing the cipher from functional mode to test mode. It leads to retrieval of secret keys although it is not part of the scan-chain [44–49].

Once the crypto-chip is fabricated, the intruder can operate the cipher with desired plaintext in mission-mode for quite a few cycles; changing it to test mode, he can scan out the result of the round register. The round register holds the interim output of the cipher operation. The intruder will then view the interim outputs of the crypto-chip and evaluate differential crypt-analysis on the findings to extract the hidden key.

The attack [44] is divided into two parts: (1) determining the scan-chain structure by identifying the bits of the round register and (2) retrieving the secret key.

1. The procedure to identify the round register bits in the scanned out response is as follows.

    (a) Rerun step 1 by inputting other plaintext with 1-bit XOR difference than previous plaintext and generating the scan-out response corresponding to current plaint text. This is termed as response f2 as illustrated in Fig. 6.
    (b) Calculate the XOR difference of these two responses f1 and f2. The flops corresponding to the round register will have the value “1” in XOR difference output.
    (c) The second and third steps need to be repeated until all the round register flops are recognized.

    These steps are required to find out each word of round registers in the scan-out response. A single-bit difference in the input plaintext changes one word as per the MixColumn operation.

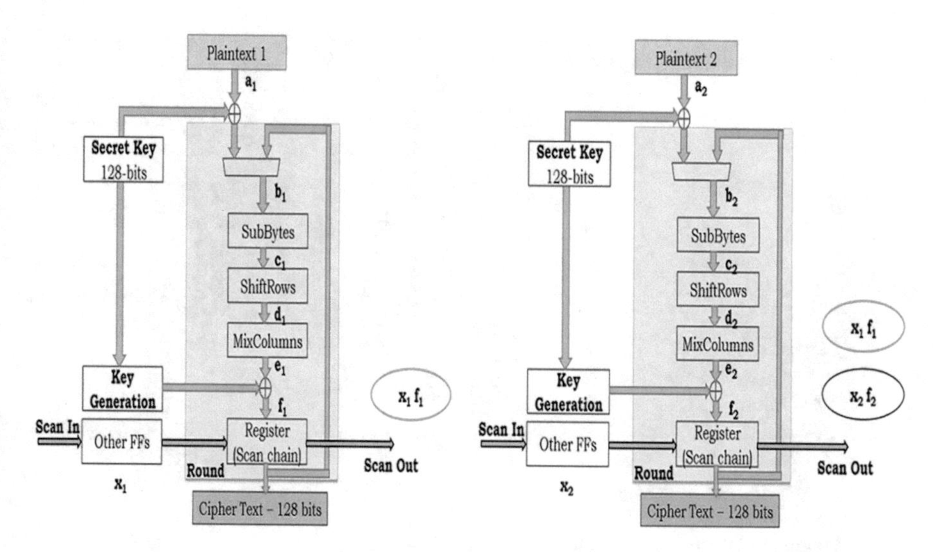

Fig. 6 Application of first plaintext during DSA

2. The second part of the attack is to retrieve the secrete key. The differential property of AES cipher is exploited by the intruder. In order to perform this, the intruder input all possible plaintext pairs with a single-bit difference in the LSB of any byte and generating the corresponding output pairs. The intruder can observe the unique hamming distance (HD) in just four-output-pair differences. The unique HD means it is produced by unique SubBytes inputs.

As illustrated in Fig. 7, the four unique HDs are 9, 12, 23, and 24. Hence, to recover the secret key, the intruder inputs every possible 128 plaintext pairs. These pairs are having a 1-bit difference in the LSB of a specific byte in the plaintext. Next, the intruder notices the HD in the generated round register response. The unique SubBytes inputs are determined to form a unique HD captured in the corresponding scan-out response pairs. As shown in Fig. 7, the SubBytes pairs are (b1, b1 XOR 1) for unique HDs.

Hence, the two secret key bytes can be guessed by only XORing applied plaintext with corresponding SubBytes input pair as per the architecture of AES cipher. Continue this process to find out for all potential key bytes. In this way, it is possible to disclose the whole secret key. In this scenario, the intruder will retrieve two possible key byte values for each input plaintext byte. So, the intruder at the end will have in total 2^{16} secret keys, and each one is 128-bit long. It is needed to apply this key with plaintext and observed the ciphertext in each case which is much fewer computations than applying all possible key bit combinations (2^{128}) using brute force attack. To perform this attack, the intruder may apply $128 \times 16 = 2048$ plaintext in the worst-case scenario. The entire 128-bit of the secret key of AES cipher is normally retrieved by applying a total of 544 plaintexts.

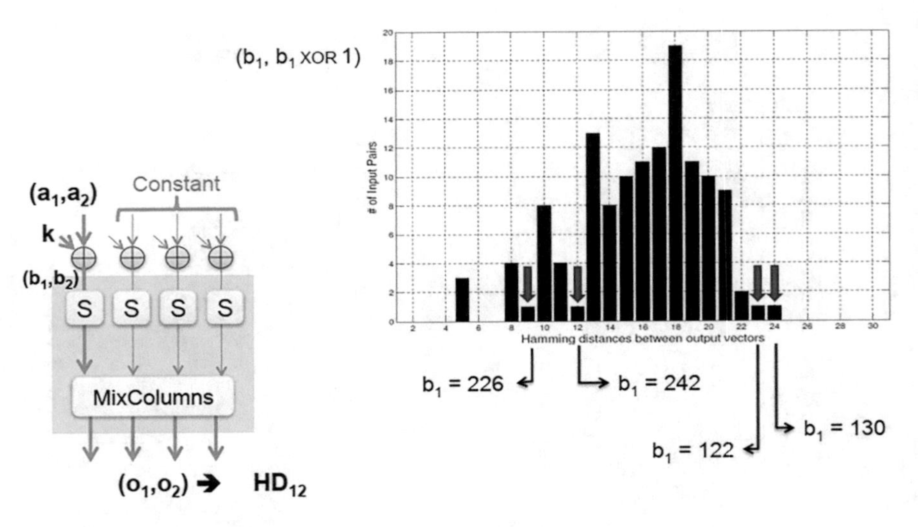

Fig. 7 Hamming Distance to consider during DSA [44, 45]

The above demonstrated scan attack can be successfully mounted on more complex scan chain structures. The authors in [45–49] further extended the attack algorithm used in [44] to mount an attack on advanced DfT architecture with feature like test decompression, mask decoder, and response compactors.

4.4.2 Test-Mode-Only (TMO) Attack

The mentioned DSA may be carried out if the intruder may switch the circuit from mission to test mode and vice versa. This is why countermeasures focused on resetting the scan chains when the circuit is shifted from the test mode to practical mode have been established. The author [50, 51] proposed a thoroughly checked scan assault on AES. It is known as TMO since the AES inputs are placed in the test mode through the chip inputs and the value of the round register is observed. This is seen in Fig. 8. The attacker is allowed to use only the highlighted blue and green operations in this attack. This is developed to overcome the mode reset countermeasures [52].

Attack Assumptions

Along with the assumptions of the DSA, the TMO is carried out upon consideration of the following assumptions.

- The attack performed by considering the user key, which is embedded on-chip or saved in memory.
- A reset can be achieved by clearing the content of the round counter which forces to run the cipher again from the first round.

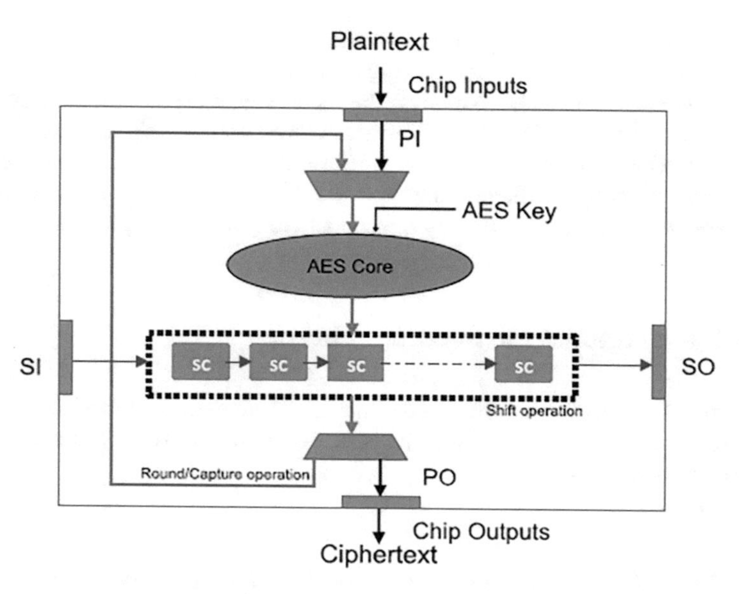

Fig. 8 Test-mode-only attack operations [50]

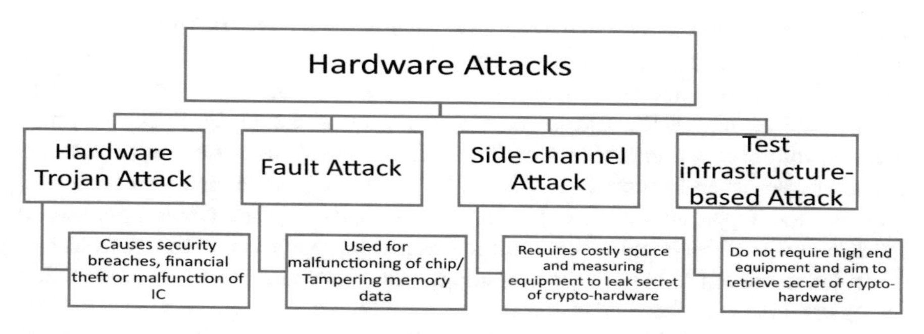

Fig. 9 Implementation attacks on crypto hardware

The DSA and TMO attacks can be mounted on different DFT architecture of AES [53–55].

Different hardware attacks and their properties are illustrated in Fig. 9. The hardware Trojans are referred to as malicious modifications made during design or fabrication process by insertion of malicious circuitry. This type of attack causes the threat of security breach, financial theft, and malfunctioning of SoCs. The side-channel attacks require the costly source and measuring equipment.

Some of the fault attacks have been deemed practically not possible due to the highly accurate timing requirement of the fault injection and the rigid requirement on the position of the injected fault. Some are applied after decapsulating the device package and using the expensive setup for the application of attack. They are mainly used to malfunctioning of chip/tampering of memory data.

The scan chain is attacked to recover the secret key of crypto-devices. Compared to earlier described attacks, the test-infrastructure-based attacks are considered as classical attacks in literature as they retrieve the data without requiring costly instruments, without malfunctioning the stored data, without depackaging the device, and without the need of very precise and accurate timing to exercise attack.

5 Defense Against the Security Threats

This section will cover the different defensive techniques/countermeasures against hardware Trojan attacks, side-channel attacks, fault injection attacks, and test-infrastructure-based attacks.

5.1 *State-of-the-Art Techniques for Hardware Trojan Detection*

State-of-the-art Trojan detection methods are classified in Fig. 10. Trojan detections approaches are categorized into two types: destructive and nondestructive [56]. Destructive approaches is very expensive and time-consuming since it requires silicon delayering of IC, taking images of metal layers using SEM (scanning electron microscope), combining of this different metal layer images to get final gate-level netlist,and comparing this netlist with the golden one. An adversary may add the Trojan in some samples of ICs as opposed to the whole populace of fabricated ICs. Subsequently, this technique isn't extremely viable for the detection of Trojan in the era of nanometer technology ICs. The nondestructive approach has five possible ways of detecting the presence of Trojan: (1) logic testing, (2) side-channel analysis, (3) IP trust verification, (4) design-for-security (DFS), and (5) runtime monitoring.

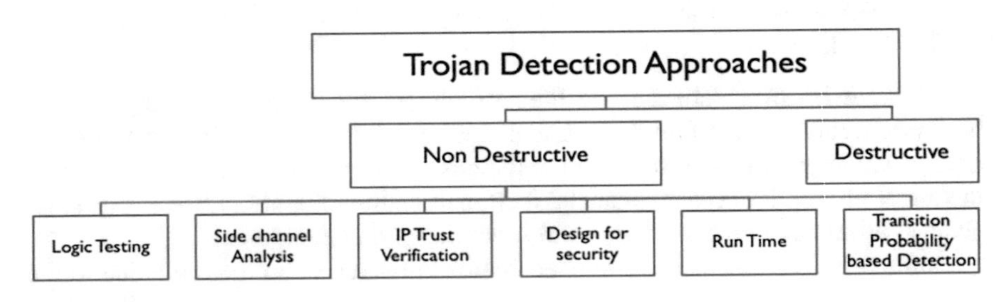

Fig. 10 Trojan detection techniques

Logic/Functional Testing After the fabrication of IC, logic testing is used to detect Trojan instances using test patterns to detect malfunction. This is also not very useful because testing is mainly carried out for detecting manufacturing defects and is not intended for detecting extra malfunctioning hardware.

Side-Channel Analysis It is mainly performed to measure side-channel parameters such as power, transient and quiescent currents, delay, and frequency. There may be a change in such parameter value due to the presence of Trojan in IC.

IP trust verification attempts to detect the existence of Trojans at an IP level using pre-silicon techniques.

Design-for-Security Here, extra hardware can be added during the design or test development phase either to make the Trojan insertion very difficult or to make Trojan detection very easy. Nevertheless, the trust verification and DFS method are not fully capable of detecting the wide variety of Trojans. Runtime monitoring is performed on IC during in-field operation to detect any malfunction for the potentially undetected Trojans. They are considered as the last defense against the Trojan attacks.

Transition Probability-Based Trojan Detection In this [57], initially the logic probabilities on the each of the golden netlist is computed. It is then used for calculation of transition probability of each nets. The nets which are having lowest transition probability are considered as vulnerable nets since the attacker exploits such low transition probability area in the netlist to insert hard-to-detect and stealthy trojan gates. The transition probability changes with the insertion of extra malicious Trojan gates traversing from primary input (PI) to primary output(PO) path in the netlist. The Trojan-affected area is identified by comparing the transition probability of golden netlist and Trojan-inserted netlist. The technique is evaluated for the different ISCAS'85 benchmark circuits to successfully detect the hardware Trojan insertion in low probability circuit nodes.

5.2 Countermeasures Against Side-Channel Attacks (SCA)

As mentioned in Sect. 3.2, the noninvasive side-channel attacks are performed by analyzing execution time, electromagnetic radiation, and power consumption of SoCs. The countermeasure against each of the side-channel attacks are discussed as follows.

Timing SCA Countermeasures There are basically two types of countermeasures to defend the SoC against timing-based side-channel attacks: (1) randomizing the execution time of various operations [58] and (2) making all the operations at constant-time [59, 60]. They will thwart the data leaks via execution time. The first countermeasure can be achieved easily by inserting different delays into the various execution path. This is done during design stages, and the designer has full control to add number of buffers to achieve the expected delay. When it is in place, it is hard to perform timing-based side channel. But it does not ensure the full security against timing-based attacks. The second countermeasure ensures full security against timing SCA, but it is hard to implement in real-time scenario.

Power SCA countermeasures To make the power traces independent of the inter-operational data of crypto algorithm of secured SoC, there are mainly three countermeasures based on primitive logic cells: (1) sense-amplified-based logic (SABL) [61], (2) wave dynamic differential logic (WDDL) [62] and (3) t-private logic circuit [63]. The first two (SABL, WDDL) countermeasures are designed to make the power consumption identical in every clock cycle. However, the third (t-private logic) countermeasure uses randomization technique which make power consumption value different in each clock cycle. The t-private logic circuit masks each bit by t-random bits. Hence, all of the countermeasures prevent the secret-key retrieval though power traces.

Electromagnetic (EM) SCA countermeasures There are several countermeasures to thwart EM-based SCA. The main countermeasure is to redesign the circuit to optimize the linking between different component which results to the minimizing of EM radiation. Another effective precaution is the addition of an extra layer of insulation to the chip, which serves to protect against EM radiation. Adding inoperable components that generate excessive EM noise-signal levels can make it difficult to acquire sensitive data because of the noise levels being generated in the similar frequency range. Through inserting fake function in the middle of the stage of crypto-operation, the attacker is restricted from determining if bits are true or fake, although after conducting EM SCA successfully. EM SCAs are easily performed through various existing technique, and crucial data can be leaked. Therefore, the EM SCA countermeasures are required to be applied immediately in the early stage of design to make the SoC more secure.

5.3 Countermeasures Against Fault Injection Attacks

In this section, we have discussed the existing countermeasures toward the fault injection attacks: duplication of critical operation, error-detection schemes, and anti-tamper protection module.

Replication (Duplication) of Critical Operations It is a widely used countermeasure where the same crypto-operations are being performed several times to calculate the two outcomes [64–67]. While comparing the outcomes with each other, it is assumed that fault is introduced if they are not the same. There are two methods to replicate (duplicate) the crucial crypto-operations: spatial and temporal. In spatial duplication, there are two identical hardware crypto blocks to perform desired crypto-operation and recalculate the outcomes. Temporal duplication utilizes the same hardware crypto block for the recalculation of the outcomes across successive timestamps. While comparing spatial duplication with temporal duplication, the former is impacting the area, and the latter is impacting the time requirements.

Error Detection Schemes (EDS) Another countermeasure against fault injection attack relies on parity detection. Due to fault injection if parity bits are not matching, the EDS deactivate the crucial crypto operations of the SoC. It can detect single-bit fault injection attack, but it is not an effective countermeasure against multi-bit fault injection attack. The area and time requirements shoot up prominently if multiple faults are required to be detected or corrected. They have less area and time overhead as compared to previously mentioned replication countermeasure [68–70].

Anti-Tamper Protection Module This countermeasure is used to protect against fault injection attacks. There is a tamper-proof module which holds the crypto hardware. The fault attacks are restricted with the help of sensors attached to the module. If there is any physical tampering with crypto hardware is made, the sensors will detect and alert the same. The anti-tamper module is applied to IBM-4764 crypto-processor [71].

5.4 Countermeasures Against Test-Infrastructure-Based Attacks

Many countermeasures have been proposed in the literature, aimed to face one or more of the test-infrastructure based attacks described in Sect. 3.4.

In this chapter, we have classified the countermeasures as follows:

1. Applicable to DSA only
2. Applicable to TMO only
3. Unified countermeasures which are applicable to DSA and TMO both

The classification of countermeasures against test infrastructure-based attacks is shown in Fig. 11.

5.4.1 Countermeasures Against DSA Only

The five countermeasures against DSA are mentioned in the literature and discussed as follows.

1. *Masking of round register and compactor output*: The attacker cannot retrieve the secret key information by masking/unmasking round register output or by using extended LFSR (eLFSR) as shown in Figs. 12 and 13. In this scheme, the round register output after a round operation is masked in AES cipher and

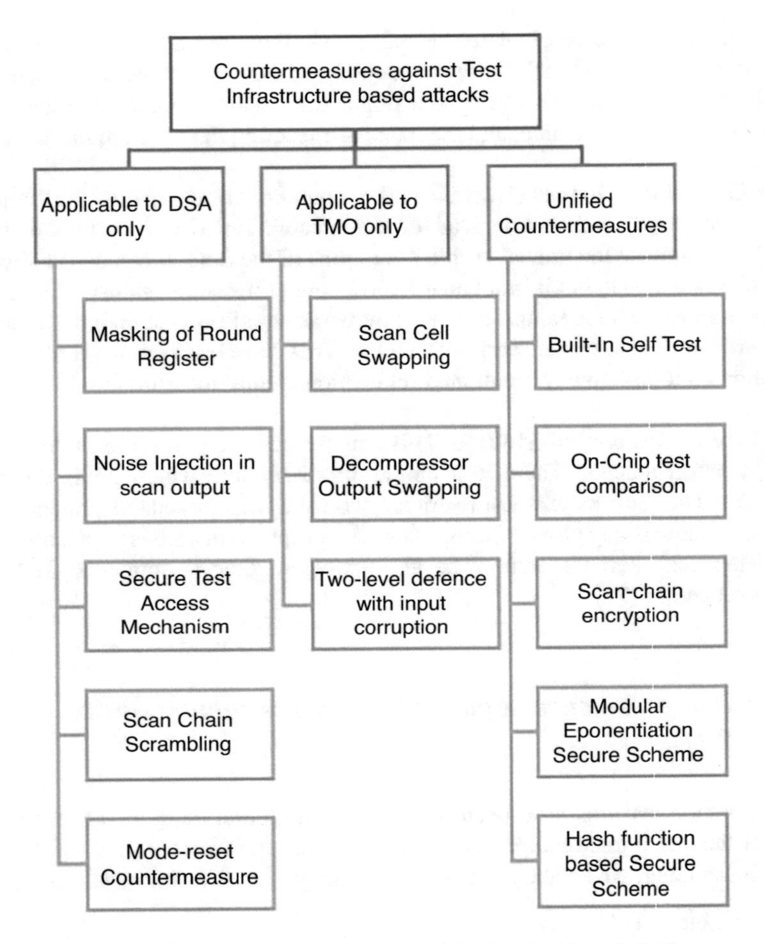

Fig. 11 Classification of countermeasures against test-infrastructure-based attack

then unmasked it before the next round starts as shown in Fig. 12. In the second scheme, an extended LFSR method is used to XOR the compacted response with pseudo-random bits as shown in Fig. 13. In both cases, an attacker cannot retrieve the secret key-related information from scan-out data after the first round of operation. Thus, conventional DSA is not applicable. This comes with the cost of area overhead as well as the longest critical path [45].

2. *Noise injection in scan output*: This method is utilizing on-chip LFSR and TRNG (true random number generator). The 50% scan-out bits are becoming noisy as TRNG conceal some of the LFSR output bits and the remaining 50% bits are not changed. This is depicted in Fig. 14. The masking needs to be applied at every clock cycle for the attack to be successful [49].

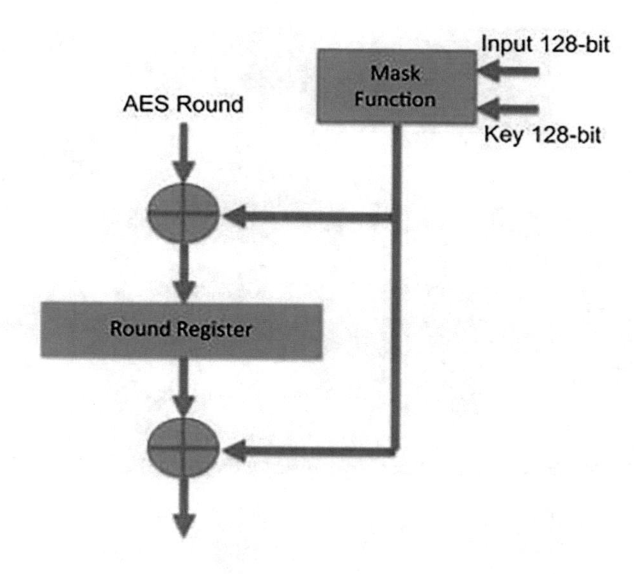

Fig. 12 Masking of round register [45]

3. *Secure test access mechanism*: In this method, the circuit works only in two modes: secure and insecure. The switching from test to functional mode is feasible during the insecure mode. However, the circuit only functions in mission mode during the secure mode. Switching from secure mode to insecure mode is only possible through power-off reset as illustrated in Fig. 15. Hence, intermediate information of cipher is not observed by the attacker, and DSA cannot be applied. The group of registers is needed, and a revision in the test controller also required to change the starting of the test session [44].
4. *Scan-chain scrambling*: In scan-chain scrambling [72], the pseudo-random choice of scan chains for loading is made with the help of LFSR. Hence, it can easily thwart DSA. However, this countermeasure seriously impacts the device area and increased power consumption in the functional mode of chip.
5. *Mode-reset countermeasure*: In this method [73], the author proposed the modification in the chip test controller such that scan-in and scan-out operations is are only possible after initialization reset procedure. When the test mode of the chip is requested, the chip is fully reset, and data that is going to be scanned out have no relation with chip- sensitive information. Hence, this can thwart DSA very easily as it requires switching from functional to test mode to get an intermediate state of the crypto-chip scanned out. However, this scheme requires test controller modification as well as XOR combinational network for reset checking.

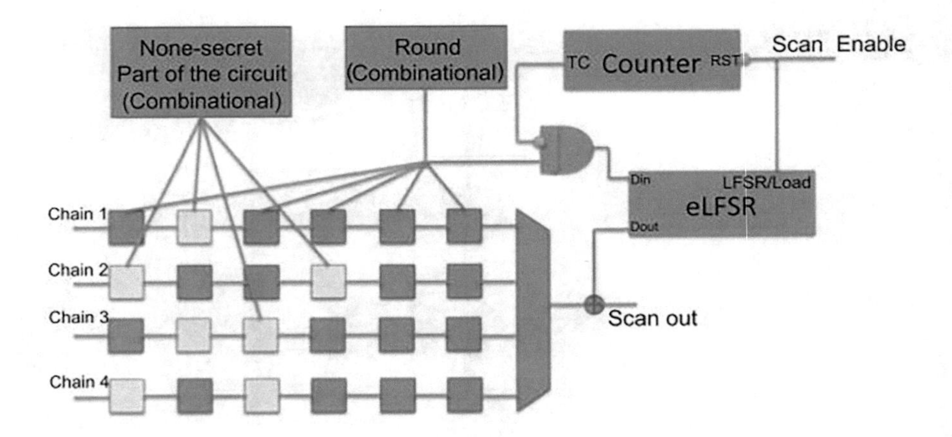

Fig. 13 Masking of response compactor output [45]

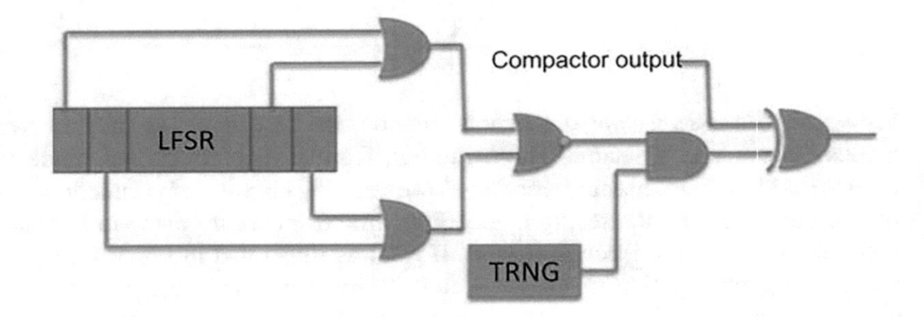

Fig. 14 Noise injection in the output [49]

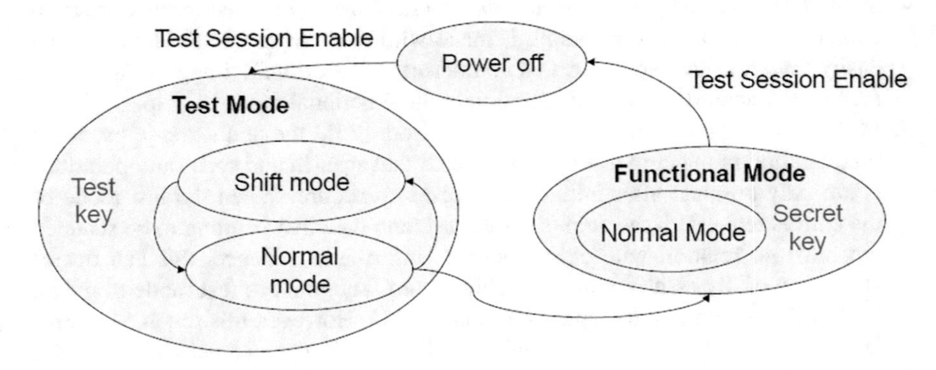

Fig. 15 Secure test access mechanism [44]

5.4.2 Countermeasures Against Test-Mode-Only Attack

The three countermeasures against TMO are discussed as follows.

1. *Scan cell swapping*: In this countermeasure [52], authors propose swapping of non-key-related flip flops with nearest key-related flip flops such that condition 1 mentioned in test-mode-only attack, for the attack to be successful, is violated. The new distributions of KFFs are such that none of the four key bytes are controllable through any value of the 28 combinations. For the correct swapping to happen, the detailed analysis of decompression structure/type and its correlation to scan cells are required. Hence, a test-mode-only attack can be prevented. This is illustrated in Fig. 16(1).
2. *Decompressor output swapping*: Another way of changing the distribution of key-related flops is done by interchanging the wires feeding data to scan chains. Hence, automatically different scan chains will be connected to decompressor output, which again needs the encoding condition for the key byte to break. Similar to the previous countermeasure, the violation of condition for one key byte occurs, and hence, the other key-byte may accidentally retrieve which satisfies that encoding condition. Hence, post-check is required, and test-mode-only attack cannot retrieve correct key bytes [52]. It is shown in Fig. 16(2).
3. *Two-level defense with input corruption*: This countermeasure [74] is applied to basic scan chain architecture without a test compression scheme. It injects the noise by flipping the scan input bits of vectors, and hence, the attacker cannot categorize scan cells into words and bytes. This is done by using LFSR which selects the location of the bit-flip of the test vector, and the newly generated bit is XORed with shifted test vector bit. The second level of defense by using another LFSR which counts specific continuous shift pulses without detecting capture pulse in between and injects the extra noise on the shifted test vector. Thus, the test-mode-only attack can be prevented. This is illustrated in Fig. 17.

5.4.3 Unified Countermeasures

The five countermeasures against both DSA and TMO are discussed as follows.

1. *Built-in self-test*: In this scheme, any kind of test output is not coming out from the chip, but the signature of the output is matched with the golden signature saved on-chip without taking it out; BIST is a countermeasure against DSA and TMO [75]. However, the scheme required additional hardware for on-chip pattern generation, response compactor to produce a signature, and a ROM to store the golden signature. Besides low fault coverage, this countermeasure compromises debug and diagnostic facilities.
2. *On-chip test comparison*: In this scheme, the generated test response is compared with golden one stored on-chip. The golden response is transferred to chip in this method using scan-out pin, and it is compared with generated test response. Apparently, it costs the extra hardware to compare both responses.

   Subsequently, the DSA and TMO is not feasible since the attacker can only see pass/fail bit at the output [76]. Test time or test coverage is not impacted by

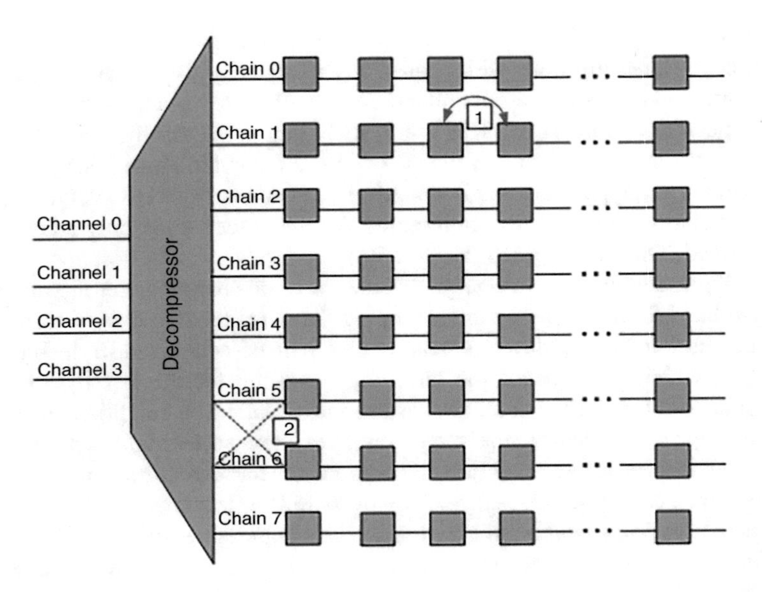

Fig. 16 (1) Scan cell swapping and (2) decompressor output swapping [52]

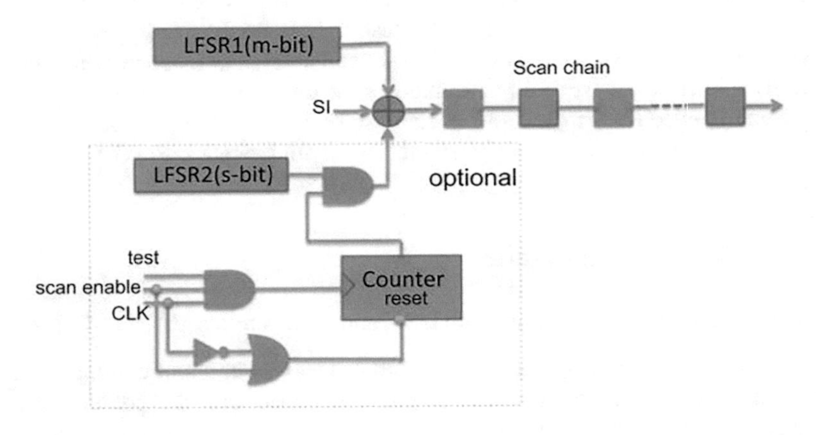

Fig. 17 Two-level defense with input corruption [74]

this countermeasure. However, diagnostic and debug facilities are impacted by the loss of observability of test response. The method is represented in Fig. 18.

3. *Scan-chain encryption*: In [77], the author has proposed on-chip encryption/decryption before and after response compaction and stimuli decompression to prevent differential scan attack on the secure circuit as shown in Fig. 19. However, the scheme required extra efforts for both input/output on-chip encryption/decryption of test-patterns on-chip as well as off-chip. With this solution, there is no impact on test diagnosis or debug facilities. Also, the test

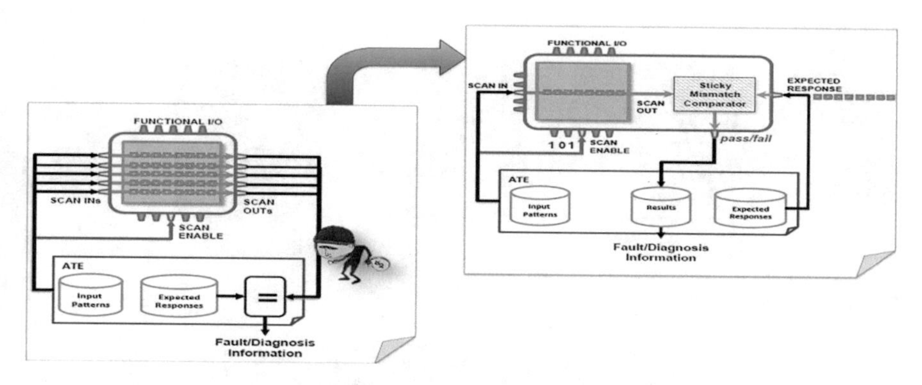

Fig. 18 On-chip test comparison [76]

coverage is not compromised. However, this comes with the cost of large-area overhead and test time increment.

4. *Modular exponentiation-based secure scheme (ME-SS)*: In this scheme [54, 55], the modular exponentiation-based countermeasure against all kinds of test infrastructure based attacks (DSA and TMO) is proposed. As shown in Fig. 20, the ME-SS countermeasure is following the AES circuit with all types of scan architecture be it a simple scan or a test compression-based scan architecture. Hence, it does not disturb today's IP core-based SoC flow and does not demand any modification to scan architecture netlist. The countermeasure completely thwart traditional DSA and TMO attacks with improving the area and test time overhead nearly 87% and 88% as compared to previous published scan-chain encryption countermeasure. However, it is not fully secure against the test infrastructure-based attack procedure in the case of statistical analysis.
5. *Hash function-based secure scheme (HSS)*: The HSS countermeasure [78] is proposed to further improve the security provided by the previous ME-SS countermeasure. It is shown that the proposed HSS countermeasure provides full security against traditional DSA and TMO attacks in the case of the AES circuit with response compactor (X-tolerant as well as MISR). As shown in Fig. 21, the modular exponentiation is performed multiple times, and final HSS encoded response is generated. The proposed HSS countermeasure with three rounds imposes 0.36% and 23% of overhead in terms of area and test time as compared to previously proposed ME-SS with improving the security of normal AES circuit with X-tolerant response compactor nearly 29% in the case of statistical analysis.

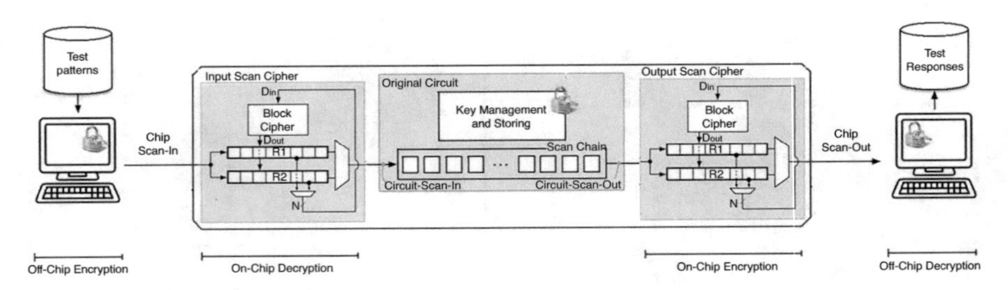

Fig. 19 On-chip encryption [77]

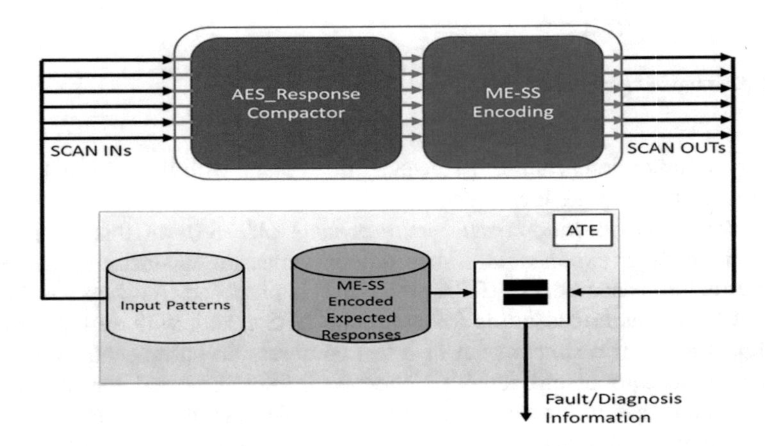

Fig. 20 Architecture of ME-SS [54]

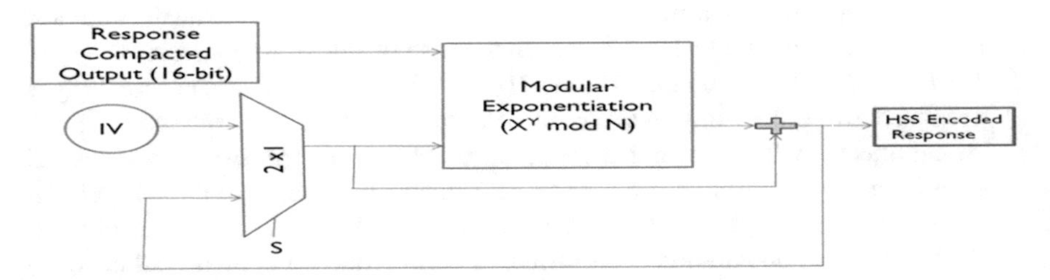

Fig. 21 HSS for AES with response compactor circuit [78]

6 Summary

This chapter presented threat models and corresponding attacks in the different stages of current-generation SoC life cycle. We also outlined existing defense approaches against the hardware Trojan attack, side-channel attack, fault injection attack, and test-infrastructure based attacks.

References

1. Ray, S., Peeters, E., Tehranipoor, M.M., Bhunia, S.: System-on-chip platform security assurance: architecture and validation. Proc. IEEE. **106**(1), 21–37 (2018)
2. Tehranipoor, M., Koushanfar, F.: A survey of hardware Trojan taxonomy and detection. IEEE Des. Test Comput. **27**(1), 10–25 (2010)
3. Tehranipoor, M., Wang, C.: Introduction to Hardware Security and Trust. Springer Science & Business Media, New York (2011)
4. Ramamoorthy, G.: Market share analysis: semiconductor design intellectual property, worldwide (2012). [Online]. Available: https://www.gartner.com/doc/2403015/market-share-analysis-semiconductordesign
5. Messmer, E.: RSA security attack demo deep-fries apple mac components (2014). [Online]. Available: http://www.networkworld.com/news/2014/022614-rsaapple-attack-279212.html
6. Nahiyan, A., Xiao, K., Yang, K., Jin, Y., Forte, D., Tehranipoor, M.: AVFSM: a framework for identifying and mitigating vulnerabilities in FSMs. In: *Proceedings of the 53rd Annual Design Automation Conference*, pp. 1–6 (2016)
7. Zhou, Y., Fang, Y., Zhang, Y.: Securing wireless sensor networks: a survey. IEEE Commun. Surv. Tutorials. **10**(3), 6–28 (2008)
8. Synopsis design compiler, https://www.synopsys.com/implementation-and-signoff/rtlsynthesis-test.html
9. Cadence genus synthesis solution, https://www.cadence.com/content/cadence-www/global/en_US/home/tools/digital-design-and-signoff/synthesis/genus-synthesis-solution.html
10. Nahiyan, A., Farahmandi, F., Mishra, P., Forte, D., Tehranipoor, M.: Security-aware FSM design flow for identifying and mitigating vulnerabilities to fault attacks. IEEE Trans. Comput. Aided Des. Integr. Circuits Syst. **38**(6), 1003–1016 (2019)
11. DARPA system security integrated through hardware and firmware (SSITH), https://www.fbo.gov/index?s=opportunity&mode=form&id=ea2550cb0c42eb91c7292377824a58b7
12. DARPA, TRUST in integrated circuits (TIC) – proposer information pamphlet (2007). [Online]. Available: http://www.darpa.mil/MTO/solicitations/baa07-24/index.html
13. Defense science board, Task force on high performance microchip supply (2005). [Online]. Available: http://www.acq.osd.mil/dsb/reports/200502HPMSReportFinal.pdf
14. Australian Government DoD-DSTO, Towards countering the rise of the silicon trojan (2008). [Online]. Available: https://www.semanticscholar.org/paper/Towards-Countering-the-Rise-of-the-Silicon-Trojan-Anderson-North/9916af435dc14416b986558910b8556e3b403855
15. Adee, S.: The Hunt for the Kill Switch. IEEE Spectr. **45**(5), 34–39 (2008)
16. Alkabani, Y., Koushanfar, F.: Designers Hardware Trojan Horse. HOST (2008)
17. King, S., et al.: Designing and Implementing Malicious Hardware. LEET (2008)
18. Abramovici, M., Bradley, P.: Integrated Circuit Security – New Threats and Solutions. CSIIR Workshop (2009)
19. Banga, M., Hsiao, M.S.: A Region Based Approach for the Identification of Hardware Trojans. HOST (2008)
20. Wolff, F., et al.: Towards Trojan-Free Trusted ICs: Problem Analysis and Detection Scheme. DATE (2008)
21. Jin, Y., Makris, Y.: Hardware Trojan Detection Using Path Delay Fingerprint. HOST (2008)
22. Chen, Z., et al.: Hardware Trojan Designs on BASYS FPGA Board (Virginia Tech). CSAW Embedded System Challenge (2008). [Online]. Available: https://www.semanticscholar.org/paper/Hardware-Trojan-Designs-on-Basys-Fpga-Board-Chen-Guo/69c85c799e9f21bd63caaa02e88fb3f572b3a609
23. Baumgarten, A., et al.: Embedded Systems Challenge (Iowa State University). CSAW Embedded System Challenge (2008)
24. Jin, Y., Kupp, N.: CSAW 2008 Team Report (Yale University). CSAW Embedded System Challenge (2008). [Online]. Available: http://www.eecs.ucf.edu/~jinyier/courses/EEE6306/files/submit%20code/CSAW%20Report%20-%20TRELA.pdf

25. Chakraborty, R.S., Wolf, F., Papachristou, C., Bhunia, S.: MERO: a statistical approach for hardware Trojan detection. In: *International Workshop on Cryptographic Hardware and Embedded Systems (CHES'09)*, pp. 369–410 (2009)
26. Dhem, J.-F., Koeune, F., Leroux, P.-A., Mestr, P., Quisquater, J.J., Willems, J.-J.: A practical implementation of the timing attack. In: Quisquater, J., Schneier, B. (eds.) Lecture Notes in Computer Science, vol. 1820, pp. 167–182. CARDIS (1998)
27. Kocher, P., Jaffe, J., Jun, B.: Differential power analysis. In: Advances in *Cryptology-CRYPTO 99*, LNCS *1666*, pp. 388–397 (1999)
28. Brier, E., Clavier, C., Olivier, F.: Correlation power analysis with a leakage model. In: *Cryptographic Hardware and Embedded Systems -CHES* 2004, pp. 16–29. Springer, Berlin Heidelberg (2004)
29. Dofe, J., Pahlevanzadeh, H., Yu, Q.: A comprehensive FPGA-based assessment on fault-resistant AES against correlation power analysis attack. J. Electron. Test. **32**(5), 611–624 (2016)
30. Quisquater, J.-J., Samyde, D.: ElectroMagnetic analysis (EMA): measures and countermeasures for smart cards. In: Attali, I., Jensen, T.P. (eds.) *E-smart, Lecture Notes in Computer Science*, vol. 2140, p. 200210 (2001)
31. Gandolfi, K., Mourtel, C., Olivier, F.: Electromagnetic analysis: concrete results. In: Ko, K., et al. (eds.) cKKNP01, pp. 251–261 (2001)
32. Peterson, I.: Chinks in digital armor: exploiting faults to break smartcard cryptosystems. Sci. News. **151**, 7879 (1997)
33. Boneh, D., DeMillo, R.A., Lipton, R.J.: On the importance of checking cryptographic protocols for faults. In: *16th Annual International Conference on Theory and Application of Cryptographic Techniques*, ser. EUROCRYPT 1997, Berlin, Heidelberg, p. 3751 (1997)
34. Skorobogatov, S.: Low temperature data remanence in static RAM. In: *Computer Laboratory*, Tech. Rep. UCAM-CL-TR-536. University of Cambridge (2002)
35. Barenghi, A., Bertoni, G., Parrinello, E., Pelosi, G.: Low voltage fault attacks on the RSA cryptosystem. In: *Proc. Workshop Fault Diagnosis Tolerance Cryptogr.*, pp. 23–31 (2009)
36. Barenghi, A., Bertoni, G.M., Breveglieri, L., Pellicioli, M., Pelosi, G.: Low voltage fault attacks to AES. In: *Proc. Int. Symp.* Hardware-Oriented Security Trust, pp. 7–12 (2010)
37. Selmane, N., Guilley, S., Danger, J.-L.: Practical setup time violation attacks on AES. In: *Proc. Eur. Dependable Comput. Conf.*, pp. 91–96 (2008)
38. Barenghi, A., Hocquet, C., Bol, D., Standaert, F.-X., Regazzoni, F., Koren, I.: Exploring the feasibility of low cost fault injection attacks on sub-threshold devices through an example of a 65 nm AES implementation. In: *Proc.* Workshop RFID Security Privacy, pp. 48–60 (2011)
39. Kommerling, O., Kuhn, M.G.: Desig Principles for Tamper-resistant Smartcard Processors. In: *Proceedings of the USENIX Workshop on Smartcard Technology*, p. 22. USENIX Association, Berkeley, CA, USA (1999)
40. Bar-El, H., Choukri, H., Naccache, D., Tunstall, M., Whelan, C.: The sorcerer's apprentice guide to fault attacks. Proc. IEEE. **94**(2), 370382 (2006)
41. Skorobogatov, S.P., Anderson, R.J.: Optical fault induction attacks. In: *International Workshop on Cryptographic Hardware and Embedded Systems-CHES 2002*, p. 212 (2002)
42. Quisquater, J.-J., Samyde, D.: Eddy current for magnetic analysis with active sensor. In: *Esmart* 2002, Nice, France (2002)
43. Schmidt, J.-M., Hutter, M.: Optical and EM fault-attacks on CRTbased RSA: concrete results. In: Karl, J.W., Posch, C. (eds.) *Austrochip* 2007*, 15th Austrian Workhop on Microelectronics*, 11 October 2007, Graz, Austria, Proceedings, pp. 61–67. Verlag der Technischen Universitat Graz (2007)
44. Yang, B., Wu, K., Karri, R.: Secure scan: a design-for-test architecture for crypto chips. IEEE Trans. Comput. Aided Des. Integr. Circuits Syst. **25**(10), 2287–2293 (2006)
45. DaRolt, J., Di Natale, G., Flottes, M.L., et al.: Scan attacks and countermeasures in presence of scan response compactors. In: 2011 *16th IEEE European Test Symp.*, Trondheim, pp. 19–24 (2011)

46. Yang, B., Wu, K., Karri, R.: Scan based side channel attack on dedicated hardware implementations of Data Encryption Standard. In: Proceedings of IEEE International Test Conference, pp. 339–344 (2004)
47. Da Rolt, J., Di Natale, G., Flottes, M.L., et al.: Are advanced DFT structures sufficient for preventing scan-attacks? In: 2012 IEEE 30th VLSI Test Symp. (VTS), Hyatt Maui, HI, pp. 246–251 (2012)
48. Ege, B., Das, A., Gosh, S., Verbauwhede, I.: September: "differential scan attack on AES with Xtolerant and X-masked test response compactor". In: Digital System Design (DSD), 2012 IEEE 15th Euromicro Conference on, pp. 545–552 (2012)
49. Das, A., Ege, B., Ghosh, S., Batina, L., Verbauwhede, I.: Security analysis of industrial test compression schemes. IEEE Trans. Comput. Aided Des. Integr. Circuits Syst. **32**(12), 1966–1977 (2013)
50. Ali, S.S., Sinanoglu, O., Saeed, S.M., Karri, R.: New scan-based attack using only the test mode. In: Very large scale integration (VLSI-SoC), 2013 IFIP/IEEE 21st international conference on, pp. 234–239. IEEE (2013)
51. Ali, S.S., Sinanoglu, O., Saeed, S.M., Karri, R.: New scan attacks against state-of-the-art countermeasures and DFT. In: 2014 IEEE International Symposium on Hardware-Oriented Security and Trust (HOST), pp. 142–147 (2014)
52. Ali, S.S., Saeed, S.M., Sinanoglu, O., Karri, R.: Novel test-mode-only scan attack and countermeasure for compression based scan architectures. IEEE Trans. Comput. Aided Des. Integr. Circuits Syst. **34**(5), 808–821 (2015)
53. Ali, S.S., Saeed, S.M., Sinanoglu, O., Karri, R.: Scan attack in presence of mode-reset countermeasure. In: 2013 IEEE 19th International On-Line Testing Symposium (IOLTS), pp. 230–231. IEEE (2013)
54. Popat, J., Mehta, U.: A novel countermeasure against differential scan attack in AES algorithm. In: VLSI Design And Test (VDAT) (2018)
55. Popat, J., Mehta, U.: Statistical security analysis of AES with X-tolerant response compactor against all types of test infrastructure attacks with/without novel unified countermeasure. IET Circuits Devices Syst. **13**(8), 1117–1124 (2019)
56. Dong, C., Xu, Y., Liu, X., Zhang, F., He, G., Chen, Y.: Hardware trojans in chips: a survey for detection and prevention. *Sensors*. ***20***(18), 5165 (2020)
57. Popat, J., Mehta, U.: Transition probabilistic approach for detection and diagnosis of Hardware Trojan in combinational circuits. In: *IEEE Annual India Conference (INDICON)*, pp. 1–6. IEEE (2016)
58. Durvaux, F., Renauld, M., Standaert, F.-X., Oldenzeel, L., Veyrat-Charvillon, N.: Cryptanalysis of the CHES 2009/2010 random delay countermeasure. In: *IACR Cryptology ePrint Archive*, p. 38 (2012)
59. Jayasinghe, D., Ragel, R., Elkaduwe, D.: Constant time encryption as a countermeasure against remote cache timing attacks. In: 2012 IEEE 6th International Conference on Information and Automation for Sustainability, Beijing, China, pp. 129–134 (2012)
60. Barthe, G., Betarte, G., Campo, J.D., et al.: System-level non-interference of constant-time cryptography. Part II: verified static analysis and stealth memory. J. Autom. Reasoning. **64**, 1685–1729 (2020)
61. Tiri, K., Akmal, M., Verbauwhede, I.: A dynamic and differential CMOS logic with signal independent power consumption to withstand differential power analysis on smart cards. In: *Solid-State Circuits Conference, 2002*. ESSCIRC 2002. Proceedings of the 28th European, pp. 403–406 (2002)
62. Tiri, K., Verbauwhede, I.: A VLSI design flow for secure side-channel attack resistant ICs. In: *Proceedings of the Conference on Design, Automation and Test in Europe* – Volume 3, DATE '05, pp. 58–63. IEEE Computer Society, Washington, DC, USA (2005)
63. Ishai, Y., Sahai, A., Wagner, D.: Private circuits: securing hardware against probing attacks. In: *Advances in Cryptology – CRYPTO* 2003, *23rd Annual International Cryptology Conference*, Santa Barbara, California, USA, August 17–21, pp. 463–481 (2003)

64. Giraud, C.: "DFA on AES", in *International Conference on Advanced Encryption Standard*. Springer, Berlin Heidelberg (2004)
65. Koren, I., Krishna, C.M.: Fault Tolerant Systems. Morgan-Kaufman, San Francisco, CA (2007)
66. Karri, R., Wu, K., Mishra, P., Kim, Y.: Fault-based side-channel cryptanalysis tolerant Rijndael symmetric block cipher architecture. In: *Proc. IEEE Int. Symp. Defect Fault Tolerance VLSI Syst.*, pp. 427–435 (2001)
67. Maistri, P., Vanhauwaert, P., Leveugle, R.: A novel double-data-rate AES architecture resistant against fault injection. In: *Proc. Workshop Fault Diagnosis Tolerance Cryptogr.*, pp. 54–61 (2007)
68. Bertoni, G., Breveglieri, L., Koren, I., Maistri, P., Piuri, V.: Error analysis and detection procedures for a hardware implementation of the advanced encryption standard. IEEE Trans. Comput. **52**(4), 492–505 (2003)
69. Bertoni, G., Breveglieri, L., Koren, I., Maistri, P.: An efficient hardware-based fault diagnosis scheme for AES: Performances and cost. In: *Proc. IEEE Int. Symp. Defect Fault Tolerance VLSI Syst.*, pp. 130–138 (2004)
70. Butter, A., Kao, C., Kuruts, J.: DES encryption and decryption unit with error checking, US Patent 5 432 848 (1995)
71. IBM, *Ibm 4764 pci-x Cryptographic Coprocessor Specifications*. [Online]. Available: http://www.ibm.com/security/cryptocards/pdfs/bs330.pdf
72. Hely, D., Flottes, M.-L., Bancel, F., Rouzeyre, B., Berard, N., Renovell, M.: Scan design and secure chip. In: IOLTS, vol. 4, pp. 219–224 (2004)
73. Hely, D., Bancel, F., Flottes, M.-L., Rouzeyre, B.: Test control for secure scan designs. In: Test Symposium, 2005. European, pp. 190–195. IEEE (2005)
74. Ali, S.S., Saeed, S.M., Sinanoglu, O., Karri, R.: New scan-based attack using only the test mode and an input corruption countermeasure. In: IFIP/IEEE International Conference on Very Large-Scale Integration-System on a Chip, pp. 48–68. Springer, Cham (2013)
75. Natale, G.D., Doulcier, M., Flottes, M.L., Rouzeyre, B.: Self-Test Techniques for Crypto-Devices. IEEE Trans. Very Large Scale Integr. VLSI Syst. **18**(2), 329–333 (2010)
76. Da Rolt, J., Di Natale, G., Flottes, M.L., Rouzeyre, B.: On-chip test comparison for protecting confidential data in secure ICS. In: 2012 17th IEEE European Test Symposium (ETS), p. 1 (2012)
77. Silva, D., Mathieu, M.-L.F., Di Natale, G., Rouzeyre, B.: Preventing scan attacks on secure circuits through scan chain encryption. IEEE Trans. Comput. Aided Des. Integr. Circuits Syst. **38**(3), 538–550 (2018)
78. Popat, J., Mehta, U., Upadhyay, M.: A Hash based secure scheme against scan-based attacks on AES cipher. In: International Test Conference India (ITC India). IEEE (2020)

Defect Diagnosis Techniques for Silicon Customer Returns

Patrick Girard, Alberto Bosio, Aymen Ladhar, and Arnaud Virazel

1 Introduction

Modern electronic systems are composed of complex systems on a chip (SoCs) which are made of various heterogeneous blocks that include memories, digital, analog and mixed-signal parts, etc. These SoCs demand a huge amount of knowledge and expertise to be designed, fabricated, and embedded on the final support with the required levels of functionality and reliability. To guarantee their correct behavior and hence fit a given quality level required by the application standard (e.g., automotive, avionic, etc.), SoCs pass through a comprehensive test flow (functional, structural, parametric, etc.) at the end of the manufacturing process. SoCs that pass the test flow are further used in the field by the target application.

Despite the high-quality level of the test flow used during manufacturing test, SoCs may still fail in the field. So, in order to identify the source of failures and avoid their reoccurrence in the next generation of products, each defective SoC (referred to as *customer return*) is always returned to the manufacturer who is in charge of analyzing the device to determine the root cause of failures. This is particularly true for safety-critical applications. A customer return is a circuit that passed the full manufacturing test flow but failed on the customer's side [1]. The two main causes of a customer return are test escape during manufacturing test and latent

P. Girard (✉) · A. Virazel
LIRMM, University of Montpellier/CNRS, Montpellier, France
e-mail: Patrick.Girard@lirmm.fr; Arnaud.Virazel@lirmm.fr

A. Bosio
Univ. of Lyon, ECL, INSA Lyon, CNRS, UCBL, CPE Lyon, INL, Ecully, France
e-mail: alberto.bosio@ec-lyon.fr

A. Ladhar
STMicroelectronics, Crolles, France
e-mail: aymen.ladhar@st.com

A. Iranmanesh (ed.), *Frontiers of Quality Electronic Design (QED)*,
https://doi.org/10.1007/978-3-031-16344-9_17

defect mechanisms during lifetime [2]. Latent defects cause two types of failures: (i) early-life failures that do not manifest during manufacturing test but that degrade over time due to electrical and thermal stress during in-field use and (ii) failures caused by wear-out mechanisms. Wear-out or aging manifesting as progressive performance degradation is induced by various mechanisms such as negative-bias temperature instability or hot-carrier injection. All these failures that occur in the field are extremely critical as they may lead to catastrophic consequences.

When a customer return is detected, it is crucial to reproduce the failure mechanism in the failure analysis lab of the manufacturer with the appropriate test conditions (temperature and voltage) and the original test flow. In the case of test escape, test engineers must generate new test patterns that will exhibit the failure in the same test conditions. In the case of latent defect, the task will often succeed, and a diagnosis program made of several routines is used to identify, step by step, the failing part, and, finally, the suspected defect(s). Each routine consists in the application of a diagnosis algorithm at a given hierarchy level (system, core, and cell levels) [3].

Diagnosis is a software-based method that analyzes the applied test sequences, the tester responses, and the circuit structure (generally with layout information) to generate a list of candidates that represent the possible locations and types of defects within the defective circuit. The quality of a diagnosis outcome is evaluated thanks to two metrics: accuracy and resolution. A diagnosis is accurate if the real defect is included in the reported list of candidates. Resolution refers to the total number of candidates reported by the diagnosis. An accurate diagnosis with perfect resolution (i.e., one candidate which is the real defect) is the ideal case.

Diagnosis is usually followed by *physical failure analysis* (PFA), a time-consuming process for exposing the defect physically (thanks to special techniques and tools such as acoustic microscopy, X-ray imaging, transmission electron microscopy, photon emission microscopy, laser-induced voltage alteration, thermal imaging, etc.) in order to characterize the failure mechanism. Due to the high cost of PFA and its destructive nature, diagnosis accuracy and resolution are of critical importance. In practice, it is very uncommon to perform PFA on any defect with more than five candidates [4]. This ensures that the likelihood for identifying the root-cause of failure is maximized when performing PFA.

With more defects inside standard cells at leading-edge technology nodes [5], *cell-aware* (CA) test, which deterministically targets defect locations inside standard cells, and CA diagnosis, which can identify the location and type of cell-internal defects at the transistor level, are quality assessment solutions widely adopted today in industry [6, 7]. CA diagnosis is valuable in the context of large and complex cells such as multipliers, adders, and multi-bit sequential elements. Even when a defect is known to be within a cell, finding defects in such a complex cell during PFA can be challenging and time-consuming, especially if the defect is exhibited for a specific test condition and undetected for some others. CA diagnosis shortens the lengthy investigation process by pinpointing a small subsection of the suspected cell. It improves results for diagnosis scenarios such as customer return analysis and volume diagnosis applications like yield analysis [7].

Unfortunately, diagnosis resolution today is typically far from ideal due to SoC complexity. In particular, with the advent of nanometer scale technologies (i.e., 7 nm), a high resolution (very few or one candidate) is not always achievable by existing CA logic diagnosis tools based on conventional methods [8]. For this reason, considerable researches have been dedicated on improving resolution by using machine learning techniques, mainly through the extraction of features that allow correct candidates (those that correctly represent defect locations) to be distinguished from incorrect ones [2, 4, 9–11].

Even though they are efficient, these techniques address volume diagnosis for yield ramp-up. This is a different problem than *defect diagnosis of customer returns*. Indeed, during volume diagnosis for yield improvement, a lot of data collected during manufacturing test and subsequent diagnosis phases are available and can be used, such as hundreds of identical failed chips with candidates correctly identified (good, bad). It is thus possible to use these data for failure diagnosis of a new failed chip. On the contrary, during fault diagnosis of a customer return, only one failed chip has to be investigated, with no information about the defective behavior of any other similar chip used in the same conditions (application, environment, workload). For this reason, approaches existing for volume diagnosis cannot be reused in a straightforward manner for customer returns.

Historically, conventional approaches based on *critical path tracing* or *fault simulation* were used in industry for defect diagnosis of customer returns. However, with the fast development and vast application of *machine learning* (ML) in recent years, ML-based techniques have been demonstrated to be quite valuable for diagnosis. Most of these techniques are based on supervised learning, because it naturally aligns with the common practice of training with labeled historical data and usually performs well in industrial diagnosis tasks [12].

This chapter reviews the latest developments in the field of customer return diagnosis based on ML. It is organized as follows. Section 2 gives some background on test and diagnosis in the context of customer return analysis. Section 3 first explains how customer returns are usually (re-)tested for diagnosis purpose and what are the limitations. Next, a proposal of best practices for customer return test pattern generation is done and discussed. Section 4 summarizes the state-of-the-art in the field of defect diagnosis for customer returns. Both conventional approaches and ML-based techniques are discussed. Section 5 presents some industrial case studies performed with one of the latest ML-based diagnosis techniques. Section 6 concludes the chapter and draws some conclusions.

2 Background on Test and Fault Diagnosis

During the manufacturing process of integrated circuits (ICs), defects may occur in the physical structure of the IC, next leading to erroneous behaviors. The role of testing is to detect ICs affected by defects and discard them from the set of ICs sent to the customers. Moreover, testing is also important to gather as much information

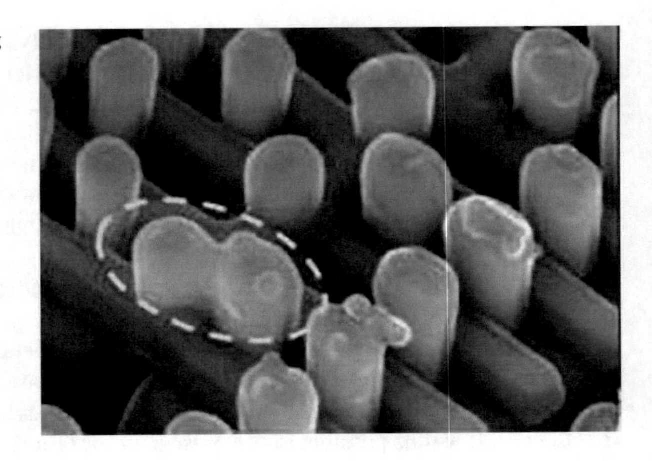

Fig. 1 Short defect affecting a logic gate

as possible and further use them during fault diagnosis, in order to understand the root causes of the observed failures. This section provides some backgrounds on defects, test, and fault diagnosis.

2.1 From Defects to Failures

Physical defects like shorts and opens may occur during any single step of the fabrication process. These defects can be randomly caused by contaminations or due to systematic process-design interactions [13]. In modern nanometer-scale technologies, defects appear not only in the cell interconnection (inter-cell defect) but also inside the cell itself (intra-cell defect) [14, 15]. This is caused by the reduced circuit sizes, the use of new process technologies, new materials, and the ever-increasing number of vias and contacts. For example, Fig. 1 depicts a short defect affecting a 4-input AOI cell made of 48 transistors in a 32-nm STMicroelectronics design [16].

The representation of a defect is a *fault model*, an anomalous physical condition that may lead to an *error*. An error is the exhibition of a fault in a system that might or might not be propagated and, in this last case, give rise to *failure* [17].

Figure 2 summarizes the above concepts through a simple example. The IC is a combinational circuit composed of three gates (G1, G2, and G3), two primary inputs (PI1 and PI2), and one primary output (PO). The circuit is affected by a defect represented by a stuck-at-0 fault model (S@0) located at the input *a* of G3 (i.e., the behavior of the defect can be modeled as a logic value always set to "0" at input *a* of G3). The circuit is stimulated by the logical values applied to PIs ("00"). The applied values lead to have "1" at input *a* of G3. However, because of the S@0, the value is actually set at "0." This situation is represented as "1/0," where "1" is the expected value and "0" is the wrong valued induced by the fault. Since the expected value is

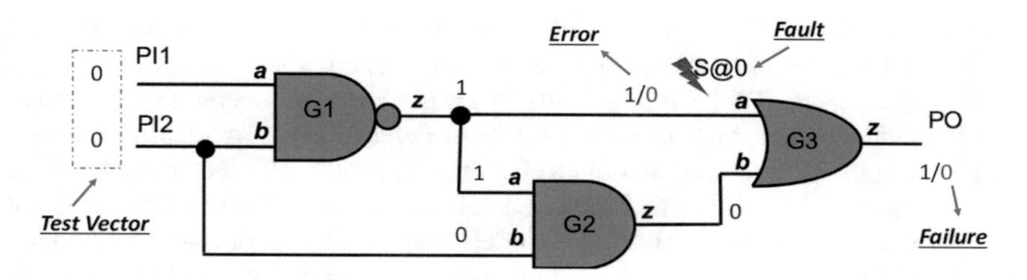

Fig. 2 Fault, error, and failure example

different from the wrong one, the fault is said to be *sensitized*, and the wrong value is the error induced by the fault. In the example, the error is propagated through G3 and reaches the PO. At this point, the error becomes observable and leads to a failure. The set of applied logic values is called *test vector* since it can detect the presence of the S@0 affecting input *a* of G3. Several fault models exist and are used in industry. Among them, the most popular are the following:

- *Stuck-at fault model* [17]: The logic value of a given net appears to be stuck at a constant logic value ("0" or "1"), referred to as stuck-at-0 or stuck-at-1.
- *Transition fault model* [17]: The transition from a given logic value V to the opposite logic value V at the output of a gate is delayed. In this case, the delay of the gate is changed and is assumed to be large enough to prevent a passing transition from reaching any output of the circuit within the clock period. Two types of transition fault are defined: slow-to-rise (slow transition from logic "0" to logic "1") and slow-to-fall (slow transition from logic "1" to logic "0").
- *Bridging fault model* [17]: Usually modeled at the gate or transistor level, it represents a short between a group of signals. The logic value of the shorted net can be modeled as a 1-dominant (OR bridge), 0-dominant (AND bridge), or indeterminate, depending upon the technology in which the circuit is implemented.
- *Cell-aware fault model* [15]: It represents a defect inside a given logic cell. The faulty behavior depends on the logic cell transistor-level structure. This fault model has to be defined every time a new technology library is implemented.

2.2 Testing

IC testing consists in applying a set of test vectors (forming a so-called test sequence) in order to detect the highest number of faults as possible. The test sequence quality is measured by using to the following metrics:

- *Fault coverage* (FC): This is the ratio between the number of detected faults and the total number of faults. Ideally, the FC has to be equal to 1.
- *Defect coverage* (DC): Similar to the FC, it gives the ratio between the number of detected defects and the total number of defects. It is important to mention that a high FC does not automatically implies a high DC. For example, it is demonstrated in [14, 15] that intra-cell defects cannot be detected by using test approaches based on classical fault models such as stuck-at or transition faults.
- *Test coverage* (TC): This is the ratio between the number of detected faults and the total number of detectable faults. Redundant faults are omitted by this metric. This is the main metric used in industry to qualify a test sequence. The higher the test coverage, the better is the quality of the test sequence.
- *Test length* (TL): The number of test vectors composing the test sequence. The higher the length, the higher is the cost of test in terms of test time and test data volume.

The test sequence is generated by using a commercial EDA tool, known as automatic test pattern generator (ATPG). In short, an ATPG aims at generating a test sequence that maximizes the FC while minimizing the TL. The description of ATPG architectures and algorithms is out the scope of this chapter. The reader can refer to [17] for detailed information.

IC testing is always executed after the manufacturing process. It allows to quantify the quality of the manufacturing process itself through the yield defined as

$$\text{Yield} = \frac{\#\text{Good}}{\#\text{Total}}$$

where #Good is the number of fault-free devices over the total number of manufactured devices. Every time a new technology node and its manufacturing process is used, the yield loss can be very high (yield <40%). The process of identifying yield losses and quantifying and improving them is referred to as *yield learning*. Testing and fault diagnosis play a crucial role for yield learning.

Even when the manufactured IC passes the testing phase and thus is used in the field, testing may be still needed. This is the case of safety-critical applications, like automotive or avionic, where all the hardware components have to be continuously tested to ensure the correct behavior. Any IC that fails in the field can be considered as a *customer return*.

2.3 Fault Diagnosis

Fault diagnosis is the process applied to a failing IC to shed light into the actual defect and then apply corrective actions to prevent failure reoccurrence in next-generation products. We can identify two main types of fault diagnosis depending on the scope:

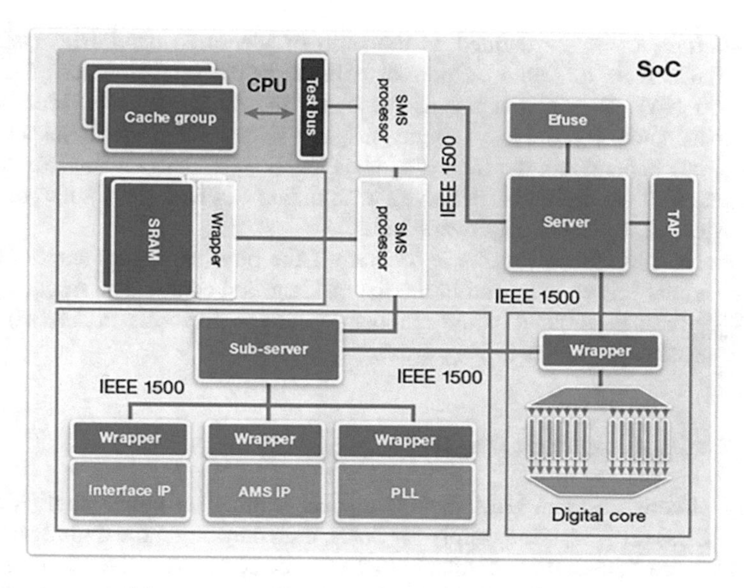

Fig. 3 Heterogeneous system-on-chip. (Courtesy of Synopsys)

- *High-volume fault diagnosis*: It is applied for yield learning. Indeed, diagnosing the sources of failures assists the designers in collecting valuable information regarding the underlying failure mechanisms, in order to enhance yield through improvement of the manufacturing process and development of new design techniques that minimize the failure rate [2].
- *Customer return fault diagnosis*: It is applied on a given IC to determine the root cause of failures that have occurred in the field. In this scenario, failures are not easy to reproduce in the manufacturer's lab as the real mission conditions, and executed workload are unknown and cannot be exhaustively modeled.

In the rest of this chapter, we mainly refer to fault diagnosis of customer returns. Fault diagnosis can be applied at different levels depending on the complexity of the IC. Todays' ICs are complex devices that consist of independent and heterogeneous blocks, and each of them may comprise memory, digital circuits, analog and mixed-signal (AMS) circuits, etc. (see Fig. 3).

The first level of fault diagnosis is thus the system level that aims at determining the failing block. The second level of fault diagnosis is the block level. Depending on the nature of the failing block, i.e., analog, digital, memory, etc., different fault diagnosis techniques (as well as test sequences) have to be applied. This chapter focuses on *digital circuit blocks* only. Memory and analog and mixed-signal (AMS) fault diagnosis is beyond the scope of this chapter, but the reader can refer to [18, 19] for more details. The third level of fault diagnosis is the cell level, called *cell-aware fault diagnosis*. It consists in identifying defects within a logic cell.

The key metrics characterizing fault diagnosis performance are as follows:

- *Resolution*: $R = \frac{\#C}{\#S}$ defined as the ratio of identified candidates ($\#C$), i.e., potential defects or faults on nets or cells, over the total number of possible suspects ($\#S$). The smaller the R, the better the fault diagnosis. Ideally, fault diagnosis should provide a single suspect. The definition of suspects and candidates depends on the fault diagnosis level. At system level, a candidate is a block, while at digital circuit level, a candidate can be a gate, a net, or even a transistor in the case of cell-aware faults.
- *Accuracy*: The fault diagnosis is accurate if the physical defect responsible for the observed failure is indeed in the list of identified candidates. Accuracy $A = \frac{\#\text{Correct}}{\#\text{Diagnosis}}$ can be defined as the number of correct diagnosis (#Correct) over a given set of experiments (#Diagnosis).

2.3.1 System-Level Fault Diagnosis

Existing effective system-level fault diagnosis techniques either apply only for boards (PCBs) [20], or, if they apply for SoCs, then they target the digital part of the SoC [21]. In particular, the individual blocks can be accessed and tested in isolation so as to identify the faulty block. The problem is exacerbated when the failure occurs during the mission (functional) mode. In this case, as the SoC operates in functional mode, no information is logged at block level, i.e., the functional stimuli applied to each block may be unknown. A promising solution is to use machine learning. In the literature, machine learning techniques have been already exploited for system-level diagnosis, but only for boards [22].

2.3.2 Digital Block-Level Fault Diagnosis

Numerous research works as and commercial tools (from EDA vendors) exist for digital block-level fault diagnosis. Almost all existing approaches are based on the "cause-effect" [23] or the "effect-cause" [24] paradigms. The "cause-effect" paradigm requires a pre-computed fault dictionary that can be obtained by simulating all targeted faults in a specific design with a given set of test vectors. During diagnosis, a search in the fault dictionary is performed to determine a set of candidates that can explain the observed errors. On the other hand, the "effect-cause" diagnosis approach determines the set of candidates by using a back-tracing algorithm [24]. This algorithm is carried out starting from each failing primary output and traces back through circuit nets to reach primary inputs. Each traced net is classified as a candidate. Compared to "cause-effect" diagnosis, the "effect-cause" approach does not require any pre-computed fault dictionary and is therefore independent of a targeted fault model. However, "effect-cause" diagnosis may require a very high computational time proportional to the circuit complexity (e.g., number of gates and test vectors). Hence, the "effect-cause" approach may not be appropriate for large designs.

Irrespective of the adopted paradigm (i.e., cause-effect or effect-cause), the result is a list of nets (e.g., connections between logic cells or flip-flops) that are declared as candidates. Even if only one candidate is included in the list, it is often not precise enough to isolate the defect causing the error. For example, one cell generally contains many transistors (usually more than 100), and one net could be extended to several metal layers. Without more accurate information of the defect location inside a cell, PFA may fail, i.e., the root cause may not be found. Hence, it is crucial to identify which components within a cell are more likely to be the defective ones so as to successfully perform PFA. This shift in the diagnosis accuracy level is obtained by applying transistor-level diagnosis inside a cell, referred to as *cell-aware fault diagnosis*.

2.3.3 Cell-Aware Fault Diagnosis

Previous works on cell-aware (CA) fault diagnosis focusing on logic cells can be classified into three approaches. The first approach converts a transistor-level netlist into an equivalent gate-level netlist by using complex transformation rules [25]. Therefore, any classical fault diagnosis approach can be applied on the equivalent gate-level netlist. The main limitation of this approach is that the set of transformation rules depends on the targeted defect, and, thereby, non-modeled defects may not be diagnosed. The second approach is based on the "cause-effect" paradigm [25, 26]. The transistor-level netlist of a cell is exploited, in order to inject the targeted defects. Therefore, a defect dictionary is created by transistor-level simulations, and the defect signatures of all the defects affecting the cells in the library are stored in this defect dictionary. Then, during fault diagnosis, the defect signature of all defects affecting a suspected cell is compared with the observed failures to obtain a list of candidates inside the cell. These approaches can be further classified depending on the "accuracy" of the injected defects and the simulation "precision." In [26], a large number of defects are simulated at transistor-level using SPICE. For a given defect, different resistance values are simulated, in order to be as accurate as possible. This approach leads to more precise results, but it requires a huge simulation time. To reduce the simulation time and the fault dictionary size while keeping a high resolution, authors in [25] propose to exploit layout information, in order to consider only realistic defects. For example, for each cell, only the realistic, potential net-bridging defects and via open defects are extracted and then simulated. Then, the identified set of realistic defects is simulated at transistor level. The third intra-cell fault diagnosis approach is based on the "effect-cause" paradigm [27]. All the existing diagnosis techniques depend on the targeted fault models or defects. In [27], the main goal is to achieve a resolution close to the transistor level. However, instead of explicitly considering defects at transistor level, the idea is to exploit the knowledge of the faulty behavior induced by the defects.

Unfortunately, due to circuit complexity of today's circuits, cell-aware fault diagnosis resolution is usually far from ideal. For this reason, a lot of efforts has been done to improve resolution by using machine learning techniques, initially through the extraction of features that allow correct candidates (those that correctly represent defect locations) to be distinguished from incorrect ones [2, 11]. Even if they are efficient, these techniques address volume diagnosis for yield improvement, which is a different problem than fault diagnosis of customer returns (as already mentioned and explained in Sect. 1). For this reason, these techniques cannot be reused for fault diagnosis of customer returns.

3 Test of Customer Returns for Diagnosis Purpose

The ultimate objective of an IC test engineer in charge of a given IC product is to reach a zero *defective parts per million* (DPPM) with a reasonable test cost (test time and test data volume). To this purpose, several test sequences made of test patterns are generated during test preparation to target all types of defects (static, dynamic) by using different (i.e., slow or fast) test clock schemes. Note that a test pattern can be composed of only one vector to deal with static faults, e.g., stuck-at faults, or of two vectors (two-vector test pattern) to deal with dynamic faults, e.g., transition faults. These test patterns are usually generated in an incremental manner to avoid multiple detection of the same defects and hence reduce test costs. In this section, we first provide an example of a typical test scenario used in production (manufacturing) industrial test. Next, we discuss some limitations of this type of scenarios in terms of diagnosis accuracy and resolution. Finally, we suggest some of the best test practices to be applied for custom return diagnosis.

3.1 Typical Test Scenario

When a customer return occurs and is sent back to the manufacturer, the first step is to reproduce the failure mechanism by reusing the initial manufacturing test program and collect the failure files to be diagnosed. Therefore, before proceeding with fault diagnosis, it is important to understand the ATPG process as well as its implementation into the test program. This knowledge is valuable and helps to enhance the diagnosis process and accelerate the time needed to retrieve the silicon failure.

Figure 4 gives an example of a multi-run ATPG flow used in industry for screening defects in logic parts of a SoC. It can be seen that several ATPG runs are implemented at various speeds (low-speed or at-speed) to get different test sequences targeting different defect categories (static and dynamic) with different fault models (stuck-at, transition, bridging, and cell-aware). The first category is *static defects*, and their detection requires one-vector test patterns, called *static test patterns*. The second category is *dynamic defects*, and their detection requires two-vector test patterns, called *dynamic test patterns*. Dynamic defects do not modify

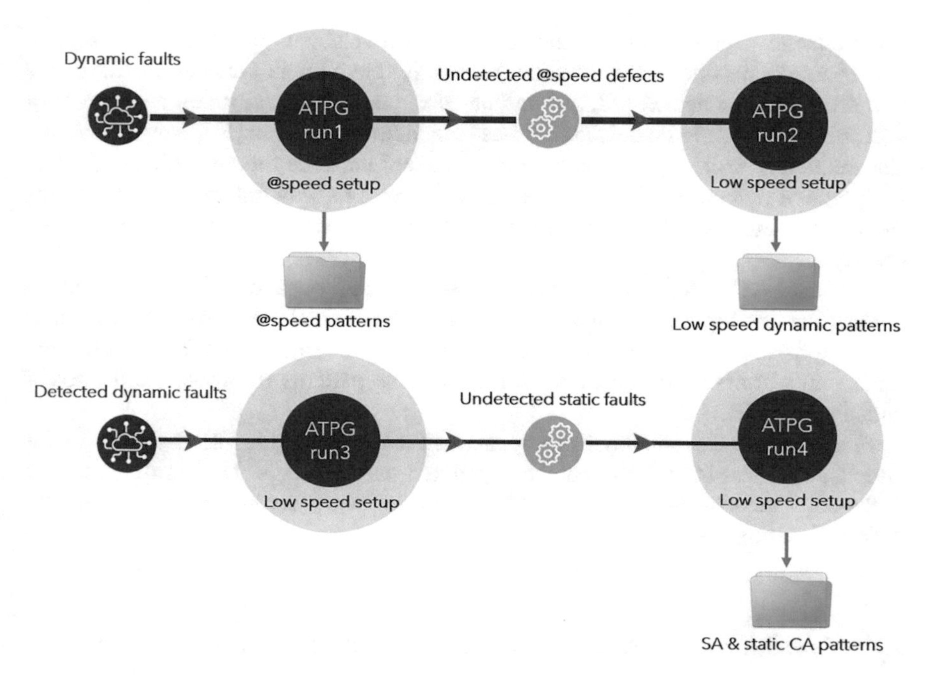

Fig. 4 Example of an industrial multi-run ATPG flow

the functional behavior of the IC but rather induce some delays that prevent the IC to operate at the desired clock frequency. With the new technologies, the incidence of dynamic defects is continuously growing, not only during the ICs manufacturing process but also during the IC lifetime where latent or wear-out defects may appear due to different stress conditions (operational, functional, environmental, etc.).

The flow sketched in Fig. 4 shows four ATPG runs implemented to generate three types of test patterns (at-speed, low-speed dynamic, and static test patterns). The first step in this ATPG flow is to generate at-speed test patterns for dynamic defect detection. These patterns target the detection of delay defects either at the interconnect wires or inside cells of the logic part of the SoC. At-speed test is made so that the PLL clock is used instead of the tester clock during the capture phase. This is usually performed by using an on-chip clock controller (OCC). An OCC is an added DFT logic to control internal clocks during manufacturing test. Since at-speed testing requires two clock pulses in capture phase with the same frequency as the internal clock, these at-speed pulses would need to be provided through I/O pads without OCC. However, these pads are limited in terms of supported maximum frequency. Conversely, an OCC uses an internal PLL for generating test clock pulses and hence allows application of at-speed tests. During static testing, the OCC ensures that only one clock pulse is generated during the capture phase.

Once the dynamic patterns are generated, the undetected dynamic defects are targeted with a second ATPG run (see Fig. 4). In fact, it has been proven that even at low frequency, some dynamic defects can be still detected. Typically, a stuck-open or a source/drain open defect results from a complete break between circuit nodes that should be connected [28, 29] and has a sequential behavior that requires two-vector test patterns to be detected. It has been demonstrated in [30] that varying the test frequency, voltage, and temperature does not enhance the test efficiency when such type of dynamic defects is targeted. This explains why this type of ATPG runs (low-speed setup) is used in industrial test flows to generate low-speed dynamic test patterns.

The third ATPG run is a preprocessing step for the fourth one. In fact, the fourth run requires as input a list of all static defects not detected by the dynamic tests. This information can be identified only through fault simulation. To this end, the static faults are fault simulated with the dynamic patterns to determine the set of undetected faults.

Finally, three test sequences are generated and then applied sequentially to the circuit under test (CUT) to achieve a targeted test coverage. During diagnosis of a customer return, the same test sequences can be reused to exhibit the failure observed during mission mode.

3.2 Limitation of Manufacturing Test for Customer Returns

The goal of an ATPG as used for manufacturing test is to detect the maximum number of faults with the minimum number of test patterns. Unfortunately, this may not be adequate for fault diagnosis for the following main reasons:

- Distinguishing between defect candidates during fault diagnosis is achievable only when test sequences are made of patterns that each sensitizes a limited number of faults. Conversely, test patterns generated by conventional ATPG sensitize as many faults as possible to avoid long test time. These two conflicting objectives between test pattern generation for testing purpose and test pattern generation for diagnosis purpose generally lead to poor diagnosability of test sequences generated by an ATPG. Table 1 shows an example where the generation of a test pattern is needed to detect two defects – internal D1 and D2 – in a NOR2 gate with A and B as inputs and Z as output. As can be seen, any ATPG tool would only generate P1 as it can detect both defects on output Z (Z_{D1} and Z_{D2} both have a value different than the fault-free value on Z). However, these patterns are unable to distinguish between the two defects, and hence any diagnostic tool would report D1 and D2 as fault candidates. As can be observed, P2 can detect only D2 and hence is able to distinguish between D1 and D2. The same is true for P3 which is able to distinguish between D1 and D2. Therefore, an ideal test sequence for diagnosis purpose would be made of (P2, P3).

Table 1 Example of distinguishing patterns between defect candidates

Pattern	AB	Z	Z_{D1}	Z_{D2}
P1	00	0	1	1
P2	01	0	0	1
P3	10	0	1	0
P4	11	1	1	1

- Generally, there is no full picture on how an actual defect behaves with regard to passing patterns (patterns declared as "pass" during test application). In fact, the purpose of a test pattern is to create a failing excitation of the defect and ensure its propagation to an observable point. This means that the behavior of the defect is known only for the failing patterns, and most of the time no information is collected for the passing patterns. Let us consider again the example in Table 1. An ATPG tool will generate P1 and eventually P2 to detect and distinguish between D1 and D2. Consequently, P3 and P4 will not be generated. However, the knowledge of this information would be important to improve the diagnosis accuracy since it would provide additional indications on how the defect behaves with a complete set of test stimulus.
- Scan chain diagnosis of ICs with an embedded test compression mechanism usually leads to low diagnostic capabilities [31]. This is the case of most ICs nowadays. In fact, compressed test patterns and test responses are broadly used in industry to reduce test data volume and scan input/output requirements. The usage of these techniques can negatively impact fault diagnosis accuracy. In fact, the defect responses are captured in scan flip-flops which are not directly observed. This means that using typical chain test patterns is not enough to distinguish between failing scan chains, and thus generating additionally distinguishing patterns is needed to improve the resolution.
- Test truncation is a widely used technique in production test. Indeed, not all the test responses of a failing CUT are collected, and this is limited to a predefined number of failing patterns. Indeed, recording the failing patterns and their observation points is a time-consuming step especially when the collected failures are huge. Proceeding with test truncation can reduce the test time since not all failing patterns are recorded. However, this procedure has a negative impact on the diagnostic quality as exploiting only a subset of the failing patterns to retrieve the fault candidates limits the diagnostic tool capabilities.

The abovementioned limitations of manufacturing test can prevent the successful failure analysis of a customer return. In the following subsection, we first explain how to adapt such a test flow for a customer return. Then, we present different techniques to generate *diagnostic patterns* (also referred to as distinguishing patterns).

3.3 Best Practices for Customer Return Test Pattern Generation

Two approaches exist to improve diagnosis quality. The first one is to improve the efficiency of diagnostic algorithms, which is usually done by CAD vendors. The second one is to improve the test sequence quality in such a way that more valuable information can be collected during test [32]. In both cases, the goal is to make the diagnostic tool easily retrieve the defect location using a minimum set of candidates. In the rest of the section, we suggest some of the best practices to be used during test for this purpose, as well as advice on how to generate new diagnostic test patterns for silicon costumer returns.

Before testing a customer returns for diagnosis purpose, two modifications on the test program must be performed. The first one is to include all test patterns generated by the ATPG into the test program and ensure that the test does not stop at the first failing pattern. The second one is to remove or to increase the truncation value. Once these two modifications have been done, the test program can be started, followed by the first run of fault diagnosis. Depending on the diagnostic results, i.e., in the case of low diagnostic resolution, a diagnostic pattern generation process can be launched. In the following, some scenarios that require the generation of new diagnostic patterns are detailed.

- *Diagnostic ATPG for cell internal defects*. In the case of a suspected cell internal defect, information provided by the application of the entire failing and passing test patterns applied at the inputs of the defective cell is crucial to efficiently find out the cell internal defect. The best way to get this information is to generate a so-called cell exhaustive test. This test applies all the possible combinations at the cell inputs. These combinations can be static or dynamic. Static combinations include logic values "0" and "1," whereas dynamic combinations include rising and falling transitions. By applying such type of test, all nonequivalent cell internal defects can be distinguished. It is recommended to apply the dynamic part of the test with different frequencies (low speed, at-speed) to get information about the failure mechanism with respect to the clock frequency.
- *Diagnostic ATPG for interconnect open defects*. In the case of a suspected interconnect open defect, it is important to generate test patterns targeting each segment composing this interconnection and ensure a different fault propagation through different primary outputs. Figure 5 shows a net with two metal layers (*M2* and *M3*) and five segments (*s1* to *s5*). An open defect located on segment *s2* disconnects two cells (*C2* and *C3*). To help the diagnostic tool to report the defective location, it must be forced to select all possible fault propagation paths during test pattern generation (*C2 only*, *C3 only*, *C4 only*, *C2 and C3*, *C2 and C4*, *C3 and C4*, *all fanout cells: C2 and C3 and C4*). With this additional information, the diagnostic tool will be able to locate the failing segment more accurately. In this example, only an open defect on s2 can propagate through C2 and C3, and not through C4.

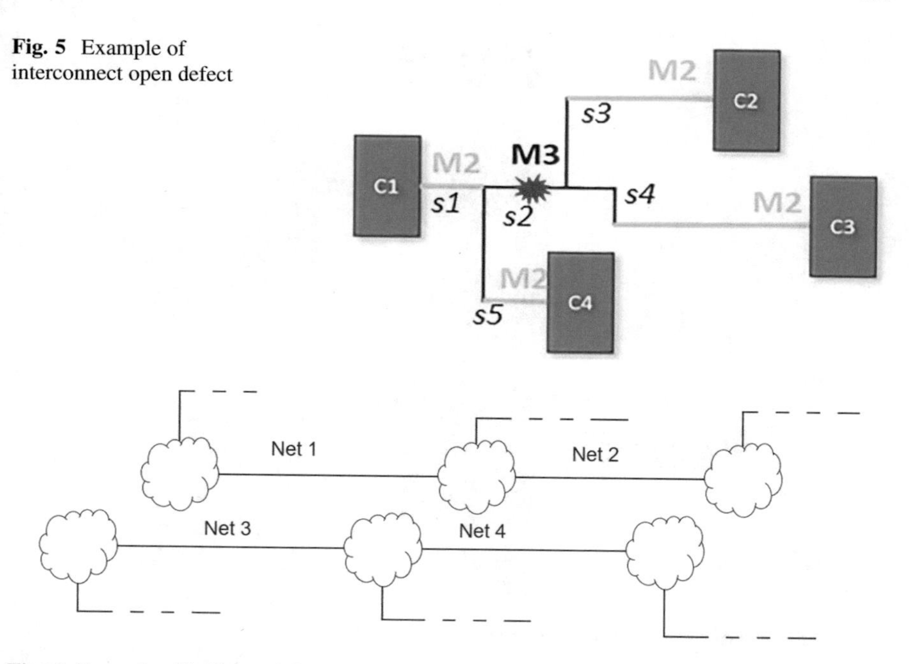

Fig. 5 Example of interconnect open defect

Fig. 6 Example of bridging defects

- *Diagnostic ATPG for bridging defects.* In the case of a suspected bridging defect, it is important to have the list of bridging pairs, extracted from the layout database, to generate additional diagnostic patterns. The goal in this case is to test an opposite value on each bridging pair separately ("0" on one net and "1" on the other one, or vice-versa) and avoid testing several bridging pairs at the same time. In Fig. 6, let us assume that two possible bridges (net1-net3 and net2-net4) are reported by the diagnostic tool. A new diagnostic pattern must be generated to distinguish between these two defects. Any ATPG tool would try to target net1-net3 and net2-net4 with the same test pattern, which is not appropriate for fault diagnosis. A diagnostic test pattern generation will try to test net1-net2 by delivering an opposite value on these two nets and ensure that net2 and net4 have the same logic value. With this additional information, a diagnostic tool will be able to differentiate these two potential bridging defects and identify the actual failing one.
- *Diagnostic ATPG for chain defects.* In the case of a scan chain failure in a design using test compression, it is crucial to generate additional patterns that can distinguish between two or more faults. These patterns are called *chain diagnostic patterns*. These patterns are not used during production (manufacturing) test but can be generated and applied for diagnostic purpose, especially in the case of a customer return. Figure 7 shows an example of a test program in which additional chain diagnostic patterns are inserted. For a fault-free CUT, the flow starts by testing the chain integrity followed by the ATPG test. In the case of a failure,

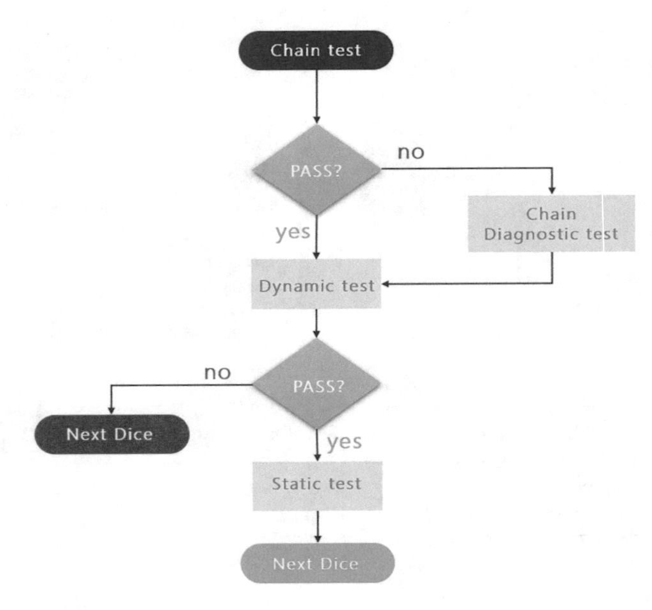

Fig. 7 Example of a manufacturing test program

additional chain diagnostic patterns are used. The main drawback of using chain diagnostic patterns, however, is the increased test time.

4 Defect Diagnosis Techniques for Customer Returns

A conventional diagnosis flow of ICs is depicted in Fig. 8. The circuit is first designed according to specifications and supported by a large set of design and verification tools. When the circuit design is completed and verified, the circuit can be manufactured. After manufacturing, all parts of the circuit (logic, memory, analog, etc.) must be tested by different methods. This step is mandatory for any manufacturer to ensure a high quality of products before delivery to the customer. If the test reveals an abnormal behavior (i.e., test fail), information from the ATE are subsequently exploited during the diagnosis step to identify the source of failure and take necessary actions to correct the design or modify the manufacturing process. The final objective of this step is to improve the yield ramp-up. Conversely, if the test is passed, the IC is sold and embedded in a system.

During the lifetime of the IC (e.g., in field), and especially during mission mode, the IC may fail by providing incorrect and unexpected responses to functional stimuli (see Fig. 8). In the case such erroneous behavior is observed, the IC is sent back to the manufacturer and is considered as a customer return. It is then tested with

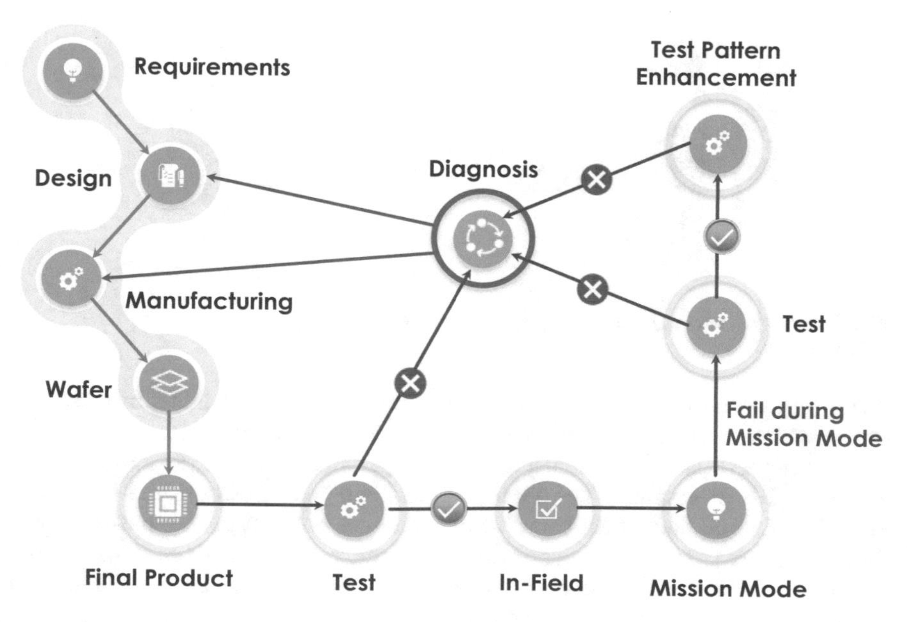

Fig. 8 Diagnosis flow of integrated circuits

the test sequences initially used after manufacturing process in order to reproduce the failure occurrence. In the case this test phase does not reveal any error, then a test pattern enhancement step is launched to produce new test patterns able to exhibit the erroneous behavior. Obtained test data log are then analyzed during the diagnosis step to finally identify the source of failure.

Fault diagnosis must be able to identify the fault location and the fault type. This information is further used to guide the failure analysis process in pinpointing the defect on silicon and identifying the root cause of failure. For example, the root cause may be a problem in the physical implementation process or a misalignment in the production masks. The results of these investigations during PFA can be further used to optimize the design and manufacturing processes and avoid reoccurrence of the failure in next-generation products.

The next subsections give details on conventional diagnosis approaches and advanced methods based on machine learning.

4.1 Conventional Approaches

Conventional diagnosis approaches aim at identifying the list of suspected nets and gates (or cells) in a digital circuit. Two main paradigms can be distinguished: cause-effect and effect-cause. The "cause-effect" paradigm needs a pre-computed fault

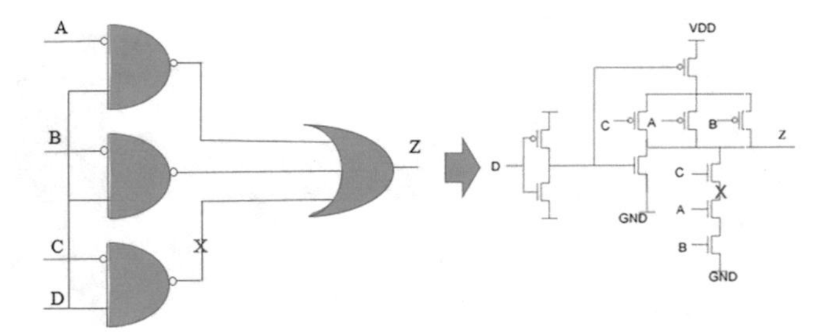

Fig. 9 Inter-cell (left) vs intra-cell (right) diagnosis results

dictionary, which can be obtained by simulating the targeted faults in a specific design with a set of test patterns [23]. From this fault dictionary and the set of failing and passing test patterns obtained after test application, a diagnosis tool is able to identify a list of suspects (i.e., fault or defect candidates). Compared to "cause-effect" diagnosis, the "effect-cause" approach does not require any pre-computed fault dictionary [33]. It is based on critical path tracing (CPT) and proceeds by back tracing sensitive paths in the circuit from every failing output identified after test application to identify the suspected faults.

For each conventional diagnosis approach, there are two levels of diagnosis: inter-cell and intra-cell. In the case of inter-cell diagnosis, each candidate is a circuit net (i.e., a connection between cells) or a cell. On the other hand, for intra-cell diagnosis, each candidate is a net inside one cell. Figure 9 gives an example of diagnosis report obtained from an inter-cell diagnosis (left part of Fig. 9) and an intra-cell diagnosis (right part of Fig. 9).

The next subsections give details on "cause-effect" and "cause-effect" diagnosis state-of-the-art approaches.

4.1.1 Diagnosis Using Fault Simulation

Inter-Cell Diagnosis

A typical cause-effect diagnosis method is illustrated in Fig. 10.

To build a fault dictionary, a specific fault model, such as the stuck-at fault model or the transition fault model, is first assumed. The dictionary, which records the responses of all test patterns for all possible faults, is generated by intensively performing fault simulations. This dictionary is referred to as the fault dictionary. Once the fault dictionary is built, the failure syndrome of the failing device is examined by using the fault dictionary. The fault(s) whose test response matches the observed failure during test application will be considered as fault candidate(s).

The time required for constructing the fault dictionary corresponds to the time for fault simulating all test patterns for all faults considered for the circuit under diagnosis. This is acceptable as it is done only once. During diagnosis, analyzing

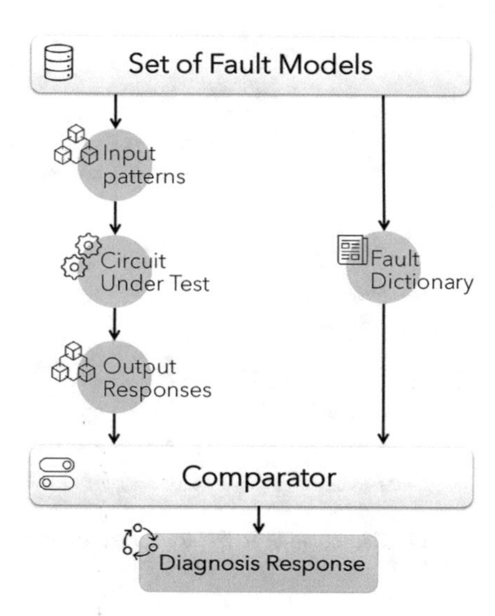

Fig. 10 Cause-effect diagnosis flow

the fault dictionary to derive fault candidates is usually quite fast. However, for practical applications, the cause-effect diagnosis approach can be limited due to some problems.

The first one is the size of the fault dictionary since it requires a large amount of storage capacity for recording all test responses for all faults against all test patterns. With the increasing size of today's design, this method thus becomes sometimes unpractical. Some methods have been proposed to reduce the size of the fault dictionary. The pass-fail dictionary is the simple way to reduce the size by using a single pass/fail bit to replace the output response of the test vector [34]. However, this solution may negatively impact the diagnosis resolution as some faults become undistinguishable by using only pass-fail bits. Another method proposed in [35] consists in constructing small fault dictionaries by recording only the test responses of the failing patterns instead of recording test responses of all test patterns for all faults. This can reduce the memory requirement without sacrificing the resolution. Authors in [36] proposed another technique to compress the fault dictionary by using a multiple input signature register (MISR) to generate compressed fault signatures. However, the problem of this method is that two different test responses may be compressed and lead to the same failing signature.

Another issue that may occur when using the cause-effect diagnosis approach is the lack of accuracy (i.e., the real defect is not in the list of suspects). In fact, if the defect is not modeled by the fault models used to compute the dictionary, it cannot be identified by the diagnosis process. To solve this problem, diagnosis must be performed using several dictionaries, one for each fault model [37].

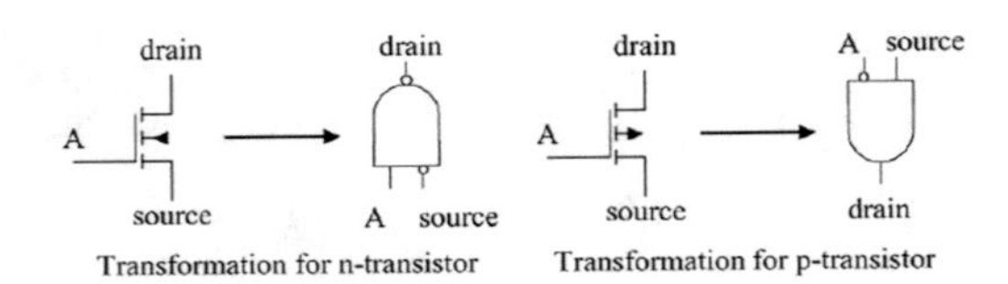

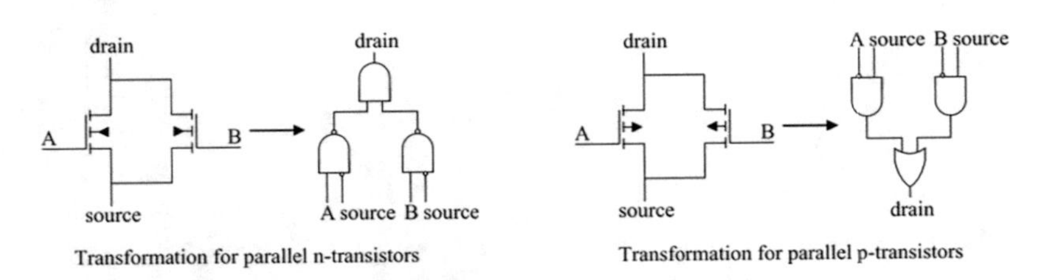

Fig. 11 Examples of transistor transformation to gate netlist

Intra-Cell Diagnosis

There are many research works focusing on intra-cell diagnosis. They can be classified into two categories.

The first category proposes a conversion from transistor-level netlist into an equivalent gate-level netlist, based on complex transformations rules [38–40]. Figure 11 gives some examples of these rules. Then, any classical inter-cell diagnosis solution can be applied on the equivalent gate-level netlist. The main limitation of this approach is that the set of transformation rules depends on the targeted defects, so that non-modeled defects may not be diagnosed.

The second category of intra-cell diagnosis techniques is based on the "cause-effect" paradigm. The transistor-level netlist of a cell is exploited in order to inject the targeted defects. Therefore, a defect dictionary is generated using results from transistor level simulations. The defect signature of all defects affecting a library cell is stored in this defect dictionary. Then, during diagnosis, the signature of all defects affecting a suspected cell, is compared to the observed failures to obtain a list of candidates internal to the cell.

These approaches can be further classified depending on the accuracy of the injected defects and the simulation precision. In [41] targeted defects are simulated by using a switch-level simulator, thus leading to a less precise defect injection, but this solution saves simulation time. In [6, 42] a large number of defects is simulated at transistor level (i.e., by using SPICE simulations). For a given defect, different sizes of resistance are simulated to be as accurate as possible. This approach leads to more precise diagnosis results, but it requires a huge simulation time. Moreover, the size of the defect dictionary is usually very high.

In order to reduce the simulation time and the dictionary size while keeping a high precision, authors in [43–45] propose to exploit layout information in order to consider only realistic defects. For example, for each cell, only the realistic potential

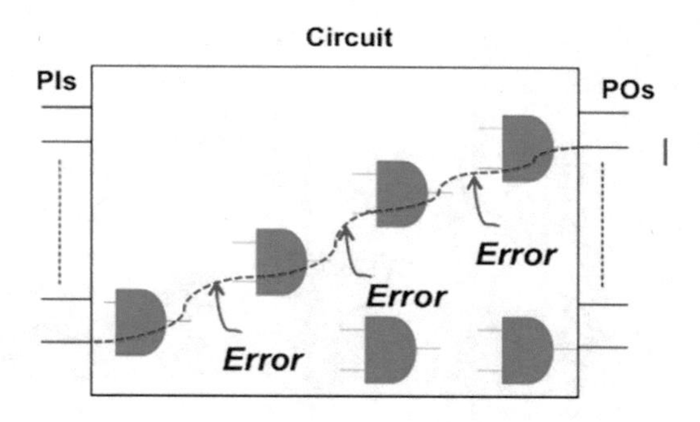

Fig. 12 Critical path tracing principle

bridging defects and via open defects are extracted and then simulated. By this way, only the identified set of realistic defects is simulated at transistor-level.

4.1.2 Diagnosis Using Critical Path Tracing

Inter-Cell Diagnosis

Conversely to the cause-effect paradigm, effect-cause diagnosis approaches directly derive the fault candidates by using a CPT algorithm as illustrated in Fig. 12, thus avoiding the construction of any fault dictionary.

A generic effect-cause diagnosis algorithm that makes the single fault assumption (only one fault at a time can occur in the circuit) is summarized as follows:

- Step 1. *Identification of initial faulty candidates*: In [46, 47], the CPT technique is applied for logic diagnosis. In both cases, a specific fault model is considered, the stuck-at fault model for [46] and the delay fault model for [47]. The process of critical path tracing consists in starting from each faulty primary output and back tracing sensitive paths up to the primary inputs of the circuit. By this way, a number of critical paths containing logic gates and nets is obtained. If a single fault is assumed, then the intersection of all the critical paths traced from all failing outputs is considered as the final candidate set (that contains suspect gates and nets). Otherwise, in the case of a multiple fault assumption, the union will be the final candidate set.
- Step 2. *Reduction of the candidate list by using passing patterns*: The initial suspect set can be reduced by using passing test patterns. To this purpose, the same critical path tracing process is done from each fault-free output of the circuit. By definition, the real defect cannot produce any faulty behavior when applying passing patterns. Therefore, a candidate (suspected gate or net) will be removed from the initial suspect set if it is contained in the set of critical paths traced from the fault-free primary outputs.

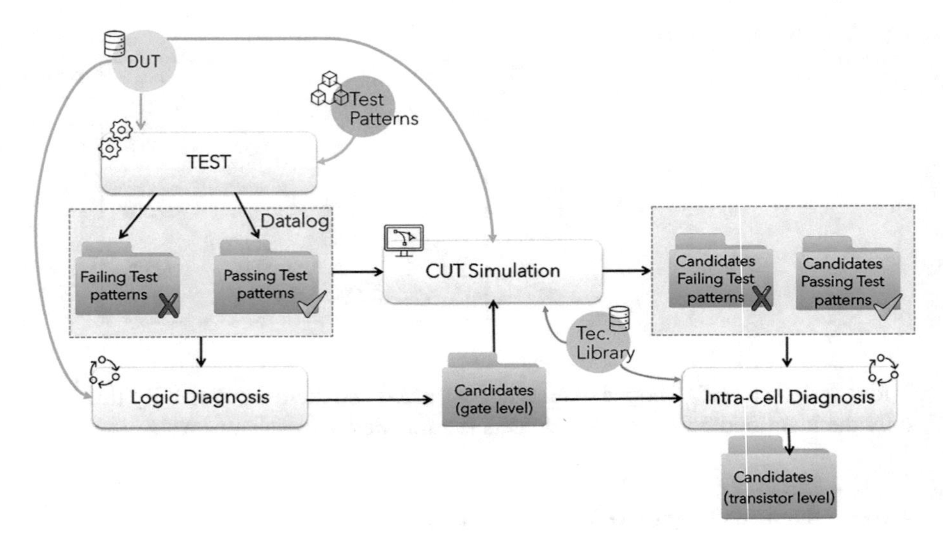

Fig. 13 Overall intra-cell diagnosis flow

Intra-Cell Diagnosis

In [48], the authors proposed an intra-cell diagnosis method based on the "effect-cause" paradigm that aims at locating the root cause of the observed failures inside a logic cell. It is based on a CPT algorithm this time applied at transistor level. The main characteristic of this approach is that it exploits the analysis of the faulty behavior induced by the actual defect. In other word, a defect is located by analyzing the effect that it induces in the circuit. Moreover, since the complexity of a single cell in terms of number of transistors is low, the proposed intra-cell diagnosis approach requires a negligible computation time.

Figure 13 sketches the overall diagnosis flow proposed in [48]. First, test patterns are applied to the circuit under diagnosis (CUD) to distinguish between the correct circuit behavior and a faulty circuit behavior caused by defects. These defects produce failing output responses for one or more input test patterns. Input test patterns leading to observed faulty behavior (i.e., failing test patterns) are stored into a file called datalog.

Then, an inter-cell fault diagnosis (i.e., logic diagnosis) algorithm exploits datalog information to determine a list of suspected logic cells. Any available commercial diagnosis tool can be adopted. Then, the CUD simulation determines the local set of failing/passing patterns for each suspected logic cell reported by the logic diagnosis tool. Finally, the intra-cell diagnosis is executed for each suspected gate and the set of pre-determined local failing/passing test patterns. The diagnosis result is a list of candidates at transistor level. For each suspect, a set of fault models able to explain the observed failures is generated.

Compare with the cause-effect methodology, effect-cause approaches have several advantages. They require less memory storage since no fault/defect dictionary has to be constructed a priori. And they do not assume any fault model. Thus, they can be used for diagnosing more realistic faults.

4.2 Advanced Methods Based on Machine Learning

Although conventional diagnosis techniques discussed in the previous section can achieve a good resolution, in some cases (e.g., complex cells, complex failure mechanisms) the number of candidates may be too high to allow an efficient PFA. This problem will be even more severe in the future with the advent of nanometer-scale technologies (i.e., 7 nm and beyond). Consequently, improving diagnosis efficiency at the transistor level (i.e., CA diagnosis) is mandatory. Achieving this goal can be done by using supervised learning algorithms to determine suspected defects. Supervised learning is now used in several classification problems where the knowledge on some data can be used to classify a new instance of such data. In this section, we summarize the latest developments in the field of CA diagnosis based on supervised learning.

4.2.1 Preliminaries

Several learning-guided solutions for CA diagnosis of mission mode failures in customer returns have been proposed recently in [16, 49–53]. All solutions are based on a Bayesian classification method for accurately identifying defect candidates in combinational standard cells of a customer return. Choosing one solution over another depends on the test scenario (test sequence, test scheme, test conditions) considered during the diagnosis phase and selected according to the types of targeted defects and failure mechanisms.

The test scenarios in [16, 49–53] are sketched in Fig. 14. In [49, 50], two *distinct* processes were developed to diagnose static and dynamic defects *separately*. In [49], a basic scan testing scheme used to apply static CA test sequences is considered, so that stuck-at faults plus static intra-cell defects are targeted during diagnosis. In [50], a fast sequential testing scheme used to apply dynamic CA test sequences is considered, so that transition faults plus dynamic intra-cell defects are targeted during diagnosis. Note that [51] is just a combination of [49, 50], i.e., two testing schemes, one static and one dynamic, and two CA diagnosis flows, one static and one dynamic, are considered independently. The main limitation of the solutions in [49–51] is the required a priori knowledge of the type of targeted defects in the customer return. In other words, a test engineer needs to know what type of defects is screening before choosing between [49] or [50].

In an attempt to deal *concurrently* with all types of defects that may occur in customer returns, without any a priori knowledge of the targeted defect type, a new implementation of the CA diagnosis flow was proposed in [16–52]. Note that [16] is a fully extended version of [52]. Authors assume a test scenario in which *two*

Test Sequence / Fault & Defect	Static (low speed)	Dynamic (at-speed)	Static + Dynamic
Stuck-at & Static CA	[49] [51]	[53]	[52] [16]
Transition & Dynamic CA		[50] [51] [53]	[52] [16]

Fig. 14 Test scenarios considered in [16, 49–53]

test (static and dynamic) sequences are successively used, each one considering an appropriate testing scheme, i.e., basic scan and fast sequential. First, a static CA test sequence produced by a commercial cell-aware ATPG tool is applied to the CUD. This sequence targets all cell-level stuck-at faults plus cell-internal static defects, considering that these defects are not covered by a standard stuck-at fault ATPG. A standard (low speed) scan-based testing scheme is used to this purpose. Next, another option of the cell-aware ATPG is used to generate a dynamic CA test sequence that targets cell-level transition faults plus intra-cell dynamic defects not covered by a standard transition fault ATPG. In this case, an at-speed *launch-on-capture* (LOC) scheme (also called *fast sequential*) is used during test application.

To construct the comprehensive flow described in [16], a new framework was set up in which specific rules were defined to obtain a high level of diagnosis effectiveness in terms of resolution and accuracy. The method was based on a Gaussian naive Bayes-trained model to predict good defect candidates. This method is summarized in the next Subsection 4.2.2.

In [53], authors proposed a new version of the CA diagnosis flow, assuming a test scenario in which *both* static and dynamic defects can be diagnosed by using a *single* dynamic CA test sequence applied at-speed. According to the test flow depicted in Fig. 4, this scenario may happen when such a test sequence has been generated to test transition faults plus cell-internal dynamic defects and also cover the required percentage of stuck-at faults plus cell-internal static defects (or, more generally, satisfies the test coverage specifications). Note that in this case, only one (dynamic) datalog is generated after test application and can further be used for diagnosis purpose. Nevertheless, both static and dynamic defects are taken into account in this scenario. As only dynamic instance tables are manipulated, the representation of training and new data is simplified, i.e., a single type of feature vector is used, without no loss of information and hence without decreasing the quality of the training and inference phases.

4.2.2 Learning-Based Cell-Aware Diagnosis Flow

Figure 15 is a general view of the learning-based CA diagnosis flow utilized in [16]. It is based on supervised learning in which a known set of input data and known

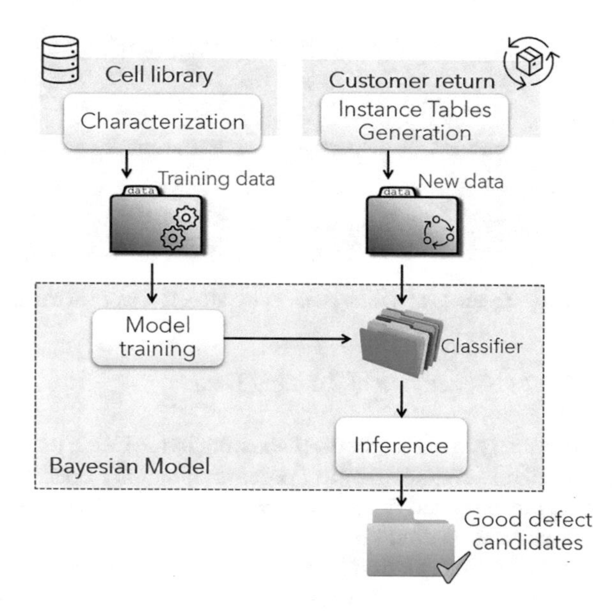

Fig. 15 Generic view of the cell-aware diagnosis flow used in [16]

responses (*labeled data*) is taken and used as training data, trains a model, and then implement a classifier based on this model to make predictions (*inferences*) for the response to new data.

After investigating several ML algorithms and observing their inference accuracies in [49], a Bayesian classification method has been chosen for the learning and inference phases in [16, 49–53]. So, the first step of the CA diagnosis flow consists in generating a naive Bayes (NB) model and to train it by using a training dataset. In this step, training data are used to improve incrementally the capability of the model to make inference. The training dataset is divided into mutually exclusive and equal subsets. For each subset, the model is trained on the union of all other subsets. Some manipulations, such as grouping data by considering equivalent defects or removing data instances of undetectable defects, are also performed during this phase. Once the training phase is complete, the accuracy (i.e., performance) of the model is evaluated by using a part of the dataset initially set aside. More details about performance evaluation as done in this framework can be found in [51]. The second main step consists in implementing the NB classifier by using a Gaussian distribution to model the *likelihood* probability functions, and use this classifier to make prediction when a new data instance has to be evaluated. The next subsections detail the various steps of the CA diagnosis flow, which is able to deal with any type of cell-internal defect (i.e., static and dynamic) that may occur in customer returns.

P1	P2	P3	P4	P5	P6	P7	P8	P9	P10	P11	P12	P13	P14	P15	P16	Pattern
00	01	10	11	0R	0F	R0	RR	RF	R1	F0	FF	FR	F1	1R	1F	Defect
1	0	0	0	0	1	0	0	0	0	1	1	0	0	0	0	D1
1	0	0	1	0	1	0	1	0	1	1	1	0	0	1	0	D2
0	1	0	0	1	0	0	0	0	0	0	0	1	1	0	0	D3
0.5	0.5	0.5	0.5	0	0	0	0	0	0	1	0	0	0	0	0	D11
0.5	0.5	0.5	0.5	1	0	0	0	0	0	0	0	1	0	0	0	D12
0.5	0.5	0.5	0.5	0	0	0	0	1	0	0	0	0	0	0	1	D13

Fig. 16 Example of training dataset for all defect types in a two-input cell as used in [16, 49–52]

4.2.2.1 Generation of Training Data

For each type of standard cell existing in the CUD, training data are generated during an off-line characterization process done only once for a given cell library. These data are extracted from CA views provided by a commercial CAD tool that contains all characterization results for a given cell type. These results are collected in the form of a fault dictionary that contain, for each defect within a cell, the cell input patterns detecting (or not) this defect. An example of training dataset, as used in [16, 49–52] and containing six instances, for example, two-input cell, is illustrated in Fig. 16. Each instance is associated with a static defect (D_1, D_2, D_3) or a dynamic defect (D_{11}, D_{12}, D_{13}). A 1 (0) indicates that defect D_i is detectable (not detectable) at the output of the cell when the cell-level test pattern P_j is applied at the inputs of the cell. Cell-level test patterns (called *cell-patterns* in the sequel) are static (one-input vector – P_1 to P_4 in Fig. 16) or dynamic (two-input vectors – P_5 to P_{16} in Fig. 16 in which R (F) indicates a rising (falling) transition at the cell input). For an n-input cell, there exists 2^n static cell-patterns and $2^n.(2^n–1)$ dynamic cell-patterns.

Dynamic defects can be detected by dynamic patterns, but not only. They can also be detected by static patterns applied using a basic scan testing scheme, provided that (i) at least one transition has been created at the cell inputs between the next-to-last scan shift cycle and the launch cycle and (ii) the delay induced by the defect is large enough to be detected (*these are the detection conditions of a dynamic defect modeled by a stuck-open or a gross delay fault*). For this reason, the value "0.5" is assigned to each dynamic defect (D_{11}, D_{12}, D_{13}) for all related static cell patterns, meaning that such a defect is detectable or not depending on whether or not the above conditions are satisfied.

As only dynamic test sequences are considered in [53], the representation of training data as done in [16, 49–52] can be simplified without any loss of information and without reducing the quality of the training phase. This comes from the observation that a static defect is a particular case of dynamic defect (e.g., a full open is a resistive open with an infinite value of the resistance) and that all static cell-patterns for a given defect are embedded in its whole set of dynamic cell-patterns. Indeed, a dynamic defect requires a two-vector test pattern (V_1V_2) in which the values of V_1 and V_2 have to be correctly defined for the defect to be detected. On the contrary, only the value of V_2 is significant for a static defect to be detected by

P1′	P2′	P3′	P4′	P5′	P6′	P7′	P8′	P9′	P10′	P11′	P12′	Pattern
0R	0F	R0	RR	RF	R1	F0	FF	FR	F1	1R	1F	Defect
0	1	0	0	0	0	1	1	0	0	0	0	D1
0	1	0	1	0	1	1	1	0	0	1	0	D2
1	0	0	0	0	0	0	0	1	1	0	0	D3
0	0	0	0	0	0	1	0	0	0	0	0	D11
1	0	0	0	0	0	0	0	1	0	0	0	D12
0	0	0	0	1	0	0	0	0	0	0	1	D13

Fig. 17 Example of training dataset for all defect types in a two-input cell as used in [53]

such pattern, irrespective of the value taken by V_1. When looking at Fig. 16, one can see that $P_1 = \{00\}$ is embedded in $P_6 = \{0F\}$, $P_{11} = \{F0\}$, and $P_{12} = \{FF\}$ and the same for P_2, P_3 and P_4. Similarly, we can see that static defect D_2 is detectable by P_1 and P_4, and hence by P_6, P_8, P_{10}, P_{11}, P_{12}, and P_{15}. So, by "compacting" a training dataset as illustrated in Fig. 17, in which only dynamic cell-patterns are considered, one can see that all meaningful information is still contained in this set, while redundant ("0" and "1" values in the first four columns of Fig. 16) or insignificant ("0.5" values in the same columns for dynamic defects) information is removed. More generally, such compact format for training data makes so that only one type of feature vector (dynamic) is used for both types of defects.

As the goal with training data is to provide a distinct feature vector for each data (defect), it is crucial to be able to distinguish between static and dynamic defects with such a new format of the training dataset. Let us consider two defects D_1 and D_{11} where D_1 is static and detectable by {00} and D_{11} is dynamic and detectable by {F0} (note that {00} is the second vector of {F0}). As can be seen in Fig. 17, these two defects can easily be distinguished since their training data instances (or *feature vectors*) are different. The consequence of using such a new format for training data (and hence for new data as will be shown later on) is not an improved accuracy or resolution, but rather a simplified manipulation of feature vectors.

4.2.2.2 Generation of Instance Tables

An instance table is a failure mapping file generated for each suspected cell by using information contained in the tester datalog. It describes the behavior (pass/fail) of the cell for each cell-pattern occurring on its inputs during test of the CUD. The generation process of instance tables is sketched in Fig. 18. First, CA test patterns are applied to the CUD. These test patterns are obtained from a commercial CA test pattern-generation tool that targets intra-cell defects. Next, a datalog containing information on the failing test patterns and corresponding failing primary outputs is obtained. From this datalog and the circuit netlist, a logic diagnosis is carried out (still using a commercial tool) and gives the list of suspected cells. From this

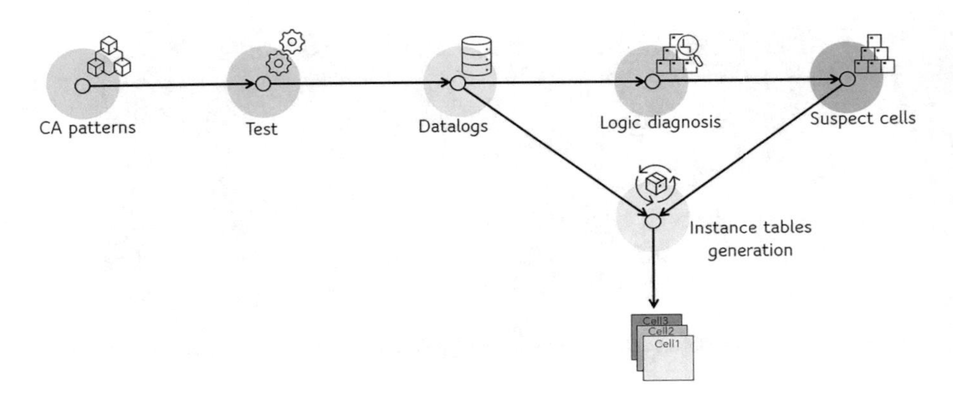

Fig. 18 Generation flow of instance tables

Fig. 19 Example of static and dynamic instance tables

NOR Cell - NR2NHVTX1

Z	Output	L412/C1381A
A	Input	U59/Z
B	Input	U28/Z

Pattern 1 PASSING Z: stuck-at-0
Z 000011111111111 - 1
A 111100000000000 - 0
B 000000000000000 - 0
Pattern 2 FAILING Z: stuck-at-1
Z 011100000000000 - 0
A 000011111111111 - 1
B 100011111111111 - 1

list and the datalog information, one can finally generate an instance table for each suspected cell. Note that in the case several test sequences, e.g., one static and one dynamic, are used for diagnosis of the CUD, the generation process is repeated so as to produce static and dynamic instance tables for all suspected cells as in the case reported in [16].

The format of a static instance table is illustrated in Fig. 19 for a given two-input NOR cell and two static cell-patterns. In this example, the first part of the file gives information on how the cell is connected to other cells in the circuit, while the second part represents, respectively, the number of patterns, the pattern status (failing, passing), and the cell output Z with the associated fault model for which exercising conditions are reported. These conditions reported right below each cell-pattern in Fig. 19 represent the stimulus arriving at the cell inputs during the shift phase (before "-") and applied during the launch cycle (after "-"). For example, cell-pattern 2 consists in applying a "1" on input A and B, and failing in detecting a stuck-at 1 on Z.

P1	P2	P3	P4	P5	P6	P7	P8	P9	P10	P11	P12	P13	P14	P15	P16	
0.5	0.5	0.5	0.5	0	0	0	0.5	0	1	0	0	0	0	0	0	D_i

Fig. 20 Format of a new data instance for a two-input cell

P1'	P2'	P3'	P4'	P5'	P6'	P7'	P8'	P9'	P10'	P11'	P12'	
0	0	0.5	1	0	1	0	0	0.5	0	0	0	Di

Fig. 21 Format of a new data instance as used in [53]

4.2.2.3 Generation of New Data

New data are generated after post-processing of instance tables. They are made of various instances, each of them being associated to one suspected cell in the CUD, and represents a feature vector that characterizes the actual behavior of the cell during test application. From each new data instance, one or more defect candidates can be extracted that have to be classified as good or bad candidate with a corresponding probability to be the root cause of failure. This classification is done by comparing the new data instance with the training data of the corresponding suspected cell, and identify those training data instances that match (or not) with the new data instance.

The formats of a new data instance as used in [16, 49–52] and [53] are illustrated in Figs. 20 and 21, respectively. This format is close to the format of a training data instance, but has a different meaning. In each instance, the value "1" (resp. "0") is associated to a failing (resp. passing) cell-pattern P_i for a given defect candidate, meaning that the candidate is *actually* detectable (resp. undetectable) by the cell-pattern P_i at the output of the cell during test of the CUD, and hence can (cannot) be the real defect. In such instance, the value "0.5" is associated to a cell-pattern for a given defect candidate when this pattern cannot appear at the inputs of a suspected cell during real test application with an ATE. The median value "0.5" was chosen to avoid missing information in new data instances while not biasing the features of these data.

4.2.2.4 Diagnosis of Defects in Sequential Cells

All the work carried out in [16, 49–53] was about diagnosis of defects occurring in combinational standard cells of a customer returns. However, defects in SoCs may also occur in sequential standard cells of logic blocks. In this section, we show how the previous diagnosis flow can handle sequential cells and related defects by adding new information to the training dataset [54].

The two main differences between a combinational cell and a sequential cell are that (i) the latter has a clock input pin and (ii) the fact that the previous logic value of a sequential cell output can affect the current output value of the cell. To take

P1	P2	P3	P4	P5	P6	P7	P8	Pattern
U00	U01	U10	U11	UR0	UR1	UF0	UF1	Defect
0	0	1	0	0	0	1	0	D1
1	1	0	0	0	0	1	1	D2
0.5	0.5	0.5	0.5	1	1	0	0	D11
0.5	0.5	0.5	0.5	1	0	0	1	D12

Fig. 22 Example of training dataset for all defect types (static and dynamic) in a sequential cell. The pin order is clock-data-previous output

this difference into account, each cell-pattern for a sequential cell is considered as a tuple in which the first value represents the input clock signal (pulsing or not), the second value is associated to the main input of the cell (e.g., D), and the third value is associated to a virtual input pin representing the previous value of the output pin of the cell (e.g., Q). Note that in the case of sequential cells with multiple real inputs (e.g., D flip-flop with a D, scan-in, scan-enable, and clock input signals), the cell-pattern representation is expanded accordingly. In each tuple, the first value is either U (i.e., pulse) or 0, depending on whether or not there is an active clock signal. The second value can be 0, 1, R, or F. The third value can only be static (i.e., 0 or 1). An example of training dataset for all defect types (static and dynamic) that may occur in a sequential cell is illustrated in Fig. 22. Note that the CA views used during the generation of training data do not contain information about cell patterns with non-pulsing clock signals (i.e., none of the cell internal defects can be detected at the cell output without clock pulse). Consequently, the training data do not include such cell-patterns as can be observed in the example of Fig. 22. Note also that instance tables of sequential cells may contain cell patterns with no transition on the main inputs of the cell. To allow the ML algorithm understanding this information, the solution consists in including static cell patterns (e.g., P1 to P4 in Fig. 22) in the training data of sequential cells.

With the above representation of training data for sequential cells, one can see that the diagnosis flow in Fig. 15 can be used in a straightforward manner without any change. The two main steps (model training by using a training dataset, implementation of the NB classifier to make inference) remain the same irrespective of the type of manipulated standard cells.

5 Industrial Case Studies

The CA diagnosis flow described in Sect. 4.2.2 and targeting defects in both combinational and sequential cells of customer returns has been implemented in a Python program. For validation purpose, authors in [16, 49–54] have tested the proposed flow in three different approaches:

Table 2 Main features of the silicon test chip

#cells	#PIs	#POs	#SFF
3.8 M	97	32	17.5 k

Table 3 Average pattern count in instance tables of the first simulated case study

#passing patterns	#unique passing patterns	#failing patterns	#unique failing patterns
43.4	24.0	15.5	8.6

- First, they performed experiments on ITC'99 benchmark circuits with defect injection campaigns targeting *combinational cells* in each circuit. Various results are reported in [16, 49–53] to show the superiority of the framework when compared to commercial diagnosis solutions.
- Next, they considered a test chip developed and designed using STMicroelectronics 28 nm FDSOI technology, and they conducted two defect injection campaigns targeting *sequential cells* [54]. Results are reported in Subsection 5.1 and also demonstrate the effectiveness of the diagnosis framework.
- Finally, they considered *a customer return* from STMicroelectronics and performed a silicon case study with a real defect subsequently analyzed and detected during PFA. Results are reported in Subsection 5.2.

5.1 Simulated Test Case Studies

Authors in [54] conducted experiments on a silicon test chip developed and designed using STMicroelectronics 28 nm FDSOI technology. The test chip contains only digital and memory blocks, as well as one PLL. The digital blocks include 3.8 million cells. Other features (number of primary inputs, primary outputs and scan flip-flops) are given in Table 2.

A first simulated case study was done for *static* defects. All possible static defects were successively injected into three scan flip-flops (SFF) of a single full-scan digital block. This block was tested with a static CA test sequence achieving a stuck-at + static CA fault coverage of 100%. The average numbers of passing and failing test patterns are given in Table 3. The CA diagnosis flow was executed and achieved good results. In fact, the accuracy was *100%* for all the injected defects (the injected defect was always identified and reported), and the resolution was *1.25*. The resolution varies between 1 and 3, and Fig. 23 shows the distribution of this resolution with respect to the total number of simulated cases. Obviously, in most of the cases, the number of suspects is equal to 1 (perfect resolution).

A second simulated case study with another defect injection campaign was performed on the same test chip. All possible dynamic defects were successively injected into three scan flip-flops of a single full-scan digital block. This time, a dynamic CA test sequence was applied and achieved a transition + dynamic CA fault coverage of 89.8%. The average number of failing test patterns was 7.9. Again,

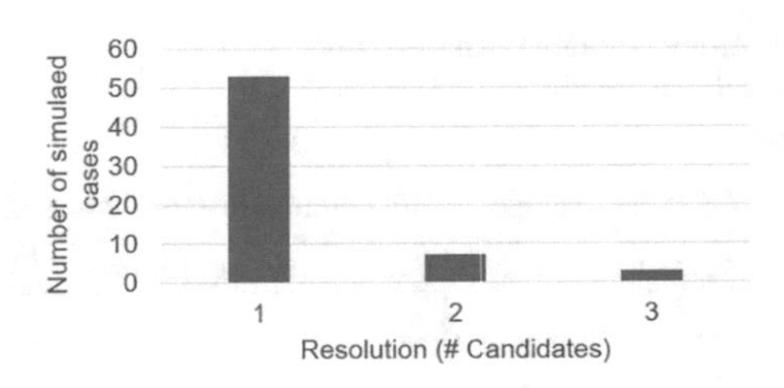

Fig. 23 Distribution of the resolution with respect to the simulated cases

CA diagnosis flow was executed and achieved good results. The accuracy was *100%* for all the injected defects. The average resolution obtained for dynamic defect injection experiments was 1.37. Again, the resolution ranged between 1 and 3, and in most of the case, the number of suspects was equal to 1.

5.2 Silicon Test Case Study

Next, a silicon case study was performed on a customer return designed with a 28 nm FDSOI technology from STMicroelectronics [54]. Experiments were performed in the following test conditions: a nominal supply voltage of 0.83 V, a tester frequency of 10 MHz, a launch-to-capture clock speed (for the dynamic CA test sequence application) aligned to the internal clock frequency, and a temperature of 25 °C. The design process was considered as typical. The CA diagnosis flow was experimented, and the following results were obtained. Initially, the static CA test sequence was failing when applied at the nominal voltage. The fails were collected in a "static" datalog. Then, a logic diagnosis reported a short list of candidates among them a cell which a six-input SFF cell made of 56 transistors and having a Reset, an Enable, a Test-Input and Test-Enable input pins. The cell includes 758 potential short or open defects. A static instance table was then produced for this suspected cell and contained 5 failing and 75 passing cell-level test patterns. This instance table was post-processed to generate the new data, and then the NB classifier identified four suspected defects among which defect D62 (a short between the gate and source of NMOS 19).

The above CA candidate were presented to the FA team of STMicroelectronics, who performed a PFA in the past on this customer return. The result obtained with the CA diagnosis flow was confirmed as defect D62 was found to be the real defect. Figure 24 shows the layout view of the suspected cell and the suspect transistors. Yellow circles indicate defect candidates and red mark indicates actual observed defect.

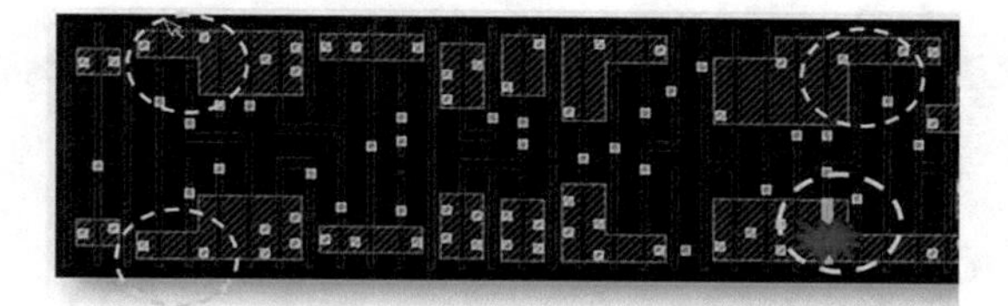

Fig. 24 GDS view of the suspected cell and the suspected transistors

6 Discussion and Conclusion

This chapter has identified the key challenges in cell-aware diagnosis of silicon customer returns and has described a full methodology and subsequent results that have emerged from pioneering research in this domain. In a comprehensive form, it proposed a compendium of solutions existing in this field. More in detail, the chapter has presented a framework for cell-aware diagnosis of customer returns based on supervised learning. The flow proposes solutions for static and dynamic defects that may appear in combinational cells or sequential cells of real circuits. A naive Bayes classifier is used to accurately pinpoint defect candidates. Experiments on silicon test cases have been performed to confirm the flow efficacy in terms of accuracy and resolution.

Experimental results proved the relevance of a learning-based method to solve the challenge of customer return diagnosis, despite the small size of the training dataset used (only one sample for one defect class). If multiple defect sizes and test conditions are used, this becomes even truer. In fact, multiple samples (one for each defect size or defect size range, one for each PVT test condition) can be associated to a given defect class, simply because the behavior of the defect differs when applying the same set of test patterns. This method has a low impact in term of timing and complexity. Indeed, the training dataset is obtained from characterized cell libraries that are generated anyway for test and diagnosis purpose. So, even with large cell libraries with a huge number of defects to be simulated (e.g., 631 cells in a library, each with 4 to 6 inputs, 951 shorts, and 749 opens on average – typical example of an ST library), the diagnosis framework will still be simple and time-efficiently relevant.

It is worth noting that among other factors, the effectiveness of the framework can be enlightened by the fact that naïve Bayes algorithm usually offers good classification performance [55]. The NB classifier needs a small amount of training data to estimate its parameters [56], which is the case in the proposed method, as only one instance per class (i.e., CA defect) is available. On the other hand, other popular ML classification algorithms, such as K-nearest neighbor (KNN) classifier or classifiers based on a support vector machine (SVM), which estimate the class of a new sample by analyzing the classes of similar training samples, cannot properly work when only one sample per class is available.

Acknowledgments This work has been funded by the French National Research Agency (ANR) under the framework of the ANR-17-CE24-0014-01 EDITSoC (Electrical Diagnosis for IoT SoCs in automotive) project.

References

1. Sumikawa, N., Drmanac, D., Wang, L.C., Abadir, M.S.: Understanding customer returns from a test perspective. In: Proc. IEEE VLSI Test Symposium, pp. 2–7 (2011)
2. Tikkanen, R.J., Siatkowski, S., Wang, L.-C., Abadir, M.S.: Yield optimization using advanced statistical correlation methods. In: Proc. IEEE International Test Conference (2014). https://doi.org/10.1109/TEST.2014.7035326
3. Benabboud, Y., Bosio, A., Dilillo, L., Girard, P., Virazel, A., Riewer, O.: A comprehensive system-on-chip logic diagnosis. In: Proc. IEEE Asian Test Symposium (2010). https://doi.org/10.1109/ATS.2010.49
4. Xue, Y., Li, X., Blanton, R.D., Lim, C., Enamul Amyeen, M.: Diagnosis resolution improvement through learning-guided physical failure analysis. In: Proc. IEEE International Test Conference (2016). https://doi.org/10.1109/TEST.2016.7805824
5. Pateras, S.: IC test solutions for the automotive market, Mentor Graphics, White Paper (2017). https://www.techonline.com/tech-papers/ic-test-solutions-for-the-automotive-market/
6. Hapke, F., et al.: Cell-aware test. IEEE Trans. Comput. Aided Des. **33**(9), 1396–1409 (2014)
7. Maxwell, P., Hapke, F., Tang, H.: Cell-aware diagnosis: defective inmates exposed in their cells. In: Proc. IEEE European Test Symposium (2016). https://doi.org/10.1109/ETS.2016.7519313
8. Xue, Y., Li, X., Blanton, R.D., Lim, C.: Improving diagnostic resolution of failing ICs through learning. IEEE Trans. Comput. Aided Des. **37**(6), 1288–1297 (2018)
9. Ren, X., Martin, M., Blanton, R.D.: Improving accuracy of on-chip diagnosis via incremental learning. In: Proc. IEEE VLSI Test Symposium, pp. 1–6 (2015)
10. Huang, Y., Yang, W., Cheng, W.: Advancements in diagnosis driven yield analysis (DDYA): a survey of state-of-the-art scan diagnosis and yield analysis technologies. In: Proc. IEEE European Test Symposium, pp. 1–10 (2015). https://doi.org/10.1109/ETS.2015.7138758
11. Xue, Y., Poku, O., Li, X., Blanton, R.D.: PADRE: physically- aware diagnostic resolution enhancement. In: Proc. IEEE International Test Conference (2013). https://doi.org/10.1109/TEST.2013.6651899
12. Pan, R., Zhang, Z., Li, X., Chakrabarty, K., Gu, X.: Unsupervised root-cause analysis for integrated systems. In: Proc. IEEE International Test Conference (2020). https://doi.org/10.1109/ITC44778.2020.9325268
13. Kruseman, B., Majhi, A., Hora, C., Eichenberger, S., Meirlevede, J.: Systematic defects in deep sub-micron technologies. In: Proc. IEEE International Test Conference, pp. 290–299 (2005)
14. Hapke, F., Krenz-Baath, R., Glowatz, A., Schloeffel, J., Hashempour, H., Eichenberger, S., Hora, C., Adolfsson, D.: Defect-oriented cell-aware ATPG and fault simulation for industrial cell libraries and designs. In: Proc. IEEE International Test Conference, pp. 1–2 (2009). https://doi.org/10.1109/TEST.2009.5355741
15. Hapke, F., Redemund, W., Schloeffel, J., Krenz-Baath, R., Glowatz, A., Wittke, M., Hashempour, H., Eichenberger, S.: Defect-oriented cell-internal testing. In: Proc. IEEE International Test Conference, paper 10.1 (2010). https://doi.org/10.1109/TEST.2010.5699229
16. Mhamdi, S., Girard, P., Virazel, A., Bosio, A., Ladhar, A.: A learning-based cell-aware diagnosis flow for industrial customer returns. In: Proc. IEEE International Test Conference (2020). https://doi.org/10.1109/ITC44778.2020.9325246
17. Bushnell, M., Agrawal, V.: Essentials of Electronic Testing for Digital, Memory and Mixed-Signal VLSI Circuits. Springer (2002)., ISBN 978–0–7923-7991-1

18. Ho, T.P., Faehn, E., Virazel, A., Bosio, A., Girard, P.: An effective intra-cell diagnosis flow for industrial SRAMs. In: Proc. IEEE International Test Conference, pp. 1–8 (2018). https://doi.org/10.1109/TEST.2018.8624799
19. Pavlidis, A., Faehn, E., Louërat, M.-M., Stratigopoulos, H.-G.: BIST-assisted analog fault diagnosis. In: Proc. IEEE European Test Symposium, pp. 1–6 (2021). https://doi.org/10.1109/ETS50041.2021.9465386
20. Ye, F., Zhang, Z., Chakrabarty, K., Gu, X.: Adaptive board-level functional fault diagnosis using incremental decision trees. IEEE Trans. Comput. Aided Des. Integr. Circuits Syst. **35**(2), 323–336 (2016)
21. Benabboud, Y., Bosio, A., Girard, P., Pravossoudovitch, S., Virazel, A., Bouzaida, L., Izaute, I.: A case study on logic diagnosis for system-on-chip. In: Proc. IEEE International Symposium on Quality Electronic Design, pp. 253–259 (2009)
22. Ye, F., Zhang, Z., Chakrabarty, K., Gu, X.: Board-level functional fault diagnosis using multikernel support vector machines and incremental learning. IEEE Trans. Comput. Aided Des. Integr. Circuits Syst. **33**(2), 279–290 (2014)
23. Waicukauski, J.A., Lindbloom, E.: Failure diagnosis of structured VLSI. IEEE Des. Test Comput. **6**(4), 49–60 (1989)
24. Bosio, A., Girard, P., Pravossoudovitch, S., Virazel, A.: A comprehensive framework for logic diagnosis of arbitrary defects. IEEE Trans. Comput. **59**(3), 289–300 (2010)
25. Fan, X., Moore, W., Hora, C., Gronthood, G.: A novel stuck-at based method for transistor stuck-open fault diagnosis. In: Proc. IEEE International Test Conference, pp. 386–395 (2005)
26. Hapke, F., Reese, M., Rivers, J., Over, A., Ravikumar, V., Redemund, W., Glowatz, A., Schloeffel, J., Rajski, J.: Cell-aware production test results from a 32-nm notebook processor. In: Proc. IEEE International Test Conference (2012). https://doi.org/10.1109/TEST.2012.6401533
27. Sun, A., Bosio, A., Dillilo, L., Girard, P., Virazel, A., Pravossoudovitch, S., Auvray, E.: Intra-cell defects diagnosis. J. Electron. Test. **30**(5), 541–555 (2014)
28. Li, J., et al.: Diagnosis for sequence dependent chips. In: Proc VLSI Test Symposium, pp. 187–192 (2002)
29. Li, J., McCluskey, E.J.: Diagnosis of resistive-open and stuck-open defects in digital CMOS ICs. IEEE Trans. Comput. Aided Des. Integr. Circuits Syst. **24**(11), 1748–1759 (2005)
30. Li, J., Tsang, C.W., McCluskey, E.J.: Testing for resistive opens and stuck opens. In: Proc. IEEE International Test Conference, pp. 1049–1058 (2001)
31. Huang, Y., Guo, R., Cheng, W.-T., Chien-Mo Li, J.: Survey of scan chain diagnosis. IEEE Des. Test Comput. **25**(3), 240–248 (June 2008)
32. Girard, P., Landrault, C., Pravossoudovitch, S., Rodriguez, B.: A diagnostic ATPG for delay faults based on genetic algorithms. In: *Proc.* IEEE International Test Conference (1996)
33. Abramovici, M., Breuer, M.A.: Fault diagnosis based on effect-cause analysis: an introduction. In: Proc. ACM Design Automation Conference, pp. 69–76 (1980)
34. Pomeranz and Reddy: On the generation of small dictionaries for fault location. In: Proc. IEEE/ACM International Conference on Computer-Aided Design, pp. 272–279 (1992)
35. Chess, B., Larrabee, T.: Creating small fault dictionaries [logic circuit fault diagnosis]. IEEE Trans. Comput. Aided Des. Integr. Circuits Syst. **18**(3), 346–356 (1999)
36. Tang, H., Liu, C., Cheng, W.-T., Reddy, S.M., Zou, W.: Improving performance of effect-cause diagnosis with minimal memory overhead. In: Proc. IEEE Asian Test, pp. 281–287 (2007)
37. Wu, J., Rudnick, E.M.: Bridge fault diagnosis using stuck-at fault simulation. IEEE Trans. Comput. Aided Des. Integr. Circuits Syst. **19**(4), 489–495 (2000)
38. Fan, X., Moore, W., Hora, C., Gronthoud, G.: A novel stuck-at based method for transistor stuck-open fault diagnosis. In: Proc. IEEE International Test Conference, pp. 378–386 (2005)
39. Fan, X., Moore, W., Hora, C., Konijnenburg, M., Gronthoud, G.: A gate-level method for transistor-level bridging fault diagnosis. In: Proc. IEEE VLSI Test Symposium, pp. 266–271 (2006)
40. Fan, X., Moore, W.R., Hora, C., Gronthoud, G.: Extending gate-level diagnosis tools to CMOS intra-gate faults. IET Comput. Digital Tech. **1**(6), 685 (2007)

41. Amyeen, M., Nayak, D., Venkataraman, S.: Improving precision using mixed-level fault diagnosis. In: Proc. IEEE International Test Conference, pp. 1–10 (2006)
42. Ladhar, A., Masmoudi, M., Bouzaida, L.: Efficient and accurate method for intra-gate defect diagnoses in nanometer technology and volume data. In: Proc. IEEE/ACM Design, Automation & Test in Europe Conference & Exhibition, pp. 988–993 (2009)
43. Ladhar, A.: Extraction and diagnosis of submicron defects. PhD thesis, Tunis University (2011)
44. Ladhar, A., Masmoudi, M., Bouzaida, L.: Extraction and simulation of potential bridging faults and open defects affecting standard cell libraries. In: Proc. IEEE International Conference on Signals, Circuits and Systems, pp. 1–6 (2008)
45. Ladhar, A., Masmoudi, M.: A novel algorithm to extract open defects from industrial designs. In: Proc. IEEE International Conference on Electronics, Circuits and Systems, pp. 675–678 (2009)
46. Abramovici, M., Menon, P.R., Miller, D.T.: Critical path tracing: an alternative to fault simulation. IEEE Des. Test Comput. **1**(1), 83–93 (1984)
47. Girard, P., Landrault, C., Pravossoudovitch, S.: Delay-fault diagnosis by critical-path tracing. IEEE Des. Test Comput. **9**(4), 27–32 (1992)
48. Sun, Z., Bosio, A., Dilillo, L., Girard, P., Todri-Sanial, A., Virazel, A., Auvray, E.: Effect-cause intra-cell diagnosis at transistor level. In: Proc. IEEE International Symposium on Quality Electronic Design, pp. 460–467 (2013)
49. Mhandi, S., Virazel, A., Girard, P., Bosio, A., Auvray, E., Faehn, E., Ladhar, A.: Towards improvement of mission mode failure diagnosis for system-on-chip. In: Proc. IEEE International on-Line Testing Symposium (2019)
50. Mhandi, S., Girard, P., Virazel, A., Bosio, A., Ladhar, A.: Cell-aware diagnosis of automotive customer returns based on supervised learning. In: presented at *IEEE Automotive Reliability and Test Workshop* (2019)
51. Mhamdi, S., Girard, P., Virazel, A., Bosio, A., Faehn, E., Ladhar, A.: Cell-aware defect diagnosis of customer returns based on supervised learning. IEEE Trans. Device Mater. Reliab. **20**(2), 329–340 (2020)
52. Mhandi, S., Girard, P., Virazel, A., Bosio, A., Ladhar, A.: Learning-based cell-aware defect diagnosis of customer returns. In: Proc. IEEE European Test Symposium (2020)
53. Mhandi, S., Girard, P., Virazel, A., Bosio, A., Ladhar, A.: Cell-aware diagnosis of customer returns using Bayesian inference. In: Proc. IEEE International Symposium on Quality Electronic Design (2021)
54. d'Hondt, P., Mhamdi, S., Girard, P., Virazel, A., Ladhar, A.: A comprehensive framework for cell-aware diagnosis of customer returns. Microelectron. Reliab. J. **135**, 114595 (2021)
55. Zhang, H.: The optimality of naive Bayes. In: Proc. 17th International Florida Artificial Intelligence Research Society Conference (2004)
56. Webpage: 1.9. Naive Bayes — Scikit-Learn 0.24.1 documentation. https://scikit-learn.org/stable/modules/naive_bayes.html

Index

A. Iranmanesh (ed.), *Frontiers of Quality Electronic Design (QED)*,
https://doi.org/10.1007/978-3-031-16344-9

Printed in the United States
by Baker & Taylor Publisher Services